Developments in Mathematics

VOLUME 34

For further volumes:
www.springer.com/series/5834

Simeon Reich • Alexander J. Zaslavski

Genericity in Nonlinear Analysis

Simeon Reich
Department of Mathematics
Technion-Israel Institute of Technology
Haifa, Israel

Alexander J. Zaslavski
Department of Mathematics
Technion-Israel Institute of Technology
Haifa, Israel

ISSN 1389-2177 ISSN 2197-795X (electronic)
Developments in Mathematics
ISBN 978-1-4939-4858-1 ISBN 978-1-4614-9533-8 (eBook)
DOI 10.1007/978-1-4614-9533-8
Springer New York Heidelberg Dordrecht London

Mathematics Subject Classification: 06-xx, 34-xx, 46-xx, 47-xx, 49-xx, 54-xx, 65-xx, 90-xx

Printed on acid-free paper

Springer is part of Springer Science+Business Media (www.springer.com)

Preface

In recent years it has become more and more evident that Nonlinear Functional Analysis is of crucial importance in the Mathematical Sciences. This is because functional analytic ideas and methods have turned out to be essential tools in the analysis of nonlinear phenomena in many areas of Mathematics and its applications. Among these areas one can mention Ordinary Differential Equations, Partial Differential Equations, the Geometry of Banach Spaces, Nonlinear Operator Theory, the Calculus of Variations, Optimal Control Theory, Optimization and Mathematical Economics.

One of the main features of the functional analytic approach is the investigation and solution of general classes of problems rather than of more specific individual ones. When one uses this approach, the following question arises:

We consider a class of problems which is identified with some functional space equipped with a natural complete metric. We know that for some elements of the functional space the corresponding problems possess a solution (or a solution with some desirable properties) and for some elements such solutions do not exist. We usually know some sufficient conditions for the existence of solutions, but often these conditions are difficult to verify or they hold for rather small subsets of the whole space. In such situations it is natural to ask if a solution (or a solution with some desirable properties) exists for most elements of the functional space in the sense of Baire category. This means that the functional space under consideration contains an everywhere dense G_δ subset such that for all its elements a solution exists.

It turns out that this generic approach is very useful and many interesting and important problems can be solved using it. The goal of our book is to demonstrate this. Although it is, of course, impossible to cover the whole spectrum of present-day trends in Nonlinear Analysis and its applications where the generic approach is used, we do present quite a few of the main topics which are of current research interest. They include fixed point theory of both single- and set-valued mappings, convergence analysis of infinite products, best approximation problems, discrete and continuous descent methods for minimization in a general Banach space, and the structure of minimal energy configurations with rational numbers in the Aubry-Mather theory.

Now we describe the structure of the book. We begin in Chap. 1 with the applications of the Baire theory to fixed point theory. A self-mapping of a complete metric space is called nonexpansive if it is Lipschitz with Lipschitz constant one. If the Lipschitz constant is less than one, then it is called a strict contraction. According to Banach's celebrated result, a strict contraction has a unique fixed point and all its iterates converge to it. It was unclear what happens when a mapping acting on a closed and convex subset of a general Banach space is just nonexpansive until the classical paper by De Blasi and Myjak of 1976 [49], where they show, using the Baire approach, that most mappings in the class of nonexpansive self-mappings of a bounded, closed and convex subset of a general Banach space possess a unique fixed point which attracts uniformly all their iterates. Note that they also show that the subclass of strict contractions is a small set in the whole class of nonexpansive mappings.

Chapter 2 is devoted to further generalizations, extensions and developments concerning this result of De Blasi and Myjak. Using the Baire approach, we establish existence and uniqueness of a fixed point for a generic mapping, convergence of iterates of a generic nonexpansive mapping, stability of the fixed point under small perturbations of a mapping, convergence of Krasnosel'skii-Mann iterations of nonexpansive mappings, generic power convergence of order preserving mappings, and existence and uniqueness of positive eigenvalues and eigenvectors of order-preserving linear operators. In this chapter we also study convergence of iterates of nonexpansive mappings in the presence of computational errors.

Chapter 3 is devoted to an important subclass of the class of nonexpansive mappings which consists of the so-called contractive mappings. A contractive mapping is obtained if in the definition of a strict contraction the constant is replaced by a monotonically decreasing function with nonnegative values which do not exceed one and which is a function of the distance between two points. This topic has recently become rather popular. In Chap. 3 we study different types of contractive mappings, existence of fixed points for such mappings, convergence of their powers to a fixed point, stability of a fixed point under small perturbations of the mapping, and use the Baire approach to show that most nonexpansive mappings are contractive.

In Chap. 4 we use the generic approach in order to study the asymptotic behavior of trajectories of a certain dynamical system which originates in a convex minimization problem. Usually, an algorithm for the minimization of an objective function on a set can be considered a self-mapping of the set for which the objective function is a Lyapunov function. In our case the set is a closed subset of a Banach space. The results presented in this chapter show that for most algorithms, the values of the objective function along all the trajectories tend to its infimum.

In Chap. 5 we generalize some of the results of Chap. 2 for mappings which are relatively nonexpansive with respect to Bregman distances. Such mappings appear in optimization theory and in studies of feasibility problems [37, 39].

Chapter 6 is devoted to the study of convergence of infinite products of different classes of mappings. The convergence of infinite products of nonexpansive mappings is of major importance because of their many applications in the study of feasibility and optimization problems. We study the convergence of typical (generic) infinite products of mappings to the set of their common fixed points, and establish weak ergodic theorems (a term which originates in population biology), which roughly mean that all trajectories generated by infinite products converge to each other. We study convergence and its stability for generic infinite products of nonexpansive mappings, uniformly continuous mappings, order-preserving mappings, order-preserving linear mappings, homogeneous order-preserving mappings, products of affine mappings, as well as products of resolvents of accretive operators.

In Chap. 7 we study best approximation problems in a general Banach space. A best approximation problem is determined by a pair consisting of a point and a closed (convex) subset of a Banach space. We consider the complete metric space of such pairs equipped with a natural complete metric and show that for most (in the sense of Baire category) pairs the corresponding best approximation problem has a unique solution. We also provide some generalizations and extensions of this result.

In Chap. 8 we study discrete and continuous descent methods for minimizing a convex (Lipschitz) function on a general Banach space. We consider a space of vector fields V such that for any point x in the Banach space, the directional derivative in the direction Vx is nonpositive. This space of vector fields is equipped with a complete metric. Each vector field generates two gradient type algorithms (discrete descent methods) and a flow which consists of the solutions of the corresponding evolution equation (continuous descent method). We show that most (in the sense of Baire category) vector fields produce algorithms for which values of the objective function tend to its infimum as t tends to infinity. Actually, we introduce the subclass of regular vector fields, show that the convergence property stated above holds for them and that a generic vector field is regular. We also show that this convergence property is stable under small perturbations of a given regular vector field.

Chapter 9 is devoted to set-valued mappings. We study approximate fixed points of such mappings, existence of fixed points, and the convergence and stability of iterates of set-valued mappings.

Chapter 10 is devoted to the Aubry-Mather theory applied to the famous Frenkel-Kontorova model, an infinite discrete model of solid-state physics related to dislocations in one-dimensional crystals. In this model a configuration of a system is a sequence of real numbers with indices from $-\infty$ to $+\infty$. We are interested in (h)-minimal configurations with respect to an energy function h. A configuration is called (h)-minimal if its total energy cannot be made less by changing its final states. Classical Aubry-Mather theory is concerned with finding and investigating h-minimal configurations with a given rotation number, where the function h is fixed. It implies that the set of all periodic h-minimal configurations of a rational rotation number p/q is totally ordered. Moreover, between any two neighboring periodic h-minimal configurations with rotation number p/q, there are (non-periodic) h-minimal heteroclinic connections having the same rotation number p/q. We consider a complete metric space of energy functions h equipped with a certain C^2

topology and show that for most energy functions in this space, there exist three different h-minimal configurations with rotation number p/q such that any other h-minimal configuration with the same rotation number p/q is a translation of one of these three.

Haifa
December 31, 2012

Simeon Reich
Alexander J. Zaslavski

Contents

Chapter 1
Introduction

Let X be a complete metric space. According to Baire's theorem, the intersection of every countable collection of open dense subsets of X is dense in X. This rather simple, yet powerful result has found many applications. In particular, given a property which elements of X may have, it is of interest to determine whether this property is generic, that is, whether the set of elements which do enjoy this property contains a countable intersection of open dense sets. Such an approach, when a certain property is investigated for the whole space X and not just for a single point in X, has already been successfully applied in many areas of Analysis. In this chapter we discuss several recent results in metric fixed point theory which exhibit these generic phenomena.

1.1 Hyperbolic Spaces

It turns out that the class of hyperbolic spaces is a natural setting for our generic results. In this section we briefly review this concept.

Let (X, ρ) be a metric space and let R^1 denote the real line. We say that a mapping $c : R^1 \to X$ is a metric embedding of R^1 into X if

$$\rho\big(c(s), c(t)\big) = |s - t|$$

for all real s and t. The image of R^1 under a metric embedding will be called a metric line. The image of a real interval $[a, b] = \{t \in R^1 : a \le t \le b\}$ under such a mapping will be called a metric segment.

Assume that (X, ρ) contains a family M of metric lines such that for each pair of distinct points x and y in X, there is a unique metric line in M which passes through x and y. This metric line determines a unique metric segment joining x and y. We denote this segment by $[x, y]$. For each $0 \le t \le 1$, there is a unique point z in $[x, y]$ such that

$$\rho(x, z) = t\rho(x, y) \quad \text{and} \quad \rho(z, y) = (1 - t)\rho(x, y).$$

This point will be denoted by $(1 - t)x \oplus ty$.

S. Reich, A.J. Zaslavski, *Genericity in Nonlinear Analysis*,
Developments in Mathematics 34, DOI 10.1007/978-1-4614-9533-8_1,

We will say that X, or more precisely (X, ρ, M), is a hyperbolic space if

$$\rho\left(\frac{1}{2}x \oplus \frac{1}{2}y, \frac{1}{2}x \oplus \frac{1}{2}z\right) \le \frac{1}{2}\rho(y, z)$$

for all x, y and z in X.

An equivalent requirement is that

$$\rho\left(\frac{1}{2}x \oplus \frac{1}{2}y, \frac{1}{2}w \oplus \frac{1}{2}z\right) \le \frac{1}{2}\big(\rho(x, w) + \rho(y, z)\big)$$

for all x, y, z and w in X. A set $K \subset X$ is called ρ-convex if $[x, y] \subset K$ for all x and y in K.

It is clear that all normed linear spaces are hyperbolic. A discussion of more examples of hyperbolic spaces and in, particular, of the Hilbert ball can be found, for instance, in [66, 68, 81, 124].

In the sequel we will repeatedly use the following fact (cf. pp. 77 and 104 of [68] and [124]): If (X, ρ, M) is a hyperbolic space, then

$$\rho\big((1-t)x \oplus tz, (1-t)y \oplus tw\big) \le (1-t)\rho(x, y) + t\rho(z, w) \tag{1.1}$$

for all x, y, z and w in X and $0 \le t \le 1$.

1.2 Successive Approximations

Let (X, ρ, M) be a complete hyperbolic space and let K be a closed ρ-convex subset of X. Denote by $\mathcal{A}$ the set of all operators $A : K \to K$ such that

$$\rho(Ax, Ay) \le \rho(x, y) \quad \text{for all } x, y \in K.$$

In other words, the set $\mathcal{A}$ consists of all the nonexpansive self-mappings of K.

Fix some $\theta \in K$ and for each $s > 0$, set

$$B(\theta, s) = B(s) = \{x \in K : \rho(x, \theta) \le s\}.$$

For the set $\mathcal{A}$ we consider the uniformity determined by the following base:

$$E(n, \varepsilon) = \{(A, B) \in \mathcal{A} \times \mathcal{A} : \rho(Ax, Bx) \le \varepsilon, x \in B(n)\},$$

where $\varepsilon > 0$ and n is a natural number. Clearly the space $\mathcal{A}$ with this uniformity is metrizable and complete. We equip the space $\mathcal{A}$ with the topology induced by this uniformity.

A mapping $A : K \to K$ is called regular if there exists a necessarily unique $x_A \in K$ such that

$$\lim_{n\to\infty} A^n x = x_A \quad \text{for all } x \in K.$$

A mapping $A : K \to K$ is called super-regular if there exists a necessarily unique $x_A \in K$ such that for each $s > 0$,

$$A^n x \to x_A \quad \text{as } n \to \infty, \text{ uniformly on } B(s).$$

Denote by I the identity operator. For each pair of operators $A, B : K \to K$ and each $t \in [0, 1]$, define an operator $tA \oplus (1-t)B$ by

$$\big(tA \oplus (1-t)B\big)(x) = tAx \oplus (1-t)Bx, \quad x \in K.$$

Note that if A and B belong to $\mathcal{A}$, then so does $tA \oplus (1-t)B$.

In Chap. 2 we establish generic existence and uniqueness of a fixed point for a generic mapping, convergence of iterates of a generic nonexpansive mapping, stability of the fixed point under small perturbations of a mapping and many other results. Among these results are the following two theorems obtained in [132].

The first result shows that in addition to (locally uniform) power convergence, super-regular mappings also provide stability, while the second result shows that most mappings in $\mathcal{A}$ are, in fact, super-regular. This is an improvement of the classical result of De Blasi and Myjak [49] who established power convergence (to a unique fixed point) for a generic nonexpansive self-mapping of a bounded closed convex subset of a Banach space.

Theorem 1.1 *Let $A : K \to K$ be super-regular and let ε, s be positive numbers. Then there exist a neighborhood U of A in $\mathcal{A}$ and an integer $n_0 \geq 2$ such that for each $B \in U$, each $x \in B(s)$ and each integer $n \geq n_0$, we have $\rho(x_A, B^n x) \leq \varepsilon$.*

Theorem 1.2 *There exists a set $\mathcal{F}_0 \subset \mathcal{A}$ which is a countable intersection of open everywhere dense sets in $\mathcal{A}$ such that each $A \in \mathcal{F}_0$ is super-regular.*

1.3 Contractive Mappings

In Chap. 3 we consider the class of contractive mappings which we now define.

Let K be a bounded, closed and convex subset of a Banach space $(X, \|\cdot\|)$.

Denote by $\mathcal{A}$ the set of all operators $A : K \to K$ such that

$$\|Ax - Ay\| \leq \|x - y\| \quad \text{for all } x, y \in K.$$

Set

$$d(K) = \sup\big\{\|x - y\| : x, y \in K\big\}.$$

We equip the set $\mathcal{A}$ with the metric $h(\cdot, \cdot)$ defined by

$$h(A, B) = \sup\big\{\|Ax - Bx\| : x \in K\big\}, \quad A, B \in \mathcal{A}.$$

Clearly, the metric space $(\mathcal{A}, h)$ is complete.

We say that a mapping $A \in \mathcal{A}$ is contractive if there exists a decreasing function $\phi^A : [0, d(K)] \to [0, 1]$ such that

$$\phi^A(t) < 1 \quad \text{for all } t \in \big(0, d(K)\big]$$

and

$$\|Ax - Ay\| \leq \phi^A\big(\|x - y\|\big)\|x - y\| \quad \text{for all } x, y \in K.$$

The notion of a contractive mapping, as well as its modifications and applications, were studied by many authors. See, for example, [114, 116] and the references mentioned there. We now quote a convergence result which is valid in all complete metric spaces [114].

Theorem 1.3 *Assume that $A \in \mathcal{A}$ is contractive. Then there exists a unique $x_A \in K$ such that $A^n x \to x_A$ as $n \to \infty$, uniformly on K.*

In Chap. 3 we show that most of the mappings in $\mathcal{A}$ (in the sense of Baire's categories) are, in fact, contractive and prove the following result obtained in [131].

Theorem 1.4 *There exists a set $\mathcal{F}$ which is a countable intersection of open everywhere dense sets in $\mathcal{A}$ such that each $A \in \mathcal{F}$ is contractive.*

Note that at least in Hilbert space the set of strict contractions is only of the first Baire category in $\mathcal{A}$ [13, 49].

In Chap. 3 we continue with a discussion of nonexpansive mappings which are contractive with respect to a given subset of their domain. We now define this class of mappings.

Let K be a closed (not necessarily bounded) ρ-convex subset of the complete hyperbolic space (X, ρ, M). Denote by $\mathcal{A}$ the set of all nonexpansive self-mappings of K.

For each $x \in K$ and each subset $E \subset K$, let $\rho(x, E) = \inf\{\rho(x, y) : y \in E\}$. For each $x \in K$ and each $r > 0$, set

$$B(x,r) = \{y \in K : \rho(x, y) \le r\}.$$

Fix $\theta \in K$. We equip the set $\mathcal{A}$ with the same uniformity and topology as in the previous section.

Let F be a nonempty, closed and ρ-convex subset of K. Denote by $\mathcal{A}^{(F)}$ the set of all $A \in \mathcal{A}$ such that $Ax = x$ for all $x \in F$. Clearly, $\mathcal{A}^{(F)}$ is a closed subset of $\mathcal{A}$. We consider the topological subspace $\mathcal{A}^{(F)} \subset \mathcal{A}$ with the relative topology.

An operator $A \in \mathcal{A}^{(F)}$ is said to be contractive with respect to F if for any natural number n, there exists a decreasing function $\phi_n^A : [0, \infty) \to [0, 1]$ such that

$$\phi_n^A(t) < 1 \quad \text{for all } t > 0$$

and

$$\rho(Ax, F) \le \phi_n^A(\rho(x, F))\rho(x, F) \quad \text{for all } x \in B(\theta, n).$$

Clearly, this definition does not depend on our choice of $\theta \in K$.

The following result, which was obtained in [131], shows that the iterates of an operator in $\mathcal{A}^{(F)}$ converge to a retraction of K onto F.

Theorem 1.5 *Let $A \in \mathcal{A}^{(F)}$ be contractive with respect to F. Then there exists $B \in \mathcal{A}^{(F)}$ such that $B(K) = F$ and $A^n x \to Bx$ as $n \to \infty$, uniformly on $B(\theta, m)$ for any natural number m.*

Finally, we present the following theorem of [131] which shows that if $\mathcal{A}^{(F)}$ contains a retraction, then almost all the mappings in $\mathcal{A}^{(F)}$ are contractive with respect to F.

Theorem 1.6 *Assume that there exists*

$$Q \in \mathcal{A}^{(F)} \quad \textit{such that} \quad Q(K) = F.$$

Then there exists a set $\mathcal{F} \subset \mathcal{A}^{(F)}$ which is a countable intersection of open everywhere dense sets in $\mathcal{A}^{(F)}$ such that each $B \in \mathcal{F}$ is contractive with respect to F.

1.4 Infinite Products

In Chap. 6 we present several results concerning the asymptotic behavior of (random) infinite products of generic sequences of nonexpansive, as well as uniformly continuous, operators on closed and convex subsets of a complete hyperbolic space.

Let $(X, \|\cdot\|)$ be a Banach space and let K be a nonempty, bounded, closed and convex subset of X with the topology induced by the norm $\|\cdot\|$.

Denote by $\mathcal{A}$ the set of all sequences $\{A_t\}_{t=1}^{\infty}$, where each $A_t : K \to K$ is a continuous operator, $t = 1, 2, \ldots$. Such a sequence will occasionally be denoted by a boldface $\mathbf{A}$.

For the set $\mathcal{A}$ we consider the metric $\rho_s : \mathcal{A} \times \mathcal{A} \to [0, \infty)$ defined by

$$\rho_s\big(\{A_t\}_{t=1}^{\infty}, \{B_t\}_{t=1}^{\infty}\big) = \sup\big\{\|A_t x - B_t x\| : x \in K, t = 1, 2, \ldots\big\},$$
$$\{A_t\}_{t=1}^{\infty}, \{B_t\}_{t=1}^{\infty} \in \mathcal{A}.$$

It is easy to see that the metric space $(\mathcal{A}, \rho_s)$ is complete. The topology generated in $\mathcal{A}$ by the metric ρ_s will be called the strong topology.

In addition to this topology on $\mathcal{A}$, we will also consider the uniformity determined by the base

$$E(N, \varepsilon) = \big\{\big(\{A_t\}_{t=1}^{\infty}, \{B_t\}_{t=1}^{\infty}\big) \in \mathcal{A} \times \mathcal{A} :$$
$$\|A_t x - B_t x\| \le \varepsilon, t = 1, \ldots, N, x \in K\big\},$$

where N is a natural number and $\varepsilon > 0$. It is easy to see that the space $\mathcal{A}$ with this uniformity is metrizable (by a metric $\rho_w : \mathcal{A} \times \mathcal{A} \to [0, \infty)$) and complete. The topology generated by ρ_w will be called the weak topology.

Define

$$\mathcal{A}_{ne} = \big\{\{A_t\}_{t=1}^{\infty} \in \mathcal{A} : A_t \text{ is nonexpansive for } t = 1, 2, \ldots\big\}.$$

Clearly, $\mathcal{A}_{ne}$ is a closed subset of $\mathcal{A}$ in the weak topology. We will consider the topological subspace $\mathcal{A}_{ne} \subset \mathcal{A}$ with both the weak and strong relative topologies.

In Theorem 2.1 of [129] we showed that for a generic sequence $\{C_t\}_{t=1}^{\infty}$ in the space $\mathcal{A}_{ne}$ with the weak topology,

$$\|C_T \cdot \cdots \cdot C_1 x - C_T \cdot \cdots \cdot C_1 y\| \to 0 \quad \text{as } T \to \infty,$$

uniformly for all $x, y \in K$. (Such results are usually called weak ergodic theorems in the population biology literature; see [43, 107].)

Here is the precise formulation of this weak ergodic theorem.

Theorem 1.7 *There exists a set $\mathcal{F} \subset \mathcal{A}_{ne}$, which is a countable intersection of open (in the weak topology) everywhere dense (in the strong topology) subsets of $\mathcal{A}_{ne}$, such that for each $\{B_t\}_{t=1}^{\infty} \in \mathcal{F}$ and each $\varepsilon > 0$, there exist a neighborhood U of $\{B_t\}_{t=1}^{\infty}$ in $\mathcal{A}_{ne}$ with the weak topology and a natural number N such that:*

For each $\{C_t\}_{t=1}^{\infty} \in U$, each $x, y \in K$, and each integer $T \geq N$,

$$\|C_T \cdot \cdots \cdot C_1 x - C_T \cdot \cdots \cdot C_1 y\| \leq \varepsilon.$$

Note that in Chap. 6 we also prove a random version of this theorem.

We will say that a set E of operators $A : K \to K$ is uniformly equicontinuous (ue) if for any $\varepsilon > 0$, there exists $\delta > 0$ such that $\|Ax - Ay\| \leq \varepsilon$ for all $A \in E$ and all $x, y \in K$ satisfying $\|x - y\| \leq \delta$.

Define

$$\mathcal{A}_{ue} = \big\{\{A_t\}_{t=1}^{\infty} \in \mathcal{A} : \{A_t\}_{t=1}^{\infty} \text{ is a (ue) set}\big\}.$$

Clearly, $\mathcal{A}_{ue}$ is a closed subset of $\mathcal{A}$ in the strong topology.

We will consider the topological subspace $\mathcal{A}_{ue} \subset \mathcal{A}$ with both the weak and strong relative topologies.

Denote by $\mathcal{A}_{ne}^*$ the set of all $\{A_t\}_{t=1}^{\infty} \in \mathcal{A}_{ne}$ which have a common fixed point and denote by $\bar{\mathcal{A}}_{ne}^*$ the closure of $\mathcal{A}_{ne}^*$ in the strong topology of the space $\mathcal{A}_{ne}$.

Let $\mathcal{A}_{ue}^*$ be the set of all $\mathbf{A} = \{A_t\}_{t=1}^{\infty} \in \mathcal{A}_{ue}$ for which there exists $x(\mathbf{A}) \in K$ such that for each integer $t \geq 1$,

$$A_t x(\mathbf{A}) = x(\mathbf{A}) \quad \text{and} \quad \big\|A_t y - x(\mathbf{A})\big\| \leq \big\|y - x(\mathbf{A})\big\| \quad \text{for all } y \in K,$$

and denote by $\bar{\mathcal{A}}_{ue}^*$ the closure of $\mathcal{A}_{ue}^*$ in the strong topology of the space $\mathcal{A}_{ue}$.

We consider the topological subspaces $\bar{\mathcal{A}}_{ne}^*$ and $\bar{\mathcal{A}}_{ue}^*$ with the relative strong topologies. In Theorem 2.4 of [129] we showed that a generic sequence $\{C_t\}_{t=1}^{\infty}$ in the space $\bar{\mathcal{A}}_{ue}^*$ has a unique common fixed point x_* and all random products of the operators $\{C_t\}_{t=1}^{\infty}$ converge to x_*, uniformly for all $x \in K$. We now quote this theorem.

Theorem 1.8 *There exists a set $\mathcal{F} \subset \bar{\mathcal{A}}_{ue}^*$, which is a countable intersection of open everywhere dense (in the strong topology) subsets of $\bar{\mathcal{A}}_{ue}^*$, such that for each $\{B_t\}_{t=1}^{\infty} \in \mathcal{F}$, there exists $x_* \in K$ for which the following assertions hold:*

1. *$B_t x_* = x_*$, $t = 1, 2, \ldots$, and*

$$\|B_t y - x_*\| \leq \|y - x_*\|, \quad y \in K, t = 1, 2, \ldots.$$

2. *For each $\varepsilon > 0$, there exist a neighborhood U of $\{B_t\}_{t=1}^{\infty}$ in $\bar{\mathcal{A}}_{ue}^*$ with the strong topology and a natural number N such that for each $\{C_t\}_{t=1}^{\infty} \in U$, each integer $T \geq N$, each mapping $r : \{1, \ldots, T\} \to \{1, 2, \ldots\}$, and each $x \in K$,*

$$\|C_{r(T)} \cdot \cdots \cdot C_{r(1)} x - x_*\| \leq \varepsilon.$$

In [129] we also proved an analog of this theorem for the space $\bar{\mathcal{A}}^*_{ne}$.

We remark in passing that one can easily construct an example of a sequence of operators $\{A_t\}_{t=1}^{\infty} \in \mathcal{A}^*_{ue}$ for which the convergence properties described in the previous theorem do not hold. Namely, they do not hold for the sequence each term of which is the identity operator.

Now assume that F is a nonempty, closed and convex subset of K and that $Q : K \to F$ is a nonexpansive operator such that

$$Qx = x, \quad x \in F.$$

Such an operator Q is usually called a nonexpansive retraction of K onto F (see [68]). Denote by $\mathcal{A}_{ne}^{(F)}$ the set of all $\{A_t\}_{t=1}^{\infty} \in \mathcal{A}_{ne}$ such that

$$A_t x = x, \quad x \in F, t = 1, 2, \ldots.$$

Clearly, $\mathcal{A}_{ne}^{(F)}$ is a closed subset of $\mathcal{A}_{ne}$ in the weak topology. We equip the topological subspace $\mathcal{A}_{ne}^{(F)} \subset \mathcal{A}_{ne}$ with both the weak and strong relative topologies.

In Theorem 3.1 of [129] we showed that for a generic sequence of operators $\{B_t\}_{t=1}^{\infty}$ in the space $\mathcal{A}_{ne}^{(F)}$ with the weak topology there exists a nonexpansive retraction $P_* : K \to F$ such that

$$B_t \cdot \cdots \cdot B_1 x \to P_* x \quad \text{as } t \to \infty,$$

uniformly for all $x \in K$. We end this section with the precise statement of this convergence theorem.

Theorem 1.9 *There exists a set $\mathcal{F} \subset \mathcal{A}_{ne}^{(F)}$, which is a countable intersection of open (in the weak topology) everywhere dense (in the strong topology) subsets of $\mathcal{A}_{ne}^{(F)}$, such that for each $\{B_t\}_{t=1}^{\infty} \in \mathcal{F}$, the following assertions hold:*

1. *There exists an operator $P_* : K \to F$ such that*

$$\lim_{t\to\infty} B_t \cdot \cdots \cdot B_1 x = P_* x \quad \text{for each } x \in K.$$

2. *For each $\varepsilon > 0$, there exist a neighborhood U of $\{B_t\}_{t=1}^{\infty}$ in $\mathcal{A}_{ne}^{(F)}$ with the weak topology and a natural number N such that for each $\{C_t\}_{t=1}^{\infty} \in U$, each integer $T \geq N$, and each $x \in K$,*

$$\|C_T \cdot \cdots \cdot C_1 x - P_* x\| \leq \varepsilon.$$

Theorem 3.2 of [129] is a random version of this theorem.

1.5 Contractive Set-Valued Mappings

In Chap. 9 we study contractive set-valued mappings.

Assume that $(X, \|\cdot\|)$ is a Banach space, K is a nonempty, bounded and closed subset of X and there exists $\theta \in K$ such that for each $x \in K$,

$$tx + (1-t)\theta \in K, \quad t \in (0, 1).$$

We consider the complete metric space K with the metric $\|x - y\|$, $x, y \in K$. Denote by $S(K)$ the set of all nonempty closed subsets of K. For $x \in K$ and $D \subset K$, set

$$\rho(x, D) = \inf\{\|x - y\| : y \in D\},$$

and for each $C, D \in S(K)$, let

$$H(C, D) = \max\left\{\sup_{x \in C} \rho(x, D), \sup_{y \in D} \rho(y, C)\right\}.$$

We equip the set $S(K)$ with the Hausdorff metric $H(\cdot, \cdot)$. It is well known that the metric space $(S(K), H)$ is complete.

Denote by $\mathcal{A}$ the set of all nonexpansive operators $T : S(K) \to S(K)$. For the set $\mathcal{A}$ we consider the metric $\rho_{\mathcal{A}}$ defined by

$$\rho_{\mathcal{A}}(T_1, T_2) = \sup\{H(T_1(D), T_2(D)) : D \in S(K)\}, \quad T_1, T_2 \in \mathcal{A}.$$

Denote by $\mathcal{N}$ the set of all mappings $T : K \to S(K)$ such that

$$H(T(x), T(y)) \le \|x - y\|, \quad x, y \in K.$$

Set

$$d(K) = \sup\{\|x - y\| : x, y \in K\}.$$

A mapping $T \in \mathcal{N}$ is called contractive if there exists a decreasing function ϕ : $[0, d(K)] \to [0, 1]$ such that

$$\phi(t) < 1 \quad \text{for all } t \in (0, d(K)]$$

and

$$H(T(x), T(y)) \le \phi(\|x - y\|)\|x - y\| \quad \text{for all } x, y \in K.$$

Assume that $T \in \mathcal{N}$. For each $D \in S(K)$, denote by $\tilde{T}(D)$ the closure of the set $\bigcup\{T(x) : x \in D\}$ in the norm topology.

It was shown in [144] that for any $T \in \mathcal{N}$, the mapping $\tilde{T}$ belongs to $\mathcal{A}$ and moreover, the mapping $\tilde{T}$ is contractive if and only if the mapping T is contractive.

We equip the set $\mathcal{N}$ with the metric $\rho_{\mathcal{N}}$ defined by

$$\rho_{\mathcal{N}}(T_1, T_2) = \sup\{H(T_1(x), T_2(x)) : x \in K\}, \quad T_1, T_2 \in \mathcal{N}.$$

It is not difficult to verify that the metric space $(\mathcal{N}, \rho_{\mathcal{N}})$ is complete.

For each $T \in \mathcal{N}$ set $P(T) = \tilde{T}$. It is easy to see that for each $T_1, T_2 \in \mathcal{N}$,

$$\rho_{\mathcal{A}}(P(T_1), P(T_2)) = \rho_{\mathcal{N}}(T_1, T_2).$$

Denote

$$\mathcal{B} = \{P(T) : T \in \mathcal{N}\}.$$

Clearly, the metric spaces $(\mathcal{B}, \rho_{\mathcal{A}})$ and $(\mathcal{N}, \rho_{\mathcal{N}})$ are isometric.

In [144] we obtained the following results.

Theorem 1.10 *Assume that the operator $T \in \mathcal{N}$ is contractive. Then there exists a unique set $A_T \in S(K)$ such that $\tilde{T}(A_T) = A_T$ and $(\tilde{T})^n(B) \to A_T$ as $n \to \infty$, uniformly for all $B \in S(K)$.*

Theorem 1.11 *There exists a set $\mathcal{F}$, which is a countable intersection of open and everywhere dense subsets of $(\mathcal{N}, \rho_{\mathcal{N}})$, such that each $T \in \mathcal{F}$ is contractive.*

1.6 Nonexpansive Set-Valued Mappings

Let $(X, \|\cdot\|)$ be a Banach space and denote by $S_{co}(X)$ the set of all nonempty, closed and convex subsets of X. For $x \in X$ and $D \subset X$, set

$$\rho(x, D) = \inf\{\|x - y\| : y \in D\},$$

and for each $C, D \in S_{co}(X)$, let

$$H(C, D) = \max\Big\{\sup_{x \in C} \rho(x, D), \sup_{y \in D} \rho(y, C)\Big\}.$$

The interior of a subset $D \subset X$ will be denoted by $\text{int}(D)$. For each $x \in X$ and each $r > 0$, set $B(x, r) = \{y \in X : \|y - x\| \le r\}$. For the set $S_{co}(X)$ we consider the uniformity determined by the following base:

$$\mathcal{G}(n) = \{(C, D) \in S_{co}(X) \times S_{co}(X) : H(C, D) \le n^{-1}\},$$

$n = 1, 2, \ldots$. It is well known that the space $S_{co}(X)$ with this uniformity is metrizable and complete. We endow the set $S_{co}(X)$ with the topology induced by this uniformity.

Assume now that K is a nonempty, closed and convex subset of X, and denote by $S_{co}(K)$ the set of all $D \in S_{co}(X)$ such that $D \subset K$. Clearly, $S_{co}(K)$ is a closed subset of $S_{co}(X)$. We equip the topological subspace $S_{co}(K) \subset S_{co}(X)$ with its relative topology.

Denote by $\mathcal{N}_{co}$ the set of all mappings $T : K \to S_{co}(K)$ such that $T(x)$ is bounded for all $x \in K$ and

$$H(T(x), T(y)) \le \|x - y\|, \quad x, y \in K.$$

In other words, the set $\mathcal{N}_{co}$ consists of those nonexpansive set-valued self-mappings of K which have nonempty, bounded, closed and convex point images.

Fix $\theta \in K$. For the set $\mathcal{N}_{co}$ we consider the uniformity determined by the following base:

$$\mathcal{E}(n) = \{(T_1, T_2) \in \mathcal{N}_{co} \times \mathcal{N}_{co} : H(T_1(x), T_2(x)) \le n^{-1} \text{ for all } x \in K \text{ satisfying } \|x - \theta\| \le n\}, \quad n = 1, 2, \ldots.$$

It is not difficult to verify that the space $\mathcal{N}_{co}$ with this uniformity is metrizable and complete.

The following result is well known [45, 102]; see also [116].

Theorem 1.12 *Assume that $T : K \to S(K)$, $\gamma \in (0, 1)$, and*

$$H(T(x), T(y)) \le \gamma \|x - y\|, \quad x, y \in K.$$

Then there exists $x_T \in K$ such that $x_T \in T(x_T)$.

The existence of fixed points for set-valued mappings which are merely nonexpansive is more delicate and was studied by several authors. See, for example, [67, 94, 119] and the references therein. We now state a result established in [145] which shows that if int(K) is nonempty, then a generic nonexpansive mapping does have a fixed point. This result will be proved in Chap. 9.

Theorem 1.13 *Assume that* int$(K) \neq \emptyset$. *Then there exists an open everywhere dense set* $\mathcal{F} \subset \mathcal{N}_{co}$ *with the following property: for each* $\widehat{S} \in \mathcal{F}$, *there exist* $\bar{x} \in K$ *and a neighborhood* $\mathcal{U}$ *of* $\widehat{S}$ *in* $\mathcal{N}_{co}$ *such that* $\bar{x} \in S(\bar{x})$ *for each* $S \in \mathcal{U}$.

1.7 Porosity

In this section we present a refinement of the classical result obtained by De Blasi and Myjak [49]. This refinement involves the notion of porosity which we now recall [51, 123, 180, 182].

Let (Y, d) be a complete metric space. We denote by $B(y, r)$ the closed ball of center $y \in Y$ and radius $r > 0$. A subset $E \subset Y$ is called porous (with respect to the metric d) if there exist $\alpha \in (0, 1)$ and $r_0 > 0$ such that for each $r \in (0, r_0]$ and each $y \in Y$, there exists $z \in Y$ for which

$$B(z, \alpha r) \subset B(y, r) \setminus E.$$

A subset of the space Y is called σ-porous (with respect to d) if it is a countable union of porous subsets of Y.

Remark 1.14 It is known that in the above definition of porosity, the point y can be assumed to belong to E.

Since porous sets are nowhere dense, all σ-porous sets are of the first Baire category. If Y is a finite-dimensional Euclidean space, then σ-porous sets are of Lebesgue measure 0. In fact, the class of σ-porous sets in such a space is much smaller than the class of sets which have Lebesgue measure 0 and are of the Baire first category. Also, every Banach space contains a set of the first Baire category which is not σ-porous.

To point out the difference between porous and nowhere dense sets, note that if $E \subset Y$ is nowhere dense, $y \in Y$ and $r > 0$, then there is a point $z \in Y$ and a number $s > 0$ such that $B(z, s) \subset B(y, r) \setminus E$. If, however, E is also porous, then for small enough r we can choose $s = \alpha r$, where $\alpha \in (0, 1)$ is a constant which depends only on E.

Let (X, ρ, M) be a complete hyperbolic space and $K \subset X$ a nonempty, bounded, closed and ρ-convex set. Once again we denote by $\mathcal{A}$ the set of all nonexpansive self-mappings of K. For each $A, B \in \mathcal{A}$ we again define

$$h(A, B) = \sup\{\rho(Ax, Bx) : x \in K\}. \tag{1.2}$$

It is easy to verify that $(\mathcal{A}, h)$ is a complete metric space.

The following result was established in [142].

Theorem 1.15 *There exists a set $\mathcal{F} \subset \mathcal{A}$ such that the complement $\mathcal{A} \setminus \mathcal{F}$ is σ-porous in $(\mathcal{A}, h)$ and for each $A \in \mathcal{F}$ the following property holds:*

There exists a unique $x_A \in K$ for which $Ax_A = x_A$ and $A^n x \to x_A$ as $n \to \infty$, uniformly on K.

Proof Set

$$d(K) = \sup\{\rho(x, y) : x, y \in K\}. \tag{1.3}$$

Fix $\theta \in K$. For each integer $n \geq 1$, denote by $\mathcal{A}_n$ the set of all $A \in \mathcal{A}$ which have the following property:

(C1) There exists a natural number $p(A)$ such that

$$\rho\big(A^{p(A)}x, A^{p(A)}y\big) \leq 1/n \quad \text{for all } x, y \in K. \tag{1.4}$$

Let $n \geq 1$ be an integer. We will show that $\mathcal{A} \setminus \mathcal{A}_n$ is porous in $(\mathcal{A}, h)$. To this end, let

$$\alpha = \big(d(K) + 1\big)^{-1}(8n)^{-1}. \tag{1.5}$$

Assume that $A \in \mathcal{A}$ and $r \in (0, 1]$. Set

$$\gamma = 2^{-1}r\big(d(K) + 1\big)^{-1} \tag{1.6}$$

and define $A_\gamma \in \mathcal{A}$ by

$$A_\gamma x = (1 - \gamma)Ax \oplus \gamma\theta, \quad x \in K. \tag{1.7}$$

It is easy to see that

$$\rho(A_\gamma x, A_\gamma y) \leq (1 - \gamma)\rho(x, y), \quad x, y \in K, \tag{1.8}$$

and

$$h(A, A_\gamma) \leq \gamma d(K). \tag{1.9}$$

Choose a natural number p for which

$$p > r^{-1}\big(d(K) + 1\big)^2 4n + 1. \tag{1.10}$$

Let $B \in \mathcal{A}$ satisfy

$$h(A_\gamma, B) \leq \alpha r, \tag{1.11}$$

and let $x, y \in K$. We will show that $\rho(B^p x, B^p y) \leq 1/n$. (We use the convention that $C^0 = I$, the identity operator.)

Assume the contrary. Then for $i = 0, \ldots, p$,

$$\rho\big(B^i x, B^i y\big) > 1/n. \tag{1.12}$$

It follows from (1.11), (1.2), (1.8) and (1.12) that for $i = 0, \ldots, p - 1$,

$$\begin{aligned}
\rho\big(B^{i+1}x, B^{i+1}y\big) &\leq \rho\big(B^{i+1}x, A_\gamma B^i x\big) + \rho\big(A_\gamma B^i x, A_\gamma B^i y\big) + \rho\big(A_\gamma B^i y, B^{i+1}y\big) \\
&\leq \alpha r + \rho\big(A_\gamma B^i x, A_\gamma B^i y\big) + \alpha r \\
&\leq 2\alpha r + (1 - \gamma)\rho\big(B^i x, B^i y\big) \leq \rho\big(B^i x, B^i y\big) + 2\alpha r - \gamma/n
\end{aligned}$$

and

$$\rho(B^i x, B^i y) - \rho(B^{i+1}x, B^{i+1}y) \geq \gamma/n - 2\alpha r.$$

When combined with (1.3), (1.6) and (1.5), this latter inequality implies that

$$\begin{aligned} d(K) &\geq \rho(x, y) - \rho(B^p x, B^p y) \\ &= \sum_{i=0}^{p-1} [\rho(B^i x, B^i y) - \rho(B^{i+1}x, B^{i+1}y)] \geq p(\gamma/n - 2\alpha r) \\ &\geq p[r(d(K)+1)^{-1}(2n)^{-1} - 2r(d(K)+1)^{-1}(8n)^{-1}] \\ &\geq pr(d(K)+1)^{-1}(4n)^{-1} \end{aligned}$$

and

$$p \leq r^{-1} d(K)(d(K)+1)4n,$$

a contradiction (see (1.10)). Thus $\rho(B^p x, B^p y) \leq 1/n$ for all $x, y \in K$. This means that

$$\{B \in \mathcal{A} : h(A_\gamma, B) \leq \alpha r\} \subset \mathcal{A}_n. \tag{1.13}$$

It now follows from (1.9), (1.6) and (1.5) that

$$\begin{aligned} \{B \in \mathcal{A} : h(A_\gamma, B) \leq \alpha r\} &\subset \{B \in \mathcal{A} : h(A, B) \leq \alpha r + \gamma d(K)\} \\ &\subset \{B \in \mathcal{A} : h(A, B) \leq r\}. \end{aligned}$$

In view of (1.13) this inclusion implies that $\mathcal{A} \setminus \mathcal{A}_n$ is porous in $(\mathcal{A}, h)$. Define $\mathcal{F} = \bigcap_{n=1}^{\infty} \mathcal{A}_n$. Then $\mathcal{A} \setminus \mathcal{F}$ is σ-porous in $(\mathcal{A}, h)$.

Let $A \in \mathcal{F}$. It follows from property (C1) that for each integer $n \geq 1$, there exists a natural number s such that $\rho(A^i x, A^j y) \leq 1/n$ for all $x, y \in K$ and all integers $i, j \geq s$. Since n is an arbitrary natural number, we conclude that for each $x \in K$, $\{A^i x\}_{i=1}^{\infty}$ is a Cauchy sequence which converges to a point $x_* \in K$ satisfying $Ax_* = x_*$ and moreover, $A^i x \to x_*$ as $i \to \infty$, uniformly on K. This completes the proof of Theorem 1.15. □

1.8 Examples

Most of the results obtained in this book are generic existence theorems. Usually, we study a certain property for a class of problems which is identified with a complete metric space and it is shown that for a typical (generic) element of this space the corresponding problem has a unique solution. Of course, such results are of interest only if there is a problem which does not possess the desired property. It should be mentioned that such problems do exist. Let us consider, for instance, the space of mappings discussed in Sect. 1.2. By Theorem 1.2, a typical element of this space is super-regular. It is easy to see that the identity operator is not super-regular. If our metric space is a Banach space, then any translation is not super-regular. Of course both of these mappings are not contractive too. In the book we also consider other examples which are more interesting and complicated.

In Sect. 3.4 we construct a contractive mapping $A : [0, 1] \to [0, 1]$ such that none of its powers is a strict contraction. Section 3.5 contains an example of a mapping $A : [0, 1] \to [0, 1]$ such that

$$|Ax - Ay| \le |x - y| \quad \text{for all } x, y \in [0, 1],$$
$$A^n x \to 0 \quad \text{as } n \to \infty, \text{ uniformly on } [0, 1],$$

and for each integer $m \ge 0$, the power A^m is not contractive. In Sect. 3.6 we construct a nonexpansive mapping with nonuniformly convergent powers.

In Sect. 2.24 we construct an example of an operator T on a complete metric space such that all of its orbits converge to its unique fixed point and for any nonsummable sequence of errors and any initial point, there exists a divergent inexact orbit with a convergent subsequence. In Sect. 2.26 we construct an example of an operator T on a certain complete metric space X (a bounded, closed and convex subset of a Banach space) such that all of its orbits converge to its unique fixed point, and for any nonsummable sequence of errors and any initial point, there exists an inexact orbit which does not converge to any compact set.

Chapter 2
Fixed Point Results and Convergence of Powers of Operators

In this chapter we establish existence and uniqueness of a fixed point for a generic mapping, convergence of iterates of a generic nonexpansive mapping, stability of the fixed point under small perturbations of a mapping and many other results.

2.1 Convergence of Iterates for a Class of Nonlinear Mappings

Let K be a nonempty, bounded, closed and convex subset of a Banach space $(X, \|\cdot\|)$. We show that the iterates of a typical element (in the sense of Baire's categories) of a class of continuous self-mappings of K converge uniformly on K to the unique fixed point of this typical element.

We consider the topological subspace $K \subset X$ with the relative topology induced by the norm $\|\cdot\|$. Set

$$\operatorname{diam}(K) = \sup\{\|x - y\| : x, y \in K\}. \tag{2.1}$$

Denote by $\mathcal{A}$ the set of all continuous mappings $A : K \to K$ which have the following property:

(P1) For each $\varepsilon > 0$, there exists $x_\varepsilon \in K$ such that

$$\|Ax - x_\varepsilon\| \le \|x - x_\varepsilon\| + \varepsilon \quad \text{for all } x \in K. \tag{2.2}$$

For each $A, B \in \mathcal{A}$, set

$$d(A, B) = \sup\{\|Ax - Bx\| : x \in K\}. \tag{2.3}$$

Clearly, the metric space $(\mathcal{A}, d)$ is complete.

We are now ready to state and prove the following result [149].

Theorem 2.1 *There exists a set $\mathcal{F} \subset \mathcal{A}$ such that the complement $\mathcal{A} \setminus \mathcal{F}$ is σ-porous in $(\mathcal{A}, d)$ and each $A \in \mathcal{F}$ has the following properties:*

S. Reich, A.J. Zaslavski, *Genericity in Nonlinear Analysis*,
Developments in Mathematics 34, DOI 10.1007/978-1-4614-9533-8_2,

(i) *There exists a unique fixed point* $x_A \in K$ *such that*

$$A^n x \to x_A \quad \text{as } n \to \infty, \text{ uniformly for all } x \in K;$$

(ii)

$$\|Ax - x_A\| \le \|x - x_A\| \quad \text{for all } x \in K;$$

(iii) *For each* $\varepsilon > 0$, *there exist a natural number* n *and a real number* $\delta > 0$ *such that for each integer* $p \ge n$, *each* $x \in K$, *and each* $B \in \mathcal{A}$ *satisfying* $d(B, A) \le \delta$,

$$\left\| B^p x - x_A \right\| \le \varepsilon.$$

The following auxiliary result will be used in the proof of Theorem 2.1.

Proposition 2.2 *Let* $A \in \mathcal{A}$ *and* $\varepsilon \in (0, 1)$. *Then there exist* $\bar{x} \in K$ *and* $B \in \mathcal{A}$ *such that*

$$d(A, B) \le \varepsilon \tag{2.4}$$

and

$$\|\bar{x} - Bx\| \le \|\bar{x} - x\| \quad \text{for all } x \in K. \tag{2.5}$$

Proof Choose a positive number

$$\varepsilon_0 < 8^{-1}\varepsilon^2\left(\text{diam}(K) + 1\right)^{-1}. \tag{2.6}$$

Since $A \in \mathcal{A}$, there exists $\bar{x} \in K$ such that

$$\|Ax - \bar{x}\| \le \|x - \bar{x}\| + \varepsilon_0 \quad \text{for all } x \in K. \tag{2.7}$$

Let $x \in K$. There are three cases:

$$\|Ax - \bar{x}\| < \varepsilon; \tag{2.8}$$

$$\|Ax - \bar{x}\| \ge \varepsilon \quad \text{and} \quad \|Ax - \bar{x}\| < \|x - \bar{x}\|; \tag{2.9}$$

$$\|Ax - \bar{x}\| \ge \varepsilon \quad \text{and} \quad \|Ax - \bar{x}\| \ge \|x - \bar{x}\|. \tag{2.10}$$

First we consider case (2.8). There exists an open neighborhood V_x of x in K such that

$$\|Ay - \bar{x}\| < \varepsilon \quad \text{for all } y \in V_x. \tag{2.11}$$

Define $\psi_x : V_x \to K$ by

$$\psi_x(y) = \bar{x}, \quad y \in V_x. \tag{2.12}$$

Clearly, for all $y \in V_x$,

$$0 = \left\|\psi_x(y) - \bar{x}\right\| \le \|y - \bar{x}\| \quad \text{and} \quad \left\|Ay - \psi_x(y)\right\| = \|Ay - \bar{x}\| < \varepsilon. \tag{2.13}$$

Consider now case (2.9). Since A is continuous, there exists an open neighborhood V_x of x in K such that

$$\|Ay - \bar{x}\| < \|y - \bar{x}\| \quad \text{for all } y \in V_x. \tag{2.14}$$

In this case we define $\psi_x : V_x \to K$ by

$$\psi_x(y) = Ay, \quad y \in V_x. \tag{2.15}$$

Finally, we consider case (2.10). Inequalities (2.10), (2.6) and (2.7) imply that

$$\|x - \bar{x}\| \geq \|Ax - \bar{x}\| - \varepsilon_0 > (7/8)\varepsilon. \tag{2.16}$$

For each $\gamma \in [0, 1]$, set

$$z(\gamma) = \gamma Ax + (1 - \gamma)\bar{x}. \tag{2.17}$$

By (2.17), (2.10) and (2.16), we have

$$\|z(0) - \bar{x}\| = 0 \quad \text{and} \quad \|z(1) - \bar{x}\| = \|Ax - \bar{x}\| \geq \|x - \bar{x}\| > (7/8)\varepsilon. \tag{2.18}$$

By (2.6) and (2.18), there exists $\gamma_0 \in (0, 1)$ such that

$$\|z(\gamma_0) - \bar{x}\| = \|x - \bar{x}\| - \varepsilon_0. \tag{2.19}$$

It now follows from (2.17), (2.19) and (2.7) that

$$\begin{aligned}\gamma_0\big(\|x - \bar{x}\| + \varepsilon_0\big) \geq \gamma_0\|Ax - \bar{x}\| &= \|\gamma_0 Ax + (1 - \gamma_0)\bar{x} - \bar{x}\| \\ &= \|z(\gamma_0) - \bar{x}\| = \|x - \bar{x}\| - \varepsilon_0\end{aligned}$$

and

$$\begin{aligned}\gamma_0 &\geq \big(\|x - \bar{x}\| - \varepsilon_0\big)\big(\|x - \bar{x}\| + \varepsilon_0\big)^{-1} = 1 - 2\varepsilon_0\big(\|x - \bar{x}\| + \varepsilon_0\big)^{-1} \\ &\geq 1 - 2\varepsilon_0\|x - \bar{x}\|^{-1}.\end{aligned} \tag{2.20}$$

Inequalities (2.20) and (2.16) imply that

$$\gamma_0 \geq 1 - 2\varepsilon_0\big((7/8)\varepsilon\big)^{-1}. \tag{2.21}$$

By (2.17), (2.1), (2.21) and (2.6),

$$\begin{aligned}\|z(\gamma_0) - Ax\| &= \|\gamma_0 Ax + (1 - \gamma_0)\bar{x} - Ax\| \\ &= (1 - \gamma_0)\|Ax - \bar{x}\| \leq (1 - \gamma_0)\operatorname{diam}(K) \leq 16\varepsilon_0(7\varepsilon)^{-1}\operatorname{diam}(K) \\ &\leq 3\varepsilon_0\operatorname{diam}(K)\varepsilon^{-1} \leq (3/8)\varepsilon\end{aligned}$$

and

$$\|z(\gamma_0) - Ax\| \leq (3/8)\varepsilon. \tag{2.22}$$

Relations (2.19) and (2.22) imply that there exists an open neighborhood V_x of x in K such that for each $y \in V_x$,

$$\|z(\gamma_0) - Ay\| < \varepsilon \quad \text{and} \quad \|z(\gamma_0) - \bar{x}\| < \|y - \bar{x}\|. \tag{2.23}$$

Define $\psi_x : V_x \to K$ by

$$\psi_x(y) = z(\gamma_0), \quad y \in V_x. \tag{2.24}$$

It is not difficult to see that in all three cases we have defined an open neighborhood V_x of x in K and a continuous mapping $\psi_x : V_x \to K$ such that for each $y \in V_x$,

$$\|Ay - \psi_x(y)\| < \varepsilon \quad \text{and} \quad \|\bar{x} - \psi_x(y)\| \le \|y - \bar{x}\|. \tag{2.25}$$

Since the metric space K with the metric induced by the norm is paracompact, there exists a continuous locally finite partition of unity $\{\phi_i\}_{i\in I}$ on K subordinated to $\{V_x\}_{x\in K}$, where each $\phi_i : K \to [0, 1]$, $i \in I$, is a continuous function such that for each $y \in K$, there is a neighborhood U of y in K such that

$$U \cap \mathrm{supp}(\phi_i) \neq \emptyset$$

only for finite number of $i \in I$;

$$\sum_{i\in I} \phi_i(x) = 1, \quad x \in K;$$

and for each $i \in I$, there is $x_i \in K$ such that

$$\mathrm{supp}(\phi_i) \subset V_{x_i}. \tag{2.26}$$

Here $\mathrm{supp}(\phi)$ is the closure of the set $\{x \in K : \phi(x) \neq 0\}$. Define

$$Bz = \sum_{i\in I} \phi_i(z)\psi_{x_i}(z), \quad z \in K. \tag{2.27}$$

Clearly, $B : K \to K$ is well defined and continuous.

Let $z \in K$. There are a neighborhood U of z in K and $i_1, \dots, i_n \in I$ such that

$$U \cap \mathrm{supp}(\phi_i) = \emptyset \quad \text{for any } i \in I \setminus \{i_1, \dots, i_n\}. \tag{2.28}$$

We may assume without any loss of generality that

$$z \in \mathrm{supp}(\phi_{i_p}), \quad p = 1, \dots, n. \tag{2.29}$$

Then

$$\sum_{p=1}^{n} \phi_{i_p}(z) = 1 \quad \text{and} \quad Bz = \sum_{p=1}^{n} \phi_{i_p}(z)\psi_{x_{i_p}}(z). \tag{2.30}$$

Relations (2.26), (2.29) and (2.25) imply that for $p = 1, \dots, n$ and $z \in V_{x_{i_p}}$,

$$\left\|Az - \psi_{x_{i_p}}(z)\right\| < \varepsilon \quad \text{and} \quad \left\|\bar{x} - \psi_{x_{i_p}}(z)\right\| \le \|\bar{x} - z\|.$$

By the equation above and (2.30),

$$\|Bz - Az\| = \left\|\sum_{p=1}^{n} \phi_{i_p}(z)\psi_{x_{i_p}}(z) - Az\right\|$$

$$\le \sum_{p=1}^{n} \phi_{i_p}(z)\left\|\psi_{x_{i_p}}(z) - Az\right\| < \varepsilon,$$

$$\|\bar{x} - Bz\| = \left\|\bar{x} - \sum_{p=1}^{n} \phi_{i_p}(z)\psi_{x_{i_p}}(z)\right\|$$

$$\le \sum_{p=1}^{n} \phi_{i_p}(z)\left\|\bar{x} - \psi_{x_{i_p}}(z)\right\| \le \|\bar{x} - z\|,$$

and

$$\|Bz - Az\| < \varepsilon, \qquad \|\bar{x} - Bz\| \le \|\bar{x} - z\|.$$

Proposition 2.2 is proved. □

Proof of Theorem 2.1 For each $C \in \mathcal{A}$ and $x \in K$, set $C^0 x = x$. For each natural number n, denote by $\mathcal{F}_n$ the set of all $A \in \mathcal{A}$ which have the following property:

(P2) There exist $\bar{x}$, a natural number q, and a positive number $\delta > 0$ such that

$$\|\bar{x} - Ax\| \le \|\bar{x} - x\| + n^{-1} \quad \text{for all } x \in K,$$

and such that for each $B \in \mathcal{A}$ satisfying $d(B, A) \le \delta$, and each $x \in K$,

$$\left\|B^q x - \bar{x}\right\| \le n^{-1}.$$

Define

$$\mathcal{F} = \bigcap_{n=1}^{\infty} \mathcal{F}_n. \tag{2.31}$$

Lemma 2.3 *Let $A \in \mathcal{F}$. Then there exists a unique fixed point $x_A \in K$ of A such that*

(i) *$A^n x \to x_A$ as $n \to \infty$, uniformly on K;*
(ii) *$\|Ax - x_A\| \le \|x - x_A\|$ for all $x \in K$;*

(iii) *For each* $\varepsilon > 0$, *there exist a natural number* q *and* $\delta > 0$ *such that for each* $B \in \mathcal{A}$ *satisfying* $d(B, A) \le \delta$, *each* $x \in K$, *and each integer* $i \ge q$,

$$\left\| B^i x - x_A \right\| \le \varepsilon.$$

Proof Let n be a natural number. Since $A \in \mathcal{F} \subset \mathcal{F}_n$, it follows from property (P2) that there exist $x_n \in K$, an integer $q_n \ge 1$, and a number $\delta_n \ge 0$ such that

$$\|x_n - Ax\| \le \|x_n - x\| + n^{-1} \quad \text{for all } x \in K; \tag{2.32}$$

(P3) For each $B \in \mathcal{A}$ satisfying $d(B, A) \le \delta_n$, and each $x \in K$,

$$\left\| B^{q_n} x - x_n \right\| \le 1/n.$$

Property (P3) implies that for each $x \in K$, $\|A^{q_n} x - x_n\| \le 1/n$. This fact implies, in turn, that for each $x \in K$,

$$\left\| A^i x - x_n \right\| \le 1/n \quad \text{for any integer } i \ge q_n. \tag{2.33}$$

Since n is any natural number, we conclude that for each $x \in K$, $\{A^i x\}_{i=1}^{\infty}$ is a Cauchy sequence and there exists $\lim_{i\to\infty} A^i x$. Inequality (2.33) implies that for each $x \in K$,

$$\left\| \lim_{i\to\infty} A^i x - x_n \right\| \le 1/n. \tag{2.34}$$

Since n is an arbitrary natural number, we conclude that $\lim_{i\to\infty} A^i x$ does not depend on x. Hence there is $x_A \in K$ such that

$$x_A = \lim_{i\to\infty} A^i x \quad \text{for all } x \in K. \tag{2.35}$$

By (2.34) and (2.35),

$$\|x_A - x_n\| \le 1/n. \tag{2.36}$$

Inequalities (2.36) and (2.32) imply that for each $x \in K$,

$$\begin{aligned} \|Ax - x_A\| &\le \|Ax - x_n\| + \|x_n - x_A\| \le 1/n + \|Ax - x_n\| \\ &\le 1/n + \|x - x_n\| + 1/n \le 2/n + \|x - x_A\| + \|x_A - x_n\| \\ &\le \|x - x_A\| + 3/n, \end{aligned}$$

so that

$$\|Ax - x_A\| \le \|x - x_A\| + 3/n.$$

Since n is an arbitrary natural number, we conclude that

$$\|Ax - x_A\| \le \|x - x_A\| \quad \text{for each } x \in K. \tag{2.37}$$

Let $\varepsilon > 0$. Choose a natural number

$$n > 8/\varepsilon. \tag{2.38}$$

Property (P3) implies that

$$\|B^i x - x_n\| \le 1/n \quad \text{for each } x \in K, \text{ each integer } i \ge q_n, \text{ and each } B \in \mathcal{A} \text{ satisfying } d(B, A) \le \delta_n. \tag{2.39}$$

Inequalities (2.39), (2.36) and (2.38) imply that for each $B \in \mathcal{A}$ satisfying $d(B, A) \le \delta_n$, each $x \in K$, and each integer $i \ge q_n$,

$$\|B^i x - x_A\| \le \|B^i x - x_n\| + \|x_n - x_A\| \le 1/n + 1/n < \varepsilon.$$

This completes the proof of Lemma 2.3. □

Completion of the proof of Theorem 2.1 In order to complete the proof of this theorem, it is sufficient, by Lemma 2.3, to show that for each natural number n, the set $\mathcal{A} \setminus \mathcal{F}_n$ is porous in $(\mathcal{A}, d)$.

Let n be a natural number. Choose a positive number

$$\alpha < (16n)^{-1} 2^{-1} \left((\mathrm{diam}(K) + 1)^2 16 \cdot 8n \right)^{-1}. \tag{2.40}$$

Let

$$A \in \mathcal{A} \quad \text{and} \quad r \in (0, 1]. \tag{2.41}$$

By Proposition 2.2, there exist $A_0 \in \mathcal{A}$ and $\bar{x} \in K$ such that

$$d(A, A_0) \le r/8 \tag{2.42}$$

and

$$\|A_0 x - \bar{x}\| \le \|x - \bar{x}\| \quad \text{for each } x \in K. \tag{2.43}$$

Set

$$\gamma = 8^{-1} r \left(\mathrm{diam}(K) + 1 \right)^{-1} \tag{2.44}$$

and choose a natural number q for which

$$1 \le q \left((\mathrm{diam}(K) + 1)^2 16n \cdot 8r^{-1} \right)^{-1} \le 2. \tag{2.45}$$

Define $\bar{A} : K \to K$ by

$$\bar{A}x = (1 - \gamma) A_0 x + \gamma \bar{x}, \quad x \in K. \tag{2.46}$$

Clearly, the mapping $\bar{A}$ is continuous and for each $x \in K$,

$$\begin{aligned} \|\bar{A}x - \bar{x}\| &= \|(1 - \gamma) A_0 x + \gamma \bar{x} - \bar{x}\| \\ &= (1 - \gamma) \|A_0 x - \bar{x}\| \le (1 - \gamma) \|x - \bar{x}\|. \end{aligned} \tag{2.47}$$

Thus $\bar{A} \in \mathcal{A}$. Relations (2.3), (2.46), (2.1), (2.44) and (2.47) imply that

$$\begin{aligned} d(\bar{A}, A_0) &= \sup\{\|\bar{A}x - A_0x\| : x \in K\} \\ &= \sup\{\gamma\|\bar{x} - A_0x\| : x \in K\} \le \gamma \operatorname{diam}(K) = r/8. \end{aligned}$$

Together with (2.42) this implies that

$$d(\bar{A}, A) \le d(\bar{A}, A_0) + d(A_0, A) \le r/4. \tag{2.48}$$

Now assume that

$$B \in \mathcal{A} \quad \text{and} \quad d(B, \bar{A}) \le \alpha r. \tag{2.49}$$

Then (2.49), (2.40) and (2.47) imply that for each $x \in K$,

$$\|Bx - \bar{x}\| \le \|Bx - \bar{A}x\| + \|\bar{A}x - \bar{x}\| \le \|x - \bar{x}\| + \alpha r \le \|x - \bar{x}\| + 1/n. \tag{2.50}$$

In addition, (2.49), (2.48) and (2.40) imply that

$$d(B, A) \le d(B, \bar{A}) + d(\bar{A}, A) \le \alpha r + r/4 \le r/2. \tag{2.51}$$

Assume that $x \in K$. We will show that there exists an integer $j \in [0, q]$ such that $\|B^j x - \bar{x}\| \le (8n)^{-1}$. Assume the contrary. Then

$$\|B^i x - \bar{x}\| > (8n)^{-1}, \quad i = 0, \ldots, q. \tag{2.52}$$

Let an integer $i \in \{0, \ldots, q-1\}$. By (2.49) and (2.47),

$$\begin{aligned} \|B^{i+1}x - \bar{x}\| &= \|B(B^i x) - \bar{x}\| \\ &\le \|B(B^i x) - \bar{A}(B^i x)\| + \|\bar{A}(B^i x) - \bar{x}\| \\ &\le d(B, \bar{A}) + \|\bar{A}(B^i x) - \bar{x}\| \\ &\le \alpha r + (1-\gamma)\|B^i x - \bar{x}\| \end{aligned}$$

and

$$\|B^{i+1}x - \bar{x}\| \le \alpha r + (1-\gamma)\|B^i x - \bar{x}\|.$$

When combined with (2.52), (2.40) and (2.44), this inequality implies that

$$\begin{aligned} \|B^i x - \bar{x}\| - \|B^{i+1}x - \bar{x}\| &\ge \|B^i x - \bar{x}\| - \alpha r - (1-\gamma)\|B^i x - \bar{x}\| \\ &= \gamma\|B^i x - \bar{x}\| - \alpha r > (8n)^{-1}\gamma - \alpha r \ge (16n)^{-1}\gamma, \end{aligned}$$

so that

$$\|B^i x - \bar{x}\| - \|B^{i+1}x - \bar{x}\| \ge (16n)^{-1}\gamma.$$

When combined with (2.1), this inequality implies that

$$\mathrm{diam}(K) \geq \|x - \bar{x}\| - \|B^q x - \bar{x}\| \geq \sum_{i=0}^{q-1} (\|B^i x - \bar{x}\| - \|B^{i+1} x - \bar{x}\|) \geq q(16n)^{-1}\gamma$$

and

$$q \leq \mathrm{diam}(K) 16n/\gamma,$$

a contradiction (see (2.45)). The contradiction we have reached shows that there exists an integer $j \in [0, \ldots, q-1]$ such that

$$\|B^j x - \bar{x}\| \leq (8n)^{-1}. \tag{2.53}$$

It follows from (2.49) and (2.47) that for each integer $i \in \{0, \ldots, q-1\}$,

$$\begin{aligned}\|B^{i+1}x - \bar{x}\| &= \|B(B^i x) - \bar{x}\| \leq \|B(B^i x) - \bar{A}(B^i x)\| + \|\bar{A}(B^i x) - \bar{x}\| \\ &\leq d(\bar{A}, B) + \|\bar{A}(B^i x) - \bar{x}\| \leq \alpha r + \|B^i x - \bar{x}\|\end{aligned}$$

and

$$\|B^{i+1}x - \bar{x}\| \leq \|B^i x - \bar{x}\| + \alpha r.$$

This implies that for each integer s satisfying $j < s \leq q$,

$$\|B^s x - \bar{x}\| \leq \|B^j x - \bar{x}\| + \alpha r(s - j) \leq \|B^j x - \bar{x}\| + \alpha r q. \tag{2.54}$$

It follows from (2.53), (2.54), (2.45) and (2.40) that

$$\|B^q x - \bar{x}\| \leq \alpha r q + (8n)^{-1} \leq (2n)^{-1}.$$

Thus we have shown that the following property holds:
For each B satisfying (2.49) and each $x \in K$,

$$\|B^q x - \bar{x}\| \leq (2n)^{-1} \quad \text{and} \quad \|Bx - \bar{x}\| \leq \|x - \bar{x}\| + 1/n$$

(see (2.50)). Thus

$$\{B \in \mathcal{A} : d(B, \bar{A}) \leq \alpha r/2\} \subset \mathcal{F}_n \cap \{B \in \mathcal{A} : d(B, A) \leq r\}.$$

In other words, we have shown that the set $\mathcal{A} \setminus \mathcal{F}_n$ is porous in $(\mathcal{A}, d)$. This completes the proof of Theorem 2.1. □

2.2 Convergence of Iterates of Typical Nonexpansive Mappings

Let $(X, \|\cdot\|)$ be a Banach space and let $K \subset X$ be a nonempty, bounded, closed and convex subset of X. In this section we show that the iterates of a typical element (in the sense of Baire category) of a class of nonexpansive mappings which take K to X converge uniformly on K to the unique fixed point of this typical element.

Denote by $\mathcal{M}_{ne}$ the set of all mappings $A : K \to X$ such that

$$\|Ax - Ay\| \le \|x - y\| \quad \text{for all } x, y \in K.$$

For each $A, B \in \mathcal{M}_{ne}$, set

$$d(A, B) = \sup\{\|Ax - Bx\| : x \in K\}. \tag{2.55}$$

It is clear that $(\mathcal{M}_{ne}, d)$ is a complete metric space. Denote by $\mathcal{M}_0$ the set of all $A \in \mathcal{M}_{ne}$ such that

$$\inf\{\|x - Ax\| : x \in K\} = 0. \tag{2.56}$$

In other words, $\mathcal{M}_0$ consists of all those nonexpansive mappings taking K into X which have approximate fixed points. Clearly, $\mathcal{M}_0$ is a closed subset of $\mathcal{M}_{ne}$.

Every nonexpansive self-mapping of K belongs to $\mathcal{M}_0$. In order to exhibit two classes of nonself-mappings of K that are also contained in $\mathcal{M}_0$, we first recall that if $x \in K$, then the inward set $I_K(x)$ of X with respect to K is defined by

$$I_K(x) := \{z \in X : z = x + \alpha(y - x) \text{ for some } y \in K \text{ and } \alpha \ge 0\}.$$

A mapping $A : K \to X$ is said to be weakly inward if Ax belongs to the closure of $I_K(x)$ for each $x \in K$. Consider now a weakly inward mapping $A \in \mathcal{M}_{ne}$. Fix a point $z \in K$ and $t \in [0, 1)$ and let the mapping $S : K \to X$ be defined by $Sx = tAx + (1 - t)z$, $x \in K$. This strict contraction is also weakly inward and therefore has a unique fixed point $x_t \in K$ by Theorem 2.4 in [118]. Since $\|x_t - Ax_t\| \to 0$ as $t \to 1^-$, we see that $A \in \mathcal{M}_0$.

If K has a nonempty interior $\mathrm{int}(K)$ and a nonexpansive mapping $A : K \to X$ satisfies the Leray-Schauder condition with respect to $w \in \mathrm{int}(K)$, that is, $Ay - w \ne m(y - w)$ for all y in the boundary of K and $m > 1$, then it also belongs to $\mathcal{M}_0$. This is because the strict contraction $S : K \to X$ defined by $Sx = tAx + (1 - t)w$, $x \in K$, also satisfies the Leray-Schauder condition with respect to $w \in \mathrm{int}(K)$ and therefore has a unique fixed point [117].

Set

$$\rho(K) = \sup\{\|z\| : z \in K\}. \tag{2.57}$$

Our purpose is to show that the iterates of a typical element (in the sense of Baire category) of $\mathcal{M}_0$ converge uniformly on K to the unique fixed point of this typical element. As a matter of fact, we are able to establish a more refined result, involving the notion of porosity.

We are now ready to formulate our result obtained in [152].

Theorem 2.4 *There exists a set $\mathcal{F} \subset (\mathcal{M}_0, d)$ such that its complement $\mathcal{M}_0 \setminus \mathcal{F}$ is a σ-porous subset of $(\mathcal{M}_0, d)$ and each $B \in \mathcal{F}$ has the following properties:*

1. *There exists a unique point $x_B \in K$ such that $Bx_B = x_B$;*

2. *For each $\varepsilon > 0$, there exist $\delta > 0$, a natural number q, and a neighborhood $\mathcal{U}$ of B in $(\mathcal{M}_{ne}, d)$ such that:*
   (a) *if $C \in \mathcal{U}$, $y \in K$, and $\|y - Cy\| \le \delta$, then $\|y - x_B\| \le \varepsilon$;*
   (b) *if $C \in \mathcal{U}$, $\{x_i\}_{i=0}^q \subset K$, and $Cx_i = x_{i+1}$, $i = 0, \dots, q-1$, then $\|x_q - x_B\| \le \varepsilon$.*

Although analogous results for the closed subspace of $(\mathcal{M}_0, d)$ comprising all nonexpansive self-mappings of K were established by De Blasi and Myjak in [49, 50], Theorem 2.4 seems to be the first generic result dealing with nonself-mappings. In this connection see also [131, 137].

We begin the proof of Theorem 2.4 with a simple lemma.

Denote by E the set of all $A \in \mathcal{M}_{ne}$ for which there exists $x \in K$ satisfying $Ax = x$. That is, E consists of all those nonexpansive mappings $A : K \to X$ which have a fixed point.

Lemma 2.5 *E is an everywhere dense subset of $(\mathcal{M}_0, d)$.*

Proof Let $A \in \mathcal{M}_0$ and $\varepsilon > 0$. By (2.56), there exists $\bar{x} \in K$ such that

$$\|\bar{x} - A\bar{x}\| < \varepsilon/2.$$

Define

$$By = Ay + \bar{x} - A\bar{x}, \quad y \in K. \tag{2.58}$$

Clearly, $B \in \mathcal{M}_{ne}$ and $B\bar{x} = \bar{x}$. Thus $B \in E$. It is easy to see that $d(A, B) = \|\bar{x} - A\bar{x}\| < \varepsilon$. This completes the proof of Lemma 2.5. □

Proof of Theorem 2.4 For each natural number n, denote by $\mathcal{F}_n$ the set of all those mappings $A \in \mathcal{M}_0$ which have the following property:

(P1) There exist a natural number q, $x_* \in K$, $\delta > 0$, and a neighborhood $\mathcal{U}$ of A in $\mathcal{M}_{ne}$ such that:
   (i) if $B \in \mathcal{U}$ and if $z \in K$ satisfies $\|z - Bz\| \le \delta$, then $\|z - x_*\| \le 1/n$;
   (ii) if $B \in \mathcal{U}$ and if $\{x_i\}_{i=0}^q \subset K$ satisfies $x_{i+1} = Bx_i$, $i = 0, \dots, q-1$, then $\|x_q - x_*\| \le 1/n$.

Set

$$\mathcal{F} = \bigcap_{n=1}^{\infty} \mathcal{F}_n.$$

We intend to prove that $\mathcal{M}_0 \setminus \mathcal{F}$ is a σ-porous subset of $(\mathcal{M}_0, d)$. To meet this goal, it is sufficient to show that for each natural number n, the set $\mathcal{M}_0 \setminus \mathcal{F}_n$ is a porous subset of $(\mathcal{M}_0, d)$.

Indeed, let n be a natural number. Choose a positive number

$$\alpha \le 2^{-11}\big(\rho(K) + 1\big)^{-1} n^{-1}. \tag{2.59}$$

Let

$$A \in \mathcal{M}_0 \quad \text{and} \quad r \in (0, 1]. \tag{2.60}$$

By Lemma 2.5, there are $A_0 \in E$ and $x_* \in K$ such that

$$d(A_0, A) < r/8 \quad \text{and} \quad A_0 x_* = x_*. \tag{2.61}$$

Set

$$\gamma = \big[32\big(\rho(K) + 1\big)\big]^{-1} r \tag{2.62}$$

and

$$\delta = (4n)^{-1}\gamma - 2\alpha r. \tag{2.63}$$

By (2.63), (2.62) and (2.56),

$$\delta > 0. \tag{2.64}$$

Now choose an integer $q \geq 4$ such that

$$(1 - \gamma)^q 2\big(\rho(K) + 1\big) < (16n)^{-1}. \tag{2.65}$$

Define

$$A_1 y = (1 - \gamma)A_0 y + \gamma x_*, \quad y \in K. \tag{2.66}$$

Clearly, $A_1 \in \mathcal{M}_{ne}$ and

$$A_1 x_* = x_*. \tag{2.67}$$

By (2.55), (2.66), (2.61) and (2.57),

$$\begin{aligned} d(A_1, A_0) &= \sup\big\{\|A_1 y - A_0 y\| : y \in K\big\} = \sup\big\{\|\gamma A_0 y - \gamma x_*\| : y \in K\big\} \\ &= \gamma \sup\big\{\|A_0 y - A_0 x_*\| : y \in K\big\} \\ &\leq \gamma \sup\big\{\|y - x_*\| : y \in K\big\} \leq 2\gamma\rho(K), \end{aligned}$$

so that

$$d(A_1, A_0) \leq 2\gamma\rho(K). \tag{2.68}$$

By (2.68), (2.61) and (2.62),

$$d(A, A_1) \leq d(A, A_0) + d(A_0, A_1) \leq r/8 + 2\gamma\rho(K) \leq r/4. \tag{2.69}$$

Assume that $B \in \mathcal{M}_{ne}$ satisfies

$$d(B, A_1) \leq 2\alpha r. \tag{2.70}$$

Assume further that

$$z \in K \quad \text{and} \quad \|z - Bz\| \leq \delta. \tag{2.71}$$

By (2.67) and (2.66),

$$\|A_1 z - x_*\| = \|A_1 z - A_1 x_*\|$$
$$= (1-\gamma)\|A_0 z - A_0 x_*\| \le (1-\gamma)\|z - x_*\|. \tag{2.72}$$

By (2.55), (2.70) and (2.72),

$$\|Bz - z\| \ge \|A_1 z - z\| - \|Bz - A_1 z\| \ge \|A_1 z - z\| - d(B, A_1)$$
$$\ge \|A_1 z - z\| - 2\alpha r \ge \|z - x_*\| - \|x_* - A_1 z\| - 2\alpha r$$
$$\ge \|z - x_*\| - (1-\gamma)\|z - x_*\| - 2\alpha r = \gamma\|z - x_*\| - 2\alpha r.$$

When combined with (2.71) and (2.63), this inequality implies that

$$\delta \ge \|Bz - z\| \ge \gamma\|z - x_*\| - 2\alpha r$$

and

$$\|z - x_*\| \le \gamma^{-1}(\delta + 2\alpha r) \le (4n)^{-1}.$$

Thus we have shown that

$$\text{if } z \in K \text{ satisfies } \|z - Bz\| \le \delta, \text{ then } \|z - x_*\| \le (4n)^{-1}. \tag{2.73}$$

Now assume that

$$\{x_i\}_{i=0}^q \subset K, \qquad Bx_i = x_{i+1}, \quad i = 0, \dots, q-1. \tag{2.74}$$

By (2.74), (2.55), (2.70), (2.66) and (2.61), for $i = 0, \dots, q-1$, there holds

$$\|x_{i+1} - x_*\| = \|Bx_i - x_*\| \le \|Bx_i - A_1 x_i\| + \|A_1 x_i - x_*\|$$
$$= \|Bx_i - A_1 x_i\| + \|A_1 x_i - A_1 x_*\|$$
$$\le d(B, A_1) + (1-\gamma)\|A_0 x_i - A_0 x_*\|$$
$$\le 2\alpha r + (1-\gamma)\|x_i - x_*\|,$$

that is,

$$\|x_{i+1} - x_*\| \le 2\alpha r + (1-\gamma)\|x_i - x_*\|.$$

In view of this inequality, which is valid for $i = 0, \dots, q-1$, we get

$$\|x_q - x_*\| \le 2\alpha r \sum_{i=0}^{q-1} (1-\gamma)^i + (1-\gamma)^q \|x_0 - x_*\|$$
$$\le 2\alpha r \gamma^{-1} + (1-\gamma)^q \|x_0 - x_*\| \le 2\alpha r \gamma^{-1} + 2\rho(K)(1-\gamma)^q.$$

When combined with (2.62), (2.65) and (2.59), this last inequality implies that

$$\|x_q - x_*\| \le (1-\gamma)^q 2\rho(K) + 2\alpha\big[32\big(\rho(K)+1\big)\big]$$
$$\le (16n)^{-1} + 64\alpha\big[\rho(K)+1\big] \le (16n)^{-1} + (32n)^{-1} < (8n)^{-1}.$$

Thus we have shown that

$$\text{if } \{x_i\}_{i=0}^{q} \subset K \text{ satisfies (2.74), then } \|x_q - x_*\| \le (8n)^{-1}. \tag{2.75}$$

By (2.75), (2.74) and (2.73), each $C \in \mathcal{M}_0$ which satisfies $d(C, A_1) \le \alpha r$ has property (P1). Therefore

$$\big\{C \in \mathcal{M}_0 : d(C, A_1) \le \alpha r\big\} \subset \mathcal{F}_n.$$

When combined with (2.59) and (2.69), this inclusion implies that

$$\big\{C \in \mathcal{M}_0 : d(C, A_1) \le \alpha r\big\} \subset \big\{B \in \mathcal{M}_0 : d(B, A) \le r\big\} \cap \mathcal{F}_n.$$

This means that $\mathcal{M}_0 \setminus \mathcal{F}_n$ is a porous set in $(\mathcal{M}_0, d)$ for all natural numbers n. Therefore $\mathcal{M}_0 \setminus \mathcal{F}$ is a σ-porous set in $(\mathcal{M}_0, d)$.

Now let $A \in \mathcal{F}$ and $\varepsilon > 0$. Choose a natural number

$$n > 8\big(\min\{1, \varepsilon\}\big)^{-1}. \tag{2.76}$$

Since $A \in \mathcal{F}_n$, property (P1) implies that there exist a natural number q_n, a number $\delta_n > 0$, a neighborhood $\mathcal{U}_n$ of A in $\mathcal{M}_{ne}$, and a point $x_n \in K$ such that the following property holds:

(P2) (i) if $B \in \mathcal{U}_n$, $z \in K$, and $\|z - Bz\| \le \delta_n$, then $\|z - x_n\| \le 1/n$;
(ii) if $B \in \mathcal{U}_n$, $\{z_i\}_{i=0}^{q_n} \subset K$, and $z_{i+1} = Bz_i$, $i = 0, \ldots, q_n - 1$, then $\|z_{q_n} - x_n\| \le 1/n$.

Since $A \in \mathcal{M}_0$, there exists a sequence $\{y_i\}_{i=1}^{\infty} \subset K$ such that

$$\lim_{i\to\infty} \|y_i - Ay_i\| = 0. \tag{2.77}$$

Hence there exists a natural number i_0 such that

$$\|y_i - Ay_i\| \le \delta_n \quad \text{for all integers } i \ge i_0.$$

When combined with (P2)(i), this implies that

$$\|x_n - y_i\| \le 1/n \quad \text{for all integers } i \ge i_0. \tag{2.78}$$

In view of (2.78), for each pair of integers $i, j \ge i_0$,

$$\|y_i - y_j\| \le \|y_i - x_n\| + \|x_n - y_j\| \le 2/n < \varepsilon.$$

Since ε is an arbitrary positive number, we conclude that $\{y_i\}_{i=1}^{\infty}$ is a Cauchy sequence and therefore there exists

$$x_A = \lim_{i \to \infty} y_i. \tag{2.79}$$

Clearly, $Ax_A = x_A$. It is easy to see that x_A is the unique fixed point of A. Indeed, if it were not unique, then we would be able to construct a nonconvergent sequence $\{y_i\}_{i=0}^{\infty}$ satisfying (2.77).

By (2.78) and (2.79),

$$\|x_A - x_n\| \le 1/n. \tag{2.80}$$

Now assume that

$$B \in \mathcal{U}_n, \qquad z \in K, \quad \text{and} \quad \|z - Bz\| \le \delta_n. \tag{2.81}$$

By (P2)(i) and (2.81),

$$\|z - x_n\| \le 1/n.$$

When combined with (2.80) and (2.76), this inequality implies that

$$\|z - x_A\| \le \|z - x_n\| + \|x_n - x_A\| \le 2/n < \varepsilon.$$

Finally, suppose that

$$B \in \mathcal{U}_n, \qquad \{z_i\}_{i=0}^{q_n} \subset K, \quad \text{and} \quad Bz_i = z_{i+1}, \quad i = 0, \dots, q_n - 1. \tag{2.82}$$

Then by (P2)(ii) and (2.82),

$$\|z_{q_n} - x_n\| \le 1/n.$$

When combined with (2.80) and (2.76), this last inequality implies that

$$\|z_{q_n} - x_A\| \le \|z_{q_n} - x_n\| + \|x_n - x_A\| \le 2/n < \varepsilon.$$

This completes the proof of Theorem 2.4. □

2.3 A Stability Result in Fixed Point Theory

Let $K \subset X$ be a nonempty, compact and convex subset of a Banach space $(X, \|\cdot\|)$. In this section, which is based on [153], we consider a complete metric space of all the continuous self-mappings of K and show that a typical element of this space (in the sense of Baire's categories) has a fixed point which is stable under small perturbations of the mapping.

Denote by $\mathcal{A}$ the set of all continuous mappings $A : K \to K$. For each $A, B \in \mathcal{A}$, set

$$d(A, B) = \sup\{\|Ax - Bx\| : x \in K\}.$$

Clearly, $(\mathcal{A}, d)$ is a complete metric space. By Schauder's fixed point theorem, for each $A \in \mathcal{A}$ there exists $x_* \in K$ such that $Ax_* = x_*$. We begin with the following simple result.

Proposition 2.6 *Let $A \in \mathcal{A}$, $\Omega = \{x \in K : Ax = x\}$, and let $\varepsilon > 0$. Then there exists a positive number δ such that for each $B \in \mathcal{A}$ satisfying $d(A, B) \leq \delta$ and each $x \in K$ satisfying $Bx = x$, there exists $y \in \Omega$ such that $\|x - y\| \leq \varepsilon$.*

Proof Assume the contrary. Then there exist a sequence $\{B_n\}_{n=1}^{\infty} \subset \mathcal{A}$ satisfying

$$d(A, B_n) \leq 1/n \quad \text{for all integers } n \geq 1, \tag{2.83}$$

and a sequence $\{x_n\}_{n=1}^{\infty} \subset K$ such that for each integer $n \geq 1$,

$$B_n x_n = x_n \quad \text{and} \quad \inf\bigl\{\|x_n - y\| : y \in \Omega\bigr\} \geq \varepsilon. \tag{2.84}$$

Since K is compact, we may assume without loss of generality that there exists

$$x_* = \lim_{n\to\infty} x_n. \tag{2.85}$$

It follows from (2.85), (2.84), (2.83) and the continuity of A that

$$\begin{aligned}\|Ax_* - x_*\| &\leq \|Ax_* - Ax_n\| + \|B_n x_n - Ax_n\| + \|B_n x_n - x_n\| + \|x_n - x_*\| \\ &\leq \|Ax_* - Ax_n\| + 1/n + \|x_n - x_*\| \to 0 \quad \text{as } n \to \infty.\end{aligned}$$

Thus $Ax_* = x_*$, $x_* \in \Omega$, and (2.85) contradicts (2.84). The contradiction we have reached proves Proposition 2.6. □

In view of this result, it is natural to ask if, given $A \in \mathcal{A}$, there is a fixed point $x_* \in K$ of A with the following property:

For each $\varepsilon > 0$ there exists $\delta > 0$ such that for each $B \in \mathcal{A}$ satisfying $d(A, B) \leq \delta$, there exists $y \in K$ such that $By = y$ and $\|y - x_*\| \leq \varepsilon$.

Example 2.7 Let $X = R^1$, $K = [0, 1]$ and $Ax = x$, $x \in K$. Clearly, the set of fixed points of A is the interval $[0, 1]$. For each integer $n \geq 1$, define

$$A_n x = (1 - 1/n)x, \qquad B_n x = \min\{x + 1/n, 1\} \quad \text{for all } x \in [0, 1].$$

Clearly, $B_n, A_n \to A$ as $n \to \infty$. It is easy to see that for each $n \geq 1$, the set of fixed points of A_n is the singleton $\{0\}$ while the set of fixed points of B_n is the interval $[1 - 1/n, 1]$.

This example shows that in general the answer to our question is negative. Nevertheless, we show in this section that for a typical $A \in \mathcal{A}$ (in the sense of Baire's categories) the answer is positive.

Let $K \subset X$ be a nonempty, closed and convex subset of a Banach space $(X, \|\cdot\|)$. Denote by $\tilde{\mathcal{A}}$ the family of all continuous mappings $A : K \to K$ such that the closure of $A(K)$ is a compact set in the norm topology. It is well known [171] that for each $A \in \tilde{\mathcal{A}}$ there is $x_A \in K$ such that $Ax_A = x_A$.

For each $A, B \in \tilde{\mathcal{A}}$ set

$$d(A, B) = \sup\{\|Ax - Bx\| : x \in K\}. \tag{2.86}$$

It is not difficult to see that $(\tilde{\mathcal{A}}, d)$ is a complete metric space.

Theorem 2.8 *There exists a subset $\mathcal{F} \subset \tilde{\mathcal{A}}$ which is a countable intersection of open everywhere dense subsets of $(\tilde{\mathcal{A}}, d)$ such that for each $A \in \mathcal{F}$, there exists $x_* \in K$ such that*

(i) $Ax_* = x_*$;
(ii) *for each $\varepsilon > 0$ there exists $\delta > 0$ such that if $B \in \tilde{\mathcal{A}}$ satisfies $d(A, B) \leq \delta$, then there is $z \in K$ which satisfies $Bz = z$ and $\|z - x_*\| \leq \varepsilon$.*

Two auxiliary propositions will precede the proof of Theorem 2.8.

Proposition 2.9 *Let $A \in \tilde{\mathcal{A}}$, $\varepsilon > 0$ and let $x_* \in K$ satisfy $Ax_* = x_*$. Then there exist $B \in \tilde{\mathcal{A}}$ and $\delta > 0$ such that $d(B, A) \leq \varepsilon$ and $Bz = x_*$ for each $z \in K$ satisfying $\|z - x_*\| \leq \delta$.*

Proof There exists $\delta > 0$ such that for each $z \in K$ satisfying $\|z - x_*\| \leq 4\delta$, the following inequality holds:

$$\|Az - x_*\| \leq \varepsilon/4. \tag{2.87}$$

By Urysohn's theorem, there exists a continuous function $\lambda : X \to [0, 1]$ such that

$$\lambda(z) = 1 \quad \text{for each } z \in X \text{ satisfying } \|z - x_*\| \leq \delta \tag{2.88}$$

and

$$\lambda(z) = 0 \quad \text{for each } z \in X \text{ satisfying } \|z - x_*\| \geq 2\delta. \tag{2.89}$$

Define

$$Bz = \lambda(z)x_* + \bigl(1 - \lambda(z)\bigr)Az \tag{2.90}$$

for all $z \in K$.

Clearly, $B : K \to K$ is continuous, $B(K)$ is contained in a compact subset of X, and

$$Bx_* = x_*. \tag{2.91}$$

By (2.90), (2.88) and (2.89), for each $z \in K$ satisfying $\|z - x_*\| \leq \delta$, we have

$$Bz = x_*, \tag{2.92}$$

and for each $z \in K$ satisfying $\|z - x_*\| \geq 2\delta$,

$$Bz = Az. \tag{2.93}$$

It follows from (2.90) and the choice of δ (see (2.87)) that for each $z \in K$ satisfying $\|z - x_*\| \leq 2\delta$,

$$\begin{aligned} \|Bz - Az\| &= \big\|\lambda(z)x_* + \big(1 - \lambda(z)\big)Az - Az\big\| \\ &\leq \|x_* - Az\| \leq \varepsilon/4. \end{aligned}$$

This completes the proof of Proposition 2.9. □

Proposition 2.10 *Let $A \in \tilde{\mathcal{A}}$, $\varepsilon > 0$, let $x_* \in K$ be a fixed point of A, and let $B \in \tilde{\mathcal{A}}$, $\delta > 0$ be as guaranteed by Proposition 2.9. Then for each $C \in \tilde{\mathcal{A}}$ satisfying $d(C, B) \leq \delta$, there is $y \in K$ such that*

$$Cy = y \quad \textit{and} \quad \|y - x_*\| \leq d(C, B).$$

Proof By Proposition 2.9,

$$d(A, B) \leq \varepsilon \tag{2.94}$$

and

$$Bz = x_* \quad \text{for each } z \in K \text{ satisfying } \|z - x_*\| \leq \delta. \tag{2.95}$$

Assume that $C \in \tilde{\mathcal{A}}$ satisfies

$$d(C, B) \leq \delta. \tag{2.96}$$

Set

$$\Omega = \big\{z \in K : \|z - x_*\| \leq d(C, B)\big\}. \tag{2.97}$$

Clearly, Ω is a closed and convex set. It follows from (2.97), (2.96) and (2.95) that for each $z \in \Omega$,

$$\|x_* - Cz\| \leq \|x_* - Bz\| + \|Bz - Cz\| = \|Bz - Cz\| \leq d(C, B)$$

and $Cz \in \Omega$. Thus $C(\Omega) \subset \Omega$. Clearly $C(\Omega) \subset C(X)$ is contained in a compact subset of X. By Schauder's theorem there is $y \in \Omega$ such that $Cy = y$. Proposition 2.10 is proved. □

Proof of Theorem 2.8 Let $A \in \tilde{\mathcal{A}}$ and $\varepsilon \in (0, 1)$. By Propositions 2.9 and 2.10, there exist

$$A_\varepsilon \in \tilde{\mathcal{A}}, \qquad x_{A,\varepsilon} \in K \quad \text{and} \quad \delta_{A,\varepsilon} \in (0, 1)$$

such that

$$d(A, A_\varepsilon) \le \varepsilon, \tag{2.98}$$

$$A_\varepsilon z = x_{A,\varepsilon} \quad \text{for each } z \in K \text{ satisfying } \|z - x_{A,\varepsilon}\| \le \delta_{A,\varepsilon}, \tag{2.99}$$

and the following property holds:

(P) For each $C \in \tilde{\mathcal{A}}$ satisfying $d(C, A_\varepsilon) \le \delta_{A,\varepsilon}$, there is $y \in K$ such that

$$Cy = y, \qquad \|y - x_{A,\varepsilon}\| \le d(C, A_\varepsilon).$$

For each integer $i \ge 1$, set

$$\mathcal{U}(A, \varepsilon, i) = \{C \in \tilde{\mathcal{A}} : d(C, A_\varepsilon) < \delta_{A,\varepsilon}/i\}. \tag{2.100}$$

Define

$$\mathcal{F} = \bigcap_{i=1}^{\infty} \bigcup \{\mathcal{U}(A, \varepsilon, i) : A \in \tilde{\mathcal{A}}, \varepsilon \in (0, 1)\}. \tag{2.101}$$

Clearly, $\mathcal{F}$ is a countable intersection of open and everywhere dense subsets of $(\tilde{\mathcal{A}}, d)$.

Let $B \in \mathcal{F}$. For each integer $i \ge 1$, there are $A_i \in \tilde{\mathcal{A}}$ and $\varepsilon_i \in (0, 1)$ such that

$$B \in \mathcal{U}(A_i, \varepsilon_i, i). \tag{2.102}$$

It follows from (2.102), (2.100) and property (P) that for each integer $i \ge 1$, there $y_i \in K$ such that

$$By_i = y_i \tag{2.103}$$

and

$$\|y_i - x_{A_i,\varepsilon_i}\| \le d(A, (A_i)_{\varepsilon_i}) \le \delta_{A_i,\varepsilon_i}/i. \tag{2.104}$$

Since $\{y_i\}_{i=1}^{\infty} \subset B(K)$, there is a subsequence $\{y_{i_k}\}_{k=1}^{\infty}$ which converges to $x_* \in K$. Clearly, $Bx_* = x_*$.

Let $\varepsilon > 0$. There exists a natural number k such that

$$i_k^{-1} < 8^{-1}\varepsilon \quad \text{and} \quad \|y_{i_k} - x_*\| \le \varepsilon/8. \tag{2.105}$$

It follows from (2.104) and (2.105) that

$$\|y_{i_k} - x_{A_{i_k},\varepsilon_{i_k}}\| \le 1/i_k < \varepsilon/8. \tag{2.106}$$

Inequalities (2.105) and (2.106) imply that

$$\|x_* - x_{A_{i_k},\varepsilon_{i_k}}\| \le \|x_* - y_{i_k}\| + \|y_{i_k} - x_{A_{i_k},\varepsilon_{i_k}}\| \le \varepsilon/4. \tag{2.107}$$

Let

$$C \in \mathcal{U}(A_{i_k}, \varepsilon_{i_k}, i_k). \tag{2.108}$$

It follows from (2.108), (2.100), (2.105) and property (P) that there exists a point $z \in K$ such that

$$Cz = z \quad \text{and} \quad \|z - x_{A_{i_k}, \varepsilon_{i_k}}\| \le d\big(C, (A_{i_k})_{\varepsilon_{i_k}}\big) \le 1/i_k \le \varepsilon/8.$$

When combined with (2.107), this implies that

$$\|z - x_*\| \le \|z - x_{A_{i_k}, \varepsilon_{i_k}}\| + \|x_{A_{i_k}, \varepsilon_{i_k}} - x_*\| \le \varepsilon/2.$$

Theorem 2.8 is proved. □

2.4 Well-Posed Null and Fixed Point Problems

The notion of well-posedness is of great importance in many areas of mathematics and its applications. In this section we consider two complete metric spaces of continuous mappings and establish generic well-posedness of certain null and fixed point problems. Our results, which were obtained in [154], are a consequence of the variational principle established in [74]. For other related results concerning the well-posedness of fixed point problems see [50, 139].

Let $(X, \|\cdot\|, \ge)$ be a Banach space ordered by a closed convex cone $X_+ = \{x \in X : x \ge 0\}$ such that $\|x\| \le \|y\|$ for each pair of points $x, y \in X_+$ satisfying $x \le y$. Let (K, ρ) be a complete metric space. Denote by $\mathcal{M}$ the set of all continuous mappings $A : K \to X$. We equip the set $\mathcal{M}$ with the uniformity determined by the following base:

$$E(\varepsilon) = \big\{(A, B) \in \mathcal{M} \times \mathcal{M} : \|Ax - Bx\| \le \varepsilon \text{ for all } x \in K\big\}, \tag{2.109}$$

where $\varepsilon > 0$. It is not difficult to see that this uniform space is metrizable (by a metric d) and complete.

Denote by $\mathcal{M}_p$ the set of all $A \in \mathcal{M}$ such that

$$Ax \in X_+ \quad \text{for all } x \in K \tag{2.110}$$

and

$$\inf\big\{\|Ax\| : x \in K\big\} = 0. \tag{2.111}$$

It is not difficult to see that $\mathcal{M}_p$ is a closed subset of $(\mathcal{M}, d)$.

We can now state and prove our first result.

Theorem 2.11 *There exists an everywhere dense G_δ subset $\mathcal{F} \subset \mathcal{M}_p$ such that for each $A \in \mathcal{F}$, the following properties hold:*

1. *There is a unique $\bar{x} \in K$ such that $A\bar{x} = 0$.*
2. *For any $\varepsilon > 0$, there exist $\delta > 0$ and a neighborhood U of A in $\mathcal{M}_p$ such that if $B \in U$ and if $x \in K$ satisfies $\|Bx\| \le \delta$, then $\rho(x, \bar{x}) \le \varepsilon$.*

Proof We obtain this theorem as a realization of the variational principle established in Theorem 2.1 of [74] with $f_A(x) = \|Ax\|$, $x \in K$. In order to prove our theorem by using this variational principle we need to prove the following assertion:

(A) For each $A \in \mathcal{M}_p$ and each $\varepsilon > 0$, there are $\bar{A} \in \mathcal{M}_p$, $\delta > 0$, $\bar{x} \in K$ and a neighborhood W of $\bar{A}$ in $\mathcal{M}_p$ such that

$$(A, \bar{A}) \in E(\varepsilon),$$

and if $B \in W$ and $z \in K$ satisfy $\|Bz\| \leq \delta$, then

$$\rho(z, \bar{x}) \leq \varepsilon.$$

Let $A \in \mathcal{M}_p$ and $\varepsilon > 0$. Choose $\bar{u} \in X_+$ such that

$$\|\bar{u}\| = \varepsilon/4, \tag{2.112}$$

and $\bar{x} \in K$ such that

$$\|A\bar{x}\| \leq \varepsilon/8. \tag{2.113}$$

Since A is continuous, there is a positive number r such that

$$r < \min\{1, \varepsilon/16\} \tag{2.114}$$

and

$$\|Ax - A\bar{x}\| \leq \varepsilon/8 \quad \text{for each } x \in K \text{ satisfying } \rho(x, \bar{x}) \leq 4r. \tag{2.115}$$

By Urysohn's theorem, there is a continuous function $\phi : K \to [0, 1]$ such that

$$\phi(x) = 1 \quad \text{for each } x \in K \text{ satisfying } \rho(x, \bar{x}) \leq r \tag{2.116}$$

and

$$\phi(x) = 0 \quad \text{for each } x \in K \text{ satisfying } \rho(x, \bar{x}) \geq 2r. \tag{2.117}$$

Define

$$\bar{A}x = \big(1 - \phi(x)\big)(Ax + \bar{u}), \quad x \in K. \tag{2.118}$$

It is clear that $\bar{A} : K \to X$ is continuous. Now (2.116)–(2.118) imply that

$$\bar{A}x = 0 \quad \text{for each } x \in K \text{ satisfying } \rho(x, \bar{x}) \leq r \tag{2.119}$$

and

$$\bar{A}x \geq \bar{u} \quad \text{for each } x \in K \text{ satisfying } \rho(x, \bar{x}) \geq 2r. \tag{2.120}$$

It is not difficult to see that $\bar{A} \in \mathcal{M}_p$. We claim that $(A, \bar{A}) \in E(\varepsilon)$.

Let $x \in K$. There are two cases: either

$$\rho(x, \bar{x}) \geq 2r \tag{2.121}$$

or

$$\rho(x, \bar{x}) < 2r. \tag{2.122}$$

Assume first that (2.121) holds. Then it follows from (2.121), (2.117), (2.118) and (2.112) that

$$\|Ax - \bar{A}x\| = \|\bar{u}\| = \varepsilon/4.$$

Now assume that (2.122) holds. Then by (2.122), (2.118) and (2.112),

$$\begin{aligned}\|\bar{A}x - Ax\| &= \big\|(1 - \phi(x))(Ax + \bar{u}) - Ax\big\| \leq \|\bar{u}\| + \|Ax\| \\ &\leq \varepsilon/4 + \|Ax\|.\end{aligned}$$

It follows from this inequality, (2.122), (2.115) and (2.113) that

$$\|\bar{A}x - Ax\| \leq \varepsilon/4 + \|Ax\| < \varepsilon/2.$$

Therefore in both cases $\|\bar{A}x - Ax\| \leq \varepsilon/2$. Since this inequality holds for any $x \in K$, we conclude that

$$(A, \bar{A}) \in E(\varepsilon). \tag{2.123}$$

Consider now an open neighborhood U of $\bar{A}$ in $\mathcal{M}_p$ such that

$$U \subset \big\{B \in \mathcal{M}_p : (\bar{A}, B) \in E(\varepsilon/16)\big\}. \tag{2.124}$$

Let

$$B \in U, \qquad z \in K \tag{2.125}$$

and

$$\|Bz\| \leq \varepsilon/16. \tag{2.126}$$

Relations (2.126), (2.125), (2.124) and (2.109) imply that

$$\|\bar{A}z\| \leq \|Bz\| + \|\bar{A}z - Bz\| \leq \varepsilon/16 + \varepsilon/16. \tag{2.127}$$

We claim that

$$\rho(z, \bar{x}) \leq \varepsilon. \tag{2.128}$$

Assume the contrary. Then by (2.114),

$$\rho(z, \bar{x}) > \varepsilon \geq 2r.$$

When combined with (2.120), this implies that

$$\bar{A}z \geq \bar{u}.$$

It follows from this inequality, the monotonicity of the norm, (2.125), (2.124), (2.109) and (2.112) that

$$\|Bz\| \geq \|\bar{A}z\| - \varepsilon/16 \geq \|\bar{u}\| - \varepsilon/16 = \varepsilon/4 - \varepsilon/16 = 3\varepsilon/16.$$

This, however, contradicts (2.126). The contradiction we have reached proves (2.128) and Theorem 2.11 itself. □

Now assume that the set K is a subset of X and

$$\rho(x, y) = \|x - y\|, \quad x, y \in K.$$

Denote by $\mathcal{M}_n$ the set of all mappings $A \in \mathcal{M}$ such that

$$Ax \geq x \quad \text{for all } x \in K$$

and

$$\inf\{\|Ax - x\| : x \in K\} = 0.$$

Clearly, $\mathcal{M}_n$ is a closed subset of $(\mathcal{M}, d)$. Define a map $J : \mathcal{M}_n \to \mathcal{M}_p$ by

$$J(A)x = Ax - x \quad \text{for all } x \in K$$

and all $A \in \mathcal{M}_n$. Clearly, there exists $J^{-1} : \mathcal{M}_p \to \mathcal{M}_n$, and both J and its inverse J^{-1} are continuous. Therefore Theorem 2.11 implies the following result regarding the generic well-posedness of the fixed point problem for $A \in \mathcal{M}_n$.

Theorem 2.12 *There exists an everywhere dense G_δ subset $\mathcal{F} \subset \mathcal{M}_n$ such that for each $A \in \mathcal{F}$, the following properties hold*:

1. *There is a unique $\bar{x} \in K$ such that $A\bar{x} = \bar{x}$.*
2. *For any $\varepsilon > 0$, there exist $\delta > 0$ and a neighborhood U of A in $\mathcal{M}_n$ such that if $B \in U$ and if $x \in K$ satisfies $\|Bx - x\| \leq \delta$, then $\|x - \bar{x}\| \leq \varepsilon$.*

2.5 Mappings in a Finite-Dimensional Euclidean Space

In this section we study the existence and stability of fixed points of continuous mappings in finite-dimensional Euclidean spaces. Our results [156] establish generic existence and stability of fixed points for a class of nonself-mappings defined on certain closed (but not necessarily either convex or bounded) subsets of a finite-dimensional Euclidean space. In these results, we endow the relevant space of mappings with two topologies, one weaker than the other. In the first result we find an open (in the weak topology) and everywhere dense (in the strong topology) set such that each mapping in it possesses a fixed point. In the second result we construct a countable intersection of open (in the weak topology) and everywhere dense (in the strong topology) sets such that each mapping in this intersection has a stable fixed point.

Let $K \subset R^n$ be a nonempty, closed subset of the n-dimensional Euclidean space $(R^n, \|\cdot\|)$. We assume that K is the closure of its nonempty interior $\mathrm{int}(K)$.

For each $x \in R^n$ and each $r > 0$, set $B(x,r) = \{y \in R^n : \|x - y\| \leq r\}$ and fix $\theta \in K$.

Denote by $\mathcal{M}$ the set of all continuous mappings $A : K \to R^n$. We equip the space $\mathcal{M}$ with the uniformity determined by the base

$$\mathcal{E}_w(N, \varepsilon) = \big\{(A, B) \in \mathcal{M} \times \mathcal{M} : \|Ax - Bx\| \leq \varepsilon \text{ for all } x \in B(\theta, N) \cap K\big\}, \tag{2.129}$$

where $N, \varepsilon > 0$.

Clearly, the space $\mathcal{M}$ with this uniformity is metrizable and complete. We equip the space $\mathcal{M}$ with the topology induced by this uniformity. This topology will be called the weak topology.

We also equip the space $\mathcal{M}$ with the uniformity determined by the base

$$\mathcal{E}_s(\varepsilon) = \big\{(A, B) \in \mathcal{M} \times \mathcal{M} : \|Ax - Bx\| \leq \varepsilon \text{ for all } x \in K\big\}, \tag{2.130}$$

where $\varepsilon > 0$. Clearly, the space $\mathcal{M}$ with this uniformity is also metrizable and complete. The topology induced by this uniformity on $\mathcal{M}$ will be called the strong topology.

Denote by $\mathcal{M}_f$ the set of all $A \in \mathcal{M}$ which have approximate fixed points. In other words, the set $\mathcal{M}_f$ consists of all $A \in \mathcal{M}$ such that

$$\inf\big\{\|x - Ax\| : x \in K\big\} = 0. \tag{2.131}$$

It is clear that $\mathcal{M}_f$ is a closed subset of $\mathcal{M}$ with the strong topology.

Note that if the set K is bounded, then $\mathcal{M}_f$ consists of all those elements of $\mathcal{M}$ which have fixed points. Every self-mapping of K which is a strict contraction, that is, has a Lipschitz constant strictly less than one, clearly belongs to $\mathcal{M}_f$.

If K is bounded and convex and a continuous mapping $A : K \to R^n$ satisfies the Leray-Schauder condition with respect to $w \in \mathrm{int}(K)$, that is, $Ay - w \neq m(y - w)$ for all y on the boundary of K and $m > 1$, then it also belongs to $\mathcal{M}_f$. If such an A is a strict contraction, then this continues to be true even if K is neither bounded nor convex.

We endow the topological subspace $\mathcal{M}_f \subset M$ with both the relative weak and strong topologies.

The following two results were obtained in [156].

Theorem 2.13 *Let $\gamma \in (0, 1)$. There exists an open (in the weak topology) and everywhere dense (in the strong topology) set $\mathcal{F}_\gamma \subset \mathcal{M}_f$ such that for each $A \in \mathcal{F}_\gamma$, there are $x_A \in \mathrm{int}(K)$, $r_A \in (0, 1)$, and a neighborhood $\mathcal{U}$ of A in $\mathcal{M}_f$ with the weak topology such that*

$$B(x_A, r_A) \subset K \quad \text{and} \quad Ax_A = x_A,$$

and for each $C \in \mathcal{U}$, there is $x_C \in K$ such that $Cx_C = x_C$ and $\|x_C - x_A\| \leq \gamma r_A$.

Theorem 2.14 *There exists a set $\mathcal{F} \subset \mathcal{M}_f$ which is a countable intersection of open (in the weak topology) and everywhere dense (in the strong topology) subsets of $\mathcal{M}_f$ such that for each $A \in \mathcal{F}$ and each $\gamma \in (0,1)$, there exist $x_A \in \mathrm{int}(K)$, $r_A \in (0,1)$, and a neighborhood $\mathcal{U}$ of A in $\mathcal{M}_f$ with the weak topology such that*

$$B(x_A, r_A) \subset K \quad \textit{and} \quad Ax_A = x_A,$$

and for each $C \in \mathcal{U}$ there is $x_C \in K$ such that $Cx_C = x_C$ and $\|x_C - x_A\| \leq \gamma r_A$.

Example 2.15 Let $n = 1$, $K = \bigcup_{j=0}^{\infty}[2j, 2j+1]$, and define, for each integer $j \geq 1$ and each $x \in [2j, 2j+1]$, $Ax = x + 2^{-j}$. Clearly, $\inf\{|x - Ax| : x \in K\} = 0$ but A is fixed point free.

In order to prove Theorem 2.13 we need two auxiliary results.

Denote by $\mathcal{E}$ the set of all $A \in \mathcal{M}_f$ for which there exist

$$x_A \in \mathrm{int}(K) \quad \text{and} \quad r_A \in (0,1) \tag{2.132}$$

such that

$$B(x_A, r_A) \subset K \quad \text{and} \quad Ay = x_A \quad \text{for all } y \in B(x_A, r_A/4). \tag{2.133}$$

Lemma 2.16 *The set $\mathcal{E}$ is an everywhere dense subset of $\mathcal{M}_f$ with the strong topology.*

Proof Let $A \in \mathcal{M}_f$ and $\varepsilon > 0$. By the definition of $\mathcal{M}_f$ (see (2.131)), there exists $x_0 \in K$ such that

$$\|Ax_0 - x_0\| < \varepsilon/16. \tag{2.134}$$

Since K is the closure of $\mathrm{int}(K)$ and A is continuous, there is $x_1 \in \mathrm{int}(K)$ such that

$$\|x_1 - x_0\| < \varepsilon/16 \quad \text{and} \quad \|Ax_1 - Ax_0\| < \varepsilon/16. \tag{2.135}$$

Set

$$A_1 y = Ay - Ax_1 + x_1, \quad y \in K. \tag{2.136}$$

Clearly, $A_1 \in \mathcal{M}$. In view of (2.136),

$$A_1 x_1 = x_1. \tag{2.137}$$

By (2.136), (2.135) and (2.134), for each $y \in K$,

$$\begin{aligned} \|Ay - A_1 y\| &= \|Ax_1 - x_1\| \leq \|Ax_1 - Ax_0\| + \|Ax_0 - x_0\| + \|x_0 - x_1\| \\ &< 3\varepsilon/16. \end{aligned} \tag{2.138}$$

Since A_1 has a fixed point (see (2.137)), it is clear that $A_1 \in \mathcal{M}_f$. Since A_1 is continuous and $x_1 \in \text{int}(K)$, there exists $r_1 \in (0, 1)$ such that

$$B(x_1, r_1) \subset K \quad \text{and} \quad \|A_1 x - A_1 x_1\| \le \varepsilon/16 \quad \text{for all } x \in B(x_1, r_1). \tag{2.139}$$

Define

$$\begin{aligned} &\psi(t) = 1, \quad t \in [0, r_1/2], \qquad \psi(t) = 0, \quad t \in [r_1, \infty), \\ &\psi(t) = 2(r_1 - t)r_1^{-1}, \quad t \in (r_1/2, r_1), \end{aligned} \tag{2.140}$$

and

$$By = \psi(\|y - x_1\|)x_1 + (1 - \psi(\|y - x_1\|))A_1 y, \quad y \in K. \tag{2.141}$$

Clearly, $B \in \mathcal{M}$. It follows from (2.141) and (2.140) that for each $y \in B(x_1, r_1/2)$,

$$By = x_1. \tag{2.142}$$

Therefore $B \in \mathcal{E}$. We will now show that

$$\|By - Ay\| \le \varepsilon \quad \text{for all } x \in K.$$

Indeed, let $y \in K$. There are two cases to be considered:

$$\|x_1 - y\| \le r_1; \tag{2.143}$$

$$\|x_1 - y\| > r_1. \tag{2.144}$$

If (2.144) holds, then (2.144), (2.141), (2.140) and (2.138) imply that

$$By = A_1 y \quad \text{and} \quad \|By - Ay\| = \|A_1 y - Ay\| < \varepsilon/4. \tag{2.145}$$

Let (2.143) hold. Then by (2.143), (2.141), (2.140), (2.137) and (2.139),

$$\|By - A_1 y\| = \|\psi(\|y - x_1\|)(x_1 - A_1 y)\| \le \|x_1 - A_1 y\| = \|A_1 x_1 - A_1 y\| < \varepsilon/16.$$

When combined with (2.138), this inequality implies that

$$\|By - Ay\| \le \|By - A_1 y\| + \|A_1 y - Ay\| \le \varepsilon/16 + 3\varepsilon/16 = \varepsilon/4.$$

Thus

$$\|By - Ay\| \le \varepsilon/4 \quad \text{for all } y \in K.$$

This completes the proof of Lemma 2.16. □

Lemma 2.17 *Let $A \in \mathcal{E}$, $x_A \in \text{int}(K)$, $r_A \in (0, 1)$ satisfy (2.133) and let $\gamma \in (0, 1)$. Then there exists a neighborhood $\mathcal{U}$ of A in $\mathcal{M}_f$ with the weak topology such that for each $B \in \mathcal{U}$, there is $x_B \in K$ such that $\|x_B - x_A\| \le \gamma r_A/4$ and $Bx_B = x_B$.*

Proof Set

$$\Delta = \gamma r_A/4 \tag{2.146}$$

and put

$$\mathcal{U} = \{B \in \mathcal{M}_f : \|Bz - Az\| \le \Delta \text{ for each } z \in B(x_A, r_A)\}. \tag{2.147}$$

Clearly, $\mathcal{U}$ is a neighborhood of A in $\mathcal{M}_f$ with the weak topology.

Let $B \in \mathcal{U}$. It follows from (2.147), (2.133) and (2.146) that for each $z \in B(x_A, \gamma r_A/4)$,

$$\|Bz - x_A\| \le \|Bz - Az\| + \|Az - x_A\| \le \Delta + \|Az - x_A\| = \Delta = \gamma r_A/4.$$

Thus

$$B\big(B(x_A, \gamma r_A/4)\big) \subset B(x_A, \gamma r_A/4).$$

Since the mapping B is continuous, there is $x_B \in B(x_A, \gamma r_A/4)$ such that

$$Bx_B = x_B.$$

Lemma 2.17 is proved. □

Proof of Theorem 2.13 Let $A \in \mathcal{E}$. There exist $x_A \in \text{int}(K)$ and $r_A \in (0, 1)$ such that (2.133) holds. By Lemma 2.17, there exists an open neighborhood $\mathcal{U}(A)$ of A in $\mathcal{M}_f$ with the weak topology such that the following property holds:

(P1) For each $B \in \mathcal{U}(f)$, there is $x_B \in K$ such that

$$Bx_B = x_B \quad \text{and} \quad \|x_B - x_A\| \le \gamma r_A/8. \tag{2.148}$$

Set

$$\mathcal{F}_\gamma = \bigcup \{\mathcal{U}(A) : A \in \mathcal{E}\}. \tag{2.149}$$

By Lemma 2.16, $\mathcal{F}_\gamma$ is an open (in the weak topology) and everywhere dense (in the strong topology) subset of $\mathcal{M}_f$.

Let $B \in \mathcal{F}_\gamma$. By (2.149), there is $A \in \mathcal{E}$ such that

$$B \in \mathcal{U}(A). \tag{2.150}$$

By property (P1), for each $C \in \mathcal{U}(A)$, there is $x_C \in K$ such that

$$Cx_C = x_C \quad \text{and} \quad \|x_C - x_A\| \le \gamma r_A/8. \tag{2.151}$$

Clearly,

$$\|x_B - x_A\| \le \gamma r_A/8. \tag{2.152}$$

It follows from (2.152) and (2.135) that

$$B(x_B, r_A/2) \subset B(x_A, r_A) \subset K. \tag{2.153}$$

By (2.151) and (2.152), for each $C \in \mathcal{U}(A)$,

$$\|x_C - x_B\| \le \|x_C - x_A\| + \|x_A - x_B\| \le \gamma r_A/8 + \gamma r_A/8 = \gamma r_A/4.$$

This completes the proof of Theorem 2.13. □

Proof of Theorem 2.14 For each integer $n \ge 1$, let $\mathcal{F}_n$ be as guaranteed in Theorem 2.13 with $\gamma = (2n)^{-1}$. Set

$$\mathcal{F} = \bigcap_{n=1}^{\infty} \mathcal{F}_n. \tag{2.154}$$

Clearly, $\mathcal{F}$ is a countable intersection of open (in the weak topology), everywhere dense (in the strong topology) subsets of $\mathcal{M}_f$.

Let $A \in \mathcal{F}$ and $\gamma \in (0, 1)$. Choose a natural number n such that

$$n^{-1} < \gamma/8. \tag{2.155}$$

Since $A \in \mathcal{F}_n$ and the assertion of Theorem 2.13 holds with $\gamma = (2n)^{-1}$ and $\mathcal{F}_\gamma = \mathcal{F}_n$, there are $x_A \in \text{int}(K)$, $r_A \in (0, 1)$, and a neighborhood $\mathcal{U}$ of A in $\mathcal{M}_f$ with the weak topology such that $B(x_A, r_A) \subset K$, $Ax_A = x_A$, and for each $C \in \mathcal{U}$, there is $x_C \in K$ such that $Cx_C = x_C$ and

$$\|x_C - x_A\| \le r_A(2n)^{-1} < r_A\gamma.$$

Thus Theorem 2.14 is also proved. □

2.6 Approximate Fixed Points

Let (K, ρ) be a complete metric space such that

$$\sup\{\rho(x, y) : x, y \in K\} = \infty,$$

and let $(X, \|\cdot\|, \ge)$ be a Banach space ordered by a closed convex cone

$$X_+ = \{x \in X : x \ge 0\}.$$

We assume that $\|x\| \le \|y\|$ for each $x, y \in X_+$ which satisfy $x \le y$.

Denote by $\mathcal{A}$ the set of all continuous mappings $A : K \to X_+$. We equip the set $\mathcal{A}$ with the uniformity determined by the following base:

$$E_s(\varepsilon) = \{(A, B) \in \mathcal{A} \times \mathcal{A} : \|Ax - Bx\| \le \varepsilon \text{ for all } x \in K\}, \tag{2.156}$$

where $\varepsilon > 0$ [80]. Clearly, the uniform space obtained in this way is metrizable and complete. The uniformity determined by (2.156) induces a topology on $\mathcal{A}$ which is called the strong topology.

Denote by $\mathcal{F}_0$ the set of all $A \in \mathcal{A}$ for which

$$\inf\{\|Ax\| : x \in K\} > 0.$$

Theorem 2.18 *The set $\mathcal{F}_0$ is an open everywhere dense subset of $\mathcal{A}$ with the strong topology.*

Proof Let $A \in \mathcal{F}_0$. There is $r > 0$ such that

$$\|Ax\| \geq r \quad \text{for all } x \in K. \tag{2.157}$$

Set

$$U = \{B \in \mathcal{A} : (B, A) \in E_s(r/4)\}. \tag{2.158}$$

Clearly, U is a neighborhood of A in $\mathcal{A}$ with the strong topology. Assume that $B \in U$. Then it follows from (2.157) and (2.158) that for each $x \in K$,

$$\begin{aligned}\|Bx\| &\geq \|Ax\| - \|Ax - Bx\| \\ &\geq r - \|Ax - Bx\| \geq r - r/4 = 3r/4.\end{aligned}$$

Thus $B \in \mathcal{F}_0$. This implies that $U \subset \mathcal{F}_0$. In other words, we have shown that $\mathcal{F}_0$ is an open subset of $\mathcal{F}_0$ with the strong topology.

Now we show that $\mathcal{F}_0$ is an everywhere dense subset of $\mathcal{A}$ with the strong topology. Let $A \in \mathcal{F}_0$ and $\varepsilon > 0$. Choose $u \in X$ such that

$$u \in X_+ \quad \text{and} \quad \|u\| = \varepsilon/2, \tag{2.159}$$

and set

$$Bx = Ax + u, \quad x \in K. \tag{2.160}$$

By (2.159) and (2.160), for each $x \in K$,

$$\|Bx\| = \|Ax + u\| \geq \|u\| = \varepsilon/2.$$

Thus $B \in \mathcal{F}_0$. In view of (2.160), (2.159) and (2.156), $(A, B) \in E_s(\varepsilon)$. Therefore $\mathcal{F}_0$ is an everywhere dense subset of $\mathcal{A}$ with the strong topology. Theorem 2.18 is proved. □

Now we equip the set $\mathcal{A}$ with a topology which will be called the weak topology. Fix $\theta \in K$. For each $\varepsilon, n > 0$, set

$$\begin{aligned}E_w(\varepsilon, n) = \{(A, B) \in \mathcal{A} \times \mathcal{A} : &\|Ax - Bx\| \leq \varepsilon \\ &\text{for each } x \in K \text{ satisfying } \rho(\theta, x) \leq n\}.\end{aligned} \tag{2.161}$$

We equip the set $\mathcal{A}$ with the uniformity determined by the base

$$E_w(\varepsilon, n), \quad \varepsilon, n > 0.$$

Clearly, the uniform space obtained in this way is metrizable and complete. The uniformity determined by (2.161) induces in the set $\mathcal{A}$ a topology which is called the weak topology.

Theorem 2.19 *There exists a set $\mathcal{F}_1 \subset \mathcal{A}$ which is a countable intersection of open everywhere dense subsets of $\mathcal{A}$ with the weak topology such that for each $A \in \mathcal{F}_1$,*

$$\inf\{\|Ax\| : x \in K\} = 0. \tag{2.162}$$

Proof Denote by $\mathcal{E}$ the set of all $A \in \mathcal{A}$ for which there is $x \in K$ such that $Ax = 0$. First we show that $\mathcal{E}$ is an everywhere dense subset of $\mathcal{A}$ with the weak topology. Let $A \in \mathcal{A}$ and $\varepsilon, n > 0$. Choose $\bar{x} \in K$ such that

$$\rho(\theta, \bar{x}) \geq 4n + 4. \tag{2.163}$$

By Urysohn's theorem there is a continuous function $\phi : K \to [0, 1]$ such that

$$\phi(x) = 1 \quad \text{if } \rho(x, \bar{x}) \leq 1$$

and

$$\phi(x) = 0 \quad \text{if } \rho(x, \bar{x}) \geq 2. \tag{2.164}$$

Set

$$Bx = \big(1 - \phi(x)\big)Ax, \quad x \in K. \tag{2.165}$$

Clearly, $B \in \mathcal{A}$. In view of (2.164) and (2.165),

$$\phi(\bar{x}) = 1 \quad \text{and} \quad B\bar{x} = 0.$$

Thus $B \in \mathcal{E}$. Let $x \in K$ satisfy

$$\rho(x, \theta) \leq n. \tag{2.166}$$

It follows from (2.166) and (2.163) that

$$\begin{aligned}\rho(\bar{x}, x) &\geq \rho(\bar{x}, \theta) - \rho(\theta, x) \\ &\geq 4n + 4 - n = 3n + 4.\end{aligned}$$

When combined with (2.164) and (2.165), this implies that

$$\phi(x) = 0 \quad \text{and} \quad Bx = Ax.$$

Thus $Bx = Ax$ for each $x \in K$ satisfying (2.166). The definition of the base E_w (see (2.161)) implies that $(A, B) \in E_w(\varepsilon, n)$. In other words we have shown that $\mathcal{E}$ is an everywhere dense subset of $\mathcal{A}$ with the weak topology.

Let $A \in \mathcal{E}$ and let $n \geq 1$ be an integer. There is $x_A \in K$ such that

$$Ax_A = 0. \tag{2.167}$$

Since A is continuous, there is $r \in (0, 1)$ such that

$$\|Ax\| \le (4n)^{-1} \quad \text{for each } x \in K \text{ satisfying } \rho(x, x_A) \le r. \tag{2.168}$$

Choose an open neighborhood $\mathcal{U}(A, n)$ of A in $\mathcal{A}$ with the weak topology such that

$$\mathcal{U}(A, n) \subset \{B \in \mathcal{A} : (A, B) \in E_w((4n)^{-1}, n + 4 + \rho(\theta, x_A))\}. \tag{2.169}$$

Let

$$B \in \mathcal{U}(A, n), \qquad x \in K, \qquad \rho(x, x_A) \le r. \tag{2.170}$$

By (2.170) and (2.168),

$$\|Ax\| \le (4n)^{-1}. \tag{2.171}$$

In view of (2.170) and since $r < 1$,

$$\begin{aligned}\rho(\theta, x) &\le \rho(\theta, x_A) + \rho(x_A, x) \\ &\le \rho(\theta, x_A) + r < \rho(\theta, x_A) + 1.\end{aligned}$$

Together with (2.169), (2.170) and (2.161), this inequality implies that

$$\|Ax - Bx\| \le (4n)^{-1}.$$

When combined with (2.171), this inequality implies that $\|Bx\| \le 1/n$. Thus we have shown that the following property holds:

(P0) For each $B \in \mathcal{U}(A, n)$, $\inf\{\|Bz\| : z \in K\} \le 1/n$.

Set

$$\mathcal{F}_1 = \bigcap_{n=1}^{\infty} \bigcup \{\mathcal{U}(A, n) : A \in \mathcal{E}\}. \tag{2.172}$$

Clearly, $\mathcal{F}_1$ is a countable intersection of open everywhere dense (in the weak topology) subsets of $\mathcal{A}$.

Let $B \in \mathcal{F}_1$ and $\varepsilon > 0$. Choose a natural number n such that

$$8/n < \varepsilon. \tag{2.173}$$

By (2.172), there is $A \in \mathcal{E}$ such that

$$B \in \mathcal{U}(A, n).$$

It follows from this inclusion, property (P0) and (2.173) that

$$\inf\{\|Bz\| : z \in K\} \le 1/n < \varepsilon.$$

Since ε is an arbitrary positive number, we conclude that

$$\inf\{\|Bz\| : z \in K\} = 0.$$

Theorem 2.19 is proved. □

Assume now that K is a subset of X and that

$$\rho(x, y) = \|x - y\|, \quad x, y \in K.$$

Denote by $\mathcal{B}$ the set of all continuous mappings $A : K \to X$ such that

$$Ax \geq x \quad \text{for all } x \in K.$$

For each $A \in \mathcal{B}$, denote by $J(A)$ the mapping defined by

$$J(A)x = Ax - x, \quad x \in K.$$

Clearly, $J(\mathcal{B}) = \mathcal{A}$, and if $A_1, A_2 \in \mathcal{B}$ are such that

$$J(A_1) = J(A_2),$$

then $A_1 = A_2$. We equip the set $\mathcal{B}$ with the uniformity determined by the following base:

$$\mathcal{E}_s(\varepsilon) = \big\{(A, B) \in \mathcal{B} \times \mathcal{B} : \|Ax - Bx\| \leq \varepsilon \text{ for all } x \in K\big\},$$

where $\varepsilon > 0$. It is not difficult to see that the space $\mathcal{B}$ with this uniformity is metrizable and complete. This uniformity induces in $\mathcal{B}$ a topology which is called the strong topology. It is easy to see that the mapping J is a homeomorphism of the spaces $\mathcal{B}$ and $\mathcal{A}$ with the strong topologies. Thus Theorem 2.18 implies the following result.

Corollary 2.20 *The set of all $A \in \mathcal{B}$ for which*

$$\inf\big\{\|Ax - x\| : x \in K\big\} > 0$$

is an open everywhere dense subset of $\mathcal{B}$ with the strong topology.

We also equip the set $\mathcal{B}$ with the uniformity determined by the following base:

$$\mathcal{E}_w(\varepsilon, n) = \big\{(A, B) \in \mathcal{B} \times \mathcal{B} : \|Ax - Bx\| \leq \varepsilon$$
$$\text{for each } x \in K \text{ satisfying } \|\theta - x\| \leq n\big\}$$

where $n, \varepsilon > 0$. It is not difficult to see that the space $\mathcal{B}$ with this uniformity is metrizable and complete. This uniformity induces in $\mathcal{B}$ a topology which is called the weak topology. It is easy to see that the mapping J is a homeomorphism of the spaces $\mathcal{B}$ and $\mathcal{A}$ with the weak topologies.

Therefore Theorem 2.19 implies the following corollary.

Corollary 2.21 *There exists a set $\mathcal{F} \subset \mathcal{B}$ which is a countable intersection of open and everywhere dense subsets of $\mathcal{B}$ with the weak topology such that for each $A \in \mathcal{F}$,*

$$\inf\big\{\|Ax - x\| : x \in K\big\} = 0.$$

The results of this section were obtained in [157].

2.7 Generic Existence of Small Invariant Sets

In this section we consider generic properties of mappings with approximate fixed points. More precisely, let K be a closed and convex subset of a Banach space $(X, \|\cdot\|)$. We consider a complete metric space of all the continuous self-mappings of K with approximate fixed points. We show that a typical element of this space (in the sense of Baire's categories) has invariant balls of arbitrarily small radii. This result was obtained in [146].

Denote by $\mathcal{A}$ the set of all mappings $A : K \to K$ such that

$$\inf\{\|x - Ax\| : x \in K\} = 0. \tag{2.174}$$

We equip the set $\mathcal{A}$ with the uniformity determined by the following base:

$$E(\varepsilon) = \{(A, B) \in \mathcal{A} \times \mathcal{A} : \|Ax - Bx\| \le \varepsilon \text{ for all } x \in K\}, \tag{2.175}$$

where $\varepsilon > 0$. It is easy to see that the uniform space $\mathcal{A}$ is metrizable (by a metric d).

We first observe that $(\mathcal{A}, d)$ is a complete metric space.

Proposition 2.22 *The metric space $(\mathcal{A}, d)$ is complete.*

Proof Let $\{A_i\}_{i=1}^{\infty} \subset \mathcal{A}$ be a Cauchy sequence. Then for any $\varepsilon > 0$, there is a natural number i_ε such that

$$\|A_i x - A_j x\| \le \varepsilon \quad \text{for all integers } i, j \ge i_\varepsilon \text{ and all } x \in K. \tag{2.176}$$

This implies that for each $x \in K$, $\{A_i x\}_{i=1}^{\infty}$ is a Cauchy sequence and there exists

$$Ax := \lim_{i \to \infty} A_i x. \tag{2.177}$$

Let $\varepsilon > 0$ and let a natural number i_ε satisfy (2.176). Relations (2.176) and (2.177) imply that for each integer $j \ge i_\varepsilon$ and each $x \in K$,

$$\|Ax - A_j x\| = \lim_{i \to \infty} \|A_i x - A_j x\| \le \varepsilon.$$

Thus

$$\|Ax - A_j x\| \le \varepsilon \quad \text{for each integer } j \ge i_\varepsilon \text{ and each } x \in K. \tag{2.178}$$

In order to complete the proof of Proposition 2.22, it is sufficient to show that the mapping A satisfies (2.174).

Let $\delta > 0$. Then in view of (2.178) there is a natural number i_0 such that

$$\|Ax - A_{i_0} x\| \le \delta/4 \quad \text{for all } x \in K. \tag{2.179}$$

Since $A_{i_0} \in \mathcal{A}$, there is $y \in K$ such that

$$\|A_{i_0} y - y\| \le \delta/4.$$

When combined with (2.179), this inequality implies that

$$\|Ay - y\| \le \|Ay - A_{i_0}y\| + \|A_{i_0}y - y\| \le \delta/4 + \delta/4 = \delta/2.$$

Since δ is any positive number, we conclude that $A \in \mathcal{A}$. This completes the proof of Proposition 2.22. □

Denote by $\mathcal{A}_c$ the set of all continuous $A \in \mathcal{A}$. Clearly, $\mathcal{A}_c$ is a closed subset of $(\mathcal{A}, d)$.

Theorem 2.23 *There exists a set $\mathcal{F} \subset \mathcal{A}_c$ which is a countable intersection of open and everywhere dense subsets of $\mathcal{A}_c$ such that each $A \in \mathcal{F}$ has the following property:*

For each $\gamma \in (0,1)$, there are $x_\gamma \in K$, $r \in (0,1]$, and a neighborhood $\mathcal{U}$ of A in $\mathcal{A}_c$ such that for each $C \in \mathcal{U}$,

$$C(\{z \in K : \|z - x_\gamma\| \le r\}) \subset \{z \in K : \|z - x_\gamma\| \le \gamma r\}. \tag{2.180}$$

Corollary 2.24 *Assume that for each $x \in K$, the set $\{z \in K : \|z - x\| \le 1\}$ is compact. Let $\mathcal{F}$ be as guaranteed by Theorem 2.23, and let $A \in \mathcal{F}$, $\gamma \in (0,1)$.*

Then there are $x_A \in K$ and a neighborhood $\mathcal{U}$ of A in $\mathcal{A}_c$ such that for each $C \in \mathcal{U}$, there is a point $z \in K$ so that $\|z - x_A\| \le \gamma$ and $Cz = z$.

Corollary 2.25 *Assume that X is finite-dimensional. Then the assertion of Corollary 2.24 holds.*

Corollary 2.26 *Assume that the assumptions of Corollary 2.24 hold, and that $A \in \mathcal{F}$ and $\varepsilon > 0$. Then there are $\bar{x} \in K$ and $r \in (0,1]$ such that*

$$A\bar{x} = \bar{x} \quad \text{and} \quad A(\{z \in K : \|z - \bar{x}\| \le r\}) \subset \{z \in K : \|z - \bar{x}\| \le \varepsilon r\}.$$

Proof Choose a positive number γ such that

$$\gamma < 1/2 \quad \text{and} \quad \gamma < \varepsilon/8. \tag{2.181}$$

By Theorem 2.23, there are $x_\gamma \in K$ and $r \in (0,1]$ such that (2.180) holds with $C = A$. By Schauder's theorem, there is $\bar{x} \in K$ such that

$$\|\bar{x} - x_\gamma\| \le \gamma r \quad \text{and} \quad A\bar{x} = \bar{x}. \tag{2.182}$$

We have, by (2.182),

$$\{z \in K : \|z - x_\gamma\| \le \gamma r\} \subset \{z \in K : \|z - \bar{x}\| \le 2\gamma r\}.$$

When combined with (2.180) (with $C = A$), this inclusion implies that

$$A(\{z \in K : \|z - x_\gamma\| \le r\}) \subset \{z \in K : \|z - \bar{x}\| \le 2\gamma r\}. \tag{2.183}$$

On the other hand, by (2.181) and (2.182),

$$\{z \in K : \|z - \bar{x}\| \le r/2\} \subset \{z \in K : \|z - x_\gamma\| \le r\}. \tag{2.184}$$

It now follows from (2.184), (2.183) and (2.181) that

$$\begin{aligned} A(\{x \in K : \|z - \bar{x}\| \le r/2\}) &\subset A(\{x \in K : \|z - x_\gamma\| \le r\}) \\ &\subset \{z \in K : \|z - \bar{x}\| \le \varepsilon r/4\}. \end{aligned}$$

Corollary 2.26 is proved. □

Corollary 2.27 *Assume that X is finite-dimensional. Then the assertion of Corollary 2.26 holds.*

Corollary 2.28 *Let K be compact. Then $\mathcal{A}_c$ is the set of all continuous mappings $A : K \to K$ and the assertion of Corollary 2.26 holds.*

We begin the proof of Theorem 2.23 with the following lemma.

Lemma 2.29 *Let $A \in \mathcal{A}_c$ and $\varepsilon > 0$. Then there are $x_* \in K$, $r > 0$, and $B \in \mathcal{A}_c$ such that*

$$\|Ax - Bx\| \le \varepsilon \quad \text{for all } x \in K,$$
$$Bx = x_* \quad \text{for all } x \in K \text{ satisfying } \|x - x_*\| \le r.$$

Proof Since $A \in \mathcal{A}_c$ (see (2.174)), there is $x_* \in K$ such that

$$\|Ax_* - x_*\| \le \varepsilon/8. \tag{2.185}$$

There also is a number $r \in (0, 1)$ such that

$$\|Ax - Ax_*\| \le \varepsilon/8 \quad \text{for each } x \in K \text{ such that } \|x - x_*\| \le 2r. \tag{2.186}$$

By Urysohn's theorem, there exists a continuous function $\phi : K \to [0, 1]$ such that

$$\phi(x) = 1, \quad x \in \{z \in K : \|z - x_*\| \le r\} \tag{2.187}$$

and

$$\phi(x) = 0, \quad x \in K \text{ and } \|x - x_*\| \ge 2r.$$

Set

$$Bx = \phi(x)x_* + (1 - \phi(x))Ax, \quad x \in K. \tag{2.188}$$

Clearly, $B : K \to K$ is continuous, and

$$Bx = x_* \quad \text{for all } x \in K \text{ such that } \|x - x_*\| \le r. \tag{2.189}$$

Now we show that

$$\|Bx - Ax\| \le \varepsilon \quad \text{for all } x \in K.$$

Let $x \in K$. There are two cases: (1) $\|x - x_*\| \le 2r$; (2) $\|x - x_*\| > 2r$.

Consider the first case. Then (2.188), (2.185) and (2.186) imply that

$$\begin{aligned}\|Ax - Bx\| &= \|Ax - \phi(x)x_* - (1 - \phi(x))Ax\| \\ &= \phi(x)\|x_* - Ax\| \le \|x_* - Ax\| \le \|x_* - Ax_*\| + \|Ax_* - Ax\| \\ &\le \varepsilon/8 + \varepsilon/8 = \varepsilon/4.\end{aligned}$$

Consider now the second case. Then by (2.188) and (2.187),

$$\|Ax - Bx\| = \|Ax - Ax\| = 0.$$

Thus $\|Ax - Bx\| \le \varepsilon$ for all $x \in K$. Lemma 2.29 is proved. □

Proof of Theorem 2.23 Denote by $\mathcal{E}$ the set of all $A \in \mathcal{A}_c$ with the following property:

There are $x_* \in K$ and $r > 0$ such that $Ax = x_*$ for all $x \in K$ satisfying $\|x - x_*\| \le r$.

By Lemma 2.29, $\mathcal{E}$ is an everywhere dense subset of $\mathcal{A}_c$.

Let $A \in \mathcal{E}$ and let n be a natural number. There are $x_A \in K$ and $r_A \in (0, 1)$ such that

$$Ax = x_A \quad \text{for all } x \in K \text{ satisfying } \|x - x_A\| \le r_A. \tag{2.190}$$

Denote by $\mathcal{U}(A, n)$ the open neighborhood of A in $\mathcal{A}_c$ such that

$$\mathcal{U}(A, n) \subset \{B \in \mathcal{A}_c : (A, B) \in E(r_A/n)\}. \tag{2.191}$$

Let $B \in \mathcal{U}(A, n)$. Clearly,

$$\|By - Ay\| \le r_A/n \le 1/n \quad \text{for all } y \in K. \tag{2.192}$$

By (2.190) and (2.192), for all $y \in K$ such that $\|y - x_A\| \le r_A$,

$$\|By - x_A\| \le \|By - Ay\| + \|Ay - x_A\| \le \|By - Ay\| \le r_A/n.$$

Thus

$$\|By - x_A\| \le r_A/n \quad \text{for all } y \in K \text{ such that } \|y - x_A\| \le r_A. \tag{2.193}$$

We have shown that the following property holds:

(P1) For each $B \in \mathcal{U}(A, n)$, (2.193) is true.

Define

$$\mathcal{F} = \bigcap_{n=1}^{\infty} \bigcup \{\mathcal{U}(A,n) : A \in \mathcal{E}\}.$$

Clearly, $\mathcal{F}$ is a countable intersection of open and everywhere dense subsets of $\mathcal{A}_c$.

Let $B \in \mathcal{F}$ and $\gamma \in (0,1)$. Choose a natural number n such that $8/n < \gamma$. By the definition of $\mathcal{F}$, there are $A \in \mathcal{E}$ such that

$$B \in \mathcal{U}(A,n). \tag{2.194}$$

It follows from property (P1) and (2.193) that for each $C \in \mathcal{U}(A,n)$,

$$C(\{z \in K : \|z - x_A\| \le r_A\}) \subset \{z \in K : \|z - x_A\| \le r_A/n\}$$
$$\subset \{z \in K : \|z - x_A\| \le \gamma r_A\}.$$

This completes the proof of Theorem 2.23. □

2.8 Many Nonexpansive Mappings Are Strict Contractions

Let K be a nonempty, bounded, closed and convex subset of a Banach space $(X, \|\cdot\|)$. In this section we consider the space of all nonexpansive self-mappings of K equipped with an appropriate complete metric d and prove that the complement of the subset of strict contractions is porous. This result was established in [150].

Set

$$\operatorname{rad}(K) = \sup\{\|x\| : x \in K\} \tag{2.195}$$

and

$$d(K) = \sup\{\|x - y\| : x, y \in K\}.$$

For each $A : K \to X$, let

$$\operatorname{Lip}(A) = \sup\{\|Ax - Ay\|/\|x - y\| : x, y \in K, x \neq y\} \tag{2.196}$$

be the Lipschitz constant of A. Denote by $\mathcal{A}$ the set of all nonexpansive mappings $A : K \to K$, that is, all self-mappings of K with $\operatorname{Lip}(A) \le 1$, or equivalently, all self-mappings of K which satisfy

$$\|Ax - Ay\| \le \|x - y\| \quad \text{for all } x, y \in K. \tag{2.197}$$

We say that a self-mapping $A : K \to K$ is a strict contraction if $\operatorname{Lip}(A) < 1$. Our new metric is defined by

$$d(A,B) = \sup\{\|Ax - Bx\| : x \in K\} + \operatorname{Lip}(A - B), \tag{2.198}$$

where $A, B \in \mathcal{A}$. It is not difficult to see that the metric space $(\mathcal{A}, d)$ is complete.

Theorem 2.30 *Denote by $\mathcal{F}$ the set of all strict contractions $A \in \mathcal{A}$. Then $\mathcal{A} \setminus \mathcal{F}$ is porous.*

Proof Fix a number $\alpha > 0$ such that

$$\alpha < \big(1 + 2\,\mathrm{rad}(K)\big)^{-1} 32^{-1} \tag{2.199}$$

and fix $\theta \in K$. Let $A \in \mathcal{A}$ and let $r \in (0, 1]$. Set

$$\gamma = \big(1 + 2\,\mathrm{rad}(K)\big)^{-1} r/8 \tag{2.200}$$

and put

$$A_\gamma x = (1 - \gamma)Ax + \gamma\theta, \quad x \in K. \tag{2.201}$$

Clearly, $A_\gamma \in \mathcal{A}$ and for each $x, y \in K$,

$$\|A_\gamma x - A_\gamma y\| = (1 - \gamma)\|Ax - Ay\| \le (1 - \gamma)\|x - y\|. \tag{2.202}$$

By (2.201), (2.195), (2.196) and (2.198), for each $x \in K$,

$$\begin{aligned}
\|A_\gamma x - Ax\| &= \big\|(1 - \gamma)Ax + \gamma\theta - Ax\big\| = \gamma\|\theta - Ax\| \\
&\le 2\gamma\,\mathrm{rad}(K), \\
\mathrm{Lip}(A_\gamma - A) &= \sup\big\{\big\|(A_\gamma - A)x - (A_\gamma - A)y\big\|/\|x - y\| : x, y \in K, x \ne y\big\} \\
&= \sup\big\{\big\|(\gamma\theta - \gamma Ax) - (\gamma\theta - \gamma Ay)\big\|/\|x - y\| : x, y \in K, x \ne y\big\} \\
&= \gamma \sup\big\{\|Ax - Ay\|/\|x - y\| : x, y \in K, x \ne y\big\} \le \gamma,
\end{aligned}$$

and

$$d(A, A_\gamma) \le 2\gamma\,\mathrm{rad}(K) + \gamma = \gamma\big(1 + 2\,\mathrm{rad}(K)\big). \tag{2.203}$$

Relations (2.200) and (2.203) imply that

$$d(A, A_\gamma) \le r/8. \tag{2.204}$$

Assume that $B \in \mathcal{A}$,

$$d(B, A_\gamma) \le \alpha r. \tag{2.205}$$

In view of (2.205), (2.198), (2.202) and (2.200), we see that

$$\begin{aligned}
\mathrm{Lip}(B) &\le \mathrm{Lip}(A_\gamma) + \mathrm{Lip}(B - A_\gamma) \le \mathrm{Lip}(A_\gamma) + d(B, A_\gamma) \\
&\le \mathrm{Lip}(A_\gamma) + \alpha r \le (1 - \gamma) + \alpha r \\
&= 1 - (r/8)\big(1 + 2\,\mathrm{rad}(K)\big)^{-1} + r(32\big(1 + 2\,\mathrm{rad}(K)\big)^{-1} \\
&\le 1 - (r/16)\big(1 + 2\,\mathrm{rad}(K)\big)^{-1} < 1
\end{aligned}$$

and so $B \in \mathcal{F}$. Clearly, by (2.205), (2.204) and (2.199),

$$d(B, A) \le d(B, A_\gamma) + d(A_\gamma, A) \le \alpha r + r/8 \le r.$$

Thus for each $B \in \mathcal{A}$ satisfying (2.205), $B \in \mathcal{F}$ and $d(B, A) \le r$. This completes the proof of Theorem 2.30. □

Now let F be a nonempty closed convex subset of K. For each $x \in K$, set

$$\rho(x, F) = \inf\{\|x - y\| : y \in F\}. \tag{2.206}$$

Assume that there exists $P \in \mathcal{A}$ such that

$$P(K) = F, \qquad Px = x, \quad x \in F. \tag{2.207}$$

Denote by $\mathcal{A}^{(F)}$ the set of all $A \in \mathcal{A}$ such that

$$Ax = x, \quad x \in F. \tag{2.208}$$

Clearly, $\mathcal{A}^{(F)}$ is a closed subset of $(\mathcal{A}, d)$.

Theorem 2.31 *Denote by $\mathcal{F}$ the set of all $A \in \mathcal{A}^{(F)}$ which have the following property*:

There is a number $q \in (0, 1)$ such that

$$\rho(Ax, F) \le q\rho(x, F) \quad \textit{for all } x \in K.$$

Then $\mathcal{A}^{(F)} \setminus \mathcal{F}$ is a porous subset of $(\mathcal{A}^{(F)}, d)$.

Proof Fix a number $\alpha > 0$ such that

$$\alpha < \left(1 + 2\,\mathrm{rad}(K)\right)^{-1} 32^{-1}. \tag{2.209}$$

Let $A \in \mathcal{A}^{(F)}$ and $r \in (0, 1]$. Set

$$\gamma = \left(1 + 2\,\mathrm{rad}(K)\right)^{-1} r/8 \tag{2.210}$$

and put

$$A_\gamma x = (1 - \gamma)Ax + \gamma Px, \quad x \in K. \tag{2.211}$$

Clearly, $A_\gamma \in \mathcal{A}$,

$$A_\gamma x = x, \quad x \in F, \quad \text{and} \quad A_\gamma \in \mathcal{A}^{(F)}. \tag{2.212}$$

For each $x \in K$ and $y \in F$, we have by (2.211),

$$\begin{aligned}\rho(A_\gamma x, F) &= \rho\big((1-\gamma)Ax + \gamma Px, F\big)\\ &\le \big\|(1-\gamma)Ax + \gamma Px - \big((1-\gamma)y + \gamma Px\big)\big\|\\ &= (1-\gamma)\|Ax - y\| \le (1-\gamma)\|x - y\|.\end{aligned}$$

Hence

$$\rho(A_\gamma x, F) \le (1-\gamma)\inf\big\{\|x-y\| : y \in F\big\} = (1-\gamma)\rho(x, F).$$

Thus

$$\rho(A_\gamma x, F) \le (1-\gamma)\rho(x, F), \quad x \in K. \tag{2.213}$$

By (2.211), (2.195), and (2.199), we have for $x \in K$,

$$\begin{aligned}\|A_\gamma x - Ax\| = \big\|(1-\gamma)Ax + \gamma Px - Ax\big\| &= \gamma\|Px - Ax\| \le 2\gamma\,\mathrm{rad}(K),\\ \mathrm{Lip}(A_\gamma - A) &= \mathrm{Lip}\big((1-\gamma)A + \gamma P - A\big)\\ &= \mathrm{Lip}(\gamma P - \gamma A) \le 2\gamma\end{aligned}$$

and

$$d(A, A_\gamma) \le 2\gamma\,\mathrm{rad}(K) + 2\gamma = 2\gamma\big(\mathrm{rad}(K) + 1\big). \tag{2.214}$$

It follows from (2.214) and (2.210) that

$$d(A, A_\gamma) \le r/4. \tag{2.215}$$

Assume now that

$$B \in \mathcal{A}^{(F)}$$

and

$$d(B, A_\gamma) \le \alpha r. \tag{2.216}$$

Then by (2.216), (2.215) and (2.209),

$$d(B, A) \le d(B, A_\gamma) + d(A_\gamma, A) \le \alpha r + r/4 \le r. \tag{2.217}$$

Let $x \in K$ and $y \in F$. It follows from (2.208), (2.212), (2.211), (2.196), (2.198), (2.216), (2.209) and (2.210) that

$$\begin{aligned}\rho(Bx, F) &\le \big\|Bx - \big((1-\gamma)y + \gamma Px\big)\big\|\\ &\le \|Bx - A_\gamma x\| + \big\|A_\gamma x - \big[(1-\gamma)y + \gamma Px\big]\big\|\\ &\le \big\|(Bx - By) - (A_\gamma x - A_\gamma y)\big\| + (1-\gamma)\|Ax - y\|\\ &\le \big\|(B - A_\gamma)x - (B - A_\gamma)y\big\| + (1-\gamma)\|x - y\|\\ &\le \mathrm{Lip}(B - A_\gamma)\|x - y\| + (1-\gamma)\|x - y\|\end{aligned}$$

$$\leq \alpha r\|x-y\| + (1-\gamma)\|x-y\| = \|x-y\|(\alpha r + 1 - \gamma)$$
$$\leq \|x-y\|\big(1-\big(1+2\,\mathrm{rad}(K)\big)^{-1}/16\big).$$

Therefore

$$\rho(Bx,F) \leq \big(1-\big(1+2\,\mathrm{rad}(K)\big)^{-1}/16\big)\inf\big\{\|x-y\| : y \in F\big\}$$
$$= \big(1-\big(1+2\,\mathrm{rad}(K)\big)^{-1}/16\big)\rho(x,F).$$

Thus $B \in \mathcal{F}$. This completes the proof of Theorem 2.31. □

2.9 Krasnosel'skii-Mann Iterations of Nonexpansive Operators

In this section we study the convergence of Krasnosel'skii-Mann iterations of nonexpansive operators on a closed and convex, but not necessarily bounded, subset of a hyperbolic space. More precisely, we show that in an appropriate complete metric space of nonexpansive operators, there exists a subset which is a countable intersection of open and everywhere dense sets such that each operator belonging to this subset has a (necessarily) unique fixed point and the Krasnosel'skii-Mann iterations of the operator converge to it.

Let (X,ρ,M) be a complete hyperbolic space and let K be a closed and ρ-convex subset of X. Denote by $\mathcal{A}$ the set of all operators $A : K \to K$ such that

$$\rho(Ax,Ay) \leq \rho(x,y) \quad \text{for all } x,y \in K. \tag{2.218}$$

Fix some $\theta \in K$ and for each $s > 0$, set

$$B(s) = \big\{x \in K : \rho(x,\theta) \leq s\big\}. \tag{2.219}$$

For the set $\mathcal{A}$ we consider the uniformity determined by the following base:

$$E(n) = \big\{(A,B) \in \mathcal{A}\times\mathcal{A} : \rho(Ax,Bx) \leq n^{-1} \text{ for all } x,y \in B(n)\big\}, \tag{2.220}$$

where n is a natural number. Clearly the uniform space $\mathcal{A}$ is metrizable and complete.

A mapping $A : K \to K$ is called regular if there exists a necessarily unique $x_A \in K$ such that

$$\lim_{n\to\infty} A^n x = x_A \quad \text{for all } x \in K.$$

A mapping $A : K \to K$ is called super-regular if there exists a necessarily unique $x_A \in K$ such that for each $s > 0$,

$$A^n x \to x_A \quad \text{as } n \to \infty \text{ uniformly on } B(s).$$

Denote by I the identity operator. For each pair of operators $A, B : K \to K$ and each $r \in [0, 1]$, define an operator $rA \oplus (1-r)B$ by

$$\big(rA \oplus (1-r)B\big)(x) = rAx \oplus (1-r)Bx, \quad x \in K.$$

In this section we prove the following three results [132].

Theorem 2.32 *Let $A : K \to K$ be super-regular and let ε, s be positive numbers. Then there exist a neighborhood U of A in $\mathcal{A}$ and an integer $n_0 \geq 2$ such that for each $B \in U$, each $x \in B(s)$ and each integer $n \geq n_0$, the following inequality holds: $\rho(x_A, B^n x) \leq \varepsilon$.*

Theorem 2.33 *There exists a set $\mathcal{F}_0 \subset \mathcal{A}$ which is a countable intersection of open and everywhere dense sets in $\mathcal{A}$ such that each $A \in \mathcal{F}_0$ is super-regular.*

Let $\{\bar{r}_n\}_{n=1}^{\infty}$ be a sequence of positive numbers from the interval $(0, 1)$ such that

$$\lim_{n\to\infty} \bar{r}_n = 0 \quad \text{and} \quad \sum_{n=1}^{\infty} \bar{r}_n = \infty.$$

Theorem 2.34 *There exists a set $\mathcal{F} \subset \mathcal{A}$ which is a countable intersection of open and everywhere dense sets in $\mathcal{A}$ such that each $A \in \mathcal{F}$ is super-regular and the following assertion holds:*

Let $x_A \in K$ be the unique fixed point of $A \in \mathcal{F}$ and let $\delta, s > 0$. Then there exist a neighborhood U of A in $\mathcal{A}$ and an integer $n_0 \geq 1$ such that for each sequence of positive numbers $\{r_n\}_{n=1}^{\infty}$ satisfying $r_n \in [\bar{r}_n, 1]$, $n = 1, 2, \ldots$, and each $B \in U$ the following relations hold:

(i)

$$\rho\big(\big(r_n B \oplus (1-r_n)I\big)\cdots\big(r_1 B \oplus (1-r_1)I\big)x,$$
$$\big(r_n B \oplus (1-r_n)I\big)\cdots\big(r_1 B \oplus (1-r_1)I\big)y\big) \leq \delta$$

for each integer $n \geq n_0$ and each $x, y \in B(s)$;

(ii) *if $B \in U$ is regular, then*

$$\rho\big(\big(r_n B \oplus (1-r_n)I\big)\cdots\big(r_1 B \oplus (1-r_1)I\big)x, x_A\big) \leq \delta$$

for each integer $n \geq n_0$ and each $x \in B(s)$.

Proof of Theorem 2.32 We may assume that $\varepsilon \in (0, 1)$. Recall that x_A is the unique fixed point of A. There exists an integer $n_0 \geq 4$ such that for each $x \in B(2s + 2 + 2\rho(x_A, \theta))$ and each integer $n \geq n_0$,

$$\rho\big(x_A, A^n x\big) \leq 8^{-1}\varepsilon. \tag{2.221}$$

Set

$$U = \{B \in \mathcal{A} : \rho(Ax, Bx) \le (8n_0)^{-1}\varepsilon, x \in B(8s + 8 + 8\rho(x_A, \theta))\}. \tag{2.222}$$

Let $B \in U$. It is easy to see that for each $x \in K$ and all integers $n \ge 1$,

$$\begin{aligned}\rho(A^n x, B^n x) &\le \rho(A^n x, AB^{n-1}x) + \rho(AB^{n-1}x, B^n x) \\ &\le \rho(A^{n-1}x, B^{n-1}x) + \rho(AB^{n-1}x, B^n x)\end{aligned} \tag{2.223}$$

and

$$\begin{aligned}\rho(B^n x, x_A) &\le \rho(B^n x, A^n x) + \rho(A^n x, x_A) \le \rho(B^n x, A^n x) + \rho(x, x_A) \\ &\le \rho(B^n x, A^n x) + \rho(x, \theta) + \rho(\theta, x_A).\end{aligned} \tag{2.224}$$

Using (2.222), (2.223) and (2.224) we can show by induction that for all $x \in B(4s + 4 + 4\rho(x_A, \theta))$, and for all $n = 1, 2, \dots, n_0$,

$$\rho(A^n x, B^n x) \le (8n_0)^{-1}\varepsilon n \tag{2.225}$$

and

$$\rho(B^n x, \theta) \le 2\rho(x_A, \theta) + \rho(x, \theta) + \frac{1}{2}.$$

Let $y \in B(s)$. We intend to show that $\rho(x_A, B^n y) \le \varepsilon$ for all integers $n \ge n_0$. Indeed, by (2.225),

$$\rho(\theta, B^m y) \le \frac{1}{2} + 2\rho(x_A, \theta) + s, \quad m = 1, \dots, n_0. \tag{2.226}$$

By (2.225) and (2.221),

$$\rho(x_A, B^{n_0} y) \le \varepsilon/2. \tag{2.227}$$

Now we are ready to show by induction that for all integers $m \ge n_0$,

$$\rho(x_A, B^m y) \le \varepsilon. \tag{2.228}$$

By (2.227), inequality (2.228) is valid for $m = n_0$.

Assume that an integer $k \ge n_0$ and that (2.228) is valid for all integers $m \in [n_0, k]$. Together with (2.226) this implies that

$$\rho(\theta, B^i y) \le \frac{1}{2} + 2\rho(x_A, \theta) + s, \quad i = 1, \dots, k. \tag{2.229}$$

Set

$$j = 1 + k - n_0 \quad \text{and} \quad x = B^j y. \tag{2.230}$$

By (2.229), (2.230), (2.221) and (2.225),

$$\rho(A^{n_0}x, B^{n_0}x) \le \varepsilon/8, \qquad \rho(x_A, A^{n_0}x) \le \varepsilon/8 \quad \text{and} \quad \rho(x_A, B^{k+1}y) \le \varepsilon/4.$$

This completes the proof of Theorem 2.32. □

Proof of Theorem 2.33 For each $A \in \mathcal{A}$ and $\gamma \in (0,1)$, define $A_\gamma : K \to K$ by

$$A_\gamma x = (1-\gamma)Ax \oplus \gamma\theta, \quad x \in K.$$

Let $A \in \mathcal{A}$ and $\gamma \in (0,1)$. Clearly,

$$\rho(A_\gamma x, A_\gamma y) \le (1-\gamma)\rho(Ax, Ay) \le (1-\gamma)\rho(x,y), \quad x, y \in K.$$

Therefore there exists $x(A,\gamma) \in K$ such that

$$A_\gamma\big(x(A,\gamma)\big) = x(A,\gamma).$$

Evidently, A_γ is super-regular and the set $\{A_\gamma : A \in \mathcal{A}, \gamma \in (0,1)\}$ is everywhere dense in $\mathcal{A}$. By Theorem 2.32, for each $A \in \mathcal{A}$, each $\gamma \in (0,1)$ and each integer $i \ge 1$, there exist an open neighborhood $U(A,\gamma,i)$ of A_γ in $\mathcal{A}$ and an integer $n(A,\gamma,i) \ge 2$ such that the following property holds:

(i) for each $B \in U(A,\gamma,i)$, each $x \in B(4^{i+1})$ and each $n \ge n(A,\gamma,i)$,

$$\rho\big(x(A,\gamma), B^n x\big) \le 4^{-i-1}.$$

Define

$$\mathcal{F}_0 = \bigcap_{q=1}^{\infty} \bigcup \big\{U(A,\gamma,i) : A \in U, \gamma \in (0,1), i = q, q+1, \ldots\big\}.$$

Clearly, $\mathcal{F}_0$ is a countable intersection of open and everywhere dense sets in $\mathcal{A}$.

Let $A \in \mathcal{F}_0$. There exist sequences $\{A_q\}_{q=1}^{\infty} \subset \mathcal{A}$, $\{\gamma_q\}_{q=1}^{\infty} \subset (0,1)$ and a strictly increasing sequence of natural numbers $\{i_q\}_{q=1}^{\infty}$ such that

$$A \in U(A_q, \gamma_q, i_q), \quad q = 1, 2, \ldots. \tag{2.231}$$

By property (i) and (2.231), for each $x \in B(4^{i_q+1})$ and each integer $n \ge n(A_q, \gamma_q, i_q)$,

$$\rho\big(x(A_q,\gamma_q), A^n x\big) \le 4^{-i_q-1}.$$

This implies that A is super-regular. Theorem 2.33 is proved. □

In order to prove Theorem 2.34 we need the following auxiliary results.

Let

$$\bar{r}_n \in (0,1), \quad n = 1, 2, \ldots, \qquad \lim_{n\to\infty} \bar{r}_n = 0, \qquad \sum_{n=1}^{\infty} \bar{r}_n = 1. \tag{2.232}$$

Lemma 2.35 *Let $A \in \mathcal{A}$, $S_1 > 0$ and let $n_0 \ge 2$ be an integer. Then there exist a neighborhood U of A in $\mathcal{A}$ and a number $S_* > S_1$ such that for each $B \in U$, each sequence $\{r_i\}_{i=1}^{n_0-1} \subset (0,1]$ and each sequence $\{x_i\}_{i=1}^{n_0} \subset K$ satisfying*

$$x_1 \in B(S_1), \qquad x_{i+1} = r_i B x_i \oplus (1-r_i)x_i, \quad i = 1, \ldots, n_0 - 1, \tag{2.233}$$

the following relations hold:

$$x_i \in B(S_*), \quad i = 1, \dots, n_0.$$

Proof Set

$$S_{i+1} = 2S_i + 2 + 2\rho(\theta, A\theta), \quad i = 1, \dots, n_0 - 1, \quad \text{and} \quad S_* = S_{n_0}. \tag{2.234}$$

Set

$$U = \{B \in \mathcal{A} : \rho(Ax, Bx) \le 1, x \in B(S_*)\}. \tag{2.235}$$

Assume that $B \in U$, $\{r_i\}_{i=1}^{n_0-1} \subset (0, 1]$, $\{x_i\}_{i=1}^{n_0} \subset K$ and that (2.233) holds. We will show that

$$\rho(\theta, x_i) \le S_i, \quad i = 1, \dots, n_0. \tag{2.236}$$

Clearly, (2.236) is valid for $i = 1$. Assume that the integer $m \in [1, n_0 - 1]$ and that (2.236) holds for all integers $i = 1, \dots, m$. Then by (2.236) with $i = m$, (2.233), (2.235) and (2.234),

$$\begin{aligned}
\rho(\theta, x_{m+1}) &= \rho(\theta, r_m B(x_m) \oplus (1 - r_m)x_m) \\
&\le \rho(r_m B(\theta) \oplus (1 - r_m)x_m, r_m B(x_m) \oplus (1 - r_m)x_m) \\
&\quad + \rho(\theta, r_m B(\theta) \oplus (1 - r_m)x_m) \\
&\le r_m \rho(\theta, x_m) + \rho(\theta, B(\theta)) + \rho(B(\theta), r_m B(\theta) \oplus (1 - r_m)x_m) \\
&\le S_m + \rho(\theta, A(\theta)) + \rho(A(\theta), B(\theta)) + \rho(B(\theta), x_m) \\
&\le S_m + \rho(\theta, A(\theta)) + 1 + \rho(x_m, \theta) + \rho(\theta, A\theta) + \rho(A(\theta), B(\theta)) \\
&\le 2S_m + 2\rho(\theta, A(\theta)) + 2 = S_{m+1}.
\end{aligned}$$

Lemma 2.35 is proved. □

For each $A \in \mathcal{A}$ and each $\gamma \in (0, 1)$, define $A_\gamma : K \to K$ by

$$A_\gamma x = (1 - \gamma)Ax \oplus \gamma\theta, \quad x \in K. \tag{2.237}$$

Let $A \in \mathcal{A}$ and $\gamma \in (0, 1)$. Clearly,

$$\rho(A_\gamma x, A_\gamma y) \le (1 - \gamma)\rho(x, y), \quad x, y \in K. \tag{2.238}$$

There exists $x(A, \gamma) \in K$ such that

$$A_\gamma(x(A, \gamma)) = x(A, \gamma).$$

Clearly, $\mathcal{A}_\gamma$ is super-regular and the set $\{A_\gamma : A \in \mathcal{A}, \gamma \in (0, 1)\}$ is everywhere dense in $\mathcal{A}$.

Lemma 2.36 *Let* $A \in \mathcal{A}$, $\gamma \in (0,1)$, $r \in (0,1]$ *and* $x, y \in X$. *Then*

$$\rho\bigl(rA_\gamma x \oplus (1-r)x, rA_\gamma y \oplus (1-r)y\bigr) \le (1-\gamma r)\rho(x,y).$$

Proof By (2.238),

$$\begin{aligned}\rho\bigl(rA_\gamma x \oplus (1-r)x, rA_\gamma y \oplus (1-r)y\bigr) &\le r\rho(A_\gamma x, A_\gamma y) + (1-r)\rho(x,y)\\ &\le (1-r)\rho(x,y) + r(1-\gamma)\rho(x,y)\\ &= \rho(x,y)(1-\gamma r).\end{aligned}$$

Lemma 2.36 is proved. □

Lemma 2.37 *Let* $A \in \mathcal{A}$, $\gamma \in (0,1)$ *and* $\delta, S > 0$. *Then there exist a neighborhood* U *of* A_γ *in* $\mathcal{A}$ *and an integer* $n_0 \ge 4$ *such that for each* $B \in U$, *each sequence of numbers* $r_i \in [\bar{r}_i, 1]$, $i = 1, \dots, n_0 - 1$, *and each* $x, y \in B(S)$, *the following inequality holds*:

$$\begin{aligned}&\rho\bigl(\bigl(r_{n_0-1}B \oplus (1-r_{n_0-1})I\bigr)\cdots\bigl(r_1 B \oplus (1-r_1)I\bigr)x,\\ &\quad \bigl(r_{n_0-1}B \oplus (1-r_{n_0-1})I\bigr)\cdots\bigl(r_1 B \oplus (1-r_1)I\bigr)y\bigr) \le \delta.\end{aligned}$$

Proof Choose a number

$$\gamma_0 \in (0,\gamma). \tag{2.239}$$

Clearly, $\prod_{i=1}^{\infty}(1-\gamma_0\bar{r}_i) \to 0$ as $n \to \infty$. Therefore there exists an integer $n_0 \ge 4$ such that

$$(2S+2)\prod_{i=1}^{n_0-1}(1-\gamma_0\bar{r}_i) < \delta/2. \tag{2.240}$$

By Lemma 2.35, there exist a neighborhood U_1 of A_γ in $\mathcal{A}$ and a number $S_* > 0$ such that for each $B \in U_1$, each sequence $\{r_i\}_{i=1}^{n_0-1} \subset (0,1]$, and each sequence $\{x_i\}_{i=1}^{n_0} \subset X$ satisfying

$$x_1 \in B(S), \qquad x_{i+1} = r_i B x_i \oplus (1-r_i)x_i, \quad i = 1, \dots, n_0 - 1, \tag{2.241}$$

the following relations hold:

$$x_i \in B(S_*), \quad i = 1, \dots, n_0. \tag{2.242}$$

Choose a natural number m_1 such that

$$m_1 > 2S_* + 2 \quad \text{and} \quad 8m_1^{-1} < \delta(\gamma-\gamma_0)r_i, \quad i = 1, \dots, n_0 - 1, \tag{2.243}$$

and define

$$U = \bigl\{B \in U_1 : \rho(A_\gamma x, Bx) < m_1^{-1}, x \in B(m_1)\bigr\}. \tag{2.244}$$

Assume that $B \in U$, $r_i \in [\bar{r}_i, 1]$, $i = 1, \ldots, n_0 - 1$, and

$$x, y \in B(S). \tag{2.245}$$

Set

$$\begin{aligned} &x_1 = x, \qquad y_1 = y, \qquad x_{i+1} = r_i B x_i \oplus (1 - r_i) x_i, \\ &y_{i+1} = r_i B y_i \oplus (1 - r_i) y_i, \quad i = 1, \ldots, n_0 - 1. \end{aligned} \tag{2.246}$$

It follows from the definition of U_1 (see (2.241) and (2.242)) that

$$y_i, x_i \in B(S_*), \quad i = 1, \ldots, n_0. \tag{2.247}$$

To prove the lemma it is sufficient to show that

$$\rho(x_{n_0}, y_{n_0}) \le \delta. \tag{2.248}$$

Assume the contrary. Then

$$\rho(x_i, y_i) > \delta, \quad i = 1, \ldots, n_0. \tag{2.249}$$

Fix $i \in \{1, \ldots, n_0 - 1\}$. It follows from (2.246), (2.247), (2.243), (2.244) and (2.237) that

$$\begin{aligned} \rho(x_{i+1}, y_{i+1}) &= \rho\big(r_i B x_i \oplus (1 - r_i) x_i, r_i B y_i \oplus (1 - r_i) y_i\big) \\ &\le \rho\big(r_i A_\gamma x_i \oplus (1 - r_i) x_i, r_i A_\gamma y_i \oplus (1 - r_i) y_i\big) \\ &\quad + \rho(A_\gamma x_i, B x_i) + \rho(A_\gamma y_i, B y_i) \\ &\le r_i \rho(A_\gamma x_i, A_\gamma y_i) + (1 - r_i) \rho(x_i, y_i) + 2m_1^{-1} \\ &\le 2m_1^{-1} + (1 - r_i) \rho(x_i, y_i) + r_i \rho(x_i, y_i)(1 - \gamma) \\ &\le 2m_1^{-1} + \rho(x_i, y_i)\big(1 - r_i + r_i(1 - \gamma)\big) \\ &= 2m_1^{-1} + \rho(x_i, y_i)(1 - \gamma r_i). \end{aligned} \tag{2.250}$$

By (2.250), (2.243) and (2.249),

$$\rho(x_{i+1}, y_{i+1}) \le \rho(x_i, y_i)(1 - \gamma_0 r_i),$$

and since this inequality holds for all $i \in \{1, \ldots, n_0 - 1\}$, it follows from (2.245) and (2.240) that

$$\rho(x_{n_0}, y_{n_0}) \le 2S \prod_{i=1}^{n_0 - 1} (1 - \gamma_0 r_i) < \delta/2.$$

This contradicts (2.249) and proves Lemma 2.37. □

Proof of Theorem 2.34 Let

$$\{\bar{r}_n\}_{n=1}^{\infty} \subset (0,1), \qquad \lim_{n\to\infty} \bar{r}_n = 0, \qquad \sum_{n=1}^{\infty} \bar{r}_n = \infty. \tag{2.251}$$

By Theorem 2.33, there exists a set $\mathcal{F}_0 \subset \mathcal{A}$ which is a countable intersection of open and everywhere dense sets such that each $A \in \mathcal{F}_0$ is super-regular.

For each $A \in \mathcal{A}$ and each $\gamma > 0$, define $A_\gamma \in \mathcal{A}$ by

$$A_\gamma x = (1-\gamma)Ax \oplus \gamma\theta, \quad x \in K.$$

Clearly, A_γ is super-regular, and for each $A \in \mathcal{A}$ and $\gamma \in (0,1)$, there exists $x(A,\gamma) \in K$ for which

$$A_\gamma\big(x(A,\gamma)\big) = x(A,\gamma). \tag{2.252}$$

Let $A \in \mathcal{A}$, $\gamma \in (0,1)$ and let $i \ge 1$ be an integer. By Lemma 2.37, there exist an open neighborhood $U_1(A,\gamma,i)$ of A_γ in $\mathcal{A}$ and an integer $n_0(A,\gamma,i) \ge 4$ such that the following property holds:

(a) for each $B \in U_1(A,\gamma,i)$, each sequence of numbers

$$r_j \in [\bar{r}_j, 1], \quad j = 1,\dots,n_0(A,\gamma,i)-1,$$

and each pair of sequences $\{x_i\}_{i=1}^{n_0(A,\gamma,i)}, \{y_i\}_{i=1}^{n_0(A,\gamma,i)} \subset X$ satisfying

$$x_1, y_1 \in B\big(8^{i+1}\big(4 + 4\rho\big(x(A,\gamma),\theta\big)\big)\big), \tag{2.253}$$

$$x_{i+1} = r_i Bx_i \oplus (1-r_i)x_i, \qquad y_{i+1} = r_i By_i \oplus (1-r_i)y_i,$$
$$i = 1,\dots,n_0(A,\gamma,i)-1, \tag{2.254}$$

the following inequality holds:

$$\rho(x_{n_0(A,\gamma,i)}, y_{n_0(A,\gamma,i)}) \le 8^{-i-1}. \tag{2.255}$$

Since A_γ is super-regular, by Theorem 2.32 there is an open neighborhood $U(A,\gamma,i)$ of A_γ in $\mathcal{A}$ and an integer $n(A,\gamma,i)$, such that

$$U(A,\gamma,i) \subset U_1(A,\gamma,i), \qquad n(A,\gamma,i) \ge n_0(A,\gamma,i), \tag{2.256}$$

and the following property holds:

(b) for each $B \in U(A,\gamma,i)$, each $x \in B(8^{i+1}(2 + 2\rho(x(A,\gamma),\theta)))$ and each integer $m \ge n(A,\gamma,i)$,

$$\rho\big(x(A,\gamma), B^m x\big) \le 8^{-1-i}. \tag{2.257}$$

Define

$$\mathcal{F} = \mathcal{F}_0 \cap \left[\bigcap_{q=1}^{\infty} \bigcup \big\{U(A,\gamma,i) : A \in \mathcal{A}, \gamma \in (0,1), i = q, q+1, \dots\big\}\right].$$

Clearly, $\mathcal{F}$ is a countable intersection of open and everywhere dense sets in $\mathcal{A}$.

Let $A \in \mathcal{F}$. Then $A \in \mathcal{F}_0$ and it is super-regular. There exists $x(A) \in K$ such that

$$A\big(x(A)\big) = x(A). \tag{2.258}$$

There also exist sequences $\{A_q\}_{q=1}^{\infty} \subset \mathcal{A}$, $\{\gamma_q\}_{q=1}^{\infty} \subset (0, 1)$ and a strictly increasing sequence of natural numbers $\{i_q\}_{q=1}^{\infty}$ such that

$$A \in U(A_q, \gamma_q, i_q), \quad q = 1, 2, \dots. \tag{2.259}$$

Let $\delta, s > 0$. Choose a natural number q such that

$$2^q > 16(s+1) \quad \text{and} \quad 2^{-q} < 8^{-1}\delta, \tag{2.260}$$

and consider the open set $U(A_q, \gamma_q, i_q)$.

Let $r_j \in [\bar{r}_j, 1]$, $j = 1, 2, \dots$, and $B \in U(A_q, \gamma_q, i_q)$. By property (a), the first part the theorem (assertion (i)) is valid.

To prove assertion (ii), assume, in addition, that B is regular. Then there is $x(B) \in K$ such that

$$B\big(x(B)\big) = x(B). \tag{2.261}$$

By property (b),

$$\rho\big(x(A_q, \gamma_q), x(A)\big), \rho\big(x(A_q, \gamma_q), x(B)\big) \le 8^{-i_q - 1}. \tag{2.262}$$

Let $x_1 \in B(s)$ and

$$x_{j+1} = r_j B x_j \oplus (1 - r_j) x_j, \quad j = 1, 2, \dots.$$

It follows from property (a) and (2.261) that

$$\rho\big(x_j, x(B)\big) \le 8^{-i_q - 1} \quad \text{for all integers } j \ge n(A_q, \gamma_q, i_q).$$

Together with (2.262) and (2.260), this implies that for all integers $j \ge n(A_q, \gamma_q, i_q)$,

$$\rho\big(x_j, x(A)\big) \le 3 \cdot 8^{-i_q - 1} < \delta.$$

This completes the proof of Theorem 2.34. □

2.10 Power Convergence of Order-Preserving Mappings

In this section we study the asymptotic behavior of the iterations of those order-preserving mappings on an interval $\langle 0, u_* \rangle$ in an ordered Banach space X for which the origin is a fixed point. Here u_* is an interior point of the cone of positive elements X_+ of the space X. Such classes of order-preserving mappings arise, for example, in mathematical economics. We show that for a generic mapping there exists a fixed point which belongs to the interior of X_+ such that the iterations of the mapping with an initial point in the interior of X_+ converge to it.

Let $(X, \|\cdot\|)$ be a Banach space ordered by a closed cone X_+ with a nonempty interior such that $\|x\| \le \|y\|$ for each $x, y \in X_+$ satisfying $x \le y$. For each $u, v \in X$ such that $u \le v$ denote

$$\langle u, v\rangle = \{x \in X : u \le x \le v\}.$$

Let u_* be an interior point of X_+. Define

$$\|x\|_* = \inf\{r \in [0, \infty) : -ru_* \le x \le ru_*\}, \quad x \in X. \tag{2.263}$$

Clearly, $\|\cdot\|_*$ is a norm on X which is equivalent to the norm $\|\cdot\|$.

An operator $A : \langle 0, u_*\rangle \to \langle 0, u_*\rangle$ is called monotone if

$$Ax \le Ay \quad \text{for each } x, y \in \langle 0, u_*\rangle \text{ such that } x \le y. \tag{2.264}$$

Denote by $\mathcal{M}$ the set of all monotone continuous operators $A : \langle 0, u_*\rangle \to \langle 0, u_*\rangle$ such that

$$A(0) = 0 \tag{2.265}$$

and

$$A(\alpha z) \ge \alpha Az \quad \text{for all } z \in \langle 0, u_*\rangle \text{ and } \alpha \in [0, 1]. \tag{2.266}$$

Geometrically, (2.266) means that the hypograph of A is star-shaped with respect to the origin.

For the space $\mathcal{M}$ we define a metric $\rho : \mathcal{M} \times \mathcal{M} \to [0, \infty)$ by

$$\rho(A, B) = \sup\{\|Ax - Bx\|_* : x \in \langle 0, u_*\rangle\}, \quad A, B \in \mathcal{M}. \tag{2.267}$$

It is easy to see that the metric space $\mathcal{M}$ is complete.

An operator $A : \langle 0, u_*\rangle \to \langle 0, u_*\rangle$ is called concave if for all $x, y \in \langle 0, u_*\rangle$ and $\alpha \in [0, 1]$,

$$A(\alpha x + (1-\alpha)y) \ge \alpha Ax + (1-\alpha)Ay. \tag{2.268}$$

We denote by $\mathcal{M}_{co}$ the set of all concave operators $A \in \mathcal{M}$. Clearly, $\mathcal{M}_{co}$ is a closed subset of $\mathcal{M}$. We consider the topological subspace $\mathcal{M}_{co} \subset \mathcal{M}$ with the relative topology.

The spaces $\mathcal{M}$ and $\mathcal{M}_{co}$ are very important, for example, from the point of view of mathematical economics. In this area of research order-preserving mappings A are usually models of economic dynamics and the condition $A(0) = 0$ means that if we have no resources, then we produce nothing. Concavity means that the combination of resources allows one to produce at least the corresponding combination of outputs and even more than this combination. Monotonicity means that a larger input leads to a larger output. A particular class of concave operators are those operators which are positively homogeneous of degree $m \le 1$. Such operators were studied by many mathematical economists in the finite dimensional case (see [105] and the references mentioned there). For more information on ordered Banach spaces, order-preserving mappings and their applications see, for example, [3, 4].

We are now ready to state and prove the main result of this section. This result was established in [164].

Theorem 2.38 *There exist a set $\mathcal{F} \subset \mathcal{M}$ which is a countable intersection of open and everywhere dense sets in $\mathcal{M}$ and a set $\mathcal{F}_{co} \subset \mathcal{F} \cap \mathcal{M}_{co}$ which is a countable intersection of open and everywhere dense sets in $\mathcal{M}_{co}$ such that for each $P \in \mathcal{F}$, there exists $x_P \in \langle 0, u_* \rangle$ for which the following two assertions hold:*

1. *The point x_P is an interior point of X_+ and $\lim_{t \to \infty} P^t x = x_P$ for each $x \in \langle 0, u_* \rangle$ which is an interior point of the cone X_+.*
2. *For each $\gamma, \varepsilon \in (0, 1)$, there exist an integer $N \geq 1$ and a neighborhood U of P in $\mathcal{M}$ such that for each $C \in U$, each $z \in \langle \gamma u_*, u_* \rangle$ and each integer $T \geq N$,*

$$\left\| C^T z - x_P \right\|_* \leq \varepsilon.$$

Proof of Theorem 2.38 For each $x, y \in X_+$ define

$$\lambda(x, y) = \sup\{r \in [0, \infty) : rx \leq y\}. \tag{2.269}$$

In the proof of Theorem 2.38 we will use several auxiliary results.

Lemma 2.39 *The function $y \to \lambda(u_*, y)$, $y \in X_+$, is continuous, concave and positively homogeneous.*

Proof All we need to show is that the function $y \to \lambda(u_*, y)$, $y \in X_+$, is continuous. To this end, assume that $y \in X_+$, $\{y_n\}_{n=1}^{\infty} \subset X_+$ and

$$\|y_n - y\|_* \to 0 \quad \text{as } n \to \infty. \tag{2.270}$$

We show that

$$\lambda(u_*, y_n) \to \lambda(u_*, y) \quad \text{as } n \to \infty. \tag{2.271}$$

It is well known that (2.271) is true if y is an interior point of X_+. Therefore we may assume that y is not an interior point of X_+.

Clearly,

$$\lambda(u_*, y) = 0. \tag{2.272}$$

We show that

$$\lim_{n \to \infty} \lambda(u_*, y_n) = 0. \tag{2.273}$$

Assume the contrary. Then there exists a subsequence $\{y_{n_k}\}_{k=1}^{\infty}$ and a number $r > 0$ such that

$$y_{n_k} \geq r u_*, \quad k = 1, 2, \ldots. \tag{2.274}$$

Together with (2.270) this implies that

$$y \geq r u_* \quad \text{and} \quad \lambda(u_*, y) \geq r.$$

Since this contradicts (2.272), we see that (2.273) does hold. This completes the proof of Lemma 2.39. □

Define now an operator $\phi : \langle 0, u_* \rangle \to X_+$ by

$$\phi(x) = \lambda(u_*, x)^{1/2} u_*, \quad x \in \langle 0, u_* \rangle. \tag{2.275}$$

By using Lemma 2.39 one can easily check that

$$\phi \in \mathcal{M}_{co}. \tag{2.276}$$

Let $A \in \mathcal{M}$ and let $i \geq 1$ be an integer. Define an operator $A^{(i)} : \langle 0, u_* \rangle \to \langle 0, u_* \rangle$ by

$$A^{(i)} x = \left(1 - 2^{-i}\right) A x + 2^{-i} \phi(x), \quad x \in \langle 0, u_* \rangle. \tag{2.277}$$

Lemma 2.40 *Let $A \in \mathcal{M}$ and let $i \geq 1$ be an integer. Then $A^{(i)} \in \mathcal{M}$. Moreover, if $A \in \mathcal{M}_{co}$, then $A^{(i)} \in \mathcal{M}_{co}$.*

It is clear that for each $A \in \mathcal{M}$ and each integer $i \geq 1$,

$$\rho\left(A^{(i)}, A\right) \leq 2^{-i}. \tag{2.278}$$

Lemma 2.41 *Let $A \in \mathcal{M}$ and let $i \geq 1$ be an integer. Then*

$$A^{(i)}\left(16^{-i} u_*\right) \geq 8^{-i} u_*. \tag{2.279}$$

Proof By (2.277) and (2.275),

$$A^{(i)}\left(16^{-i} u_*\right) \geq 2^{-i} \phi\left(16^{-i} u_*\right) \geq 2^{-i}\left(16^{-i}\right)^{1/2} u_* \geq 8^{-i} u_*.$$ □

For each $A \in \mathcal{M}$ and each integer $i \geq 1$, we now define the operator $B^{(A,i)} : \langle 0, u_* \rangle \to \langle 0, u_* \rangle$ by

$$B^{(A,i)}(x) = \left(1 - 16^{-i}\right) A^{(i)} x + \min\left\{\lambda(u_*, x), 16^{-i}\right\} u_*, \quad x \in \langle 0, u_* \rangle. \tag{2.280}$$

Lemma 2.42 *Let $A \in \mathcal{M}$ and let $i \geq 1$ be an integer. Then*

$$B^{(A,i)}\left(16^{-i} u_*\right) \geq \left(8^{-i} + 2^{-1} \cdot 16^{-i}\right) u_* \tag{2.281}$$

and $B^{(A,i)} \in \mathcal{M}$. Moreover, if $A \in \mathcal{M}_{co}$, then $B^{(A,i)} \in \mathcal{M}_{co}$.

Proof It follows from (2.280) and (2.279) that

$$B^{(A,i)}(16^{-i}u_*) \geq (1-16^{-i})(8^{-i}u_*) + 16^{-i}u_* \geq (8^{-i} + 2^{-1} \cdot 16^{-i})u_*.$$

Therefore (2.271) is valid. By Lemma 2.40, $B^{(A,i)}(0) = 0$ and the operator $B^{(A,i)}$ is monotone. Lemmas 2.39, 2.40 and (2.280) imply that $B^{(A,i)}$ is a continuous operator. It follows from Lemma 2.39 that the operator

$$x \to \min\{\lambda(u_*, x), 16^{-i}\}u_*, \quad x \in \langle 0, u_* \rangle,$$

is concave. When combined with (2.280), Lemma 2.40 and (2.264), this implies that

$$B^{(A,i)}(\alpha z) \geq \alpha B^{(A,i)} z \quad \text{for each } z \in \langle 0, u_* \rangle \text{ and each } \alpha \in [0,1],$$

and that if $A \in \mathcal{M}_{co}$, then $B^{(A,i)}$ is concave. This completes the proof of Lemma 2.42. □

It follows from (2.280), (2.278) and (2.267) that for each $A \in \mathcal{M}$ and each integer $i \geq 1$,

$$\rho(A, B^{(A,i)}) \leq 2^{-i} + 16^{-i}. \tag{2.282}$$

Lemma 2.43 *Let $A \in \mathcal{M}$ and let $i \geq 1$ be an integer. Then*

$$\lim_{t\to\infty} \lambda((B^{(A,i)})^t(u_*), (B^{(A,i)})^t(16^{-i}u_*)) = 1. \tag{2.283}$$

Proof Clearly,

$$(B^{(A,i)})^{t+1}(u_*) \leq (B^{(A,i)})^t(u_*), \quad t = 1, 2, \ldots \tag{2.284}$$

and

$$(B^{(A,i)})^t(16^{-i}u_*) \leq (B^{(A,i)})^t(u_*), \quad t = 1, 2, \ldots.$$

Lemma 2.42 (see 2.361)) implies that for each integer $t \geq 1$,

$$(B^{(A,i)})^{t+1}(16^{-i}u_*) \geq (B^{(A,i)})^t(16^{-i}u_*) \geq (8^{-i} + 2^{-1} \cdot 16^{-i})u_*. \tag{2.285}$$

For $t = 0, 1, \ldots$ we set

$$\lambda_t = \lambda((B^{(A,i)})^t(u_*), (B^{(A,i)})^t(16^{-i}u_*)). \tag{2.286}$$

By (2.284),

$$\lambda_t \leq 1, \quad t = 0, 1, \ldots. \tag{2.287}$$

Let $t \geq 0$ be an integer. It follows from (2.280), (2.286), (2.269), (2.285), Lemma 2.40 and (2.287) that

$$\begin{aligned}
\left(B^{(A,i)}\right)^{t+1}\left(16^{-i}u_*\right) &= B^{(A,i)}\left(\left(B^{(A,i)}\right)^t\left(16^{-i}u_*\right)\right) \\
&= \left(1-16^{-i}\right)A^{(i)}\left(\left(B^{(A,i)}\right)^t\left(16^{-i}u_*\right)\right) \\
&\quad + \min\left\{\lambda\left(u_*, \left(B^{(A,i)}\right)^t\left(16^{-i}u_*\right)\right), 16^{-i}\right\}u_* \\
&\ge \left(1-16^{-i}\right)A^{(i)}\left(\lambda_t\left(B^{(A,i)}\right)^t(u_*)\right) + 16^{-i}u_* \\
&\ge \left(1-16^{-i}\right)\lambda_t A^{(i)}\left(\left(B^{(A,i)}\right)^t(u_*)\right) + 16^{-i}u_* \\
&= \lambda_t\left[\left(1-16^{-i}\right)A^{(i)}\left(\left(B^{(A,i)}\right)^t(u_*)\right) + 16^{-i}u_*\right] \\
&\quad + (1-\lambda_t)16^{-i}u_* \\
&= \lambda_t\left[\left(1-16^{-i}\right)A^{(i)}\left(B^{(A,i)}\right)^t(u_*)\right. \\
&\quad \left. + \min\left\{\lambda\left(u_*, \left(B^{(A,i)}\right)^t(u_*)\right), 16^{-i}\right\}u_*\right] + (1-\lambda_t)16^{-i}u_* \\
&= \lambda_t\left(B^{(A,i)}\right)^{t+1}(u_*) + (1-\lambda_t)16^{-i}u_* \\
&\ge \left(\lambda_t + (1-\lambda_t)16^{-i}\right)\left(B^{(A,i)}\right)^{t+1}(u_*).
\end{aligned}$$

This implies that

$$\lambda_{t+1} \ge \lambda_t + (1-\lambda_t)16^{-i}. \tag{2.288}$$

Combining (2.287) and (2.288), we see that

$$\Lambda = \lim_{t\to\infty} \lambda_t \tag{2.289}$$

exists. By (2.289) and (2.288), $\Lambda \ge \Lambda + (1-\Lambda)16^{-1}$. By (2.287) this implies that $\Lambda = 1$. Lemma 2.43 is proved. □

Lemma 2.44 *Let $A \in \mathcal{M}$ and let $i \ge 1$ be an integer. Then there exists $x^{(A,i)} \in \langle 0, u_*\rangle$ such that*

$$x^{(A,i)} \ge \left(8^{-i} + 2^{-1}\cdot 16^{-i}\right)u_* \tag{2.290}$$

and

$$\lim_{t\to\infty}\left(B^{(A,i)}\right)^t\left(16^{-i}u_*\right) = \lim_{t\to\infty}\left(B^{(A,i)}\right)^t(u_*) = x^{(A,i)}. \tag{2.291}$$

Proof It is clear that inequalities (2.284) hold. Lemma 2.42 implies that for each integer $t \ge 1$, inequality (2.283) is also valid. By Lemma 2.43, (2.284) and (2.285),

$$\lim_{t\to\infty}\left[\left(B^{(A,i)}\right)^t u_* - \left(B^{(A,i)}\right)^t\left(16^{-i}u_*\right)\right] = 0, \tag{2.292}$$

and $\{(B^{(A,i)})^t u_*\}_{t=1}^{\infty}$, as well as $\{(B^{(A,i)})^t(16^{-i}u_*)\}_{t=1}^{\infty}$, are Cauchy sequences.

Therefore there exist $x_1, x_2 \in \langle 0, u_*\rangle$ such that

$$x_1 = \lim_{t\to\infty}\left(B^{(A,i)}\right)^t\left(16^{-i}u_*\right) \quad \text{and} \quad x_2 = \lim_{t\to\infty}\left(B^{(A,i)}\right)^t u_*.$$

By (2.292) and (2.285), $x_1 = x_2 \geq (8^{-i} + 2^{-1} \cdot 16^{-i})u_*$. This completes the proof of Lemma 2.44. □

Lemma 2.45 *Let $A \in \mathcal{M}$, $\varepsilon > 0$, $z \in \langle 0, u_* \rangle$ and let $n \geq 1$ be an integer. Then there exists a neighborhood U of A in $\mathcal{M}$ such that for each $C \in U$,*

$$\left\|C^n z - A^n z\right\|_* < \varepsilon.$$

Proof We prove the lemma by induction. It is clear that the assertion of the lemma is valid for $n = 1$. Assume that it is valid for an integer $n \geq 1$. There exists

$$\delta \in \left(0, 8^{-1}\varepsilon\right) \tag{2.293}$$

such that

$$\left\|Ay - A\left(A^n z\right)\right\|_* \leq 8^{-1}\varepsilon \tag{2.294}$$

for each $y \in \langle 0, u_* \rangle$ satisfying $\|y - A^n z\|_* \leq \delta$. Since the assertion of the lemma is assumed to be valid for n, there exists a neighborhood U_0 of A in $\mathcal{M}$ such that for each $C \in U_0$,

$$\left\|C^n z - A^n z\right\|_* < \delta. \tag{2.295}$$

Set

$$U = \left\{C \in U_0 : \rho(C, A) < 8^{-1}\varepsilon\right\}, \tag{2.296}$$

and let $C \in U$. The definition of U implies that

$$\begin{aligned} \left\|A^{n+1}z - C^{n+1}z\right\|_* &\leq \left\|A^{n+1}z - AC^n z\right\|_* + \left\|AC^n z - C^{n+1}z\right\|_* \\ &\leq \left\|A^{n+1}z - AC^n z\right\|_* + 8^{-1}\varepsilon. \end{aligned} \tag{2.297}$$

By (2.295),

$$\left\|A^n z - C^n z\right\|_* < \delta.$$

It follows from this inequality and the choice of δ (see (2.293) and (2.294)) that

$$\left\|AC^n z - A\left(A^n z\right)\right\|_* \leq 8^{-1}\varepsilon.$$

Together with (2.297) this implies that

$$\left\|A^{n+1}z - C^{n+1}z\right\|_* \leq 4^{-1}\varepsilon.$$

This completes the proof of Lemma 2.45. □

Let $A \in \mathcal{M}$ and let $i \geq 1$ be an integer. By Lemma 2.44, there exists an integer $N(A, i) \geq 4$ such that

$$\left\|\left(B^{(A,i)}\right)^{N(A,i)}\left(16^{-i}u_*\right) - \left(B^{(A,i)}\right)^{N(A,i)}(u_*)\right\|_* \leq 16^{-i-1}. \tag{2.298}$$

By Lemma 2.45, there exists an open neighborhood $U(A,i)$ of $B^{(A,i)}$ in $\mathcal{M}$ such that

$$U(A,i) \subset \{C \in \mathcal{M} : \rho(C, B^{(A,i)}) \le 16^{-i-2}\}, \tag{2.299}$$

and for each $C \in U(A,i)$,

$$\begin{aligned} \|C^{N(A,i)}(16^{-i}u_*) - (B^{(A,i)})^{N(A,i)}(16^{-i}u_*)\|_* &\le 16^{-i-2}, \\ \|C^{N(A,i)}(u_*) - (B^{(A,i)})^{N(A,i)}(u_*)\|_* &\le 16^{-i-2}. \end{aligned} \tag{2.300}$$

Lemma 2.46 *Let $A \in \mathcal{M}$ and let $i \ge 1$ be an integer. Assume that $C \in U(A,i)$. Then*

$$C^t(16^{-i}u_*) \ge 8^{-i}u_*, \quad t = 1, 2, \ldots, \tag{2.301}$$

and for each $z \in \langle 16^{-i}u_*, u_*\rangle$ and each integer $T \ge N(A,i)$, the following inequality holds:

$$\|C^T z - x(A,i)\|_* \le 16^{-i-1} + 16^{-i-2}. \tag{2.302}$$

Proof By the definition of $U(A,i)$ (see (2.299)) and Lemma 2.42 (see (2.281)),

$$\|C(16^{-i}u_*) - B^{(A,i)}(16^{-i}u_*)\|_* \le 16^{-i-2}$$

and

$$\begin{aligned} C(16^{-i}u_*) &\ge B^{(A,i)}(16^{-i}u_*) - 16^{-i-2}u_* \\ &\ge (8^{-i} + 2^{-1} \cdot 16^{-i})u_* - 16^{-i-2}u_* \ge 8^{-i}u_*. \end{aligned} \tag{2.303}$$

Since the operator C is monotone, (2.303) implies that

$$C^{t+1}(16^{-i}u_*) \ge C^t(16^{-i}u_*), \quad t = 0, 1, \ldots. \tag{2.304}$$

Inequalities (2.304) and (2.303) imply (2.301), as claimed.

Assume that $z \in \langle 16^{-i}u_*, u_*\rangle$ and let $T \ge N(A,i)$ be an integer. Since the operator C is monotone, it follows from (2.304) and the definition of $U(A,i)$ (see (2.300)) that

$$\begin{aligned} C^T z &\in \langle C^T(16^{-i}u_*), C^T(u_*)\rangle \subset \langle C^{N(A,i)}(16^{-i}u_*), C^{N(A,i)}u_*\rangle \\ &\subset \langle (B^{(A,i)})^{N(A,i)}(16^{-i}u_*) \\ &\quad - 16^{-i-2}u_*, (B^{(A,i)})^{N(A,i)}(u_*) + 16^{-i-2}u_*\rangle. \end{aligned} \tag{2.305}$$

By Lemma 2.44, (2.281), (2.305) and (2.298),

$$\begin{aligned} C^T z - x(A,i) \in \langle &(B^{(A,i)})^{N(A,i)}(16^{-i}u_*) - 16^{-i-2}u_* - x(A,i), \\ &(B^{(A,i)})^{N(A,i)}(u_*) + 16^{-i-2}u_* - x(A,i)\rangle \end{aligned}$$

and

$$\begin{aligned}
&x(A,i) - C^T z, -x(A,i) + C^T z \\
&\le \left(B^{(A,i)}\right)^{N(A,i)}(u_*) - \left(B^{(A,i)}\right)^{N(A,i)}\left(16^{-i}u_*\right) + 16^{-i-2}u_* \\
&\le \left(16^{-i-1} + 16^{-i-2}\right)u_*.
\end{aligned}$$

This implies (2.302) and completes the proof of Lemma 2.46. □

Completion of the proof of Theorem 2.38 Define

$$\mathcal{F} = \bigcap_{q=1}^{\infty} \bigcup \{U(A,i) : A \in \mathcal{M}, i = q, q+1, \ldots\}$$

and

$$\mathcal{F}_{co} = \bigcap_{q=1}^{\infty} \bigcup \{U(A,i) \cap \mathcal{M}_{co} : A \in \mathcal{M}_{co}, i = q, q+1, \ldots\}.$$

It is easy to see that $\mathcal{F}_{co} \subset \mathcal{F} \cap \mathcal{M}_{co}$, $\mathcal{F}$ is a countable intersection of open and everywhere dense sets in $\mathcal{M}$, and that $\mathcal{F}_{co}$ is a countable intersection of open and everywhere dense sets in $\mathcal{M}_{co}$. Assume that $P \in \mathcal{F}$ and $\varepsilon, \gamma \in (0,1)$. Choose a natural number q for which

$$64 \cdot 2^{-q} < 64^{-1} \min\{\varepsilon, \gamma\}. \tag{2.306}$$

There exist $A \in \mathcal{M}$ and a natural number $i \ge q$ such that

$$P \in U(A,i). \tag{2.307}$$

By Lemma 2.46,

$$C^t\left(16^{-i}u_*\right) \ge 8^{-i}u_* \quad \text{for all integers } t \ge 1 \text{ and all } C \in U(A,i), \tag{2.308}$$

and

$$\begin{aligned}
&\left\|C^T z - x(A,i)\right\|_* \le 16^{-i-1} + 16^{-i-2} \quad \text{for all } C \in U(A,i), \\
&\quad \text{each integer } T \ge N(A,i) \text{ and each } z \in \left\langle 16^{-i}u_*, u_*\right\rangle.
\end{aligned} \tag{2.309}$$

Now (2.309), (2.306) and (2.307) imply that

$$\begin{aligned}
&\left\|P^T z - x(A,i)\right\|_* \le \varepsilon \quad \text{for each integer } T \ge N(A,i) \\
&\quad \text{and each } z \in \langle \gamma u_*, u_*\rangle.
\end{aligned} \tag{2.310}$$

Since ε is an arbitrary number in the interval $(0, 1)$, we conclude that for each $z \in \langle \gamma u_*, u_* \rangle$, there exists $\lim_{t\to\infty} P^t z$. By (2.310),

$$\left\| \lim_{t\to\infty} P^T z - x(A, i) \right\|_* \le \varepsilon \quad \text{for each } z \in \langle \gamma u_*, u_* \rangle. \tag{2.311}$$

Hence

$$\lim_{t\to\infty} P^t z_1 = \lim_{t\to\infty} P^t z_2$$

for each $z_1, z_2 \in \langle \gamma u_*, u_* \rangle$.

Since $\gamma \in (0, 1)$ is also arbitrary, we conclude that

$$\lim_{t\to\infty} P^t z = x_P \tag{2.312}$$

for each $z \in \langle 0, u_* \rangle$ which is an interior point of X_+. By (2.308), x_P is an interior point of X_+. Now (2.309) implies that

$$\left\| x_P - x(A, i) \right\|_* \le 16^{-i-1} + 16^{-i-2}. \tag{2.313}$$

Assume that $C \in U(A, i)$, $z \in \langle \gamma u_*, u_* \rangle$, and let $T \ge N(A, i)$ be an integer. It follows from (2.309), (2.313) and (2.306) that

$$\begin{aligned} \left\| C^T z - x_P \right\|_* &\le \left\| x_P - x(A, i) \right\|_* + \left\| x(A, i) - C^T z \right\|_* \\ &\le 16^{-i-1} + 16^{-i-2} + \left\| x(A, i) - C^T z \right\|_* \\ &\le 2\left(16^{-i-1} + 16^{-i-2}\right) < \varepsilon. \end{aligned}$$

This completes the proof of Theorem 2.38. □

2.11 Positive Eigenvalues and Eigenvectors

In this section we consider a closed cone of positive operators on an ordered Banach space and prove that a generic element of this cone has a unique positive eigenvalue and a unique (up to a positive multiple) positive eigenvector. Moreover, the normalized iterations of such a generic element converge to its unique eigenvector. This section is based on [140].

Let $(X, \|\cdot\|)$ be a Banach space which is ordered by a closed convex cone X_+. For each $u, v \in X$ such that $u \le v$, we define $\langle u, v \rangle = \{z \in X : u \le z \le v\}$.

We assume that the cone X_+ has a nonempty interior and that for each $x, y \in X_+$ satisfying $x \le y$, the inequality $\|x\| \le \|y\|$ holds. We denote by $\mathrm{int}(X_+)$ the set of all interior points of X_+.

Fix an interior point η of the cone X_+ and define

$$\|x\|_\eta = \inf\{r \in [0, \infty) : -r\eta \le x \le r\eta\}, \quad x \in X. \tag{2.314}$$

Clearly, $\|\cdot\|_\eta$ is a norm on X which is equivalent to the original norm $\|\cdot\|$.

Let X' be the space of all linear continuous functionals $f : X \to R^1$ and let

$$X'_+ = \{f \in X' : f(x) \geq 0 \text{ for all } x \in X_+\}.$$

Denote by $\mathcal{A}$ the set of all linear operators $A : X \to X$ such that $A(X_+) \subset X_+$. Such operators are called positive. For the set $\mathcal{A}$ we define a metric $\rho(\cdot, \cdot)$ by

$$\rho(A, B) = \sup\{\|Ax - Bx\|_\eta : x \in \langle 0, \eta\rangle\}, \quad A, B \in \mathcal{A}.$$

This metric ρ is equivalent to the metrics induced by the operator norms derived from $\|\cdot\|$ and $\|\cdot\|_\eta$. It is clear that the metric space $(\mathcal{A}, \rho)$ is complete. Since many linear operators between Banach spaces arising in classical and modern analysis are, in fact, positive operators, the theory of positive linear operators and its applications have drawn the attention of more and more mathematicians. See, for example, [3, 86, 170] and the references cited therein.

In this section we study the asymptotic behavior of powers of positive linear operators on the ordered Banach space X. We obtain generic convergence to an operator of the form $f(\cdot)\eta$, where f is a bounded linear functional and η is a unique (up to a positive multiple) eigenvector.

We denote by $\mathcal{A}_*$ the set of all $A \in \mathcal{A}$ such that $A\xi = \xi$ for some $\xi \in \text{int}(X_+)$ and by $\bar{\mathcal{A}}_*$ the closure of $\mathcal{A}_*$ in $(\mathcal{A}, \rho)$. We equip the subspace $\bar{\mathcal{A}}_* \subset \mathcal{A}$ with the same metric ρ.

In our paper [125] we established the following result.

Theorem 2.47 *There exists a set $\mathcal{F} \subset \bar{\mathcal{A}}_*$ which is a countable intersection of open and everywhere dense sets in $\bar{\mathcal{A}}*$ such that for each $B \in \mathcal{F}$, there exists an interior point ξ_B of X_+ satisfying $B\xi_B = \xi_B$, $\|\xi_B\|_\eta = 1$, and the following two assertions hold*:

1. *There exists $f_B \in X'_+$ such that $\lim_{T\to\infty} B^T x = f_B(x)\xi_B$, $x \in X$.*
2. *For each $\varepsilon > 0$, there exists a neighborhood $\mathcal{U}$ of B in $\bar{\mathcal{A}}_*$ and a natural number N such that for each $C \in \mathcal{U} \cap \mathcal{A}_*$, each integer $T \geq N$ and each $x \in \langle -\eta, \eta\rangle$,*

$$\|C^T x - f_B(x)\xi_B\| \leq \varepsilon.$$

Since the existence of fixed points and the convergence of iterates is of fundamental importance, it is of interest to look for a larger subset of $\mathcal{A}$ for which such a result continues to hold. To this end, we introduce the set $\mathcal{A}_{q*}$ of all $A \in \mathcal{A}$ for which there exist $c_0 \in (0, 1)$ and $c_1 > 1$ such that

$$c_0\eta \leq A^n\eta \leq c_1\eta \quad \text{for all integers } n \geq 1. \tag{2.315}$$

Note that our definition of $\mathcal{A}_{q*}$ does not depend on our choice of η. Since $\mathcal{A}_* \subset \mathcal{A}_{q*}$, it is natural to ask if there is also a generic result for the closure $\bar{\mathcal{A}}_{q*}$ of $\mathcal{A}_{q*}$. Note that in contrast with $\bar{\mathcal{A}}_*$, it is not clear *a priori* if $\mathcal{A}_*$ is dense in $\bar{\mathcal{A}}_{q*}$. However, as we show in our first result that this is indeed the case.

Theorem 2.48 $\bar{\mathcal{A}}_{q*} = \bar{\mathcal{A}}_*$.

Combining Theorems 2.47 and 2.48, we see that a generic element in $\bar{\mathcal{A}}_{q*}$ has a unique (up to a positive multiple) positive fixed point and all its iterations converge to some multiple of this fixed point.

Since the existence of positive eigenvectors which are not necessarily fixed points is even more important, we devote most of the section to this problem.

Known results about the existence of positive fixed points and eigenvectors include the classical Perron-Frobenius and Krein-Rutman theorems. For a survey of more recent results of the linear theory, see Sect. 2 in [106].

We begin with the following definition.

We say that an operator $A \in \mathcal{A}$ is regular if there exist $x_A \in \text{int}(X_+)$ satisfying $\|x_A\|_\eta = 1$, $\alpha_A > 0$ and $f_A \in X'_+ \setminus \{0\}$ such that

$$Ax_A = \alpha_A x_A, \qquad \alpha_A^{-n} A^n x \to f_A(x) x_A \quad \text{as } n \to \infty,$$

uniformly for all $x \in \langle -\eta, \eta \rangle$.

Note that in the definition above, x_A, α_A and f_A are all uniquely defined and that if $x \in \text{int}(X_+)$, then $\|A^n x\|_\eta^{-1} A^n x \to x_A$ as $n \to \infty$.

We denote by $\mathcal{A}_{reg}$ the set of all regular operators in $\mathcal{A}$ and by $\bar{\mathcal{A}}_{reg}$ its closure in the space $(\mathcal{A}, \rho)$. We endow the subspace $\bar{\mathcal{A}}_{reg} \subset \mathcal{A}$ with the same metric ρ.

We continue with two theorems on regular operators.

Theorem 2.49 *Let $A \in \mathcal{A}_{reg}$ and $\varepsilon > 0$. Then there exist an integer $N \geq 1$ and a neighborhood $\mathcal{U}$ of A in $\mathcal{A}$ such that for each $B \in \mathcal{A}_{reg} \cap \mathcal{U}$,*

$$\|x_A - x_B\|_\eta \leq \varepsilon, \qquad |\alpha_A - \alpha_B| \leq \varepsilon$$

and for each $x \in \langle -\eta, \eta \rangle$ and each integer $n \geq N$,

$$\left\|\alpha_B^{-n} B^n x - f_A(x) x_A\right\|_\eta \leq \varepsilon.$$

Theorem 2.50 *Let $A \in \mathcal{A}_{reg}$, $\varepsilon > 0$ and $\Delta \in (0, 1)$. Then there exist an integer $N \geq 1$ and a neighborhood $\mathcal{U}$ of A in $\mathcal{A}$ such that the following assertion holds:*

Assume that $B \in \mathcal{U}$, $x_0 \in X_+$, $\alpha_0 > 0$, $\Delta\eta \leq x_0 \leq \eta$ and $\alpha_0 x_0 = Bx_0$. Then

$$\|x_A - x_0\|_\eta \leq \varepsilon, \qquad |\alpha_A - \alpha_0| \leq \varepsilon$$

and for each $x \in \langle -\eta, \eta \rangle$ and each integer $n \geq N$,

$$\left\|\alpha_0^{-n} B^n x - f_A(x) x_A\right\|_\eta \leq \varepsilon.$$

These theorems bring out the importance of regular operators. Such operators not only have a unique positive eigenvector but also enjoy certain convergence and stability properties. Therefore we would like to show that most operators in an appropriate space are indeed regular. Moreover, in analogy with the definition of $\mathcal{A}_{q*}$ we will also consider quasiregular operators.

We say that an operator $A \in \mathcal{A}$ is quasiregular if there exist $\alpha > 0$, $c_0 \in (0, 1)$ and $c_1 > 1$ such that

$$c_0\alpha^n \eta \le A^n \eta \le c_1\alpha^n \eta \quad \text{for all integers } n \ge 1.$$

Denote by $\mathcal{A}_{qreg}$ the set of all quasiregular $A \in \mathcal{A}$ and by $\bar{\mathcal{A}}_{qreg}$ the closure of $\mathcal{A}_{qreg}$ in $(\mathcal{A}, \rho)$. We endow the subspace $\bar{\mathcal{A}}_{qreg} \subset \mathcal{A}$ with the same metric ρ.

Theorem 2.51 *$\bar{\mathcal{A}}_{qreg} = \bar{\mathcal{A}}_{reg}$ and there exists a set $\mathcal{F} \subset \mathcal{A}_{reg}$ which is a countable intersection of open and everywhere dense subsets of $\bar{\mathcal{A}}_{reg}$.*

Theorems 2.48–2.51 were obtained in [140].

2.12 Proof of Theorem 2.48

In this section we are going to present the proof of Theorem 2.48. We precede this proof by a few preliminary results.

As usual, we set $A^0 = I$ (the identity) for each $A \in \mathcal{A}$. We denote by $g \cdot B$ the composition of $g \in X'$ and a linear operator $B : X \to X$.

Proposition 2.52 *Let $A \in \mathcal{A}$ and assume that there exist $c_0 \in (0, 1)$ and $c_1 > 1$ such that*

$$c_0\eta \le A^n\eta \le c_1\eta \quad \textit{for all integers } n \ge 1. \tag{2.316}$$

Then there exists $f_A \in X'_+$ such that

$$f_A(\eta) > 0 \quad \textit{and} \quad f_A \cdot A = f_A.$$

Proof There exists $g \in X'_+$ such that $g(\eta) = 1$. Denote by S the convex hull of the set $\{g \cdot A^n : n = 0, 1, \ldots\}$. Clearly for each $h \in S$,

$$c_0 \le h(\eta) \le c_1. \tag{2.317}$$

Denote by $\bar{S}$ the closure of S in the weak-star topology $\sigma(X', X)$. Clearly (2.317) holds for all $h \in \bar{S}$ and $\bar{S} \subset X'_+$. The set $\bar{S}$ is convex and by (2.317) compact in the weak-star topology. The operator $A' : f \to f \cdot A$, $f \in X'$, is weakly-star continuous and $A'(\bar{S}) \subset \bar{S}$. By Tychonoff's fixed point theorem, there exists $f_A \in \bar{S}$ for which $f_A \cdot A = f_A$. Since (2.317) holds for all $h \in \bar{S}$, $f_A(\eta) \ge c_0$. Proposition 2.52 is proved. □

Corollary 2.53 *Assume that $A \in \mathcal{A}$, $c_0 \in (0, 1)$, $c_1 > 1$, $\alpha > 0$ and*

$$\alpha^n c_0\eta \le A^n\eta \le \alpha^n c_1\eta \quad \textit{for all integers } n \ge 1. \tag{2.318}$$

Then there exists $f_A \in X'_+$ such that $f_A(\eta) > 0$ and $f_A \cdot A = \alpha f_A$.

Lemma 2.54 *Assume that $A \in \mathcal{A}$, there exist $c_1 > 1$ and $\alpha > 0$ such that*

$$A^n \eta \le \alpha^n c_1 \eta \quad \textit{for all integers } n \ge 1, \tag{2.319}$$

and that there exists $f_A \in X'_+$ such that

$$f_A \cdot A = \alpha f_A \quad \textit{and} \quad f_A(\eta) = 1. \tag{2.320}$$

Let $\gamma \in (0, 1)$. Define $A_\gamma \in \mathcal{A}$ by

$$A_\gamma x = (1-\gamma)Ax + \gamma\alpha f_A(x)\eta, \quad x \in X. \tag{2.321}$$

Then $f_A \cdot A_\gamma = \alpha f_A$ and for each integer $n \ge 1$, there exist positive constants $c_i^{(n)}$, $i = 0, \dots, n-1$, such that

$$\sum_{i=0}^{n-1} c_i^{(n)} = 1 - (1-\gamma)^n \tag{2.322}$$

and

$$(A_\gamma)^n x = (1-\gamma)^n A^n x + \alpha^n f_A(x) \sum_{i=0}^{n-1} \left(\alpha^{-i} c_i^{(n)} A^i \eta\right), \quad x \in X. \tag{2.323}$$

Proof We will prove this lemma by induction. Clearly $f_A \cdot A_\gamma = f_A$ and (2.322) and (2.323) hold for $n = 1$, $c_0 = \gamma$.

Assume that $k \ge 1$ is an integer and there exist positive constants $c_i^{(k)}$, $i = 0, \dots, k-1$, such that (2.322) and (2.323) hold with $n = k$. It then follows from (2.322) and (2.323) with $n = k$ and (2.321) that for each $x \in X$,

$$\begin{aligned}
(A_\gamma)^{k+1} x &= A_\gamma\left(A_\gamma^k x\right) \\
&= (1-\gamma)A\left[(A_\gamma)^k x\right] + \alpha\gamma f_A\left((A_\gamma)^k x\right)\eta \\
&= \alpha\gamma\alpha^k f_A(x)\eta \\
&\quad + (1-\gamma)A\left[(1-\gamma)^k A^k x + \alpha^k f_A(x)\left(\sum_{i=0}^{k-1} \alpha^{-i} c_i^{(k)} A^i \eta\right)\right] \\
&= \gamma\alpha^{k+1} f_A(x)\eta + (1-\gamma)^{k+1} A^{k+1} x \\
&\quad + \alpha^k f_A(x)(1-\gamma)\left(\sum_{i=0}^{k-1} \alpha^{-i} c_i^{(k)} A^{i+1} \eta\right) \\
&= \gamma\alpha^{k+1} f_A(x)\eta + (1-\gamma)^{k+1} A^{k+1} x
\end{aligned}$$

$$
\begin{aligned}
&+\alpha^{k+1} f_A(x)(1-\gamma)\left(\sum_{i=1}^{k} \alpha^{-i} c_{i-1}^{(k)} A^i \eta\right)\\
&=(1-\gamma)^{k+1} A^{k+1} x+\alpha^{k+1} f_A(x)\left(\gamma \eta+\sum_{i=1}^{k}\left((1-\gamma) \alpha^{-i} c_{i-1}^{(k)} A^i \eta\right)\right)
\end{aligned}
$$

and

$$
\gamma+\sum_{i=1}^{k}\left((1-\gamma) c_{i-1}^{(k)}\right)=\gamma+(1-\gamma)\left(1-(1-\gamma)^{(k)}\right)=1-(1-\gamma)^{k+1}.
$$

Therefore (2.322) and (2.321) are true for $n=k+1$ with $c_0^{(k+1)}=\gamma$ and $c_i^{(k+1)}=(1-\gamma) c_{i-1}^{(k)}$, $i=1, \ldots, k$. This completes the proof of Lemma 2.54. □

Lemma 2.55 *Assume that $A \in \mathcal{A}$, there exist $c_0 \in(0,1)$, $c_1>1$ and $\alpha>0$ such that*

$$
\alpha^n c_0 \eta \leq A^n \eta \leq \alpha^n c_1 \eta \quad \text{for all integers } n \geq 1, \tag{2.324}
$$

and that there exists $f_A \in X_{+}^{\prime}$ such that (2.320) holds. Let $\gamma \in(0,1)$ and let $A_\gamma \in \mathcal{A}$ be defined by (2.321). Then there exists $x_A \in\langle c_0 \eta, c_1 \eta\rangle$ such that

$$
\alpha^{-n}\left(A_\gamma\right)^n x-f_A(x) x_A \rightarrow 0 \quad \text{as } n \rightarrow \infty,
$$

uniformly for all $x \in\langle 0, \eta\rangle$. Moreover, $A_\gamma x_A=\alpha x_A$.

Proof By Lemma 2.54 and (2.324), for each integer $n \geq 1$ there exists

$$
z_n \in\langle c_0 \eta, c_1 \eta\rangle \tag{2.325}
$$

such that

$$
\left(A_\gamma\right)^n x=(1-\gamma)^n A^n x+\alpha^n\left(1-(1-\gamma)^n\right) f_A(x) z_n, \quad x \in X. \tag{2.326}
$$

For each integer $n \geq 1$, by (2.320), (2.324) and (2.325),

$$
\begin{aligned}
\left(A_\gamma\right)^n \eta &=(1-\gamma)^n A^n \eta+\alpha^n\left(1-(1-\gamma)^n\right) z_n \\
&\in(1-\gamma)^n\left\langle\alpha^n c_0 \eta, \alpha^n c_1 \eta\right\rangle+\alpha^n\left(1-(1-\gamma)^n\right)\left\langle c_0 \eta, c_1 \eta\right\rangle \\
&\subset \alpha^n\left\langle c_0 \eta, c_1 \eta\right\rangle .
\end{aligned} \tag{2.327}
$$

Let $\varepsilon>0$. By (2.326), there exists an integer $n(\varepsilon) \geq 1$ such that for each $x \in\langle c_0 \eta, c_1 \eta\rangle$ and each integer $n \geq n(\varepsilon)$,

$$
\left\|\alpha^{-n}\left(A_\gamma\right)^n x-f_A(x) z_n\right\| \leq \varepsilon .
$$

Since $\{\alpha^{-i}(A_\gamma)^i\eta\}_{i=0}^\infty \subset \langle c_0\eta, c_1\eta\rangle$ and $f_A \cdot A_\gamma = \alpha f_A$, we conclude that for each integer $n \geq n(\varepsilon)$ and each integer $i \geq 0$,

$$\begin{aligned}\varepsilon &\geq \left\|\alpha^{-n}(A_\gamma)^n\left(\alpha^{-i}(A_\gamma)^i\eta\right) - f_A\left(\alpha^{-i}(A_\gamma^i\eta)\right)z_n\right\| \\ &= \left\|\alpha^{-n-i}(A_\gamma)^{n+i}\eta - z_n\right\|\end{aligned}$$

and therefore $\|z_n - z_{n+i}\| \leq 2\varepsilon$. This implies that $\{z_n\}_{n=1}^\infty$ is a Cauchy sequence. Hence there exists a vector $x_A \in \langle c_0\eta, c_1\eta\rangle$ such that $\lim_{i\to\infty}\|z_i - x_A\| = 0$. Let $\varepsilon > 0$. There exists an integer $n_0 \geq 1$ such that $\|z_i - x_A\| \leq \varepsilon/2$ for all integers $i \geq n_0$. By (2.326) and (2.324), there exists an integer $n_1 > n_0$ such that for each integer $n \geq n_1$ and each $x \in \langle 0, \eta\rangle$,

$$\left\|\alpha^{-n}(A_\gamma)^n x - f_A(x)z_n\right\| \leq 2^{-1}\varepsilon.$$

It follows from this last inequality and the definition of n_0 that for each $x \in \langle 0, \eta\rangle$ and each integer $n \geq n_1$,

$$\left\|\alpha^{-n}(A_\gamma)^n x - f_A(x)x_A\right\| \leq \varepsilon.$$

This completes the proof of Lemma 2.55. □

Proof of Theorem 2.48 It is, of course, sufficient to show that $\mathcal{A}_{q*} \subset \bar{\mathcal{A}}_*$. Towards this end, let $A \in \mathcal{A}_{q*}$. Then there exist $c_0 \in (0, 1)$ and $c_1 > 1$ such that

$$c_0\eta \leq A^n\eta \leq c_1\eta \quad \text{for all integers } n \geq 1.$$

By Proposition 2.52, there exists $f_A \in X'_+ \setminus \{0\}$ such that $f_A \cdot A = f_A$ and $f_A(\eta) = 1$.

For each $\gamma \in (0, 1)$, define $A_\gamma \in \mathcal{A}$ by

$$A_\gamma x = (1-\gamma)Ax + \gamma f_A(x)\eta, \quad x \in X.$$

By Lemma 2.55, A_γ belongs to $\mathcal{A}_*$. On the other hand, $\lim_{\gamma\to 0^+} A_\gamma = A$. Thus $\mathcal{A}_{q*} \subset \bar{\mathcal{A}}_*$ and Theorem 2.48 is proved. □

2.13 Auxiliary Results for Theorems 2.49–2.51

For each $x, y \in X_+$, define

$$\begin{aligned}\lambda(x, y) &= \sup\{\lambda \in [0, \infty) : \lambda x \leq y\}, \\ r(x, y) &= \inf\{r \in [0, \infty) : y \leq rx\}.\end{aligned} \tag{2.328}$$

Here we use the usual convention that the infimum of the empty set is ∞.

Lemma 2.56 *Assume that $A \in \mathcal{A}$, $n \geq 1$ is an integer and $\varepsilon > 0$. Then there exists a neighborhood $\mathcal{U}$ of A in $\mathcal{A}$ such that for each $B \in \mathcal{U}$ and each $x \in \langle -\eta, \eta \rangle$,*

$$\left\| A^n x - B^n x \right\|_\eta \leq \varepsilon.$$

Proof We prove the lemma by induction. Clearly for $n = 1$ the lemma is true. Assume that $k \geq 1$ is an integer and that the lemma holds for $n = k, \dots, 1$. There is a number $c_0 > 0$ such that $\|Ax\|_\eta \leq c_0$ for each $x \in \langle -\eta, \eta \rangle$. Since the lemma is true for $n = k$, there exists a neighborhood $\mathcal{U}_1$ of A in $\mathcal{A}$ such that $\|A^k x - B^k x\|_\eta \leq (4 + 4c_0)^{-1}\varepsilon$ for each $B \in \mathcal{U}_1$ and for each $x \in \langle -\eta, \eta \rangle$. It follows that there exists $c_1 > 1$ such that $\|B^k x\|_\eta \leq c_1$ for each $B \in \mathcal{U}_1$ and each $x \in \langle -\eta, \eta \rangle$. Since the lemma holds for $n = 1$, there exists a neighborhood $\mathcal{U} \subset \mathcal{U}_1$ of A in $\mathcal{A}$ such that for each $B \in \mathcal{U}$ and each $x \in \langle -\eta, \eta \rangle$, $\|Ax - Bx\|_\eta \leq (4c_1)^{-1}\varepsilon$.

Assume now that $B \in \mathcal{U}$ and $x \in \langle -\eta, \eta \rangle$. Then

$$\left\| A^{k+1} x - B^{k+1} x \right\|_\eta \leq \left\| A^{k+1} x - AB^k x \right\|_\eta + \left\| AB^k x - B^{k+1} x \right\|_\eta. \tag{2.329}$$

It follows from the definition of c_0 and $\mathcal{U}_1$ that

$$\left\| A^{k+1} x - AB^k x \right\|_\eta \leq \varepsilon/4. \tag{2.330}$$

By the definition of $\mathcal{U}$ and c_1, $\|AB^k x - B^{k+1} x\|_\eta \leq \varepsilon/4$. Together with (2.329) and (2.330), this implies that $\|A^{k+1} x - B^{k+1} x\|_\eta \leq \varepsilon$. In other words, the lemma also holds for $n = k + 1$. This completes the proof of Lemma 2.56. □

Let $A \in \mathcal{A}$ be regular,

$$\begin{aligned} x_A \in \operatorname{int}(X_+), \qquad \|x_A\|_\eta = 1, \qquad \alpha_A > 0, \\ f_A \in X'_+ \setminus \{0\}, \qquad Ax_A = \alpha_A x_A, \\ \alpha_A^{-n} A^n x \to f_A(x) x_A \quad \text{as } n \to \infty, \text{ uniformly on } \langle -\eta, \eta \rangle. \end{aligned} \tag{2.331}$$

Assumptions (2.331) and Lemma 2.56 imply the following result.

Lemma 2.57 *Let $\varepsilon > 0$. Then there exists an integer $N(\varepsilon) \geq 1$ such that for each integer $N > N(\varepsilon)$, there exists a neighborhood $\mathcal{U}$ of A in $\mathcal{A}$ such that for each $B \in \mathcal{U}$ and each $x \in \langle -\eta, \eta \rangle$,*

$$\left\| \alpha_A^{-n} B^n x - f_A x_A \right\|_\eta \leq \varepsilon, \quad n = N(\varepsilon), \dots, N.$$

Corollary 2.58 *Assume that $0 < \Delta_1 < 1 < \Delta_2$ and $\theta > 1$. Then there exists an integer $N_0 \geq 1$ such that for each integer $N > N_0$, there exists a neighborhood $\mathcal{U}$ of A in $\mathcal{A}$ such that for each $x \in \langle \Delta_1 \eta, \Delta_2 \eta \rangle$, each $B \in \mathcal{U}$ and each integer $n \in [N_0, N]$,*

$$B^n x \in \left\langle \theta^{-1} \alpha_A^n f_A(x) x_A, \theta \alpha_A^n f_A(x) x_A \right\rangle.$$

Lemma 2.59 *Assume that* $0 < \Delta_1 < 1 < \Delta_2$ *and* $\theta > 1$. *Then there exist an integer* $N_0 \geq 1$ *and a neighborhood* $\mathcal{U}$ *of* A *in* $\mathcal{A}$ *such that for each* $B \in \mathcal{U}$, $x \in \langle \Delta_1 \eta, \Delta_2 \eta \rangle$ *and each integer* $n \geq N_0$,

$$r(x_A, B^n x) \leq \theta \lambda(x_A, B^n x). \tag{2.332}$$

Proof We may assume that

$$\Delta_2 > \theta \quad \text{and} \quad \theta \Delta_1 < \lambda(\eta, x_A). \tag{2.333}$$

Choose $\theta_0 > 1$ such that

$$\theta_0^2 < \theta. \tag{2.334}$$

By Corollary 2.58, there exist an integer $N_0 \geq 1$ and a neighborhood $\mathcal{U}$ of A in $\mathcal{A}$ such that for each $x \in \langle \Delta_1 \eta, \Delta_2 \eta \rangle$, each $B \in \mathcal{U}$ and each integer $n \in [N_0, 8N_0 + 8]$,

$$B^n x \in \langle \theta_0^{-1} \alpha_A^n f_A(x) x_A, \theta_0 \alpha_A^n f_A(x) x_A \rangle. \tag{2.335}$$

Assume that $B \in \mathcal{U}$ and $x \in \langle \Delta_1 \eta, \Delta_2 \eta \rangle$. By the definition of $\mathcal{U}$ and N_0, the inclusion (2.335) is valid for each integer $n \in [N_0, 8N_0 + 8]$. The relations (2.335) and (2.334) imply that for each integer $n \in [N_0, 8N_0 + 8]$,

$$r(x_A, B^n x) \leq \theta_0 \alpha_A^n f_A(x), \qquad \lambda(x_A, B^n x) \geq \theta_0^{-1} \alpha_A^n f_A(x)$$

and

$$r(x_A, B^n x) \leq \theta_0^2 \lambda(x_A, B^n x) \leq \theta \lambda(x_A, B^n x).$$

It remains to be shown that (2.332) is valid for all integers $n > 8N_0 + 8$.

Assume the contrary. Then there exists an integer

$$N_1 > 8N_0 + 8 \tag{2.336}$$

such that

$$r(x_A, B^n x) \leq \theta \lambda(x_A, B^n x) \quad \text{for all integers } n \in [N_0, N_1 - 1] \tag{2.337}$$

and

$$r(x_A, B^{N_1} x) > \theta \lambda(x_A, B^{N_1} x). \tag{2.338}$$

Consider the vector $B^{N_1 - N_0} x$. By (2.336) and (2.337), we see that

$$r(x_A, B^{N_1 - N_0} x) \leq \theta \lambda(x_A, B^{N_1 - N_0} x) \tag{2.339}$$

and

$$\theta^{-1} r(x_A, B^{N_1 - N_0} x) x_A \leq B^{N_1 - N_0} x \leq r(x_A, B^{N_1 - N_0} x) x_A.$$

By (2.338),

$$r\big(x_A, B^{N_1-N_0}x\big) > 0. \tag{2.340}$$

It follows from (2.339), (2.340), (2.331) and (2.333) that

$$r\big(x_A, B^{N_1-N_0}x\big)^{-1}B^{N_1-N_0}x \in \big\langle\theta^{-1}x_A, x_A\big\rangle \subset \big\langle\theta^{-1}\lambda(\eta, x_A)\eta, \eta\big\rangle \subset \langle\Delta_1\eta, \Delta_2\eta\rangle.$$

It follows from this relation and the definition of $\mathcal{U}$ and N_0 (see (2.335)) that

$$r\big(x_A, B^{N_1-N_0}x\big)^{-1}B^{N_1}x \in \big\langle\theta_0^{-1}\alpha_A^{N_0}f_A(x)x_A, \theta_0\alpha_A^{N_0}f_a(x)x_A\big\rangle,$$

$$r\big(x_A, B^{N_1}x\big) \leq \theta_0\alpha_A^{N_0}f_A(x)r\big(x_A, B^{N_1-N_0}x\big),$$

$$\lambda\big(x_A, B^{N_1}x\big) \geq \theta_0^{-1}\alpha_A^{N_0}f_A(x)r\big(x_A, B^{N_1-N_0}x\big),$$

and by (2.333),

$$r\big(x_A, B^{N_1}x\big) \leq \theta\lambda\big(x_A, B^{N_1}x\big),$$

an inequality which contradicts (2.338). Thus (2.332) is indeed valid for all $n \geq N_0$ and Lemma 2.59 is proved. □

Lemma 2.60 *Let $\gamma > 1$. Then there exists a neighborhood $\mathcal{U}$ of A in $\mathcal{A}$ such that for each $B \in \mathcal{A}_{reg} \cap \mathcal{U}$, the inequalities $\gamma^{-1}x_A \leq x_B \leq \gamma x_A$ hold.*

Proof Choose a positive number $\theta > 1$ such that

$$\theta^2 < \gamma. \tag{2.341}$$

By Lemma 2.59, there exists an integer $N_0 \geq 1$ and a neighborhood $\mathcal{U}$ of A in $\mathcal{A}$ such that for each $B \in \mathcal{U}$ and each integer $n \geq N_0$,

$$r\big(x_A, B^n\eta\big) \leq \theta\lambda\big(x_A, B^n\eta\big). \tag{2.342}$$

Assume that $B \in \mathcal{A}_{reg} \cap \mathcal{U}$. Then

$$\lim_{n\to\infty} \alpha_B^{-n}B^n\eta = f_B(\eta)x_B. \tag{2.343}$$

By the definition of $\mathcal{U}$ and N_0, (2.342) is valid for each integer $n \geq N_0$. This implies that for each integer $n \geq N_0$,

$$\alpha_B^{-n}\lambda\big(x_A, B^n\eta\big)x_A \leq \alpha_B^{-n}B^n\eta \leq \alpha_B^{-n}r\big(x_A, B^n\eta\big)x_A$$

and

$$r\big(x_A, \alpha_B^{-n}B^n\eta\big) \leq \theta\lambda\big(x_A, \alpha_B^{-n}B^n\eta\big).$$

When combined with (2.343), this implies that

$$r\big(x_A, f_B(\eta)x_B\big) \leq \theta^2\lambda\big(f_B(\eta)x_B, x_A\big) \quad \text{and} \quad r(x_A, x_B) \leq \theta^2\lambda(x_A, x_B). \tag{2.344}$$

It follows from (2.331), (2.334) and (2.341) that

$$\lambda(x_A, x_B)x_A \leq x_B \leq r(x_A, x_B)x_A \leq r(x_A, x_B)\eta,$$
$$x_A \leq \lambda(x_A, x_B)^{-1}x_B \leq \lambda(x_A, x_B)^{-1}\eta, \qquad r(x_A, x_B) \geq 1, \qquad \lambda(x_A, x_B)^{-1} \geq 1,$$
$$r(x_A, x_B) \leq \theta^2, \qquad \lambda(x_A, x_B) \geq \theta^{-2}$$

and finally, that

$$\gamma^{-1}x_A \leq \theta^{-2}x_A \leq x_B \leq \theta^2 x_A \leq \gamma x_A.$$

Lemma 2.60 is proved. □

Lemma 2.61 *Let $\theta > 1$ and $\Delta \in (0, 1)$. Then there exists a neighborhood $\mathcal{U}$ of A in $\mathcal{A}$ such that for each $B \in \mathcal{U}$, $z \in X_+$ and $\alpha > 0$ satisfying*

$$\|z\|_\eta = 1, \qquad z \geq \Delta\eta \quad \text{and} \quad Bz = \alpha z, \tag{2.345}$$

the following inequalities hold: $\theta^{-1}x_A \leq z \leq \theta x_A$.

Proof By Lemma 2.59, there exists an integer $N_0 \geq 1$ and a neighborhood $\mathcal{U}$ of A in $\mathcal{A}$ such that for each $B \in \mathcal{U}$, each integer $n \geq N_0$ and for each $x \in \langle 4^{-1}\Delta\eta, 4\eta\rangle$,

$$r\big(x_A, B^n x\big) \leq \theta\lambda\big(x_A, B^n x\big). \tag{2.346}$$

Assume that $B \in \mathcal{U}$, $z \in X_+$, $\alpha > 0$ and that (2.345) is valid. By (2.345) and the definition of $\mathcal{U}$ and N_0 (see (2.346)), for each integer $n \geq N_0$,

$$\begin{aligned}\alpha^n r(x_A, z) &= r\big(x_A, \alpha^n z\big) = r\big(x_A, B^n z\big) \leq \theta\lambda\big(x_A, B^n z\big)\\ &= \theta\lambda\big(x_A, \alpha^n z\big) = \alpha^n\theta\lambda(x_A, z) \quad \text{and} \quad r(x_A, z) \leq \theta\lambda(x_A, z).\end{aligned} \tag{2.347}$$

It follows from (2.345), (2.331) and (2.347) that

$$\lambda(x_A, z)x_A \leq z \leq r(x_A, z)x_A \leq r(x_A, z)\eta, \qquad r(x_A, z) \geq 1,$$
$$x_A \leq \lambda(x_A, z)^{-1}z \leq \lambda(x_A, z)^{-1}\eta, \qquad \lambda(x_A, z) \leq 1,$$
$$r(x_A, z) \leq \theta, \qquad \lambda(x_A, z) \geq \theta^{-1}$$

and finally, that $\theta^{-1}x_A \leq z \leq \theta x_A$. This completes the proof of Lemma 2.61. □

Lemma 2.62 *Let $\varepsilon \in (0, 1)$ and $\Delta \in (0, 1)$. Then there exists a neighborhood $\mathcal{U}$ of A in $\mathcal{A}$ such that for each $B \in \mathcal{U}$, $z \in X_+$ and $\alpha > 0$ satisfying*

$$\|z\|_\eta = 1, \qquad z \geq \Delta\eta \quad \text{and} \quad Bz = \alpha z, \tag{2.348}$$

we have $|\alpha - \alpha_A| \leq \varepsilon$.

Proof Choose a number $\gamma > 1$ for which

$$(\alpha_A + 1)(\gamma - 1) \le \varepsilon/8.$$

By Lemma 2.61, there exists a neighborhood $\mathcal{U}_1$ of A in $\mathcal{A}$ such that for each $B \in \mathcal{U}_1$, $z \in X_+$ and $\alpha > 0$ satisfying (2.348), the following inequalities hold:

$$\gamma^{-1} x_A \le z \le \gamma x_A. \tag{2.349}$$

There exists a neighborhood $\mathcal{U} \subset \mathcal{U}_1$ of A in $\mathcal{A}$ such that for each $B \in \mathcal{U}$,

$$\|Ay - By\|_\eta \le \varepsilon/8 \quad \text{for all } y \in \gamma\langle -\eta, \eta\rangle. \tag{2.350}$$

Assume that $B \in \mathcal{U}$, $z \in X_+$, $\alpha > 0$ and that (2.348) is true. Then by the definition of $\mathcal{U}_1$, (2.349) holds.

It follows from (2.348) and (2.331) that

$$\begin{aligned}|\alpha - \alpha_A| &= \big|\|\alpha z\|_\eta - \|\alpha_A x_A\|_\eta\big| \le \|\alpha z - \alpha_A x_A\|_\eta = \|Bz - Ax_A\|_\eta \\ &\le \|Ax_A - Az\|_\eta + \|Az - Bz\|_\eta.\end{aligned} \tag{2.351}$$

By our choice of γ, (2.349) and (2.331),

$$\begin{aligned}(1-\gamma)\alpha_A \eta &\le (1-\gamma)\alpha_A x_A = A(1-\gamma)x_A \le Ax_A - Az \\ &\le (1-\gamma^{-1})Ax_A \le (\gamma - 1)\alpha_A \eta\end{aligned}$$

and

$$\|Ax_A - Az\|_\eta \le \varepsilon/8. \tag{2.352}$$

It follows from (2.349) and (2.350) that

$$z \le \gamma x_A \le \gamma\eta \quad \text{and} \quad \|Az - Bz\|_\eta \le 8^{-1}\varepsilon.$$

When combined with (2.351) and (2.352), this implies that $|\alpha_A - \alpha| \le \varepsilon$. Lemma 2.62 is proved. □

Lemmas 2.62 and 2.60 imply the following result.

Lemma 2.63 *Let $\varepsilon \in (0, 1)$. Then there exists a neighborhood $\mathcal{U}$ of A in $\mathcal{A}$ such that for each $B \in \mathcal{A}_{reg} \cap \mathcal{U}$ we have $|\alpha_B - \alpha_A| \le \varepsilon$.*

2.14 Proofs of Theorems 2.49 and 2.50

In this section we prove Lemma 2.64. Theorem 2.50 follows when this lemma is combined with Lemmas 2.61 and 2.62. Theorem 2.49 is a consequence of Lemmas 2.60, 2.63 and 2.64.

Lemma 2.64 *Let $A \in \mathcal{A}$ be regular and let ε and Δ belong to the interval* $(0, 1)$. *Then there exist an integer $N \geq 1$ and a neighborhood $\mathcal{U}$ of A in $\mathcal{A}$ such that the following assertion holds*:

If

$$\begin{aligned} &B \in \mathcal{U}, \qquad x_0 \in \operatorname{int}(X_+), \\ &\Delta\eta \leq x_0 \leq \eta, \qquad \alpha_0 > 0 \quad \textit{and} \quad \alpha_0 x_0 = B x_0, \end{aligned} \tag{2.353}$$

then for each $x \in \langle -\eta, \eta \rangle$ and each integer $n \geq N$,

$$\left\| \alpha_0^{-n} B^n x - f_A(x) x_A \right\|_\eta \leq \varepsilon. \tag{2.354}$$

Proof Choose a positive number ε_0 for which

$$8\varepsilon_0 < 4^{-1} \varepsilon \Delta.$$

By Lemma 2.56, there exist a neighborhood $\mathcal{U}_1$ of A in $\mathcal{A}$ and an integer $N \geq 1$ such that for each $B \in \mathcal{U}_1$,

$$\left\| \alpha_A^{-N} B^N x - f_A(x) x_A \right\|_\eta \leq 16^{-1} \varepsilon_0 \quad \text{for all } x \in \langle -\eta, \eta \rangle. \tag{2.355}$$

There exists a number $c_1 > 1$ such that

$$\left\| B^N x \right\|_\eta \leq c_1 \quad \text{for } x \in \langle -\eta, \eta \rangle \quad \text{and} \quad B \in \mathcal{U}_1, \quad \text{and} \quad f_A(\eta) \leq c_1. \tag{2.356}$$

There exists a number $\delta_1 \in (0, \min\{1, \alpha_A/8\})$ such that

$$\left| \alpha^{-N} - \alpha_A^{-N} \right| c_1 \leq 16^{-1} \varepsilon_0 \quad \text{for each } \alpha \text{ satisfying } |\alpha - \alpha_A| \leq \delta_1. \tag{2.357}$$

By Lemmas 2.62 and 2.61 there exists a neighborhood $\mathcal{U}_2$ of A in $\mathcal{A}$ such that for each $B \in \mathcal{U}_2$, $z \in X_+$ and $\alpha > 0$ satisfying $\Delta\eta \leq z \leq \eta$ and $Bz = \alpha z$, the following inequalities are true:

$$|\alpha - \alpha_N| \leq \delta_1 \quad \text{and} \quad \|z - x_A\|_\eta \leq 16^{-1} \varepsilon_0 c_1^{-1}. \tag{2.358}$$

Set

$$\mathcal{U} = \mathcal{U}_1 \cap \mathcal{U}_2. \tag{2.359}$$

Assume that $B \in \mathcal{U}$, $x_0 \in X_+$, $\alpha_0 > 0$ and that (2.353) holds. By the definition of $\mathcal{U}_1$ and N, (2.355) holds. It follows from the definition of $\mathcal{U}_2$ (see (2.358)) and (2.353) that $|\alpha_0 - \alpha_N| \leq \delta_1$. By the latter inequality, (2.357), (2.356) and (2.355),

$$\left\| \alpha_0^{-N} B^N x - f_A(x) x_A \right\|_\eta \leq 8^{-1} \varepsilon_0 \quad \text{for all } x \in \langle -\eta, \eta \rangle. \tag{2.360}$$

By the definition of $\mathcal{U}_2$ (see (2.358)) and (2.353),

$$\|x_0 - x_A\|_\eta \leq 16^{-1} \varepsilon_0 c_1^{-1}. \tag{2.361}$$

This inequality, when combined with (2.360) and (2.356), implies that

$$\left\|\alpha_0^{-N} B^N x - f_A(x)x_0\right\|_\eta \le 8^{-1}\varepsilon_0 + 16^{-1}\varepsilon_0 \quad \text{for all } x \in \langle -\eta, \eta\rangle. \tag{2.362}$$

By (2.362) and (2.353), we have

$$\alpha_0^{-N} B^N x - f_A(x)x_0 \in \varepsilon_0\left(8^{-1} + 16^{-1}\right)\langle -\eta, \eta\rangle \subset \varepsilon_0\left(8^{-1} + 16^{-1}\right)\Delta^{-1}\langle -x_0, x_0\rangle$$

for all $x \in \langle -\eta, \eta\rangle$.

It follows from this relation and (2.353) that for each $x \in \langle -\eta, \eta\rangle$ and each integer $n \ge N$,

$$\begin{aligned}
\alpha_0^{-n} B^n x - f_A(x)x_0 &= \alpha_0^{-n+N} B^{n-N}\left[\alpha_0^{-N} B^N x - f_A(x)x_0\right] \\
&\subset \varepsilon_0\left(8^{-1} + 16^{-1}\right)\Delta^{-1}\alpha_0^{N-n} B^{n-N}\langle -x_0, x_0\rangle \\
&\subset \varepsilon_0\left(8^{-1} + 16^{-1}\right)\Delta^{-1}\langle -x_0, x_0\rangle \\
&\subset \varepsilon_0\left(8^{-1} + 16^{-1}\right)\Delta^{-1}\langle -\eta, \eta\rangle
\end{aligned}$$

and

$$\left\|\alpha_0^{-n} B^n x - f_A(x)x_0\right\|_\eta \le \Delta^{-1}\varepsilon_0/4.$$

When combined with (2.361), (2.356) and (2.354), this implies that for each $x \in \langle -\eta, \eta\rangle$ and each integer $n \ge N$,

$$\left\|\alpha_0^{-n} B^n x - f_A(x)x_A\right\|_\eta \le \Delta^{-1}\varepsilon_0/4 + 16^{-1}\varepsilon_0 < \varepsilon.$$

Lemma 2.64 is proved. □

2.15 Proof of Theorem 2.51

It follows from Lemma 2.55 and Corollary 2.53 that $\mathcal{A}_{qreg} \subset \bar{\mathcal{A}}_{reg}$. This clearly implies that $\bar{\mathcal{A}}_{reg} = \bar{\mathcal{A}}_{qreg}$.

To construct the set $\mathcal{F}$ we let $A \in \mathcal{A}_{reg}$,

$$\begin{aligned}
&x_A \in \operatorname{int}(X_+), \qquad f_A \in X'_+ \setminus \{0\}, \qquad \alpha_A > 0, \\
&Ax_A = \alpha_A x_A, \qquad f_A \cdot A = \alpha_A \cdot f_A, \\
&\alpha_A^{-n} A^n x \to f_A(x)x_A \quad \text{as } n \to \infty, \text{ uniformly on } \langle \eta, \eta\rangle.
\end{aligned} \tag{2.363}$$

Let $i \ge$ be an integer. By Lemmas 2.60 and 2.63, Theorem 2.49, Lemmas 2.61 and 2.62, and Theorem 2.50, there exist a number $r(A, i) \in (0, 4^{-i})$ and an integer $N(A, i) \ge 1$ such that the following two assertions hold:

1. Assume that $B \in \mathcal{A}_{reg}$ and $\rho(A, B) < r(A, i)$. Then

$$(1 - 4^{-i})x_A \le x_B \le (1 + 4^{-i})x_A, \qquad |\alpha_A - \alpha_B| \le 4^{-i}\min\{1, \alpha_A\}$$

and

$$\left\|\alpha_B^{-n} B^n x - f_A(x)x_A\right\|_\eta \le 4^{-i} \quad \text{for all } x \in \langle -\eta, \eta\rangle \text{ and each integer } n \ge N.$$

2. Assume that $B \in \mathcal{A}$, $\rho(A, B) < r(A, i)$, $x_0 \in X_+$, $\alpha_0 > 0$, $\alpha_0 x_0 = Bx_0$ and $4^{-1}x_A \le x_0 \le \eta$. Then

$$(1 - 4^{-i})x_A \le x_0 \le (1 + 4^{-i})x_A, \qquad |\alpha_A - \alpha_0| \le 4^{-i}\min\{1, \alpha_A\}$$

and

$$\left\|\alpha_0^{-n} B^n x - f_A(x)x_A\right\|_\eta \le 4^{-i}$$

for all $x \in \langle -\eta, \eta\rangle$ and each integer $n \ge N(A, i)$.

Now set

$$\mathcal{U}(A, i) = \{B \in \mathcal{A} : \rho(B, A) < r(A, i)\} \tag{2.364}$$

and define

$$\mathcal{F} = \left[\bigcap_{i=1}^{\infty} \bigcup \{\mathcal{U}(A, i) : A \in \mathcal{A}_{reg}\}\right] \cap \bar{\mathcal{A}}_{reg}. \tag{2.365}$$

Evidently, $\mathcal{F}$ is a countable intersection of open and everywhere dense subsets of $\bar{\mathcal{A}}_{reg}$.

It remains to be shown that $\mathcal{F} \subset \mathcal{A}_{reg}$. To this end, assume that $B \in \mathcal{F}$. There exist $\{A_k\}_{k=1}^{\infty} \subset \mathcal{A}_{reg}$ and a strictly increasing sequence of natural numbers $\{i_k\}_{k=1}^{\infty}$ such that

$$B \in \mathcal{U}(A_k, i_k) \quad \text{and} \quad \mathcal{U}(A_{k+1}, i_{k+1}) \subset \mathcal{U}(A_k, i_k), \quad k = 1, 2, \ldots. \tag{2.366}$$

Let $k \ge 1$. It follows from assertion 1 and (2.366) that for each integer $j \ge 1$,

$$(1 - 4^{-i_k})x_{A_k} \le x_{A_{k+j}} \le (1 + 4^{-i_k})x_{A_k} \tag{2.367}$$

and

$$|\alpha_{A_k} - \alpha_{A_{k+j}}| \le 4^{-i_k}\min\{1, \alpha_{A_k}\}.$$

It is clear that both $\{x_{A_p}\}_{p=1}^{\infty}$ and $\{\alpha_{A_p}\}_{p=1}^{\infty}$ are Cauchy sequences. Therefore there exist the limits

$$x_* = \lim_{s\to\infty} x_{A_s}, \qquad \alpha_* = \lim_{s\to\infty} \alpha_{A_s}. \tag{2.368}$$

Set

$$\lambda_* = \inf\{\lambda(x_{A_k}, \eta) : k = 1, 2, \ldots\}. \tag{2.369}$$

By (2.367), λ_* is positive. By (2.367) and (2.368),

$$\begin{aligned}&(1-4^{-i_k})x_{A_k} \le x_* \le (1+4^{-i_k})x_{A_k},\\&|\alpha_{A_k}-\alpha_*| \le 4^{-i_k}\min\{1,\alpha_{A_k}\}, \quad x_* \le \eta.\end{aligned} \tag{2.370}$$

By (2.368) and (2.366),

$$Bx_* = B\Big(\lim_{k\to\infty} x_{A_k}\Big) = \lim_{k\to\infty} A_k x_{A_k} = \lim_{k\to\infty} \alpha_{A_k} x_{A_k} = \alpha_* x_*. \tag{2.371}$$

Let $k \ge 1$ be an integer. It follows from assertion 2, (2.366), (2.370) and (2.371) that

$$\begin{aligned}&\left\|\alpha_*^{-n} B^n x - f_{A_k}(x)x_{A_k}\right\|_\eta \le 4^{-i_k} \quad \text{for all } x \in \langle -\eta, \eta\rangle\\&\text{and each integer } n \ge N(A_k, i_k).\end{aligned} \tag{2.372}$$

Note that (see (2.363) and (2.369))

$$x_{A_k} = f_{A_k}(x_{A_k})x_{A_k}, \qquad f_{A_k}(x_{A_k}) = 1$$

and

$$f_{A_k}(\eta) \le f_{A_k}(x_{A_k}) \cdot \lambda_*^{-1} = \lambda_*^{-1}.$$

When combined with (2.372) and (2.370), this implies that

$$\left\|\alpha_*^{-n} B^n x - f_{A_k}(x)x_A\right\|_\eta \le 4^{-i_k} + 4^{-i_k}\lambda_*^{-1} \tag{2.373}$$

for all $x \in \langle -\eta, \eta\rangle$ and each integer $n \ge N(A_k, i_k)$. Since k is an arbitrary natural number, we obtain that for each $x \in X$, there exists

$$\lim_{n\to\infty} \alpha_*^{-n} B^n x = f_B(x)x_*, \tag{2.374}$$

where $f_B \in X'_+$. It follows from (2.373) and (2.374) that for each integer $k \ge 1$, each integer $n \ge N(A_k, i_k)$ and each $x \in \langle -\eta, \eta\rangle$,

$$\left\|f_B(x)x_* - f_{A_k}(x)x_*\right\|_\eta \le 4^{-i_k} + 4^{-i_k}\lambda_*^{-1}$$

and

$$\left\|\alpha_*^{-n} B^n x - f_B(x)x_*\right\|_\eta \le 2(4^{-i_k} + 4^{-i_k}\lambda_*^{-1}).$$

Therefore $B \in \mathcal{A}_{reg}$ and Theorem 2.51 is established.

2.16 Convergence of Inexact Orbits for a Class of Operators

In this section we exhibit a class of nonlinear operators with the property that their iterates converge to their unique fixed points even when computational errors are present. We also show that most (in the sense of Baire category) elements in an appropriate complete metric space of operators do, in fact, possess this property.

Assume that (X, ρ) is a complete metric space and let the operator $A : X \to X$ have the following properties:

(A1) there exists a unique $x_A \in X$ such that $Ax_A = x_A$;
(A2) $A^n x \to x_A$ as $n \to \infty$, uniformly on all bounded subsets of X;
(A3) A is uniformly continuous on bounded subsets of X;
(A4) A is bounded on bounded subsets of X.

Many operators with these properties can be found, for example, in [23, 33, 50, 85, 108, 114, 126, 127, 137]. We mention, in particular, the classes of operators introduced by Rakotch [114] and Browder [23]. Note that if X is either a closed and convex subset of a Banach space or a closed and ρ-convex subset of a complete hyperbolic metric space [124], then (A4) follows from (A3).

In view of (A2), it is natural to ask if the convergence of the orbits of A will be preserved even in the presence of computational errors. In this section we provide affirmative answers to this question. More precisely, we have the following results which were obtained in [35].

Theorem 2.65 *Let K be a nonempty, bounded subset of X and let $\varepsilon > 0$ be given. Then there exist $\delta = \delta(\varepsilon, K) > 0$ and a natural number N such that for each natural number $n \geq N$, and each sequence $\{x_i\}_{i=0}^n \subset X$ which satisfies*

$$x_0 \in K \quad and \quad \rho(Ax_i, x_{i+1}) \leq \delta, \quad i = 0, \ldots, n-1,$$

the following inequality holds:

$$\rho(x_i, x_A) \leq \varepsilon, \quad i = N, \ldots, n.$$

Corollary 2.66 *Assume that $\{x_i\}_{i=0}^\infty \subset X$, $\{x_i\}_{i=0}^\infty$ is bounded, and that*

$$\lim_{i \to \infty} \rho(Ax_i, x_{i+1}) = 0.$$

Then $\rho(x_i, x_A) \to 0$ as $i \to \infty$.

Theorem 2.67 *Let $\varepsilon > 0$ be given. Then there exists $\delta = \delta(\varepsilon) > 0$ such that for each sequence $\{x_i\}_{i=0}^\infty \subset X$ which satisfies*

$$\rho(x_0, x_A) \leq \delta \quad and \quad \rho(x_{i+1}, Ax_i) \leq \delta, \quad i = 0, 1, \ldots,$$

the following inequality holds:

$$\rho(x_i, x_A) \leq \varepsilon, \quad i = 0, 1, \ldots.$$

These results show that, roughly speaking, in order to achieve an ε-approximation of x_A, it suffices to compute *inexact orbits* of A, that is, sequences $\{x_i\}_{i=0}^\infty$ such that

$$x_0 \in X \quad \text{and} \quad \rho(x_{i+1}, Ax_i) \le \delta \quad \text{for any } i \ge 0,$$

where δ is a sufficiently small positive number.

However, sometimes the operator A is not given explicitly and only some approximation of it, B_i, is available at each step i of the inexact orbit computing procedure. The next result shows that for certain operators A, the procedure of approximating x_A by inexact orbits is stable in the sense that, even in this case, the orbits determined by the sequence of operators B_i approach x_A provided that each B_i is a sufficiently accurate approximation of A in the topology of uniform convergence on bounded subsets of X. To be precise, we set, for each $x \in X$ and $E \subset X$,

$$\rho(x, E) = \inf\{\rho(x, y) : y \in E\}.$$

Denote by $\mathcal{A}$ the set of all self-mappings $A : X \to X$ which have properties (A3) and (A4). Fix $\theta \in X$. For each natural number n, set

$$E_n = \{(A, B) \in \mathcal{A} \times \mathcal{A} : \rho(Ax, Bx) \le 1/n \text{ for all } x \in B(\theta, n)\}. \tag{2.375}$$

We equip the set $\mathcal{A}$ with the uniformity determined by the base E_n, $n = 1, 2, \ldots$. This uniformity is metrizable by a complete metric.

Denote by $\mathcal{A}_{reg}$ the set of all mappings $A \in \mathcal{A}$ which satisfy (A1) and (A2), and by $\bar{\mathcal{A}}_{reg}$ the closure of $\mathcal{A}_{reg}$ in $\mathcal{A}$.

Theorem 2.68 *Assume that $A \in \mathcal{A}_{reg}$ and x_A is a fixed point of A. Let $m, \varepsilon > 0$ be given. Then there exist a neighborhood $\mathcal{U}$ of A in $\mathcal{A}$ and a natural number N such that for each $x \in B(\theta, m)$, each integer $n \ge N$, and each sequence $\{B_i\}_{i=1}^n \subset \mathcal{U}$,*

$$\rho(B_i \cdots B_1 x, x_A) \le \varepsilon \quad \text{for } i = N, \ldots, n.$$

As a matter of fact, it turns out that the stability property established in this theorem is generic. That is, it holds for most (in the sense of Baire category) operators in the closure of $\mathcal{A}_{reg}$.

Theorem 2.69 *The set $\mathcal{A}_{reg}$ contains an everywhere dense G_δ subset of $\bar{\mathcal{A}}_{reg}$.*

2.17 Proofs of Theorem 2.65 and Corollary 2.66

We first prove Theorem 2.65. To this end, set, for $x \in X$ and $r > 0$,

$$B(x, r) = \{y \in X : \rho(x, y) \le r\}.$$

We may assume without loss of generality that

$$\varepsilon \le 1 \quad \text{and} \quad B(x_A, 4) \subset K. \tag{2.376}$$

By (A2), there exists a natural number $N \geq 4$ such that

$$\rho(A^n x, x_A) \leq \varepsilon/4 \quad \text{for all integers } n \geq N \text{ and all } x \in K. \tag{2.377}$$

By (A4), the set $A^m(K)$ is bounded for all natural numbers m. Hence there exists a positive number $S > 0$ such that

$$A^i(K) \subset B(x_A, S), \quad i = 0, \ldots, 2N. \tag{2.378}$$

(Here we use the convention that A^0 is the identity operator.) By induction and (A3), we define a finite sequence of positive numbers $\{\gamma_i\}_{i=0}^{2N}$ so that

$$\gamma_{2N} = \varepsilon/4$$

and, for each $i = 0, 1, \ldots, 2N-1$,

$$\gamma_i \leq \gamma_{i+1} \tag{2.379}$$

and

$$\rho(Ax, Ay) \leq 2^{-1}\gamma_{i+1} \quad \text{for all } x, y \in B(x_A, S+4) \quad \text{with} \quad \rho(x, y) \leq \gamma_i. \tag{2.380}$$

Set

$$\delta = \gamma_0/2. \tag{2.381}$$

First, we prove the following auxiliary result.

Lemma 2.70 *Suppose that* $\{z_i\}_{i=0}^{2N} \subset X$ *satisfies*

$$z_0 \in K \quad \textit{and} \quad \rho(z_{i+1}, Az_i) \leq \delta, \quad i = 0, \ldots, 2N-1. \tag{2.382}$$

Then

$$\rho(z_i, x_A) \leq \varepsilon, \quad i = N, \ldots, 2N.$$

Proof We will show that for $i = 1, \ldots, 2N$,

$$\rho(z_i, A^i z_0) \leq \gamma_i. \tag{2.383}$$

Clearly, (2.383) holds for $i = 1$ by (2.382) and (2.381).

Assume that $i \in \{2, \ldots, 2N\}$ and

$$\rho(z_{i-1}, A^{i-1} z_0) \leq \gamma_{i-1}. \tag{2.384}$$

Then (2.382) implies that

$$\begin{aligned} \rho(z_i, A^i z_0) &\leq \rho(z_i, Az_{i-1}) + \rho(Az_{i-1}, A(A^{i-1} z_0)) \\ &\leq \delta + \rho(Az_{i-1}, A(A^{i-1} z_0)). \end{aligned} \tag{2.385}$$

It follows from the definition of γ_{i-1} (see (2.379)), (2.384), (2.382) and (2.378) that

$$A^{i-1}z_0, z_{i-1} \in B(x_A, S+1).$$

By these inclusions, the definition of γ_{i-1} (see (2.380) with $j = i - 1$) and (2.384),

$$\rho\big(A\big(A^{i-1}z_0\big), Az_{i-1}\big) \le \gamma_i/2.$$

When combined with (2.385) and (2.381), this inequality implies that

$$\rho\big(z_i, A^i z_0\big) \le \delta + \gamma_i/2 \le \gamma_i.$$

Therefore (2.383) is valid for all $i \in \{1, \dots, 2N\}$. Together with (2.377), (2.379), (2.382) and (2.383), this last inequality implies that for all $i \in \{N, \dots, 2N\}$, we have

$$\rho(z_i, x_A) \le \rho\big(z_i, A^i z_0\big) + \rho\big(A^i z_0, x_A\big) \le \gamma_i + \varepsilon/4 \le \varepsilon/2.$$

Lemma 2.70 is proved. □

Now we are ready to complete the proof of Theorem 2.65.

To this end, assume that $n \ge N$ is a natural number and that the sequence $\{x_i\}_{i=0}^{n} \subset X$ satisfies

$$x_0 \in K \quad \text{and} \quad \rho(Ax_i, x_{i+1}) \le \delta, \quad i = 0, \dots, n-1.$$

We will show that

$$\rho(x_i, x_A) \le \varepsilon, \quad i = N, \dots, n. \tag{2.386}$$

If $n \le 2N$, then (2.386) follows from Lemma 2.70. Therefore we may confine our attention to the case where $n > 2N$. Again by Lemma 2.70,

$$\rho(x_i, x_A) \le \varepsilon, \quad i = N, \dots, 2N. \tag{2.387}$$

Assume by way of contradiction that there exists an integer $q \in (2N, n]$ such that

$$\rho(x_q, x_A) > \varepsilon. \tag{2.388}$$

In view of (2.387), we may assume without loss of generality that

$$\rho(x_i, x_A) \le \varepsilon, \quad i \in \{2N, \dots, q-1\}. \tag{2.389}$$

Define $\{z_i\}_{i=0}^{2N} \subset X$ by

$$z_i = x_{i+q-N}, \quad i = 0, \dots, N, \qquad z_{i+1} = Az_i, \quad i = N, \dots, 2N-1. \tag{2.390}$$

We will show that the sequence $\{z_i\}_{i=0}^{2N}$ satisfies (2.382). To meet this goal, we only need to show that $z_0 \in K$. By (2.390), (2.389) and (2.387),

$$z_0 = x_{q-N} \quad \text{and} \quad \rho(z_0, x_A) \le \varepsilon.$$

The last inequality and (2.376) imply that $z_0 \in K$. Therefore (2.382) holds. It now follows from Lemma 2.70 and (2.390) that

$$\rho(x_A, x_q) = \rho(x_A, z_N) \leq \varepsilon.$$

This, however, contradicts (2.388). The contradiction we have reached proves (2.386) and this completes the proof of Theorem 2.65.

Finally, we are going to prove Corollary 2.66.

Set $K = \{x_n : n = 0, 1, \ldots\}$ and let $\varepsilon > 0$ we given. Let $\delta > 0$ and a natural number N be as guaranteed by Theorem 2.65. There exists a natural number j such that for each integer $i \geq j$, we have $\rho(Ax_i, x_{i+1}) \leq \delta$. It follows from the last inequality and the choice of δ that $\rho(x_i, x_A) \leq \varepsilon$ for all integers $i \geq j + N$. Since ε is an arbitrary positive number, this implies that $\lim_{i\to\infty} x_i = x_A$. The proof of Corollary 2.66 is complete.

Corollary 2.66 provides a partial answer to a question raised in [77] in the wake of Theorem 1 of [75], which is also concerned with the stability of iterations.

2.18 Proof of Theorem 2.67

We may assume without loss of generality that $\varepsilon \leq 1$. By Theorem 2.65, there exist a natural number N and a real number $\delta_0 \in (0, \varepsilon)$ such that the following property holds.

(P1) For each natural number $n \geq N$ and each sequence $\{y_i\}_{i=0}^n \subset X$ which satisfies

$$y_0 \in B(x_A, 4) \quad \text{and} \quad \rho(y_{i+1}, Ay_i) \leq \delta_0, \quad i = 0, \ldots, n-1, \tag{2.391}$$

the following inequality holds:

$$\rho(y_i, x_A) \leq \varepsilon, \quad i = N, \ldots, n. \tag{2.392}$$

By property (A4), the set $A^i(B(x_A, 4))$ is bounded for any integer $i \geq 1$. Choose a number $s > 1$ such that

$$\bigcup_{i=0}^{N} A^i\big(B(x_A, 4)\big) \subset B(x_A, s). \tag{2.393}$$

By induction and (A3), we define a finite sequence of positive numbers $\{\gamma_i\}_{i=0}^N$ so that

$$\begin{gathered} \gamma_i \leq 1, \quad i = 0, \ldots, N, \\ \gamma_N \leq \delta_0/4, \qquad \gamma_i \leq \gamma_{i+1}, \quad i = 0, \ldots, N-1, \end{gathered} \tag{2.394}$$

and for each $j \in \{0, \ldots, N-1\}$,

$$\begin{gathered} \rho(Ax, Ay) \leq 2^{-1}\gamma_{j+1} \quad \text{for all } x, y \in B(x_A, s+4) \\ \text{with} \quad \rho(x, y) \leq \gamma_j. \end{gathered} \tag{2.395}$$

Set

$$\delta = \gamma_0/4. \tag{2.396}$$

Assume that $\{x_i\}_{i=0}^{\infty} \subset X$,

$$\rho(x_0, x_A) \le \delta \quad \text{and} \quad \rho(x_{i+1}, Ax_i) \le \delta, \quad i = 0, 1, \ldots. \tag{2.397}$$

We will show that

$$\rho(x_i, x_A) \le \varepsilon \tag{2.398}$$

for all integers $i \ge 0$. By (2.397), (2.396) and (P1), inequality (2.398) holds for all integers $i \ge N$. Therefore we only need to prove (2.398) for $i < N$. Clearly, (2.398) holds for $i = 0$.

We will show that for $i = 0, \ldots, N$, we have

$$\rho(x_i, x_A) = \rho(x_i, A^i x_A) \le \gamma_i. \tag{2.399}$$

By (2.397) and (2.396), this is true for $i = 0$. Assume that $i \in \{1, \ldots, N\}$ and

$$\rho(x_{i-1}, A^{i-1} x_A) = \rho(x_{i-1}, x_A) \le \gamma_{i-1}. \tag{2.400}$$

Then (2.397) implies that

$$\rho(x_i, x_A) \le \rho(x_i, Ax_{i-1}) + \rho(Ax_{i-1}, x_A) \le \delta + \rho(Ax_{i-1}, x_A). \tag{2.401}$$

It follows from (2.400) and (2.394) that

$$x_{i-1} \in B(x_A, s). \tag{2.402}$$

By (2.402), (2.400) and the definition of γ_{i-1} (see (2.395) with $j = i - 1$),

$$\rho(Ax_{i-1}, x_A) \le 2^{-1}\gamma_i. \tag{2.403}$$

Using (2.401), (2.403), (2.396) and (2.394), we obtain

$$\rho(x_i, x_A) \le \delta + 2^{-1}\gamma_i \le \gamma_i.$$

Thus (2.399) indeed holds for all $i \in \{0, \ldots, N\}$. This fact, when combined with (2.394), implies that (2.398) is true for all $i \in \{0, \ldots, N\}$. This completes the proof of Theorem 2.67.

2.19 Proof of Theorem 2.68

We may assume, without any loss of generality, that $\varepsilon < 1$ and that $m \ge 1$ is an integer such that

$$m \ge \rho(x_A, \theta) + 4. \tag{2.404}$$

By Theorem 2.65, there exist $\delta \in (0, \varepsilon)$ and a natural number N such that the following property holds.

(P2) For each natural number $n \geq N$ and each sequence $\{x_i\}_{i=0}^n \subset X$ which satisfies

$$x_0 \in B(\theta, m) \quad \text{and} \quad \rho(Ax_i, x_{i+1}) \leq \delta, \quad i = 0, \ldots, n-1, \tag{2.405}$$

the following inequality holds:

$$\rho(x_i, x_A) \leq \varepsilon, \quad i = N, \ldots, n. \tag{2.406}$$

Set

$$K_0 = B(\theta, m) \quad \text{and} \quad K_{i+1} = \{z \in X : \rho(z, A(K_i)) \leq 1\}, \quad i = 0, 1, \ldots. \tag{2.407}$$

Clearly, the set K_i is bounded for any integer $i \geq 0$. Choose a natural number $q \geq 8$ such that

$$\bigcup_{i=0}^{2N} K_i \subset B(\theta, q) \quad \text{and} \quad 1/q < \delta/8. \tag{2.408}$$

We are going to use the following technical result.

Lemma 2.71 *Assume that*

$$z \in B(\theta, m) \quad \textit{and} \quad \{B_i\}_{i=1}^{2N} \subset \{C \in \mathcal{A} : (C, A) \in E_q\}, \tag{2.409}$$

where E_q is given by (2.375). *Then*

$$\rho(B_i \cdots B_1 z, x_A) \leq \varepsilon, \quad i = N, \ldots, 2N. \tag{2.410}$$

Proof Set

$$z_0 = z \quad \text{and} \quad z_i = B_i z_{i-1}, \quad i = 1, \ldots, 2N. \tag{2.411}$$

We will show that

$$z_i \in K_i \tag{2.412}$$

for $i = 0, \ldots, 2N$. Clearly, (2.412) holds for $i = 0$. Assume that $i \in \{0, \ldots, 2N-1\}$ and (2.412) is valid. Inclusions (2.412) and (2.408) imply that

$$z_i \in K_i \subset B(\theta, q). \tag{2.413}$$

When combined with (2.409), (2.375) and (2.411), this last inclusion implies that

$$\rho(Az_i, z_{i+1}) = \rho(Az_i, B_{i+1} z_i) \leq 1/q. \tag{2.414}$$

Consequently, (2.414), (2.413) and (2.407) imply that $z_{i+1} \in K_{i+1}$. Therefore (2.412) is true for all $i = 0, \dots, 2N$. This implies (see (2.408)) that

$$\{z_i\}_{i=0}^{2N} \subset B(\theta, q).$$

It follows from this inclusion, (2.408), (2.409) and (2.411) that for $i = 0, \dots, 2N - 1$,

$$\rho(z_{i+1}, Az_i) = \rho(B_{i+1}z_i, Az_i) \le 1/q < \delta.$$

By (P2), we see that

$$\rho(B_i \cdots B_1 z, x_A) = \rho(z_i, x_A) \le \varepsilon, \quad i = N, \dots, 2N.$$

Lemma 2.71 is proved. □

Now we are ready to complete the proof of Theorem 2.68. To this end, set

$$\mathcal{U} = \{C \in \mathcal{A} : (C, A) \in E_q\}. \tag{2.415}$$

Let $n \ge N$ be an integer, $x \in B(\theta, m)$, and $\{B_i\}_{i=1}^n \subset \mathcal{U}$. We will show that

$$\rho(B_i \cdots B_1 x, x_A) \le \varepsilon \quad \text{for } i = N, \dots, n. \tag{2.416}$$

If $n \le 2N$, then (2.416) follows from Lemma 2.71. Therefore we may restrict our attention to the case $n > 2N$. By Lemma 2.71,

$$\rho(B_i \cdots B_1 x, x_A) \le \varepsilon, \quad i = N, \dots, 2N. \tag{2.417}$$

Suppose now that there exists an integer $p > 2N$, $p \le n$, such that

$$\rho(B_p \cdots B_1 x, x_A) > \varepsilon. \tag{2.418}$$

According to (2.417), we may assume, without loss of generality, that

$$\rho(B_i \cdots B_1 x, x_A) \le \varepsilon, \quad i = 2N, \dots, p - 1. \tag{2.419}$$

Define $\{D_i\}_{i=0}^{2N} \subset \mathcal{A}$ by

$$D_i = B_{i+p-N}, \quad i = 0, \dots, N, \qquad D_i = A, \quad i = N+1, \dots, 2N, \tag{2.420}$$

and let

$$z = B_{p-N} \cdots B_1 x.$$

It follows from (2.417), (2.419), (2.420) and (2.404) that

$$\rho(z, x_A) \le \varepsilon \quad \text{and} \quad z \in B(\theta, m).$$

Applying now Lemma 2.71 to the mappings $\{D_i\}_{i=0}^{2N}$ defined by (2.420), we deduce that

$$\varepsilon \geq \rho(D_N \cdots D_1 z, x_A) = \rho(x_A, B_p \cdots B_{p-N+1} z) = \rho(x_A, B_p \cdots B_1 x),$$

which contradicts (2.418). Hence (2.416) is true and Theorem 2.68 is established.

2.20 Proof of Theorem 2.69

Let $A \in \mathcal{A}_{reg}$ and let $k \geq 1$ be an integer. There is $x_A \in K$ such that

$$Ax_A = x_A. \tag{2.421}$$

According to Theorem 2.68, there exist a natural number $N(A,k)$ and an open neighborhood $\mathcal{U}(A,k)$ of A in $\mathcal{A}$ such that the following property holds.

(P3) For each $x \in B(\theta, k)$, each natural number $n \geq N(A,k)$ and each $B \in \mathcal{U}(A,k)$, we have $\rho(B^n, x_A) \leq 1/k$.

Define

$$\mathcal{F} = \left[\bigcap_{q=1}^{\infty} \bigcup \{\mathcal{U}(A,k) : A \in \mathcal{A}_{reg}, k \geq q \text{ an integer}\} \right] \cap \bar{\mathcal{A}}_{reg}. \tag{2.422}$$

Clearly, $\mathcal{F}$ is an everywhere dense G_δ subset of $\bar{\mathcal{A}}_{reg}$.

Let $B \in \mathcal{F}$. We claim that $B \in \mathcal{A}_{reg}$. Indeed, let q be a natural number. There exists a mapping $A_q \in \mathcal{A}_{reg}$ with a fixed point x_{A_q} and a natural number $k_q \geq q$ such that

$$B \in \mathcal{U}(A_q, k_q). \tag{2.423}$$

This inclusion together with (P3) imply that the following property holds.

(P4) For each point $x \in B(\theta, q) \subset B(\theta, k_q)$ and each natural number $n \geq N(A_q, k_q)$,

$$\rho(B^n x, x_{A_q}) \leq k_q^{-1} \leq 1/q.$$

Since q is an arbitrary natural number, we obtain that for any $x \in X$, the sequence $\{B^n x\}_{n=1}^{\infty}$ is a Cauchy sequence and its limit is the unique fixed point x_B of B. Thus

$$\lim_{n \to \infty} B^N z = x_B \quad \text{for any } z \in X.$$

Property (P4) implies that

$$\rho(x_{A_q}, x_B) \leq 1/q. \tag{2.424}$$

Finally, it follows from property (P4) and (2.424) that for any $x \in B(\theta, q)$ and any $n \geq N(A_q, k_q)$,

$$\rho\big(B^n x, x_B\big) \leq 2/q.$$

This implies that $B^n x \to x_B$ as $n \to \infty$, uniformly on any bounded subset of X. This completes the proof of Theorem 2.69.

2.21 Inexact Orbits of Nonexpansive Operators

Let (X, ρ) be a complete metric space, $A : X \to X$ be a continuous mapping, and let $F(A)$ be the set of all fixed points of A. We assume that $F(A) \neq \emptyset$ and that for each $x, y \in X$,

$$\rho(Ax, Ay) \leq \rho(x, y). \tag{2.425}$$

By A^0 we denote the identity self-mapping of A. We assume that for each $x \in X$, the sequence $\{A^n x\}_{n=1}^{\infty}$ converges in (X, ρ). (Clearly, its limit belongs to $F(A)$.)

The following result was obtained in [34].

Theorem 2.72 *Let $x_0 \in X$, $\{r_n\}_{n=0}^{\infty} \subset (0, \infty)$, $\sum_{n=0}^{\infty} r_n < \infty$,*

$$\{x_n\}_{n=0}^{\infty} \subset X, \quad \rho(x_{n+1}, Ax_n) \leq r_n, \quad n = 0, 1, \ldots. \tag{2.426}$$

Then the sequence $\{x_n\}_{n=1}^{\infty}$ converges to a fixed point of A in (X, ρ).

Proof Fix a natural number k and consider the sequence $\{A^n x_k\}_{n=0}^{\infty}$. This sequence converges to $y_k \in F(A)$. By induction we will show that for each integer $i \geq 0$,

$$\rho\big(A^i x_k, x_{k+1}\big) \leq \sum_{j=k-1}^{i+k-1} r_j - r_{k-1}. \tag{2.427}$$

Clearly, for $i = 0$ (2.427) is valid. Assume that (2.427) is valid for an integer $i \geq 0$. By (2.426), (2.425) and (2.427),

$$\begin{aligned}\rho\big(x_{k+i+1}, A^{i+1} x_k\big) &\leq \rho(x_{k+i+1}, Ax_{k+i}) + \rho\big(Ax_{k+i}, A\big(A^i x_k\big)\big)\\ &\leq r_{k+i} + \rho\big(x_{k+i}, A^i x_k\big) \leq \sum_{j=k-1}^{i+k} r_i - r_{k-1}.\end{aligned}$$

Therefore (2.427) holds for all integers $i \geq 0$.

By (2.427), we have for each integer $i \geq 0$,

$$\rho(x_{k+i}, y_k) \leq \rho\big(x_{k+i}, A^i x_k\big) + \rho\big(A^i x_k, y_k\big) \leq \sum_{j=k}^{\infty} r_j + \rho\big(A^i x_k, y_k\big). \tag{2.428}$$

Since $A^i x_k$ converges to y_k in (X, ρ), there is an integer $i_0 \geq 1$ such that for each integer $i \geq i_0$,

$$\rho(A^i x_k, y_k) \leq \sum_{j=k}^{\infty} r_j/4. \tag{2.429}$$

By (2.429) and (2.428), for each pair of integers $i_1, i_2 \geq i_0$,

$$\rho(x_{k+i_1}, x_{k+i_2}) \leq \rho(x_{k+i_1}, y_k) + \rho(y_k, x_{k+i_2}) \leq 3\sum_{j=k}^{\infty} r_j.$$

Thus we have shown that for each natural number k, there is an integer $i_0 \geq 1$ such that for each pair of integers $i_1, i_2 \geq i_0$,

$$\rho(x_{k+i_1}, x_{k+i_2}) \leq 3\sum_{j=k}^{\infty} r_j.$$

Since $\sum_{j=1}^{\infty} r_j < \infty$, we see that $\{x_n\}_{n=1}^{\infty}$ is a Cauchy sequence and there exists $\bar{x} = \lim_{n\to\infty} x_n$. Together with (2.428), this equality implies that

$$\rho(\bar{x}, y_k) \leq \sum_{j=k}^{\infty} r_j.$$

Since $\sum_{j=1}^{\infty} r_j < \infty$, this inequality implies that

$$\bar{x} = \lim_{k\to\infty} y_k$$

and $A\bar{x} = \bar{x}$. Theorem 2.72 is proved. □

Now we present another result which was obtained in [34].

Let X be a nonempty closed subset of a Banach space $(E, \|\cdot\|)$ with a dual space $(E^*, \|\cdot\|_*)$ and let $A : X \to X$ satisfy

$$\|Ax - Ay\| \leq \|x - y\| \quad \text{for each } x, y \in X. \tag{2.430}$$

As usual, we denote by A^0 the identity self-mapping of X. Consider the following assumptions.

(A1) For each $x \in X$, the sequence $\{A^n x\}_{n=1}^{\infty}$ converges weakly in X.

(A2) For each $x \in X$, the sequence $\{A^n x\}_{n=1}^{\infty}$ converges weakly in X to a fixed point of A.

Theorem 2.73 *Assume that* (A1) *holds. Let* $x_0 \in X$,

$$\{r_n\}_{n=0}^{\infty} \subset (0,\infty), \sum_{n=0}^{\infty} r_n < \infty, \tag{2.431}$$

$$\{x_n\}_{n=0}^{\infty} \subset X, \qquad \|x_{n+1} - Ax_n\| \le r_n, \quad n = 0, 1, \ldots. \tag{2.432}$$

Then the sequence $\{x_n\}_{n=1}^{\infty}$ converges weakly in X. Moreover, if (A2) *holds, then its limit is a fixed point of A.*

Proof Fix a natural number k and consider a sequence $\{A^n x_k\}_{n=0}^{\infty}$. This sequence converges weakly to $y_k \in X$. (Note that if (A2) holds, then $Ay_k = y_k$.) By induction we will show that for each integer $i \ge 0$,

$$\|A^i x_k - x_{k+i}\| \le \sum_{j=k-1}^{i+k-1} r_j - r_{k-1}. \tag{2.433}$$

It is clear that (2.433) is valid for $i = 0$. Assume that $i \ge 0$ is an integer and that (2.433) is valid. By (2.432) and (2.430),

$$\begin{aligned}\|x_{k+i+1} - A^{i+1}x_k\| &\le \|x_{k+i+1} - Ax_{k+i}\| + \|Ax_{k+i} - A(A^i x_k)\| \\ &\le r_{k+i} + \|x_{k+i} - A^i x_k\| \\ &\le r_{k+i} + \sum_{j=k-1}^{i+k-1} r_j - r_{k-1} = \sum_{j=k-1}^{i+k} r_j - r_{k-1}.\end{aligned}$$

Therefore (2.433) holds for all integers $i \ge 0$. Fix an integer $q \ge 1$. By (2.433), we have

$$\|A^q x_k - x_{k+q}\| \le \sum_{j=k}^{\infty} r_j. \tag{2.434}$$

By (2.430) and (2.434), we have for each integer $i \ge 0$,

$$\|A^{q+i} x_k - A^i x_{k+q}\| \le \|A^q x_k - x_{k+q}\| \le \sum_{j=k}^{\infty} r_j. \tag{2.435}$$

In view of (2.435) and the definition of y_k and y_{k+q},

$$\|y_k - y_{k+q}\| \le \sum_{j=k}^{\infty} r_j. \tag{2.436}$$

Since the above inequality holds for each pair of natural numbers q and k and since $\sum_{j=0}^{\infty} r_j < \infty$, we conclude that $\{y_k\}_{k=1}^{\infty}$ is a Cauchy sequence and there exists

$$y_* = \lim_{k\to\infty} y_k \tag{2.437}$$

in the norm topology of E. (Note that if (A2) holds, then $Ay_* = y_*$.) By (2.437) and (2.436),

$$\|y_k - y_*\| \le \sum_{j=k}^{\infty} r_j \quad \text{for all integers } k \ge 1. \tag{2.438}$$

In order to complete the proof it is sufficient to show that $\lim_{k\to\infty} x_k = y_*$ in the weak topology.

Let $f \in E^*$ be a continuous linear functional on E such that $\|f\|_* \le 1$ and let $\varepsilon > 0$ be given. It is sufficient to show that $|f(y_* - x_i)| \le \varepsilon$ for all large enough integers i.

There is an integer $k \ge 1$ such that

$$\sum_{j=k}^{\infty} r_j < \varepsilon/4. \tag{2.439}$$

By (2.438) and (2.434), for each integer $i \ge 1$,

$$\begin{aligned} \left|f(y_* - x_{k+i})\right| &\le \left|f(y_* - y_k)\right| + \left|f\left(y_k - A^i x_k\right)\right| + \left|f\left(A^i x_k - x_{k+i}\right)\right| \\ &\le \|y_* - y_k\| + \left|f\left(y_k - A^i x_k\right)\right| + \left\|A^i x_k - x_{k+i}\right\| \\ &\le \sum_{j=k}^{\infty} r_j + \left|f\left(y_k - A^i x_k\right)\right| + \sum_{j=k}^{\infty} r_j. \end{aligned} \tag{2.440}$$

Since $y_k = \lim_{i\to\infty} A^i x_k$ in the weak topology of X, there is a natural number i_0 such that

$$\left|f\left(y_k - A^i x_k\right)\right| \le \varepsilon/4 \quad \text{for all natural numbers } i \ge i_0. \tag{2.441}$$

By (2.440), (2.439), (2.441), we have for each integer $i \ge i_0$,

$$\left|f(y_* - x_{k+i})\right| \le \varepsilon/4 + \varepsilon/4 + \varepsilon/4 = 3\varepsilon/4.$$

Theorem 2.73 is proved. □

2.22 Convergence to Attracting Sets

In this section we continue to study the influence of errors on the convergence of orbits of nonexpansive mappings in either metric or Banach spaces.

Let (X, ρ) be a metric space. For each $x \in X$ and each closed nonempty subset $A \subset X$, put

$$\rho(x, A) = \inf\{\rho(x, y) : y \in A\}.$$

Theorem 2.74 *Let $T : X \to X$ satisfy*

$$\rho(Tx, Ty) \le \rho(x, y) \quad \text{for all } x, y \in X. \tag{2.442}$$

Suppose that F is a nonempty closed subset of X such that for each $x \in X$,

$$\lim_{i \to \infty} \rho\big(T^i x, F\big) = 0.$$

Assume that $\{\gamma_n\}_{n=0}^{\infty} \subset (0, \infty)$, $\sum_{n=0}^{\infty} \gamma_n < \infty$,

$$\{x_n\}_{n=0}^{\infty} \subset X \quad \text{and} \quad \rho(x_{n+1}, Tx_n) \le \gamma_n, \quad n = 0, 1, \dots. \tag{2.443}$$

Then

$$\lim_{n \to \infty} \rho(x_n, F) = 0.$$

Proof Let $\varepsilon > 0$. Then there is an integer $k \ge 1$ such that

$$\sum_{i=k}^{\infty} \gamma_i < \varepsilon. \tag{2.444}$$

Define a sequence $\{y_i\}_{i=k}^{\infty}$ by

$$\begin{aligned} y_k &= x_k, \\ y_{i+1} &= Ty_i \quad \text{for all integers } i \ge k. \end{aligned} \tag{2.445}$$

By (2.443) and (2.445),

$$\rho(x_{k+1}, y_{k+1}) \le \gamma_k. \tag{2.446}$$

Assume that $q \ge k + 1$ is an integer and that for $i = k + 1, \dots, q$,

$$\rho(x_i, y_i) \le \sum_{j=k}^{i-1} \gamma_j. \tag{2.447}$$

(Note that in view of (2.446), inequality (2.447) is valid when $q = k + 1$.)

By (2.442) and (2.447),

$$\rho(Ty_q, Tx_q) \le \rho(y_q, x_q) \le \sum_{j=k}^{q-1} \gamma_j.$$

When combined with (2.445) and (2.443), this implies that

$$\rho(x_{q+1}, y_{q+1}) \le \rho(x_{q+1}, Tx_q) + \rho(Tx_q, Ty_q) \le \gamma_q + \sum_{j=k}^{q-1} \gamma_j = \sum_{j=k}^{q} \gamma_j,$$

so that (2.447) also holds for $i = q + 1$. Thus we have shown that for all integers $q \geq k + 1$,

$$\rho(y_q, x_q) \leq \sum_{j=k}^{q-1} \gamma_j < \sum_{j=k}^{\infty} \gamma_j < \varepsilon, \tag{2.448}$$

by (2.444). In view of (2.445) and the hypotheses of the theorem we note that

$$\lim_{i \to \infty} \rho(y_i, F) = 0. \tag{2.449}$$

By (2.448) and (2.449),

$$\limsup_{i \to \infty} \rho(x_i, F) \leq \varepsilon.$$

Since ε is an arbitrary positive number, we conclude that

$$\lim_{i \to \infty} \rho(x_i, F) = 0,$$

as asserted. □

Theorem 2.75 *Let X be a nonempty and closed subset of a reflexive Banach space $(E, \|\cdot\|)$ and let $T : X \to X$ be such that*

$$\|Tx - Ty\| \leq \|x - y\| \quad \text{for all } x, y \in X. \tag{2.450}$$

Let F be a nonempty and closed subset of X such that for each $x \in X$, the sequence $\{T^n x\}_{n=1}^{\infty}$ is bounded and all its weak limit points belong to F.

Assume that $\{\gamma_i\}_{i=0}^{\infty} \subset (0, \infty)$, $\sum_{i=0}^{\infty} \gamma_i < \infty$, $\{x_i\}_{i=0}^{\infty} \subset X$ and

$$\|x_{i+1} - Tx_i\| \leq \gamma_i \quad \text{for all integers } i \geq 0. \tag{2.451}$$

Then the sequence $\{x_i\}_{i=0}^{\infty} \subset X$ is bounded and all its weak limit points also belong to F.

Proof Let $\varepsilon > 0$ be given. There is an integer $k \geq 1$ such that

$$\sum_{i=k}^{\infty} \gamma_i < \varepsilon. \tag{2.452}$$

Define a sequence $\{y_i\}_{i=k}^{\infty}$ by

$$y_k = x_k, \qquad y_{i+1} = Ty_i \quad \text{for all integers } i \geq k. \tag{2.453}$$

Arguing as in the proof of Theorem 2.74, we can show that for all integers $q \geq k+1$,

$$\|y_q - x_q\| \leq \sum_{j=k}^{q-1} \gamma_j < \varepsilon. \tag{2.454}$$

Obviously, (2.454) implies that the sequence $\{x_k\}_{k=0}^{\infty}$ is bounded.

Assume now that z is a weak limit point of the sequence $\{x_k\}_{k=0}^{\infty}$. There exists a subsequence $\{x_{i_p}\}_{p=1}^{\infty}$ which weakly converges to z. We may assume without loss of generality that $\{y_{i_p}\}_{p=1}^{\infty}$ weakly converges to $\tilde{z} \in F$. By (2.454) and the weak lower semicontinuity of the norm,

$$\|\tilde{z} - z\| \leq \varepsilon.$$

Since ε is an arbitrary positive number, we conclude that

$$z \in F.$$

Theorem 2.75 is proved. □

Both Theorems 2.74 and 2.75 were obtained in [111].

2.23 Nonconvergence to Attracting Sets

In this section, which is based on [111], we show that both Theorems 2.72 and 2.74 cannot, in general, be improved. We begin with Theorem 2.72.

Proposition 2.76 *For any normed space X, there exists an operator $T : X \to X$ such that $\|Tx - Ty\| \leq \|x - y\|$ for all $x, y \in X$, the sequence $\{T^n x\}_{n=1}^{\infty}$ converges for each $x \in X$ and, for any sequence of positive numbers $\{\gamma_n\}_{n=0}^{\infty}$, there exists a sequence $\{x_n\}_{n=0}^{\infty} \subset X$ with $\|x_{n+1} - Tx_n\| \leq \gamma_n$ for all nonnegative integers n, which converges if and only if the sequence $\{\gamma_n\}_{n=0}^{\infty}$ is summable, i.e., $\sum_{n=0}^{\infty} \gamma_n < \infty$.*

Proof This is a simple fact because we may take T to be the identity operator: $Tx = x$, $\forall x$. Then we may take x_0 to be an arbitrary element of X with $\|x_0\| = 1$, and define by induction

$$x_{n+1} = Tx_n + \gamma_n x_0, \quad n = 0, 1, 2, \ldots.$$

Evidently, $\|x_{n+1} - Tx_n\| = \gamma_n$ and $x_{n+1} = x_0(1 + \sum_{i=0}^{n} \gamma_i)$ for all integers $n \geq 0$, so that the convergence of $\{x_n\}_{n=0}^{\infty}$ is equivalent to the summability of the sequence $\{\gamma_n\}_{n=0}^{\infty}$. □

Counterexamples to possible improvements of Theorem 2.74 are more difficult to construct because this theorem deals with convergence to attracting sets. For simplicity, we assume that the non-summable sequence $\{\gamma_n\}_{n=0}^{\infty}$ decreases to 0 and that $\gamma_1 \leq 1$.

Proposition 2.77 *Let X be an arbitrary (but not one-dimensional) normed space and let a non-summable sequence of positive numbers $\{\gamma_n\}_{n=0}^{\infty}$ decrease to 0. Then there exist a subspace $F \subset X$ and a nonexpansive (with respect to an equivalent norm on X) operator $T : X \to X$ such that $\rho(T^n u, F) \to 0$ as $n \to \infty$ for any*

$u \in X$, and there exists a sequence $\{u_n\}_{n=0}^{\infty} \subset X$ such that $\|u_{n+1} - Tu_n\| \le \gamma_n$ for all integers $n \ge 0$, but $\rho(u_n, F)$ does not tend to 0 as $n \to \infty$.

Proof We take any 2-dimensional subspace of X, identify it with R^2 (with coordinates (x, y)), and perform all constructions and proofs only in this subspace, taking as F the one-dimensional space $L := \{(x, y) \in R^2 : y = 0\}$. The same counterexample may be then applied to the whole space X if we take F to be an algebraic complement of the one-dimensional space $\{(x, y) \in R^2 : x = 0\}$ which contains L.

So, consider a plane with orthogonal axes x, y and the norm $\|u\| = \|(x, y)\| = \max(|x|, |y|)$ (recall that in a finite dimensional space all norms are equivalent). At the first stage, we only consider the case where $\gamma_{n+1}/\gamma_n \ge 1/2$ for all n and we define a decreasing function $y = \gamma(x)$ which equals γ_n at $x = 2n$, $n = 1, 2, \dots$, and is linear on the intermediate segments. Finally, we define the operator T as the superposition $T = T_4T_3T_2T_1$ of the following four mappings: (a) $T_1 : (x, y) \mapsto (|x|, |y|)$; (b) $T_2 : (x, y) \mapsto (x, \min(1, y))$; (c) $T_3 : (x, y) \mapsto (x + 2, y)$; (d) $T_4 : (x, y) \mapsto (x, [1 - \gamma(x)]y)$.

The principal point of the proof is to show that the operator T is nonexpansive.

This is obviously true for the first three mappings T_1, T_2 and T_3, so we need only consider the fourth operator T_4. For simplicity, we may assume from the very beginning that $T = T_4$.

For arbitrary $x_1 < x_2$, let $u_1 = (x_1, y_1)$ and $u_2 = (x_2, y_2)$. Then $Tu_1 = (x_1, [1 - \gamma(x_1)]y_1)$ and $Tu_2 = (x_2, [1 - \gamma(x_2)]y_2)$. Our aim is to show that $\|Tu_1 - Tu_2\| \le \|u_1 - u_2\|$, where $\|u_1 - u_2\| = \max(x_2 - x_1, |y_2 - y_1|)$ and $\|Tu_1 - Tu_2\| = \max(x_2 - x_1, |[1 - \gamma(x_2)]y_2 - [1 - \gamma(x_1)]y_1|)$. Since after the application of the first two mappings T_1 and T_2, the second coordinate y already belongs to $[0, 1]$, the case where $x_2 - x_1 \ge 1$ is trivial, because then $\|Tu_1 - Tu_2\| = \|u_1 - u_2\| = x_2 - x_1$. Hence we may assume in what follows that $x_2 - x_1 < 1$ and thus we need only consider one of the following two possibilities: either both x_1 and x_2 belong to the same interval $[2n, 2(n + 1)]$ or they belong to two adjoining intervals $[2n, 2(n + 1)]$ and $[2(n + 1), 2(n + 2)]$ for some $n = 1, 2, \dots$. We claim that in both cases,

$$\gamma(x_1) - \gamma(x_2) \le (x_2 - x_1)\gamma(x_1). \tag{2.455}$$

If $2n \le x_1 < x_2 \le 2(n + 1)$, then the points u_1 and u_2 lie on the straight line connecting the points $(2n, 1 - \gamma_n)$ and $(2(n+1), 1 - \gamma_{n+1})$, so that the ratio $(\gamma(x_1) - \gamma(x_2))/(x_2 - x_1)$ coincides with the slope of this line:

$$k_n = (\gamma_n - \gamma_{n+1})/2 \le \gamma_n/2 \le \gamma_{n+1} \le \gamma(x_1).$$

In the second case the same ratio is less than or equal to $\max(k_n, k_{n+1})$, where

$$k_{n+1} = (\gamma_{n+1} - \gamma_{n+2})/2 \le \gamma_{n+1} \le \gamma(x_1),$$

and therefore inequality (2.455) is proved in both cases.

Note that in order to compare the distances between u_1 and u_2, and between Tu_1 and Tu_2, it is enough to show that

$$\left|y_2\left[1 - \gamma(x_2)\right] - y_1\left[1 - \gamma(x_1)\right]\right| \le \max\left(x_2 - x_1, |y_2 - y_1|\right). \tag{2.456}$$

If $y_1 \geq y_2$, then

$$y_1\big[1-\gamma(x_1)\big]-y_2\big[1-\gamma(x_2)\big]=(y_1-y_2)-\big[y_1\gamma(x_1)-y_2\gamma(x_2)\big]\leq y_1-y_2,$$

because $\gamma(x_1)\geq\gamma(x_2)$. On the other hand,

$$\begin{aligned}y_1\big[1-\gamma(x_1)\big]-y_2\big[1-\gamma(x_2)\big]&=(y_1-y_2)\big[1-\gamma(x_2)\big]+y_1\big[\gamma(x_2)-\gamma(x_1)\big]\\&\geq-(x_2-x_1)\gamma(x_1)y_1\end{aligned}$$

by (2.455). Now inequality (2.456) follows because $\gamma(x_1)y_1<1$.

If $y_2-y_1\geq 0$, then also $y_2[1-\gamma(x_2)]-y_1[1-\gamma(x_1)]\geq 0$ and it suffices to estimate this difference only from above. Bearing in mind that all $y\leq 1$, we obtain by (2.455) that

$$\begin{aligned}&y_2\big[1-\gamma(x_2)\big]-y_1\big[1-\gamma(x_1)\big]\\&\quad=(y_2-y_1)\big[1-\gamma(x_1)\big]+y_2\big[\gamma(x_1)-\gamma(x_2)\big]\\&\quad\leq(y_2-y_1)\big[1-\gamma(x_1)\big]+\gamma(x_1)(x_2-x_1)\leq\max(x_2-x_1,y_2-y_1),\end{aligned}$$

as needed.

Let $u=(x,y)$ be an arbitrary point in R^2. Then $T_2T_1u\in\{(x,y):x\geq 0, 0\leq y\leq 1\}$ and thereafter the operators T_1 and T_2 coincide with the identity mapping. Defining the integer k by $2k\leq x<2(k+1)$, we see that

$$\rho\big(T^nu,F\big)=y\prod_{i=1}^{n}[1-\gamma(x+2i)]\leq y\prod_{i=k+1}^{k+n}(1-\gamma_i)\longrightarrow 0$$

as $n\to\infty$, because the series $\sum_{i=1}^{\infty}\gamma_i$ is divergent.

To finish the proof for the case where $\gamma_{n+1}/\gamma_n\geq 1/2$ for all natural numbers n, we define $u_n=(2(n-1),1)$ for $n=1,2,\ldots$. Then $Tu_n=T_4T_3u_n=(2n,1-\gamma_n)$ and $\|u_{n+1}-Tu_n\|=\gamma_n$. At the same time, $\rho(u_n,F)=1$ for all n and does not tend to 0.

We now proceed to the general case where the given sequence $\{\gamma_n\}_{n=0}^{\infty}$ does not satisfy the condition $\gamma_{n+1}/\gamma_n\geq 1/2$ for all $n\geq 0$. We then define by induction a new sequence:

$$\gamma_1'=\gamma_1,\qquad \gamma_{n+1}'=\max\big\{\gamma_{n+1},\gamma_n'/2\big\},\quad n=1,2,\ldots,$$

so that $\gamma_{n+1}'/\gamma_n'\geq 1/2$. Using the new sequence $\{\gamma_n'\}_{n=0}^{\infty}$, we construct the operator T as before, replacing each γ_n by γ_n'. The sequence $\{u_n\}_{n=0}^{\infty}$ will be defined by induction. Let $u_1=(0,1)$. If the point $u_n=(x_n,y_n)$ has already been defined, then to obtain the next point $u_{n+1}=(x_{n+1},y_{n+1})$, we put $x_{n+1}=x_n+2$, $y_{n+1}=y_n$ if $\gamma_n'=\gamma_n$, and $y_{n+1}=y_n[1-\gamma_n']$ if $\gamma_n'>\gamma_n$. Since $Tu_n=(x_{n+1},y_n[1-\gamma_n'])$ for each n, we find that $\|u_{n+1}-Tu_n\|\leq\gamma_n$ for all n, as needed.

It is easy to see that

$$y_{n+1} = \prod_{k=1}^{n} (1 - \sigma_k \gamma_k'),$$

where $\sigma_k = 1$ when $\gamma_n' > \gamma_n$ and $\sigma_k = 0$ otherwise. But the series $\sum_{k=1}^{\infty} \sigma_k \gamma_k'$ converges, since the ratio of any two consecutive nonzero terms here is not greater than $1/2$. Therefore

$$\rho(u_n, F) \geq \prod_{k=1}^{\infty} (1 - \sigma_k \gamma_k') > 0.$$

That is, the sequence $\{\rho(u_n, F)\}$ again does not tend to zero, as claimed. □

2.24 Convergence and Nonconvergence to Fixed Points

In Sect. 2.23 we have shown that Theorems 2.72 and 2.74 cannot be, in general, improved. However in Proposition 2.76 every point of the space is a fixed point of the operator T and the inexact orbits tend to infinity. In Proposition 2.77 the attracting set F is unbounded and the operator T depends on the sequence of errors. In this section we construct an operator T on a complete metric space X such that all of its orbits converge to its unique fixed point, and for any nonsummable sequence of errors and any initial point, there exists a divergent inexact orbit with a convergent subsequence. On the other hand, we emphasize that while the example of the present section is for a particular subset of an infinite-dimensional Banach space, the examples in Sect. 2.23 apply to general normed spaces, even finite-dimensional ones.

Let X be the set of all sequences $x = \{x_i\}_{i=1}^{\infty}$ of nonnegative numbers such that $\sum_{i=1}^{\infty} x_i \leq 1$. For $x = \{x_i\}_{i=1}^{\infty}$, $y = \{y_i\}_{i=1}^{\infty} \in X$, set

$$\rho(\{x_i\}_{i=1}^{\infty}, \{y_i\}_{i=1}^{\infty}) = \sum_{i=1}^{\infty} |x_i - y_i|. \tag{2.457}$$

Clearly, (X, ρ) is a complete metric space.

Define a mapping $T : X \to X$ as follows:

$$T(\{x_i\}_{i=1}^{\infty}) = (x_2, x_3, \ldots, x_i, \ldots), \quad \{x_i\}_{i=1}^{\infty} \in X. \tag{2.458}$$

In other words, for any $\{x_i\}_{i=1}^{\infty} \in X$,

$$T(\{x_i\}_{i=1}^{\infty}) = \{y_i\}_{i=1}^{\infty}, \quad \text{where } y_i = x_{i+1} \text{ for all integers } i \geq 1. \tag{2.459}$$

Set $T^0 x = x$ for all $x \in X$. Clearly,

$$\rho(Tx, Ty) \leq \rho(x, y) \quad \text{for all } x, y \in X \tag{2.460}$$

and

$$T^n x \text{ converges to } (0, 0, \dots, \dots) \quad \text{as } n \to \infty \tag{2.461}$$

for all $x \in X$.

The following result was obtained in [111].

Theorem 2.78 *Let* $\{r_i\}_{i=0}^{\infty} \subset [0, \infty)$,

$$\sum_{i=0}^{\infty} r_i = \infty, \tag{2.462}$$

and $x = \{x_i\}_{i=1}^{\infty} \in X$. *Then there exists a sequence* $\{y^{(i)}\}_{i=0}^{\infty} \subset X$ *such that*

$$y^{(0)} = x, \qquad \rho\big(Ty^{(i)}, y^{(i+1)}\big) \le r_i, \quad i = 0, 1, \dots,$$

the sequence $\{y^{(i)}\}_{i=0}^{\infty}$ *does not converge in* (X, ρ), *but* $(0, 0, \dots)$ *is a limit point of* $\{y^{(i)}\}_{i=0}^{\infty}$.

In the proof of this theorem we may assume without loss of generality that

$$r_i \le 16^{-1} \quad \text{for all integers } i \ge 0. \tag{2.463}$$

We precede the proof of Theorem 2.78 with the following lemma.

Lemma 2.79 *Let* $z^{(0)} = \{z_i^{(0)}\}_{i=1}^{\infty} \in X$ *and let* $k \ge 0$ *be an integer. Then there exist an integer* $n \ge 4$ *and a sequence* $\{z^{(i)}\}_{i=0}^{n} \subset X$ *such that*

$$\rho\big(z^{(i+1)}, Tz^{(i)}\big) \le r_{k+i}, \quad i = 0, \dots, n-1,$$

and

$$\rho\big(z^{(n)}, (0, 0, 0, \dots)\big) \ge 4^{-1}.$$

Proof There is a natural number $m > 4$ such that

$$\sum_{i=m}^{\infty} z_i^{(0)} < 16^{-1}. \tag{2.464}$$

Set

$$z^{(i+1)} = Tz^{(i)}, \quad i = 0, \dots, m-1. \tag{2.465}$$

Clearly,

$$z^{(m)} = \big(z_{m+1}^{(0)}, z_{m+2}^{(0)}, \dots, z_i^{(0)}, \dots\big). \tag{2.466}$$

By (2.462), there is a natural number $n > m$ such that

$$\sum_{j=k+m}^{k+n} r_j \geq 2^{-1}. \tag{2.467}$$

By (2.467) and (2.463), $n \geq m + 7$ and we may assume without loss of generality that

$$\sum_{j=k+m}^{k+n-1} r_j < 1/2. \tag{2.468}$$

In view of (2.457) and (2.463)

$$\sum_{j=k+m}^{k+n-1} r_j = \sum_{j=k+m}^{k+n} r_j - r_{k+n} \geq 2^{-1} - 16^{-1}. \tag{2.469}$$

For $i = m + 1, \ldots, n$, define $z^{(i)} = \{z_j^{(i)}\}_{j=1}^{\infty}$ as follows:

$$\begin{aligned} z_j^{(i)} &= z_{j+i}^{(0)}, \quad j \in \{1, 2, \ldots\} \setminus \{n + 1 - i\}, \\ z_{n+1-i}^{(i)} &= z_{n+1}^{(0)} + \sum_{j=k+m}^{k+i-1} r_j. \end{aligned} \tag{2.470}$$

Clearly, for $i = m + 1, \ldots, n$, $z^{(i)}$ is well-defined and by (2.470), (2.464) and (2.468),

$$\sum_{j=1}^{\infty} z_j^{(i)} = \sum_{j=i+1}^{\infty} z_j^{(0)} + \sum_{j=k+m}^{k+i-1} r_j \leq \sum_{j=m}^{\infty} z_j^{(0)} + \sum_{j=k+m}^{k+n-1} r_j \leq 16^{-1} + 2^{-1} < 1.$$

Thus $z^{(i)} \in X$, $i = m + 1, \ldots, n$.

Let $i \in \{m, \ldots, n - 1\}$. In order to estimate $\rho(z^{(i+1)}, Tz^{(i)})$, we first set

$$\{\tilde{z}_j\}_{j=1}^{\infty} = Tz^{(i)}. \tag{2.471}$$

In view of (2.471), (2.458) and (2.459), $\tilde{z}_j = z_{j+1}^{(i)}$ for all integers $j \geq 1$. When combined with (2.470), this implies that

$$\tilde{z}_j = z_{j+1+i}^{(0)} \quad \text{for all } j \in \{1, 2, \ldots\} \setminus \{n - i\} \tag{2.472}$$

and

$$\tilde{z}_{n-i} = z_{n+1-i}^{(i)} = z_{n+1}^{(0)} + \sum_{j=k+m}^{k+i-1} r_j.$$

By (2.472), $\tilde{z}_j = z_j^{(i+1)}$ for all $j \in \{1, 2, \ldots\} \setminus \{n-i\}$. Together with (2.473), (2.457), (2.472) and (2.470), this equality implies that

$$\rho\big(z^{(i+1)}, Tz^{(i)}\big) = \rho\big(z^{(i+1)}, \{\tilde{z}_j\}_{j=1}^{\infty}\big) = \big|z_{n-i}^{(i+1)} - \tilde{z}_{n-i}\big| = r_{k+i}.$$

It follows from this relation, which holds for all $i \in \{m, \ldots, n-1\}$, and from (2.465) that

$$\rho\big(z^{(i+1)}, Tz^{(i)}\big) \le r_{k+i}, \quad i = 0, \ldots, n-1.$$

By (2.457), (2.470) and (2.469),

$$\rho\big(z^{(n)}, (0, 0, 0, \ldots)\big) \ge z_1^{(n)} = z_{n+1}^{(0)} + \sum_{j=k+m}^{k+n-1} r_j \ge 2^{-1} - 16^{-1}.$$

This completes the proof of Lemma 2.79. □

Proof of Theorem 2.78 In order to prove the theorem, we construct by induction, using Lemma 2.79, sequences of nonnegative integers $\{t_k\}_{k=0}^{\infty}$ and $\{s_k\}_{k=0}^{\infty}$, and a sequence $\{y^{(i)}\}_{i=0}^{\infty} \subset X$ such that

$$y^{(0)} = x, \tag{2.473}$$

$$\rho\big(y^{(i+1)}, Ty^{(i)}\big) \le r_i \quad \text{for all integers } i \ge 0, \tag{2.474}$$

$$t_0 = s_0 = 0, \qquad s_k < s_{k+1} < t_{k+1} \quad \text{for all integers } k \ge 0, \tag{2.475}$$

and for all integers $k \ge 1$,

$$\rho\big(y^{(s_k)}, (0, 0, 0 \ldots)\big) \le 1/k \quad \text{and} \quad \rho\big(y^{(t_k)}, (0, 0, 0 \ldots)\big) \ge 1/4. \tag{2.476}$$

In the sequel we use the notation $y^{(i)} = \{y_j^{(i)}\}_{j=1}^{\infty}$, $i = 0, 1, \ldots$.

Set

$$y^{(0)} = x \quad \text{and} \quad t_0, s_0 = 0. \tag{2.477}$$

Assume that $q \ge 0$ is an integer and that we have already defined two sequences of nonnegative numbers $\{t_k\}_{k=0}^{q}$ and $\{s_k\}_{k=0}^{q}$, and a sequence $\{y^{(i)}\}_{i=0}^{t_q} \subset X$ such that (2.474) holds for all integers i satisfying $0 \le i < s_q$, (2.477) holds,

$$t_k < s_{k+1} < t_{k+1} \quad \text{for all integers } k \text{ satisfying } 0 \le k < q,$$

and (2.476) holds for all integers k satisfying $0 < k \le q$. (Note that for $q = 0$ this assumption does hold.)

Now we show that this assumption also holds for $q + 1$.

Indeed, there is a natural number $s_{q+1} > t_q + 1$ such that

$$\sum_{j=s_{q+1}-1-t_q}^{\infty} y_j^{(t_q)} < (q+1)^{-1}. \tag{2.478}$$

Set

$$y^{(i+1)} = Ty^{(i)}, \quad i = t_q, \dots, s_{q+1} - 1. \tag{2.479}$$

By (2.479), (2.457), (2.458), (2.459) and (2.478),

$$\rho\big(y^{(s_{q+1})}, (0, 0, \dots)\big) = \sum_{j=1}^{\infty} y_j^{(s_{q+1})} = \sum_{j=s_{q+1}-t_q+1}^{\infty} y_j^{(t_q)} < (q+1)^{-1}. \tag{2.480}$$

Applying Lemma 2.79 with

$$z^{(0)} = y^{(s_{q+1})} \quad \text{and} \quad k = s_{q+1}, \tag{2.481}$$

we obtain that there exist an integer $n \geq 4$ and a sequence $\{y^{(i)}\}_{i=s_{q+1}}^{s_{q+1}+n} \subset X$ such that

$$\rho\big(y^{(i+1)}, Ty^{(i)}\big) \leq r_i, \quad i = s_{q+1}, \dots, s_{q+1} + n - 1, \tag{2.482}$$

and

$$\rho\big(y^{(s_{q+1}+n)}, (0, 0, 0 \dots)\big) \geq 1/4. \tag{2.483}$$

Put

$$t_{q+1} = s_{q+1} + n.$$

In this way we have constructed a sequence $\{y^{(i)}\}_{i=0}^{t_{q+1}} \subset X$ and two sequences of nonnegative integers $\{t_k\}_{k=0}^{q+1}$ and $\{s_k\}_{k=0}^{q+1}$ such that (2.477) holds, (2.474) holds for all integers i satisfying $0 \leq i < t_{q+1}$ (see (2.479) and (2.482)), $t_k < s_{k+1} < t_{k+1}$ for all integers k satisfying $0 \leq k < q+1$, and (2.476) holds for all integers k satisfying $0 < k \leq q+1$ (see (2.480), (2.482) and (2.483)).

In other words, the assumption made concerning q also holds for $q+1$. It follows that we have indeed constructed two sequences of nonnegative integers $\{t_k\}_{k=0}^{\infty}$ and $\{s_k\}_{k=0}^{\infty}$, and a sequence $\{y^{(i)}\}_{i=0}^{\infty} \subset X$ which satisfy (2.473)–(2.476). This completes the proof of Theorem 2.78. □

2.25 Convergence to Compact Sets

In this section, we study the influence of computational errors on the convergence to compact sets of orbits of nonexpansive mappings in Banach and metric spaces.

Let (X, ρ) be a complete metric space. For each $x \in X$ and each nonempty closed subset $A \subset X$, put

$$\rho(x, A) = \inf\{\rho(x, y) : y \in A\}.$$

For each mapping $T : X \to X$, set $T^0 x = x$ for all $x \in X$.

The following result was obtained in [112].

Theorem 2.80 *Let $T : X \to X$ satisfy*

$$\rho(Tx, Ty) \le \rho(x, y) \quad \text{for all } x, y \in X. \tag{2.484}$$

Suppose that for each $x \in X$, there exists a nonempty compact set $E(x) \subset X$ such that

$$\lim_{i\to\infty} \rho\big(T^i x, E(x)\big) = 0. \tag{2.485}$$

Assume that $\{\gamma_n\}_{n=0}^{\infty} \subset (0, \infty)$, $\sum_{n=0}^{\infty} \gamma_n < \infty$,

$$\{x_n\}_{n=0}^{\infty} \subset X \quad \text{and} \quad \rho(x_{n+1}, Tx_n) \le \gamma_n, \quad n = 0, 1, \dots. \tag{2.486}$$

Then there exists a nonempty compact subset F of X such that

$$\lim_{n\to\infty} \rho(x_n, F) = 0.$$

Proof In order to prove the theorem it is sufficient to show that any subsequence of $\{x_n\}_{n=0}^{\infty}$ has a convergent subsequence.

To see this, it is sufficient to show that for any $\varepsilon > 0$, the following assertion holds:

(P1) Any subsequence of $\{x_n\}_{n=0}^{\infty}$ possesses a subsequence which is contained in a ball with radius ε.

Indeed, there is an integer $k \ge 1$ such that

$$\sum_{i=k}^{\infty} \gamma_i < \varepsilon/8. \tag{2.487}$$

Define a sequence $\{y_i\}_{i=k}^{\infty}$ by

$$\begin{aligned} y_k &= x_k, \\ y_{i+1} &= Ty_i \quad \text{for all integers } i \ge k. \end{aligned} \tag{2.488}$$

There exists a nonempty compact set $E \subset X$ such that

$$\lim_{i\to\infty} \rho(y_i, E) = 0. \tag{2.489}$$

By (2.486) and (2.488),

$$\rho(x_{k+1}, y_{k+1}) \le \gamma_k. \tag{2.490}$$

Assume that $q \ge k+1$ is an integer and that for $i = k+1, \dots, q$,

$$\rho(x_i, y_i) \le \sum_{j=k}^{i-1} \gamma_j. \tag{2.491}$$

(Note that in view of (2.490), inequality (2.491) is valid when $q = k + 1$.)

By (2.484) and (2.491),

$$\rho(Ty_q, Tx_q) \le \rho(y_q, x_q) \le \sum_{j=k}^{q-1} \gamma_j.$$

When combined with (2.486), this implies that

$$\rho(x_{q+1}, y_{q+1}) \le \rho(x_{q+1}, Tx_q) + \rho(Tx_q, Ty_q) \le \gamma_q + \sum_{j=k}^{q-1} \gamma_j = \sum_{j=k}^{q} \gamma_j,$$

so that (2.491) also holds for $i = q + 1$. Thus we have shown that for all integers $q \ge k + 1$,

$$\rho(y_q, x_q) \le \sum_{j=k}^{q-1} \gamma_j < \sum_{j=k}^{\infty} \gamma_j < \varepsilon/8 \tag{2.492}$$

by (2.487). In view of (2.489), for all large enough natural numbers q, we have

$$\rho(x_q, E) < \varepsilon/4. \tag{2.493}$$

By (2.493), there exist an integer $q_0 > k$ and a sequence $\{z_i\}_{i=q_0}^{\infty} \subset K$ such that

$$\rho(x_i, z_i) < \varepsilon/3 \quad \text{for all integers } i \ge q_0. \tag{2.494}$$

Consider any subsequence $\{x_{q_i}\}_{i=1}^{\infty}$ of $\{x_n\}_{n=0}^{\infty}$. Since the set E is compact, the sequence $\{z_{q_i}\}_{i=1}^{\infty}$ possesses a convergent subsequence $\{z_{q_{i_j}}\}_{j=1}^{\infty}$.

We may assume without loss of generality that all elements of this convergent subsequence belong to $B(u, \varepsilon/16)$ for some $u \in X$.

In view of (2.494),

$$x_{q_{i_j}} \in B(u, \varepsilon/2) \quad \text{for all sufficiently large natural numbers } j.$$

Thus (P1) holds and this completes the proof of the theorem. □

Note that Theorem 2.80 is an extension of Theorem 2.72.

The following result, which was obtained in [112], shows that both Theorems 2.72 and 2.80 cannot, in general, be improved (cf. Proposition 2.77).

Proposition 2.81 *For any normed space X, there exists an operator $T : X \to X$ such that $\|Tx - Ty\| \le \|x - y\|$ for all $x, y \in X$, the sequence $\{T^n x\}_{n=1}^{\infty}$ converges for each $x \in X$ and, for any sequence of positive numbers $\{\gamma_n\}_{n=0}^{\infty}$, there exists a sequence $\{x_n\}_{n=0}^{\infty} \subset X$ with $\|x_{n+1} - Tx_n\| \le \gamma_n$ for all nonnegative integers n, which converges to a compact set if and only if the sequence $\{\gamma_n\}_{n=0}^{\infty}$ is summable, i.e., $\sum_{n=0}^{\infty} \gamma_n < \infty$.*

Proof This is a simple fact because we may take T to be the identity operator: $Tx = x$, $\forall x$. Then we may take as x_0 to be an arbitrary element of X with $\|x_0\| = 1$ and define by induction

$$x_{n+1} = Tx_n + \gamma_n x_0, \quad n = 0, 1, 2, \ldots.$$

Evidently, $\|x_{n+1} - Tx_n\| = \gamma_n$ and $x_{n+1} = x_0(1 + \sum_{i=0}^{n} \gamma_i)$ for all integers $n \geq 0$, so that the convergence of $\{x_n\}_{n=0}^{\infty}$ to a compact set is equivalent to the summability of the sequence $\{\gamma_n\}_{n=0}^{\infty}$. Proposition 2.81 is proved. □

2.26 An Example of Nonconvergence to Compact Sets

In the previous section, we have shown that Theorems 2.72 and 2.80 cannot, in general, be improved. However, in Proposition 2.81 every point of the space is a fixed point of the operator T and the inexact orbits tend to infinity. In this section, we construct an operator T on a certain complete metric space X (a bounded, closed and convex subset of a Banach space) such that all of its orbits converge to its unique fixed point, and for any nonsummable sequence of errors and any initial point, there exists an inexact orbit which does not converge to any compact set. This example is based on [112].

Let X be the set of all sequences $x = \{x_i\}_{i=1}^{\infty}$ of nonnegative numbers such that $\sum_{i=1}^{\infty} x_i \leq 1$. For $x = \{x_i\}_{i=1}^{\infty}$ and $y = \{y_i\}_{i=1}^{\infty}$ in X, set

$$\rho\left(\{x_i\}_{i=1}^{\infty}, \{y_i\}_{i=1}^{\infty}\right) = \sum_{i=1}^{\infty} |x_i - y_i|. \tag{2.495}$$

Clearly, (X, ρ) is a complete metric space.

Define a mapping $T : X \to X$ as follows:

$$T\left(\{x_i\}_{i=1}^{\infty}\right) = (x_2, x_3, \ldots, x_i, \ldots), \quad \{x_i\}_{i=1}^{\infty} \in X. \tag{2.496}$$

In other words, for any $\{x_i\}_{i=1}^{\infty} \in X$,

$$T\left(\{x_i\}_{i=1}^{\infty}\right) = \{y_i\}_{i=1}^{\infty}, \quad \text{where } y_i = x_{i+1} \text{ for all integers } i \geq 1. \tag{2.497}$$

Set $T^0 x = x$ for all $x \in X$. Clearly,

$$\rho(Tx, Ty) \leq \rho(x, y) \quad \text{for all } x, y \in X \tag{2.498}$$

and

$$T^n x \text{ converges to } (0, 0, \ldots, \ldots) \quad \text{as } n \to \infty \tag{2.499}$$

for all $x \in X$.

Theorem 2.82 *Let* $\{r_i\}_{i=0}^{\infty} \subset [0, \infty)$,

$$\sum_{i=0}^{\infty} r_i = \infty, \tag{2.500}$$

and $x = \{x_i\}_{i=1}^{\infty} \in X$. *Then there exists a sequence* $\{y^{(i)}\}_{i=0}^{\infty} \subset X$ *such that*

$$y^{(0)} = x, \qquad \rho\big(Ty^{(i)}, y^{(i+1)}\big) \le r_i, \quad i = 0, 1, \ldots, \tag{2.501}$$

and that the following property holds:
there is no nonempty compact set $E \subset X$ *such that*

$$\lim_{i \to \infty} \rho\big(y^{(i)}, E\big) = \emptyset.$$

In the proof of this theorem, we may assume without any loss of generality that

$$r_i \le 16^{-1} \quad \text{for all integers } i \ge 0. \tag{2.502}$$

We precede the proof of Theorem 2.82 with the following lemma.

Lemma 2.83 *Let* $z^{(0)} = \{z_i^{(0)}\}_{i=1}^{\infty} \in X$, *let* $k \ge 0$ *be an integer and let* j_0 *be a natural number. Then there exist an integer* $n \ge 4$ *and a sequence* $\{z^{(i)}\}_{i=0}^{n} \subset X$ *such that*

$$\rho\big(z^{(i+1)}, Tz^{(i)}\big) \le r_{k+i}, \quad i = 0, \ldots, n-1,$$

and

$$z^{(n)} = \big(z_1^{(n)}, \ldots, z_i^{(n)}, \ldots\big) = \big\{z_i^{(n)}\big\}_{i=1}^{\infty}$$

with $z_{j_0+1}^{(n)} \ge 4^{-1}$.

Proof There is a natural number $m > 4$ such that

$$m > j_0 + 4,$$
$$\sum_{i=m}^{\infty} z_i^{(0)} < 16^{-1}. \tag{2.503}$$

Set

$$z^{(i+1)} = Tz^{(i)}, \quad i = 0, \ldots, m-1. \tag{2.504}$$

Then

$$z^{(m)} = \big(z_{m+1}^{(0)}, z_{m+2}^{(0)}, \ldots, z_i^{(0)}, \ldots\big). \tag{2.505}$$

By (2.500), there is a natural number $n > m$ such that

$$\sum_{j=k+m}^{k+n} r_j \geq 2^{-1}. \tag{2.506}$$

By (2.506) and (2.502),

$$n \geq m + 7 \tag{2.507}$$

and we may assume without loss of generality that

$$\sum_{j=k+m}^{k+n-1} r_j < 1/2. \tag{2.508}$$

In view of (2.506) and (2.502),

$$\sum_{j=k+m}^{k+n-1} r_j = \sum_{j=k+m}^{k+n} r_j - r_{k+n} \geq 2^{-1} - 16^{-1}. \tag{2.509}$$

For $i = m+1, \dots, n$, define $z^{(i)} = \{z_j^{(i)}\}_{j=1}^{\infty}$ as follows:

$$\begin{aligned} z_j^{(i)} &= z_{j+i}^{(0)}, \quad j \in \{1, 2, \dots\} \setminus \{n + 1 + j_0 - i\}, \\ z_{n+1+j_0-i}^{(i)} &= z_{n+1+j_0}^{(0)} + \sum_{j=k+m}^{k+i-1} r_j. \end{aligned} \tag{2.510}$$

Clearly, for $i = m+1, \dots, n$, $z^{(i)}$ is well-defined and by (2.510), (2.503) and (2.508),

$$\sum_{j=1}^{\infty} z_j^{(i)} = \sum_{j=i+1}^{\infty} z_j^{(0)} + \sum_{j=k+m}^{k+i-1} r_j \leq \sum_{j=m}^{\infty} z_j^{(0)} + \sum_{j=k+m}^{k+n-1} r_j \leq 16^{-1} + 2^{-1} < 1.$$

Thus $z^{(i)} \in X$, $i = m+1, \dots, n$.

Let $i \in \{m, \dots, n-1\}$. We now estimate $\rho(z^{(i+1)}, Tz^{(i)})$. If $i = m$, then by (2.496), (2.497), (2.505) and (2.514),

$$\rho\big(z^{(i+1)}, Tz^{(i)}\big) \leq r_{k+i}. \tag{2.511}$$

Let $i > m$. We first set

$$\{\tilde{z}_j\}_{j=1}^{\infty} = Tz^{(i)}. \tag{2.512}$$

In view of (2.506), (2.496) and (2.497), $\tilde{z}_j = z_{j+1}^{(i)}$ for all integers $j \geq 1$. When combined with (2.510), this implies that

$$\tilde{z}_j = z_{j+1+i}^{(0)} \quad \text{for all } j \in \{1, 2, \ldots\} \setminus \{n - i + j_0\},$$

$$\tilde{z}_{n+j_0-i} = z_{n+1+j_0-i}^{(i)} = z_{n+1+j_0}^{(0)} + \sum_{j=k+m}^{k+i-1} r_j. \tag{2.513}$$

By (2.510) and (2.513),

$$\tilde{z}_j = z_j^{(i+1)} \tag{2.514}$$

for all $j \in \{1, 2, \ldots\} \setminus \{n + j_0 - i\}$. It now follows from (2.512), (2.514), (2.510) and (2.513) that

$$\rho\big(z^{(i+1)}, Tz^{(i)}\big) = \rho\big(z^{(i+1)}, \{\tilde{z}_j\}_{j=1}^{\infty}\big) = \big|z_{n+j_0-i}^{(i+1)} - \tilde{z}_{n+j_0-i}\big|$$

$$= \left| z_{n+1+j_0}^{(0)} + \sum_{j=k+m}^{k+i} r_j - \left(z_{n+1+j_0}^{(0)} + \sum_{j=k+m}^{k+i-1} r_j \right) \right| < r_{k+i}.$$

When combined with (2.504), this implies that

$$\rho\big(z^{(i+1)}, Tz^{(i)}\big) \leq r_{k+i}, \quad i = 0, \ldots, n-1.$$

By (2.509) and (2.510),

$$z_{j_0+1}^{(n)} = z_{n+1+j_0-n}^{(n)} \geq \sum_{j=k+m}^{k+n-1} r_j \geq 4^{-1}.$$

This completes the proof of Lemma 2.83. □

Proof of Theorem 2.82 In order to prove the theorem, we construct by induction, using Lemma 2.83, a sequence of nonnegative integers $\{s_k\}_{k=0}^{\infty}$ and a sequence $\{y^{(i)}\}_{i=0}^{\infty} \subset X$ such that

$$y^{(0)} = x,$$

$$\rho\big(y^{(i+1)}, Ty^{(i)}\big) \leq r_i \quad \text{for all integers } i \geq 0, \tag{2.515}$$

$$s_0 = 0, \qquad s_k < s_{k+1} \quad \text{for all integers } k \geq 0, \tag{2.516}$$

and for all integers $k \geq 1$,

$$y_{k+1}^{(s_k)} \geq 1/4. \tag{2.517}$$

In the sequel we use the notation $y^{(i)} = \{y_j^{(i)}\}_{j=1}^{\infty}$, $i = 0, 1, \ldots$.

Set

$$y^{(0)} = x, \qquad s_0 = 0. \tag{2.518}$$

Assume that $q \geq 0$ is an integer and we have already defined a (finite) sequence of nonnegative integers $\{s_k\}_{k=0}^{q}$ and a (finite) sequence $\{y^{(i)}\}_{i=0}^{s_q} \subset X$ such that (2.518) is valid, (2.515) holds for all integers i satisfying $0 \leq i < s_q$,

$$s_i < s_{i+1} \quad \text{for all integers } i \text{ satisfying } 0 \leq i < q,$$

and that (2.517) holds for all integers k satisfying $0 < k \leq q$. (Note that for $q = 0$ this assumption does hold.)

Now we show that this assumption also holds for $q + 1$.

Indeed, applying Lemma 2.83 with

$$z^{(0)} = y^{(s_q)} \quad \text{and} \quad j_0 = q + 1, \qquad k = s_q,$$

we obtain that there exist an integer $s_{q+1} \geq 4 + s_q$ and a sequence $\{y^{(i)}\}_{i=s_q}^{s_{q+1}} \subset X$ such that

$$\rho\big(y^{(i+1)}, Ty^{(i)}\big) \leq r_i, \quad i = s_q, \ldots, s_{q+1} - 1,$$

and

$$y_{q+2}^{(s_{q+1})} \geq 1/4.$$

Thus the assumption made for q also holds for $q + 1$. Therefore we have constructed by induction a sequence $\{y^{(i)}\}_{i=0}^{\infty} \subset X$ and a sequence of nonnegative integers $\{s_k\}_{k=0}^{\infty}$ which satisfy (2.515) and (2.516) for all integers $i, k \geq 0$, respectively, and (2.517) for all integers $k \geq 1$.

Finally, we show that there is no nonempty compact set $E \subset X$ such that

$$\lim_{i \to \infty} \rho\big(y^{(i)}, E\big) = 0.$$

Assume the contrary. Then there does exist a nonempty compact set $E \subset X$ such that

$$\lim_{i \to \infty} \rho\big(y^{i)}, E\big) = 0.$$

This implies that any subsequence of $\{y^{(k)}\}_{k=0}^{\infty}$ possesses a convergent subsequence.

Consider such a subsequence $\{y^{(s_q)}\}_{q=1}^{\infty}$. This subsequence has a convergent subsequence $\{y^{s_{q_p}}\}_{p=1}^{\infty}$. There are, therefore, a point $z = \{z_i\}_{i=0}^{\infty} \in X$ such that

$$z = \lim_{p \to \infty} y^{(s_{q_p})}$$

and a natural number p_0 such that

$$\rho\big(z, y^{(s_{q_p})}\big) \leq 16^{-1} \quad \text{for all integers } p \geq p_0. \tag{2.519}$$

By (2.518) and (2.519), we have for all integers $p \geq p_0$,

$$\left|z_{q_p+1} - y^{(s_{q_p})}_{q_p+1}\right| \leq \rho\left(z, y^{(s_{q_p})}\right) \leq 16^{-1}$$

and

$$z_{q_p+1} \geq y^{(s_{q_p})}_{q_p+1} - 16^{-1} \geq 8^{-1}.$$

This, of course, contradicts the inequality $\sum_{i=1}^{\infty} z_i \leq 1$. The contradiction we have reached completes the proof of Theorem 2.82. □

Chapter 3
Contractive Mappings

In this chapter we consider the class of contractive mappings and show that a typical nonexpansive mapping (in the sense of Baire's categories) is contractive. We also study nonexpansive mappings which are contractive with respect to a given subset of their domain.

3.1 Many Nonexpansive Mappings Are Contractive

Assume that $(X, \|\cdot\|)$ is a Banach space and let K be a bounded, closed and convex subset of X. Denote by $\mathcal{A}$ the set of all operators $A : K \to K$ such that

$$\|Ax - Ay\| \le \|x - y\| \quad \text{for all } x, y \in K. \tag{3.1}$$

In other words, the set $\mathcal{A}$ consists of all the nonexpansive self-mappings of K. Set

$$d(K) = \sup\{\|x - y\| : x, y \in K\}. \tag{3.2}$$

We equip the set $\mathcal{A}$ with the metric $h(\cdot, \cdot)$ defined by

$$h(A, B) = \sup\{\|Ax - Bx\| : x \in K\}, \quad A, B \in \mathcal{A}.$$

Clearly, the metric space $(\mathcal{A}, h)$ is complete.

We say that a mapping $A \in \mathcal{A}$ is contractive if there exists a decreasing function $\phi^A : [0, d(K)] \to [0, 1]$ such that

$$\phi^A(t) < 1 \quad \text{for all } t \in (0, d(K)] \tag{3.3}$$

and

$$\|Ax - Ay\| \le \phi^A(\|x - y\|)\|x - y\| \quad \text{for all } x, y \in K. \tag{3.4}$$

The notion of a contractive mapping, as well as its modifications and applications, were studied by many authors. See, for example, [85]. We now quote a convergence result which is valid in all complete metric spaces [114].

S. Reich, A.J. Zaslavski, *Genericity in Nonlinear Analysis*, Developments in Mathematics 34, DOI 10.1007/978-1-4614-9533-8_3,

Theorem 3.1 *Assume that $A \in \mathcal{A}$ is contractive. Then there exists $x_A \in K$ such that $A^n x \to x_A$ as $n \to \infty$, uniformly on K.*

In [131] we prove that a generic element in the space of all nonexpansive mappings is contractive. In [137] we show that the set of all noncontractive mappings is not only of the first category, but also σ-porous. Namely, the following result was obtained there.

Theorem 3.2 *There exists a set $\mathcal{F} \subset \mathcal{A}$ such that $\mathcal{A} \setminus \mathcal{F}$ is σ-porous in $(\mathcal{A}, h)$ and each $A \in \mathcal{F}$ is contractive.*

Proof For each natural number n, denote by $\mathcal{A}_n$ the set of all $A \in \mathcal{A}$ which have the following property:

(P1) There exists $\kappa \in (0, 1)$ such that $\|Ax - Ay\| \le \kappa \|x - y\|$ for all $x, y \in K$ satisfying $\|x - y\| \ge d(K)(2n)^{-1}$.

Let $n \ge 1$ be an integer. We will show that the set $\mathcal{A} \setminus \mathcal{A}_n$ is porous in $(\mathcal{A}, h)$. Set

$$\alpha = 8^{-1} \min\{d(K), 1\}(2n)^{-1}\big(d(K) + 1\big)^{-1}. \tag{3.5}$$

Fix $\theta \in K$. Let $A \in \mathcal{A}$ and $r \in (0, 1]$. Set

$$\gamma = 2^{-1} r\big(d(K) + 1\big)^{-1} \tag{3.6}$$

and define

$$A_\gamma x = (1 - \gamma)Ax + \gamma\theta, \quad x \in K. \tag{3.7}$$

Clearly, $A_\gamma \in \mathcal{A}$,

$$h(A_\gamma, A) \le \gamma d(K), \tag{3.8}$$

and for all $x, y \in K$,

$$\|A_\gamma x - A_\gamma y\| \le (1 - \gamma)\|Ax - Ay\| \le (1 - \gamma)\|x - y\|. \tag{3.9}$$

Assume that $B \in \mathcal{A}$ and

$$h(B, A_\gamma) \le \alpha r. \tag{3.10}$$

We will show that $B \in \mathcal{A}_n$.

Let

$$x, y \in K \quad \text{and} \quad \|x - y\| \ge (2n)^{-1} d(K). \tag{3.11}$$

It follows from (3.9) and (3.11) that

$$\|x - y\| - \|A_\gamma x - A_\gamma y\| \ge \gamma \|x - y\| \ge \gamma d(K)(2n)^{-1}. \tag{3.12}$$

By (3.10),

$$\|Bx - By\| \le \|Bx - A_\gamma x\| + \|A_\gamma x - A_\gamma y\| + \|A_\gamma y - By\| \le \|A_\gamma x - A_\gamma y\| + 2\alpha r.$$

When combined with (3.12), (3.6), and (3.5), this implies that

$$\begin{aligned}\|x-y\| - \|Bx - By\| &\ge \|x-y\| - \|A_\gamma x - A_\gamma y\| - 2\alpha r \\ &\ge \gamma d(K)(2n)^{-1} - 2\alpha r \\ &= 2^{-1} r\big[(2n)^{-1} d(K)\big(d(K)+1\big)^{-1} - 4\alpha\big] \\ &\ge 2^{-1} r d(K)(4n)^{-1}\big(d(K)+1\big)^{-1}.\end{aligned}$$

Thus

$$\begin{aligned}\|Bx - By\| &\le \|x-y\| - r d(K)\big(d(K)+1\big)^{-1}(8n)^{-1} \\ &\le \|x-y\|\big(1 - r(8n)^{-1}\big(d(K)+1\big)^{-1}\big).\end{aligned}$$

Since this holds for all $x, y \in K$ satisfying (3.11), we conclude that $B \in \mathcal{A}_n$. Thus each $B \in \mathcal{A}$ satisfying (3.10) belongs to $\mathcal{A}_n$. In other words,

$$\big\{B \in \mathcal{A} : h(B, A_\gamma) \le \alpha r\big\} \subset \mathcal{A}_n. \tag{3.13}$$

If $B \in \mathcal{A}$ satisfies (3.10), then by (3.8), (3.5) and (3.6), we have

$$h(A, B) \le h(B, A_\gamma) + h(A_\gamma, A) \le \alpha r + \gamma d(K) \le 8^{-1} r + 2^{-1} r \le r.$$

Thus

$$\big\{B \in \mathcal{A} : h(B, A_\gamma) \le \alpha r\big\} \subset \big\{B \in \mathcal{A} : h(B, A) \le r\big\}.$$

When combined with (3.13), this inclusion implies that $\mathcal{A} \setminus \mathcal{A}_n$ is porous in $(\mathcal{A}, h)$. Set $\mathcal{F} = \bigcap_{n=1}^{\infty} \mathcal{A}_n$. Clearly, $\mathcal{A} \setminus \mathcal{F}$ is σ-porous in $(\mathcal{A}, h)$. By property (P1), each $A \in \mathcal{F}$ is contractive. □

3.2 Attractive Sets

In this section, we study nonexpansive mappings which are contractive with respect to a given subset of their domain.

Assume that $(X, \|\cdot\|)$ is a Banach space and that K is a closed, bounded and convex subset of X. Once again, denote by $\mathcal{A}$ the set of all mappings $A : K \to K$ such that

$$\|Ax - Ay\| \le \|x - y\| \quad \text{for all } x, y \in K. \tag{3.14}$$

For each $x \in K$ and each subset $E \subset K$, let

$$\rho(x, E) = \inf\{\|x - y\| : y \in E\}. \tag{3.15}$$

Let F be a nonempty, closed and convex subset of K. Denote by $\mathcal{A}^{(F)}$ the set of all $A \in \mathcal{A}$ such that $Ax = x$ for all $x \in F$. Clearly, $\mathcal{A}^{(F)}$ is a closed subset of $(\mathcal{A}, h)$. In what follows we consider the complete metric space $(\mathcal{A}^{(F)}, h)$.

An operator $A \in \mathcal{A}^{(F)}$ is said to be contractive with respect to F if there exists a decreasing function $\phi^A : [0, d(K)] \to [0, 1]$ such that

$$\phi^A(t) < 1 \quad \text{for all } t \in (0, d(K)] \tag{3.16}$$

and

$$\rho(Ax, F) \le \phi^A(\rho(x, F))\rho(x, F) \quad \text{for all } x \in K. \tag{3.17}$$

We now show that if $\mathcal{A}^{(F)}$ contains a retraction, then the complement of the set of contractive mappings (with respect to F) in $\mathcal{A}^{(F)}$ is σ-porous. This result was also obtained in [137].

Theorem 3.3 *Assume that there exists $Q \in \mathcal{A}^{(F)}$ such that*

$$Q(K) = F. \tag{3.18}$$

Then there exists a set $\mathcal{F} \subset \mathcal{A}^{(F)}$ such that $\mathcal{A}^{(F)} \setminus \mathcal{F}$ is σ-porous in $(\mathcal{A}^{(F)}, h)$ and each $B \in \mathcal{F}$ is contractive with respect to F.

Proof For each natural number n, denote by $\mathcal{A}_n$ the set of all $A \in \mathcal{A}^{(F)}$ which have the following property:

(P2) There exists $\kappa \in (0, 1)$ such that $\rho(Ax, F) \le \kappa\rho(x, F)$ for all $x \in K$ such that $\rho(x, F) \ge \min\{d(K), 1\}/n$. Define

$$\mathcal{F} = \bigcap_{n=1}^{\infty} \mathcal{A}_n. \tag{3.19}$$

Clearly, each element of $\mathcal{F}$ is contractive with respect to F. We need to show that $\mathcal{A}^{(F)} \setminus \mathcal{A}_n$ is porous in $(\mathcal{A}^{(F)}, h)$ for all integers $n \ge 1$. To this end, let $n \ge 1$ be an integer and set

$$\alpha = (d(K) + 1)^{-1} \min\{d(K), 1\}(16n)^{-1}. \tag{3.20}$$

Let $A \in \mathcal{A}^{(F)}$ and $r \in (0, 1]$. Set

$$\gamma = 2^{-1}r(d(K) + 1)^{-1} \tag{3.21}$$

and define

$$A_\gamma x = (1 - \gamma)Ax + \gamma Qx, \quad x \in K. \tag{3.22}$$

It is obvious that $A_\gamma \in \mathcal{A}^{(F)}$. By (3.22),

$$h(A, A_\gamma) \le \sup\{\|A_\gamma x - Ax\| : x \in K\}$$
$$\le \gamma \sup\{\|Ax - Qx\| : x \in K\} \le \gamma d(K). \qquad (3.23)$$

Let $B \in \mathcal{A}^{(F)}$ be such that

$$h(A_\gamma, B) \le \alpha r. \qquad (3.24)$$

Then by (3.24), (3.23), (3.21), and (3.20),

$$h(A, B) \le h(A, A_\gamma) + h(A_\gamma, B) \le \gamma d(K) + \alpha r$$
$$< 1/2r + r/2 \le r.$$

Thus (3.24) implies that $h(A, B) \le r$ and

$$\{C \in \mathcal{A}^{(F)} : h(A_\gamma, C) \le \alpha r\}$$
$$\subset \{C \in \mathcal{A}^{(F)} : h(A, C) \le r\}. \qquad (3.25)$$

Let $x \in K$ with

$$\rho(x, F) \ge \min\{d(K), 1\}/n. \qquad (3.26)$$

For each $\varepsilon > 0$, there exists $z \in F$ such that $\rho(x, F) + \varepsilon \ge \|x - z\|$, and by (3.22) and (3.18),

$$\rho(A_\gamma x, F) = \rho\big((1-\gamma)Ax + \gamma Qx, F\big)$$
$$\le \big((1-\gamma)Ax + Qx\big) - \big((1-\gamma)z + \gamma Qx\big) \le (1-\gamma)\|Ax - z\|$$
$$\le (1-\gamma)\|x - z\| \le (1-\gamma)\rho(x, F) + \varepsilon(1-\gamma).$$

Since ε is an arbitrary positive number, we conclude that

$$\rho(A_\gamma x, F) \le (1-\gamma)\rho(x, F).$$

Since $|\rho(y_1, F) - \rho(y_2, F)| \le \|y_1 - y_2\|$ for all $y_1, y_2 \in K$, it follows from (3.24) that

$$\rho(Bx, F) \le \|A_\gamma x - Bx\| + \rho(A_\gamma x, F) \le \alpha r + \rho(A_\gamma x, F)$$
$$\le \alpha r + (1-\gamma)\rho(x, F),$$

and

$$\rho(Bx, F) \le (1-\gamma)\rho(x, F) + \alpha r.$$

It now follows from this inequality, (3.26), (3.20) and (3.21) that

$$\begin{aligned}\rho(Bx,F) &\le \rho(x,F)\big(1-\gamma+\alpha r\big(\rho(x,F)\big)^{-1}\big)\\ &\le \rho(x,F)\big[1-2^{-1}r\big(d(K)+1\big)^{-1}+\alpha r\big(\min\{d(K),1\}/n\big)^{-1}\big]\\ &\le \rho(x,F)\big[1-r2^{-1}\big(d(K)+1\big)^{-1}+r\big(16\big(d(K)+1\big)\big)^{-1}\big]\\ &\le \rho(x,F)\big(1-r4^{-1}d(K+1)^{-1}\big).\end{aligned}$$

Thus

$$\rho(Bx,F)\le \rho(x,F)\big(1-r4^{-1}\big(d(K)+1\big)^{-1}\big)$$

for each $x\in K$ satisfying (3.26). This fact implies that $B\in \mathcal{A}_n$. Since this inclusion holds for any B satisfying (3.24), combining it with (3.25) we obtain that

$$\big\{C\in\mathcal{A}^{(F)}: h(A_\gamma,C)\le \alpha r\big\}\subset\big\{C\in\mathcal{A}^{(F)}: h(A,C)\le r\big\}\cap\mathcal{A}_n.$$

This shows that $\mathcal{A}^{(F)}\setminus\mathcal{A}_n$ is indeed porous in $(\mathcal{A}^{(F)},h)$. □

3.3 Attractive Subsets of Unbounded Spaces

In this section we continue to study nonexpansive mappings which are contractive with respect to a given subset of their domain.

Assume that (X,ρ) is a hyperbolic complete metric space and that K is a closed (not necessarily bounded) and ρ-convex subset of X. Denote by $\mathcal{A}$ the set of all mappings $A: K\to K$ such that

$$\rho(Ax,Ay)\le\rho(x,y)\quad\text{for all } x,y\in K. \tag{3.27}$$

For each $x\in K$ and each subset $E\subset K$, let $\rho(x,E)=\inf\{\rho(x,y): y\in E\}$. For each $x\in K$ and each $r>0$, set

$$B(x,r)=\big\{y\in K:\rho(x,y)\le r\big\}. \tag{3.28}$$

Fix $\theta\in K$. For the set $\mathcal{A}$ we consider the uniformity determined by the following base:

$$E(n,\varepsilon)=\big\{(A,B)\in\mathcal{A}\times\mathcal{A}:\rho(Ax,Bx)\le\varepsilon, x\in B(\theta,n)\big\}, \tag{3.29}$$

where $\varepsilon>0$ and n is a natural number. Clearly the space $\mathcal{A}$ with this uniformity is metrizable and complete. We equip the space $\mathcal{A}$ with the topology induced by this uniformity.

Let F be a nonempty, closed and ρ-convex subset of K. Denote by $\mathcal{A}^{(F)}$ the set of all $A\in\mathcal{A}$ such that $Ax=x$ for all $x\in F$. Clearly, $\mathcal{A}^{(F)}$ is a closed subset of $\mathcal{A}$. We consider the topological subspace $\mathcal{A}^{(F)}\subset\mathcal{A}$ with the relative topology.

An operator $A \in \mathcal{A}^{(F)}$ is said to be contractive with respect to F if for any natural number n there exists a decreasing function $\phi_n^A : [0, \infty) \to [0, 1]$ such that

$$\phi_n^A(t) < 1 \quad \text{for all } t > 0 \tag{3.30}$$

and

$$\rho(Ax, F) \leq \phi_n^A\big(\rho(x, F)\big)\rho(x, F) \quad \text{for all } x \in B(\theta, n). \tag{3.31}$$

Clearly, this definition does not depend on our choice of θ.

We begin our discussion of such mappings by proving that the set F attracts all the iterates of A. This result was obtained in [131].

Theorem 3.4 *Let $A \in \mathcal{A}^{(F)}$ be contractive with respect to F. Then there exists $B \in \mathcal{A}^{(F)}$ such that $B(K) = F$ and $A^n x \to Bx$ as $n \to \infty$, uniformly on $B(\theta, m)$ for any natural number m.*

Proof We may assume without loss of generality that $\theta \in F$. Then for each real $r > 0$,

$$C\big(B(\theta, r)\big) \subset B(\theta, r) \quad \text{for all } C \in \mathcal{A}^{(F)}. \tag{3.32}$$

Let r be a natural number. To prove the theorem, it is sufficient to show that there exists $B : B(\theta, r) \to F$ such that

$$A^n x \to Bx \quad \text{as } n \to \infty, \text{ uniformly on } B(\theta, r). \tag{3.33}$$

There exists a decreasing function $\phi_r^A : [0, \infty) \to [0, 1]$ such that

$$\phi_r^A(t) < 1 \quad \text{for all } t > 0 \tag{3.34}$$

and

$$\rho(Ax, F) \leq \phi_r^A\big(\rho(x, F)\big)\rho(x, F) \quad \text{for all } x \in B(\theta, r). \tag{3.35}$$

Let $\varepsilon \in (0, 1)$. Choose a natural number $m \geq 4$ such that

$$\phi_r^A(\varepsilon r)^m < 8^{-1}\varepsilon. \tag{3.36}$$

Let $x \in B(\theta, r)$. We will show that

$$\rho\big(A^m x, F\big) < \varepsilon r. \tag{3.37}$$

Assume the contrary. Then for each $i = 0, \ldots, m$, $\rho(A^i x, F) \geq \varepsilon r$, and by (3.35) and (3.32),

$$A^i x \in B(\theta, r), \quad \rho\big(A^{i+1}x, F\big) \leq \phi_r^A\big(\rho\big(A^i x, F\big)\big)\rho\big(A^i x, F\big)$$
$$\leq \phi_r^A(\varepsilon r)\rho\big(A^i x, F\big).$$

When combined with (3.36), these inequalities imply that

$$\rho\big(A^m x, F\big) \le \phi_r^A(\varepsilon r)^m \rho(x, F) \le 8^{-1}\varepsilon\rho(x,\theta) \le 8^{-1}\varepsilon r,$$

a contradiction. Therefore (3.27) is valid and for each $x \in B(\theta, r)$, there exists $C_\varepsilon(x) \in F$ such that $\rho(A^m x, C_\varepsilon x) < \varepsilon r$. This implies that for each $x \in B(\theta, r)$,

$$\rho\big(A^i x, C_\varepsilon x\big) < \varepsilon r \quad \text{for all integers } i \ge m. \tag{3.38}$$

Since ε is an arbitrary number in (0, 1), we conclude that for each $x \in B(\theta, r)$, $\{A^i x\}_{i=1}^\infty$ is a Cauchy sequence and there exists $Bx = \lim_{i\to\infty} A^i x$. Clearly,

$$\rho\big(Bx, C_\varepsilon(x)\big) \le \varepsilon r \quad \text{for all } x \in B(\theta, r). \tag{3.39}$$

Since (3.39) is true for any ε in (0, 1), we conclude that $B(B(\theta, r)) \subset F$.

By (3.39) and (3.38), for each $x \in B(\theta, r)$,

$$\rho\big(A^i x, Bx\big) \le 2\varepsilon r \quad \text{for all integers } i \ge m.$$

Finally, since $\varepsilon \in (0, 1)$ is arbitrary, we conclude that (3.33) is valid. This completes the proof of Theorem 3.4. □

Proposition 3.5 *Assume that $A, B \in \mathcal{A}^{(F)}$ and that A is contractive with respect to F. Then AB and BA are also contractive with respect to F.*

Proof We may assume that $\theta \in F$. Then for each real $r > 0$,

$$C\big(B(\theta, r)\big) \subset B(\theta, r) \quad \text{for all } C \in \mathcal{A}^{(F)}. \tag{3.40}$$

Fix $r > 0$. There exists a decreasing function $\phi_r^A : [0, \infty) \to [0, 1]$ such that

$$\phi_r^A(t) < 1 \quad \text{for all } t > 0 \tag{3.41}$$

and

$$\rho(Ax, F) \le \phi_r^A\big(\rho(x, F)\big)\rho(x, F) \quad \text{for all } x \in B(\theta, r). \tag{3.42}$$

By (3.42), for each $x \in B(\theta, r)$,

$$\begin{aligned}\rho(BAx, F) &= \inf\big\{\rho(BAx, y) : y \in F\big\} \le \inf\big\{\rho(Ax, y) : y \in F\big\}\\ &= \rho(Ax, F) \le \phi_r^A\big(\rho(x, F)\big)\rho(x, F).\end{aligned}$$

Therefore BA is contractive with respect to F.

Let now x belong to $B(\theta, r)$. By (3.42) and (3.40), $Bx \in B(\theta, r)$ and

$$\rho(ABx, F) \le \phi_r^A\big(\rho(Bx, F)\big)\rho(Bx, F). \tag{3.43}$$

There are two cases: (1) $\rho(Bx, F) \geq 2^{-1}\rho(x, F)$; (2) $\rho(Bx, F) < 2^{-1}\rho(x, F)$. In the first case, we have by (3.43),

$$\rho(ABx, F) \leq \phi_r^A\big(2^{-1}\rho(x, F)\big)\rho(Bx, F) \leq \phi_r^A\big(2^{-1}\rho(x, F)\big)\rho(x, F),$$

and in the second case, (3.43) implies that

$$\rho(ABx, F) \leq \rho(Bx, F) \leq 2^{-1}\rho(x, F).$$

Thus in both cases we obtain that

$$\begin{aligned}\rho(ABx, F) &\leq \max\big\{\phi_r^A\big(2^{-1}\rho(x, F)\big), 2^{-1}\big\}\rho(x, F)\\ &= \psi\big(\rho(x, F)\big)\rho(x, F),\end{aligned}$$

where $\psi(t) = \max\{\phi_r^A(2^{-1}t), 2^{-1}\}$, $t \in [0, \infty)$. Therefore AB is also contractive with respect to F. Proposition 3.5 is proved. □

We now show that if $\mathcal{A}^{(F)}$ contains a retraction, then almost all the mappings in $\mathcal{A}^{(F)}$ are contractive with respect to F.

Theorem 3.6 *Assume that there exists*

$$Q \in \mathcal{A}^{(F)} \quad \textit{such that} \quad Q(K) = F. \tag{3.44}$$

Then there exists a set $\mathcal{F} \subset \mathcal{A}^{(F)}$ which is a countable intersection of open and everywhere dense sets in $\mathcal{A}^{(F)}$ such that each $B \in \mathcal{F}$ is contractive with respect to F.

Proof We may assume that $\theta \in F$. Then for each real $r > 0$,

$$C\big(B(\theta, r)\big) \subset B(\theta, r) \quad \text{for all } C \in \mathcal{A}^{(F)}. \tag{3.45}$$

For each $A \in \mathcal{A}^{(F)}$ and each $\gamma \in (0, 1)$, define $A_\gamma \in \mathcal{A}^{(F)}$ by

$$A_\gamma x = (1-\gamma)Ax \oplus \gamma Qx, \quad x \in K. \tag{3.46}$$

Clearly, for each $A \in \mathcal{A}^{(F)}$, $A_\gamma \to A$ as $\gamma \to 0^+$ in $\mathcal{A}^{(F)}$. Therefore the set $\{A_\gamma : A \in \mathcal{A}^{(F)}, \gamma \in (0, 1)\}$ is everywhere dense in $\mathcal{A}^{(F)}$.

Let $A \in \mathcal{A}^{(F)}$ and $\gamma \in (0, 1)$. Evidently,

$$\begin{aligned}\rho(A_\gamma x, F) &= \inf_{y\in F}\big\{\rho\big((1-\gamma)Ax \oplus \gamma Qx, y\big)\big\}\\ &\leq \inf_{y\in F}\big\{\rho\big((1-\gamma)Ax \oplus \gamma Qx, (1-\gamma)y \oplus \gamma Qx\big)\big\}\\ &\leq \inf_{y\in F}\big\{(1-\gamma)\rho(Ax, y)\big\} \leq (1-\gamma)\rho(x, F)\end{aligned}$$

for all $x \in K$. Thus

$$\rho(A_\gamma x, F) \le (1-\gamma)\rho(x, F) \quad \text{for all } x \in K. \tag{3.47}$$

For each integer $i \ge 1$, denote by $U(A, \gamma, i)$ an open neighborhood of A_γ in $\mathcal{A}^{(F)}$ for which

$$U(A, \gamma, i) \subset \{B \in \mathcal{A}^{(F)} : (B, A_\gamma) \in E(2^i, 8^{-i}\gamma)\} \tag{3.48}$$

(see (3.29)).

We will show that for each $A \in \mathcal{A}^{(F)}$, each $\gamma \in (0, 1)$ and each integer $i \ge 1$, the following property holds:

P(2) For each $B \in U(A, \gamma, i)$ and each $x \in B(\theta, 2^i)$ satisfying $\rho(x, F) \ge 4^{-i}$, the inequality $\rho(Bx, F) \le (1 - 2^{-1}\gamma)\rho(x, F)$ is true.

Indeed, let $A \in \mathcal{A}^{(F)}$, $\gamma \in (0, 1)$ and let $i \ge 1$ be an integer. Assume that

$$B \in U(A, \gamma, i), \qquad x \in B(\theta, 2^i) \quad \text{and} \quad \rho(x, F) \ge 4^{-i}. \tag{3.49}$$

Using (3.47), (3.48) and (3.49), we see that

$$\begin{aligned} \rho(Bx, F) &\le \rho(A_\gamma x, F) + 8^{-i}\gamma \le (1-\gamma)\rho(x, F) + 8^{-i}\gamma \\ &\le (1-\gamma)\rho(x, F) + 2^{-1}\gamma\rho(x, F) \le (1 - 2^{-1}\gamma)\rho(x, F). \end{aligned}$$

Thus property P(2) holds for each $A \in \mathcal{A}^{(F)}$, each $\gamma \in (0, 1)$ and each integer $i \ge 1$. Define

$$\mathcal{F} = \bigcap_{q=1}^{\infty} \bigcup \{U(A, \gamma, i) : A \in \mathcal{A}^{(F)}, \gamma \in (0, 1), i \ge q\}.$$

Clearly, $\mathcal{F}$ is a countable intersection of open and everywhere dense sets in $\mathcal{A}^{(F)}$.

Let $B \in \mathcal{F}$. To show that B is contractive with respect to F, it is sufficient to show that for each $r > 0$ and each $\varepsilon \in (0, 1)$, there is $\kappa \in (0, 1)$ such that

$$\rho(Bx, F) \le \kappa\rho(x, F) \quad \text{for each } x \in B(\theta, r) \text{ satisfying } \rho(x, F) \ge \varepsilon.$$

Let $r > 0$ and $\varepsilon \in (0, 1)$. Choose a natural number q such that

$$2^q > 8r \quad \text{and} \quad 2^{-q} < 8^{-1}\varepsilon.$$

There exist $A \in \mathcal{A}^{(F)}$, $\gamma \in (0, 1)$ and an integer $i \ge q$ such that $B \in U(A, \gamma, i)$. By property P(2), for each $x \in B(\theta, r) \subset B(\theta, 2^i)$ satisfying $\rho(x, F) \ge \varepsilon > 2^{-i}$, the following inequality holds:

$$\rho(Bx, F) \le (1 - 2^{-1}\gamma)\rho(x, F).$$

Thus B is contractive with respect to F. This completes the proof of Theorem 3.6. □

3.4 A Contractive Mapping with no Strictly Contractive Powers

Let

$$X = [0, 1] \quad \text{and} \quad \rho(x, y) = |x - y| \quad \text{for each } x, y \in X.$$

In this section, which is based on [155], we construct a contractive mapping $A : [0, 1] \to [0, 1]$ such that none of its powers is a strict contraction.

We begin by setting

$$A(0) = 0. \tag{3.50}$$

Next, we define, for each natural number n, the mapping A on the interval $[(n + 1)^{-1}, n^{-1}]$ by

$$A\big((n+1)^{-1} + t\big) = (n+2)^{-1} + t\big(n^{-1} - (n+1)^{-1}\big)^{-1}\big((n+1)^{-1} - (n+2)^{-1}\big)$$
$$\text{for all } t \in \big[0, n^{-1} - (n+1)^{-1}\big]. \tag{3.51}$$

It is clear that for each natural number n,

$$A\big(n^{-1}\big) = (n+1)^{-1}, \tag{3.52}$$

the restriction of A to the interval $[(n + 1)^{-1}, n^{-1}]$ is affine, and that the mapping $A : [0, 1] \to [0, 1]$ is well defined.

First, we show that A is nonexpansive, that is, $|Ax - Ay| \le |x - y|$ for all $x, y \in [0, 1]$.

Indeed, if $x \in [0, 1]$, then

$$\big|Ax - A(0)\big| \le |x|. \tag{3.53}$$

Assume now that n is a natural number and that

$$x, y \in \big[(n+1)^{-1}, n^{-1}\big]. \tag{3.54}$$

By (3.51) and (3.54),

$$\begin{aligned}
&|Ax - Ay| \\
&\quad = \big|(n+2)^{-1} + \big(x - (n+1)^{-1}\big)\big(n^{-1} - (n+1)^{-1}\big)^{-1}\big((n+1)^{-1} - (n+2)^{-1}\big) \\
&\qquad - \big[(n+2)^{-1} + \big(y - (n+1)^{-1}\big)\big(n^{-1} - (n+1)^{-1}\big)^{-1} \\
&\qquad \times \big((n+1)^{-1} - (n+2)^{-1}\big)\big]\big| \\
&\quad = |x - y|\big(n^{-1} - (n+1)^{-1}\big)^{-1}\big((n+1)^{-1} - (n+2)^{-1}\big) \\
&\quad = |x - y| n(n+1)\big((n+1)(n+2)\big)^{-1} = |x - y| n(n+2)^{-1}.
\end{aligned}$$

Thus for each natural number n and each $x, y \in [(n+1)^{-1}, n^{-1}]$,

$$|Ax - Ay| \le |x - y| n(n+2)^{-1}. \tag{3.55}$$

Together with (3.53) this last inequality implies that

$$|Ax - Ay| \le |x - y| \quad \text{for all } x, y \in [0, 1], \tag{3.56}$$

as claimed.

Next, we show that the power A^m is not a strict contraction for any integer $m \ge 1$. Assume the converse. Then there would exist a natural number m and $c \in (0, 1)$ such that for each $x, y \in [0, 1]$,

$$\left|A^m x - A^m y\right| \le c|x - y|. \tag{3.57}$$

Since

$$(m+i)(m+i+1)i^{-1}(i+1)^{-1} \to 1 \quad \text{as } i \to \infty,$$

there is an integer $p \ge 4$ such that

$$p(p+1) > (p+m)(p+m+1)c. \tag{3.58}$$

By (3.52), (3.50) and (3.58),

$$\begin{aligned} &A^m\left(p^{-1}\right) - A^m\left((p+1)^{-1}\right) \\ &\quad = (p+m)^{-1} - (p+m+1)^{-1} = (p+m)^{-1}(p+m+1)^{-1} \\ &\quad > cp^{-1}(p+1)^{-1} = c\left(p^{-1} - (p+1)^{-1}\right), \end{aligned}$$

which contradicts (3.57).

The contradiction we have reached proves that A^m is not a strict contraction for any integer $m \ge 1$.

Finally, we show that A is contractive. Let $\varepsilon \in (0, 1)$. We claim that there exists $c \in (0, 1)$ such that

$$|Ax - Ay| \le c|x - y| \quad \text{for each } x, y \in [0, 1] \text{ satisfying } |x - y| \ge \varepsilon. \tag{3.59}$$

Indeed, choose a natural number $p \ge 4$ such that

$$p > 18\varepsilon^{-2}, \tag{3.60}$$

and assume that

$$x, y \in [0, 1] \quad \text{and} \quad |x - y| \ge \varepsilon. \tag{3.61}$$

We may assume without loss of generality that

$$y > x. \tag{3.62}$$

There are two cases:

$$x < (4p)^{-1}; \tag{3.63}$$

$$x \geq (4p)^{-1}. \tag{3.64}$$

Assume that (3.63) holds. There exists a natural number n such that

$$(1+n)^{-1} < y \leq n^{-1}. \tag{3.65}$$

By (3.65), (3.62) and (3.61),

$$\varepsilon \leq y \leq 1/n, \qquad (n+2)^{-1} \geq (3n)^{-1} \geq \varepsilon/3. \tag{3.66}$$

By (3.65) and (3.51),

$$\begin{aligned} Ay &= (n+2)^{-1} + \big(y - (n+1)^{-1}\big)\big(n^{-1} - (n+1)^{-1}\big)^{-1}\big((n+1)^{-1} - (n+2)^{-1}\big) \\ &= (n+2)^{-1} + \big(y - (n+1)^{-1}\big)n(n+1)(n+1)^{-1}(n+2)^{-1} \\ &\leq y - (n+1)^{-1} + (n+2)^{-1} \end{aligned}$$

and

$$y - Ay \geq (n+1)^{-1}(n+2)^{-1}.$$

When combined with (3.66), the above inequality implies that

$$Ay - Ax \leq Ay \leq y - (n+1)^{-1}(n+2)^{-1} \leq y - (n+2)^{-2} \leq y - \varepsilon^2/9. \tag{3.67}$$

By (3.63), (3.60) and (3.67),

$$\begin{aligned} \big(1 - 18^{-1}\varepsilon^2\big)(y-x) &\geq \big(1 - 18^{-1}\varepsilon^2\big)y - x \geq \big(1 - 18^{-1}\varepsilon^2\big)y - (4p)^{-1} \\ &\geq y - \varepsilon^2/18 - (4p)^{-1} \geq y - \varepsilon^2/18 - \varepsilon^2/18 \\ &\geq Ay - Ax. \end{aligned}$$

Thus we have shown that if (3.63) holds, then

$$|Ax - Ay| \leq \big(1 - \varepsilon^2/18\big)|x - y|. \tag{3.68}$$

Now assume that (3.64) holds. By (3.64) and (3.62),

$$x, y \in \big[(4p)^{-1}, 1\big].$$

In view of (3.55), the Lipschitz constant of the restriction of A to the interval $[(4p)^{-1}, 1]$ does not exceed $(4p+2)(4p+4)^{-1}$ and therefore we have

$$|Ax - Ay| \leq (4p+2)(4p+4)^{-1}|x - y|.$$

By this inequality and (3.68), we see that, in both cases,

$$|Ax - Ay| \le \max\{(1 - \varepsilon^2/18), (4p + 2)(4p + 4)^{-1}\}|x - y|.$$

Since this inequality holds for each $x, y \in X$ satisfying (3.61), we conclude that (3.59) is satisfied and therefore A is contractive.

3.5 A Power Convergent Mapping with no Contractive Powers

Let $X = [0, 1]$ and let $\rho(x, y) = |x - y|$ for all $x, y \in X$. In this section, which is based on [155], we construct a mapping $A : [0, 1] \to [0, 1]$ such that

$$|Ax - Ay| \le |x - y| \quad \text{for all } x, y \in [0, 1],$$

$$A^n x \to 0 \quad \text{as } n \to \infty, \text{ uniformly on } [0, 1],$$

and for each integer $m \ge 0$, the power A^m is not contractive.

To this end, let

$$A(0) = 0 \tag{3.69}$$

and for $t \in [2^{-1}, 1]$, set

$$A(t) = t - 1/4. \tag{3.70}$$

Clearly,

$$A(1) = 3/4 \quad \text{and} \quad A(1/2) = 1/4. \tag{3.71}$$

For $t \in [4^{-1}, 2^{-1})$, set

$$A(t) = 4^{-1} - 16^{-1} + (t - 4^{-1})4^{-1}. \tag{3.72}$$

Clearly, A is continuous on $[4^{-1}, 1]$ and

$$A(4^{-1}) = 4^{-1} - 16^{-1}. \tag{3.73}$$

Now let $n \ge 2$ be a natural number. We define the mapping A on the interval $[2^{-2^n}, 2^{-2^{n-1}}]$ as follows. For each $t \in [2^{-2^n+1}, 2^{-2^{n-1}}]$, set

$$A(t) = t - 2^{-2^n}. \tag{3.74}$$

Clearly,

$$A(2^{-2^n+1}) = 2^{-2^n} \quad \text{and} \quad A(2^{-2^{n-1}}) = 2^{-2^{n-1}} - 2^{-2^n}. \tag{3.75}$$

For $t \in [2^{-2^n}, 2^{-2^n+1})$, set

$$A(t) = 2^{-2^n} - 2^{-2^{n+1}} + \left(t - 2^{-2^n}\right)2^{2^n}\left(2^{-2^{n+1}}\right)$$
$$= 2^{-2^n} - 2^{-2^{n+1}} + 2^{-2^n}\left(t - 2^{-2^n}\right). \tag{3.76}$$

It is clear that

$$A\left(2^{-2^n}\right) = 2^{-2^n} - 2^{-2^{n+1}}$$

and

$$\lim_{t \to (2^{-2^n+1})+} A(t) = 2^{-2^n} - 2^{-2^{n+1}} + 2^{-2^n}\left(2^{-2^n+1} - 2^{-2^n}\right) = 2^{-2^n}. \tag{3.77}$$

It follows from (3.74)–(3.77) that the mapping A is continuous on each one of the intervals $[2^{-2^n}, 2^{-2^{n-1}}]$, $n = 2, 3, \ldots$. It is not difficult to check that A is well defined on $[0, 1]$ and that it is increasing.

By (3.70) and (3.72), for each $x \in [1/4, 1]$ we have $Ax < x$. We will now show that this inequality holds for all $x \in (0, 1]$.

Let $n \geq 2$ be an integer and let $x \in [2^{-2^n}, 2^{-2^{n-1}}]$. It is clear that $Ax < x$ if $x \in [2^{-2^n+1}, 2^{-2^{n-1}}]$. If $x \in [2^{-2^n}, 2^{-2^n+1})$, then by (3.74) and (3.75),

$$Ax < A\left(2^{-2^n+1}\right) \leq 2^{-2^n} \leq x.$$

Thus $Ax < x$ for all $x \in [2^{-2^n}, 2^{-2^{n-1}}]$ and for any integer $n \geq 2$. Therefore we have indeed shown that

$$Ax < x \quad \text{for all } x \in (0, 1], \tag{3.78}$$

as claimed.

Next, we will show that

$$|Ax - Ay| \leq |x - y| \quad \text{for each } x, y \in [0, 1]. \tag{3.79}$$

If $x = 0$ and $y > 0$, then

$$|Ay - Ax| = Ay \leq y = |y - x|. \tag{3.80}$$

Assume that $x, y \in (0, 1]$. Note that the restrictions of the mapping A to the interval $[1/4, 1]$ and to all of the intervals $[2^{-2n}, 2^{-2^{n-1}}]$, where $n \geq 2$ is an integer, are Lipschitz with Lipschitz constant one. This obviously implies that the mapping A is 1-Lipschitz on all of $(0, 1]$. Therefore (3.79) is true.

Let $x \in (0, 1]$. By (3.78), the sequence $\{A^n x\}_{n=1}^{\infty}$ is decreasing and there exists the limit

$$x_* = \lim_{n \to \infty} A^n x.$$

Clearly, $Ax_* = x_*$. If $x_* > 0$, then by (3.78), $Ax_* < x_*$, a contradiction. Thus $x_* = 0$ and $\lim_{n \to \infty} A^n(1) = 0$. Since the mapping A is increasing, this implies that

$$A^n x \to 0 \quad \text{as } n \to \infty, \text{ uniformly on } [0, 1].$$

Finally, we will show that for each integer $m \geq 1$, the power A^m is not contractive.

Indeed, let $m \geq 1$ be an integer. It is sufficient to show that there exist $x, y \in [0, 1]$ such that

$$x \neq y \quad \text{and} \quad |A^m - A^m y| = |x - y|.$$

To this end, choose a natural number $n \geq m + 4$ such that

$$2^{2^{n-1}} - 3 \geq m + 2. \tag{3.81}$$

Using induction and (3.74), we show that for each integer $i \in \{1, \dots, 2^{2^{n-1}} - 2\}$,

$$A^i(2^{-2^{n-1}}) = 2^{-2^{n-1}} - i2^{-2^n} \geq 2^{-2^n+1}$$

and

$$A^i(2^{-2^{n-1}}) \in [2^{-2^n+1}, 2^{-2^{n-1}}].$$

Put

$$x = 2^{-2^{n-1}} \quad \text{and} \quad y = A(2^{-2^{n-1}}).$$

Then for $i = 1, \dots, 2^{2^{n-1}} - 3$, we have

$$|A^i x - A^i y| = |x - y|,$$

and in view of (3.81),

$$|A^m x - A^m y| = |x - y|.$$

Thus the power A^m is not contractive, as asserted.

3.6 A Mapping with Nonuniformly Convergent Powers

In [155] we proved the following result.

Theorem 3.7 *Let (X, ρ) be a compact metric space, let a mapping $A : X \to X$ satisfy*

$$\rho(Ax, Ay) \leq \rho(x, y) \quad \textit{for each } x, y \in X, \tag{3.82}$$

and let $x_A \in X$ satisfy

$$A^n x \to x_A \quad \textit{as } n \to \infty, \textit{ for each } x \in X.$$

Then $A^n x \to x_A$ as $n \to \infty$, uniformly on X.

Proof Let $\varepsilon > 0$. For each $x \in X$, there is a natural number $n(x)$ such that

$$\rho(A^n x, x_A) \leq \varepsilon/2 \quad \text{for all integers } n \geq n(x). \tag{3.83}$$

Let

$$x, y \in X \quad \text{with} \quad \rho(x, y) < \varepsilon/2. \tag{3.84}$$

By (3.83) and (3.84), for each integer $n \geq n(x)$,

$$\rho(A^n y, x_A) \leq \rho(A^n y, A^n x) + \rho(A^n x, x_A) < \varepsilon/2 + \varepsilon/2.$$

Thus the following property holds:

(P) For each $x \in X$, each integer $n \geq n(x)$, and each $y \in X$ satisfying $\rho(x, y) < \varepsilon/2$, we have

$$\rho(A^n y, x_A) < \varepsilon.$$

Since X is compact, there exist finitely many points $x_1, \dots, x_q \in X$ such that

$$\bigcup_{i=1}^{q} \{y \in X : \rho(y, x_i) < \varepsilon/2\} = X.$$

Assume that $y \in X$ and that the integer $n \geq \max\{n(x_i) : i = 1, \dots, q\}$. Then there is $j \in \{1, \dots, q\}$ such that $\rho(y, x_j) < \varepsilon/2$. By property (P),

$$\rho(A^n y, x_A) < \varepsilon.$$

This completes the proof of Theorem 3.7. □

The following example was constructed in [155].

Let X be the set of all sequences $(x_1, x_2, \dots, x_n, \dots)$ such that $\sum_{i=1}^{\infty} |x_i| \leq 1$ and set

$$\rho(x, y) = \rho((x_i), (y_i)) = \sum_{i=1}^{\infty} |x_i - y_i|.$$

In other words, (X, ρ) is the closed unit ball of ℓ_1. Clearly, (X, ρ) is a complete metric space. Define

$$A(x_1, x_2, \dots, x_n, \dots) = (x_2, x_2, \dots, x_n, \dots), \quad x = (x_1, x_2, \dots) \in X.$$

Then the mapping A is nonexpansive, and for each $x \in X$, $A^n x \to 0$ as $n \to \infty$.

However, if n is a natural number and e_n is the n-th unit vector of X, then $\rho(A^n e_{n+1}, 0) = 1$.

3.7 Two Results in Metric Fixed Point Theory

In this section, which is based on [115], we establish two fixed point theorems for certain mappings of contractive type. The first result is concerned with the case where such mappings take a nonempty and closed subset of a complete metric space X into X, and the second with an application of the continuation method to the case where they satisfy the Leray-Schauder boundary condition in Banach spaces.

The following result was obtained in [115].

Theorem 3.8 *Let K be a nonempty and closed subset of a complete metric space (X, ρ). Assume that $T : K \to X$ satisfies*

$$\rho(Tx, Ty) \le \phi\big(\rho(x, y)\big)\rho(x, y) \quad \textit{for each } x, y \in K, \tag{3.85}$$

where $\phi : [0, \infty) \to [0, 1]$ is a monotonically decreasing function such that $\phi(t) < 1$ for all $t > 0$.

Assume that $K_0 \subset K$ is a nonempty and bounded set with the following property:

(P1) *For each natural number n, there exists $x_n \in K_0$ such that $T^i x_n$ is defined for all $i = 1, \dots, n$.*

Then

(A) *the mapping T has a unique fixed point $\bar{x}$ in K;*
(B) *For each $M, \varepsilon > 0$, there exist $\delta > 0$ and a natural number k such that for each integer $n \ge k$ and each sequence $\{x_i\}_{i=0}^n \subset K$ satisfying*

$$\rho(x_0, \bar{x}) \le M \quad \textit{and} \quad \rho(x_{i+1}, Tx_i) \le \delta, \quad i = 0, \dots, n-1,$$

we have

$$\rho(x_i, \bar{x}) \le \varepsilon, \quad i = k, \dots, n. \tag{3.86}$$

Proof of Theorem 3.8(A) The uniqueness of $\bar{x}$ is obvious. To establish its existence, let $x_n \in K_0$ be, for each natural number n, the point provided by property (P1). Fix $\theta_0 \in K$. Since K_0 is bounded, there is $c_0 > 0$ such that

$$\rho(\theta, z) \le c_0 \quad \text{for all } z \in K_0. \tag{3.87}$$

Let $\varepsilon > 0$ be given. We will show that there exists a natural number k such that the following property holds:

(P2) If $n > k$ is an integer and if an integer i satisfies $k \le i < n$, then

$$\rho\big(T^i x_n, T^{i+1} x_n\big) \le \varepsilon. \tag{3.88}$$

Assume the contrary. Then for each natural number k, there exist natural numbers n_k and i_k such that

$$k \le i_k < n_k \quad \text{and} \quad \rho\big(T^{i_k} x_{n_k}, T^{i_k+1} x_{n_k}\big) > \varepsilon. \tag{3.89}$$

Choose a natural number k such that

$$k > \big(\varepsilon\big(1-\phi(\varepsilon)\big)\big)^{-1}\big(2c_0+\rho(\theta,T\theta)\big). \tag{3.90}$$

By (3.89) and (3.85),

$$\rho\big(T^i x_{n_k}, T^{i+1}x_{n_k}\big) > \varepsilon, \quad i=0,\dots,i_k. \tag{3.91}$$

(Here we use the notation that $T^0 z = z$ for all $z \in K$.) It follows from (3.85), (3.91) and the monotonicity of ϕ that for all $i=0,\dots,i_k-1$,

$$\begin{aligned}\rho\big(T^{i+2}x_{n_k}, T^{i+1}x_{n_k}\big) &\le \phi\big(\rho\big(T^{i+1}x_{n_k}, T^i x_{n_k}\big)\big)\rho\big(T^{i+1}x_{n_k}, T^i x_{n_k}\big)\\ &\le \phi(\varepsilon)\rho\big(T^{i+1}x_{n_k}, T^i x_{n_k}\big)\end{aligned}$$

and

$$\begin{aligned}&\rho\big(T^{i+2}x_{n_k}, T^{i+1}x_{n_k}\big) - \rho\big(T^{i+1}x_{n_k}, T^i x_{n_k}\big)\\ &\quad\le \big(\phi(\varepsilon)-1\big)\rho\big(T^{i+1}x_{n_k}, T^i x_{n_k}\big) < -\big(1-\phi(\varepsilon)\big)\varepsilon.\end{aligned} \tag{3.92}$$

Inequalities (3.92) and (3.89) imply that

$$\begin{aligned}-\rho(x_{n_k}, Tx_{n_k}) &\le \rho\big(T^{i_k+1}x_{n_k}, T^{i_k}x_{n_k}\big) - \rho(x_{n_k}, Tx_{n_k})\\ &= \sum_{i=0}^{i_k-1}\big[\rho\big(T^{i+2}x_{n_k}, T^{i+1}x_{n_k}\big) - \rho\big(T^{i+1}x_{n_k}, T^i x_{n_k}\big)\big]\\ &\le -\big(1-\phi(\varepsilon)\varepsilon\big)i_k \le -k\big(1-\phi(\varepsilon)\big)\varepsilon\end{aligned}$$

and

$$k\big(1-\phi(\varepsilon)\big)\varepsilon \le \rho(x_{n_k}, Tx_{n_k}). \tag{3.93}$$

In view of (3.93), (3.85) and (3.87),

$$\begin{aligned}k\big(1-\phi(\varepsilon)\big)\varepsilon &\le \rho(x_{n_k}, Tx_{n_k})\\ &\le \rho(x_{n_k},\theta)+\rho(\theta,T\theta)+\rho(T\theta,Tx_{n_k}) \le c_0+\rho(\theta,T\theta)+c_0\end{aligned}$$

and

$$k \le \big(\varepsilon\big(1-\phi(\varepsilon)\big)\big)^{-1}\big(2c_0+\rho(\theta,T\theta)\big).$$

This contradicts (3.90). The contradiction we have reached proves that for each $\varepsilon > 0$, there exists a natural number k such that (P2) holds.

Now let $\delta > 0$ be given. We show that there exists a natural number k such that the following property holds:

(P3) If $n > k$ is an integer and if integers i, j satisfy $k \le i, j < n$, then

$$\rho\big(T^i x_n, T^j x_n\big) \le \delta.$$

To this end, choose a positive number

$$\varepsilon < 4^{-1}\delta\big(1-\phi(\delta)\big). \tag{3.94}$$

We have already shown that there exists a natural number k such that (P2) holds.

Assume that the natural numbers n, i and j satisfy

$$n > k \quad \text{and} \quad k \le i, j < n. \tag{3.95}$$

We claim that $\rho(T^i x_n, T^j x_n) \le \delta$.

Assume the contrary. Then

$$\rho\big(T^i x_n, T^j x_n\big) > \delta. \tag{3.96}$$

By (P2), (3.95), (3.85), (3.96) and the monotonicity of ϕ,

$$\begin{aligned}\rho\big(T^i x_n, T^j x_n\big) &\le \rho\big(T^i x_n, T^{i+1} x_n\big) + \rho\big(T^{i+1} x_n, T^{j+1} x_n\big) + \rho\big(T^{j+1} x_n, T^j x_n\big)\\ &\le \varepsilon + \rho\big(T^{i+1} x_n, T^{j+1} x_n\big) + \varepsilon\\ &\le 2\varepsilon + \phi\big(\rho\big(T^i x_n, T^j x_n\big)\big)\rho\big(T^i x_n, T^j x_n\big)\\ &\le 2\varepsilon + \phi(\delta)\rho\big(T^i x_n, T^j x_n\big).\end{aligned}$$

Together with (3.94) this implies that

$$\rho\big(T^i x_n, T^j x_n\big) \le 2\varepsilon\big(1-\phi(\delta)\big)^{-1} < \delta,$$

a contradiction. Thus we have shown that for each $\delta > 0$, there exists a natural number k such that (P3) holds.

Let $\varepsilon > 0$ be given. We will show that there exists a natural number k such that the following property holds:

(P4) If $n_1, n_2 \ge k$ are integers, then $\rho(T^k x_{n_1}, T^k x_{n_2}) \le \varepsilon$.

Choose a natural number k such that

$$k > \big(\big(1-\phi(\varepsilon)\big)(\varepsilon)\big)^{-1} 4c_0 \tag{3.97}$$

and assume that the integers n_1 and n_2 satisfy

$$n_1, n_2 \ge k. \tag{3.98}$$

We claim that $\rho(T^k x_{n_1}, T^k x_{n_2}) \le \varepsilon$. Assume the contrary. Then

$$\rho\big(T^k x_{n_1}, T^k x_{n_2}\big) > \varepsilon.$$

Together with (3.85) this implies that

$$\rho\big(T^i x_{n_1}, T^i x_{n_2}\big) > \varepsilon, \quad i = 0, \ldots, k. \tag{3.99}$$

By (3.85), (3.99) and the monotonicity of ϕ, we have for $i = 0, \ldots, k-1$,

$$\begin{aligned}\rho(T^{i+1}x_{n_1}, T^{i+1}x_{n_2}) &\le \phi(\rho(T^i x_{n_1}, T^i x_{n_2}))\rho(T^i x_{n_1}, T^i x_{n_2}) \\ &\le \phi(\varepsilon)\rho(T^i x_{n_1}, T^i x_{n_2})\end{aligned}$$

and

$$\begin{aligned}&\rho(T^{i+1}x_{n_1}, T^{i+1}x_{n_2}) - \rho(T^i x_{n_1}, T^i x_{n_2}) \\ &\quad \le (\phi(\varepsilon) - 1)\rho(T^i x_{n_1}, T^i x_{n_2}) \le -(1 - \phi(\varepsilon))\varepsilon.\end{aligned}$$

This implies that

$$\begin{aligned}-\rho(x_{n_1}, x_{n_2}) &\le \rho(T^k x_{n_1}, T^k x_{n_2}) - \rho(x_{n_1}, x_{n_2}) \\ &= \sum_{i=0}^{k-1}[\rho(T^{i+1}x_{n_1}, T^{i+1}x_{n_2}) - \rho(T^i x_{n_1}, T^i x_{n_2})] \le -k(1 - \phi(\varepsilon))\varepsilon.\end{aligned}$$

Together with (3.87) this implies that

$$k(1 - \phi(\varepsilon))\varepsilon \le \rho(x_{n_1}, x_{n_2}) \le \rho(x_{n_1}, \theta) + \rho(\theta, x_{n_2}) \le 2c_0.$$

This contradicts (3.97). Thus we have shown that

$$\rho(T^k x_{n_1}, T^k x_{n_2}) \le \varepsilon.$$

In other words, there exists a natural number k for which (P4) holds.

Let $\varepsilon > 0$ be given. By (P4), there exists a natural number k_1 such that

$$\rho(T^{k_1}x_{n_1}, T^{k_1}x_{n_2}) \le \varepsilon/4 \quad \text{for all integers } n_1, n_2 \ge k_1. \tag{3.100}$$

By (P3), there exists a natural number k_2 such that

$$\rho(T^i x_n, T^j x_n) \le \varepsilon/4 \quad \text{for all natural numbers } n, j, i \text{ satisfying } k_2 \le i, j < n. \tag{3.101}$$

Assume now that the natural numbers n_1, n_2, i and j satisfy

$$n_1, n_2 > k_1 + k_2, \qquad i, j \ge k_1 + k_2, \qquad i < n_1, \qquad j < n_2. \tag{3.102}$$

We claim that

$$\rho(T^i x_{n_1}, T^j x_{n_2}) \le \varepsilon.$$

By (3.100), (3.102) and (3.85),

$$\rho(T^{k_1+k_2}x_{n_1}, T^{k_1+k_2}x_{n_2}) \le \rho(T^{k_1}x_{n_1}, T^{k_1}x_{n_2}) \le \varepsilon/4. \tag{3.103}$$

In view of (3.102) and (3.101),

$$\rho(T^{k_1+k_2}x_{n_1}, T^i x_{n_1}) \le \varepsilon/4 \quad \text{and} \quad \rho(T^{k_1+k_2}x_{n_2}, T^j x_{n_2}) \le \varepsilon/4.$$

Together with (3.103) these inequalities imply that

$$\begin{aligned}&\rho\big(T^i x_{n_1}, T^j x_{n_2}\big)\\&\quad\le \rho\big(T^i x_{n_1}, T^{k_1+k_2} x_{n_1}\big) + \rho\big(T^{k_1+k_2} x_{n_1}, T^{k_1+k_2} x_{n_2}\big) + \rho\big(T^{k_1+k_2} x_{n_2}, T^j x_{n_2}\big)\\&\quad< \varepsilon.\end{aligned}$$

Thus we have shown that the following property holds:

(P5) For each $\varepsilon > 0$, there exists a natural number $k(\varepsilon)$ such that

$$\rho\big(T^i x_{n_1}, T^j x_{n_2}\big) \le \varepsilon$$

for all natural numbers $n_1, n_2 \ge k(\varepsilon)$, $i \in [k(\varepsilon), n_1)$ and $j \in [k(\varepsilon), n_2)$.

Consider the two sequences $\{T^{n-2} x_n\}_{n=2}^{\infty}$ and $\{T^{n-1} x_n\}_{n=2}^{\infty}$. Property (P5) implies that both of them are Cauchy and that

$$\lim_{n\to\infty} \rho\big(T^{n-1} x_n, T^{n-2} x_n\big) = 0.$$

Therefore there exists $\bar{x} \in K$ such that

$$\lim_{n\to\infty} \rho\big(\bar{x}, T^{n-2} x_n\big) = \lim_{n\to\infty} \rho\big(\bar{x}, T^{n-1} x_n\big) = 0.$$

Since the mapping T is continuous, $T\bar{x} = \bar{x}$ and assertion (A) is proved. □

Proof of Theorem 3.8(B) For each $x \in X$ and $r > 0$, set

$$B(x, r) = \big\{y \in X : \rho(x, y) \le r\big\}. \tag{3.104}$$

Choose $\delta_0 > 0$ such that

$$\delta_0 < M\big(1 - \phi(M/2)\big)/4. \tag{3.105}$$

Assume that

$$y \in K \cap B(\bar{x}, M), \qquad z \in X \quad \text{and} \quad \rho(z, Ty) \le \delta_0. \tag{3.106}$$

By (3.106) and (3.85),

$$\begin{aligned}\rho(\bar{x}, z) &\le \rho(\bar{x}, Ty) + \rho(Ty, z) \le \rho(T\bar{x}, Ty) + \delta_0\\&\le \phi\big(\rho(\bar{x}, y)\big)\rho(\bar{x}, y) + \delta_0.\end{aligned} \tag{3.107}$$

There are two cases:

$$\rho(y, \bar{x}) \le M/2; \tag{3.108}$$

$$\rho(y, \bar{x}) > M/2. \tag{3.109}$$

Assume that (3.108) holds. By (3.107), (3.108) and (3.105),

$$\rho(\bar{x}, z) \le \rho(\bar{x}, y) + \delta_0 \le M/2 + \delta_0 < M. \tag{3.110}$$

If (3.109) holds, then by (3.107), (3.106), (3.109) and the monotonicity of ϕ,

$$\begin{aligned}\rho(\bar{x}, z) &\le \delta_0 + \phi(M/2)\rho(\bar{x}, y) \le \delta_0 + \phi(M/2)M \\ &< (M/4)\big(1 - \phi(M/2)\big) + \phi(M/2)M \le M.\end{aligned}$$

Thus $\rho(\bar{x}, z) \le M$ in both cases.

We have shown that

$$\begin{aligned}&\rho(\bar{x}, z) \le M \quad \text{for each } z \in X \text{ and } y \in K \cap B(\bar{x}, M) \\ &\quad \text{satisfying } \rho(z, Ty) \le \delta_0.\end{aligned} \tag{3.111}$$

Since M is any positive number, we conclude that there is $\delta_1 > 0$ such that

$$\begin{aligned}&\rho(\bar{x}, z) \le \varepsilon \quad \text{for each } z \in X \text{ and } y \in K \cap B(\bar{x}, \varepsilon) \\ &\quad \text{satisfying } \rho(z, Ty) \le \delta_1.\end{aligned} \tag{3.112}$$

Choose a positive number δ such that

$$\delta < \min\big\{\delta_0, \delta_1, \varepsilon\big(1 - \phi(\varepsilon)\big)4^{-1}\big\} \tag{3.113}$$

and a natural number k such that

$$k > 4(M+1)\big(1 - \phi(\varepsilon)\varepsilon\big)^{-1} + 4. \tag{3.114}$$

Let $n \ge k$ be a natural number and assume that $\{x_i\}_{i=0}^n \subset K$ satisfies

$$\rho(x_0, \bar{x}) \le M \quad \text{and} \quad \rho(x_{i+1}, Tx_i) \le \delta, \quad i = 0, \dots, n-1. \tag{3.115}$$

We claim that (3.86) holds. By (3.111), (3.115) and the inequality $\delta < \delta_0$ (see (3.113)),

$$\{x_i\}_{i=0}^k \subset B(\bar{x}, M). \tag{3.116}$$

Assume that (3.86) does not hold. Then there is an integer j such that

$$j \in \{k, n\} \quad \text{and} \quad \rho(x_j, \bar{x}) > \varepsilon. \tag{3.117}$$

By (3.117), (3.115), (3.112) and (3.113),

$$\rho(x_i, \bar{x}) > \varepsilon, \quad i = 0, \dots, j. \tag{3.118}$$

Let $i \in \{0, \dots, j-1\}$. By (3.115), (3.118), the monotonicity of ϕ, (3.113) and (3.85),

$$\rho(x_{i+1},\bar{x}) \le \rho(x_{i+1},Tx_i) + \rho(Tx_i,T\bar{x}) \le \delta + \phi\big(\rho(x_i,\bar{x})\big)\rho(x_i,\bar{x})$$
$$\le \delta + \phi(\varepsilon)\rho(x_i,\bar{x})$$

and

$$\rho(x_{i+1},\bar{x}) - \rho(x_i,\bar{x}) \le \delta - \big(1-\phi(\varepsilon)\big)\rho(x_i,\bar{x}) \le \delta - \big(1-\phi(\varepsilon)\big)\varepsilon$$
$$\le -\big(1-\phi(\varepsilon)\big)\varepsilon/2.$$

By (3.115) and (3.117) and the above inequalities,

$$-M \le -\rho(x_0,\bar{x}) \le \rho(x_j,\bar{x}) - \rho(x_0,\bar{x})$$
$$= \sum_{i=0}^{j-1}\big[\rho(x_{i+1},\bar{x}) - \rho(x_i,\bar{x})\big] \le -j\big(1-\phi(\varepsilon)\varepsilon/2\big) \le -k\big(1-\phi(\varepsilon)\big)\varepsilon/2.$$

This contradicts (3.114). The contradiction we have reached proves (3.86) and assertion (B). □

Let G be a nonempty subset of a Banach space $(Y, \|\cdot\|)$. In [64] J. A. Gatica and W. A. Kirk proved that if $T : \overline{G} \to Y$ is a strict contraction, then T must have a unique fixed point x_1, under the additional assumptions that the origin is in the interior $\text{Int}(G)$ of G and that T satisfies a certain boundary condition known as the Leray-Schauder condition:

$$Tx \ne \lambda x \quad \forall x \in \partial G, \forall \lambda > 1. \tag{L-S}$$

Here G is not necessarily convex or bounded. Their proof was nonconstructive. Later, M. Frigon, A. Granas and Z. E. A. Guennoun [61], and M. Frigon [60] proved that if x_t is the unique fixed point of tT, then, in fact, the mapping $t \to x_t$ is Lipschitz, so it gives a partial way to approximate x_1. Our second result in this section, which was also obtained in [115], extends these theorems to the case where T merely satisfies (3.85).

Theorem 3.9 *Let G be a nonempty subset of a Banach space Y with $0 \in \text{Int}(G)$. Suppose that $T : \overline{G} \to X$ is nonexpansive and that it satisfies condition* (L-S). *Then for each $t \in [0, 1)$, the mapping $tT : \overline{G} \to X$ has a unique fixed point $x_t \in \text{Int}(G)$ and the mapping $t \to x_t$ is Lipschitz on $[0, b]$ for any $0 < b < 1$. If, in addition, T satisfies* (3.85), *then it has a unique fixed point $x_1 \in \overline{G}$ and the mapping $t \to x_t$ is continuous on $[0, 1]$. In particular, $x_1 = \lim_{t\to 1^-} x_t$.*

Proof In the first part of the proof we assume that T is nonexpansive, i.e., it satisfies (3.85) with ϕ identically equal to one.

Let $S \subset [0, 1)$ be the following set:

$$S = \big\{t \in [0, 1) : tT \text{ has a unique fixed point } x_t \in \text{Int}(G)\big\}.$$

Since tT is a strict contraction for each $t \in [0, 1)$, it has at most one fixed point. In order to prove the first part of this theorem, we have to show that $S = [0, 1)$. Since $0 \in S$ by assumption and since $[0, 1)$ is connected, it is enough to show that S is both open and closed.

1. S is open: Let $t_0 \in S$. From the definition of S it is clear that $t_0 < 1$, so there is a real number q such that $t_0 < q < 1$. Let $x_{t_0} \in \mathrm{Int}(G)$ be the unique fixed point of $t_0 T$.

Since $\mathrm{Int}(G)$ is open, there is $r > 0$ such that the closed ball $B[x_{t_0}, r]$ of radius r and center x_{t_0} is contained in $\mathrm{Int}(G)$. We have, for all $x \in B[x_{t_0}, r]$ and $t \in [0, 1)$,

$$\begin{aligned}\|tTx - x_{t_0}\| &\leq \|tTx - tTx_{t_0}\| + |t - t_0|\|Tx_{t_0}\| + \|t_0Tx_{t_0} - x_{t_0}\| \\ &\leq t\|x - x_{t_0}\| + |t - t_0|\|Tx_{t_0}\| \leq tr + |t - t_0|\big(\|Tx_{t_0}\| + 1\big). \end{aligned} \tag{3.119}$$

Suppose that $t \in [0, 1)$ satisfies

$$|t - t_0| < \min\left\{\frac{r(1-q)}{1 + \|Tx_{t_0}\|}, q - t_0\right\}. \tag{3.120}$$

Then $t < q$ and

$$|t - t_0| \leq \frac{r(1-t)}{1 + \|Tx_{t_0}\|},$$

so $\|tTx - x_{t_0}\| \leq r$ by (3.119). Consequently, the closed ball $B[x_{t_0}, r]$ is invariant under tT, and the Banach fixed point theorem ensures that tT has a unique fixed point $x_t \in B[x_{t_0}, r] \subset \mathrm{Int}(G)$. Thus $t \in S$ for all $t \in [0, 1)$ satisfying (3.120).

2. S is closed: Suppose $t_0 \in [0, 1)$ is a limit point of S. We have to prove that $t_0 \in S$, and since $0 \in S$ we can assume that $t_0 > 0$. There is a sequence $(t_n)_n$ in $[0, 1)$ such that $t_0 = \lim_{n\to\infty} t_n$, and since $t_0 < 1$, there is $0 < q < 1$ such that $t_n < q$ for n large enough. Define

$$A_0 := \{x_t : t \in S \cap [0, q]\}.$$

The set A_0 is not empty since $0 \in A_0$. In addition, if $t \in S \cap [0, q]$, then

$$\|x_t\| = \|tTx_t\| \leq q\big(\|Tx_t - T0\| + \|T0\|\big) \leq q\phi\big(\|x_t - 0\|\big)\|x_t - 0\| + q\|T0\|.$$

Therefore

$$\|x_t\| \leq \frac{q\|T0\|}{1 - \phi(\|x_t\|)q} \leq \frac{\|T0\|}{1-q}, \tag{3.121}$$

so A_0 is a bounded set, and since T is Lipschitz, $T(A_0)$ is also bounded, say by M. We will show that $(x_{t_n})_n$ is a Cauchy sequence which converges to the fixed point x_{t_0} of t_0T. Indeed, since x_{t_n} and x_{t_m} are the fixed points of t_nT and t_mT, respectively, it follows that

$$\begin{aligned}\|x_{t_n} - x_{t_m}\| &= \|t_nTx_{t_n} - t_mTx_{t_m}\| \leq |t_n - t_m|\|Tx_{t_n}\| + \|t_mTx_{t_n} - t_mTx_{t_m}\| \\ &\leq |t_n - t_m|M + t_m\phi\big(\|x_{t_n} - x_{t_m}\|\big)\|x_{t_n} - x_{t_m}\|.\end{aligned}$$

Hence

$$\|x_{t_n} - x_{t_m}\| \le \frac{|t_n - t_m|M}{1 - t_m\phi(\|x_{t_n} - x_{t_m}\|)} \le \frac{|t_n - t_m|M}{1-q}. \tag{3.122}$$

Since $t_n \to t_0$ as $n \to \infty$, we see that $(x_{t_n})_n$ is indeed Cauchy and hence converges to $x_{t_0} \in \overline{G}$. Using again the equality $t_n T x_{t_n} = x_{t_n}$, we obtain

$$\begin{aligned}\|t_0 T x_{t_0} - x_{t_0}\| &\le \|t_0 T x_{t_0} - t_0 T x_{t_n}\| + \|t_0 T x_{t_n} - t_n T x_{t_n}\| + \|t_n T x_{t_n} - x_{t_0}\| \\ &= t_0\|T x_{t_0} - T x_{t_n}\| + |t_0 - t_n|\|T x_{t_n}\| + \|x_{t_n} - x_{t_0}\| \\ &\le \|x_{t_0} - x_{t_n}\| + |t_0 - t_n|M + \|x_{t_n} - x_{t_0}\| \to 0,\end{aligned}$$

so $t_0 T x_{t_0} = x_{t_0}$, i.e., x_{t_0} is indeed a fixed point of $t_0 T$. It remains to show that $x_{t_0} \in \operatorname{Int}(G)$, and this follows from the (L-S) condition: since $T x_{t_0} = \frac{1}{t_0} x_{t_0}$, so (L-S) implies that $x_{t_0} \notin \partial G$ (recall that $0 < t_0 < 1$). Hence S is closed, as claimed.

The fact that the mapping $t \to x_t$ is Lipschitz on the interval $[0, b]$ for any $0 < b < 1$ follows from (3.122).

Suppose now that T satisfies (3.85) with $\phi(t) < 1$ for all positive t. Let $(t_n)_n$ be a sequence in $[0, 1)$ such that $t_n \to t_0 = 1$. The set A_0 (and hence the set $T(A_0)$) remain bounded also when $q = 1$, because if $\|x_t\| \ge 1$, then in (3.121) we get $\|x_t\| \le \frac{\|T0\|}{1-\phi(1)}$, so in any case $\|x_t\| \le \max\left(1, \frac{\|T0\|}{1-\phi(1)}\right)$ (recall that $\phi(t) < 1$). Now, in order to prove that $x_1 := \lim_{t \to 1^{-1}} x_t$ exists, note first that $(x_{t_n})_n$ is Cauchy if $t_n \to 1$, because otherwise there is $\varepsilon > 0$ and a subsequence (call it again t_n) such that $\|x_{t_{2n+1}} - x_{t_{2n+2}}\| \ge \varepsilon$, but from (3.122) we obtain

$$\|x_{t_{2n+1}} - x_{t_{2n+2}}\| \le \frac{|t_{2n+1} - t_{2n+2}|M}{1 - t_{2n+2}\phi(\varepsilon)} \to 0,$$

a contradiction. Now, all these sequences approach the same limit because for any two such sequences

$$(x_{t_n})_n, \qquad (x_{s_n})_n,$$

the interlacing sequence $(t_1, s_1, t_2, s_2, \ldots) \to 1$, so $(x_{t_1}, x_{s_1}, x_{t_2}, x_{s_2}, \ldots)$ is also Cauchy. The fact that x_1 is a fixed point of T is proved as above (here, however, one cannot use (L-S) to conclude that $x_1 \in \operatorname{Int}(G)$, and indeed it may happen that $x_1 \in \partial G$ as the mapping $T : [-1, \infty) \to R$, defined by $Tx = \frac{x-1}{2}$, shows). □

3.8 A Result on Rakotch Contractions

In this section, which is based on [160], we establish fixed point and convergence theorems for certain mappings of contractive type which take a closed subset of a complete metric space X into X.

Let K be a nonempty and closed subset of a complete metric space (X, ρ). For each $x \in X$ and $r > 0$, set

$$B(x, r) = \{y \in X : \rho(x, y) \le r\}.$$

In the following result, which was obtained in [160], we provide a new sufficient condition for the existence and approximation of the unique fixed point of a contractive mapping which maps a nonempty and closed subset of a complete metric space X into X.

Theorem 3.10 *Assume that $T : K \to X$ satisfies*

$$\rho(Tx, Ty) \le \phi\big(\rho(x, y)\big)\rho(x, y) \quad \text{for all } x, y \in K, \tag{3.123}$$

where $\phi : [0, \infty) \to [0, 1]$ is a monotonically decreasing function such that $\phi(t) < 1$ for all $t > 0$.

Assume that there exists a sequence $\{x_n\}_{n=1}^{\infty} \subset K$ such that

$$\lim_{n\to\infty} \rho(x_n, Tx_n) = 0. \tag{3.124}$$

Then there exists a unique $\bar{x} \in K$ such that $T\bar{x} = \bar{x}$.

Proof The uniqueness of $\bar{x}$ is obvious. To establish its existence, let $\varepsilon \in (0, 1)$ be given and choose a positive number γ such that

$$\gamma < \big(1 - \phi(\varepsilon)\big)\varepsilon/8. \tag{3.125}$$

By (3.124), there is a natural number n_0 such that

$$\rho(x_n, Tx_n) < \gamma \quad \text{for all integers } n \ge n_0. \tag{3.126}$$

Assume that the integers $m, n \ge n_0$. We claim that $\rho(x_m, x_n) \le \varepsilon$. Assume the contrary. Then

$$\rho(x_m, x_n) > \varepsilon. \tag{3.127}$$

By (3.125), (3.123), (3.127), the monotonicity of ϕ, and (3.126),

$$\begin{aligned}
\rho(x_m, x_n) &\le \rho(x_m, Tx_m) + \rho(Tx_m, Tx_n) + \rho(Tx_n, x_n) \\
&\le 2\gamma + \phi\big(\rho(x_m, x_n)\big)\rho(x_m, x_n) \le 2\gamma + \phi(\varepsilon)\rho(x_m, x_n) \\
&= \rho(x_m, x_n) - \big(1 - \phi(\varepsilon)\big)\rho(x_m, x_n) + 2\gamma \\
&< \rho(x_m, x_n) - \big(1 - \phi(\varepsilon)\big)\rho(x_m, x_n) + \big(1 - \phi(\varepsilon)\big)\varepsilon/4 \\
&\le \rho(x_m, x_n) - \big(1 - \phi(\varepsilon)\big)\rho(x_m, x_n)(3/4) \\
&= \rho(x_m, x_n)\big[(1/4) + \phi(\varepsilon)(3/4)\big] < \rho(x_m, x_n),
\end{aligned}$$

a contradiction.

The contradiction we have reached proves that $\rho(x_m, x_n) \le \varepsilon$ for all integers $m, n \ge n_0$, as claimed.

Since ε is an arbitrary number in $(0, 1)$, we conclude that $\{x_n\}_{n=1}^{\infty}$ is a Cauchy sequence and there exists $\bar{x} \in X$ such that $\lim_{n\to\infty} x_n = \bar{x}$. By (3.123), for all integers $n \ge 1$,

$$\begin{aligned}\rho(T\bar{x}, \bar{x}) &\le \rho(T\bar{x}, Tx_n) + \rho(Tx_n, x_n) + \rho(x_n, \bar{x}) \\ &\le 2\rho(x_n, \bar{x}) + \rho(Tx_n, x_n) \to 0 \quad \text{as } n \to \infty.\end{aligned}$$

This concludes the proof of Theorem 3.10. □

In the following result, which was also obtained in [160], we present another proof of the fixed point theorem established in Theorem 1(A) of [115]. This proof is based on Theorem 3.10.

Theorem 3.11 *Let $T : K \to X$ satisfy*

$$\rho(Tx, Ty) \le \phi\big(\rho(x, y)\big)\rho(x, y) \quad \textit{for all } x, y \in K,$$

where $\phi : [0, \infty) \to [0, 1]$ is a monotonically decreasing function such that $\phi(t) < 1$ for all $t > 0$.

Assume that $K_0 \subset K$ is a nonempty and bounded set with the following property:

For each natural number n, there exists $y_n \in K_0$ such that $T^i y_n$ is defined for all $i = 1, \ldots, n$.

Then the mapping T has a unique fixed point $\bar{x}$ in K.

Proof By Theorem 3.10, it is sufficient to show that for each $\varepsilon \in (0, 1)$, there is $x \in K$ such that $\rho(x, Tx) < \varepsilon$. Indeed, let $\varepsilon \in (0, 1)$. There is $M > 0$ such that

$$\rho(y_0, y_i) \le M, \quad i = 1, 2, \ldots. \tag{3.128}$$

By (3.123) and (3.128), for each integer $i \ge 1$,

$$\rho(y_i, Ty_i) \le \rho(y_i, y_0) + \rho(y_0, Ty_0) + \rho(Ty_0, Ty_i) \le 2M + \rho(y_0, Ty_0). \tag{3.129}$$

Choose a natural number $q \ge 4$ such that

$$(q - 1)\varepsilon\big(1 - \phi(\varepsilon)\big) > 4M + 2\rho(y_0, Ty_0). \tag{3.130}$$

Set $T^0 z = z$, $z \in K$.

We claim that $\rho(T^{q-1} y_q, T^q y_q) < \varepsilon$. Assume the contrary. Then by (3.123),

$$\rho\big(T^i y_q, T^{i+1} y_q\big) \ge \varepsilon, \quad i = 0, \ldots, q - 1. \tag{3.131}$$

In view of (3.123), (3.131) and the monotonicity of ϕ, we have for $i = 0, \ldots, q - 2$,

$$\begin{aligned}\rho\big(T^{i+1} y_q, T^{i+2} y_q\big) &\le \phi\big(\rho\big(T^i y_q, T^{i+1} y_q\big)\big)\rho\big(T^i y_q, T^{i+1} y_q\big) \\ &\le \phi(\varepsilon)\rho\big(T^i y_q, T^{i+1} y_q\big)\end{aligned}$$

and

$$\rho(T^i y_q, T^{i+1} y_q) - \rho(T^{i+1} y_q, T^{i+2} y_q) \geq (1 - \phi(\varepsilon))\rho(T^i y_q, T^{i+1} y_q)$$
$$\geq (1 - \phi(\varepsilon))\varepsilon. \tag{3.132}$$

By (3.129) and (3.132),

$$\begin{aligned} 2M + \rho(y_0, Ty_0) &\geq \rho(y_q, Ty_q) - \rho(T^{q-1} y_q, T^q y_q) \\ &\geq \sum_{i=0}^{q-2} [\rho(T^i y_q, T^{i+1} y_q) - \rho(T^{i+1} y_q, T^{i+2} y_q)] \\ &\geq (q-1)(1 - \phi(\varepsilon))\varepsilon \end{aligned}$$

and

$$2M + \rho(y_0, Ty_0) \geq (q-1)(1 - \phi(\varepsilon))\varepsilon.$$

This contradicts (3.130). The contradiction we have reached shows that

$$\rho(T^{q-1} y_q, T^q y_q) < \varepsilon,$$

as claimed. Theorem 3.11 is proved. □

In the following result, also obtained in [160], we establish a convergence result for (unrestricted) infinite products of mappings which satisfy a weak form of condition (3.123).

Theorem 3.12 *Let* $\phi : [0, \infty) \to [0, 1]$ *be a monotonically decreasing function such that* $\phi(t) < 1$ *for all* $t > 0$.

Let

$$\bar{x} \in K, \qquad T_i : K \to X, \quad i = 0, 1, \ldots, \qquad T_i \bar{x} = \bar{x}, \quad i = 0, 1, \ldots, \tag{3.133}$$

and assume that

$$\rho(T_i x, \bar{x}) \leq \phi(\rho(x, \bar{x}))\rho(x, \bar{x}) \quad \textit{for each } x \in K, i = 0, 1, \ldots. \tag{3.134}$$

Then for each $M, \varepsilon > 0$, *there exist* $\delta > 0$ *and a natural number* k *such that for each integer* $n \geq k$, *each mapping* $r : \{0, 1, \ldots, n-1\} \to \{0, 1, \ldots\}$, *and each sequence* $\{x_i\}_{i=0}^{n-1} \subset K$ *satisfying*

$$\rho(x_0, \bar{x}) \leq M \quad \textit{and} \quad \rho(x_{i+1}, T_{r(i)} x_i) \leq \delta, \quad i = 0, \ldots, n-1,$$

we have

$$\rho(x_i, \bar{x}) \leq \varepsilon, \quad i = k, \ldots, n. \tag{3.135}$$

Proof Choose $\delta_0 > 0$ such that

$$\delta_0 < M\big(1 - \phi(M/2)\big)/4. \tag{3.136}$$

Assume that

$$y \in K \cap B(\bar{x}, M), \qquad i \in \{0, 1, \ldots\}, \qquad z \in X \quad \text{and} \quad \rho(z, T_i y) \le \delta_0. \tag{3.137}$$

By (3.137) and (3.134),

$$\rho(\bar{x}, z) \le \rho(\bar{x}, T_i y) + \rho(T_i, z) \le \phi\big(\rho(\bar{x}, y)\big)\rho(\bar{x}, y) + \delta_0. \tag{3.138}$$

There are two cases:

$$\rho(y, \bar{x}) \le M/2 \tag{3.139}$$

and

$$\rho(y, \bar{x}) > M/2. \tag{3.140}$$

Assume that (3.139) holds. Then by (3.138), (3.139) and (3.136),

$$\rho(\bar{x}, z) \le \rho(\bar{x}, y) + \delta_0 \le M/2 + \delta_0 < M. \tag{3.141}$$

If (3.140) holds, then by (3.138), (3.137), (3.136) and the monotonicity of ϕ,

$$\begin{aligned}\rho(\bar{x}, z) &\le \delta_0 + \phi(M/2)\rho(\bar{x}, y) \le \delta_0 + \phi(M/2)M \\ &< (M/4)\big(1 - \phi(M/2)\big) + \phi(M/2)M \le M.\end{aligned}$$

Thus $\rho(\bar{x}, z) \le M$ in both cases.

We have shown that

$$\text{if } y \in K \cap B(\bar{x}, M), i \in \{0, 1, \ldots\}, z \in X, \rho(z, T_i y) \le \delta_0, \text{ then } \rho(\bar{x}, z) \le M. \tag{3.142}$$

Since M is any positive number, we conclude that there is $\delta_1 > 0$ such that

$$\text{if } y \in K \cap B(\bar{x}, \varepsilon), i \in \{0, 1, \ldots\}, z \in X, \rho(z, T_i y) \le \delta_1, \text{ then } \rho(\bar{x}, z) \le \varepsilon. \tag{3.143}$$

Now choose a positive number δ such that

$$\delta < \min\big\{\delta_0, \delta_1, \varepsilon\big(1 - \phi(\varepsilon)\big)4^{-1}\big\} \tag{3.144}$$

and a natural number k such that

$$k > 4(M + 1)\big(\big(1 - \phi(\varepsilon)\big)\varepsilon\big)^{-1} + 4. \tag{3.145}$$

Let $n \geq k$ be a natural number. Assume that $r : \{0, \ldots, n-1\} \to \{0, 1, \ldots\}$ and that

$$\{x_i\}_{i=0}^{n-1} \subset K$$

satisfies

$$\rho(x_0, \bar{x}) \leq M \quad \text{and} \quad \rho(x_{i+1}, T_{r(i)}x_i) \leq \delta, \quad i = 0, \ldots, n-1. \tag{3.146}$$

We claim that (3.135) holds. By (3.142), (3.146) and the inequality $\delta < \delta_0$,

$$\{x_i\}_{i=0}^{n} \subset B(\bar{x}, M). \tag{3.147}$$

Assume to the contrary that (3.135) does not hold. Then there is an integer j such that

$$j \in \{k, \ldots, n\} \quad \text{and} \quad \rho(x_j, \bar{x}) > \varepsilon. \tag{3.148}$$

By (3.148) and (3.134),

$$\rho(x_i, \bar{x}) > \varepsilon, \quad i = 0, \ldots, j. \tag{3.149}$$

Let $i \in \{0, \ldots, j-1\}$. By (3.146), (3.134) and the monotonicity of ϕ,

$$\begin{aligned}\rho(x_{i+1}, \bar{x}) &\leq \rho(x_{i+1}, T_{r(i)}x_i) + \rho(T_{r(i)}x_i, \bar{x}) \leq \delta + \phi\big(\rho(x_i, \bar{x})\big)\rho(x_i, \bar{x}) \\ &\leq \delta + \phi(\varepsilon)\rho(x_i, \bar{x}).\end{aligned}$$

When combined with (3.144) and (3.49), this implies that

$$\begin{aligned}\rho(x_{i+1}, \bar{x}) - \rho(x_i, \bar{x}) &\leq \delta - \big(1 - \phi(\varepsilon)\big)\rho(x_i, \bar{x}) \leq \delta - \big(1 - \phi(\varepsilon)\big)\varepsilon \\ &< -\big(1 - \phi(\varepsilon)\big)\varepsilon/2.\end{aligned} \tag{3.150}$$

Finally, by (3.146), (3.150) and (3.148),

$$\begin{aligned}-M &\leq -\rho(x_0, \bar{x}) \leq \rho(x_j, \bar{x}) - \rho(x_0, \bar{x}) \\ &= \sum_{i=0}^{j-1}\big[\rho(x_{i+1}, \bar{x}) - \rho(x_i, \bar{x})\big] \leq -j\big(1 - \phi(\varepsilon)\big)\varepsilon/2 \leq -k\big(1 - \phi(\varepsilon)\big)\varepsilon/2.\end{aligned}$$

This contradicts (3.145). The contradiction we have reached proves (3.135) and Theorem 3.12 itself. □

3.9 Asymptotic Contractions

In this section, which is based on [8], we provide sufficient conditions for the iterates of an asymptotic contraction on a complete metric space X to converge to its unique fixed point, uniformly on each bounded subset of X.

Let (X, d) be a complete metric space. The following theorem is the main result of Chen [40]. It improves upon Kirk's original theorem [83]. In this connection, see also [6] and [76].

Theorem 3.13 *Let $T : X \to X$ be such that*

$$d(T^n x, T^n y) \leq \phi_n(d(x, y))$$

for all $x, y \in X$ and all natural numbers n, where $\phi_n : [0, \infty) \to [0, \infty)$ and $\lim_{n\to\infty} \phi_n = \phi$, uniformly on any bounded interval $[0, b]$. Suppose that ϕ is upper semicontinuous and that $\phi(t) < t$ for all $t > 0$. Furthermore, suppose that there exists a positive integer n_* such that ϕ_{n_*} is upper semicontinuous and $\phi_{n_*}(0) = 0$. If there exists $x_0 \in X$ which has a bounded orbit $O(x_0) = \{x_0, Tx_0, T^2x_0, \dots\}$, then T has a unique fixed point $x_* \in X$ and $\lim_{n\to\infty} T^n x = x_*$ for all $x \in X$.

Note that Theorem 3.13 does not provide us with uniform convergence of the iterates of T on bounded subsets of X, although this does hold for many classes of mappings of contractive type (e.g., [23, 114]). This property is important because it yields stability of the convergence of iterates even in the presence of computational errors [35]. In this section we show that this conclusion can be derived in the setting of Theorem 3.13 if for each natural number n, the function ϕ_n is assumed to be bounded on any bounded interval. To this end, we first prove a somewhat more general result (Theorem 3.14) which, when combined with Theorem 3.13, yields our strengthening of Chen's result (Theorem 3.15).

Theorem 3.14 *Let $x_* \in X$ be a fixed point of $T : X \to X$. Assume that*

$$d(T^n x, x_*) \leq \phi_n(d(x, x_*)) \quad \text{for all } x \in X \text{ and all natural numbers } n, \tag{3.151}$$

where $\phi_n : [0, \infty) \to [0, \infty)$ and $\lim_{n\to\infty} \phi_n = \phi$, uniformly on any bounded interval $[0, b]$. Suppose that ϕ is upper semicontinuous and $\phi(t) < t$ for all $t > 0$. Then $T^n x \to x_*$ as $n \to \infty$, uniformly on each bounded subset of X.

Theorem 3.15 *Let $T : X \to X$ be such that*

$$d(T^n x, T^n y) \leq \phi_n(d(x, y))$$

for all $x, y \in X$ and all natural numbers n, where $\phi_n : [0, \infty) \to [0, \infty)$ and $\lim_{n\to\infty} \phi_n = \phi$, uniformly on any bounded interval $[0, b]$. Suppose that ϕ is upper semicontinuous and $\phi(t) < t$ for all $t > 0$. Furthermore, suppose that there exists a positive integer n_* such that ϕ_{n_*} is upper semicontinuous and $\phi_{n_*}(0) = 0$. If there exists $x_0 \in X$ which has a bounded orbit $O(x_0) = \{x_0, Tx_0, T^2x_0, \dots\}$, then T has a unique fixed point $x_* \in X$ and $\lim_{n\to\infty} T^n x = x_*$, uniformly on each bounded subset of X.

Proof of Theorem 3.14 We may assume without loss of generality that $\phi(0) = 0$ and $\phi_n(0) = 0$ for all integers $n \geq 1$.

For each $x \in X$ and each $r > 0$, set

$$B(x,r) = \{y \in X : d(x, y) \leq r\}.$$

We first prove three lemmata.

Lemma 3.16 *Let $K > 0$. Then there exists a natural number $\bar{q}$ such that for all integers $s \geq \bar{q}$,*

$$T^s(B(x_*, K)) \subset B(x_*, K + 1).$$

Proof There exists a natural number $\bar{q}$ such that for all integers $s \geq \bar{q}$,

$$|\psi_s(t) - \phi(t)| < 1 \quad \text{for all } t \in [0, K].$$

Let $s \geq \bar{q}$ be an integer. Then for all $x \in B(x_*, K)$,

$$d(T^s x, x_*) \leq \phi_s(d(x, x_*)) < \phi(d(x, x_*)) + 1 < d(x, x_*) + 1 < K + 1.$$

Lemma 3.16 is proved. □

Lemma 3.17 *Let $0 < \varepsilon_1 < \varepsilon_0$. Then there exists a natural number q such that for each integer $j \geq q$,*

$$T^j(B(x_*, \varepsilon_1)) \subset B(x_*, \varepsilon_0).$$

Proof There exists an integer $q \geq 1$ such that for each integer $j \geq q$,

$$|\phi_j(t) - \phi(t)| < (\varepsilon_0 - \varepsilon_1)/2 \quad \text{for all } t \in [0, \varepsilon_0]. \tag{3.152}$$

Assume that

$$j \in \{q, q + 1, \ldots\} \quad \text{and} \quad x \in B(x_*, \varepsilon_1).$$

By (3.151) and (3.152),

$$\begin{aligned} d(T^j x, x_*) &\leq \phi_j(d(x, x_*)) < \phi(d(x, x_*)) + (\varepsilon_0 - \varepsilon_1)/2 \\ &\leq \varepsilon_1 + (\varepsilon_0 - \varepsilon_1)/2 = (\varepsilon_0 + \varepsilon_1)/2. \end{aligned}$$

Lemma 3.17 is proved. □

Lemma 3.18 *Let $K, \varepsilon > 0$ be given. Then there exists a natural number q such that for each $x \in B(x_*, K)$,*

$$\min\{d(T^j x, x_*) : j = 1, \ldots, q\} \leq \varepsilon.$$

Proof By Lemma 3.16, there is a natural number $\bar{q}$ such that

$$T^n\big(B(x_*, K)\big) \subset B(x_*, K+1) \quad \text{for all natural numbers } n \geq \bar{q}. \tag{3.153}$$

We may assume without loss of generality that $\varepsilon < K/8$. Since the function $t - \phi(t)$, $t \in (0, \infty)$, is lower semicontinuous and positive, there is

$$\delta \in (0, \varepsilon/8) \tag{3.154}$$

such that

$$t - \phi(t) \geq 2\delta \quad \text{for all } t \in [\varepsilon/2, K+1]. \tag{3.155}$$

There is a natural number $s \geq \bar{q}$ such that

$$\big|\phi(t) - \phi_s(t)\big| \leq \delta \quad \text{for all } t \in [0, K+1]. \tag{3.156}$$

By (3.155) and (3.156), we have, for all $t \in [\varepsilon/2, K+1]$,

$$\phi_s(t) \leq \phi(t) + \delta \leq t - 2\delta + \delta = t - \delta. \tag{3.157}$$

In view of (3.156) and (3.154), we have, for all $t \in [0, \varepsilon/2]$,

$$\phi_s(t) \leq \phi(t) + \delta \leq t + \delta \leq \varepsilon/2 + \delta < (3/4)\varepsilon. \tag{3.158}$$

Choose a natural number p such that

$$p > 4 + \delta^{-1}(K+1). \tag{3.159}$$

Let

$$x \in B(x_*, K). \tag{3.160}$$

We will show that

$$\min\big\{d\big(T^j x, x_*\big) : j = 1, 2, \ldots, ps\big\} \leq \varepsilon. \tag{3.161}$$

Assume the contrary. Then

$$d\big(T^j x, x_*\big) > \varepsilon \quad \text{for all } j = s, \ldots, ps. \tag{3.162}$$

By (3.160) and(3.153),

$$T^j x \in B(x_*, K+1), \quad j = s, \ldots, ps. \tag{3.163}$$

Let a natural number i satisfy $i \leq p - 1$. By (3.162) and (3.163),

$$d\big(T^{is} x, x_*\big) > \varepsilon \quad \text{and} \quad d\big(T^{is} x, x_*\big) \leq K + 1. \tag{3.164}$$

It follows from (3.151), (3.164) and (3.157) that

$$d\big(T^s\big(T^{is} x\big), x_*\big) \leq \phi_s\big(d\big(T^{is} x, x_*\big)\big) \leq d\big(T^{is} x, x_*\big) - \delta.$$

Thus for each natural number $i \le p - 1$,

$$d(T^{(i+1)s}x, x_*) \le d(T^{is}x, x_*) - \delta.$$

This inequality implies that

$$d(T^{ps}x, x_*) \le d(T^{(p-1)s}x, x_*) - \delta \le \cdots \le d(T^{s}x, x_*) - (p-1)\delta.$$

When combined with (3.163) and (3.159), this implies, in turn, that

$$d(T^{ps}x, x_*) \le K + 1 - (p-1)\delta < 0.$$

The contradiction we have reached proves (3.161) and completes the proof of Lemma 3.18. □

Completion of the proof of Theorem 3.14 Let $K, \varepsilon > 0$ be given. Choose $\varepsilon_1 \in (0, \varepsilon)$. By Lemma 3.17, there exists a natural number q_1 such that

$$T^j(B(x_*, \varepsilon_1)) \subset B(x_*, \varepsilon) \quad \text{for all integers } j \ge q_1. \tag{3.165}$$

By Lemma 3.18, there exists a natural number q_2 such that

$$\min\{d(T^j x, x_*) : j = 1, \ldots, q_2\} \le \varepsilon_1 \quad \text{for all } x \in B(x_*, K). \tag{3.166}$$

Assume that

$$x \in B(x_*, K).$$

By (3.166), there is a natural number $j_1 \le q_2$ such that

$$d(T^{j_1}x, x_*) \le \varepsilon_1. \tag{3.167}$$

In view of (3.167) and (3.165),

$$T^j(T^{j_1}x) \in B(x_*, \varepsilon) \quad \text{for all integers } j \ge q_1. \tag{3.168}$$

Inclusion (3.168) and the inequality $j_1 \le q_2$ now imply that

$$T^i x \in B(x_*, \varepsilon) \quad \text{for all integers } i \ge q_1 + q_2.$$

Theorem 3.14 is proved. □

3.10 Uniform Convergence of Iterates

Let (X, d) be a complete metric space. The following theorem [9] is the main result of this section. In contrast with Theorem 3.14, here we only assume that a subsequence of $\{\phi_n\}_{n=1}^{\infty}$ converges to ϕ.

Theorem 3.19 *Let $x_* \in X$ be a fixed point of $T : X \to X$. Assume that*

$$d(T^n x, x_*) \le \phi_n(d(x, x_*)) \tag{3.169}$$

for all $x \in X$ and all natural numbers n, where the functions $\phi_n : [0, \infty) \to [0, \infty)$, $n = 1, 2, \ldots$, satisfy the following conditions:

(i) *For each $b > 0$, there is a natural number n_b such that*

$$\sup\{\phi_n(t) : t \in [0, b] \text{ and all } n \ge n_b\} < \infty; \tag{3.170}$$

(ii) *there exist an upper semicontinuous function $\phi : [0, \infty) \to [0, \infty)$ satisfying $\phi(t) < t$ for all $t > 0$ and a strictly increasing sequence of natural numbers $\{m_k\}_{k=1}^{\infty}$ such that $\lim_{k\to\infty} \phi_{m_k} = \phi$, uniformly on any bounded interval $[0, b]$.*

Then $T^n x \to x_*$ as $n \to \infty$, uniformly on any bounded subset of X.

Proof Set $T^0 x = x$ for all $x \in X$. For each $x \in X$ and each $r > 0$, set

$$B(x, r) = \{z \in X : d(x, z) \le r\}. \tag{3.171}$$

Let $M > 0$ and $\varepsilon \in (0, 1)$ be given. By (i), there are $M_1 > M$ and an integer $n_1 \ge 1$ such that

$$\phi_i(t) \le M_1 \quad \text{for all } t \in [0, M+1] \text{ and all integers } i \ge n_1. \tag{3.172}$$

In view of (3.169) and (3.172), for each $x \in B(x_*, M)$ and each integer $n \ge n_1$,

$$d(T_n x, x_*) \le \phi_n(d(x, x_*)) \le M_1. \tag{3.173}$$

Since the function $t - \phi(t)$ is lower semicontinuous, there is $\delta > 0$ such that

$$\delta < \varepsilon/8 \tag{3.174}$$

and

$$t - \phi(t) \ge 2\delta, \quad t \in [\varepsilon/8, 4M_1 + 4]. \tag{3.175}$$

By (ii), there is an integer $n_2 \ge 2n_1 + 2$ such that

$$|\phi_{n_2}(t) - \phi(t)| \le \delta, \quad t \in [0, 4M_1 + 4]. \tag{3.176}$$

Assume that

$$x \in B(x_*, M_1 + 4). \tag{3.177}$$

If $d(x, x_*) \le \varepsilon/8$, then it follows from (3.169), (3.174), (3.176) and (3.177) that

$$d(T^{n_2} x, x_*) \le \phi_{n_2}(d(x, x_*)) \le \phi(d(x, x_*)) + \delta \le d(x, x_*) + \delta < \varepsilon/4.$$

If $d(x, x_*) \geq \varepsilon/8$, then relations (3.169), (3.175), (3.176) and (3.177) imply that

$$d\big(T^{n_2}x, x_*\big) \leq \phi_{n_2}\big(d(x, x_*)\big) \leq \phi\big(d(x, x_*)\big) + \delta \leq d(x, x_*) - 2\delta + \delta = d(x, x_*) - \delta.$$

Thus in both cases we have

$$d\big(T^{n_2}x, x_*\big) \leq \max\big\{d(x, x_*) - \delta, \varepsilon/4\big\}. \tag{3.178}$$

Now choose a natural number $q > 2$ such that

$$q > (8 + 2M_1)\delta^{-1}. \tag{3.179}$$

Assume that

$$x \in B(x_*, M_1 + 4) \quad \text{and} \quad T^{in_2}x \in B(x_*, M_1 + 4), \quad i = 1, \ldots, q - 1. \tag{3.180}$$

We claim that

$$\min\big\{d\big(T^{jn_2}x, x_*\big) : j = 1, \ldots, q\big\} \leq \varepsilon/4. \tag{3.181}$$

Assume the contrary. Then by (3.178) and (3.180), for each $j = 1, \ldots, q$, we have

$$d\big(T^{jn_2}x, x_*\big) \leq d\big(T^{(j-1)n_2}x, x_*\big) - \delta$$

and

$$d\big(T^{qn_2}x, x_*\big) \leq d\big(T^{(q-1)n_2}x, x_*\big) - \delta \leq \cdots \leq d(x, x_*) - q\delta \leq M_1 + 4 - q\delta.$$

This contradicts (3.179). The contradiction we have reached proves (3.181).

Assume that an integer j satisfies $1 \leq j \leq q - 1$ and

$$d\big(T^{jn_2}x, x_*\big) \leq \varepsilon/4.$$

When combined with (3.178) and (3.180), this implies that

$$d\big(T^{(j+1)n_2}x, x_*\big) \leq \max\big\{d\big(T^{jn_2}x, x_*\big) - \delta, \varepsilon/4\big\} \leq \varepsilon/4.$$

It follows from this inequality and (3.181) that

$$d\big(T^{qn_2}x, x_*\big) \leq \varepsilon/4 \tag{3.182}$$

for all points x satisfying (3.177).

Assume now that $x \in B(x_*, M)$ and let an integer s be such that $s \geq n_1 + qn_2$. By (3.173),

$$T^i x \in B(x_*, M_1) \quad \text{for all integers } i \geq n_1$$

and

$$T^{s-qn_2}x \in B(x_*, M_1). \tag{3.183}$$

Since $T^s x = T^{qn_2}(T^{s-qn_2}x)$, it follows from (3.182) and (3.183) that

$$d(T^s x, x_*) = d\big(T^{qn_2}\big(T^{s-qn_2}x\big), x_*\big) < \varepsilon/4.$$

This completes the proof of Theorem 3.19. □

The following result, which was also obtained in [9], is an extension of Theorem 3.19.

Theorem 3.20 *Let $x_* \in X$ be a fixed point of $T : X \to X$. Assume that $\{m_k\}_{k=1}^{\infty}$ is a strictly increasing sequence of natural numbers such that*

$$d\big(T^{m_k}x, x_*\big) \le \phi_{m_k}\big(d(x, x_*)\big)$$

for all $x \in X$ and all natural numbers k, where T and the functions $\phi_{m_k} : [0, \infty) \to [0, \infty)$, $k = 1, 2, \ldots$, satisfy the following conditions:

(i) *For each $M > 0$, there is $M_1 > 0$ such that*

$$T^i\big(B(x_*, M)\big) \subset B(x_*, M_1) \quad \textit{for each integer } i \ge 0;$$

(ii) *there exists an upper semicontinuous function $\phi : [0, \infty) \to [0, \infty)$ satisfying $\phi(t) < t$ for all $t > 0$ such that $\lim_{k\to\infty} \phi_{m_k} = \phi$, uniformly on any bounded interval $[0, b]$.*

Then $T^n x \to x_*$ as $n \to \infty$, uniformly on any bounded subset of X.

Proof Let i be a natural number such that $i \ne m_k$ for all natural numbers k. For each $t \ge 0$, set

$$\phi_i(t) = \sup\big\{d\big(T^i x, x_*\big) : x \in B(x_*, t)\big\}.$$

Clearly, $\phi_i(t)$ is finite for all $t \ge 0$. It is easy to see that all the assumptions of Theorem 3.19 hold. Therefore Theorem 3.19 implies that $T^n x \to x_*$ as $n \to \infty$, uniformly on all bounded subsets of X. Theorem 3.20 is proved. □

Now we show that Theorem 3.19 has a converse.

Assume now that $T : X \to X$, $x_* \in X$, $T^n x \to x_*$ as $n \to \infty$, uniformly on all bounded subsets of X, and that $T(C)$ is bounded for any bounded $C \subset X$. We claim that T necessarily satisfies all the hypotheses of Theorem 3.19 with an appropriate sequence $\{\phi_n\}_{n=1}^{\infty}$.

Indeed, fix a natural number n and for all $t \ge 0$, set

$$\phi_n(t) = \sup\big\{d\big(T^n x, x_*\big) : x \in B(x_*, t)\big\}.$$

Clearly, $\phi_n(t)$ is finite for all $t \ge 0$ and all natural numbers n, and

$$d\big(T^n x, x_*\big) \le \phi_n\big(d(x, x_*)\big)$$

for all $x \in X$ and all natural numbers n. It is also obvious that $\phi_n \to 0$ as $n \to \infty$, uniformly on any bounded subinterval of $[0, \infty)$, and that for any $b > 0$,

$$\sup\{\phi_n(t) : t \in [0, b], n \geq 1\} < \infty.$$

Thus all the assumptions of Theorem 3.19 hold with $\phi(t) = 0$ identically.

3.11 Well-Posedness of Fixed Point Problems

Let (K, ρ) be a bounded complete metric space. We say that the fixed point problem for a mapping $A : K \to K$ is well posed if there exists a unique $x_A \in K$ such that $Ax_A = x_A$ and the following property holds:

if $\{x_n\}_{n=1}^{\infty} \subset K$ and $\rho(x_n, Ax_n) \to 0$ as $n \to \infty$, then $\rho(x_n, x_A) \to 0$ as $n \to \infty$.

The notion of well-posedness is of central importance in many areas of Mathematics and its applications. In our context this notion was studied in [50], where generic well-posedness of the fixed point problem is established for the space of nonexpansive self-mappings of K.

In this section, which is based on [139], we first show (Theorem 3.21) that the fixed point problem is well posed for any contractive self-mapping of K. Since it is known that in Banach spaces (see Theorem 3.2) almost all nonexpansive mappings are contractive in the sense of Baire's categories, the generic well-posedness of the fixed point problem for the space of nonexpansive self-mappings of K follows immediately in this case. In our second result (Theorem 3.22) we show that the fixed point problem is well posed as soon as the uniformly continuous self-mapping of K has a unique fixed point which is the uniform limit of every sequence of iterates.

Let (K, ρ) be a bounded complete metric space. Define

$$d(K) = \sup\{\rho(x, y) : x, y \in K\}. \tag{3.184}$$

Recall that a mapping $A : K \to K$ is contractive if there exists a decreasing function $\phi : [0, d(K)] \to [0, 1]$ such that

$$\phi(t) < 1, \quad t \in (0, d(K)] \tag{3.185}$$

and

$$\rho(Ax, Ay) \leq \phi(\rho(x, y))\rho(x, y) \quad \text{for all } x, y \in K. \tag{3.186}$$

Theorem 3.21 *Assume that a mapping $A : K \to K$ is contractive. Then the fixed point problem for A is well posed.*

Proof Since the mapping A is contractive, there exists a decreasing function $\phi : [0, d(K)] \to [0, 1]$ such that (3.185) and (3.186) hold. By Theorem 3.1, there exists a unique $x_A \in K$ such that

$$Ax_A = x_A. \tag{3.187}$$

Let $\{x_n\}_{n=1}^{\infty} \subset K$ satisfy

$$\lim_{n\to\infty} \rho(x_n, Ax_n) = 0. \tag{3.188}$$

We claim that $x_n \to x_A$ as $n \to \infty$. Assume the contrary. By extracting a subsequence, if necessary, we may assume without loss of generality that there exists $\varepsilon > 0$ such that

$$\rho(x_n, x_A) \ge \varepsilon \quad \text{for all integers } n \ge 1. \tag{3.189}$$

Then it follows from (3.187), (3.186), (3.189) and the monotonicity of the function ϕ that for all integers $n \ge 1$,

$$\begin{aligned}\rho(x_A, x_n) &\le \rho(x_A, Ax_n) + \rho(Ax_n, x_n) \le \rho(Ax_n, x_n) + \phi\big(\rho(x_n, x_A)\big)\rho(x_n, x_A)\\ &\le \rho(Ax_n, x_n) + \phi(\varepsilon)\rho(x_A, x_n).\end{aligned} \tag{3.190}$$

Inequalities (3.190) and (3.189) imply that for all integers $n \ge 1$,

$$\varepsilon\big(1 - \phi(\varepsilon)\big) \le \big(1 - \phi(\varepsilon)\big)\rho(x_A, x_n) \le \rho(Ax_n, x_n),$$

a contradiction (see (3.188)). The contradiction we have reached proves Theorem 3.21. □

Theorem 3.22 *Assume that $A : K \to K$ is a uniformly continuous mapping, $x_A \in K$, $Ax_A = x_A$, and that $A^n x \to x_A$ as $n \to \infty$, uniformly on K. Then the fixed point problem for the mapping A is well posed.*

Proof Let $\varepsilon > 0$ be given. In order to prove this theorem, it is sufficient to show that there exists $\delta > 0$ such that for each $y \in K$ satisfying $\rho(y, Ay) < \delta$, the inequality $\rho(y, x_A) < \varepsilon$ is true.

There exists a natural number $n_0 \ge 3$ such that

$$\rho\big(A^n x, x_A\big) \le \varepsilon/8 \quad \text{for any } x \in K \text{ and any integer } n \ge n_0. \tag{3.191}$$

Set

$$\delta_0 = \varepsilon(8n_0)^{-1}. \tag{3.192}$$

Using induction, we define a sequence of positive numbers $\{\delta_i\}_{i=0}^{\infty}$ such that for any integer $i \ge 0$,

$$\delta_{i+1} < \delta_i \tag{3.193}$$

and

$$\text{if } x, y \in K \text{ and } \rho(x, y) \le \delta_{i+1}, \text{ then } \rho(Ax, Ay) \le \delta_i. \tag{3.194}$$

We now show that if $y \in K$ satisfies $\rho(y, Ay) < \delta_{n_0}$, then $\rho(y, x_A) < \varepsilon/2$. Indeed, let $y \in K$ satisfy

$$\rho(y, Ay) < \delta_{n_0}. \tag{3.195}$$

It follows from the definition of the sequence $\{\delta_i\}_{i=0}^{\infty}$ (see (3.193), (3.194)) and (3.195) that for any integer $j \in [1, n_0]$,

$$\rho(A^j y, A^{j+1} y) \le \delta_{n_0 - j}. \tag{3.196}$$

Relations (3.196), (3.193) and (3.192) imply that

$$\rho(y, A^{n_0+1} y) \le \sum_{j=0}^{n_0} \rho(A^j y, A^{j+1} y) \le (n_0 + 1)\delta_0 < \varepsilon/4. \tag{3.197}$$

(Here we use the notation $A^0 x = x$ for all $x \in K$.) It follows from (3.197) and the definition of n_0 (see (3.191)) that

$$\rho(y, x_A) \le \rho(y, A^{n_0+1} y) + \rho(A^{n_0+1} y, x_A) < \varepsilon/4 + \varepsilon/8 < \varepsilon/2.$$

Thus we have indeed shown that if $y \in K$ satisfies $\rho(y, Ay) < \delta_{n_0}$, then $\rho(y, x_A) < \varepsilon/2$. This completes the proof of Theorem 3.22. □

3.12 A Class of Mappings of Contractive Type

Let (X, ρ) be a complete metric space. In this section, which is based on [158], we present a sufficient condition for the existence and approximation of the unique fixed point of a contractive mapping which maps a nonempty, closed subset of X into X.

Theorem 3.23 *Let K be a nonempty and closed subset of a complete metric space (X, ρ). Assume that $T : K \to X$ satisfies*

$$\rho(Tx, Ty) \le \phi(\rho(x, y)) \quad \textit{for each } x, y \in K, \tag{3.198}$$

where $\phi : [0, \infty) \to [0, \infty)$ is upper semicontinuous and satisfies $\phi(t) < t$ for all $t > 0$.

Assume further that $K_0 \subset K$ is a nonempty and bounded set with the following property:

(P1) *For each natural number n, there exists $x_n \in K_0$ such that $T^n x_n$ is defined.*

Then the following assertions hold.

(A) *There exists a unique $\bar{x} \in K$ such that $T\bar{x} = \bar{x}$.*
(B) *Let $M, \varepsilon > 0$. Then there exist $\delta > 0$ and a natural number k such that for each integer $n \ge k$ and each sequence $\{x_i\}_{i=0}^{n} \subset K$ satisfying*

$$\rho(x_0, \bar{x}) \le M$$

and

$$\rho(x_{i+1}, Tx_i) \le \delta, \quad i = 0, \dots, n-1,$$

the inequality $\rho(x_i, \bar{x}) \le \varepsilon$ *holds for* $i = k, \dots, n$.

Proof (A) The uniqueness of $\bar{x}$ is obvious. To establish its existence, we may and shall assume that $\phi(0) = 0$.

For each natural number n, let x_n be as guaranteed by (P1). Fix $\theta \in K$. Since K_0 is bounded, there is $c_0 > 0$ such that

$$\rho(\theta, z) \le c_0 \quad \text{for all } z \in K_0. \tag{3.199}$$

Let $\varepsilon > 0$ be given. We will show that there exists a natural number k such that the following property holds:

(P2) If n and i are integers such that $k \le i < n$, then

$$\rho\big(T^i x_n, T^{i+1} x_n\big) \le \varepsilon.$$

Assume the contrary. Then for each natural number k, there exist natural numbers n_k and i_k such that

$$k \le i_k < n_k \quad \text{and} \quad \rho\big(T^{i_k} x_{n_k}, T^{i_k+1} x_{n_k}\big) > \varepsilon. \tag{3.200}$$

Since the function $t - \phi(t)$ is positive for all $t > 0$ and lower semicontinuous, there is $\gamma > 0$ such that

$$t - \phi(t) \ge \gamma \quad \text{for all } t \in \big[\varepsilon/2, 2c_0 + \rho(\theta, T\theta) + \varepsilon\big]. \tag{3.201}$$

Choose a natural number k such that

$$k > \gamma^{-1}\big(2c_0 + \rho(\theta, T\theta)\big). \tag{3.202}$$

Then (3.200) holds. By (3.200) and (3.198),

$$\rho\big(T^i x_{n_k}, T^{i+1} x_{n_k}\big) > \varepsilon, \quad i = 0, \dots, i_k. \tag{3.203}$$

(Here we use the convention that $T^0 z = z$ for all $z \in K$.) By (3.198),

$$\rho(x_{n_k}, Tx_{n_k}) \ge \rho\big(T^i x_{n_k}, T^{i+1} x_{n_k}\big) \quad \text{for each integer } i \text{ satisfying } 0 \le i < i_k. \tag{3.204}$$

By (P1), (3.199) and (3.198),

$$\begin{aligned} \rho(x_{n_k}, Tx_{n_k}) &\le \rho(x_{n_k}, \theta) + \rho(\theta, T\theta) + \rho(T\theta, Tx_{n_k}) \\ &\le c_0 + \rho(\theta, T\theta) + c_0. \end{aligned} \tag{3.205}$$

Together with (3.203) and (3.204) this implies that

$$\varepsilon < \rho\big(T^i x_{n_k}, T^{i+1} x_{n_k}\big) \le 2c_0 + \rho(\theta, T\theta) \quad \text{for all } i = 0, \ldots, i_k. \tag{3.206}$$

It follows from (3.198), (3.206) and (3.201) that for all $i = 0, \ldots, i_k - 1$,

$$\rho\big(T^{i+2} x_{n_k}, T^{i+1} x_{n_k}\big) \le \phi\big(\rho\big(T^{i+1} x_{n_k}, T^i x_{n_k}\big)\big) \le \rho\big(T^{i+1} x_{n_k}, T^i x_{n_k}\big) - \gamma.$$

When combined with (3.205) and (3.200), this implies that

$$\begin{aligned}
-\rho(\theta, T\theta) - 2c_0 &\le -\rho(x_{n_k}, T x_{n_k}) \le \rho\big(T^{i_k+1} x_{n_k}, T^{i_k} x_{n_k}\big) - \rho(x_{n_k}, T x_{n_k}) \\
&= \sum_{i=0}^{i_k-1} \big[\rho\big(T^{i+2} x_{n_k}, T^{i+1} x_{n_k}\big) - \rho\big(T^{i+1} x_{n_k}, T^i x_{n_k}\big)\big] \\
&\le -\gamma i_k \le -k\gamma
\end{aligned}$$

and

$$k\gamma \le 2c_0 + \rho(\theta, T\theta).$$

This contradicts (3.202). The contradiction we have reached proves the existence of a natural number k such that property (P2) holds.

Now let $\delta > 0$ be given. We will show that there exists a natural number k such that the following property holds:

(P3) If n, i and j are integers such that $k \le i, j < n$, then

$$\rho\big(T^i x_n, T^j x_n\big) \le \delta.$$

Assume to the contrary that there is no natural number k for which (P3) holds.

Then for each natural number k, there exist natural numbers n_k, i_k and j_k such that

$$k \le i_k < j_k < n_k \tag{3.207}$$

and

$$\rho\big(T^{i_k} x_{n_k}, T^{j_k} x_{n_k}\big) > \delta.$$

We may assume without loss of generality that for each natural number k, the following property holds:

If an integer j satisfies $i_k \le j < j_k$, then

$$\rho\big(T^{i_k} x_{n_k}, T^j x_{n_k}\big) \le \delta. \tag{3.208}$$

We have already shown that there exists a natural number k_0 such that (P2) holds with $k = k_0$ and $\varepsilon = \delta$.

Assume now that k is a natural number. It follows from (3.207) and (3.208) that

$$\delta < \rho(T^{i_k}x_{n_k}, T^{j_k}x_{n_k}) \le \rho(T^{j_k}x_{n_k}, T^{j_k-1}x_{n_k}) + \rho(T^{j_k-1}x_{n_k}, T^{i_k}x_{n_k})$$
$$\le \rho(T^{j_k}x_{n_k}, T^{j_k-1}x_{n_k}) + \delta. \tag{3.209}$$

By property (P2),

$$\lim_{k\to\infty} \rho(T^{j_k}x_{n_k}, T^{j_k-1}x_{n_k}) = 0.$$

When combined with (3.209), this implies that

$$\lim_{k\to\infty} \rho(T^{i_k}x_{n_k}, T^{j_k}x_{n_k}) = \delta. \tag{3.210}$$

By (3.207), for each integer $k \ge 1$,

$$\delta < \rho(T^{i_k}x_{n_k}, T^{j_k}x_{n_k})$$
$$\le \rho(T^{i_k}x_{n_k}, T^{i_k+1}x_{n_k}) + \rho(T^{i_k+1}x_{n_k}, T^{j_k+1}x_{n_k}) + \rho(T^{j_k+1}x_{n_k}, T^{j_k}x_{n_k})$$
$$\le \rho(T^{i_k}x_{n_k}, T^{i_k+1}x_{n_k}) + \rho(T^{j_k+1}x_{n_k}, T^{j_k}x_{n_k}) + \phi(\rho(T^{i_k}x_{n_k}, T^{j_k}x_{n_k})). \tag{3.211}$$

Since by (P2),

$$\lim_{k\to\infty} \rho(T^{i_k}x_{n_k}, T^{i_k+1}x_{n_k}) = \lim_{k\to\infty} \rho(T^{j_k}x_{n_k}, T^{j_k+1}x_{n_k}) = 0,$$

(3.210) and (3.211) imply that $\delta \le \phi(\delta)$, a contradiction.

The contradiction we have reached proves that there exists a natural number k such that (P3) holds.

Let $\varepsilon > 0$ be given. We will show that there exists a natural number k such that the following property holds:

(P4) If the integers $n_1, n_2 > k$, then $\rho(T^k x_{n_1}, T^k x_{n_2}) \le \varepsilon$.

Assume the contrary. Then for each integer $k \ge 1$, there are integers $n_1^{(k)}, n_2^{(k)} > k$ such that

$$\rho(T^k x_{n_1^{(k)}}, T^k x_{n_2^{(k)}}) > \varepsilon. \tag{3.212}$$

By (P1), (3.198) and (3.199), the sequence

$$\{\rho(T^k x_{n_1^{(k)}}, T^k x_{n_2^{(k)}})\}_{k=1}^{\infty}$$

is bounded. Set

$$\delta = \limsup_{k\to\infty} \rho(T^k x_{n_1^{(k)}}, T^k x_{n_2^{(k)}}). \tag{3.213}$$

By definition, there exists a strictly increasing sequence of natural numbers $\{k_i\}_{i=1}^{\infty}$ such that

$$\delta = \lim_{i\to\infty} \rho(T^{k_i} x_{n_1^{(k_i)}}, T^{k_i} x_{n_2^{(k_i)}}). \tag{3.214}$$

By (3.212) and (3.213),

$$\delta \geq \varepsilon. \tag{3.215}$$

By (3.198), for each natural number i,

$$\begin{aligned}
\rho\big(T^{k_i}x_{n_1^{(k_i)}}, T^{k_i}x_{n_2^{(k_i)}}\big) &\leq \rho\big(T^{k_i+1}x_{n_1^{(k_i)}}, T^{k_i}x_{n_1^{(k_i)}}\big) \\
&\quad + \rho\big(T^{k_i+1}x_{n_1^{(k_i)}}, T^{k_i+1}x_{n_2^{(k_i)}}\big) + \rho\big(T^{k_i+1}x_{n_2^{(k_i)}}, T^{k_i}x_{n_2^{(k_i)}}\big) \\
&\leq \rho\big(T^{k_i+1}x_{n_1^{(k_i)}}, T^{k_i}x_{n_1^{(k_i)}}\big) + \rho\big(T^{k_i+1}x_{n_2^{(k_i)}}, T^{k_i}x_{n_2^{(k_i)}}\big) \\
&\quad + \phi\big(\rho\big(T^{k_i}x_{n_1^{(k_i)}}, T^{k_i}x_{n_2^{(k_i)}}\big)\big).
\end{aligned} \tag{3.216}$$

By property (P2),

$$\lim_{i\to\infty} \rho\big(T^{k_i+1}x_{n_j^{(k_i)}}, T^{k_i}x_{n_j^{(k_i)}}\big) = 0, \quad j = 1, 2. \tag{3.217}$$

Now it follows from (3.216), (3.217), (3.204) and (3.215) that $\varepsilon \leq \delta \leq \phi(\delta)$, a contradiction. This contradiction implies that there is indeed a natural number k such that (P4) holds, as claimed.

Let $\varepsilon > 0$ be given. By (P4), there exists a natural number k_1 such that

$$\rho\big(T^{k_1}x_{n_1}, T^{k_1}x_{n_2}\big) \leq \varepsilon/4 \quad \text{for all integers } n_1, n_2 \geq k_1. \tag{3.218}$$

By (P3), there exists a natural number k_2 such that

$$\rho\big(T^i x_n, T^j x_n\big) \leq \varepsilon/4 \quad \text{for all natural numbers } n, i, j \text{ satisfying } k_2 \leq i, j < n. \tag{3.219}$$

Assume that the natural numbers n_1, n_2, i and j satisfy

$$n_1, n_2 > k_1 + k_2, \qquad i, j \geq k_1 + k_2, \qquad i < n_1, \qquad j < n_2. \tag{3.220}$$

We claim that $\rho(T^i x_{n_1}, T^j x_{n_2}) \leq \varepsilon$. By (3.198), (3.218) and (3.220),

$$\rho\big(T^{k_1+k_2}x_{n_1}, T^{k_1+k_2}x_{n_2}\big) \leq \rho\big(T^{k_1}x_{n_1}, T^{k_1}x_{n_2}\big) \leq \varepsilon/4. \tag{3.221}$$

In view of (3.219) and (3.220),

$$\rho\big(T^{k_1+k_2}x_{n_1}, T^i x_{n_1}\big) \leq \varepsilon/4 \quad \text{and} \quad \rho\big(T^{k_1+k_2}x_{n_2}, T^j x_{n_2}\big) \leq \varepsilon/4. \tag{3.222}$$

Inequalities (3.222) and (3.221) imply that

$$\begin{aligned}
\rho\big(T^i x_{n_1}, T^j x_{n_2}\big) &\leq \rho\big(T^i x_{n_1}, T^{k_1+k_2}x_{n_1}\big) + \rho\big(T^{k_1+k_2}x_{n_1}, T^{k_1+k_2}x_{n_2}\big) \\
&\quad + \rho\big(T^{k_1+k_2}x_{n_2}, T^j x_{n_2}\big) < \varepsilon.
\end{aligned}$$

Thus we have shown that the following property holds:

(P5) For each $\varepsilon > 0$, there exists a natural number $k(\varepsilon)$ such that

$$\rho\big(T^i x_{n_1}, T^j x_{n_2}\big) \le \varepsilon \quad \text{for all natural numbers } n_1, n_2, i \text{ and } j$$

such that

$$n_1, n_2 > k(\varepsilon), \qquad i \in \big[k(\varepsilon), n_1\big) \quad \text{and} \quad j \in \big[k(\varepsilon), n_2\big).$$

Consider now the sequences $\{T^{n-2}x_n\}_{n=3}^{\infty}$ and $\{T^{n-1}x_n\}_{n=3}^{\infty}$. Property (P5) implies that both of them are Cauchy sequences and that

$$\lim_{n\to\infty} \rho\big(T^{n-2}x_n, T^{n-1}x_n\big) = 0.$$

Hence there exists $\bar{x} \in K$ such that

$$\lim_{n\to\infty} \rho\big(\bar{x}, T^{n-2}x_n\big) = \lim_{t\to\infty} \rho\big(\bar{x}, T^{n-1}x_n\big) = 0.$$

Since the mapping T is continuous, it follows that $T\bar{x} = \bar{x}$. Thus part (A) of our theorem is proved.

We now turn to the proof of part (B). Clearly,

$$\inf\big\{t - \phi(t) : t \in [M/2, M]\big\} > 0.$$

Choose a positive number δ_0 such that

$$\delta_0 < \min\big\{M/2, \inf\big\{t - \phi(t) : t \in [M/2, M]\big\}/4\big\}. \tag{3.223}$$

For each $x \in X$ and $r > 0$, set

$$B(x, r) = \big\{y \in X : \rho(x, y) \le r\big\}.$$

Assume that

$$y \in K \cap B(\bar{x}, M), \qquad z \in X \quad \text{and} \quad \rho(z, Ty) \le \delta_0. \tag{3.224}$$

By (3.224) and (3.198),

$$\rho(\bar{x}, z) \le \rho(\bar{x}, Ty) + \rho(Ty, z) \le \rho(T\bar{x}, Ty) + \delta_0 \le \phi\big(\rho(\bar{x}, y)\big) + \delta_0. \tag{3.225}$$

There are two cases:

$$\rho(y, \bar{x}) \le M/2; \tag{3.226}$$

$$\rho(y, \bar{x}) > M/2. \tag{3.227}$$

Assume that (3.226) holds. By (3.225), (3.226), (3.198) and (3.223),

$$\rho(\bar{x}, z) \le \rho(\bar{x}, y) + \delta_0 \le M/2 + \delta_0 < M.$$

Assume that (3.227) holds. Then by (3.223), (3.225), (3.224) and (3.227),

$$\rho(\bar{x}, z) \le \delta_0 + \phi(\rho(\bar{x}, y)) < [\rho(\bar{x}, y) - \phi(\rho(\bar{x}, y))]4^{-1} + \phi(\rho(\bar{x}, y))$$
$$< \rho(\bar{x}, y) \le M.$$

Thus $\rho(\bar{x}, z) \le M$ in both cases.

We have shown that

$$\rho(\bar{x}, z) \le M \quad \text{for each } z \in X \text{ such that}$$
$$\text{there exists } y \in K \cap B(\bar{x}, M) \text{ satisfying } \rho(z, Ty) \le \delta_0. \tag{3.228}$$

Since M is an arbitrary positive number, we may conclude that there is $\delta_1 > 0$ so that

$$\rho(\bar{x}, z) \le \varepsilon \quad \text{for each } z \in X \text{ such that}$$
$$\text{there exists } y \in K \cap B(\bar{x}, \varepsilon) \text{ satisfying } \rho(z, Ty) \le \delta_1. \tag{3.229}$$

Choose a positive number δ such that

$$\delta < \min\{\delta_0, \delta_1, 4^{-1} \inf\{t - \phi(t) : t \in [\varepsilon, M + \varepsilon + 1]\}\} \tag{3.230}$$

and a natural number k such that

$$k > 2(M+1)\delta^{-1} + 2. \tag{3.231}$$

Assume that n is a natural number such that $n \ge k$ and that $\{x_i\}_{i=0}^n \subset K$ satisfies

$$\rho(x_0, \bar{x}) \le M, \qquad \rho(x_{i+1}, Tx_i) \le \delta, \quad i = 0, \dots, n-1. \tag{3.232}$$

We claim that

$$\rho(x_i, \bar{x}) \le \varepsilon, \quad i = k, \dots, n. \tag{3.233}$$

By (3.228), (3.230) and (3.232),

$$\{x_i\}_{i=0}^n \subset B(\bar{x}, M). \tag{3.234}$$

Assume that (3.233) does not hold. Then there is an integer j such that

$$j \in \{k, \dots, n\} \quad \text{and} \quad \rho(x_j, \bar{x}) > \varepsilon. \tag{3.235}$$

By (3.229), (3.230) and (3.232),

$$\rho(x_i, \bar{x}) > \varepsilon, \quad i = 0, \dots, j. \tag{3.236}$$

Let $i \in \{0, \dots, j-1\}$. By (3.232), (3.198), (3.234), (3.236) and (3.230),

$$\rho(x_{i+1}, \bar{x}) \le \rho(x_{i+1}, Tx_i) + \rho(Tx_i, T\bar{x}) \le \delta + \phi(\rho(x_i, \bar{x}))$$
$$< \phi(\rho(x_i, \bar{x})) + 4^{-1}(\rho(x_i, \bar{x}) - \phi(\rho(x_i, \bar{x})))$$

$$< \phi(\rho(x_i, \bar{x})) + 2^{-1}(\rho(x_i, \bar{x}) - \phi(\rho(x_i, \bar{x}))) - \delta$$
$$\leq \rho(x_i, \bar{x}) - \delta.$$

When combined with (3.232) and (3.235), this implies that

$$-M \leq -\rho(x_0, \bar{x}) \leq \rho(x_j, \bar{x}) - \rho(x_0, \bar{x})$$
$$= \sum_{i=0}^{j-1} [\rho(x_{i+1}, \bar{x}) - \rho(x_i, \bar{x})] \leq -j\delta \leq -k\delta.$$

Thus

$$k\delta \leq M$$

which contradicts (3.231).

Hence (3.233) is true, as claimed, and part (B) of our theorem is also proved. □

3.13 A Fixed Point Theorem for Matkowski Contractions

Let (X, ρ) be a complete metric space. In this section, which is based on [159], we present a sufficient condition for the existence and approximation of the unique fixed point of a Matkowski contraction [99] which maps a nonempty and closed subset of X into X.

Theorem 3.24 *Let K be a nonempty and closed subset of a complete metric space (X, ρ). Assume that $T : K \to X$ satisfies*

$$\rho(Tx, Ty) \leq \phi(\rho(x, y)) \quad \text{for each } x, y \in K, \tag{3.237}$$

where $\phi : [0, \infty) \to [0, \infty)$ is increasing and satisfies $\lim_{n\to\infty} \phi^n(t) = 0$ for all $t > 0$. Assume that $K_0 \subset K$ is a nonempty and bounded set with the following property:

(P1) *For each natural number n, there exists $x_n \in K_0$ such that $T^n x_n$ is defined.*

Then the following assertions hold.

(A) *There exists a unique $\bar{x} \in K$ such that $T\bar{x} = \bar{x}$.*
(B) *Let $M, \varepsilon > 0$. Then there exists a natural number k such that for each sequence $\{x_i\}_{i=0}^n \subset K$ with $n \geq k$ satisfying*

$$\rho(x_0, \bar{x}) \leq M \quad \text{and} \quad Tx_i = x_{i+1}, \quad i = 0, \dots, n-1,$$

the inequality $\rho(x_i, \bar{x}) \leq \varepsilon$ holds for all $i = k, \dots, n$.

Proof For each $x \in X$ and $r > 0$, set

$$B(x,r) = \{y \in X : \rho(x, y) \le r\}. \tag{3.238}$$

(A) Since $\phi^n(t) \to 0$ as $n \to \infty$ for all $t > 0$, and since ϕ is increasing, we have

$$\phi(t) < t \quad \text{for all } t > 0. \tag{3.239}$$

This implies the uniqueness of $\bar{x}$. Clearly, $\phi(0) = 0$.

For each natural number n, let x_n be as guaranteed by property (P1). Fix $\theta \in K$. Since K_0 is bounded, there is $c_0 > 0$ such that

$$\rho(\theta, z) \le c_0 \quad \text{for all } z \in K_0. \tag{3.240}$$

Let $\varepsilon > 0$ be given. We will show that there exists a natural number k such that the following property holds:

(P2) If the integers i and n satisfy $k \le i < n$, then

$$\rho(T^i x_n, T^{i+1} x_n) \le \varepsilon.$$

By (3.236) and (3.240), for each $z \in K_0$,

$$\begin{aligned}\rho(z, Tz) &\le \rho(z, \theta) + \rho(\theta, T\theta) + \rho(T\theta, Tz) \\ &\le 2\rho(z, \theta) + \rho(\theta, T\theta) \le 2c_0 + \rho(\theta, T\theta).\end{aligned} \tag{3.241}$$

Clearly, there is a natural number k such that

$$\phi^k(2c_0 + \rho(\theta, T\theta)) < \varepsilon. \tag{3.242}$$

Assume now that the integers i and n satisfy $k \le i < n$.

By (3.236), (3.239), (3.241), the choice of x_n, and (3.242),

$$\begin{aligned}\rho(T^i x_n, T^{i+1} x_n) &\le \rho(T^k x_n, T^{k+1} x_n) \le \phi^k(\rho(x_n, Tx_n)) \\ &\le \phi^k(2c_0 + \rho(\theta, T\theta)) < \varepsilon.\end{aligned}$$

Thus property (P2) holds for this k.

Let $\delta > 0$ be given. We claim that there exists a natural number k such that the following property holds:

(P3) If the integers i, j and n satisfy $k \le i < j < n$, then

$$\rho(T^i x_n, T^j x_n) \le \delta.$$

Indeed, by (3.239),

$$\phi(\delta) < \delta. \tag{3.243}$$

By (P2) and (3.243), there is a natural number k such that (P2) holds with $\varepsilon = \delta - \phi(\delta)$.

Assume now that the integers i and n satisfy $k \le i < n$. In view of the choice of k and property (P2) with $\varepsilon = \delta - \phi(\delta)$, we have

$$\rho(T^i x_n, T^{i+1} x_n) \le \delta - \phi(\delta). \tag{3.244}$$

Now let

$$x \in K \cap B(T^i x_n, \delta). \tag{3.245}$$

It follows from (3.236), (3.244) and (3.245) that

$$\begin{aligned}\rho(Tx, T^i x_n) &\le \rho(Tx, T^{i+1} x_n) + \rho(T^{i+1} x_n, T^i x_n) \le \phi(\rho(x, T^i x_n)) + \delta - \phi(\delta) \\ &\le \delta.\end{aligned}$$

Thus

$$T(K \cap B(T^i x_n, \delta)) \subset B(T^i x_n, \delta),$$

and if an integer j satisfies $i < j < n$, then $\rho(T^i x_n, T^j x_n) \le \delta$. Hence property (P3) does hold, as claimed.

Let $\varepsilon > 0$ be given. We will show that there exists a natural number k such that the following property holds:

(P4) If the integers n_1, n_2 and i satisfy $k \le i \le \min\{n_1, n_2\}$, then

$$\rho(T^i x_{n_1}, T^i x_{n_2}) \le \varepsilon.$$

Indeed, there exists a natural number k such that

$$\phi^i(2c_0) < \varepsilon \quad \text{for all integers } i \ge k. \tag{3.246}$$

Assume now that the natural numbers n_1, n_2 and i satisfy

$$k \le i \le \min\{n_1, n_2\}. \tag{3.247}$$

By (3.236), (3.240) and (3.246),

$$\rho(T^i x_{n_1}, T^i x_{n_2}) \le \phi^i(\rho(x_{n_1}, x_{n_2})) \le \phi^i(2c_0) < \varepsilon.$$

Thus property (P4) indeed holds.

Let $\varepsilon > 0$ be given. By (P4), there exists a natural number k_1 such that

$$\begin{aligned}\rho(T^i x_{n_1}, T^i x_{n_2}) \le \varepsilon/4 \quad &\text{for all integers } n_1, n_2 \ge k_1 \\ &\text{and all integers } i \text{ satisfying } k_1 \le i \le \min\{n_1, n_2\}.\end{aligned} \tag{3.248}$$

By property (P3), there exists a natural number k_2 such that

$$\rho(T^i x_n, T^j x_n) \le \varepsilon/4 \quad \text{for all natural numbers } n, i, j \text{ satisfying } k_2 \le i, j < n. \tag{3.249}$$

Assume that the natural numbers n_1, n_2, i and j satisfy

$$n_1, n_2 > k_1 + k_2, \qquad i, j \ge k_1 + k_2, \qquad i < n_1, \qquad j < n_2. \tag{3.250}$$

We claim that

$$\rho(T^i x_{n_1}, T^j x_{n_2}) \le \varepsilon.$$

By (3.238), (3.243), (3.248) and (3.250),

$$\rho(T^{k_1+k_2} x_{n_1}, T^{k_1+k_2} x_{n_2}) \le \rho(T^{k_1} x_{n_1}, T^{k_1} x_{n_2}) \le \varepsilon/4. \tag{3.251}$$

In view of (3.249) and (3.250),

$$\rho(T^{k_1+k_2} x_{n_1}, T^i x_{n_1}) \le \varepsilon/4 \quad \text{and} \quad \rho(T^{k_1+k_2} x_{n_2}, T^j x_{n_2}) \le \varepsilon/4.$$

When combined with (3.251), this implies that

$$\begin{aligned}\rho(T^i x_{n_1}, T^j x_{n_2}) &\le \rho(T^i x_{n_1}, T^{k_1+k_2} x_{n_1}) + \rho(T^{k_1+k_2} x_{n_1}, T^{k_1+k_2} x_{n_2}) \\ &\quad + \rho(T^{k_1+k_2} x_{n_2}, T^j x_{n_2}) \\ &\le \varepsilon/4 + \varepsilon/4 + \varepsilon/4 < \varepsilon.\end{aligned}$$

Thus we have shown that the following property holds:

(P5) For each $\varepsilon > 0$, there exists a natural number $k(\varepsilon)$ such that

$$\rho(T^i x_{n_1}, T^j x_{n_2}) \le \varepsilon$$

for all natural numbers $n_1, n_2 > k(\varepsilon)$, $i \in [k(\varepsilon), n_1)$ and $j \in [k(\varepsilon), n_2)$.

Consider now the sequences $\{T^{n-2} x_n\}_{n=3}^{\infty}$ and $\{T^{n-1} x_n\}_{n=3}^{\infty}$. Property (P5) implies that these sequences are Cauchy sequences and that

$$\lim_{n\to\infty} \rho(T^{n-2} x_n, T^{n-1} x_n) = 0.$$

Hence there exists $\bar{x} \in K$ such that

$$\lim_{n\to\infty} \rho(\bar{x}, T^{n-2} x_n) = \lim_{n\to\infty} \rho(\bar{x}, T^{n-1} x_n) = 0.$$

Since the mapping T is continuous, $T\bar{x} = \bar{x}$ and part (A) is proved.

(B) Since T is a Matkowski contraction, there is a natural number k such that $\phi^k(M) < \varepsilon$.

Assume that a point $x_0 \in B(\bar{x}, M)$, an integer $n \ge k$, and that $T^i x_0$ is defined for all $i = 0, \dots, n$. Then $T^i x_0 \in K$, $i = 0, \dots, n-1$, and by (3.236),

$$\rho(T^k x_0, \bar{x}) \le \phi^k(\rho(x_0, \bar{x})) \le \phi^k(M) < \varepsilon.$$

By (3.236) and (3.239), we have for $i = k, \dots, n$,

$$\rho(T^i x_0, \bar{x}) \le \rho(T^k x_0, \bar{x}) \le \varepsilon.$$

Thus part (B) of our theorem is also proved. □

3.14 Jachymski-Schröder-Stein Contractions

Suppose that (X, d) is a complete metric space, N_0 is a natural number, and $\phi : [0, \infty) \to [0, \infty)$ is a function which is upper semicontinuous from the right and satisfies $\phi(t) < t$ for all $t > 0$. We call a mapping $T : X \to X$ for which

$$\min\{d(T^i x, T^i y) : i \in \{1, \dots, N_0\}\} \le \phi(d(x, y)) \quad \text{for all } x, y \in X \tag{3.252}$$

a Jachymski-Schröder-Stein contraction (with respect to ϕ).

Condition (3.252) was introduced in [78]. Such mappings with $\phi(t) = \gamma t$ for some $\gamma \in (0, 1)$ have recently been of considerable interest [10, 78, 79, 100, 101, 174]. In this section, which is based on [161], we study general Jachymski-Schröder-Stein contractions and prove two fixed point theorems for them (Theorems 3.25 and 3.26 below). In our first result we establish convergence of iterates to a fixed point, and in the second this conclusion is strengthened to obtain uniform convergence on bounded subsets of X. This last type of convergence is useful in the study of inexact orbits [35]. Our theorems contain the (by now classical) results in [23] as well as Theorem 2 in [78]. In contrast with that theorem, in Theorem 3.25 we only assume that ϕ is upper semicontinuous from the right and we do not assume that $\liminf_{t\to\infty}(t - \phi(t)) > 0$. Moreover, our arguments are completely different from those presented in [78], where the Cantor Intersection Theorem was used. We remark in passing that Cantor's theorem was also used in this context in [65] (cf. also [68]).

Theorem 3.25 *Let (X, d) be a complete metric space and let $T : X \to X$ be a Jachymski-Schröder-Stein contraction. Assume there is $x_0 \in X$ such that T is uniformly continuous on the orbit $\{T^i x_0 : i = 1, 2, \dots\}$. Then there exists $\bar{x} = \lim_{i\to\infty} T^i x_0$ in (X, d). Moreover, if T is continuous at $\bar{x}$, then $\bar{x}$ is the unique fixed point of T.*

Proof Set

$$T^0 x = x, \quad x \in X. \tag{3.253}$$

We are going to define a sequence of nonnegative integers $\{k_i\}_{i=0}^{\infty}$ by induction. Set $k_0 = 0$. Assume that $i \ge 0$ is an integer, and that the integer $k_i \ge 0$ has already been defined. Clearly, there exists an integer k_{i+1} such that

$$1 \le k_{i+1} - k_i \le N_0 \tag{3.254}$$

and

$$d(T^{k_{i+1}} x_0, T^{k_{i+1}+1} x_0) = \min\{d(T^{j+k_i} x_0, T^{j+k_i+1} x_0) : j = 1, \dots, N_0\}. \tag{3.255}$$

By (3.252), (3.254) and (3.255), the sequence $\{d(T^{k_j} x_0, T^{k_j+1} x_0)\}_{j=0}^{\infty}$ is decreasing. Set

$$r = \lim_{j\to\infty} d(T^{k_j} x_0, T^{k_j+1} x_0). \tag{3.256}$$

Assume that $r > 0$. Then by (3.252), (3.254) and (3.255), for each integer $j \geq 0$,

$$d\big(T^{k_{j+1}}x_0, T^{k_{j+1}+1}x_0\big) \leq \phi\big(d\big(T^{k_j}x_0, T^{k_j+1}x_0\big)\big).$$

When combined with (3.256), the monotonicity of the sequence

$$\big\{d\big(T^{k_j}x_0, T^{k_j+1}x_0\big)\big\}_{j=0}^{\infty},$$

and the upper semicontinuity from the right of ϕ, this inequality implies that

$$r \leq \limsup_{j\to\infty} \phi\big(d\big(T^{k_j}x_0, T^{k_j+1}x_0\big)\big) \leq \phi(r),$$

a contradiction. Thus $r = 0$ and

$$\lim_{j\to\infty} d\big(T^{k_j}x_0, T^{k_j+1}x_0\big) = 0. \tag{3.257}$$

We claim that, in fact,

$$\lim_{i\to\infty} d\big(T^{i}x_0, T^{i+1}x_0\big) = 0.$$

Indeed, let $\varepsilon > 0$ be given. Since T is uniformly continuous on the set

$$\Omega := \big\{T^i x_0 : i = 1, 2, \dots\big\}, \tag{3.258}$$

there is

$$\varepsilon_0 \in (0, \varepsilon) \tag{3.259}$$

such that

$$\text{if } x, y \in \Omega, i \in \{1, \dots, N_0\}, d(x, y) \leq \varepsilon_0, \text{ then } d\big(T^i x, T^i y\big) \leq \varepsilon. \tag{3.260}$$

By (3.257), there is a natural number j_0 such that

$$d\big(T^{k_j}x_0, T^{k_j+1}x_0\big) \leq \varepsilon_0 \quad \text{for all integers } j \geq j_0. \tag{3.261}$$

Let p be an integer such that

$$p \geq k_{j_0} + N_0.$$

Then by (3.254) there is an integer $j \geq j_0$ such that

$$k_j < p \leq k_j + N_0. \tag{3.262}$$

By (3.261) and the inequality $j \geq j_0$,

$$d\big(T^{k_j}x_0, T^{k_j+1}x_0\big) \leq \varepsilon_0.$$

Together with (3.262) and (3.261), this implies that

$$d(T^p x_0, T^{p+1} x_0) \le \varepsilon.$$

Thus this inequality holds for any integer $p \ge k_{j_0} + N_0$ and we conclude that

$$\lim_{p\to\infty} d(T^p x_0, T^{p+1} x_0) = 0, \tag{3.263}$$

as claimed.

Now we show that $\{T^i x_0\}_{i=1}^{\infty}$ is a Cauchy sequence. Assume the contrary. Then there exists $\varepsilon > 0$ such that for each natural number p, there exist integers $m_p > n_p \ge p$ such that

$$d(T^{m_p} x_0, T^{n_p} x_0) \ge \varepsilon. \tag{3.264}$$

We may assume without loss of generality that for each natural number p,

$$d(T^i x_0, T^{n_p} x_0) < \varepsilon \quad \text{for all integers } i \text{ satisfying } n_p < i < m_p. \tag{3.265}$$

By (3.264) and (3.265), for any integer $p \ge 1$,

$$\begin{aligned}\varepsilon &\le d(T^{m_p} x_0, T^{n_p} x_0) \le d(T^{m_p} x_0, T^{m_p-1} x_0) + d(T^{m_p-1} x_0, T^{n_p} x_0)\\ &\le d(T^{m_p} x_0, T^{m_p-1} x_0) + \varepsilon.\end{aligned}$$

When combined with (3.263), this implies that

$$\lim_{p\to\infty} d(T^{m_p} x_0, T^{n_p} x_0) = \varepsilon. \tag{3.266}$$

Let $\delta > 0$ be given. By (3.263), there is an integer $p_0 \ge 1$ such that

$$d(T^{i+1} x_0, T^i x_0) \le \delta(4N_0)^{-1} \quad \text{for all integers } i \ge p_0. \tag{3.267}$$

Let $p \ge p_0$ be an integer. By (3.263), there is $j \in \{1, \dots, N_0\}$ such that

$$d(T^{m_p+j} x_0, T^{n_p+j} x_0) \le \phi(d(T^{m_p} x_0, T^{n_p} x_0)). \tag{3.268}$$

By the inequalities $m_p > n_p \ge p$, (3.267) and (3.268),

$$\begin{aligned}d(T^{m_p} x_0, T^{n_p} x_0) &\le \sum_{i=0}^{j-1} d(T^{m_p+i} x_0, T^{m_p+i+1} x_0) + d(T^{m_p+j} x_0, T^{n_p+j} x_0)\\ &\quad + \sum_{i=0}^{j-1} d(T^{n_p+i} x_0, T^{n_p+i+1} x_0)\\ &\le 2j\delta(4N_0)^{-1} + \phi(d(T^{m_p} x_0, T^{n_p} x_0))\\ &< \delta + \phi(d(T^{m_p} x_0, T^{n_p} x_0)).\end{aligned} \tag{3.269}$$

By (3.266), (3.269), (3.264), and the upper semicontinuity from the right of ϕ,

$$\varepsilon = \lim_{p\to\infty} d\big(T^{m_p}x_0, T^{n_p}x_0\big) \le \delta + \limsup_{p\to\infty} \phi\big(d\big(T^{m_p}x_0, T^{n_p}x_0\big)\big) \le \delta + \phi(\varepsilon).$$

Since δ is an arbitrary positive number, we conclude that $\varepsilon \le \phi(\varepsilon)$. The contradiction we have reached proves that $\{T^i x_0\}_{i=1}^{\infty}$ is indeed a Cauchy sequence. Set

$$\bar{x} = \lim_{i\to\infty} T^i x_0.$$

Clearly, if T is continuous, then $T\bar{x} = \bar{x}$ and $\bar{x}$ is the unique fixed point of T. Theorem 3.25 is proved. □

For each $x \in X$ and $r > 0$, set

$$B(x,r) = \big\{z \in X : \rho(x,z) \le r\big\}.$$

Theorem 3.26 *Let (X,d) be a complete metric space and let $T : X \to X$ be a Jachymski-Schröder-Stein contraction with respect to the function $\phi : [0,\infty) \to [0,\infty)$. Assume that ϕ is upper semicontinuous, T is uniformly continuous on the set $\{T^i x : i = 1,2,\dots\}$ for each $x \in X$, and that T is continuous on X. Then there exists a unique fixed point $\bar{x}$ of T such that $T^n x \to \bar{x}$ as $n \to \infty$, uniformly on bounded subsets of X.*

Proof By Theorem 3.25, T has a unique fixed point $\bar{x}$ and

$$T^n x \to \bar{x} \quad \text{as } n \to \infty \text{ for all } x \in X. \tag{3.270}$$

Let $r > 0$ be given. We claim that $T^n x \to \bar{x}$ as $n \to \infty$, uniformly on $B(\bar{x}, r)$.

Indeed, let

$$\varepsilon \in (0, r). \tag{3.271}$$

Since T is continuous, there is

$$\varepsilon_0 \in (0, \varepsilon) \tag{3.272}$$

such that

$$\text{if } x \in X, d(x,\bar{x}) \le \varepsilon_0, i \in \{1,\dots,N_0\}, \text{ then } d\big(T^i x, \bar{x}\big) \le \varepsilon. \tag{3.273}$$

Since ϕ is upper semicontinuous, there is

$$\delta \in (0, \varepsilon_0) \tag{3.274}$$

such that

$$\text{if } t \in [\varepsilon_0, r], \text{ then } t - \phi(t) \ge \delta. \tag{3.275}$$

Choose a natural number N_1 such that

$$N_1\delta > 2r. \tag{3.276}$$

Assume that

$$x \in X, \quad d(\bar{x}, x) \le r. \tag{3.277}$$

We will show that

$$d(\bar{x}, T^i x) \le \varepsilon \quad \text{for all integers } i \ge N_0 + N_0 N_1. \tag{3.278}$$

To this end, set $k_0 = 0$. Define by induction an increasing sequence of integers $\{k_i\}_{i=1}^{\infty}$ such that

$$k_{i+1} - k_i \in [1, N_0], \quad d(T^{k_i+1}x, \bar{x}) = \min\{d(T^{j+k_i}x, \bar{x}) : j \in \{1, \dots, N_0\}\}. \tag{3.279}$$

By (3.252) and (3.279), the sequence $\{d(T^{k_i}x, \bar{x})\}_{i=0}^{\infty}$ is decreasing. We claim that $d(T^{k_{N_1}}x, \bar{x}) \le \varepsilon_0$.

Assume the contrary. Then by (3.277) and (3.252),

$$r \ge d(T^{k_j}x, \bar{x}) > \varepsilon_0, \quad j = 0, \dots, N_1. \tag{3.280}$$

By (3.279), (3.252), (3.280) and (3.275), we have for $j = 0, \dots, N_1$,

$$d(T^{k_j}x, \bar{x}) - d(T^{k_j+1}x, \bar{x}) \ge d(T^{k_j}x, \bar{x}) - \phi(d(T^{k_j}x, \bar{x})) \ge \delta. \tag{3.281}$$

Together with (3.277), this implies that

$$r \ge d(T^{k_0}x, \bar{x}) - d(T^{k_{N_1+1}}x, \bar{x}) \ge \delta(N_1 + 1),$$

which contradicts (3.276). The contradiction we have reached and the monotonicity of the sequence $\{d(T^{k_j}x, \bar{x})\}_{j=0}^{\infty}$ show that there is $p \in \{0, 1, \dots, N_1\}$ such that

$$d(T^{k_j}x, \bar{x}) \le \varepsilon_0 \quad \text{for all integers } j \ge p. \tag{3.282}$$

Assume that $i \ge N_0 + N_0 N_1$ is an integer. By (3.279), there is an integer $j \ge 0$ such that

$$k_j \le i < k_{j+1}. \tag{3.283}$$

By (3.279), (3.283) and the choice of p,

$$(j+1)N_0 > i,$$

$$j + 1 > i/N_0 \ge N_1 + 1,$$

and

$$j > N_1 \ge p. \tag{3.284}$$

By (3.284) and (3.282), $d(T^{k_j}x, \bar{x}) \leq \varepsilon_0$. Together with (3.283), (3.279), (3.272) and (3.273), this inequality implies that

$$d(\bar{x}, T^i x) \leq \varepsilon,$$

as claimed. Theorem 3.26 is proved. □

3.15 Two Results on Jachymski-Schröder-Stein Contractions

Suppose that (X, d) is a complete metric space, N_0 is a natural number, and $\phi : [0, \infty) \to [0, \infty)$ is a function. In this section we continue to study Jachymski-Schröder-Stein contractions (with respect to ϕ) $T : X \to X$ for which

$$\min\{d(T^i x, T^i y) : i \in \{1, \ldots, N_0\}\} \leq \phi(d(x, y)) \quad \text{for all } x, y \in X. \tag{3.285}$$

In the previous section we studied general Jachymski-Schröder-Stein contractions, where ϕ is upper semicontinuous from the right and satisfies $\phi(t) < 1$ for all positive t. In this section, which is based on [162], we study the case where ϕ is increasing and satisfies

$$\lim_{n\to\infty} \phi(t)^n = 0 \tag{3.286}$$

for all $t > 0$. Here $\phi^n = \phi^{n-1} \circ \phi$ for all integers $n \geq 1$. This condition on ϕ originates in Matkowski's fixed point theorem [99].

More precisely, we establish two fixed point theorems (Theorems 3.27 and 3.28 below). In our first result we prove convergence of iterates to a fixed point, and in the second this conclusion is strengthened to obtain uniform convergence on bounded subsets of X.

Theorem 3.27 *Let (X, d) be a complete metric space and $T : X \to X$ be a Jachymski-Schröder-Stein contraction such that ϕ is increasing and satisfies (3.286). Let $x_0 \in X$. Assume there is $x_0 \in X$ such that T is uniformly continuous on the orbit $\{T^i x_0 : i = 1, 2, \ldots\}$. Then there exists $\bar{x} = \lim_{i\to\infty} T^i x_0$. Moreover, if T is continuous at $\bar{x}$, then $\bar{x}$ is the unique fixed point of T.*

Proof Since $\phi^n(t) \to 0$ s $n \to \infty$ for $t > 0$,

$$\phi(\varepsilon) < \varepsilon \quad \text{for any } \varepsilon > 0. \tag{3.287}$$

Set $T^0 x = x$, $x \in X$. Using induction, we now define a sequence of nonnegative integers $\{k_i\}_{i=0}^{\infty}$. Set $k_0 = 0$. Assume that $i \geq 0$ is an integer and that the integer $k_i \geq 0$ has already been defined. Clearly, by (3.286) there exists an integer k_{i+1} such that

$$1 \leq k_{i+1} - k_i \leq N_0 \tag{3.288}$$

and

$$d\big(T^{k_{i+1}}x_0, T^{k_{i+1}+1}x_0\big) = \min\big\{d\big(T^{j+k_i}x_0, T^{j+k_i+1}x_0\big) : i = 1, \dots, N_0\big\}. \quad (3.289)$$

By (3.285), (3.287), (3.288) and (3.289), the sequence $\{d(T^{k_j}x_0, T^{k_j+1}x_0)\}_{j=0}^{\infty}$ is decreasing and for any integer $i \ge 0$,

$$d\big(T^{k_{i+1}}x_0, T^{k_{i+1}+1}x_0\big) \le \phi\big(d\big(T^{k_i}x_0, T^{k_i+1}x_0\big)\big). \quad (3.290)$$

Since ϕ is indecreasing, it follows from (3.290) and (3.285) that for any integer $j \ge 1$,

$$d\big(T^{k_j}x_0, T^{k_j+1}x_0\big) \le \phi^j\big(d(x_0, Tx_0)\big) \to 0 \quad \text{as } j \to \infty.$$

Thus

$$\lim_{j\to\infty} d\big(T^{k_j}x_0, T^{k_j+1}x_0\big) = 0. \quad (3.291)$$

We claim that

$$\lim_{i\to\infty} d\big(T^i x_0, T^{i+1}x_0\big) = 0.$$

Let $\varepsilon > 0$ be given. Since T is uniformly continuous on the set

$$\Omega := \big\{T^i x_0 : i = 1, 2, \dots\big\}, \quad (3.292)$$

there is

$$\varepsilon_0 \in (0, \varepsilon) \quad (3.293)$$

such that

$$\text{if } x, y \in \Omega, i \in \{1, \dots, N_0\}, d(x, y) \le \varepsilon_0, \text{ then } d\big(T^i x, T^i y\big) \le \varepsilon. \quad (3.294)$$

By (3.291), there is a natural number j_0 such that

$$d\big(T^{k_j}x_0, T^{k_j+1}x_0\big) \le \varepsilon_0 \quad \text{for all integers } j \ge j_0. \quad (3.295)$$

Consider an integer

$$p \ge k_{j_0} + N_0. \quad (3.296)$$

Then by (3.288) and (3.296), there is an integer $j \ge j_0$ such that

$$k_j < p \le k_j + N_0. \quad (3.297)$$

By (3.295) and the inequality $j \ge j_0$, we have

$$d\big(T^{k+j}x_0, T^{k_j+1}x_0\big) \le \varepsilon_0.$$

Together with (3.294) and (3.297) this implies

$$d\big(T^p x_0, T^{p+1}x_0\big) \le \varepsilon.$$

Since this inequality holds for any integer $p \ge k_{j_0} + N_0$, we conclude that

$$\lim_{p\to\infty} d\big(T^p x_0, T^{p+1} x_0\big) = 0, \tag{3.298}$$

as claimed.

Next we show that $\{T^i x_0\}_{i=1}^{\infty}$ is a Cauchy sequence. To this end, let $\varepsilon > 0$ be given. By (3.287),

$$\phi(\varepsilon) < \varepsilon. \tag{3.299}$$

By (3.299), there exists $\varepsilon_0 > 0$ such that

$$\varepsilon_0 < \big(\varepsilon - \phi(\varepsilon)\big)4^{-1}. \tag{3.300}$$

By (3.298), there exists a natural number n_0 such that

$$\text{if the integers } i, j \ge n_0, |i - j| \le 2N_0 + 2, \text{ then } d\big(T^i x_0, T^j x_0\big) \le \varepsilon_0. \tag{3.301}$$

We show that for each pair of integers $i, j \ge n_0$,

$$d\big(T^i x_0, T^j x_0\big) \le \varepsilon.$$

Assume the contrary. Then there exist integers $p, q \ge n_0$ such that

$$d\big(T^p x_0, T^q x_0\big) > \varepsilon. \tag{3.302}$$

We may assume without loss of generality that

$$p < q.$$

We also may assume without loss of generality that

$$\text{if an integer } i \text{ satisfies } p \le i < q, \text{ then } d\big(T^i x_0, T^p x_0\big) \le \varepsilon. \tag{3.303}$$

By (3.302), (3.301) and (3.300),

$$q - p > 2N_0 + 2$$

and

$$q - N_0 > p + N_0 + 2. \tag{3.304}$$

By (3.303) and (3.304),

$$d\big(T^{q-N_0} x_0, T^p x_0\big) \le \varepsilon. \tag{3.305}$$

There is $s \in \{1, \dots, N_0\}$ such that

$$d\big(T^{q-N_0+s} x_0, T^{p+s} x_0\big) = \min\big\{d\big(T^{q-N_0+j} x_0, T^{p+j} x_0\big) : j \in \{1, \dots, N_0\}\big\}. \tag{3.306}$$

By (3.285), (3.305) and (3.306),

$$d\big(T^{q-N_0+s}x_0, T^{p+s}x_0\big) \le \phi\big(d\big(T^{q-N_0}x_0, T^p x_0\big)\big) \le \phi(\varepsilon). \tag{3.307}$$

Hence,

$$\begin{aligned} d\big(T^q x_0, T^p x_0\big) &\le d\big(T^p x_0, T^{p+s}x_0\big) \\ &\quad + d\big(T^{p+s}x_0, T^{q-N_0+s}x_0\big) + d\big(T^{q-N_0+s}x_0, T^q x_0\big) \\ &\le d\big(T^p x_0, T^{p+s}x_0\big) + \phi(\varepsilon) + d\big(T^{q-N_0+s}x_0, T^q x_0\big). \end{aligned} \tag{3.308}$$

By (3.301) and (3.304) and the choice of s,

$$d\big(T^p x_0, T^{p+s}x_0\big), d\big(T^{q-N_0+s}, T^q x_0\big) \le \varepsilon_0. \tag{3.309}$$

By (3.299), (3.300), (3.308) and (3.309),

$$d\big(T^q x_0, T^p x_0\big) \le 2\varepsilon_0 + \phi(\varepsilon) \le 2^{-1}\varepsilon + 2^{-1}\phi(\varepsilon) < \varepsilon.$$

However, the inequality above contradicts (3.302). The contradiction we have reached proves that

$$d\big(T^i x_0, T^j x_0\big) \le \varepsilon \quad \text{for all integers } i, j \ge n_0.$$

Since ε is an arbitrary positive number, we conclude that $\{T^i x_0\}_{i=1}^{\infty}$ is indeed a Cauchy sequence and there exists $\bar{x} = \lim_{i\to\infty} T^i x_0$.

Clearly, if T is continuous, then $\bar{x}$ is a fixed point of T and it is the unique fixed point of T.

This completes the proof of Theorem 3.27. □

Theorem 3.28 *Let (X, d) be a complete metric space and $T : X \to X$ be a Jachymski-Schröder-Stein contraction such that ϕ is increasing and satisfies* (3.286). *Assume that T is continuous on X and uniformly continuous on the orbit $\{T^i x : i = 1, 2, \dots\}$ for each $x \in X$. Then there exists a unique fixed point $\bar{x}$ of T and $T^n x \to \bar{x}$ as $n \to \infty$, uniformly on all bounded subsets of X.*

Proof By Theorem 3.27, there exists a unique fixed point of T. Let $r > 0$ be given. We claim that $T^n x \to \bar{x}$ as $n \to \infty$, uniformly on the ball $B(\bar{x}, r) = \{y \in X : \rho(\bar{x}, y) \le r\}$.

Indeed, let $\varepsilon \in (0, r)$. Clearly, there exists a number $\varepsilon_0 \in (0, \varepsilon)$ such that

$$\text{if } x \in X, d(x, \bar{x}) \le \varepsilon_0, i \in \{1, \dots, N_0\}, \text{ then } d\big(T^i x, \bar{x}\big) \le \varepsilon. \tag{3.310}$$

By (3.286), there is a natural number n_0 such that

$$\phi^{n_0}(r) < \varepsilon_0. \tag{3.311}$$

Let $x \in X$ satisfy $d(x, \bar{x}) \le r$. Set $k_0 = 0$. We now define by induction an increasing sequence of integers $\{k_i\}_{i=0}^{\infty}$ such that for all integers $i \ge 0$,

$$k_{i+1} - k_i \in [1, N_0],$$
$$d\big(T^{k_{i+1}}x, \bar{x}\big) = \min\big\{d\big(T^{k_i+j}x, \bar{x}\big) : j \in \{1, \dots, N_0\}\big\}. \tag{3.312}$$

By (3.312), (3.285) and (3.287), the sequence $\{d(T^{k_i}x, \bar{x})\}_{i=1}^{\infty}$ is decreasing.

For each integer $i \ge 0$,

$$d\big(T^{k_{i+1}}x, \bar{x}\big) \le \phi\big(d\big(T^{k_i}x, \bar{x}\big)\big). \tag{3.313}$$

By (3.313) and the choice of x, for each integer $m \ge 1$,

$$d\big(T^{k_m}x, \bar{x}\big) \le \phi^m\big(d(x, \bar{x})\big) \le \phi^m(r).$$

By (3.287) and (3.311), for each integer $m \ge n_0$,

$$d\big(T^{k_m}x, \bar{x}\big) \le \phi^m(r) \le \phi^{n_0}(r) < \varepsilon_0. \tag{3.314}$$

Assume now that $i \ge N_0(n_0 + 2)$ is an integer. By (3.312), there is an integer $j \ge 0$ such that

$$k_j \le i < k_{j+1}. \tag{3.315}$$

By (3.312) and (3.315),

$$(j+1)N_0 > i, \qquad j + 1 > iN_0^{-1} \ge n_0 + 2, \qquad j > n_0.$$

Together with (3.314) this implies that

$$d\big(T^{k_j}x, \bar{x}\big) < \varepsilon_0.$$

When combined with (3.315), (3.312) and (3.310), this implies that

$$d\big(T^i x, \bar{x}\big) > \varepsilon.$$

Theorem 3.28 is proved. □

Chapter 4
Dynamical Systems with Convex Lyapunov Functions

4.1 Minimization of Convex Functionals

In this section, which is based on [128], we consider a metric space of sequences of continuous mappings acting on a bounded, closed and convex subset of a Banach space, which share a common convex Lyapunov function. We show that for a generic sequence taken from that space the values of the Lyapunov function along all trajectories tend to its infimum.

Assume that $(X, \|\cdot\|)$ is a Banach space with norm $\|\cdot\|$, $K \subset X$ is a bounded, closed and convex subset of X, and $f : K \to R^1$ is a convex and uniformly continuous function. Set

$$\inf(f) = \inf\{f(x) : x \in K\}.$$

Observe that this infimum is finite because K is bounded and f is uniformly continuous. We consider the topological subspace $K \subset X$ with the relative topology. Denote by $\mathcal{A}$ the set of all continuous self-mappings $A : K \to K$ such that

$$f(Ax) \le f(x) \quad \text{for all } x \in K. \tag{4.1}$$

Later in this chapter (see Sect. 4.4), we construct many such mappings.

For the set $\mathcal{A}$ we define a metric $\rho : \mathcal{A} \times \mathcal{A} \to R^1$ by

$$\rho(A, B) = \sup\{\|Ax - Bx\| : x \in K\}, \quad A, B \in \mathcal{A}. \tag{4.2}$$

Clearly, the metric space $\mathcal{A}$ is complete. Denote by $\mathcal{M}$ the set of all sequences $\{A_t\}_{t=1}^{\infty} \subset \mathcal{A}$. Members $\{A_t\}_{t=1}^{\infty}$, $\{B_t\}_{t=1}^{\infty}$ and $\{C_t\}_{t=1}^{\infty}$ of $\mathcal{M}$ will occasionally be denoted by boldface **A**, **B** and **C**, respectively. For the set $\mathcal{M}$ we consider the uniformity determined by the following base:

$$E(N, \varepsilon) = \left\{\left(\{A_t\}_{t=1}^{\infty}, \{B_t\}_{t=1}^{\infty}\right) \in \mathcal{M} \times \mathcal{M} : \rho(A_t, B_t) \le \varepsilon, t = 1, \dots, N\right\},$$

S. Reich, A.J. Zaslavski, *Genericity in Nonlinear Analysis*,
Developments in Mathematics 34, DOI 10.1007/978-1-4614-9533-8_4,

where N is a natural number and $\varepsilon > 0$. Clearly the uniform space $\mathcal{M}$ is metrizable (by a metric $\rho_w : \mathcal{M} \times \mathcal{M} \to R^1$) and complete (see [80]).

From the point of view of the theory of dynamical systems, each element of $\mathcal{M}$ describes a nonstationary dynamical system with a Lyapunov function f. Also, some optimization procedures in Banach spaces can be represented by elements of $\mathcal{M}$ (see the first example in Sect. 4.4 and [97, 98]).

In this section we intend to show that for a generic sequence taken from the space $\mathcal{M}$ the values of the Lyapunov function along all trajectories tend to its infimum.

We now present the two main results of this section. They were obtained in [128]. Theorem 4.1 deals with sequences of operators (the space $\mathcal{M}$), while Theorem 4.2 is concerned with the stationary case (the space $\mathcal{A}$).

Theorem 4.1 *There exists a set $\mathcal{F} \subset \mathcal{M}$, which is a countable intersection of open and everywhere dense sets in $\mathcal{M}$, such that for each $\mathbf{B} = \{B_t\}_{t=1}^{\infty} \in \mathcal{F}$ the following assertion holds:*

For each $\varepsilon > 0$, there exist a neighborhood U of $\mathbf{B}$ in $\mathcal{M}$ and a natural number N such that for each $\mathbf{C} = \{C_t\}_{t=1}^{\infty} \in U$ and each $x \in K$,

$$f(C_N \cdots C_1 x) \leq \inf(f) + \varepsilon.$$

Theorem 4.2 *There exists a set $\mathcal{G} \subset \mathcal{A}$, which is a countable intersection of open and everywhere dense sets in $\mathcal{A}$, such that for each $B \in \mathcal{G}$ the following assertion holds:*

For each $\varepsilon > 0$, there exist a neighborhood U of B in $\mathcal{A}$ and a natural number N such that for each $C \in U$ and each $x \in K$,

$$f\left(C^N x\right) \leq \inf(f) + \varepsilon.$$

The following proposition is the key auxiliary result which will be used in the proofs of these two theorems.

Proposition 4.3 *There exists a mapping $A_* \in \mathcal{A}$ with the following property:*

Given $\varepsilon > 0$, there is $\delta(\varepsilon) > 0$ such that for each $x \in K$ satisfying $f(x) \geq \inf(f) + \varepsilon$, the inequality

$$f(A_* x) \leq f(x) - \delta(\varepsilon)$$

is true.

Remark 4.4 If there is $x_{min} \in K$ for which $f(x_{min}) = \inf(f)$, then we can set $A_*(x) = x_{min}$ for all $x \in K$.

Section 4.2 contains the proof of Proposition 4.3. Proofs of Theorems 4.1 and 4.2 are given in Sect. 4.3. Section 4.4 is devoted to two examples.

4.2 Proof of Proposition 4.3

By Remark 4.4, we may assume that

$$\{x \in K : f(x) = \inf(f)\} = \emptyset. \tag{4.3}$$

For each $x \in K$, define an integer $p(x) \geq 1$ by

$$p(x) = \min\{i : i \text{ is a natural number and } f(x) \geq \inf(f) + 2^{-i}\}. \tag{4.4}$$

By (4.3), the function $p(x)$ is well defined for all $x \in K$. Now we will define an open covering $\{V_x : x \in K\}$ of K. For each $x \in K$, there is an open neighborhood V_x of x in K such that:

$$|f(y) - f(x)| \leq 8^{-p(x)-1} \quad \text{for all } y \in V_x \tag{4.5}$$

and

$$\text{if } p(x) > 1 \text{ then } f(y) < \inf(f) + 2^{-p(x)+1} \text{ for all } y \in V_x. \tag{4.6}$$

For each $x \in K$, choose $a_x \in K$ such that

$$f(a_x) \leq \inf(f) + 2^{-p(x)-9}. \tag{4.7}$$

Clearly, $\bigcup\{V_x : x \in K\} = K$ and $\{V_x : x \in K\}$ is an open covering of K.

Lemma 4.5 *Let $x \in K$. Then for all $y \in V_x$,*

$$f(y) \geq \inf(f) + 2^{-p(x)-1} \tag{4.8}$$

and

$$|p(y) - p(x)| \leq 1. \tag{4.9}$$

Proof Let $y \in V_x$. Then (4.8) follows from (4.5) and (4.4). The definition of $p(x)$ (see (4.4)) and (4.8) imply that $p(y) \leq p(x) + 1$. Now we will show that $p(y) \geq p(x) - 1$. It is sufficient to consider the case $p(x) > 1$. Then by the definition of V_x (see (4.6)) and (4.4), $f(y) < \inf(f) + 2^{-p(x)+1}$ and $p(y) \geq p(x)$. This completes the proof of the lemma. □

Since metric spaces are paracompact, there is a continuous locally finite partition of unity $\{\phi_x\}_{x\in K}$ on K subordinated to $\{V_x\}_{x\in K}$ (namely, $\operatorname{supp}\phi_x \subset V_x$ for all $x \in K$ and $\sum_{x\in K} \phi_x(y) = 1$ for all $y \in K$).

For $y \in K$, define

$$A_* y = \sum_{x\in K} \phi_x(y) a_x. \tag{4.10}$$

Clearly, the mapping A_* is well defined, $A_*(K) \subset K$ and A_* is continuous.

Lemma 4.6 *For each* $y \in K$,

$$f(A_* y) \le f(y) - 2^{-p(y)-1}. \tag{4.11}$$

Proof Let $y \in K$. There is an open neighborhood U of y in K and $x_1, \dots, x_n \in K$ such that

$$\{x \in K : \operatorname{supp} \phi_x \cap U \neq \emptyset\} = \{x_i\}_{i=1}^n. \tag{4.12}$$

We have

$$A_* y = \sum_{i=1}^n \phi_{x_i}(y) a_{x_i}. \tag{4.13}$$

We may assume that there is an integer $m \in \{1, \dots, n\}$ such that

$$\phi_{x_i}(y) > 0 \quad \text{if and only if} \quad 1 \le i \le m. \tag{4.14}$$

By (4.12) and (4.14), $\sum_{i=1}^m \phi_{x_i}(y) = 1$. When combined with (4.13) and (4.14), this implies that

$$f(A_* y) \le \max\{f(a_{x_i}) : i = 1, \dots, m\}. \tag{4.15}$$

Let $i \in \{1, \dots, m\}$. It follows from (4.14) and Lemma 4.5 that

$$y \in \operatorname{supp} \phi_{x_i} \subset V_{x_i} \quad \text{and} \quad |p(y) - p(x_i)| \le 1. \tag{4.16}$$

By (4.7) and (4.16),

$$f(a_{x_i}) \le \inf(f) + 2^{-p(x_i)-9} \le \inf(f) + 2^{-p(y)-8}.$$

Thus, by (4.15),

$$f(A_* y) \le \inf(f) + 2^{-p(y)-8}. \tag{4.17}$$

On the other hand, by (4.4), $f(y) \ge \inf(f) + 2^{-p(y)}$. Together with (4.17) this implies (4.11). The lemma is proved. □

Completion of the proof of Proposition 4.3 Clearly, $A_* \in \mathcal{A}$. Let $\varepsilon > 0$ be given. Choose an integer $j \ge 1$ such that $2^{-j} < \varepsilon$.

Let $x \in K$ satisfy $f(x) \ge \inf(f) + \varepsilon$. Then by (4.4), $p(x) \le j$ and by Lemma 4.6,

$$f(A_* x) \le f(x) - 2^{-p(x)-1} \le f(x) - 2^{-j-1}.$$

This completes the proof of the proposition (with $\delta(\varepsilon) = 2^{-j-1}$). □

Remark 4.7 As a matter of fact, if $\varepsilon \in (0, 1)$, then the proof of Proposition 4.3 shows that it holds with $\delta(\varepsilon) = \varepsilon/4$.

4.3 Proofs of Theorems 4.1 and 4.2

Set

$$r_K = \sup\{\|x\| : x \in K\} \quad \text{and} \quad d_0 = \sup\{|f(x)| : x \in K\}. \tag{4.18}$$

Let $A_* \in \mathcal{A}$ be one of the mappings the existence of which is guaranteed by Proposition 4.3. For each $\{A_t\}_{t=1}^{\infty} \in \mathcal{M}$ and each $\gamma \in (0, 1)$, we define a sequence of mappings $A_t^{\gamma} : K \to K$, $t = 1, 2, \dots$, by

$$A_t^{\gamma} x = (1 - \gamma)A_t x + \gamma A_* x, \quad x \in K, t = 1, 2, \dots. \tag{4.19}$$

It is easy to see that for each $\{A_t\}_{t=1}^{\infty} \in \mathcal{M}$ and each $\gamma \in (0, 1)$,

$$\{A_t^{\gamma}\}_{t=1}^{\infty} \in \mathcal{M} \quad \text{and} \quad \rho(A_t^{\gamma}, A_t) \le 2\gamma r_K, \quad t = 1, 2, \dots. \tag{4.20}$$

We may assume that the function $\delta(\varepsilon)$ of Proposition 4.3 satisfies $\delta(\varepsilon) < \varepsilon$ for all $\varepsilon > 0$.

Lemma 4.8 *Assume that $\varepsilon, \gamma \in (0, 1)$, $\{A_t\}_{t=1}^{\infty} \in \mathcal{M}$ and let an integer $N \ge 4$ satisfy*

$$2^{-1}N\gamma\delta(\varepsilon) > 2d_0 + 1. \tag{4.21}$$

Then there exists a number $\Delta > 0$ such that for each sequence $\{B_t\}_{t=1}^{N} \subset \mathcal{A}$ satisfying

$$\rho(B_t, A_t^{\gamma}) \le \Delta, \quad t = 1, \dots, N, \tag{4.22}$$

it follows that, for each $x \in K$,

$$f(B_N \cdots B_1 x) \le \inf(f) + \varepsilon. \tag{4.23}$$

Proof Since the function f is uniformly continuous, there is $\Delta \in (0, 16^{-1}\delta(\varepsilon))$ such that

$$|f(y_1) - f(y_2)| \le 16^{-1}\gamma\delta(\varepsilon) \tag{4.24}$$

for each $y_1, y_2 \in K$ satisfying $\|y_1 - y_2\| \le \Delta$.

Assume that $\{B_t\}_{t=1}^{N} \subset \mathcal{A}$ satisfies (4.22) and that $x \in K$. We now show that (4.23) holds.

Assume the contrary. Then

$$f(x) > \inf(f) + \varepsilon \quad \text{and} \quad f(B_n \cdots B_1 x) > \inf(f) + \varepsilon, \quad n = 1, \dots, N. \tag{4.25}$$

Set

$$x_0 = x, \qquad x_{t+1} = B_{t+1}x_t, \quad t = 0, 1, \dots, N - 1. \tag{4.26}$$

For each $t \geq 0$ satisfying $t \leq N-1$, it follows from (4.22), (4.26) and the definition of Δ (see (4.24)) that

$$\|B_{t+1}x_t - A^{\gamma}_{t+1}x_t\| \leq \Delta \tag{4.27}$$

and

$$\begin{aligned}|f(x_{t+1}) - f(A^{\gamma}_{t+1}x_t)| &= |f(B_{t+1}x_t) - f(A^{\gamma}_{t+1}x_t)| \\ &\leq 16^{-1}\gamma\delta(\varepsilon).\end{aligned} \tag{4.28}$$

By (4.19), (4.25), (4.26), the definition of $\delta(\varepsilon)$ and the properties of the mapping A_*, we have for each $t = 0, \ldots, N-1$,

$$\begin{aligned}f(A^{\gamma}_{t+1}x_t) &= f((1-\gamma)A_{t+1}x_t + \gamma A_* x_t) \\ &\leq (1-\gamma)f(A_{t+1}x_t) + \gamma f(A_* x_t) \leq (1-\gamma)f(x_t) + \gamma\big(f(x_t) - \delta(\varepsilon)\big) \\ &= f(x_t) - \gamma\delta(\varepsilon).\end{aligned}$$

Together with (4.28) this implies that for $t = 0, \ldots, N-1$,

$$f(x_{t+1}) \leq 16^{-1}\gamma\delta(\varepsilon) + f(x_t) - \gamma\delta(\varepsilon).$$

By induction we can show that for all $t = 1, \ldots, N$,

$$f(x_t) \leq f(x_0) - 2^{-1}\gamma\delta(\varepsilon)t.$$

Together with (4.21) and (4.18) this implies that

$$\begin{aligned}f(B_N \cdots B_1 x) = f(x_N) &\leq f(x_0) - 2^{-1}N\gamma\delta(\varepsilon) \\ &\leq d_0 - 2^{-1}N\gamma\delta(\varepsilon) \leq -d_0 - 1 \leq \inf(f) - 1.\end{aligned}$$

This obvious contradiction proves (4.23) and the lemma itself. □

By Lemma 4.8, for each $\mathbf{A} = \{A_t\}_{t=1}^{\infty} \in \mathcal{M}$, each $\gamma \in (0,1)$ and each integer $q \geq 1$, there exist an integer $N(\mathbf{A}, \gamma, q) \geq 4$ and an open neighborhood $U(\mathbf{A}, \gamma, q)$ of $\{A^{\gamma}_t\}_{t=1}^{\infty}$ in $\mathcal{M}$ such that the following property holds:

(a) For each $\{B_t\}_{t=1}^{\infty} \in U(\mathbf{A}, \gamma, q)$ and each $x \in K$,

$$f(B_{N(\mathbf{A},\gamma,q)} \cdots B_1 x) \leq \inf(f) + 4^{-q}.$$

Proof of Theorem 4.1 It follows from (4.20) that the set

$$\big\{\{A^{\gamma}_t\}_{t=1}^{\infty} : \{A_t\}_{t=1}^{\infty} \in \mathcal{M}, \gamma \in (0,1)\big\}$$

is everywhere dense in $\mathcal{M}$. Define

$$\mathcal{F} = \bigcap_{q=1}^{\infty} \bigcup \big\{U(\mathbf{A}, \gamma, q) : \mathbf{A} \in \mathcal{M}, \gamma \in (0,1)\big\}.$$

Clearly, $\mathcal{F}$ is a countable intersection of open and everywhere dense sets in $\mathcal{M}$.

Assume that $\{B_t\}_{t=1}^{\infty} \in \mathcal{F}$ and that $\varepsilon > 0$. Choose an integer $q \geq 1$ such that

$$4^{-q} < \varepsilon. \tag{4.29}$$

There exist $\{A_t\}_{t=1}^{\infty} \in \mathcal{M}$ and $\gamma \in (0, 1)$ such that

$$\{B_t\}_{t=1}^{\infty} \in U\big(\{A_t\}_{t=1}^{\infty}, \gamma, q\big). \tag{4.30}$$

It follows from (4.29) and property (a) that for each $\{C_t\}_{t=1}^{\infty} \in U(\mathbf{A}, \gamma, q)$ and each $x \in K$,

$$f(C_{N(\mathbf{A},\gamma,q)} \cdots C_1 x) \leq \inf(f) + 4^{-q} < \inf(f) + \varepsilon.$$

This completes the proof of Theorem 4.1. □

Proof of Theorem 4.2 For each $A \in \mathcal{A}$, define

$$\widehat{A}_t = A, \quad t = 1, 2, \ldots. \tag{4.31}$$

Clearly, $\{\widehat{A}_t\}_{t=1}^{\infty} \in \mathcal{M}$ for $A \in \mathcal{A}$, and for each $A \in \mathcal{A}$ and each $\gamma \in (0, 1)$,

$$\widehat{A}_t^{\gamma} x = (1 - \gamma)Ax + \gamma A_* x, \quad x \in K, t = 1, 2, \ldots \tag{4.32}$$

(see (4.19)). By property (a) (which follows from Lemma 4.8), for each $A \in \mathcal{A}$, each $\gamma \in (0, 1)$ and each integer $q \geq 1$, there exist an integer $N(A, \gamma, q) \geq 4$ and an open neighborhood $U(A, \gamma, q)$ of the mapping $(1 - \gamma)A + \gamma A_*$ in $\mathcal{A}$ such that the following property holds:

(b) For each $B \in U(A, \gamma, q)$ and each $x \in K$,

$$f\big(B^{N(A,\gamma,q)} x\big) \leq \inf(f) + 4^{-q}.$$

Clearly, the set

$$\big\{(1 - \gamma)A + \gamma A_* : A \in \mathcal{A}, \gamma \in (0, 1)\big\}$$

is everywhere dense in $\mathcal{A}$. Define

$$\mathcal{G} = \bigcap_{q=1}^{\infty} \bigcup \big\{U(A, \gamma, q) : A \in \mathcal{A}, \gamma \in (0, 1)\big\}.$$

It is clear that $\mathcal{G}$ is a countable intersection of open and everywhere dense sets in $\mathcal{A}$. Assume that $B \in \mathcal{G}$ and $\varepsilon > 0$. Choose an integer $q \geq 1$ such that (4.29) is valid. There exist $A \in \mathcal{A}$ and $\gamma \in (0, 1)$ such that $B \in U(A, \gamma, q)$. It now follows from (4.29) and property (b) that for each $C \in U(A, \gamma, q)$ and each $x \in K$,

$$f\big(C^{N(A,\gamma,q)} x\big) \leq \inf(f) + 4^{-q} < \inf(f) + \varepsilon.$$

Theorem 4.2 is established. □

4.4 Examples

Let $(X, \|\cdot\|)$ be a Banach space. In this section we consider examples of continuous mappings $A : K \to K$ satisfying $f(Ax) \le f(x)$ for all $x \in K$, where K is a bounded, closed and convex subset of X and $f : K \to R^1$ is a convex function.

Example 4.9 Let $f : X \to R^1$ be a convex uniformly continuous function satisfying

$$f(x) \to \infty \quad \text{as } \|x\| \to \infty.$$

Evidently, the function f is bounded from below. For each real number c, let $K_c = \{x \in X : f(x) \le c\}$. Fix a real number c such that $K_c \ne \emptyset$. Clearly, the set K_c is bounded, closed and convex. We assume that the function f is strictly convex on K_c, namely,

$$f\big(\alpha x + (1-\alpha)y\big) < \alpha f(x) + (1-\alpha) f(y)$$

for all $x, y \in K_c$, $x \ne y$, and all $\alpha \in (0, 1)$.

Let $V : K_c \to X$ be any continuous mapping. For each $x \in K_c$, there is a unique solution of the following minimization problem:

$$f(z) \to \min, \quad z \in \big\{x + \alpha V(x) : \alpha \in [0, 1]\big\}.$$

This solution will be denoted by Ax. Since $f(Ax) \le f(x)$ for all $x \in K_c$, we conclude that $A(K_c) \subset K_c$.

We will show that the mapping $A : K_c \to K_c$ is continuous. To this end, consider a sequence $\{x_n\}_{n=1}^\infty \subset K_c$ such that $\lim_{n\to\infty} x_n = x_*$. We intend to show that $\lim_{n\to\infty} Ax_n = Ax_*$. For each integer $n \ge 1$, there is $\alpha_n \in [0, 1]$ such that $Ax_n = x_n + \alpha_n V x_n$. There is also $\alpha_* \in [0, 1]$ such that $Ax_* = x_* + \alpha_* V(x_*)$. We may assume without loss of generality that the limit $\bar{\alpha} = \lim_{n\to\infty} \alpha_n$ exists. By the definition of A,

$$f(Ax_*) \le f\big(x_* + \bar{\alpha} V(x_*)\big).$$

Since the function f is strictly convex, to complete the proof it is sufficient to show that

$$f(Ax_*) = f\big(x_* + \alpha_* V(x_*)\big) = f\big(x_* + \bar{\alpha} V(x_*)\big). \tag{4.33}$$

Assume the contrary. Then

$$\begin{aligned}\lim_{n\to\infty} f\big(x_n + \alpha_* V(x_n)\big) &= f\big(x_* + \alpha_* V(x_*)\big) \\ &< f\big(x_* + \bar{\alpha} V(x_*)\big) = \lim_{n\to\infty} f\big(x_n + \alpha_n V(x_n)\big),\end{aligned}$$

and for all large enough n,

$$f\big(x_n + \alpha_* V(x_n)\big) < f\big(x_n + \alpha_n V(x_n)\big) = f(Ax_n).$$

This contradicts the definition of A. Hence (4.33) is true and the mapping A is indeed continuous.

Example 4.10 Let K be a bounded, closed and convex subset of X and $f : K \to R^1$ be a convex continuous function which is bounded from below. For each $x_0, x_1 \in K$ satisfying $f(x_0) > f(x_1)$, we will construct a continuous mapping $A : K \to K$ such that $f(Ax) \le f(x)$ for all $x \in K$ and $Ax = x_1$ for all x in a neighborhood of x_0.

Indeed, let $x_0, x_1 \in K$ with $f(x_0) > f(x_1)$. There are numbers r_0, ε_0 such that

$$f(x) - \varepsilon_0 > f(x_1) \quad \text{for all } x \in K \text{ satisfying } \|x - x_0\| \le r_0. \tag{4.34}$$

Now we define an open covering $\{V_x : x \in K\}$ of K. Let $x \in K$. If $\|x - x_0\| < r_0$ we set

$$V_x = \left\{y \in K : \|y - x_0\| < r_0\right\} \quad \text{and} \quad a_x = x_1.$$

If $\|x - x_0\| \ge r_0$, then there is $r_x \in (0, 4^{-1}r_0)$ and $a_x \in K$ such that

$$f(a_x) \le f(y) \quad \text{for all } y \in \left\{z \in K : \|z - x\| \le r_x\right\}. \tag{4.35}$$

In this case we set

$$V_x = \left\{y \in K : \|y - x\| < r_x\right\}.$$

Clearly, $\bigcup\{V_x : x \in K\} = K$. There is a continuous locally finite partition of unity $\{\phi_x\}_{x \in K}$ on K subordinated to $\{V_x\}_{x \in K}$ (namely, $\operatorname{supp}\phi_x \subset V_x$ for all $x \in K$). For $y \in K$, define

$$Ay = \sum_{x \in K} \phi_x(y) a_x.$$

Evidently, the mapping A is well defined, $A : K \to X$ and A is continuous. Since $\sum_{x \in K} \phi_x(y) = 1$ for all $y \in K$ and K is convex, we see that $A(K) \subset K$.

We will now show that $f(Ay) \le f(y)$ for all $y \in K$ and that $Ay = x_1$ if $\|y - x_0\| \le 4^{-1}r_0$.

Let $y \in K$. There are $z_1, \dots, z_n \in K$ and a neighborhood U of y in K such that

$$\{z \in K : U \cap \operatorname{supp}\phi_z \ne \emptyset\} = \{z_1, \dots, z_n\}.$$

We have

$$Ay = \sum_{i=1}^{n} \phi_{z_i}(y) a_{z_i}, \qquad \sum_{i=1}^{n} \phi_{z_i}(y) = 1, \qquad f(Ay) \le \sum_{i=1}^{n} \phi_{z_i}(y) f(a_{z_i}). \tag{4.36}$$

We may assume without loss of generality that there is $p \in \{1, \dots, n\}$ such that

$$\phi_{z_i}(y) > 0 \quad \text{if and only if} \quad 1 \le i \le p. \tag{4.37}$$

Let $1 \le i \le p$. Then

$$y \in \operatorname{supp}\phi_{z_i} \subset V_{z_i} \tag{4.38}$$

and by the definition of V_{z_i} and a_{z_i} (see (4.34) and (4.35)), $f(y) \geq f(a_{z_i})$. When combined with (4.36) and (4.37), this implies that $f(Ay) \leq f(y)$.

Assume in addition that $\|y - x_0\| \leq 4^{-1}r_0$. Then it follows from the definition of $\{V_z : z \in K\}$ and (4.38) that $\|z_i - x_0\| < r_0$ and $a_{z_i} = x_1$ for each $i = 1, \dots, p$. By (4.36) and (4.37), $Ay = x_1$. Thus we have indeed constructed a continuous mapping $A : K \to K$ such that $f(Ay) \leq f(y)$ for all $y \in K$, and $Ay = x_1$ for all $y \in K$ satisfying $\|y - x_0\| \leq 4^{-1}r_0$.

4.5 Normal Mappings

Assume that $(X, \|\cdot\|)$ is a Banach space with norm $\|\cdot\|$, $K \subset X$ is a nonempty, bounded, closed and convex subset of X, and $f : K \to R^1$ is a convex and uniformly continuous function. Set

$$\inf(f) = \inf\{f(x) : x \in K\}.$$

Observe that this infimum is finite because K is bounded and f is uniformly continuous. We consider the topological subspace $K \subset X$ with the relative topology. Denote by $\mathcal{A}$ the set of all self-mappings $A : K \to K$ such that

$$f(Ax) \leq f(x) \quad \text{for all } x \in K \tag{4.39}$$

and by $\mathcal{A}_c$ the set of all continuous mappings $A \in \mathcal{A}$. In Sect. 4.4 we constructed many mappings which belong to $\mathcal{A}_c$.

We equip the set $\mathcal{A}$ with a metric $\rho : \mathcal{A} \times \mathcal{A} \to R^1$ defined by

$$\rho(A, B) = \sup\{\|Ax - Bx\| : x \in K\}, \quad A, B \in \mathcal{A}. \tag{4.40}$$

Clearly, the metric space $\mathcal{A}$ is complete and $\mathcal{A}_c$ is a closed subset of $\mathcal{A}$. In the sequel we will consider the metric space $(\mathcal{A}_c, \rho)$. Denote by $\mathcal{M}$ the set of all sequences $\{A_t\}_{t=1}^{\infty} \subset \mathcal{A}$ and by $\mathcal{M}_c$ the set of all sequences $\{A_t\}_{t=1}^{\infty} \subset \mathcal{A}_c$. Members $\{A_t\}_{t=1}^{\infty}$, $\{B_t\}_{t=1}^{\infty}$ and $\{C_t\}_{t=1}^{\infty}$ of $\mathcal{M}$ will occasionally be denoted by boldface **A**, **B** and **C**, respectively. For the set $\mathcal{M}$ we will consider two uniformities and the topologies induced by them. The first uniformity is determined by the following base:

$$E_w(N, \varepsilon) = \{(\{A_t\}_{t=1}^{\infty}, \{B_t\}_{t=1}^{\infty}) \in \mathcal{M} \times \mathcal{M} : \\ \rho(A_t, B_t) \leq \varepsilon, t = 1, \dots, N\}, \tag{4.41}$$

where N is a natural number and $\varepsilon > 0$. Clearly the uniform space $\mathcal{M}$ with this uniformity is metrizable (by a metric $\rho_w : \mathcal{M} \times \mathcal{M} \to R^1$) and complete (see [80]). We equip the set $\mathcal{M}$ with the topology induced by this uniformity. This topology will be called weak and denoted by τ_w. Clearly $\mathcal{M}_c$ is a closed subset of $\mathcal{M}$ with the weak topology.

The second uniformity is determined by the following base:

$$E_s(\varepsilon) = \left\{ \left(\{A_t\}_{t=1}^{\infty}, \{B_t\}_{t=1}^{\infty} \right) \in \mathcal{M} \times \mathcal{M} : \rho(A_t, B_t) \le \varepsilon, t \ge 1 \right\}, \quad (4.42)$$

where $\varepsilon > 0$. Clearly this uniformity is metrizable (by a metric $\rho_s : \mathcal{M} \times \mathcal{M} \to R^1$) and complete (see [80]). Denote by τ_s the topology induced by this uniformity in $\mathcal{M}$. Since τ_s is clearly stronger than τ_w, it will be called strong. We consider the topological subspace $\mathcal{M}_c \subset \mathcal{M}$ with the relative weak and strong topologies.

In Sects. 4.1–4.3 we showed that for a generic sequence taken from the space $\mathcal{M}_c$, the sequence of values of the Lyapunov function f along any trajectory tends to the infimum of f.

A mapping $A \in \mathcal{A}$ is called normal if given $\varepsilon > 0$, there is $\delta(\varepsilon) > 0$ such that for each $x \in K$ satisfying $f(x) \ge \inf(f) + \varepsilon$, the inequality

$$f(Ax) \le f(x) - \delta(\varepsilon)$$

is true.

A sequence $\{A_t\}_{t=1}^{\infty} \in \mathcal{M}$ is called normal if given $\varepsilon > 0$, there is $\delta(\varepsilon) > 0$ such that for each $x \in K$ satisfying $f(x) \ge \inf(f) + \varepsilon$ and each integer $t \ge 1$, the inequality

$$f(A_t x) \le f(x) - \delta(\varepsilon)$$

holds.

In this chapter we show that a generic element taken from the spaces $\mathcal{A}$, $\mathcal{A}_c$, $\mathcal{M}$ and $\mathcal{M}_c$ is normal. This is important because it turns out that the sequence of values of the Lyapunov function f along any (unrestricted) trajectory of such an element tends to the infimum of f on K.

For $\alpha \in (0, 1)$, $\mathbf{A} = \{A_t\}_{t=1}^{\infty}$, $\mathbf{B} = \{B_t\}_{t=1}^{\infty} \in \mathcal{M}$ define $\alpha\mathbf{A} + (1-\alpha)\mathbf{B} = \{\alpha A_t + (1-\alpha) B_t\}_{t=1}^{\infty} \in \mathcal{M}$.

We can easily prove the following fact.

Proposition 4.11 *Let $\alpha \in (0, 1)$, $\mathbf{A}, \mathbf{B} \in \mathcal{M}$ and let $\mathbf{A}$ be normal. Then $\alpha\mathbf{A} + (1-\alpha)\mathbf{B}$ is also normal.*

In this chapter we will prove the following results obtained in [63].

Theorem 4.12 *Let $\mathbf{A} = \{A_t\}_{t=1}^{\infty} \in \mathcal{M}$ be normal and let $\varepsilon > 0$. Then there exists a neighborhood U of $\mathbf{A}$ in $\mathcal{M}$ with the strong topology and a natural number N such that for each $\mathbf{C} = \{C_t\}_{t=1}^{\infty} \in U$, each $x \in K$ and each $r : \{1, 2, \dots\} \to \{1, 2, \dots\}$,*

$$f(C_{r(N)} \cdots C_{r(1)} x) \le \inf(f) + \varepsilon.$$

Theorem 4.13 *Let $\mathbf{A} = \{A_t\}_{t=1}^{\infty} \in \mathcal{M}$ be normal and let $\varepsilon > 0$. Then there exists a neighborhood U of $\mathbf{A}$ in $\mathcal{M}$ with the weak topology and a natural number N such that for each $\mathbf{C} = \{C_t\}_{t=1}^{\infty} \in U$ and each $x \in K$,*

$$f(C_N \cdots C_1 x) \le \inf(f) + \varepsilon.$$

Theorem 4.14 *There exists a set $\mathcal{F} \subset \mathcal{M}$ which is a countable intersection of open and everywhere dense sets in $\mathcal{M}$ with the strong topology and a set $\mathcal{F}_c \subset \mathcal{F} \cap \mathcal{M}_c$ which is a countable intersection of open and everywhere dense sets in $\mathcal{M}_c$ with the strong topology such that each $\mathbf{A} \in \mathcal{F}$ is normal.*

Theorem 4.15 *There exists a set $\mathcal{F} \subset \mathcal{A}$ which is a countable intersection of open and everywhere dense sets in $\mathcal{A}$ and a set $\mathcal{F}_c \subset \mathcal{F} \cap \mathcal{A}_c$, which is a countable intersection of open and everywhere dense sets in $\mathcal{A}_c$ such that each $\mathbf{A} \in \mathcal{F}$ is normal.*

4.6 Existence of a Normal $A \in \mathcal{A}_c$

If there is $x_{min} \in K$ for which $f(x_{min}) = \inf(f)$, then we can set $A(x) = x_{min}$ for all $x \in K$ and this A is normal. Therefore in order to show the existence of a normal $A \in \mathcal{A}_c$ we may assume that

$$\{x \in K : f(x) = \inf(f)\} = \emptyset. \tag{4.43}$$

The existence of a normal $A \in \mathcal{A}_c$ follows from Michael's selection theorem.

Proposition 4.16 *There exists a normal $A_* \in \mathcal{A}_c$.*

Proof We may assume that (4.43) is true. Define a set-valued map $a : K \to 2^K$ as follows: for each $x \in K$, denote by $a(x)$ the closure (in the norm topology of X) of the set

$$\{y \in K : f(y) < 2^{-1}(f(x) + \inf(f))\}. \tag{4.44}$$

It is clear that for each $x \in K$, the set $a(x)$ is nonempty, closed and convex. We will show that a is lower semicontinuous.

Let $x_0 \in K$, $y_0 \in a(x_0)$ and let $\varepsilon > 0$ be given. In order to prove that a is lower semicontinuous, we need to show that there exists a positive number δ such that for each $x \in K$ satisfying $\|x - x_0\| < \delta$,

$$a(x) \cap \{y \in K : \|y - y_0\| < \varepsilon\} \neq \emptyset.$$

By the definition of $a(x_0)$, there exists a point $y_1 \in K$ such that

$$f(y_1) < 2^{-1}(f(x_0) + \inf(f)) \quad \text{and} \quad \|y_1 - y_0\| < \varepsilon/2.$$

Since the function f is continuous, there is a number $\delta > 0$ such that for each $x \in K$ satisfying $\|x - x_0\| < \delta$,

$$f(y_1) < 2^{-1}(f(x) + \inf(f)).$$

Hence $y_1 \in a(x)$ by definition. Therefore a is indeed lower semicontinuous. By Michael's selection theorem, there exists a continuous mapping $A_* : K \to K$ such that $A_* x \in a(x)$ for all $x \in K$. It follows from the definition of a (see (4.44)) that for each $x \in K$,

$$f(A_* x) \leq 2^{-1}\big(f(x) + \inf(f)\big).$$

This implies that A_* is normal. This completes the proof of Proposition 4.16. □

4.7 Auxiliary Results

By Proposition 4.16, there exists a normal mapping $A_* \in \mathcal{A}_c$. For each $\{A_t\}_{t=1}^{\infty} \in \mathcal{M}$ and each $\gamma \in (0, 1)$, we define a sequence of mappings $\mathbf{A}^{\gamma} = \{A_t^{\gamma}\}_{t=1}^{\infty} \in \mathcal{M}$ by

$$A_t^{\gamma} x = (1-\gamma) A_t x + \gamma A_* x, \quad x \in K, t = 1, 2, \ldots. \tag{4.45}$$

Clearly, for each $\mathbf{A} = \{A_t\}_{t=1}^{\infty} \in \mathcal{M}_c$ and each $\gamma \in (0, 1)$, $\mathbf{A}^{\gamma} \in \mathcal{M}_c$. By (4.45) and Proposition 4.11, $\mathbf{A}^{\gamma}$ is normal for each $\mathbf{A} \in \mathcal{M}$ and each $\gamma \in (0, 1)$. It is obvious that for each $\mathbf{A} \in \mathcal{M}$,

$$\mathbf{A}^{\gamma} \to \mathbf{A} \quad \text{as } \gamma \to 0^+ \text{ in the strong topology.} \tag{4.46}$$

Lemma 4.17 *Let $\mathbf{A} = \{A_t\}_{t=1}^{\infty} \in \mathcal{M}$ be normal and let $\varepsilon > 0$ be given. Then there exist a neighborhood U of $\mathbf{A}$ in $\mathcal{M}$ with the strong topology and a number $\delta > 0$ such that for each $\mathbf{B} = \{B_t\}_{t=1}^{\infty} \in U$, each $x \in K$ satisfying*

$$f(x) \geq \inf(f) + \varepsilon \tag{4.47}$$

and each integer $t \geq 1$,

$$f(B_t x) \leq f(x) - \delta.$$

Proof Since $\mathbf{A}$ is normal, there is $\delta_0 > 0$ such that for each integer $t \geq 1$ and each $x \in K$ satisfying (4.47),

$$f(A_t x) \leq f(x) - \delta_0. \tag{4.48}$$

Since f is uniformly continuous, there is $\delta \in (0, 4^{-1}\delta_0)$ such that

$$\big|f(y) - f(z)\big| \leq 4^{-1}\delta_0 \tag{4.49}$$

for each $y, z \in K$ satisfying $\|y - z\| \leq 2\delta$. Set

$$U = \big\{\mathbf{B} \in \mathcal{M} : (\mathbf{A}, \mathbf{B}) \in E_s(\delta)\big\}. \tag{4.50}$$

Assume that $\mathbf{B} = \{B_t\}_{t=1}^{\infty} \in U$, let $t \geq 1$ be an integer and let $x \in K$ satisfy (4.47). By (4.47) and the definition of δ_0, (4.48) is true. The definitions of δ and U (see (4.49) and (4.50)) imply that

$$\|A_t x - B_t x\| \leq \delta \quad \text{and} \quad \big|f(A_t x) - f(B_t x)\big| \leq \delta_0/4.$$

When combined with (4.48), this implies that

$$f(B_t x) \le f(x) + 4^{-1}\delta_0 - \delta_0 \le f(x) - \delta.$$

This completes the proof of the lemma. □

4.8 Proof of Theorem 4.12

Assume that $\mathbf{A} = \{A_t\}_{t=1}^{\infty} \in \mathcal{M}$ is normal and let $\varepsilon > 0$ be given. By Lemma 4.17, there exist a neighborhood U of $\mathbf{A}$ in $\mathcal{M}$ with the strong topology and a number $\delta > 0$ such that the following property holds:

(Pi) For each $\{B_t\}_{t=1}^{\infty} \in U$, each integer $t \ge 1$ and each $x \in K$ satisfying (4.47), the inequality

$$f(B_t x) \le f(x) - \delta \tag{4.51}$$

holds.

Choose a natural number $N \ge 4$ such that

$$\delta N > 2(\varepsilon + 1) + 2\sup\{|f(z)| : z \in K\}. \tag{4.52}$$

Assume that

$$\mathbf{C} = \{C_t\}_{t=1}^{\infty} \in U, \qquad x \in K \quad \text{and} \quad r : \{1, 2, \ldots\} \to \{1, 2, \ldots\}. \tag{4.53}$$

We claim that

$$f(C_{r(N)} \cdots C_{r(1)} x) \le \inf(f) + \varepsilon. \tag{4.54}$$

Assume the contrary. Then

$$f(x) > \inf(f) + \varepsilon, \qquad f(C_{r(n)} \cdots C_{r(1)} x) > \inf(f) + \varepsilon, \quad n = 1, \ldots, N. \tag{4.55}$$

It follows from (4.55), (4.53) and property (Pi) that

$$f(C_{r(1)} x) \le f(x) - \delta,$$
$$f(C_{r(n+1)} C_{r(n)} \cdots C_{r(1)} x) \le f(C_{r(n)} \cdots C_{r(1)} x) - \delta, \quad n = 1, \ldots, N-1.$$

This implies that

$$f(C_{r(n)} \cdots C_{r(1)} x) \le f(x) - N\delta \le -2 - \sup\{|f(z)| : z \in K\},$$

a contradiction. Therefore (4.54) is valid and Theorem 4.12 is proved.

4.9 Proof of Theorem 4.13

Assume that $\mathbf{A} = \{A_t\}_{t=1}^{\infty} \in \mathcal{M}$ is normal and let $\varepsilon > 0$ be given. Since $\mathbf{A}$ is normal, there is $\delta \in (0, 1)$ such that for each integer $t \geq 1$ and each $x \in K$ satisfying

$$f(x) \geq \inf(f) + \varepsilon, \tag{4.56}$$

the following inequality is valid:

$$f(A_t x) \leq f(x) - \delta. \tag{4.57}$$

Choose a natural number $N > 4$ for which

$$N > 4\delta^{-1} + 4\delta^{-1} \sup\{|f(z)| : z \in K\}. \tag{4.58}$$

Since f is uniformly continuous, there is $\Delta \in (0, 4^{-1}\delta)$ such that

$$|f(z) - f(y)| \leq 8^{-1}\delta \tag{4.59}$$

for each $y, z \in K$ satisfying $\|z - y\| \leq 4\Delta$. Set

$$U = \{\mathbf{B} \in \mathcal{M} : (\mathbf{A}, \mathbf{B}) \in E_w(N, \Delta)\}. \tag{4.60}$$

Assume that

$$\mathbf{C} = \{C_t\}_{t=1}^{\infty} \in U \quad \text{and} \quad x \in K. \tag{4.61}$$

We claim that

$$f(C_N \cdots C_1 x) \leq \inf(f) + \varepsilon. \tag{4.62}$$

Assume the contrary. Then

$$f(x) > \inf(f) + \varepsilon, \qquad f(C_n \cdots C_1 x) > \inf(f) + \varepsilon, \quad n = 1, \ldots, N. \tag{4.63}$$

Define $C_0 : K \to K$ by $C_0 x = x$ for all $x \in K$. Let $t \in \{0, \ldots, N - 1\}$. It follows from (4.63) and the definition of δ (see (4.56) and (4.57)) that

$$f(A_{t+1} C_t \cdots C_0 x) \leq f(C_t \cdots C_0 x) - \delta. \tag{4.64}$$

The definition of U (see (4.60)) and (4.61) imply that $\|A_{t+1} C_t \cdots C_0 x - C_{t+1} C_t \cdots C_0 x\| \leq \Delta$. By this inequality and the definition of Δ (see (4.59)),

$$|f(A_{t+1} C_t \cdots C_0 x) - f(C_{t+1} C_t \cdots C_0 x)| \leq 8^{-1}\delta.$$

When combined with (4.64), this implies that

$$f(C_{t+1} C_t \cdots C_0 x) \leq f(C_t \cdots C_0 x) - 2^{-1}\delta.$$

Since this inequality is true for all $t \in \{0, \ldots, N-1\}$, we conclude that

$$f(C_N \cdots C_1 x) \le f(x) - 2^{-1} N\delta.$$

Together with (4.58) this implies that

$$\begin{aligned} -\sup\{|f(z)| : z \in K\} &\le \sup\{|f(z)| : z \in K\} - 2^{-1}\delta N \\ &\le -2 - \sup\{|f(z)| : z \in K\}, \end{aligned}$$

a contradiction. Therefore (4.62) does hold and Theorem 4.13 is proved.

4.10 Proof of Theorem 4.14

Let $\mathbf{A} \in \mathcal{M}$, $\gamma \in (0, 1)$ and let $i \ge 1$ be an integer. Consider the sequence $\mathbf{A}^\gamma \in \mathcal{M}$ defined by (4.45). By Proposition 4.11, $\mathbf{A}^\gamma$ is normal. By Lemma 4.17, there exists an open neighborhood $U(\mathbf{A}, \gamma, i)$ of $\mathbf{A}^\gamma$ in $\mathcal{M}$ with the strong topology and a number $\delta(\mathbf{A}, \gamma, i) > 0$ such that the following property holds:

(Pii) For each $\mathbf{B} = \{B_t\}_{t=1}^\infty \in U(\mathbf{A}, \gamma, i)$, each integer $t \ge 1$ and each $x \in K$ satisfying $f(x) \ge \inf(f) + 2^{-i}$,

$$f(B_t x) \le f(x) - \delta(\mathbf{A}, \gamma, i).$$

Define

$$\mathcal{F} = \bigcap_{i=1}^{\infty} \bigcup \{U(\mathbf{A}, \gamma, i) : \mathbf{A} \in \mathcal{M}, \gamma \in (0, 1)\} \tag{4.65}$$

and

$$\mathcal{F}_c = \left[\bigcap_{i=1}^{\infty} \bigcup \{U(\mathbf{A}, \gamma, i) : \mathbf{A} \in \mathcal{M}_c, \gamma \in (0, 1)\} \right] \cap \mathcal{M}_c.$$

Clearly, $\mathcal{F}_c \subset \mathcal{F}$, $\mathcal{F}$ is a countable intersection of open and everywhere dense sets in $\mathcal{M}$ with the strong topology, and $\mathcal{F}_c$ is a countable intersection of open and everywhere dense sets in $\mathcal{M}_c$ with the strong topology.

Assume that $\mathbf{B} = \{B_t\}_{t=1}^\infty \in \mathcal{F}$. We will show that $\mathbf{B}$ is normal.

Let $\varepsilon > 0$ be given. Choose an integer $i \ge 1$ such that

$$2^{-i} < \varepsilon/8. \tag{4.66}$$

By (4.65), there exist $\mathbf{A} \in \mathcal{M}$ and $\gamma \in (0, 1)$ such that

$$\mathbf{B} \in U(\mathbf{A}, \gamma, i). \tag{4.67}$$

Let $t \geq 1$ be an integer, $x \in K$, and $f(x) \geq \inf(f) + \varepsilon$. Then by (4.66), (4.67) and property (Pii),

$$f(B_t x) \leq f(x) - \delta(\mathbf{A}, \gamma, i).$$

Thus $\mathbf{B}$ is indeed normal and Theorem 4.14 is proved.

The proof of Theorem 4.15 is analogous to that of Theorem 4.14.

4.11 Normality and Porosity

In this section, which is based on [133], we continue to consider a complete metric space of sequences of mappings acting on a bounded, closed and convex subset K of a Banach space which share a common convex Lyapunov function f. In previous sections, we introduced the concept of normality and showed that a generic element taken from this space is normal. The sequence of values of the Lyapunov uniformly continuous function f along any (unrestricted) trajectory of such an element tends to the infimum of f on K. In the present section, we first present a convergence result for perturbations of such trajectories. We then show that if f is Lipschitzian, then the complement of the set of normal sequences is σ-porous.

Assume that $(X, \|\cdot\|)$ is a Banach space with norm $\|\cdot\|$, $K \subset X$ is a nonempty, bounded, closed and convex subset of X, and $f : K \to R^1$ is a convex and uniformly continuous function. Observe that the function f is bounded because K is bounded and f is uniformly continuous. Set

$$\inf(f) = \inf\{f(x) : x \in K\} \quad \text{and} \quad \sup(f) = \sup\{f(x) : x \in K\}.$$

We consider the topological subspace $K \subset X$ with the relative topology. Denote by $\mathcal{A}$ the set of all self-mappings $A : K \to K$ such that

$$f(Ax) \leq f(x) \quad \text{for all } x \in K$$

and by $\mathcal{A}_c$ the set of all continuous mappings $A \in \mathcal{A}$.

For the set $\mathcal{A}$ we define a metric $\rho : \mathcal{A} \times \mathcal{A} \to R^1$ by

$$\rho(A, B) = \sup\{\|Ax - Bx\| : x \in K\}, \quad A, B \in \mathcal{A}.$$

It is clear that the metric space $\mathcal{A}$ is complete and $\mathcal{A}_c$ is a closed subset of $\mathcal{A}$. We will study the metric space $(\mathcal{A}_c, \rho)$. Denote by $\mathcal{M}$ the set of all sequences $\{A_t\}_{t=1}^{\infty} \subset \mathcal{A}$ and by $\mathcal{M}_c$ the set of all sequences $\{A_t\}_{t=1}^{\infty} \subset \mathcal{A}_c$. For the set $\mathcal{M}$ we define a metric $\rho_{\mathcal{M}} : \mathcal{M} \times \mathcal{M} \to R^1$ by

$$\rho_{\mathcal{M}}(\{A_t\}_{t=1}^{\infty}, \{B_t\}_{t=1}^{\infty}) = \sup\{\rho(A_t, B_t) : t = 1, 2, \ldots\}, \quad \{A_t\}_{t=1}^{\infty}, \{B_t\}_{t=1}^{\infty} \in \mathcal{M}.$$

Clearly, the metric space $\mathcal{M}$ is complete and $\mathcal{M}_c$ is a closed subset of $\mathcal{M}$. We will also study the metric space $(\mathcal{M}_c, \rho_{\mathcal{M}})$.

We recall the following definition of normality.

A mapping $A \in \mathcal{A}$ is called normal if given $\varepsilon > 0$, there is $\delta(\varepsilon) > 0$ such that for each $x \in K$ satisfying $f(x) \geq \inf(f) + \varepsilon$, the inequality

$$f(Ax) \leq f(x) - \delta(\varepsilon)$$

is true.

A sequence $\{A_t\}_{t=1}^{\infty} \in \mathcal{M}$ is called normal if given $\varepsilon > 0$, there is $\delta(\varepsilon) > 0$ such that for each $x \in K$ satisfying $f(x) \geq \inf(f) + \varepsilon$ and each integer $t \geq 1$, the inequality

$$f(A_t x) \leq f(x) - \delta(\varepsilon)$$

holds.

We now present two theorems which were obtained in [133]. Their proofs are given in the next two sections.

Theorem 4.18 *Let $\{A_t\}_{t=1}^{\infty} \in \mathcal{M}$ be normal and let ε be positive. Then there exist a natural number n_0 and a number $\gamma > 0$ such that for each integer $n \geq n_0$, each mapping $r : \{1, \dots, n\} \to \{1, 2, \dots\}$ and each sequence $\{x_i\}_{i=0}^{n} \subset K$ which satisfies*

$$\|x_{i+1} - A_{r(i+1)} x_i\| \leq \gamma, \quad i = 0, \dots, n-1,$$

the inequality $f(x_i) \leq \inf(f) + \varepsilon$ holds for $i = n_0, \dots, n$.

Theorem 4.19 *Let $\mathcal{F}$ be the set of all normal sequences in the space $\mathcal{M}$ and let*

$$F = \big\{A \in \mathcal{A} : \{A_t\}_{t=1}^{\infty} \in \mathcal{F} \text{ where } A_t = A, t = 1, 2, \dots\big\}.$$

Assume that the function f is Lipschitzian. Then the complement of the set $\mathcal{F}$ is a σ-porous subset of $\mathcal{M}$ and the complement of the set $\mathcal{F} \cap \mathcal{M}_c$ is a σ-porous subset of $\mathcal{M}_c$. Moreover, the complement of the set F is a σ-porous subset of $\mathcal{A}$ and the complement of the set $F \cap \mathcal{A}_c$ is a σ-porous subset of $\mathcal{A}_c$.

4.12 Proof of Theorem 4.18

We may assume that $\varepsilon < 1$. Since $\{A_t\}_{t=1}^{\infty}$ is normal, there exists a function $\delta : (0, \infty) \to (0, \infty)$ such that for each $s > 0$, each $x \in K$ satisfying $f(x) \geq \inf(f) + s$ and each integer $t \geq 1$,

$$f(A_t x) \leq f(x) - \delta(s). \tag{4.68}$$

We may assume that $\delta(s) < s$, $s \in (0, \infty)$. Choose a natural number

$$n_0 > 4\big(1 + \sup(f) - \inf(f)\big)\delta\big(8^{-1}\varepsilon\big)^{-1}. \tag{4.69}$$

Since f is uniformly continuous, there exists a number $\gamma > 0$ such that for each $y_1, y_2 \in K$ satisfying $\|y_1 - y_2\| \leq \gamma$, the following inequality holds:

$$\left|f(y_1) - f(y_2)\right| \leq \delta\left(8^{-1}\varepsilon\right)8^{-1}(n_0+1)^{-1}. \tag{4.70}$$

We claim that the following assertion is true:

(A) Suppose that

$$\{x_i\}_{i=0}^{n_0} \subset K, r : \{1, \ldots, n_0\} \to \{1, 2, \ldots\},$$
$$\|x_{i+1} - A_{r(i+1)}x_i\| \leq \gamma, \quad i = 0, \ldots, n_0 - 1. \tag{4.71}$$

Then there exists an integer $n_1 \in \{1, \ldots, n_0\}$ such that

$$f(x_{n_1}) \leq \inf(f) + \varepsilon/8. \tag{4.72}$$

Assume the contrary. Then

$$f(x_i) > \inf(f) + \varepsilon/8, \quad i = 1, \ldots, n_0. \tag{4.73}$$

By (4.73) and the definition of $\delta : (0,\infty) \to (0,\infty)$ (see (4.68)), for each $i = 1, \ldots, n_0 - 1$, we have

$$f(A_{r(i+1)}x_i) \leq f(x_i) - \delta\left(8^{-1}\varepsilon\right). \tag{4.74}$$

It follows from (4.71) and the definition of γ (see (4.70)) that for $i = 1, \ldots, n_0 - 1$,

$$\left|f(x_{i+1}) - f(A_{r(i+1)}x_i)\right| \leq \delta\left(8^{-1}\varepsilon\right)8^{-1}(n_0+1)^{-1}.$$

When combined with (4.74), this inequality implies that for $i = 1, \ldots, n_0 - 1$,

$$f(x_{i+1}) - f(x_i) \leq f(x_{i+1}) - f(A_{r(i+1)}x_i) + f(A_{r(i+1)}x_i) - f(x_i)$$
$$\leq \delta\left(8^{-1}\varepsilon\right)8^{-1}(n_0+1)^{-1} - \delta\left(8^{-1}\varepsilon\right) \leq (-1/2)\delta\left(8^{-1}\varepsilon\right).$$

This, in turn, implies that

$$\inf(f) - \sup(f) \leq f(x_{n_0}) - f(x_1) \leq (n_0 - 1)(-1/2)\delta\left(8^{-1}\varepsilon\right),$$

a contradiction (see (4.69)). Thus there exists an integer $n_1 \in \{1, \ldots, n_0\}$ such that (4.72) is true. Therefore assertion (A) is valid, as claimed.

Assume now that we are given an integer $n \geq n_0$, a mapping

$$r : \{1, \ldots, n\} \to \{1, 2, \ldots\} \tag{4.75}$$

and a finite sequence

$$\{x_i\}_{i=0}^{n} \subset K \quad \text{such that} \quad \|x_{i+1} - A_{r(i+1)}x_i\| \leq \gamma, \quad i = 0, \ldots, n-1. \tag{4.76}$$

It follows from assertion (A) that there exists a finite sequence of natural numbers $\{j_p\}_{p=1}^q$ such that

$$\begin{aligned} &1 \le j_1 \le n_0, \qquad 1 \le j_{p+1} - j_p \le n_0 \quad \text{if } 1 \le p \le q-1, n - j_q < n_0, \\ &f(x_{j_p}) \le \inf(f) + \varepsilon/8, \quad p = 1, \dots, q. \end{aligned} \tag{4.77}$$

Let $i \in \{n_0, \dots, n\}$. We will show that $f(x_i) \le \inf(f) + \varepsilon/2$. There exists $p \in \{1, \dots, q\}$ such that

$$0 \le i - j_p \le n_0.$$

If $i = j_p$, then by (4.77), $f(x_i) = f(x_{j_p}) \le \inf(f) + \varepsilon/8$. Thus we may assume that $i > j_p$. For all integers $j_p \le s < i$, it follows from (4.76) and the definition of γ (see (4.70)) that

$$\begin{aligned} &f(A_{r(s+1)}x_s) \le f(x_s), \\ &\left| f(x_{s+1}) - f(A_{r(s+1)}x_s) \right| \le \delta(8^{-1}\varepsilon)8^{-1}(n_0+1)^{-1} \end{aligned}$$

and

$$\begin{aligned} f(x_{s+1}) &\le f(A_{r(s+1)}x_s) + \delta(8^{-1}\varepsilon)8^{-1}(n_0+1)^{-1} \\ &\le f(x_s) + \delta(8^{-1}\varepsilon)8^{-1}(n_0+1)^{-1}. \end{aligned}$$

Thus

$$f(x_{s+1}) - f(x_s) \le \delta(8^{-1}\varepsilon)8^{-1}(n_0+1)^{-1}, \quad j_p \le s < i.$$

This implies that

$$\begin{aligned} f(x_i) &\le f(x_{j_p}) + \delta(8^{-1}\varepsilon)8^{-1}(n_0+1)^{-1}(n_0+1) \\ &\le \inf(f) + \varepsilon/8 + 8^{-1}\delta(8^{-1}\varepsilon) \le \inf(f) + \varepsilon/2. \end{aligned}$$

Therefore $f(x_i) \le \inf(f) + \varepsilon/2$ for all integers $i \in [n_0, n]$ and Theorem 4.18 is proved.

4.13 Proof of Theorem 4.19

Since $f : K \to R^1$ is assumed to be Lipschitzian, there exists a constant $L(f) > 0$ such that

$$\left| f(x) - f(y) \right| \le L(f)\|x - y\| \quad \text{for all } x, y \in K. \tag{4.78}$$

By Proposition 4.16, there exist a normal continuous mapping $A_* : K \to K$ and a function $\phi : (0, \infty) \to (0, \infty)$ such that for each $\varepsilon > 0$ and each $x \in K$ satisfying $f(x) \ge \inf(f) + \varepsilon$, the inequality $f(A_*x) \le f(x) - \phi(\varepsilon)$ holds.

Let $\varepsilon > 0$ be given. We say that a sequence $\{A_t\}_{t=1}^{\infty} \in \mathcal{M}$ is (ε)-quasinormal if there exists $\delta > 0$ such that if $x \in K$ satisfies $f(x) \geq \inf(f) + \varepsilon$, then $f(A_t x) \leq f(x) - \delta$ for all integers $t \geq 1$.

Recall that $\mathcal{F}$ is defined to be the set of all normal sequences in $\mathcal{M}$. For each integer $n \geq 1$, denote by $\mathcal{F}_n$ the set of all (n^{-1})-quasinormal sequences in $\mathcal{M}$. Clearly,

$$\mathcal{F} = \bigcap_{n=1}^{\infty} \mathcal{F}_n. \tag{4.79}$$

Set

$$d(K) = \sup\{\|z\| : z \in K\}. \tag{4.80}$$

Let $n \geq 1$ be an integer. Choose $\alpha \in (0, 1)$ such that

$$2L(f)\alpha < (1-\alpha)\phi(n^{-1})8^{-1}(d(K)+1)^{-1}. \tag{4.81}$$

Assume that $0 < r \leq 1$ and $\{A_t\}_{t=1}^{\infty} \in \mathcal{M}$. Set

$$\gamma = (1-\alpha)r8^{-1}(d(K)+1)^{-1} \tag{4.82}$$

and define for each integer $t \geq 1$, the mapping $A_{t\gamma} : K \to K$ by

$$A_{t\gamma}x = (1-\gamma)A_t x + \gamma A_* x, \quad x \in K. \tag{4.83}$$

It is clear that $\{A_{t\gamma}\}_{t=1}^{\infty} \in \mathcal{M}$ and

$$\rho_{\mathcal{M}}\big(\{A_t\}_{t=1}^{\infty}, \{A_{t\gamma}\}_{t=1}^{\infty}\big) \leq 2\gamma \sup\{\|z\| : z \in K\} \leq 2\gamma d(K). \tag{4.84}$$

Note that $\{A_{t\gamma}\}_{t=1}^{\infty} \in \mathcal{M}_c$ if $\{A_t\}_{t=1}^{\infty} \in \mathcal{M}_c$ and that $A_{t\gamma} = A_{1\gamma}$, $t = 1, 2, \ldots$, if $A_t = A_1$, $t = 1, 2, \ldots$.

Assume that

$$\{C_t\}_{t=1}^{\infty} \in \mathcal{M} \quad \text{and} \quad \rho_{\mathcal{M}}\big(\{A_{t\gamma}\}_{t=1}^{\infty}, \{C_t\}_{t=1}^{\infty}\big) \leq \alpha r. \tag{4.85}$$

Then by (4.85), (4.84) and (4.82),

$$\begin{aligned}\rho_{\mathcal{M}}\big(\{A_t\}_{t=1}^{\infty}, \{C_t\}_{t=1}^{\infty}\big) &\leq \alpha r + 2\gamma d(K) \leq \alpha r + (1-\alpha)r/2 \\ &= r(1+\alpha)/2 < r. \end{aligned} \tag{4.86}$$

Assume now that $x \in K$ satisfies

$$f(x) \geq \inf(f) + n^{-1} \tag{4.87}$$

and that $t \geq 1$ is an integer. By (4.87), the properties of A_* and ϕ, and (4.83),

$$\begin{aligned} &f(A_* x) \leq f(x) - \phi(n^{-1}), \\ f(A_{t\gamma}x) &\leq (1-\gamma)f(A_t x) + \gamma f(A_* x) \\ &\leq (1-\gamma)f(x) + \gamma\big(f(x) - \phi(n^{-1})\big) = f(x) - \gamma\phi(n^{-1}). \end{aligned} \tag{4.88}$$

By (4.85), $\|C_t x - A_{t\gamma} x\| \leq \alpha r$. Together with (4.78) this inequality yields

$$\left|f(C_t x) - f(A_{t\gamma} x)\right| \leq L(f)\alpha r.$$

By the latter inequality, (4.88), (4.82) and (4.81),

$$\begin{aligned} f(C_t x) &\leq f(A_{t\gamma} x) + L(f)\alpha r \\ &\leq L(f)\alpha r + f(x) - \gamma\phi\left(n^{-1}\right) \\ &\leq f(x) - \phi\left(n^{-1}\right)(1-\alpha) r 8^{-1}\left(d(K)+1\right)^{-1} + L(f)\alpha r \\ &\leq f(x) - L(f)\alpha r. \end{aligned}$$

Thus for each $\{C_t\}_{t=1}^{\infty} \in \mathcal{M}$ satisfying (4.85), inequalities (4.86) hold and $\{C_t\}_{t=1}^{\infty} \in \mathcal{F}_n$. Summing up, we have shown that for each integer $n \geq 1$, $\mathcal{M} \setminus \mathcal{F}_n$ is porous in $\mathcal{M}$, $\mathcal{M}_c \setminus \mathcal{F}_n$ is porous in $\mathcal{M}_c$, the complement of the set

$$\left\{A \in \mathcal{A} : \{A_t\}_{t=1}^{\infty} \in \mathcal{F}_n \text{ with } A_t = A \text{ for all integers } t \geq 1\right\}$$

is porous in $\mathcal{A}$ and the complement of the set

$$\left\{A \in \mathcal{A}_c : \{A_t\}_{t=1}^{\infty} \in \mathcal{F}_n \text{ with } A_t = A \text{ for all integers } t \geq 1\right\}$$

is porous in $\mathcal{A}_c$.

Combining these facts with (4.79), we conclude that $\mathcal{M} \setminus \mathcal{F}$ is σ-porous in $\mathcal{M}$, $\mathcal{M}_c \setminus \mathcal{F}$ is σ-porous in $\mathcal{M}_c$, $\mathcal{A} \setminus F$ is σ-porous in $\mathcal{A}$ and $\mathcal{A}_c \setminus F$ is σ-porous in $\mathcal{A}_c$. This completes the proof of Theorem 4.19.

4.14 Convex Functions Possessing a Sharp Minimum

In this section, which is based on the paper [7], we are given a convex, Lipschitz function f, defined on a bounded, closed and convex subset K of a Banach space X, which possesses a sharp minimum. A minimization algorithm is a self-mapping $A : K \to K$ such that $f(Ax) \leq f(x)$ for all $x \in K$. We show that for most of these algorithms A, the sequences $\{A^n x\}_{n=1}^{\infty}$ tend to this sharp minimum (at an exponential rate) for all initial values $x \in K$.

Let $K \subset X$ be a nonempty, bounded, closed and convex subset of a Banach space X. For each $A : K \to X$, set

$$\operatorname{Lip}(A) = \sup\left\{\|Ax - Ay\|/\|x - y\| : x, y \in K \text{ such that } x \neq y\right\}. \tag{4.89}$$

Assume that $f : K \to R^1$ is a convex, Lipschitz function such that $\operatorname{Lip}(f) > 0$. We have

$$\left|f(x) - f(y)\right| \leq \operatorname{Lip}(f)\|x - y\| \quad \text{for all } x, y \in K.$$

Assume further that there exists a point $x_* \in K$ and a number $c_0 > 0$ such that

$$\inf(f) := \inf\{f(x) : x \in K\} = f(x_*)$$

and

$$f(x) \geq f(x_*) + c_0\|x - x_*\| \quad \text{for all } x \in K. \tag{4.90}$$

In other words, we assume that the function f possesses a sharp minimum (cf. [26, 109]).

Denote by $\mathcal{A}$ the set of all self-mappings $A : K \to K$ such that $\mathrm{Lip}(A) < \infty$ and

$$f(Ax) \leq f(x) \quad \text{for all } x \in K. \tag{4.91}$$

We equip the set $\mathcal{A}$ with the uniformity determined by the base

$$\mathcal{E}(\varepsilon) = \{(A, B) \in \mathcal{A} \times \mathcal{A} : \|Ax - Bx\| \leq \varepsilon \text{ for all } x \in K \text{ and } \mathrm{Lip}(A - B) \leq \varepsilon\},$$

where $\varepsilon > 0$. Clearly, the uniform space $\mathcal{A}$ is metrizable and complete.

Theorem 4.20 *There exists an open and everywhere dense subset $\mathcal{B} \subset \mathcal{A}$ such that for each $B \in \mathcal{B}$, there exist an open neighborhood $\mathcal{U}$ of B in $\mathcal{A}$ and a number $\lambda_0 \in (0, 1)$ such that for each $C \in \mathcal{U}$, each $x \in K$, and each natural number n,*

$$\|C^n x - x_*\| \leq c_0^{-1}\lambda^n(f(x) - f(x_*)).$$

Proof Let $\gamma \in (0, 1)$ and $A \in \mathcal{A}$ be given. Set

$$A_\gamma x = (1 - \gamma)Ax + \gamma x_*, \quad x \in K. \tag{4.92}$$

Clearly, for all $x \in K$,

$$f(A_\gamma x) \leq (1 - \gamma)f(Ax) + \gamma f(x_*) \tag{4.93}$$

and

$$A_\gamma \in \mathcal{A}. \tag{4.94}$$

Next, we prove the following lemma.

Lemma 4.21 *Let $A \in \mathcal{A}$, $\gamma \in (0, 1)$ and $B \in \mathcal{A}$. Then for each $x \in K$,*

$$f(Bx) - f(x_*) \leq [(1 - \gamma) + \mathrm{Lip}(f)\,\mathrm{Lip}(B - A_\gamma)c_0^{-1}](f(x) - f(x_*)).$$

Proof Let $x \in K$. By (4.93), the relations $A_\gamma x_* = Bx_* = x_*$ and (4.90),

$$\begin{aligned} f(Bx) - f(x_*) &= f(A_\gamma x) - f(x_*) + f(Bx) - f(A_\gamma x) \\ &\leq (1 - \gamma)(f(x) - f(x_*)) + \mathrm{Lip}(f)\|Bx - A_\gamma x\| \end{aligned}$$

$$\begin{aligned}
&\leq (1-\gamma)\big(f(x)-f(x_*)\big)+\mathrm{Lip}(f)\,\mathrm{Lip}(B-A_\gamma)\|x-x_*\| \\
&\leq (1-\gamma)\big(f(x)-f(x_*)\big) \\
&\quad +\mathrm{Lip}(f)\,\mathrm{Lip}(B-A_\gamma)c_0^{-1}\big(f(x)-f(x_*)\big) \\
&\leq \big[(1-\gamma)+\mathrm{Lip}(f)\,\mathrm{Lip}(B-A_\gamma)c_0^{-1}\big]\big(f(x)-f(x_*)\big).
\end{aligned}$$

The lemma is proved. □

Completion of the proof of Theorem 4.20 Let $A \in \mathcal{A}$ and $\gamma \in (0, 1)$ be given. Choose $r(\gamma) > 0$ such that

$$\lambda_\gamma := (1-\gamma)+\mathrm{Lip}(f)r(\gamma)c_0^{-1} < 1. \tag{4.95}$$

Denote by $\mathcal{U}(A, \gamma)$ the open neighborhood of A_γ in $\mathcal{A}$ such that

$$\mathcal{U}(A,\gamma) \subset \big\{B \in \mathcal{A} : (A_\gamma, B) \in \mathcal{E}\big(r(\gamma)\big)\big\}. \tag{4.96}$$

Set

$$\mathcal{B} = \bigcup\big\{\mathcal{U}(A,\gamma) : A \in \mathcal{A}, \gamma \in (0,1)\big\}. \tag{4.97}$$

Clearly, we have for each $A \in \mathcal{A}$,

$$A_\gamma \to A \quad \text{as } \gamma \to 0^+.$$

Therefore $\mathcal{B}$ is an everywhere dense, open subset of $\mathcal{A}$. Let $B \in \mathcal{A}$. There are $A \in \mathcal{A}$ and $\gamma \in (0, 1)$ such that

$$B \in \mathcal{U}(A,\gamma). \tag{4.98}$$

Assume that

$$C \in \mathcal{U}(A,\gamma) \quad \text{and} \quad x \in K. \tag{4.99}$$

By Lemma 4.21, (4.99), (4.96) and (4.95),

$$\begin{aligned}
f(Cx)-f(x_*) &\leq \big[(1-\gamma)+\mathrm{Lip}(f)\,\mathrm{Lip}(C-A_\gamma)c_0^{-1}\big]\big(f(x)-f(x_*)\big) \\
&\leq \lambda_\gamma\big(f(x)-f(x_*)\big).
\end{aligned}$$

This implies that for each $x \in K$ and each natural number n,

$$f\big(C^n x\big)-f(x_*) \leq \lambda_\gamma^n\big(f(x)-f(x_*)\big).$$

When combined with (4.90), this last inequality implies, in its turn, that for each $x \in K$ and each integer $n \geq 1$,

$$\big\|C^n x - x_*\big\| \leq c_0^{-1}\big(f\big(C^n x\big)-f(x_*)\big) \leq c_0^{-1}\lambda_\gamma^n\big(f(x)-f(x_*)\big).$$

This completes the proof of Theorem 4.20. □

Chapter 5
Relatively Nonexpansive Operators with Respect to Bregman Distances

5.1 Power Convergence of Operators in Banach Spaces

The following problem often occurs in functional analysis and optimization theory, as well as in other fields of pure and applied mathematics: given a nonempty, closed and convex subset K of a Banach space X and an operator $T : K \to K$, do the sequences iteratively generated in K by the rule $x^{k+1} = Tx^k$ converge to a fixed point of T no matter how the initial point $x^0 \in K$ is chosen? It is well known that this indeed happens, in some sense, for "standard" classes of operators (e.g., certain nonexpansive operators and operators of contractive type which were studied in Chaps. 2 and 3, and in [24, 68]). Note that in [27, 38, 122] it was shown that the question asked above has an affirmative answer even if the operator T is not contractive in any standard sense, but still satisfies some requirements which make the orbits of T behave like the orbits of contractive operators. A careful analysis shows that the operators discussed in these papers share the following property:

There exists a convex function $f : X \to R^1 \cup \{\infty\}$ such that K is a subset of the interior of

$$\mathrm{dom}(f) = \{x \in X : f(x) < \infty\},$$

and for some $z_T \in K$, we have

$$D_f(z_T, Tx) \le D_f(z_T, x) \tag{5.1}$$

for all $x \in K$, where $D_f : \mathcal{D} \times \mathcal{D}^0 \to [0, \infty)$ denotes the *Bregman distance* [37, 39] with respect to f (here $\mathcal{D} = \mathrm{dom}(f)$ and $\mathcal{D}^0$ is the interior of $\mathcal{D}$) defined by

$$D_f(y, x) = f(y) - f(x) + f^0(x, x - y), \tag{5.2}$$

where

$$f^0(x, v) = \lim_{t \to 0^+} t^{-1}\big(f(x + tv) - f(x)\big). \tag{5.3}$$

Operators satisfying (5.1) will be called nonexpansive with respect to f in the sequel. In general, operators which are nonexpansive with respect to some totally

S. Reich, A.J. Zaslavski, *Genericity in Nonlinear Analysis*, Developments in Mathematics 34, DOI 10.1007/978-1-4614-9533-8_5,

convex function f are not nonexpansive in the usual sense of the term, that is, they do not necessarily satisfy the condition

$$\|Tx - Ty\| \le \|x - y\|, \tag{5.4}$$

or even the condition $D_f(Tx, Ty) \le D_f(x, y)$ for all $x, y \in K$. Examples of this phenomenon can be found in [28]. Also, it may happen that the orbits of an operator T which is nonexpansive with respect to some convex function f are not convergent or do not converge to fixed points of T, although such an operator T must have fixed points (z_T is a fixed point of T because of (5.1)). Moreover, even if all the orbits of T converge to fixed points of T, it may happen that the limits of these orbits are not equal to the point z_T in (5.1). For instance, take $X = R^1$, $f(x) = x^2$, $K = [0, 1]$ and $Tx = x^2$. Then T is nonexpansive with respect to f and $z_T = 0$ satisfies (5.1). However, the orbit of T starting at $x^0 = 1$ converges to 1 (a fixed point of T which does not satisfy (5.1)).

The convergence of orbits of significant classes of operators satisfying (5.1) was studied because of its importance in optimization theory and in other fields. Our aim in this chapter is to show that strong convergence is not the exception, but the rule. More precisely, we show that in appropriate complete metric spaces of operators which are nonexpansive with respect to a uniformly convex function f, there exists a subset which is a countable intersection of open and everywhere dense sets such that for any operator belonging to this subset, all its orbits converge strongly.

The practical meaning of our results is that whenever one applies iterative algorithms of the form $x^{k+1} = Tx^k$ to compute a fixed point of an operator T, then there is a good chance that the convergence of the resulting sequence $\{x^k\}$ is actually strong. This conclusion is consistent with many computational experiments despite the fact that the study of particular classes of operators T satisfying (5.1) has not yet produced general strong convergence theorems.

5.2 Power Convergence for a Class of Continuous Mappings

Let $(X, \|\cdot\|)$ be a Banach space, $K \subset X$ a nonempty, closed and convex subset of X, and let $f : X \to R^1 \cup \{\infty\}$ be convex. Let $\mathcal{D}$ be the domain of f and let $D_f : \mathcal{D} \times \mathcal{D}^0 \to [0, \infty)$ denote the Bregman distance with respect to f defined by (5.2). We assume in the sequel that $K \subset \mathcal{D}^0$.

Denote by $\mathcal{M}$ the set of all mappings $T : K \to K$ which are bounded on bounded subsets of K. For the set $\mathcal{M}$ we consider the uniformity determined by the following base:

$$E(N, \varepsilon) = \big\{(T_1, T_2) \in \mathcal{M} \times \mathcal{M} : \|T_1x - T_2x\| \le \varepsilon \text{ for all } x \in K \text{ satisfying } \|x\| \le N\big\}, \tag{5.5}$$

where $N, \varepsilon > 0$. Clearly, this uniform space is metrizable and complete. We equip the space $\mathcal{M}$ with the topology induced by this uniformity. Denote by $\mathcal{M}_c$ the set of all continuous $T \in \mathcal{M}$. Clearly, $\mathcal{M}_c$ is a closed subset of $\mathcal{M}$. We consider the topological subspace $\mathcal{M}_c \subset \mathcal{M}$ with the relative topology.

Denote by $\mathcal{M}_0$ the set of all $T \in \mathcal{M}_c$ for which there is $z_T \in K$ such that the following assumptions hold:

A(i)

$$Tz_T = z_T, \qquad D_f(z_T, \cdot) : K \to R^1 \text{ is convex,}$$

$$D_f(z_T, Tx) \leq D_f(z_T, x) \quad \text{for all } x \in K;$$

A(ii) for any $\varepsilon > 0$, there exists $\delta > 0$ such that if $x \in K$ and $D_f(z_T, x) \leq \delta$, then $\|z_T - x\| \leq \varepsilon$;

A(iii) $D_f(z_T, \cdot) : K \to R^1$ is Lipschitzian in a neighborhood of z_T.

Denote by $\bar{\mathcal{M}}_0$ the closure of $\mathcal{M}_0$ in $\mathcal{M}$. We consider the topological subspace $\bar{\mathcal{M}}_0 \subset \mathcal{M}$ with the relative topology.

Note that **A**(iii) holds if the function $D_f(z_T, \cdot) : \mathcal{D}^0 \to R^1$ is convex. Note also that **A**(ii) holds if the function f is uniformly convex. Examples of such functions f can be found in [28]. Let $\xi \in K$ be given. Denote by $\mathcal{M}_{0,\xi}$ the set of all $T \in \mathcal{M}_0$ such that Assumption **A** holds with $z_T = \xi$ and denote by $\bar{\mathcal{M}}_{0,\xi}$ the closure of $\mathcal{M}_{0,\xi}$ in $\mathcal{M}$. We consider the topological subspace $\bar{\mathcal{M}}_{0,\xi} \subset \mathcal{M}$ with the relative topology.

In this chapter we prove the following six results, which were obtained in [30].

Theorem 5.1 *Let $x_j \in K$, $j = 1, \dots, p$, where p is a natural number. Then there exists a set $\mathcal{F} \subset \bar{\mathcal{M}}_0$, which is a countable intersection of open and everywhere dense sets in $\bar{\mathcal{M}}_0$ such that for each $T \in \mathcal{F}$, the following assertions hold:*

1. *There exists $z_* \in K$ such that $T^n x_j \to z_*$ as $n \to \infty$ for each $j = 1, \dots, p$.*
2. *For each $\varepsilon > 0$, there exist an integer $N \geq 1$, a neighborhood $\mathcal{U}$ of T in $\mathcal{M}$ and neighborhoods V_j of x_j in K for $j = 1, \dots, p$ such that for $j = 1, \dots, p$,*

$$\left\|S^n y - z_*\right\| \leq \varepsilon \quad \text{for each } S \in \mathcal{U}, \text{ each } y \in V_j \text{ and each integer } n \geq N.$$

Theorem 5.2 *Let $\xi \in K$ and $x_j \in K$, $j = 1, \dots, p$, where p is a natural number. Then there exists a set $\mathcal{F}_\xi \subset \bar{\mathcal{M}}_{0,\xi}$ which is a countable intersection of open and everywhere dense sets in $\bar{\mathcal{M}}_{0,\xi}$ such that for each $T \in \mathcal{F}_\xi$,*

$$\lim_{n \to \infty} T^n x_j = \xi, \quad j = 1, \dots, p,$$

and the following assertion holds:

For each $\varepsilon > 0$, there exist an integer $N \geq 1$, a neighborhood $\mathcal{U}$ of T in $\mathcal{M}$ and neighborhoods V_j of x_j in K for $j = 1, \dots, p$ such that for $j = 1, \dots, p$,

$$\left\|S^n y - \xi\right\| \leq \varepsilon \quad \text{for each } S \in \mathcal{U}, \text{ each } y \in V_j \text{ and each integer } n \geq N.$$

Let $\xi \in K$ be given. We equip the topological spaces $K \times \mathcal{M}$, $K \times \bar{\mathcal{M}}_0$ and $K \times \bar{\mathcal{M}}_{0,\xi}$ with the appropriate product topologies.

Theorem 5.3 *There exists a set $\mathcal{F} \subset K \times \bar{\mathcal{M}}_0$ which is a countable intersection of open and everywhere dense sets in $K \times \bar{\mathcal{M}}_0$ such that for each $(x, T) \in \mathcal{F}$, there exists $\lim_{n\to\infty} T^n x$ and the following assertion holds:*

For each $\varepsilon > 0$, there exists an integer $N \geq 1$ and a neighborhood $\mathcal{U}$ of (x, T) in $K \times \mathcal{M}$ such that for each $(y, S) \in \mathcal{U}$ and each integer $i \geq N$,

$$\left\| S^i y - \lim_{n\to\infty} T^n x \right\| \leq \varepsilon.$$

Our next theorem is an analog of Theorem 5.3 for the space $\bar{\mathcal{M}}_{0,\xi} \subset \bar{\mathcal{M}}_0$.

Theorem 5.4 *Let $\xi \in K$ be given. Then there exists a set $\mathcal{F}_\xi \subset K \times \bar{\mathcal{M}}_{0,\xi}$ which is a countable intersection of open and everywhere dense sets in $K \times \bar{\mathcal{M}}_{0,\xi}$ such that for each $(x, T) \in \mathcal{F}_\xi$, $\lim_{n\to\infty} T^n x = \xi$ and the following assertion holds:*

For each $\varepsilon > 0$, there exists an integer $N \geq 1$ and a neighborhood $\mathcal{U}$ of (x, T) in $K \times \mathcal{M}$ such that for each $(y, S) \in \mathcal{U}$ and each integer $i \geq N$,

$$\| S^i y - \xi \| \leq \varepsilon.$$

Theorem 5.5 *Let K_0 be a nonempty, separable and closed subset of K. Then there exists a set $\mathcal{F} \subset \bar{\mathcal{M}}_0$ which is a countable intersection of open and everywhere dense sets in $\bar{\mathcal{M}}_0$ such that for each $B \in \mathcal{F}$, there exist $x_B \in K$ and a set $\mathcal{K}_B \subset K_0$ which is a countable intersection of open and everywhere dense sets in K_0 with the relative topology such that the following assertions hold:*

1. *$\lim_{n\to\infty} B^n x = x_B$ for each $x \in \mathcal{K}_B$.*
2. *For each $x \in \mathcal{K}_B$ and each $\varepsilon > 0$, there exist an integer $N \geq 1$ and a neighborhood $\mathcal{U}$ of (x, B) in $K \times \mathcal{M}$ such that for each $(y, S) \in \mathcal{U}$ and each integer $i \geq N$,*

$$\| S^i y - x_B \| \leq \varepsilon.$$

Theorem 5.6 *Let K_0 be a nonempty, separable and closed subset of K, and let $\xi \in K$ be given. Then there exists a set $\mathcal{F}_\xi \subset \bar{\mathcal{M}}_{0,\xi}$ which is a countable intersection of open and everywhere dense sets in $\bar{\mathcal{M}}_{0,\xi}$ such that for each $B \in \mathcal{F}_\xi$, there exists a set $\mathcal{K}_B \subset K_0$ which is a countable intersection of open and everywhere dense sets in K_0 with the relative topology such that the following assertions hold:*

1. *$\lim_{n\to\infty} B^n x = \xi$ for each $x \in \mathcal{K}_B$.*
2. *For each $x \in \mathcal{K}_B$ and each $\varepsilon > 0$, there exist an integer $N \geq 1$ and a neighborhood $\mathcal{U}$ of (x, B) in $K \times \mathcal{M}$ such that for each $(y, S) \in \mathcal{U}$ and each integer $i \geq N$,*

$$\| S^i y - \xi \| \leq \varepsilon.$$

5.3 Preliminary Lemmata for Theorems 5.1–5.6

With the notions and notations of Sects. 5.1 and 5.2, assume that $T \in \mathcal{M}$, $z_T \in K$,

$$\begin{aligned} Tz_T = z_T, \qquad D_f(z_T, \cdot) : K \to R^1 \text{ is convex}, \\ D_f(z_T, Tx) \le D_f(z_T, x) \quad \text{for all } x \in K, \end{aligned} \tag{5.6}$$

and that

$$\begin{aligned} &\text{for any } \varepsilon > 0, \text{ there exists } \delta > 0 \text{ such that if } x \in K \text{ and} \\ &D_f(z_T, x) \le \delta, \text{ then } \|z_T - x\| \le \varepsilon. \end{aligned} \tag{5.7}$$

For any $\gamma \in (0, 1)$, define a mapping $T_\gamma : K \to K$ by

$$T_\gamma x = \gamma z_T + (1 - \gamma)Tx, \quad x \in K. \tag{5.8}$$

Clearly, for each $\gamma \in (0, 1)$,

$$T_\gamma \in \mathcal{M} \text{ and if } T \in \mathcal{M}_c, \text{ then } T_\gamma \in \mathcal{M}_c, \tag{5.9}$$

and

$$T_\gamma \to T \quad \text{in } \mathcal{M} \text{ as } \gamma \to 0^+. \tag{5.10}$$

Lemma 5.7 *Let $\gamma \in (0, 1)$ be given. Then $T_\gamma z_T = z_T$ and*

$$D_f(z_T, T_\gamma x) \le (1 - \gamma)D_f(z_T, x) \quad \textit{for all } x \in K. \tag{5.11}$$

Proof Evidently, $T_\gamma z_T = z_T$. Assume that $x \in K$. Then by (5.8) and (5.6),

$$\begin{aligned} D_f(z_T, T_\gamma x) &= D_f\big(z_T, \gamma z_T + (1 - \gamma)Tx\big) \\ &\le \gamma D_f(z_T, z_T) + (1 - \gamma)D_f(z_T, Tx) \le (1 - \gamma)D_f(z_T, x), \end{aligned}$$

as claimed. □

Lemma 5.8 *Assume that the function $D_f(z_T, \cdot) : K \to R^1$ is Lipschitzian in a neighborhood of z_T. Let $\varepsilon, \gamma \in (0, 1)$. Then there exist a number $\delta \in (0, \varepsilon)$ and a neighborhood $\mathcal{U}$ of T_γ in $\mathcal{M}$ such that for each $S \in \mathcal{U}$ and each $x \in K$ satisfying $\|x - z_T\| \le \delta$, the inequality $\|S^n x - z_T\| \le \varepsilon$ holds for all integers $n \ge 0$ (note that $S^0 x = x$).*

Proof We may assume without loss of generality that there is $c_0 > 1$ such that

$$\big|D_f(z_T, y_1) - D_f(z_T, y_2)\big| \le c_0\|y_1 - y_2\| \tag{5.12}$$

for each y_1 and $y_2 \in K$ satisfying

$$\|y_i - z_T\| \le 8\varepsilon, \quad i = 1, 2. \tag{5.13}$$

By (5.7), there exists $\Delta \in (0, \varepsilon)$ such that

$$\|z_T - y\| \le \varepsilon \quad \text{for each } y \in K \text{ satisfying } D_f(z_T, y) \le \Delta. \tag{5.14}$$

Choose a positive number

$$\delta < (4c_0)^{-1}(1-\gamma)\gamma\Delta \tag{5.15}$$

and set

$$\mathcal{U} = \{S \in \mathcal{M} : \|Sy - T_\gamma y\| \le \delta \text{ for all } y \in K \\ \text{satisfying } \|y - z_T\| \le 4\}. \tag{5.16}$$

Assume that

$$S \in \mathcal{U}, \qquad x \in K \quad \text{and} \quad \|x - z_T\| \le \delta. \tag{5.17}$$

We intend to show that

$$\|S^n x - z_T\| \le \varepsilon \quad \text{for all integers } n \ge 1. \tag{5.18}$$

By (5.14), in order to prove (5.18), it is sufficient to show that

$$D_f(z_T, S^n x) \le \Delta \quad \text{for all integers } n \ge 1. \tag{5.19}$$

It follows from (5.17), (5.12), (5.13) and (5.15) that

$$D_f(z_T, x) \le D_f(z_T, z_T) + c_0\|z_T - x\| \le c_0\delta < \Delta. \tag{5.20}$$

Assume that (5.19) is not true. Then by (5.20), there exists an integer $m \ge 0$ such that

$$D_f(z_T, S^i x) \le \Delta, \quad i = 0, \ldots, m, \quad \text{and} \quad D_f(z_T, S^{m+1}x) > \Delta. \tag{5.21}$$

Inequalities (5.21) and (5.14) imply that

$$\|z_T - S^m x\| \le \varepsilon. \tag{5.22}$$

By Lemma 5.7,

$$D_f(z_T, T_\gamma(S^m x)) \le (1-\gamma)D_f(z_T, S^m x). \tag{5.23}$$

It follows from (5.23) and (5.21) that $D_f(z_T, T_\gamma(S^m x)) \le \Delta$. Together with (5.14) this inequality implies that

$$\|z_T - T_\gamma(S^m x)\| \le \varepsilon. \tag{5.24}$$

Note that (5.22), (5.16) and (5.17) imply that

$$\|T_\gamma(S^m x) - S^{m+1}x\| \le \delta. \tag{5.25}$$

When combined with (5.24) and (5.15), this inequality implies that

$$\left\| z_T - S^{m+1}x \right\| \le \varepsilon + \delta \le 2\varepsilon. \tag{5.26}$$

By (5.24), (5.26), (5.12), (5.13) and (5.25),

$$\left| D_f\left(z_T, T_\gamma\left(S^m x\right)\right) - D_f\left(z_T, S^{m+1}x\right) \right| \le c_0 \left\| T_\gamma\left(S^m x\right) - S^{m+1}x \right\| \le c_0\delta. \tag{5.27}$$

It follows from (5.27) and (5.23) that

$$D_f\left(z_T, S^{m+1}x\right) \le c_0\delta + D_f\left(z_T, T_\gamma\left(S^m x\right)\right) \le c_0\delta + (1-\gamma) D_f\left(z_T, S^m x\right). \tag{5.28}$$

There are two cases: (i) $D_f(z_T, S^m x) \le 2^{-1}\Delta$; (ii) $D_f(z_T, S^m x) > 2^{-1}\Delta$.

Consider first case (i). Then by (5.28) and (5.15),

$$D_f\left(z_T, S^{m+1}x\right) \le c_0\delta + 2^{-1}(1-\gamma)\Delta < \Delta,$$

a contradiction (see (5.21)).

Consider now case (ii). Then by (5.28) and (5.15),

$$D_f\left(z_T, S^m x\right) - D_f\left(z_T, S^{m+1}x\right) \ge \gamma D_f\left(z_T, S^m x\right) - c_0\delta \ge 2^{-1}\gamma\Delta - c_0\delta > 0,$$

$$\text{so that} \quad D_f\left(z_T, S^{m+1}x\right) < D_f\left(z_T, S^m x\right),$$

a contradiction (see (5.21)). Thus in both cases we have reached a contradiction. Therefore (5.19) is valid and (5.18) is also true. This completes the proof of Lemma 5.8. □

Lemma 5.9 *Assume that the mapping $T: K \to K$ is continuous and that $\gamma > 0$. Then for each $x \in K$, each $\varepsilon > 0$ and each integer $n \ge 1$, there exist a number $\delta > 0$ and a neighborhood $\mathcal{U}$ of T_γ in $\mathcal{M}$ such that for each $S \in \mathcal{U}$ and each $y \in K$ satisfying $\|y - x\| \le \delta$, the inequality $\|(T_\gamma)^n x - S^n y\| \le \varepsilon$ holds.*

Proof We prove this lemma by induction. It is clear that for $n = 1$ it is valid. Assume that $m \ge 1$ is an integer and that the lemma is true for $n = m$. We will show that it is also true for $n = m + 1$.

Let $x \in K$ and $\varepsilon > 0$ be given. Since the lemma is true for $n = 1$, there are a neighborhood $\mathcal{U}_0$ of T_γ in $\mathcal{M}$ and a number $\delta_0 > 0$ such that for each $y \in K$ satisfying $\|y - (T_\gamma)^m x\| \le \delta_0$ and each $S \in \mathcal{U}_0$, the following inequality holds:

$$\left\| Sy - T_\gamma\left((T_\gamma)^m x\right) \right\| \le 4^{-1}\varepsilon. \tag{5.29}$$

Since we assume that the lemma is true for $n = m$, there exist a number $\delta > 0$ and a neighborhood $\mathcal{U}$ of T_γ in $\mathcal{M}$ such that $\mathcal{U} \subset \mathcal{U}_0$ and for each $y \in K$ satisfying $\|y - x\| \le \delta$ and each $S \in \mathcal{U}$, the inequality

$$\left\| S^m y - (T_\gamma)^m x \right\| \le 2^{-1}\delta_0 \tag{5.30}$$

is true.

Assume that

$$S \in \mathcal{U}, \quad y \in K \quad \text{and} \quad \|y - x\| \le \delta. \tag{5.31}$$

By (5.31) and the definition of $\mathcal{U}$, (5.30) is true. By (3.30) and the definition of $\mathcal{U}_0$ and δ_0 (see (5.29)),

$$\big\|S\big(S^m y\big) - T_\gamma\big((T_\gamma)^m y\big)\big\| \le 4^{-1}\varepsilon. \tag{5.32}$$

Therefore for each $S \in \mathcal{U}$ and each $y \in K$ satisfying (5.31), inequality (5.32) holds. Thus the lemma is true for $n = m + 1$. This completes the proof of Lemma 5.9. □

Lemma 5.10 *Assume that the function $D_f(z_T, \cdot) : K \to R^1$ is Lipschitzian in a neighborhood of z_T and that the mapping $T : K \to K$ is continuous. Let $\gamma, \varepsilon \in (0, 1)$ and $x \in K$ be given. Then there exist a neighborhood $\mathcal{U}$ of T_γ in $\mathcal{M}$, a number $\delta > 0$ and an integer $N \ge 1$ such that for each $y \in K$ satisfying $\|y - x\| \le \delta$, each $S \in \mathcal{U}$ and each integer $n \ge N$,*

$$\big\|S^n y - z_T\big\| \le \varepsilon.$$

Proof By Lemma 5.8, there are a number $\Delta \in (0, \varepsilon)$ and a neighborhood $\mathcal{U}_0$ of T_γ in $\mathcal{M}$ such that the following property holds:

A(i) For each $S \in \mathcal{U}_0$ and each $y \in K$ satisfying $\|y - z_T\| \le \Delta$, the following relation holds:

$$\big\|S^n y - z_T\big\| \le \varepsilon \quad \text{for all integers } n \ge 1.$$

By (5.7), there is $\delta_0 > 0$ such that

$$\|z_T - y\| \le 4^{-1}\Delta \quad \text{if } y \in K \text{ and } D_f(z_T, y) \le \delta_0. \tag{5.33}$$

By Lemma 5.7, there exists an integer $N \ge 1$ such that

$$D_f\big(z_T, (T_\gamma)^N x\big) \le \delta_0.$$

When combined with (5.33), this inequality implies that

$$\big\|z_T - (T_\gamma)^N x\big\| \le 4^{-1}\Delta. \tag{5.34}$$

By Lemma 5.9, there exist a neighborhood $\mathcal{U} \subset \mathcal{U}_0$ of T_γ in $\mathcal{M}$ and a number $\delta > 0$ such that for each $y \in K$ satisfying $\|y - x\| \le \delta$ and each $S \in \mathcal{U}$,

$$\big\|S^N y - (T_\gamma)^N x\big\| \le 4^{-1}\Delta.$$

By the definition of $\mathcal{U}$, δ and (5.34), the following property holds:

A(ii) For each $y \in K$ satisfying $\|y - x\| \le \delta$ and each $S \in \mathcal{U}$,

$$\big\|S^N y - z_T\big\| \le 2^{-1}\Delta.$$

By properties A(i) and A(ii), for each $S \in \mathcal{U}$ and each $y \in K$ satisfying $\|y - x\| \le \delta$,

$$\|S^n y - z_T\| \le \varepsilon \quad \text{for all integers } n \ge N.$$

Lemma 5.10 is proved. □

5.4 Proofs of Theorems 5.1–5.6

Proofs of Theorems 5.1 and 5.2 Let $\xi, x_j \in K$, $j = 1, \ldots, p$, where p is a natural number. With each $T \in \mathcal{M}_0$, we associate a point $z_T \in K$ satisfying Assumption **A**. If $T \in \mathcal{M}_{0,\xi}$, then $z_T = \xi$.

By Lemma 5.7, for each $T \in \mathcal{M}_0$ and each $\gamma \in (0, 1)$, $T_\gamma \in \mathcal{M}_0$ and $T_\gamma \in \mathcal{M}_{0,\xi}$ if $T \in \mathcal{M}_{0,\xi}$.

By Lemma 5.10, for each $T \in \mathcal{M}_0$, each $\gamma \in (0, 1)$ and each integer $i \ge 1$, there exist a natural number $N(T, \gamma, i)$, a real number $\delta(T, \gamma, i) > 0$ and an open neighborhood $\mathcal{U}(T, \gamma, i)$ of T_γ in $\mathcal{M}$ such that the following property holds:

C(i) For each $S \in \mathcal{U}(T, \gamma, i)$, each $j \in \{1, \ldots, p\}$, each $y \in K$ satisfying $\|y - x_j\| \le \delta(N, \gamma, i)$ and each integer $n \ge N(T, \gamma, i)$,

$$\|S^n y - z_T\| \le 2^{-i}.$$

Define

$$\mathcal{F} = \left[\bigcap_{q=1}^{\infty} \bigcup \{\mathcal{U}(T, \gamma, i) : T \in \mathcal{M}_0, \gamma \in (0, 1), i = q, q + 1, \ldots\}\right] \cap \bar{\mathcal{M}}_0,$$

$$\mathcal{F}_\xi = \left[\bigcap_{q=1}^{\infty} \bigcup \{\mathcal{U}(T, \gamma, i) : T \in \mathcal{M}_{0,\xi}, \gamma \in (0, 1), i = q, q + 1, \ldots\}\right] \cap \bar{\mathcal{M}}_{0,\xi}.$$

It is clear that $\mathcal{F}$ (respectively, $\mathcal{F}_\xi$) is a countable intersection of open and everywhere dense sets in $\bar{\mathcal{M}}_0$ (respectively, $\bar{\mathcal{M}}_{0,\xi}$), and that $\mathcal{F}_\xi \subset \mathcal{F}$.

Let $B \in \mathcal{F}$ and $\varepsilon > 0$ be given. Choose an integer $q \ge 1$ such that

$$2^{-q} < 4^{-1}\varepsilon. \tag{5.35}$$

There exist $T \in \mathcal{M}_0$ ($T \in \mathcal{M}_{0,\xi}$ if $B \in \mathcal{F}_\xi$), $\gamma \in (0, 1)$ and an integer $i \ge q$ such that

$$B \in \mathcal{U}(T, \gamma, i). \tag{5.36}$$

It follows from property C(i) and (5.35) that the following property holds:

C(ii) For each $S \in \mathcal{U}(T, \gamma, i)$, each $j \in \{1, \ldots, p\}$, each $y \in K$ satisfying $\|y - x_j\| \le \delta(N, \gamma, i)$ and each integer $n \ge N(T, \gamma, i)$,

$$\|S^n y - z_T\| \le 4^{-1}\varepsilon. \tag{5.37}$$

Since ε is an arbitrary positive number, we conclude that for each $j = 1, \dots, p$, $\{B^n x_j\}_{n=1}^{\infty}$ is a Cauchy sequence and there exists $\lim_{n\to\infty} B^n x_j$. Inequality (5.37) implies that

$$\left\| \lim_{n\to\infty} B^n x_j - z_T \right\| \le 4^{-1}\varepsilon, \quad j = 1, \dots, p. \tag{5.38}$$

(If $B \in \mathcal{F}_\xi$, then $T \in \mathcal{M}_{0,\xi}$, $z_T = \xi$ and since (5.38) holds for any $\varepsilon > 0$, we see that $\lim_{n\to\infty} B^n x_j = \xi$, $j = 1, \dots, p$.)

Since (5.38) holds for any $\varepsilon > 0$, we conclude that

$$\lim_{n\to\infty} B^n x_j = \lim_{n\to\infty} B^n x_1, \quad j = 1, \dots, p.$$

It follows from property C(ii) and (5.38) that for each $S \in \mathcal{U}(T, \gamma, i)$, each $j = 1, \dots, p$, each $y \in K$ satisfying $\|y - x_j\| \le \delta(N, \gamma, i)$ and each integer $r \ge N(T, \gamma, i)$,

$$\left\| S^r y - \lim_{n\to\infty} B^n x_j \right\| \le 2^{-1}\varepsilon.$$

This completes the proofs of Theorems 5.1 and 5.2. □

Now we are going to show that Theorems 5.3 and 5.4 are also true.

Proofs of Theorems 5.3 and 5.4 With each $T \in \mathcal{M}_0$ we associate a point $z_T \in K$ satisfying Assumption **A**. If $T \in \mathcal{M}_{0,\xi}$, then $z_T = \xi$. By Lemma 5.7, for each $T \in \mathcal{M}_0$ and each $\gamma \in (0, 1)$, $T_\gamma \in \mathcal{M}_0$ and $T_\gamma \in \mathcal{M}_{0,\xi}$ if $T \in \mathcal{M}_{0,\xi}$. By Lemma 5.10, for each $(x, T) \in K \times \mathcal{M}_0$, each $\gamma \in (0, 1)$ and each integer $i \ge 1$, there exist an integer $N(x, T, \gamma, i) \ge 1$ and an open neighborhood $\mathcal{U}(x, T, \gamma, i)$ of (x, T_γ) in $K \times \mathcal{M}$ such that the following property holds:

C(iii) For each $(y, S) \in \mathcal{U}(x, T, \gamma, i)$ and each integer $n \ge N(x, T, \gamma, i)$,

$$\|S^n y - z_T\| \le 2^{-i}.$$

Define

$$\mathcal{F} = \left[\bigcap_{q=1}^{\infty} \bigcup \{\mathcal{U}(x, T, \gamma, i) : x \in K, T \in \mathcal{M}_0, \right.$$

$$\left. \gamma \in (0, 1), i = q, q+1, \dots\} \right] \cap (K \times \bar{\mathcal{M}}_0),$$

$$\mathcal{F}_\xi = \left[\bigcap_{q=1}^{\infty} \bigcup \{\mathcal{U}(x, T, \gamma, i) : x \in K, T \in \mathcal{M}_{0,\xi}, \right.$$

$$\left. \gamma \in (0, 1), i = q, q+1, \dots\} \right] \cap (K \times \bar{\mathcal{M}}_{0,\xi}).$$

Clearly, $\mathcal{F}$ (respectively, $\mathcal{F}_\xi$) is a countable intersection of open and everywhere dense sets in $K \times \bar{\mathcal{M}}_0$ (respectively, $K \times \bar{\mathcal{M}}_{0,\xi}$) and $\mathcal{F}_\xi \subset \mathcal{F}$.

Let $(z, B) \in \mathcal{F}$ and $\varepsilon > 0$ be given. Choose an integer $q \geq 1$ such that

$$2^{-q} < 4^{-1}\varepsilon. \tag{5.39}$$

There exist $x \in K$, $T \in \mathcal{M}_0$ ($T \in \mathcal{M}_{0,\xi}$ if $(z, B) \in \mathcal{F}_\xi$), $\gamma \in (0, 1)$ and an integer $i \geq q$ such that

$$(z, B) \in \mathcal{U}(x, T, \gamma, i). \tag{5.40}$$

It follows from property C(iii) and (5.39) that the following property also holds:

C(iv) For each $(y, S) \in \mathcal{U}(x, T, \gamma, i)$ and each integer $n \geq N(x, T, \gamma, i)$,

$$\left\| S^n y - z_T \right\| \leq 4^{-1}\varepsilon. \tag{5.41}$$

Note that $z_T = \xi$ if $(z, B) \in \mathcal{F}_\xi$. Since ε is an arbitrary positive number, we conclude that $\{B^n z\}_{n=1}^\infty$ is a Cauchy sequence and there exists $\lim_{n\to\infty} B^n z$. Inequality (5.41) implies that

$$\left\| \lim_{n\to\infty} B^n z - z_T \right\| \leq 4^{-1}\varepsilon. \tag{5.42}$$

(If $(z, B) \in \mathcal{F}_\xi$, then $z_T = \xi$ and since (5.42) holds for any $\varepsilon > 0$, we conclude that $\lim_{n\to\infty} B^n z = \xi$.)

It follows from property C(iv) and (5.42) that for each $(y, S) \in \mathcal{U}(x, T, \gamma, i)$ and each integer $j \geq N(x, T, \gamma, i)$,

$$\left\| S^j y - \lim_{n\to\infty} B^n z \right\| \leq 2^{-1}\varepsilon.$$

This completes the proofs of Theorems 5.3 and 5.4. □

Proof of Theorem 5.5 Assume that K_0 is a nonempty, closed and separable subset of K. Let the sequence $\{x_j\}_{j=1}^\infty \subset K_0$ be dense in K_0 and let p be a natural number. By Theorem 5.1, there exists a set $\mathcal{F}_p \subset \bar{\mathcal{M}}_0$, which is a countable intersection of open and everywhere dense sets in $\bar{\mathcal{M}}_0$ such that, for each $T \in \mathcal{F}_p$, the following two properties hold:

C(v) For $j = 1, \ldots, p$ there exists $\lim_{n\to\infty} T^n x_j$ and

$$\lim_{n\to\infty} T^n x_j = \lim_{n\to\infty} T^n x_1, \quad j = 1, \ldots, p;$$

C(vi) For each $\varepsilon > 0$, there exist a neighborhood $\mathcal{U}$ of T in $\mathcal{M}$, a real number $\delta > 0$ and a natural number $N \geq 1$ such that, for each $S \in \mathcal{U}$, each $j = 1, \ldots, p$, each $y \in K$ satisfying $\|y - x_j\| \leq \delta$ and each integer $m \geq N$,

$$\left\| S^m y - \lim_{i\to\infty} T^i x_j \right\| \leq \varepsilon.$$

Set

$$\mathcal{F}=\bigcap_{p=1}^{\infty}\mathcal{F}_p. \tag{5.43}$$

It is clear that $\mathcal{F}$ is a countable intersection of open and everywhere dense sets in $\bar{\mathcal{M}}_0$.

Assume next that $T\in\mathcal{F}$. By (5.43) and C(v), there exists $x_T\in K$ such that

$$\lim_{n\to\infty}T^n x_j=x_T,\quad j=1,2,\ldots. \tag{5.44}$$

Now we construct the set $\mathcal{K}_T\subset K_0$. To this end, observe that property C(vi), (5.44) and (5.43) imply that for each pair of natural numbers (q,i) there exist a neighborhood $\mathcal{U}(q,i)$ of T in $\mathcal{M}$, a number $\delta(q,i)>0$ and a natural number $N(q,i)$ such that the following property holds:

C(vii) For each $S\in\mathcal{U}(q,i)$, each $y\in K$ satisfying $\|y-x_q\|\le\delta(q,i)$ and each integer $m\ge N(q,i)$,

$$\left\|S^m y-x_T\right\|\le 2^{-i}.$$

Define

$$\mathcal{K}_T=\bigcap_{n=1}^{\infty}\bigcup\left\{\left\{y\in K_0:\|y-x_q\|<\delta(q,i)\right\}:q\ge 1, i\ge n\right\}. \tag{5.45}$$

Clearly, $\mathcal{K}_T$ is a countable intersection of open and everywhere dense sets in K_0.

Assume that $x\in\mathcal{K}_T$ and $\varepsilon>0$ are given. Choose an integer $n\ge 1$ such that

$$2^{-n}<4^{-1}\varepsilon. \tag{5.46}$$

By (5.45), there exist a natural number q and an integer $i\ge n$ such that

$$\|x-x_q\|<\delta(q,i). \tag{5.47}$$

Combining (5.46) with property C(vii), we see that the following property is also true:

C(viii) For each $S\in\mathcal{U}(q,i)$, each $y\in K$ satisfying $\|y-x_q\|\le\delta(q,i)$ and each integer $m\ge N(q,i)$,

$$\left\|S^m y-x_T\right\|\le 4^{-1}\varepsilon.$$

Using this fact and (5.47), we get

$$\left\|T^m x-x_T\right\|\le 4^{-1}\varepsilon \quad \text{for all integers } m\ge N(q,i).$$

Since ε is an arbitrary positive number we conclude that $\{T^m x\}_{m=1}^{\infty}$ is a Cauchy sequence and

$$\lim_{m\to\infty}T^m x=x_T. \tag{5.48}$$

Applying property C(viii) and (5.35), it follows that for each $S \in \mathcal{U}(q,i)$, each $y \in K$ satisfying $\|y - x\| < \delta(q,i) - \|x - x_q\|$ and each integer $m \geq N(q,i)$,

$$\|S^m y - x_T\| \leq 4^{-1}\varepsilon.$$

This completes the proof of Theorem 5.5. □

We omit the proof of Theorem 5.6 because it is analogous to that of Theorem 5.5.

5.5 A Class of Uniformly Continuous Mappings

Denote by $\mathcal{A}$ the set of all mappings $T : K \to K$ which are uniformly continuous on bounded subsets of K. Clearly any $T \in \mathcal{A}$ is bounded on bounded subsets of K and $\mathcal{A}$ is a closed subset of the complete uniform space $\mathcal{M}$ defined in Sect. 5.2. We consider the topological subspace $\mathcal{A} \subset \mathcal{M}$ with the relative topology.

Denote by $\mathcal{A}_*$ the set of all $T \in \mathcal{A}$ for which there is a point $z_T \in K$ such that the following three properties hold:

B(i)

$$Tz_T = z_T, \qquad D_f(z_T, \cdot) : K \to R^1 \text{ is convex,}$$
$$D_f(z_T, Tx) \leq D_f(z_T, x) \quad \text{for all } x \in K;$$

B(ii) The function $D_f(z_T, \cdot)$ is bounded from above on any bounded subset of K;
B(iii) For any $\varepsilon > 0$ there exists $\delta > 0$ such that if $x \in K$ and $D_f(z_T, x) \leq \delta$, then $\|z_T - x\| \leq \varepsilon$.

Denote by $\bar{\mathcal{A}}_*$ the closure of $\mathcal{A}_*$ in the space $\mathcal{A}$. We consider the topological subspace $\bar{\mathcal{A}}_* \subset \mathcal{A}$ with the relative topology.

We note that **B**(iii) holds if the function f is uniformly convex.

Theorem 5.11 *There exists a set $\mathcal{F} \subset \bar{\mathcal{A}}_*$, which is a countable intersection of open and everywhere dense sets in $\bar{\mathcal{A}}_*$, such that for each $T \in \mathcal{F}$, the following two assertions hold*:

1. *There exists $z_* \in K$ such that $T^n x \to z_*$ as $n \to \infty$ for all $x \in K$.*
2. *For each $\varepsilon > 0$ and each bounded set $C \subset K$, there exist an integer $N \geq 1$ and a neighborhood $\mathcal{U}$ of T in $\mathcal{A}$ such that for each $S \in \mathcal{U}$ and each $x \in C$,*

$$\|S^n x - z_*\| \leq \varepsilon \quad \textit{for all integers } n \geq N.$$

Let $\xi \in K$ be given. Denote by $\mathcal{A}_\xi$ the set of all $T \in \mathcal{A}$ which satisfy Property **B** with $z_T = \xi$ and denote by $\bar{\mathcal{A}}_\xi$ the closure of $\mathcal{A}_\xi$ in $\mathcal{A}$. We consider the topological subspace $\bar{\mathcal{A}}_\xi \subset \mathcal{A}$ with the relative topology.

Theorem 5.12 *There exists a set $\mathcal{F}_\xi \subset \bar{\mathcal{A}}_\xi$, which is a countable intersection of open and everywhere dense sets in $\bar{\mathcal{A}}_\xi$, such that for each $T \in \mathcal{F}_\xi$, the following two assertions hold:*

1. *$T^n x \to \xi$ as $n \to \infty$ for all $x \in K$.*
2. *For each $\varepsilon > 0$ and each bounded set $C \subset K$, there exists an integer $N \geq 1$ and a neighborhood $\mathcal{U}$ of T in $\mathcal{A}$ such that for each $S \in \mathcal{U}$ and each $x \in C$,*

$$\left\| S^n x - \xi \right\| \leq \varepsilon \quad \textit{for all integers } n \geq N.$$

5.6 An Auxiliary Result

This section is devoted to an auxiliary result which will be used in the next section.

Proposition 5.13 *Let K_0 be a bounded subset of K, $T \in \mathcal{A}$, $\varepsilon > 0$ and let $n \geq 1$ be an integer. Then there exists a neighborhood $\mathcal{U}$ of T in $\mathcal{A}$ such that for each $S \in \mathcal{U}$ and each $x \in K_0$, the inequality $\|T^n x - S^n x\| \leq \varepsilon$ holds.*

Proof We prove this proposition by induction. Clearly, it is valid for $n = 1$. Assume that $m \geq 1$ is an integer and that the proposition is true for $n = m$. We now show that it is also true for $n = m + 1$.

Since the proposition is true for $n = m$, there is a neighborhood $\mathcal{U}_0$ of T in $\mathcal{A}$ such that

$$\Delta_0 = \sup\left\{ \left\| S^m x \right\| : S \in \mathcal{U}_0, x \in K_0 \right\} < \infty. \tag{5.49}$$

Set

$$K_1 = \left\{ x \in K : \|x\| \leq \Delta_0 + 1 \right\} \tag{5.50}$$

and define

$$\mathcal{U}_1 = \left\{ S \in \mathcal{U}_0 : \|Tx - Sx\| \leq 8^{-1}\varepsilon \text{ for all } x \in K_1 \right\}. \tag{5.51}$$

Since the mapping T is uniformly continuous on K_1, there is $\delta > 0$ such that

$$\|Tx - Ty\| \leq 8^{-1}\varepsilon \tag{5.52}$$

for each $x, y \in K_1$ satisfying $\|x - y\| \leq \delta$. Since the proposition is true for $n = m$, there is a neighborhood $\mathcal{U}$ of T in $\mathcal{A}$ such that

$$\mathcal{U} \subset \mathcal{U}_1 \tag{5.53}$$

and for each $S \in \mathcal{U}$ and each $x \in K_0$, the following inequality holds:

$$\left\| T^m x - S^m x \right\| \leq \delta. \tag{5.54}$$

Assume that $S \in \mathcal{U}$ and $x \in K_0$. Then

$$\left\| T^{m+1}x - S^{m+1}x \right\| \le \left\| T^{m+1}x - T\left(S^m x\right) \right\| + \left\| T\left(S^m x\right) - S^{m+1}x \right\|. \tag{5.55}$$

By the definition of $\mathcal{U}$, inequality (5.54) is true. Now (5.49), (5.53), (5.51) and (5.50) imply that

$$T^m x, S^m x \in K_1. \tag{5.56}$$

It follows from (5.54), (5.56) and the definition of δ (see (5.52)) that

$$\left\| T\left(S^m x\right) - T^{m+1}x \right\| \le 8^{-1}\varepsilon. \tag{5.57}$$

By (5.53), (5.51) and (5.56),

$$\left\| T\left(S^m x\right) - S^{m+1}x \right\| \le 8^{-1}\varepsilon. \tag{5.58}$$

Combining (5.57), (5.58) and (5.55), we obtain that $\|T^{m+1}x - S^{m+1}x\| \le 2^{-1}\varepsilon$. Proposition 5.13 is proved. □

5.7 Proofs of Theorems 5.11 and 5.12

Let $T \in \mathcal{A}_*$, $\gamma \in (0,1)$ and let $z_T \in K$ satisfy Property **B**. Define a mapping $T_\gamma : K \to K$ by

$$T_\gamma x = \gamma z_T + (1-\gamma)Tx, \quad x \in K. \tag{5.59}$$

Clearly, $T_\gamma \in \mathcal{A}$. By Lemma 5.7, $T_\gamma \in \mathcal{A}_*$ with $z_{T(\gamma)} = z_T$ and

$$D_f(z_T, T_\gamma x) \le (1-\gamma)D_f(z_T, x) \quad \text{for all } x \in K. \tag{5.60}$$

It is clear that for each $T \in \mathcal{A}_*$,

$$T_\gamma \to T \quad \text{as } \gamma \to 0 \text{ in } \mathcal{A}. \tag{5.61}$$

We precede the proof of Theorems 5.11 and 5.12 by the following lemma.

Lemma 5.14 *Let $T \in \mathcal{A}_*$, $\varepsilon, \gamma \in (0,1)$ and let $z_T \in K$ satisfy Property* **B**. *Let K_0 be a nonempty and bounded subset of K. Then there exist a natural number N and a neighborhood $\mathcal{U}$ of T_γ in $\mathcal{A}$ such that for each $S \in \mathcal{U}$, each $x \in K_0$ and each integer $n \ge N$,*

$$\left\| S^n x - z_T \right\| \le \varepsilon. \tag{5.62}$$

Proof We may assume without any loss of generality that

$$\left\{x \in K : \|x - z_T\| \le 1\right\} \subset K_0. \tag{5.63}$$

According to Property **B**, there exists $\delta \in (0, \varepsilon)$ such that

$$\text{if } x \in K \text{ and } D_f(z_T, x) \le 2\delta, \text{ then } \|z_T - x\| \le 2^{-1}\varepsilon. \tag{5.64}$$

Also by Property **B**, there exists a number $c_0 > 0$ such that

$$D_f(z_T, x) \le c_0 \quad \text{for all } x \in K_0. \tag{5.65}$$

Choose a natural number N such that

$$(1-\gamma)^N (c_0 + 1) \le 2^{-1}\delta. \tag{5.66}$$

It follows from (5.66), (5.65) and (5.61) that for each $x \in K_0$, and each integer $n \ge N$,

$$D_f\big(z_T, T_\gamma^n x\big) \le (1-\gamma)^n D_f(z_T, x) \le (1-\gamma)^N c_0 \le 2^{-1}\delta.$$

This inequality and (5.64) imply that for each $x \in K_0$ and each integer $n \ge N$, we have

$$\big\|z_T - T_\gamma^n x\big\| \le 2^{-1}\varepsilon. \tag{5.67}$$

Proposition 5.13 guarantees that there exists a neighborhood $\mathcal{U}$ of T_γ in $\mathcal{A}$ such that for each $x \in K_0$, all integers $n = N, N+1, \dots, 4N$, and each $S \in \mathcal{U}$,

$$\big\|T_\gamma^n x - S^n x\big\| \le 4^{-1}\varepsilon. \tag{5.68}$$

Assume that $S \in \mathcal{U}$ and $x \in K_0$. We claim that for all integers $n \ge N$, we have

$$\big\|S^n x - z_T\big\| \le \varepsilon. \tag{5.69}$$

In order to show this, suppose, by way of contradiction, that the claim is false. Then there is an integer $q \ge N$ such that

$$\big\|S^q x - z_T\big\| > \varepsilon. \tag{5.70}$$

It follows from the definition of $\mathcal{U}$ (see (5.68)) and (5.67) that

$$\big\|z_T - S^n y\big\| \le 3 \cdot 4^{-1}\varepsilon \quad \text{for all } y \in K_0 \text{ and all } n = N, N+1, \dots, 4N. \tag{5.71}$$

Inequalities (5.70) and (5.71) imply that $q > 4N$. Note that we may assume without loss of generality that

$$\big\|z_T - S^i x\big\| \le \varepsilon \quad \text{for all } i = N, \dots, q-1. \tag{5.72}$$

Together with (5.63) this implies that $S^{q-N}x \in K_0$. Combining this with (5.71), we see that

$$\big\|z_T - S^q x\big\| = \big\|z_T - S^N\big(S^{q-N}\big)x\big\| \le 3 \cdot 4^{-1}\varepsilon,$$

which contradicts (5.70). Thus (5.69) is true for all integers $n \geq N$. This completes the proof of Lemma 5.14. □

Now we proceed to prove Theorems 5.11 and 5.12.

Proofs of Theorems 5.11 and 5.12 Fix $\theta \in K$. For each natural number i, set

$$K_i = \{x \in K : \|x - \theta\| \leq i\}. \tag{5.73}$$

With each $T \in \mathcal{A}_*$, we associate a point $z_T \in K$ satisfying Property **B**. If $T \in \mathcal{A}_\xi$, then $z_T = \xi$.

By Lemma 5.14, for each $T \in \mathcal{A}_*$, $\gamma \in (0, 1)$ and for each integer $i \geq 1$, there exist a natural number $N(T, \gamma, i)$ and an open neighborhood $\mathcal{U}(T, \gamma, i)$ of T_γ in $\mathcal{A}$ such that the following property holds:

P(i) for each $S \in \mathcal{U}(T, \gamma, i)$, each $x \in K_{2i}$ and each integer $n \geq N(T, \gamma, i)$,

$$\|S^n x - z_T\| \leq 2^{-i}.$$

Define

$$\mathcal{F} = \left[\bigcap_{q=1}^{\infty} \bigcup \{\mathcal{U}(T, \gamma, i) : T \in \mathcal{A}_*, \gamma \in (0, 1), i = q, q+1, \dots\}\right] \cap \bar{\mathcal{A}}_*,$$

$$\mathcal{F}_\xi = \left[\bigcap_{q=1}^{\infty} \bigcup \{\mathcal{U}(T, \gamma, i) : T \in \mathcal{A}_\xi, \gamma \in (0, 1), i = q, q+1, \dots\}\right] \cap \bar{\mathcal{A}}_\xi.$$

Evidently, $\mathcal{F}$ (respectively, $\mathcal{F}_\xi$) is a countable intersection of open and everywhere dense sets in $\bar{\mathcal{A}}_*$ (respectively, $\bar{\mathcal{A}}_\xi$) and $\mathcal{F}_\xi \subset \mathcal{F}$.

Let C be a bounded subset of K and let $B \in \mathcal{F}$, $\varepsilon > 0$ be given. There exists an integer $q \geq 1$ such that

$$C \subset K_{2q} \quad \text{and} \quad 2^{-q} < 4^{-1}\varepsilon. \tag{5.74}$$

There exist $T \in \mathcal{A}_*$ ($T \in \mathcal{A}_\xi$ if $B \in \mathcal{F}_\xi$), $\gamma \in (0, 1)$ and an integer $i \geq q$ such that

$$B \in \mathcal{U}(T, \gamma, i). \tag{5.75}$$

It follows from property P(i), (5.75) and (5.74) that the following property holds:

P(ii) For each $x \in C$, each $S \in \mathcal{U}(T, \gamma, i)$ and each integer $n \geq N(T, \gamma, i)$,

$$\|S^n x - z_T\| \leq 4^{-1}\varepsilon.$$

Note that $z_T = \xi$ if $B \in \mathcal{F}_\xi$.

Relation (5.75) and property P(ii) imply that for each $x \in C$ and each integer $n \geq N(T, \gamma, i)$,

$$\|B^n x - z_T\| \leq 4^{-1}\varepsilon. \tag{5.76}$$

Since ε is an arbitrary positive number and C is an arbitrary bounded set in K, we conclude that for each $x \in K$, $\{B^n x\}_{n=1}^{\infty}$ is a Cauchy sequence. Therefore for each $x \in K$, there exists $\lim_{n\to\infty} B^n x$. Inequality (5.76) implies that

$$\left\| \lim_{n\to\infty} B^n x - z_T \right\| \le 4^{-1}\varepsilon \quad \text{for all } x \in C. \tag{5.77}$$

Again, since ε is an arbitrary positive number and C is an arbitrary bounded subset of K, (5.77) implies that there is $z_* \in K$ such that

$$z_* = \lim_{n\to\infty} B^n x \quad \text{for all } x \in K. \tag{5.78}$$

By (5.78) and (5.77),

$$\|z_* - z_T\| \le 4^{-1}\varepsilon. \tag{5.79}$$

(If $B \in \mathcal{F}_\xi$, then $T \in \mathcal{A}_\xi$, $z_T = \xi$ and since the inequality above is true for any $\varepsilon > 0$, we obtain that $z_* = \xi$.) It follows from property P(ii) and (5.79) that for each $S \in \mathcal{U}(T, \gamma, i)$, each $x \in C$ and each integer $n \ge N(T, \gamma, i)$, $\|S^n x - z_*\| \le 2^{-1}\varepsilon$. This completes the proofs of Theorems 5.11 and 5.12. □

5.8 Mappings with a Uniformly Continuous Bregman Function

In this section we use the definitions and notations from Sects. 5.1 and 5.2 and the complete uniform spaces $\mathcal{M}$ and $\mathcal{M}_c$ introduced there.

Denote by $\mathcal{M}_*$ the set of all $T \in \mathcal{M}$ for which there is $z_T \in K$ such that the following assumptions hold:

C(i)

$$Tz_T = z_T, \qquad D_f(z_T, \cdot) : K \to R^1 \text{ is convex,}$$
$$D_f(z_T, Tx) \le D_f(z_T, x) \quad \text{for all } x \in K;$$

C(ii) The function $D_f(z_T, \cdot)$ is uniformly continuous on any bounded subset of K;

C(iii) For any $\varepsilon > 0$, there exists $\delta > 0$ such that if $x \in K$ and $D_f(z_T, x) \le \delta$, then $\|z_T - x\| \le \varepsilon$;

C(iv) For each $\alpha > 0$, the level set $\{y \in K : D_f(z_T, y) \le \alpha\}$ is bounded.

Set $\mathcal{M}_{*c} = \mathcal{M}_c \cap \mathcal{M}_*$. Denote by $\bar{\mathcal{M}}_*$ the closure of $\mathcal{M}_*$ in the space $\mathcal{M}$ and by $\bar{\mathcal{M}}_{*c}$ the closure of $\mathcal{M}_{*c}$ in the space $\mathcal{M}$. We consider the topological subspaces $\bar{\mathcal{M}}_{*c}$ and $\bar{\mathcal{M}}_* \subset \mathcal{M}$ with the relative topologies.

Again we note that **C(iii)** holds if the function f is uniformly convex.

Theorem 5.15 *There exists a set $\mathcal{F} \subset \bar{\mathcal{M}}_*$, which is a countable intersection of open and everywhere dense sets in $\bar{\mathcal{M}}_*$, and a set $\mathcal{F}_c \subset \mathcal{F} \cap \bar{\mathcal{M}}_{c*}$, which is a countable intersection of open and everywhere dense sets in $\bar{\mathcal{M}}_{*c}$, such that for each $T \in \mathcal{F}$, the following assertions hold:*

1. *There exists $z_* \in K$ such that $T^n x \to z_*$ as $n \to \infty$ for all $x \in K$.*
2. *For each $\varepsilon > 0$, and each bounded set $C \subset K$, there exist an integer $N \geq 1$ and a neighborhood $\mathcal{U}$ of T in $\mathcal{M}$ such that for each $S \in \mathcal{U}$ and each $x \in C$,*

$$\|S^n x - z_*\| \leq \varepsilon \quad \textit{for all integers } n \geq N.$$

Let $\xi \in K$. Denote by $\mathcal{M}_\xi$ the set of all $T \in \mathcal{M}$ which satisfy Assumption **C** with $z_T = \xi$. Set $\mathcal{M}_{\xi c} = \mathcal{M}_\xi \cap \mathcal{M}_c$. Denote by $\bar{\mathcal{M}}_\xi$ the closure of $\mathcal{M}_\xi$ in $\mathcal{M}$ and by $\bar{\mathcal{M}}_{\xi c}$ the closure of $\mathcal{M}_{\xi c}$ in $\mathcal{M}$. We consider the topological subspaces $\bar{\mathcal{M}}_\xi$ and $\bar{\mathcal{M}}_{\xi c} \subset \mathcal{M}$ with the relative topologies.

Theorem 5.16 *There exists a set $\mathcal{F}_\xi \subset \bar{\mathcal{M}}_\xi$ (respectively, $\mathcal{F}_{\xi c} \subset \bar{\mathcal{M}}_{\xi c} \cap \mathcal{F}_\xi$), which is a countable intersection of open and everywhere dense sets in $\bar{\mathcal{M}}_\xi$ (respectively, in $\bar{\mathcal{M}}_{\xi c}$), such that for each $T \in \mathcal{F}_\xi$, the following assertions hold:*

1. *$T^n x \to \xi$ as $n \to \infty$ for all $x \in K$.*
2. *For each $\varepsilon > 0$ and each bounded set $C \subset K$, there exist an integer $N \geq 1$ and a neighborhood $\mathcal{U}$ of T in $\mathcal{M}$ such that for each $S \in \mathcal{U}$ and each $x \in C$,*

$$\|S^n x - \xi\| \leq \varepsilon \quad \textit{for all integers } n \geq N.$$

5.9 Proofs of Theorems 5.15 and 5.16

Let $T \in \mathcal{M}_*$, $\gamma \in (0, 1)$ and let $z_T \in K$ satisfy Assumption **C**. Define a mapping $T_\gamma : K \to K$ by

$$T_\gamma x = (1 - \gamma)Tx + \gamma z_T, \quad x \in K. \tag{5.80}$$

Clearly,

$$T_\gamma \in \mathcal{M} \text{ and if } T \in \mathcal{M}_c, \text{ then } T_\gamma \in \mathcal{M}_c. \tag{5.81}$$

By Lemma 5.7, $T_\gamma \in \mathcal{M}_*$ with $z(T_\gamma) = z(T)$ and

$$D_f(z_T, T_\gamma x) \leq (1 - \gamma)D_f(z_T, x) \quad \text{for all } x \in K. \tag{5.82}$$

Evidently, for each $T \in \mathcal{M}_*$,

$$T_\gamma \to T \quad \text{as } \gamma \to 0 \text{ in } \mathcal{M}. \tag{5.83}$$

Lemma 5.17 *Let $T \in \mathcal{M}_*$, $\varepsilon, \gamma \in (0, 1)$ and let z_T satisfy Assumption* **C**. *Let K_0 be a bounded subset of K. Then there exist a natural number N and a neighborhood $\mathcal{U}$ of T_γ in $\mathcal{M}$ such that for each $S \in \mathcal{U}$, each $x \in K_0$ and each integer $n \geq N$,*

$$\|S^n x - z_T\| \leq \varepsilon.$$

Proof By Assumption **C**, there exist $\delta \in (0, \varepsilon)$, $c_0 > 1$ and $c_1 > 0$ such that

$$\text{if } x \in K \text{ and } D_f(z_T, x) \le 2\delta, \text{ then } \|z_T - x\| \le 2^{-1}\varepsilon, \tag{5.84}$$

$$D_f(z_T, x) \le c_0 \quad \text{for all } x \in K_0 \tag{5.85}$$

and

$$\|x\| \le c_1 \quad \text{for all } x \in K \text{ satisfying } D_f(z_T, x) \le c_0 + 2. \tag{5.86}$$

Set

$$K_1 = \{x \in K : \|x\| \le c_1\} \quad \text{and} \quad K_2 = \{x \in K : \|x\| \le c_1 + 2\}. \tag{5.87}$$

Clearly, $K_0 \subset K_1$. By Assumption **C**, the function $D_f(z_T, \cdot)$ is uniformly continuous on K_2. Therefore there is $\delta_0 \in (0, 4^{-1}\delta)$ such that

$$\|D_f(z_T, x_1) - D_f(z_T, x_2)\| \le \gamma\delta 8^{-1} \tag{5.88}$$

for each $x_1, x_2 \in K_2$ satisfying $\|x_1 - x_2\| \le \delta_0$.

Choose a natural number N such that

$$8^{-1}N\gamma\delta > c_0 + 2 \tag{5.89}$$

and define

$$\mathcal{U} = \{S \in \mathcal{M} : \|Sx - T_\gamma x\| \le \delta_0 \text{ for all } x \in K_1\}. \tag{5.90}$$

Assume that $S \in \mathcal{U}$ and $x \in K_0$. We claim that

$$\|S^n x - z_T\| \le \varepsilon \quad \text{for all integers } n \ge N. \tag{5.91}$$

By the definition of δ (see (5.84)), in order to prove (5.91), it is sufficient to show that

$$D_f(z_T, S^n x) \le 2\delta \quad \text{for all integers } n \ge N. \tag{5.92}$$

First we will show by induction that for all integers $n \ge 0$,

$$D_f(z_T, S^n x) \le c_0 \tag{5.93}$$

and

$$D_f(z_T, S^{n+1}x) \le (1-\gamma)D_f(z_T, S^n x) + 8^{-1}\gamma\delta \quad \text{for all integers } n \ge 0. \tag{5.94}$$

Clearly, by (5.85), inequality (5.93) is valid for $n = 0$. (Note that $S^0 x = x$.)

Assume that (5.93) is true for some integer $n \ge 0$. We will show that (5.94) is also true. By (5.87), (5.86) and (5.93),

$$S^n x \in K_1 \quad \text{and} \quad \|S^n x\| \le c_1. \tag{5.95}$$

Combining (5.95) and (5.90), we see that

$$\left\| S^{n+1}x - T_\gamma S^n x \right\| \le \delta_0. \tag{5.96}$$

By (5.82) and (5.93),

$$D_f\left(z_T, T_\gamma\left(S^n x\right)\right) \le (1-\gamma) D_f\left(z_T, S^n x\right) \le (1-\gamma) c_0. \tag{5.97}$$

Together with (5.86) this implies that

$$\left\| T_\gamma\left(S^n x\right) \right\| \le c_1. \tag{5.98}$$

Combining (5.96), (5.98) and (5.87), we see that

$$T_\gamma\left(S^n x\right) \in K_1 \quad \text{and} \quad S^{n+1}x \in K_2. \tag{5.99}$$

It follows from (5.99), (5.96) and the definition of δ_0 (see (5.88)) that

$$\left\| D_f\left(z_T, S^{n+1}x\right) - D_f\left(z_T, T_\gamma S^n x\right) \right\| \le 8^{-1}\gamma\delta. \tag{5.100}$$

By (5.100) and (5.97),

$$\begin{aligned} D_f\left(z_T, S^{n+1}x\right) &\le 8^{-1}\gamma\delta + (1-\gamma) D_f\left(z_T, S^n x\right) \\ &\le 8^{-1}\gamma\delta + (1-\gamma)c_0 \le \gamma + (1-\gamma)c_0 \le c_0. \end{aligned}$$

Thus (5.94) is true and $D_f(z_T, S^{n+1}x) \le c_0$. Therefore both inequalities (5.93) and (5.94) are valid for all integers $n \ge 0$.

Let $n \ge 0$ be an integer. If $D_f(z_T, S^n x) \le \delta$, then by (5.94) we have

$$D_f\left(z_T, S^{n+1}x\right) \le (1-\gamma)\delta + 8^{-1}\gamma\delta \le \delta.$$

Therefore in order to prove (5.92), it is sufficient to show that $D_f(z_T, S^n x) \le \delta$ for some integer $n \in [0, N]$.

If this were not true, then it would follow that $D_f(z_T, S^n x) > \delta$, $n = 0, \ldots, N$. Thus according to (5.94), for $n = 0, \ldots, N$, we would get

$$D_f\left(z_T, S^n x\right) - D_f\left(z_T, S^{n+1}x\right) \ge \gamma D_f\left(z_T, S^n x\right) - 8^{-1}\gamma\delta \ge 2^{-1}\gamma\delta.$$

When combined with (5.89), this would yield

$$D_f(z_T, x) \ge D_f(z_T, x) - D_f\left(z_T, S^{N+1}x\right) \ge 2^{-1}\gamma\delta N > c_0 + 2,$$

which contradicts (5.93). Hence (5.92) and therefore (5.91) are valid for all integers $n \ge N$. This completes the proof of Lemma 5.17. □

Proofs of Theorems 5.15 and 5.16 The proofs of Theorems 5.15 and 5.16 follows the pattern of the proofs of Theorems 5.11 and 5.12. The main difference is that we use Lemma 5.17 instead of Lemma 5.14. □

5.10 Generic Power Convergence to a Retraction

We continue to consider the problem of whether and under what conditions, relatively nonexpansive operators T defined on, and with values in, a nonempty, closed and convex subset K of a Banach space $(X, \|\cdot\|)$ have the property that the sequences $\{T^k x\}_{k=1}^{\infty}$ converge strongly to fixed points of T, whenever $x \in K$. For a given nonempty, closed and convex subset F of K, we consider complete metric spaces of self-mappings of K which fix all the points of F and are relatively nonexpansive with respect to a given convex function f on X. We show (under certain assumptions on f) that the iterates of a generic mapping in these spaces converge strongly to a retraction onto F.

These results were obtained in [33].

We say that an operator $T : K \to K$ is relatively nonexpansive with respect to the convex function $f : X \to R^1 \cup \{\infty\}$ if K is a subset of the algebraic interior $\mathcal{D}^0$ of the domain of f,

$$\mathcal{D} := \mathrm{dom}(f) = \{x \in X : f(x) < \infty\},$$

the function f is lower semicontinuous on K and there exists a point $z \in K$ such that, for any $x \in K$, we have

$$D_f(z, Tx) \le D_f(z, x), \tag{5.101}$$

where $D_f : X \times \mathcal{D}^0 \to [0, \infty)$ stands for the Bregman distance given by

$$D_f(y, x) = f(y) - f(x) + f^0(x, x - y), \tag{5.102}$$

and $f^0(x, d)$ denotes the right-hand derivative of f at x in the direction d. In this case, the point z is called a pole of T with respect to f.

Let $\mathcal{M} = \mathcal{M}(f, K, F)$ be the set of all operators $T : K \to K$ which are relatively nonexpansive with respect to the same convex function $f : X \to R^1 \cup \{\infty\}$ and which have a nonempty, closed and convex set F of common poles. We assume that the function f satisfies the following conditions:

A(i) For any nonempty bounded set $E \subset K$ and any $\varepsilon > 0$, there exists $\delta > 0$ such that

$$\text{if } x \in E,\, z \in F \text{ and } D_f(z, x) \le \delta, \text{ then } \|z - x\| \le \varepsilon. \tag{5.103}$$

A(ii) There exists $\theta \in F$ such that the restriction to K of the function $g(\cdot) := D_f(\theta, \cdot)$ has the following property: for any subset $E \subset K$, $g(E)$ is bounded if and only if E is bounded.

A(iii) For any $z \in F$, the function $D_f(z, \cdot) : K \to R^1$ is convex and lower semicontinuous.

A(iv) For any $x \in K$, there exists a vector $Px \in F$ such that

$$D_f(Px, x) \le D_f(z, x) \quad \text{for all } z \in F. \tag{5.104}$$

In practical situations one also uses the following stronger version of A(i):

For any nonempty and bounded set $E \subset K$, $\inf\{\nu_f(x,t) : x \in E\}$ is positive for all $t > 0$, where

$$\nu_f(x,t) = \inf\{D_f(y,x) : y \in X \text{ and } \|y-x\| = t\}. \tag{5.105}$$

In [28] this condition is termed sequential compatibility of the function f with the relative topology of the set K. We will show (see Lemma 5.18 below) that sequential compatibility implies A(i). In its turn, condition A(i) implies that all $z \in F$ are common fixed points of the operators in $\mathcal{M}$. Condition A(ii) guarantees that any operator $T \in \mathcal{M}$ is bounded on bounded subsets of K (a feature which is essential in our proofs) because, for any bounded set $E \subset K$, we have

$$D_f(\theta, Tx) \le D_f(\theta, x), \tag{5.106}$$

where, according to condition A(ii), the function $D_f(\theta,\cdot)$ is bounded on E, and therefore so is the set $\{Tx : x \in E\}$. Condition A(ii), even taken in conjunction with A(i), is satisfied by many useful functions and, among them, by many functions which are sequentially compatible with the relative topology of K. In contrast, condition A(iii) is quite restrictive. However, it does hold for many functions f which are of interest in current applications (see the examples below). The vector Px satisfying (5.104) was termed the Bregman projection with respect to f of x onto F in [38].

Condition A(iv) is automatically satisfied when X is reflexive and f is totally convex on K (in particular, when f is sequentially compatible with the relative topology of K) as follows from Proposition 2.1.5(i) of [28]. In this case, if f is differentiable on the algebraic interior of its domain, then, for each $x \in K$, there exists a unique vector Px in F which satisfies (5.104). We now mention four typical situations in which all the conditions A(i)–A(iv) are satisfied simultaneously.

(i) (cf. [28]) X is a Hilbert space, K and F are nonempty closed convex subsets of X such that $F \subset K$ and $f(x) = \|x\|^2$;
(ii) (cf. [29]) $F \subset K \subset R^n_{++}$ and f is the negentropy;
(iii) (cf. [31]) X is a Lebesgue space L^p or l^p, $1 < p \le 2$, $f(x) = \|x\|^p$ and K consists of either nonnegative or nonpositive functions;
(iv) (cf. [32]) X is smooth and uniformly convex, F is a singleton $\{z\}$, and $f(x) = \|x-z\|^r$ with $r > 1$.

We provide the set $\mathcal{M} = \mathcal{M}(f,K,F)$ with the uniformity determined by the following base:

$$E(N,\varepsilon) = \{(T_1,T_2) \in \mathcal{M} \times \mathcal{M} : \|T_1x - T_2x\| \le \varepsilon \text{ for all } x \in K \text{ satisfying } \|x\| \le N\},$$

where $N, \varepsilon > 0$. Clearly, this uniform space is metrizable and complete. We equip the space $\mathcal{M}$ with the topology induced by this uniformity. Let $\mathcal{M}_c$ be the set of all operators in $\mathcal{M}$ which are continuous on K. This is a closed subset of $\mathcal{M}$ and we endow it with the relative topology. The subset of $\mathcal{M}_c$ consisting of those operators which are uniformly continuous on bounded subsets of K is denoted by $\mathcal{M}_u$. Again, this set is closed in $\mathcal{M}$ and we endow it with the relative topology. We will show that the sequence of powers of a generic mapping T in $\mathcal{M}_u$, $\mathcal{M}_c$ and $\mathcal{M}$, respectively, converges in the uniform topology to a relatively nonexpansive operator which belongs to the same space and is a retraction onto F. Consequently, the sequences $\{T^k x\}_{k=1}^{\infty}$ generated by a generic mapping T are strongly convergent to points in F, i.e., to fixed points of T.

In this chapter we have shown that the iterates of a generic operator in certain other spaces of relatively nonexpansive operators converge strongly to its unique fixed point. As we have just noted above, in the different situation considered now, the iterates of a generic operator converge to a retraction onto its fixed point set F.

5.11 Two Lemmata

This section is devoted to two lemmata. The first one shows that sequential compatibility implies condition A(i), while the second shows that the retraction, the existence of which is stipulated in condition A(iv), belongs to $\mathcal{M}$.

Lemma 5.18 *If the convex function f is sequentially compatible with the relative topology of K, then it satisfies condition* A(i).

Proof Let the convex function f be sequentially compatible with the relative topology of K. For any nonempty set $E \subset K$ and any $t \geq 0$, set

$$\nu_f(E,t) = \inf\{D_f(y,x) : x \in E,\, y \in X \text{ and } \|y-x\| = t\}.$$

Since f is assumed to be sequentially compatible with the relative topology of K, $\nu_f(E,t) > 0$ for any nonempty and bounded set $E \subset K$, and any $t > 0$, and the function $\nu_f(x,\cdot)$ is strictly increasing (see Proposition 1.2.2 of [28]).

Assume now that we are given a nonempty and bounded subset M of K and an $\varepsilon > 0$. Let $\delta = \nu_f(M,\varepsilon)$. If $x \in M$, $y \in F$ and $D_f(y,x) \leq \delta$, then

$$\nu_f(x, \|y-x\|) \leq D_f(y,x) \leq \delta \leq \nu_f(x,\varepsilon).$$

Since the function $\nu_f(x,\cdot)$ is strictly increasing, we conclude that $\|y-x\| \leq \varepsilon$. Lemma 5.18 is proved. □

Note that the functions in the examples (i)–(iv) listed in the previous section are all sequentially compatible with the relative topology of any closed and convex subset of their respective domains.

Lemma 5.19 *Let an operator $P : K \to F$ be as guaranteed in condition* A(iv). *Then for any $x \in K$ and for any $z \in F$, we have*

$$D_f(z,Px) \leq D_f(z,x). \tag{5.107}$$

Proof Fix $x \in K$ and $z \in F$. Denote $\hat{x} = Px$ and let

$$u(\alpha) = \hat{x} + \alpha(z - \hat{x}) \tag{5.108}$$

for any $\alpha \in [0, 1]$. Observe that $D_f(\cdot, x)$ and f are convex and, therefore, the following limits exist, and for all $y \in K$ and $d \in X$,

$$\begin{aligned}
&\big[D_f(\cdot, x)\big]^0(y, d) \\
&\quad = \lim_{t \to 0^+} \big[D_f(y + td, x) - D_f(y, x)\big]/t \\
&\quad = \lim_{t \to 0^+} \big[f(y + td) - f(x) + f^0(x, x - y - td) \\
&\qquad - \big(f(y) - f(x) + f^0(x, x - y)\big)\big]/t \\
&\quad = \lim_{t \to 0^+} \big[f(y + td) - f(y)\big]/t + \lim_{t \to 0^+} \big[f^0(x, x - y - td) - f^0(x, x - y)\big]/t \\
&\quad = f^0(y, d) + \lim_{t \to 0^+} \big[f^0(x, x - y - td) - f^0(x, x - y)\big]/t.
\end{aligned}$$

The function $f^0(x, \cdot)$ is subadditive and positively homogeneous because f is convex. Consequently, we have

$$f^0(x, x - y) \le f^0(x, x - y - td) + t f^0(x, d).$$

Combining this inequality and the previous formula, we get

$$\big[D_f(\cdot, x)\big]^0(y, d) \ge f^0(y, d) - f^0(x, d). \tag{5.109}$$

Now since $\hat{x} = Px$, we have by (5.104) and (5.109) that for any $\alpha \in (0, 1]$,

$$\begin{aligned}
0 &\ge D_f(\hat{x}, x) - D_f\big(u(\alpha), x\big) \ge \big[D_f(\cdot, x)\big]^0\big(u(\alpha), \hat{x} - u(\alpha)\big) \\
&= \big[D_f(\cdot, x)\big]^0\big(u(\alpha), -\alpha(z - \hat{x})\big) = \alpha\big[D_f(\cdot, x)\big]^0\big(u(\alpha), \hat{x} - z\big) \\
&\ge \alpha\big[f^0\big(u(\alpha), \hat{x} - z\big) - f^0(x, \hat{x} - z)\big].
\end{aligned}$$

Hence, for any $\alpha \in (0, 1]$, we get

$$f^0(x, \hat{x} - z) \ge f^0\big(u(\alpha), \hat{x} - z\big). \tag{5.110}$$

Note that by A(iii), the function $\phi(x) = f^0(x, x - z)$, $x \in K$, is lower semicontinuous. Hence the function $\phi(u(\alpha))$, $\alpha \in [0, 1]$, is also lower semicontinuous. Since

$$\phi\big(u(\alpha)\big) = f^0\big(u(\alpha), u(\alpha) - z\big) = (1 - \alpha) f^0\big(u(\alpha), \hat{x} - z\big), \quad \alpha \in [0, 1),$$

the function $\alpha \to f^0(u(\alpha), \hat{x} - z)$, $\alpha \in [0, 1)$, is lower semicontinuous too.

Applying $\liminf_{\alpha \to 0^+}$ to both sides of inequality (5.110), we see that

$$f^0(x, \hat{x} - z) \geq f^0(\hat{x}, \hat{x} - z).$$

This, in turn, implies that

$$f(z) - f(\hat{x}) + f^0(x, \hat{x} - z) \geq D_f(z, \hat{x}).$$

Since $f^0(x, \cdot)$ is sublinear, it follows that

$$f(z) - f(\hat{x}) + f^0(x, \hat{x} - x) + f^0(x, x - z) \geq D_f(z, \hat{x}).$$

Hence

$$D_f(z, x) + \left[f(z) - f(\hat{x}) - f(z) + f(x) + f^0(x, \hat{x} - x)\right] \geq D_f(z, \hat{x}). \quad (5.111)$$

Note that the quantity between square brackets is exactly

$$-\left[f(\hat{x}) - f(x) - f^0(x, \hat{x} - x)\right] \leq 0$$

because f is convex. This inequality and (5.111) imply (5.107). The proof of Lemma 5.19 is complete. □

In the remaining sections of this chapter we use the following notation.

For each $x \in K$ and each nonempty $G \subset K$, set

$$\rho_f(x, G) := \inf\left\{D_f(z, x) : z \in G\right\}. \quad (5.112)$$

5.12 Convergence of Powers of Uniformly Continuous Mappings

We assume that the operator P, the existence of which is stipulated in condition A(iv), belongs to $\mathcal{M}_u$, and that the following condition holds:

$$\text{For each bounded set } K_0 \subset K \text{ and each } \varepsilon > 0, \text{ there is } \delta > 0$$
$$\text{such that if } x \in K_0, z \in F \text{ and } \|z - x\| \leq \delta, \text{ then } D_f(z, x) \leq \varepsilon. \quad (5.113)$$

Remark 5.20 Note that condition (5.113) indeed holds if the function f is Lipschitzian on each bounded subset of K.

Theorem 5.21 *There exists a set $\mathcal{F} \subset \mathcal{M}_u$, which is a countable intersection of open and everywhere dense subsets of $\mathcal{M}_u$, such that for each $B \in \mathcal{F}$, the following assertions hold:*

(i) *There exists $P_B \in \mathcal{M}_u$ such that $P_B(K) = F$ and $B^n x \to P_B x$ as $n \to \infty$, uniformly on bounded subsets of K;*

(ii) *for each* $\varepsilon > 0$ *and each bounded set* $C \subset K$, *there exist a neighborhood* U *of* B *in* $\mathcal{M}_u$ *and an integer* $N \geq 1$ *such that for each* $S \in U$, *each* $x \in C$ *and each integer* $n \geq N$,

$$\|S^n x - P_B x\| \leq \varepsilon.$$

This theorem is established in Sect. 5.15.

5.13 Convergence to a Retraction

In this section we assume that the function $D_f(\cdot, \cdot) : F \times K \to R^1$ is uniformly continuous on bounded subsets of $F \times K$ and state two theorems, the proofs of which will be given in Sect. 5.16.

Theorem 5.22 *There exists a set* $\mathcal{F} \subset \mathcal{M}$, *which is a countable intersection of open and everywhere dense subsets of* $\mathcal{M}$, *such that for each* $B \in \mathcal{F}$, *the following assertions hold*:

1. *There exists* $P_B \in \mathcal{M}$ *such that* $P_B(K) = F$ *and* $B^n x \to P_B x$ *as* $n \to \infty$, *uniformly on bounded subsets of* K; *if* $B \in \mathcal{M}_c$, *then* $P_B \in \mathcal{M}_c$.
2. *For each* $\varepsilon > 0$ *and each nonempty bounded set* $C \subset K$, *there exist a neighborhood* U *of* B *in* $\mathcal{M}$ *and a natural number* N *such that for each* $S \in U$ *and each* $x \in C$, *there is* $z(S, x) \in F$ *such that* $\|S^n x - z(S, x)\| \leq \varepsilon$ *for all integers* $n \geq N$.

Moreover, if $P \in \mathcal{M}_c$, *then there exists a set* $\mathcal{F}_c \subset \mathcal{F} \cap \mathcal{M}_c$, *which is a countable intersection of open and everywhere dense subsets of* $\mathcal{M}_c$.

Theorem 5.23 *Let the set* $\mathcal{F} \subset \mathcal{M}$ *be as guaranteed in Theorem 5.22,* $B \in \mathcal{F} \cap \mathcal{M}_c$, $P_B z = \lim_{n\to\infty} B^n z$, $z \in K$, *and let* $x \in K$, $\varepsilon > 0$ *be given. Then there exist a neighborhood* U *of* B *in* $\mathcal{M}$, *a number* $\delta > 0$ *and a natural number* N *such that for each* $y \in K$ *satisfying* $\|x - y\| \leq \delta$, *each* $S \in U$ *and each integer* $n \geq N$, $\|S^n y - P_B x\| \leq \varepsilon$.

5.14 Auxiliary Results

In this section we prove two lemmata which will be used in the proofs of our theorems. We use the convention that $S^0 x = x$ for each $x \in K$ and each $S \in \mathcal{M}$.

For each $\gamma \in (0, 1)$ and each $T \in \mathcal{M}$, define a mapping $T_\gamma : K \to K$ by

$$T_\gamma x = \gamma P x + (1 - \gamma) T x, \quad x \in K \tag{5.114}$$

(see condition A(iv)).

Lemma 5.24 *Let $T \in \mathcal{M}$ and $\gamma \in (0, 1)$. Then $T_\gamma \in \mathcal{M}$. If $T, P \in \mathcal{M}_u$ (respectively, $T, P \in \mathcal{M}_c$), then $T_\gamma \in \mathcal{M}_u$ (respectively, $T_\gamma \in \mathcal{M}_c$).*

Proof Clearly $T_\gamma \in \mathcal{M}$ and $T_\gamma x = x$ for all $x \in F$. By (5.114), A(iii), (5.101), A(iv) and Lemma 5.19, for each $z \in F$ and each $x \in K$,

$$\begin{aligned} D_f(z, T_\gamma x) &= D_f\big(z, \gamma Px + (1-\gamma)Tx\big) \\ &\le \gamma D_f(z, Px) + (1-\gamma) D_f(z, Tx) \le D_f(z, x). \end{aligned}$$

Thus $T_\gamma \in \mathcal{M}$. Clearly, $T_\gamma \in \mathcal{M}_u$ if $T, P \in \mathcal{M}_u$ and $T_\gamma \in \mathcal{M}_c$ if $T, P \in \mathcal{M}_c$. Lemma 5.24 is proved. □

It is obvious that for each $T \in \mathcal{M}$,

$$T_\gamma \to T \quad \text{as } \gamma \to 0^+ \text{ in } \mathcal{M}. \tag{5.115}$$

Lemma 5.25 *Let $T \in \mathcal{M}$, $\gamma \in (0, 1)$ and let $x \in K$. Then*

$$\rho_f(T_\gamma x, F) \le \big(1 - \gamma^2\big) \rho_f(x, F). \tag{5.116}$$

Proof Let $\varepsilon > 0$ be given. There exists $y \in F$ such that (see (5.112))

$$D_f(y, x) \le \rho_f(x, F) + \varepsilon. \tag{5.117}$$

It follows from (5.114), A(iv), Lemma 5.19, A(iii) and (5.101) that

$$\begin{aligned} \rho_f(T_\gamma x, F) &= \rho_f\big(\gamma Px + (1-\gamma)Tx, F\big) \\ &\le D_f\big(\gamma Px + (1-\gamma)y, (1-\gamma)Tx + \gamma Px\big) \\ &\le \gamma D_f\big(Px, \gamma Px + (1-\gamma)Tx\big) + (1-\gamma) D_f\big(y, \gamma Px + (1-\gamma)Tx\big) \\ &\le \gamma^2 D_f(Px, Px) + \gamma(1-\gamma) D_f(Px, Tx) \\ &\quad + (1-\gamma)\gamma D_f(y, Px) + (1-\gamma)^2 D_f(y, Tx) \\ &\le \gamma(1-\gamma) D_f(Px, x) + (1-\gamma)\gamma D_f(y, Px) + (1-\gamma)^2 D_f(y, Tx). \end{aligned} \tag{5.118}$$

It follows from (5.118), A(iv), Lemma 5.19 and (5.117) that

$$\begin{aligned} \rho_f(T_\gamma x, F) &\le \gamma(1-\gamma)\rho_f(x, F) + (1-\gamma)\gamma D_f(y, x) + (1-\gamma)^2 D_f(y, x) \\ &\le \varepsilon + \big(1 - \gamma^2\big)\rho_f(x, F). \end{aligned}$$

Since ε is an arbitrary positive number, we conclude that (5.116) holds. This completes the proof of Lemma 5.25. □

5.15 Proof of Theorem 5.21

We begin with the following lemma.

Lemma 5.26 *Let $T \in \mathcal{M}_u$, $\gamma \in (0,1)$, $\varepsilon > 0$ and let K_0 be a nonempty and bounded subset of K. Then there exist a neighborhood U of T_γ in $\mathcal{M}_u$ and a natural number N such that for each $x \in K_0$, there exists $Qx \in F$ such that for each integer $n \geq N$ and each $S \in U$,*

$$\|S^n x - Qx\| \leq \varepsilon.$$

Proof Set

$$K_1 = \bigcup \{S^i(K_0) : S \in \mathcal{M}, i \geq 0\}. \tag{5.119}$$

Assumption A(ii) and (5.101) imply that the set K_1 is bounded. Evidently,

$$S(K_1) \subset K_1 \quad \text{for all } S \in \mathcal{M}^{(F)}. \tag{5.120}$$

By A(i), there exists $\varepsilon_0 \in (0, \varepsilon)$ such that

$$\text{if } x \in K_1, z \in F \text{ and } D_f(z,x) \leq \varepsilon_0, \text{ then } \|z - x\| \leq 4^{-1}\varepsilon. \tag{5.121}$$

By (5.113), there is $\varepsilon_1 \in (0, 2^{-1}\varepsilon_0)$ such that

$$\text{if } x \in K_1, z \in F \text{ and } \|x - z\| \leq 2\varepsilon_1, \text{ then } D_f(z,x) \leq 2^{-1}\varepsilon_0. \tag{5.122}$$

By A(i), there is $\varepsilon_2 \in (0, 2^{-1}\varepsilon_1)$ such that

$$\text{if } x \in K_1, z \in F \text{ and } D_f(z,x) \leq 2\varepsilon_2, \text{ then } \|x - z\| \leq 2^{-1}\varepsilon_1. \tag{5.123}$$

Set

$$c_0 = \sup\{\rho_f(x, F) : x \in K_1\}. \tag{5.124}$$

By A(ii), $c_0 < \infty$. Choose a natural number $N \geq 4$ such that

$$(1 - \gamma^2)^N (c_0 + 1) \leq 2^{-1}\varepsilon_2. \tag{5.125}$$

It follows from Lemma 5.25, (5.124) and (5.125) that for each $x \in K_1$,

$$\rho_f(T_\gamma^N x, F) \leq (1 - \gamma^2)^N \rho_f(x, F) \leq (1 - \gamma^2)^N c_0 < 2^{-1}\varepsilon_2.$$

Thus for each $x \in K_1$, there is $Qx \in F$ such that $D_f(Qx, T_\gamma^N x) \leq 2^{-1}\varepsilon_2$. When combined with (5.120) and (5.123), the last inequality implies that

$$\|T_\gamma^N x - Qx\| \leq 2^{-1}\varepsilon_1 \quad \text{for all } x \in K_1. \tag{5.126}$$

By Proposition 5.13, there exists a neighborhood U of T_γ in $\mathcal{M}_u$ such that for each $x \in K_1$ and each $S \in U$,

$$\|S^N x - T_\gamma^N x\| \le 4^{-1}\varepsilon_1. \tag{5.127}$$

Assume that $x \in K_0$ and $S \in U$. Evidently, $\{S^i x\}_{i=0}^\infty \subset K_1$. By (5.126) and (5.127), $\|S^N x - Qx\| \le 3 \cdot 4^{-1}\varepsilon_1$. It follows from this inequality and (5.122) that $D_f(Qx, S^N x) \le 2^{-1}\varepsilon_0$. Since $S \in \mathcal{M}_u$, it follows from the last inequality that $D_f(Qx, S^n x) \le 2^{-1}\varepsilon_0$ for all integers $n \ge N$. When combined with (5.121), this implies that $\|Qx - S^n x\| \le \varepsilon$ for all integers $n \ge N$. Lemma 5.26 is proved. □

Proof of Theorem 5.21 By (5.115), the set $\{T_\gamma : T \in \mathcal{M}_u, \gamma \in (0,1)\}$ is an everywhere dense subset of $\mathcal{M}_u$. For each natural number i, set

$$K_i = \{x \in K : \|x - \theta\| \le i\}. \tag{5.128}$$

By Lemma 5.26, for each $T \in \mathcal{M}_u$, each $\gamma \in (0,1)$ and each integer $i \ge 1$, there exist an open neighborhood $\mathcal{U}(T,\gamma,i)$ of T_γ in $\mathcal{M}_u$ and a natural number $N(T,\gamma,i)$ such that the following property holds:

P(i) For each $x \in K_{2^i}$, there is $Qx \in F$ such that

$$\|S^n x - Qx\| \le 2^{-i} \quad \text{for all integers } n \ge N(T,\gamma,i) \text{ and all } S \in \mathcal{U}(T,\gamma,i).$$

Define

$$\mathcal{F} := \bigcap_{q=1}^{\infty} \bigcup \{\mathcal{U}(T,\gamma,q) : T \in \mathcal{M}_u, \gamma \in (0,1)\}.$$

Clearly, $\mathcal{F}$ is a countable intersection of open and everywhere dense subsets of $\mathcal{M}_u$.

Let $B \in \mathcal{F}$, $\varepsilon > 0$ and let C be a bounded subset of K. There exists an integer $q \ge 1$ such that

$$C \subset K_{2^q} \quad \text{and} \quad 2^{-q} < 4^{-1}\varepsilon. \tag{5.129}$$

There also exist $T \in \mathcal{M}_u$ and $\gamma \in (0,1)$ such that

$$B \in \mathcal{U}(T,\gamma,q). \tag{5.130}$$

It now follows from Property P(i), (5.129) and (5.130) that the following property also holds:

P(ii) For each $x \in C$, there is $Qx \in F$ such that

$$\|S^n x - Qx\| \le 4^{-1}\varepsilon$$

for each integer $n \ge N(T,\gamma,q)$ and each $S \in \mathcal{U}(T,\gamma,q)$.

Property P(ii) and (5.130) imply that for each $x \in C$ and each integer $n \geq N(T, \gamma, q)$,

$$\left\| B^n x - Qx \right\| \leq 4^{-1}\varepsilon. \tag{5.131}$$

Since ε is an arbitrary positive number and C is an arbitrary bounded subset of K, we conclude that for each $x \in K$, $\{B^n x\}_{n=1}^{\infty}$ is a Cauchy sequence. Therefore for each $x \in K$, there exists

$$P_B x = \lim_{n \to \infty} B^n x. \tag{5.132}$$

By (5.131) and (5.132), for each $x \in C$,

$$\| P_B x - Qx \| \leq 4^{-1}\varepsilon. \tag{5.133}$$

Once again, since ε is an arbitrary positive number and C is an arbitrary bounded subset of K, we conclude that

$$P_B(K) = F. \tag{5.134}$$

It now follows from property (Pii) and (5.133) that for each $x \in C$, each $S \in \mathcal{U}(T, \gamma, q)$ and each integer $n \geq N(T, \gamma, q)$,

$$\left\| S^n x - P_B x \right\| \leq 2^{-1}\varepsilon.$$

This completes the proof of Theorem 5.21. □

5.16 Proofs of Theorems 5.22 and 5.23

We begin with four lemmata.

Lemma 5.27 *Let K_0 be a nonempty and bounded subset of K, and let β be a positive number. Then the set $\{(z, y) \in F \times K_0 : D_f(z, y) \leq \beta\}$ is bounded.*

Proof If this assertion were not true, then there would exist a sequence $\{(z_i, x_i)\}_{i=1}^{\infty} \subset F \times K_0$ such that

$$D_f(z_i, x_i) \leq \beta, \quad i = 1, 2, \ldots, \quad \text{and} \quad \|z_i\| \to \infty \quad \text{as } i \to \infty. \tag{5.135}$$

By (5.101), $D_f(z_i, Px_i) \leq \beta$, $i = 1, 2, \ldots$. Clearly, the sequence $\{Px_i\}_{i=1}^{\infty}$ is bounded. We may assume that $\|z_i - Px_i\| \geq 16$, $i = 1, 2, \ldots$. For each integer $i \geq 1$, there exists $\alpha_i > 0$ such that

$$\left\| \left[(1 - \alpha_i) Px_i + \alpha_i z_i \right] - Px_i \right\| = 1. \tag{5.136}$$

Clearly, $\alpha_i \to 0$ as $i \to \infty$. It is easy to see that for each integer $i \geq 1$,

$$D_f\big((1 - \alpha_i) Px_i + \alpha_i z_i, Px_i\big) \leq \alpha_i D_f(z_i, Px_i) \leq \alpha_i \beta \to 0 \quad \text{as } i \to \infty.$$

When combined with A(i), this implies that $\|Px_i - [(1-\alpha_i)Px_i + \alpha_i z_i]\| \to 0$ as $i \to \infty$. Since this contradicts (5.136), Lemma 5.27 follows. □

Lemma 5.28 *Let $T \in \mathcal{M}$, $\gamma, \varepsilon \in (0,1)$ and let K_0 be a nonempty and bounded subset of K. Then there exists a neighborhood U of T_γ in $\mathcal{M}$ such that for each $S \in U$ and each $x \in K_0$ satisfying $\rho_f(x,F) > \varepsilon$, the following inequality holds*:

$$\rho_f(Sx, F) \le \rho_f(x,F) - \varepsilon\gamma^2/4. \tag{5.137}$$

Proof Set

$$K_1 = \bigcup \{S^i(K_0) : S \in \mathcal{M}, i \ge 0\}. \tag{5.138}$$

Assumption A(ii) and (5.101) imply that the set K_1 is bounded. Evidently, $S(K_1) \subset K_1$ for all $S \in \mathcal{M}$. By A(ii), there exists $c_0 > 0$ such that

$$4 + \sup\{D_f(\theta, x) : x \in K_1\} < c_0. \tag{5.139}$$

By Lemma 5.27, there exists a number $c_1 > 0$ such that

$$\text{if } (z,x) \in F \times K_1 \text{ and } D_f(z,x) \le c_0 + 2, \text{ then } \|z\| \le c_1. \tag{5.140}$$

We may assume without loss of generality that

$$c_1 > \sup\{\|Px\| : x \in K_1\}. \tag{5.141}$$

Since $D_f(\cdot,\cdot)$ is uniformly continuous on bounded subsets of $F \times K$, there exists a number $\delta \in (0, 2^{-1})$ such that for each pair of points,

$$(z, x_1), (z, x_2) \in \{\xi \in F : \|\xi\| \le c_1\} \times K_1$$

satisfying $\|x_1 - x_2\| \le \delta$, the following inequality holds:

$$|D_f(z,x_1) - D_f(z,x_2)| \le 4^{-1}\varepsilon\gamma^2. \tag{5.142}$$

Set

$$U = \{S \in \mathcal{M} : \|Sx - T_\gamma x\| \le \delta \text{ for all } x \in K_1\}. \tag{5.143}$$

It is clear that U is a neighborhood of T_γ in $\mathcal{M}$.

Assume that

$$S \in U, \quad x \in K_0 \quad \text{and} \quad \rho_f(x,F) > \varepsilon. \tag{5.144}$$

We claim that (5.137) is valid. By Lemma 5.25,

$$\rho_f(T_\gamma x, F) \le (1-\gamma^2)\rho_f(x,F). \tag{5.145}$$

Let

$$\Delta \in (0, 4^{-1}\gamma^2\varepsilon). \tag{5.146}$$

There is $z \in F$ such that

$$D_f(z, T_\gamma x) \le (1 - \gamma^2)\rho_f(x, F) + \Delta. \tag{5.147}$$

By (5.147), (5.146), (5.139) and (5.140),

$$D_f(z, T_\gamma x) \le c_0 \quad \text{and} \quad \|z\| \le c_1. \tag{5.148}$$

By (5.143) and (5.144),

$$\|T_\gamma x - Sx\| \le \delta. \tag{5.149}$$

By (5.148) and (5.142),

$$(z, T_\gamma x), (z, Sx) \in \{\xi \in F : \|\xi\| \le c_1\} \times K_1. \tag{5.150}$$

By (5.150), (5.149) and the definition of δ (see (5.142)),

$$|D_f(z, T_\gamma x) - D_f(z, Sx)| \le 4^{-1}\varepsilon\gamma^2.$$

When combined with (5.147) and (5.146), this implies that

$$\begin{aligned}\rho_f(Sx, F) &\le D_f(z, Sx) \le 4^{-1}\varepsilon\gamma^2 + D_f(z, T_\gamma x)\\ &\le 4^{-1}\varepsilon\gamma^2 + (1 - \gamma^2)\rho_f(x, F) + \Delta\\ &\le (1 - \gamma^2)\rho_f(x, F) + 2^{-1}\varepsilon\gamma^2.\end{aligned}$$

Thus

$$\rho_f(Sx, F) \le (1 - \gamma^2)\rho_f(x, F) + 2^{-1}\varepsilon\gamma^2.$$

Inequality (5.137) follows from this inequality and (5.144). Lemma 5.28 is proved. □

Lemma 5.29 *Let $T \in \mathcal{M}$, $\gamma, \varepsilon \in (0, 1)$ and let K_0 be a nonempty and bounded subset of K. Then there exist a neighborhood U of T_γ in $\mathcal{M}$ and a natural number N such that for each $S \in U$ and each $x \in K_0$,*

$$\rho_f(S^N x, F) \le \varepsilon. \tag{5.151}$$

Proof Define the set K_1 by (5.138). Assumption A(ii) and (5.101) imply that the set K_1 is bounded. Clearly, $S(K_1) \subset K_1$ for all $S \in \mathcal{M}_u$. By A(ii), there is a positive number c_0 such that (5.139) is valid. By Lemma 5.28, there exists a neighborhood U of T_γ in $\mathcal{M}$ such that for each $S \in U$ and each $x \in K_1$ satisfying $\rho_f(x, F) > \varepsilon$, the following inequality holds:

$$\rho_f(Sx, F) \le \rho_f(x, F) - \varepsilon\gamma^2/4. \tag{5.152}$$

Choose a natural number N for which

$$8^{-1}\varepsilon\gamma^2 N > c_0 + 1. \tag{5.153}$$

Assume that $S \in U$ and $x \in K_0$. We claim that inequality (5.151) is valid. If it were not, then we would have $\rho(S^i x, F) > \varepsilon$ for all $i = 0, \dots, N$. When combined with the definition of U (see (5.152)), these inequalities imply that for all $i = 0, \dots, N-1$,

$$\rho_f\big(S^{i+1}x, F\big) \le \rho_f\big(S^i x, F\big) - \varepsilon\gamma^2/4.$$

Therefore

$$\rho_f\big(S^N x, F\big) \le \rho_f(x, F) - \varepsilon\gamma^2 N/4.$$

By this inequality, (5.139) and (5.153),

$$0 \le \rho_f\big(S^n x, F\big) \le c_0 - 4^{-1}\varepsilon\gamma^2 N \le -1.$$

This contradiction proves (5.151) and Lemma 5.29 follows. □

Lemma 5.30 *Let $T \in \mathcal{M}$, $\gamma, \varepsilon \in (0,1)$ and let K_0 be a nonempty and bounded subset of K. Then there exist a neighborhood U of T_γ in $\mathcal{M}$ and a natural number N such that for each $S \in U$ and each $x \in K_0$, there is $z(S,x) \in F$ such that*

$$\big\|S^i x - z(S,x)\big\| \le \varepsilon \quad \textit{for all integers } i \ge N. \tag{5.154}$$

Proof Define K_1 by (5.138). Assumption A(ii) and (5.101) imply that K_1 is bounded. By Assumption A(i), there exists $\delta \in (0,1)$ such that

$$\text{if } x \in K_1, z \in F \text{ and } D_f(z,x) \le \delta, \text{ then } \|x - z\| \le 2^{-1}\varepsilon. \tag{5.155}$$

By Lemma 5.29, there exists a neighborhood U of T_γ in $\mathcal{M}$ and a natural number N such that

$$\rho_f\big(S^N x, F\big) \le \delta/2 \quad \text{for each } S \in U \text{ and } x \in K_1.$$

This implies that for each $x \in K_0$ and each $S \in U$, there is $z(S,x) \in F$ for which $D_f(z(S,x), S^N x) < \delta$. When combined with (5.155) this implies that for each $x \in K_0$, each $S \in U$, and each integer $i \ge N$,

$$D_f\big(z(S,x), S^i x\big) < \delta \quad \text{and} \quad \big\|S^i x - z(S,x)\big\| \le 2^{-1}\varepsilon.$$

Lemma 5.30 is proved. □

Proof of Theorem 5.22 By (5.115), the set $\{T_\gamma : T \in \mathcal{M}, \gamma \in (0,1)\}$ is an everywhere dense subset of $\mathcal{M}$ and if $P \in \mathcal{M}_c$, then $\{T_\gamma : T \in \mathcal{M}_c, \gamma \in (0,1)\}$ is an everywhere dense subset of $\mathcal{M}_c$. For each natural number i, set

$$K_i = \big\{x \in K : \|x - \theta\| \le i\big\}. \tag{5.156}$$

By Lemma 5.30, for each $T \in \mathcal{M}$, each $\gamma \in (0,1)$, and each integer $i \geq 1$, there exist an open neighborhood $\mathcal{U}(T,\gamma,i)$ of T_γ in $\mathcal{M}$ and a natural number $N(T,\gamma,i)$ such that the following property holds:

P(iii) For each $x \in K_{2^i}$ and each $S \in \mathcal{U}(T,\gamma,i)$, there is $z(S,x) \in F$ such that

$$\|S^n x - z(S,x)\| \leq 2^{-i} \quad \text{for all integers } n \geq N(T,\gamma,i).$$

Define

$$\mathcal{F} := \bigcap_{q=1}^{\infty} \bigcup \{\mathcal{U}(T,\gamma,q) : T \in \mathcal{M}, \gamma \in (0,1)\}.$$

Clearly, $\mathcal{F}$ is a countable intersection of open and everywhere dense subsets of $\mathcal{M}$. If $P \in \mathcal{M}_c$, then we define

$$\mathcal{F}_c := \left[\bigcap_{q=1}^{\infty} \bigcup \{\mathcal{U}(T,\gamma,q) : T \in \mathcal{M}_c, \gamma \in (0,1)\}\right] \cap \mathcal{M}_c.$$

In this case, $\mathcal{F}_c \subset \mathcal{F}$ and $\mathcal{F}_c$ is a countable intersection of open and everywhere dense subsets of $\mathcal{M}_c$.

Let $B \in \mathcal{F}$, $\varepsilon > 0$, and let C be a bounded subset of K. There exists an integer $q \geq 1$ such that

$$C \subset K_{2^q} \quad \text{and} \quad 2^{-q} < 4^{-1}\varepsilon. \tag{5.157}$$

There also exist $T \in \mathcal{M}$ and $\gamma \in (0,1)$ such that

$$B \in \mathcal{U}(T,\gamma,q). \tag{5.158}$$

Note that if $P \in \mathcal{M}_c$ and $B \in \mathcal{F}_c$, then $T \in \mathcal{M}_c$.

It follows from Property P(iii), (5.157) and (5.158) that the following property also holds:

P(iv) For each $S \in \mathcal{U}(T,\gamma,q)$ and each $x \in C$, there is $z(S,x) \in F$ such that $\|S^n x - z(S,x)\| \leq 4^{-1}\varepsilon$ for each integer $n \geq N(T,\gamma,q)$.

Relation (5.158) and property P(iv) imply that for each $x \in C$ and each integer $n \geq N(T,\gamma,q)$,

$$\|B^n x - z(B,x)\| \leq 4^{-1}\varepsilon. \tag{5.159}$$

Since ε is an arbitrary positive number and C is an arbitrary bounded subset of K, we conclude that for each $x \in K$, $\{B^n x\}_{n=1}^{\infty}$ is a Cauchy sequence. Therefore for each $x \in K$, there exists

$$P_B x = \lim_{n\to\infty} B^n x.$$

Now (5.159) implies that for each $x \in C$,

$$\|P_B x - z(B,x)\| \leq 4^{-1}\varepsilon. \tag{5.160}$$

Once again, since ε is an arbitrary positive number and C is an arbitrary bounded subset of K, we conclude that

$$P_B(K) = F.$$

It follows from (5.159) and (5.160) that for each $x \in C$ and each integer $n \geq N(T, \gamma, q)$,

$$\|B^n x - P_B x\| \leq 2^{-1}\varepsilon.$$

This implies that $P_B \in \mathcal{M}$ and if $B \in \mathcal{M}_c$, then $P_B \in \mathcal{M}_c$. Theorem 5.22 is established. □

We will use the next lemma in the proof of Theorem 5.23.

Lemma 5.31 *Let $B \in \mathcal{M}_c$, $x \in K$, $\varepsilon \in (0, 1)$ and let $N \geq 1$ be an integer. Then there exist a neighborhood U of B in $\mathcal{M}$ and a number $\delta > 0$ such that for each $S \in U$ and each $y \in K$ satisfying $\|y - x\| \leq \delta$, the following inequality holds:*

$$\|S^n y - B^n x\| \leq \varepsilon.$$

This lemma is proved by induction on n.

Proof of Theorem 5.23 By Theorem 5.22, there exist a natural number N and a neighborhood U_0 of B in $\mathcal{M}$ such that

$$\|P_B y - B^n y\| \leq 8^{-1}\varepsilon \quad \text{for each } y \in K \text{ satisfying } \|y - x\| \leq 1 \text{ and each } n \geq N; \tag{5.161}$$

and for each $S \in U_0$ and each $y \in K$ satisfying $\|y - x\| \leq 1$, there is $z(S, y) \in F$ such that

$$\|S^n y - z(S, y)\| \leq 8^{-1}\varepsilon \quad \text{for all integers } n \geq N. \tag{5.162}$$

By Lemma 5.31, there exist a number $\delta \in (0, 1)$ and a neighborhood U of B in $\mathcal{M}$ such that $U \subset U_0$ and

$$\|S^N y - B^N x\| \leq 8^{-1}\varepsilon \quad \text{for each } S \in U \text{ and each } y \in K \text{ for which } \|y - x\| \leq \delta. \tag{5.163}$$

Assume that

$$y \in K, \qquad \|x - y\| \leq \delta \quad \text{and} \quad S \in U. \tag{5.164}$$

By (5.164), (5.163) and (5.161),

$$\|S^N y - B^N x\| \leq 8^{-1}\varepsilon, \qquad \|S^N y - z(S, y)\| \leq 8^{-1}\varepsilon \quad \text{and}$$
$$\|P_B x - B^N x\| \leq 8^{-1}\varepsilon.$$

These inequalities imply that

$$\|z(S, y) - P_B x\| \leq 3 \cdot 8^{-1} \varepsilon.$$

When combined with (5.162), the last inequality implies that

$$\|S^n y - P_B x\| \leq 2^{-1} \varepsilon \quad \text{for all integers } n \geq N.$$

This completes the proof of Theorem 5.23. □

5.17 Convergence of Powers for a Class of Continuous Operators

In this section we assume that $P \in \mathcal{M}_c$ and that the function

$$D_f(z, \cdot) : K \to R^1 \text{ is continuous} \quad \text{for all } z \in F. \tag{5.165}$$

Theorem 5.32 *Let $x \in K$. Then there exists a set $\mathcal{F} \subset \mathcal{M}_c$, which is a countable intersection of open and everywhere dense subsets of $\mathcal{M}_c$, such that for each $B \in \mathcal{F}$, the following assertions hold:*

1. *There exists* $\lim_{n \to \infty} B^n x \in F$.
2. *For each $\varepsilon > 0$, there exist a neighborhood U of B in $\mathcal{M}_c$, a natural number N and a number $\delta > 0$ such that for each $S \in U$, each $y \in K$ satisfying $\|y - x\| \leq \delta$ and each integer $n \geq N$, $\|S^n y - \lim_{i \to \infty} B^i x\| \leq \varepsilon$.*

We equip the space $K \times \mathcal{M}_c$ with the product topology.

Theorem 5.33 *There exists a set $\mathcal{F} \subset K \times \mathcal{M}_c$, which is a countable intersection of open and everywhere dense subsets of $K \times \mathcal{M}_c$, such that for each $(z, B) \in \mathcal{F}$, the following assertions hold:*

1. *There exists* $\lim_{n \to \infty} B^n z \in F$.
2. *For each $\varepsilon > 0$, there exist a neighborhood U of (z, B) in $K \times \mathcal{M}_c$ and a natural number N such that for each $(y, S) \in U$ and each integer $n \geq N$,*

$$\left\| S^n y - \lim_{i \to \infty} B^i z \right\| \leq \varepsilon.$$

Theorem 5.34 *Assume that the set K_0 is a nonempty, separable and closed subset of K. Then there exists a set $\mathcal{F} \subset \mathcal{M}_c$, which is a countable intersection of open and everywhere dense subsets of $\mathcal{M}_c$, such that for each $T \in \mathcal{F}$, there exists a set $\mathcal{K}_T \subset K_0$, which is a countable intersection of open and everywhere dense subsets of K_0 with the relative topology, such that the following assertions hold:*

1. *For each $x \in \mathcal{K}_T$, there exists* $\lim_{n \to \infty} T^n x \in F$.
2. *For each $x \in \mathcal{K}_T$ and each $\varepsilon > 0$, there exist an integer $N \geq 1$ and a neighborhood U of (x, T) in $\mathcal{K} \times \mathcal{M}_c$ such that for each $(y, S) \in U$ and each integer $i \geq N$, $\|S^i y - \lim_{n \to \infty} T^n x\| \leq \varepsilon$.*

5.18 Proofs of Theorems 5.32–5.34

We precede the proofs of Theorems 5.32 and 5.33 by the following lemma.

Lemma 5.35 *Let $T \in \mathcal{M}_c$, $\gamma, \varepsilon \in (0, 1)$ and let $x \in K$. Then there exist a neighborhood U of T_γ in $\mathcal{M}_c$, a natural number N, a point $\widehat{z} \in F$ and a number $\delta > 0$ such that for each $S \in U$, each $y \in K$ satisfying $\|y - x\| \le \delta$ and each integer $n \ge N$,*

$$\left\|S^n y - \widehat{z}\right\| \le \varepsilon. \tag{5.166}$$

Proof Define

$$K_1 := \bigcup \left\{S^i\left(\{y \in K : \|y - x\| \le 1\}\right) : S \in \mathcal{M}, i = 0, 1, \ldots\right\}. \tag{5.167}$$

By A(ii) and (5.101), the set K_1 is bounded. By A(i), there is $\varepsilon_0 \in (0, \varepsilon/2)$ such that

$$\text{if } z \in F, y \in K_1 \text{ and } D_f(z, y) \le 2\varepsilon_0, \text{ then } \|z - y\| \le \varepsilon/2. \tag{5.168}$$

Choose a natural number N for which

$$\left(1 - \gamma^2\right)^N \left(\rho_f(x, F) + 1\right) < \varepsilon_0/8. \tag{5.169}$$

By Lemma 5.25, this implies that

$$\rho_f\left(T_\gamma^N x, F\right) \le \left(1 - \gamma^2\right)^N \rho_f(x, F) < \varepsilon_0/8.$$

Therefore there exists $\widehat{z} \in F$ for which

$$D_f\left(\widehat{z}, T_\gamma^N x\right) < \varepsilon_0/8. \tag{5.170}$$

Since the function $D_f(\widehat{z}, \cdot) : K \to R^1$ is continuous (see (5.165)), there exists $\varepsilon_1 \in (0, \varepsilon_0/2)$ such that

$$D_f(\widehat{z}, \xi) < \varepsilon_0/8 \quad \text{for all } \xi \in K \text{ satisfying } \left\|\xi - T_\gamma^N x\right\| \le \varepsilon_1. \tag{5.171}$$

It follows from the continuity of T_γ that there exist a neighborhood U of T_γ in $\mathcal{M}_c$ and a number $\delta \in (0, 1)$ such that for each $S \in U$ and each $y \in K$ satisfying $\|y - x\| \le \delta$,

$$\left\|S^N y - T_\gamma^N x\right\| \le \varepsilon_1 \tag{5.172}$$

(see Lemma 5.31).

Assume that

$$S \in U, \qquad y \in K, \quad \text{and} \quad \|y - x\| \le \delta.$$

By the definition of U and δ, inequality (5.172) is valid. By (5.172) and (5.173), $D_f(\widehat{z}, S^N y) < \varepsilon_0/8$. This implies that $D_f(\widehat{z}, S^n y) < \varepsilon_0/8$ for all integers $n \ge N$.

When combined with (5.168), this implies that $\|\widehat{z} - S^n y\| \le \varepsilon$ for all integers $n \ge N$. Lemma 5.35 is proved. □

Proof of Theorem 5.32 Let $x \in K$ be given. By Lemma 5.35, for each $T \in \mathcal{M}_c$, each $\gamma \in (0, 1)$ and each integer $i \ge 1$, there exist an open neighborhood $\mathcal{U}(T, \gamma, i)$ of T_γ in $\mathcal{M}_c$, a natural number $N(T, \gamma, i)$, a point $z(T, \gamma, i) \in F$ and a number $\delta(T, \gamma, i) > 0$ such that the following property holds:

(Pv) For each $S \in \mathcal{U}(T, \gamma, i)$, each $y \in K$ satisfying $\|x - y\| \le \delta(T, \gamma, i)$ and each integer $n \ge N(T, \gamma, i)$,

$$\|S^n y - z(T, \gamma, i)\| \le 2^{-i}.$$

Define

$$\mathcal{F} := \bigcap_{q=1}^{\infty} \bigcup \{\mathcal{U}(T, \gamma, q) : T \in \mathcal{M}_c, \gamma \in (0, 1)\}.$$

Clearly, $\mathcal{F}$ is a countable intersection of open and everywhere dense subsets of $\mathcal{M}_c$.

Let $B \in \mathcal{F}$ and $\varepsilon > 0$ be given. There exists an integer $q \ge 1$ such that

$$2^{-q} < 4^{-1}\varepsilon. \tag{5.173}$$

There also exist $T \in \mathcal{M}_c$ and $\gamma \in (0, 1)$ such that

$$B \in \mathcal{U}(T, \gamma, q). \tag{5.174}$$

It follows from property (Pv) and (5.173) that the following property also holds:

(Pvi) For each $S \in \mathcal{U}(T, \gamma, q)$, each $y \in K$ satisfying $\|y - x\| \le \delta(T, \gamma, q)$ and each integer $n \ge N(T, \gamma, q)$,

$$\|S^n y - z(T, \gamma, q)\| \le 4^{-1}\varepsilon. \tag{5.175}$$

Since ε is an arbitrary positive number, we conclude that $\{B^n x\}_{n=1}^{\infty}$ is a Cauchy sequence and there exists $\lim_{n\to\infty} B^n x$. Inequality (5.175) implies that

$$\left\| \lim_{n\to\infty} B^n x - z(T, \gamma, q) \right\| \le 4^{-1}\varepsilon.$$

Since ε is an arbitrary positive number, we conclude that $\lim_{n\to\infty} B^n x$ belongs to F. It follows from this inequality and property (Pvi) that for each $S \in \mathcal{U}(T, \gamma, q)$, each $y \in K$ satisfying $\|y - x\| \le \delta(T, \gamma, q)$, and each integer $n \ge N(T, \gamma, q)$,

$$\left\| S^n y - \lim_{i\to\infty} B^i x \right\| \le 2^{-1}\varepsilon.$$

Theorem 5.32 is proved. □

Proof of Theorem 5.33 By Lemma 5.35, for each $(x, T) \in K \times \mathcal{M}_c$, each $\gamma \in (0, 1)$, and each integer $i \geq 1$, there exist an open neighborhood $\mathcal{U}(x, T, \gamma, i)$ of (x, T_γ) in $K \times \mathcal{M}_c$, a natural number $N(x, T, \gamma, i)$ and a point $z(x, T, \gamma, i) \in F$ such that the following property holds:

(Pvii) For each $(y, S) \in \mathcal{U}(x, T, \gamma, i)$ and each integer $n \geq N(x, T, \gamma, i)$,

$$\left\| S^n y - z(x, T, \gamma, i) \right\| \leq 2^{-i}.$$

Define

$$\mathcal{F} := \bigcap_{q=1}^{\infty} \bigcup \left\{ \mathcal{U}(x, T, \gamma, q) : (x, T) \in K \times \mathcal{M}_c, \gamma \in (0, 1) \right\}.$$

Clearly, $\mathcal{F}$ is a countable intersection of open and everywhere dense subsets of $K \times \mathcal{M}_c$.

Let $(z, B) \in \mathcal{F}$ and $\varepsilon > 0$ be given. There exists an integer $q \geq 1$ such that

$$2^{-q} < 4^{-1}\varepsilon. \tag{5.176}$$

There exist $x \in K$, $T \in \mathcal{M}_c$, and $\gamma \in (0, 1)$ such that

$$(z, B) \in \mathcal{U}(x, T, \gamma, q). \tag{5.177}$$

By (5.176) and property (Pvii), the following property also holds:

(Pviii) For each $(y, S) \in \mathcal{U}(x, T, \gamma, q)$ and each integer $n \geq N(x, T, \gamma, q)$,

$$\left\| S^n y - z(x, T, \gamma, q) \right\| \leq 4^{-1}\varepsilon. \tag{5.178}$$

Since ε is an arbitrary positive number, we conclude that $\{B^n z\}_{n=1}^{\infty}$ is a Cauchy sequence and there exists $\lim_{n\to\infty} B^n z$. Property (Pviii) and (5.177) now imply that

$$\left\| \lim_{n\to\infty} B^n z - z(x, T, \gamma, q) \right\| \leq 4^{-1}\varepsilon. \tag{5.179}$$

Since ε is an arbitrary positive number, we conclude that $\lim_{n\to\infty} B^n z \in F$. It follows from (5.179) and property (Pviii) that for each $(y, S) \in \mathcal{U}(x, T, \gamma, q)$ and each integer $n \geq N(x, T, \gamma, q)$,

$$\left\| S^n y - \lim_{i\to\infty} B^i z \right\| \leq 2^{-1}\varepsilon.$$

This completes the proof of Theorem 5.33. □

Proof of Theorem 5.34 Assume that K_0 is a nonempty, closed and separable subset of K. Let $\{x_j\}_{j=1}^{\infty} \subset K_0$ be a sequence such that K_0 is the closure of $\{x_j\}_{j=1}^{\infty}$. For each integer $p \geq 1$, there exists by Theorem 5.32 a set $\mathcal{F}_p \subset \mathcal{M}_c$ which is a countable intersection of open and everywhere dense subsets of $\mathcal{M}_c$ such that for each $T \in \mathcal{F}_p$, the following properties hold:

C(i) There exists $\lim_{n\to\infty} T^n x_p \in F$.

C(ii) For each $\varepsilon > 0$, there exist a neighborhood U of T in $\mathcal{M}_c$, a number $\delta > 0$ and a natural number N such that for each $S \in U$, each $y \in K$ satisfying $\|y - x_p\| \le \delta$ and each integer $m \ge N$,

$$\left\| S^m y - \lim_{n\to\infty} T^n x_p \right\| \le \varepsilon.$$

Set

$$\mathcal{F} = \bigcap_{p=1}^{\infty} \mathcal{F}_p. \tag{5.180}$$

Clearly, $\mathcal{F}$ is a countable intersection of open and everywhere dense subsets of $\mathcal{M}_c$.

Assume that $T \in \mathcal{F}$. Then for each $p \ge 1$, there exists $\lim_{n\to\infty} T^n x_p \in F$.

Now we construct the set $\mathcal{K}_T \subset K_0$. By property C(ii), for each pair of natural numbers q, i, there exist a neighborhood $\mathcal{U}(q,i)$ of T in $\mathcal{M}_c$, a number $\delta(q,i) > 0$ and a natural number $N(q,i)$ such that the following property holds:

C(iii) For each $S \in \mathcal{U}(q,i)$, each $y \in K$ satisfying $\|y - x_q\| \le \delta(q,i)$, and each integer $m \ge N(q,i)$,

$$\left\| S^m y - \lim_{n\to\infty} T^n x_q \right\| \le 2^{-i}.$$

Define

$$\mathcal{K}_T := \bigcap_{n=1}^{\infty} \bigcup \{\{y \in K_0 : \|y - x_q\| < \delta(q,i)\} : q \ge 1, i \ge n\}. \tag{5.181}$$

Clearly, $\mathcal{K}_T$ is a countable intersection of open and everywhere dense subsets of K_0.

Assume that $x \in \mathcal{K}_T$ and $\varepsilon > 0$ are given. There exists an integer $n \ge 1$ such that

$$2^{-n} < 4^{-1}\varepsilon. \tag{5.182}$$

By (5.181), there exist a natural number q and an integer $i \ge n$ such that

$$\|x - x_q\| < \delta(q,i). \tag{5.183}$$

It follows from (5.182) and C(iii) that the following property also holds:

C(iv) For each $S \in \mathcal{U}(q,i)$, each $y \in K$ satisfying $\|y - x_q\| \le \delta(q,i)$, and each integer $m \ge N(q,i)$,

$$\left\| S^m y - \lim_{j\to\infty} T^j x_q \right\| \le 4^{-1}\varepsilon.$$

By property C(iv) and (5.183),

$$\left\| T^m x - \lim_{j\to\infty} T^j x_q \right\| \le 4^{-1}\varepsilon$$

for all integers $m \geq N(q, i)$. Since ε is an arbitrary positive number, we conclude that $\{T^m x\}_{m=1}^{\infty}$ is a Cauchy sequence and there exists $\lim_{m\to\infty} T^m x$. We also have

$$\left\| \lim_{m\to\infty} T^m x - \lim_{m\to\infty} T^m x_q \right\| \leq 4^{-1}\varepsilon. \tag{5.184}$$

Since $\lim_{m\to\infty} T^m x_q \in F$, we conclude that $\lim_{m\to\infty} T^m x$ also belongs to F. By (5.184) and property C(iv), for each $S \in \mathcal{U}(q, i)$, each $y \in K$ satisfying $\|y - x\| < \delta(q, i) - \|x - x_q\|$, and each integer $m \geq N(q, i)$, we have

$$\left\| S^m y - \lim_{j\to\infty} T^j x \right\| \leq 2^{-1}\varepsilon.$$

Theorem 5.34 is proved. □

Chapter 6
Infinite Products

6.1 Nonexpansive and Uniformly Continuous Operators

In this section we discuss several results concerning the asymptotic behavior of (random) infinite products of generic sequences of nonexpansive as well as uniformly continuous operators on bounded, closed and convex subsets of a Banach space. These results were obtained in [129]. In addition to weak ergodic theorems, we also study convergence to a unique common fixed point and more generally, to a nonexpansive retraction. Infinite products of operators find application in many areas of mathematics (see, for example, [17, 18, 38, 57] and the references mentioned there). More precisely, we show that in appropriate spaces of sequences of operators there exists a subset which is a countable intersection of open and everywhere dense sets such that for each sequence belonging to this subset, the corresponding infinite product converges.

Let X be a Banach space normed by $\|\cdot\|$ and let K be a nonempty, bounded, closed and convex subset of X with the topology induced by the norm $\|\cdot\|$.

Denote by $\mathcal{A}$ the set of all sequences $\{A_t\}_{t=1}^{\infty}$, where each $A_t : K \to K$ is a continuous operator, $t = 1, 2, \ldots$. Such a sequence will occasionally be denoted by a boldface $\mathbf{A}$.

For the set $\mathcal{A}$ we consider the metric $\rho_s : \mathcal{A} \times \mathcal{A} \to [0, \infty)$ defined by

$$\begin{aligned}&\rho_s\big(\{A_t\}_{t=1}^{\infty}, \{B_t\}_{t=1}^{\infty}\big) = \sup\big\{\|A_t x - B_t x\| : x \in K, t = 1, 2, \ldots\big\},\\&\{A_t\}_{t=1}^{\infty}, \{B_t\}_{t=1}^{\infty} \in \mathcal{A}.\end{aligned} \tag{6.1}$$

It is easy to see that the metric space $(\mathcal{A}, \rho_s)$ is complete. The topology generated in $\mathcal{A}$ by the metric ρ_s will be called the strong topology.

In addition to this topology on $\mathcal{A}$, we will also consider the uniformity determined by the base

$$\begin{aligned}E(N, \varepsilon) = \big\{&\big(\{A_t\}_{t=1}^{\infty}, \{B_t\}_{t=1}^{\infty}\big) \in \mathcal{A} \times \mathcal{A} :\\&\|A_t x - B_t x\| \le \varepsilon, t = 1, \ldots, N, x \in K\big\},\end{aligned} \tag{6.2}$$

S. Reich, A.J. Zaslavski, *Genericity in Nonlinear Analysis*,
Developments in Mathematics 34, DOI 10.1007/978-1-4614-9533-8_6,

where N is a natural number and $\varepsilon > 0$. It is easy to see that the space $\mathcal{A}$ with this uniformity is metrizable (by a metric $\rho_w : \mathcal{A} \times \mathcal{A} \to [0, \infty)$) and complete. The topology generated by ρ_w will be called the weak topology.

An operator $A : K \to K$ is called nonexpansive if

$$\|Ax - Ay\| \le \|x - y\| \quad \text{for all } x, y \in K.$$

Define

$$\mathcal{A}_{ne} = \big\{\{A_t\}_{t=1}^{\infty} \in \mathcal{A} : A_t \text{ is nonexpansive for } t = 1, 2, \ldots\big\}. \tag{6.3}$$

Clearly, $\mathcal{A}_{ne}$ is a closed subset of $\mathcal{A}$ in the weak topology. We will consider the topological subspace $\mathcal{A}_{ne} \subset \mathcal{A}$ with both the relative weak and strong topologies.

We will show (Theorem 6.1) that for a generic sequence $\{C_t\}_{t=1}^{\infty}$ in the space $\mathcal{A}_{ne}$ with the weak topology,

$$\|C_T \cdot \cdots \cdot C_1 x - C_T \cdot \cdots \cdot C_1 y\| \to 0,$$

uniformly for all $x, y \in K$. We will also prove Theorem 6.2 which shows that for a generic sequence $\{C_t\}_{t=1}^{\infty}$ in $\mathcal{A}_{ne}$ with the strong topology, this type of uniform convergence holds for random products of the operators $\{C_t\}_{t=1}^{\infty}$. (Such results are usually called weak ergodic theorems in the population biology literature [43].)

We will say that a set E of operators $A : K \to K$ is uniformly equicontinuous (ue) if for any $\varepsilon > 0$, there exists $\delta > 0$ such that $\|Ax - Ay\| \le \varepsilon$ for all $A \in E$ and all $x, y \in K$ satisfying $\|x - y\| \le \delta$.

Define

$$\mathcal{A}_{ue} = \big\{\{A_t\}_{t=1}^{\infty} \in \mathcal{A} : \{A_t\}_{t=1}^{\infty} \text{ is a (ue) set}\big\}. \tag{6.4}$$

It is clear that $\mathcal{A}_{ue}$ is a closed subset of $\mathcal{A}$ in the strong topology.

We will consider the topological subspace $\mathcal{A}_{ue} \subset \mathcal{A}$ with the relative weak and strong topologies.

Denote by $\mathcal{A}_{ne}^*$ the set of all $\{A_t\}_{t=1}^{\infty} \in \mathcal{A}_{ne}$ such that

$$\bigcap_{t=1}^{\infty} \{x \in K : A_t x = x\} \ne \emptyset,$$

and denote by $\bar{\mathcal{A}}_{ne}^*$ the closure of $\mathcal{A}_{ne}^*$ in the strong topology of the space $\mathcal{A}_{ne}$.

Denote by $\mathcal{A}_{ue}^*$ the set of all $\mathbf{A} = \{A_t\}_{t=1}^{\infty} \in \mathcal{A}_{ue}$ for which there exists $x(\mathbf{A}) \in K$ such that for each integer $t \ge 1$,

$$A_t x(\mathbf{A}) = x(\mathbf{A}), \qquad \big\|A_t y - x(\mathbf{A})\big\| \le \big\|y - x(\mathbf{A})\big\| \quad \text{for all } y \in K,$$

and denote by $\bar{\mathcal{A}}_{ue}^*$ the closure of $\mathcal{A}_{ue}^*$ in the strong topology of the space $\mathcal{A}_{ue}$.

We will consider the topological subspaces $\bar{\mathcal{A}}_{ne}^*$ and $\bar{\mathcal{A}}_{ue}^*$ with the relative strong topologies and show (Theorems 6.3 and 6.4) that for a generic sequence $\{C_t\}_{t=1}^{\infty}$ in the space $\bar{\mathcal{A}}_{ne}^*$ ($\bar{\mathcal{A}}_{ue}^*$, respectively), there exists a unique common fixed point x_*

and all random products of the operators $\{C_t\}_{t=1}^{\infty}$ converge to x_*, uniformly for all $x \in K$.

Assume that F is a nonempty, closed and convex subset of K, and $Q : K \to F$ is a nonexpansive operator such that

$$Qx = x, \quad x \in F. \tag{6.5}$$

(Such an operator Q is usually called a nonexpansive retraction of K onto F [68].)

Denote by $\mathcal{A}_{ne}^{(F)}$ the set of all $\{A_t\}_{t=1}^{\infty} \in \mathcal{A}_{ne}$ such that

$$A_t x = x, \quad x \in F, t = 1, 2, \ldots. \tag{6.6}$$

It is clear that $\mathcal{A}_{ne}^{(F)}$ is a closed subset of $\mathcal{A}_{ne}$ in the weak topology.

We will consider the topological subspace $\mathcal{A}_{ne}^{(F)} \subset \mathcal{A}_{ne}$ with both the relative weak and strong topologies.

We will show (see Theorem 6.5) that for a generic sequence of operators $\{B_t\}_{t=1}^{\infty}$ in the space $\mathcal{A}_{ne}^{(F)}$ with the weak topology, there exists a nonexpansive retraction $P_* : K \to F$ such that

$$B_t \cdot \cdots \cdot B_1 x \to P_* x \quad \text{as } t \to \infty,$$

uniformly for all $x \in K$. We will also prove Theorem 6.6, which shows that for a generic sequence of operators $\{B_t\}_{t=1}^{\infty}$ in the space $\mathcal{A}_{ne}^{(F)}$ with the strong topology, all its random products

$$B_{r(t)} \cdot \cdots \cdot B_{r(1)} x$$

also converge to a nonexpansive retraction $P_r : K \to F$, uniformly for all $x \in K$, where $r : \{1, 2, \ldots\} \to \{1, 2, \ldots\}$. Finally, we will prove Theorem 6.7, which extends Theorem 6.6 to a larger class of operators described in Sect. 6.3.

In Sect. 6.4 we also point out that our results can, in fact, be extended to all hyperbolic spaces.

6.2 Asymptotic Behavior

In this section we will first formulate precisely our weak ergodic theorems [129].

Theorem 6.1 *There exists a set $\mathcal{F} \subset \mathcal{A}_{ne}$, which is a countable intersection of open (in the weak topology) everywhere dense (in the strong topology) subsets of $\mathcal{A}_{ne}$, such that for each $\{B_t\}_{t=1}^{\infty} \in \mathcal{F}$ and each $\varepsilon > 0$, there exist a neighborhood U of $\{B_t\}_{t=1}^{\infty}$ in $\mathcal{A}_{ne}$ with the weak topology and a natural number N such that:*

For each $\{C_t\}_{t=1}^{\infty} \in U$, each $x, y \in K$ and each integer $T \geq N$,

$$\|C_T \cdot \cdots \cdot C_1 x - C_T \cdot \cdots \cdot C_1 y\| \leq \varepsilon.$$

Theorem 6.2 *There exists a set $\mathcal{F} \subset \mathcal{A}_{ne}$, which is a countable intersection of open everywhere dense (in the strong topology) subsets of $\mathcal{A}_{ne}$, such that for each*

$\{B_t\}_{t=1}^{\infty} \in \mathcal{F}$ *and each* $\varepsilon > 0$, *there exist a neighborhood* U *of* $\{B_t\}_{t=1}^{\infty}$ *in* $\mathcal{A}_{ne}$ *with the strong topology and a natural number* N *such that*:

For each $\{C_t\}_{t=1}^{\infty} \in U$, *each* $x, y \in K$, *each integer* $T \geq N$ *and each mapping* $r : \{1, \ldots, T\} \to \{1, 2, \ldots\}$,

$$\|C_{r(T)} \cdot \cdots \cdot C_{r(1)}x - C_{r(T)} \cdot \cdots \cdot C_{r(1)}y\| \leq \varepsilon.$$

The following theorems [129] establish generic convergence to a unique fixed point.

Theorem 6.3 *There exists a set* $\mathcal{F} \subset \bar{\mathcal{A}}_{ne}^*$, *which is a countable intersection of open everywhere dense (in the strong topology) subsets of* $\bar{\mathcal{A}}_{ne}^*$, *such that for each* $\{B_t\}_{t=1}^{\infty} \in \mathcal{F}$, *there exists* $x_* \in K$ *for which the following assertions hold*:

1. $B_t x_* = x_*$, $t = 1, 2, \ldots$.
2. *For each* $\varepsilon > 0$, *there exist a neighborhood* U *of* $\{B_t\}_{t=1}^{\infty}$ *in* $\bar{\mathcal{A}}_{ne}^*$ *with the strong topology and a natural number* N *such that for each* $\{C_t\}_{t=1}^{\infty} \in U$, *each integer* $T \geq N$, *each mapping* $r : \{1, \ldots, T\} \to \{1, 2, \ldots\}$ *and each* $x \in K$,

$$\|C_{r(T)} \cdot \cdots \cdot C_{r(1)}x - x_*\| \leq \varepsilon.$$

Theorem 6.4 *There exists a set* $\mathcal{F} \subset \bar{\mathcal{A}}_{ue}^*$, *which is a countable intersection of open everywhere dense (in the strong topology) subsets of* $\bar{\mathcal{A}}_{ue}^*$, *such that for each* $\{B_t\}_{t=1}^{\infty} \in \mathcal{F}$, *there exists* $x_* \in K$ *for which the following assertions hold*:

1. $B_t x_* = x_*$, $t = 1, 2, \ldots$,

$$\|B_t y - x_*\| \leq \|y - x_*\|, \quad y \in K, t = 1, 2, \ldots.$$

2. *For each* $\varepsilon > 0$, *there exist a neighborhood* U *of* $\{B_t\}_{t=1}^{\infty}$ *in* $\bar{\mathcal{A}}_{ue}^*$ *with the strong topology and a natural number* N *such that for each* $\{C_t\}_{t=1}^{\infty} \in U$, *each integer* $T \geq N$, *each mapping* $r : \{1, \ldots, T\} \to \{1, 2, \ldots\}$ *and each* $x \in K$,

$$\|C_{r(T)} \cdot \cdots \cdot C_{r(1)}x - x_*\| \leq \varepsilon.$$

One can easily construct an example of a sequence of operators $\{A_t\}_{t=1}^{\infty} \in \mathcal{A}_{ue}^*$ for which the convergence properties described in Theorems 6.1–6.3 do not hold. Namely, they do not hold for a sequence each term of which is the identity operator.

6.3 Nonexpansive Retractions

In this section we assume that F is a nonempty, closed and convex subset of K, and that $Q : K \to F$ is a nonexpansive retraction, namely

$$Qx = x, \quad x \in F, \tag{6.7}$$

$$\|Qx - Qy\| \leq \|x - y\|, \quad x, y \in K. \tag{6.8}$$

The following two theorems [129] establish generically uniform convergence of (random) infinite products to nonexpansive retractions.

Theorem 6.5 *There exists a set $\mathcal{F} \subset \mathcal{A}_{ne}^{(F)}$, which is a countable intersection of open (in the weak topology) everywhere dense (in the strong topology) subsets of $\mathcal{A}_{ne}^{(F)}$, such that for each $\{B_t\}_{t=1}^{\infty} \in \mathcal{F}$, the following two assertions hold:*

1. *There exists an operator $P_* : K \to F$ such that*

$$\lim_{t\to\infty} B_t \cdot \dots \cdot B_1 x = P_* x \quad \text{for each } x \in K.$$

2. *For each $\varepsilon > 0$, there exist a neighborhood U of $\{B_t\}_{t=1}^{\infty}$ in $\mathcal{A}_{ne}^{(F)}$ with the weak topology and a natural number N such that for each $\{C_t\}_{t=1}^{\infty} \in U$, each integer $T \geq N$ and each $x \in K$,*

$$\|C_T \cdot \dots \cdot C_1 x - P_* x\| \leq \varepsilon.$$

Theorem 6.6 *There exists a set $\mathcal{F} \subset \mathcal{A}_{ne}^{(F)}$, which is a countable intersection of open everywhere dense subsets of $\mathcal{A}_{ne}^{(F)}$ (in the strong topology), such that for each $\{B_t\}_{t=1}^{\infty} \in \mathcal{F}$, the following two assertions hold:*

1. *For each $r : \{1, 2, \dots\} \to \{1, 2, \dots\}$, there exists an operator $P_r : K \to F$ such that*

$$\lim_{T\to\infty} B_{r(T)} \cdot \dots \cdot B_{r(1)} x = P_r x \quad \text{for each } x \in K.$$

2. *For each $\varepsilon > 0$, there exist a neighborhood U of $\{B_t\}_{t=1}^{\infty}$ in the space $\mathcal{A}_{ne}^{(F)}$ with the strong topology and a natural number N such that for each $\{C_t\}_{t=1}^{\infty} \in U$, each mapping $r : \{1, 2, \dots\} \to \{1, 2, \dots\}$, each integer $T \geq N$ and each $x \in K$,*

$$\|C_{r(T)} \cdot \dots \cdot C_{r(1)} x - P_r x\| \leq \varepsilon.$$

In our next result [129] we extend Theorem 6.6 to a subspace of $\mathcal{A}_{ue}$ consisting of sequences of quasi-nonexpansive operators. More precisely, we now assume that F is a nonempty, closed and convex subset of K and $Q : K \to F$ is a uniformly continuous operator such that

$$Qx = x, \quad x \in F, \qquad \|Qy - x\| \leq \|y - x\|, \quad y \in K, x \in F. \tag{6.9}$$

Denote by $\mathcal{A}_{ue}^{(F)}$ the set of all $\{A_t\}_{t=1}^{\infty} \in \mathcal{A}_{ue}$ such that for each integer $t \geq 1$,

$$A_t x = x, \quad x \in F, \qquad \|A_t y - x\| \leq \|y - x\|, \quad y \in K, x \in F. \tag{6.10}$$

It is clear that $\mathcal{A}_{ue}^{(F)}$ is a closed subset of $\mathcal{A}_{ue}$ in the strong topology.

We will consider the topological subspace $\mathcal{A}_{ue}^{(F)}$ with the relative strong topology and establish the following result.

Theorem 6.7 *There exists a set $\mathcal{F} \subset \mathcal{A}_{ue}^{(F)}$, which is a countable intersection of open everywhere dense subsets of $\mathcal{A}_{ue}^{(F)}$ (in the strong topology), such that for each $\{B_t\}_{t=1}^{\infty} \in \mathcal{F}$, the following two assertions hold:*

1. *For each mapping $r : \{1, 2, \ldots\} \to \{1, 2, \ldots\}$, there exists a uniformly continuous operator $P_r : K \to F$ such that*

$$\lim_{T\to\infty} B_{r(T)} \cdot \cdots \cdot B_{r(1)}x = P_r x \quad \textit{for each } x \in K.$$

2. *For each $\varepsilon > 0$, there exist a neighborhood U of $\{B_t\}_{t=1}^{\infty}$ in the space $\mathcal{A}_{ue}^{(F)}$ with the strong topology and a natural number N such that for each $\{C_t\}_{t=1}^{\infty} \in U$, each mapping $r : \{1, 2, \ldots\} \to \{1, 2, \ldots\}$, each integer $T \geq N$ and each $x \in K$,*

$$\|C_{r(T)} \cdot \cdots \cdot C_{r(1)}x - P_r x\| \leq \varepsilon.$$

6.4 Preliminary Results

In this section we will prove three auxiliary lemmas which will be used in the proofs of Theorems 6.1–6.7.

For each bounded operator $A : K \to X$, we set

$$\|A\| = \sup\{\|Ax\| : x \in K\}. \tag{6.11}$$

For each $x \in K$ and each $E \subset X$, we set

$$d(x, E) = \inf\{\|x - y\| : y \in E\}, \qquad \mathrm{rad}(E) = \sup\{\|y\| : y \in E\}. \tag{6.12}$$

Lemma 6.8 *Assume that F is a nonempty, closed and convex subset of K, $Q : K \to F$ and $A : K \to K$ are continuous operators such that*

$$\begin{aligned} &Qx = x, \quad x \in F, \qquad \|Qy - x\| \leq \|y - x\| \quad \textit{for all } y \in K \textit{ and } x \in F,\\ &Ax = x, \quad x \in F, \qquad \|Ay - x\| \leq \|y - x\| \quad \textit{for all } y \in K \textit{ and } x \in F, \end{aligned} \tag{6.13}$$

and $\gamma \in (0, 1)$. Define an operator $B : K \to K$ by

$$Bx = (1 - \gamma)Ax + \gamma Qx, \quad x \in K.$$

Then

$$Bx = x, \quad x \in F, \qquad \|By - x\| \leq \|y - x\| \quad \textit{for all } y \in K \textit{ and } x \in F,$$

and

$$d(Bx, F) \leq (1 - \gamma)d(x, F), \quad x \in K. \tag{6.14}$$

Moreover, if A and Q are nonexpansive, then B is nonexpansive.

Proof It is sufficient to show that (6.14) is valid. Let $x \in K$ and $\varepsilon > 0$ be given. There exists $z \in F$ such that

$$\|x - z\| \leq d(x, F) + \varepsilon.$$

It is easy to verify that $\gamma Qx + (1 - \gamma)Az \in F$. Hence

$$\begin{aligned} d(Bx, F) &\leq \big\| \big((1-\gamma)Ax + \gamma Qx\big) - \big(\gamma Qx + (1-\gamma)Az\big) \big\| \\ &\leq (1-\gamma)\|x - z\| \leq (1-\gamma)d(x, F) + (1-\gamma)\varepsilon. \end{aligned}$$

Since ε is any positive number, we conclude that (6.14) holds. The lemma is proved. □

Lemma 6.9 *Assume that E is a nonempty uniformly continuous set of operators $A : K \to K$, N is a natural number and ε is a positive number. Then there exists a number $\delta > 0$ such that for each sequence $\{A_t\}_{t=1}^N \subset E$, each sequence $\{B_t\}_{t=1}^N$, where the (not necessarily continuous) operators $B_t : K \to K$, $t = 1, \dots, N$, satisfy*

$$\|B_t - A_t\| \leq \delta, \quad t = 1, \dots, N, \tag{6.15}$$

and each $x \in K$, the following inequality holds:

$$\|B_N \cdot \dots \cdot B_1 x - A_N \cdot \dots \cdot A_1 x\| \leq \varepsilon. \tag{6.16}$$

Proof Set

$$\varepsilon_N = (4N)^{-1}\varepsilon. \tag{6.17}$$

By induction we define a sequence of positive numbers $\{\varepsilon_i\}_{i=0}^N$ such that for each $i \in \{1, \dots, N\}$,

$$\varepsilon_{i-1} < (4N)^{-1}\varepsilon_i, \tag{6.18}$$

and for each $A \in E$ and each $x, y \in K$ satisfying $\|x - y\| \leq \varepsilon_{i-1}$, the following inequality holds:

$$\|Ax - Ay\| \leq 2^{-1}\varepsilon_i. \tag{6.19}$$

Set $\delta = \varepsilon_0$.

Assume that $\{A_t\}_{t=1}^N \subset E$, $B_t : K \to K$, $t = 1, \dots, N$, and that (6.15) holds. We will show that (6.16) is valid for each $x \in K$.

Let $x \in K$. We will show by induction that for $t = 1, \dots, N$,

$$\|B_t \cdot \dots \cdot B_1 x - A_t \cdot \dots \cdot A_1 x\| \leq \varepsilon_t. \tag{6.20}$$

Inequalities (6.15) and (6.18) imply that $\|B_1 x - A_1 x\| < \varepsilon_1$.

Assume that $t \in \{1, \dots, N\}$, $t < N$, and

$$\|B_t \cdot \dots \cdot B_1 x - A_t \cdot \dots \cdot A_1 x\| \leq \varepsilon_t.$$

It follows from the definition of ε_t (see (6.18), (6.19)) and (6.15) that

$$\|A_{t+1}B_t \cdot \dots \cdot B_1x - A_{t+1}A_t \cdot \dots \cdot A_1x\| \le 2^{-1}\varepsilon_{t+1},$$
$$\|A_{t+1}B_t \cdot \dots \cdot B_1x - B_{t+1}B_t \cdot \dots \cdot B_1x\| \le \delta$$

and

$$\|A_{t+1}A_t \cdot \dots \cdot A_1x - B_{t+1}B_t \cdot \dots \cdot B_1x\| \le \varepsilon_{t+1}.$$

Thus we have shown by induction that (6.20) holds for $t = 1, \dots, N$. This implies that (6.16) is valid and the lemma is proved. □

Lemma 6.10 *Assume that F is a nonempty, closed and convex subset of K, $Q : K \to F$ is a uniformly continuous operator such that*

$$Qx = x, \quad x \in F, \qquad \|Qy - x\| \le \|y - x\| \quad \text{for all } y \in K \text{ and } x \in F,$$

$\varepsilon > 0$, $\gamma \in (0, 1)$ and E is a nonempty uniformly continuous set of operators $A : K \to K$ such that for each $A \in E$, the following relations hold:

$$Ax = x, \quad x \in F, \qquad \|Ay - x\| \le \|y - x\| \quad \text{for all } y \in K \text{ and } x \in F.$$

Let $N \ge 1$ be an integer such that

$$(1 - \gamma)^N \big(\mathrm{rad}(K)\big) < 16^{-1}\varepsilon. \tag{6.21}$$

For each $A \in E$, define an operator $A_\gamma : K \to K$ by

$$A_\gamma x = (1 - \gamma)Ax + \gamma Qx, \quad x \in K.$$

Then the set $\{A_\gamma : A \in E\}$ is uniformly continuous and there exists a number $\delta > 0$ such that for each sequence $\{C_t\}_{t=1}^N \subset \{A_\gamma : A \in E\}$, each sequence of (not necessarily continuous) operators $B_t : K \to K$, $t = 1, \dots, N$, satisfying

$$\|B_t - C_t\| \le \delta, \quad t = 1, \dots, N, \tag{6.22}$$

the following inequality holds:

$$d(B_N \cdot \dots \cdot B_1x, F) \le \varepsilon, \quad x \in K.$$

Proof Evidently, the set $\{A_\gamma : A \in E\}$ is uniformly continuous. By Lemma 6.8 and (6.21), for each sequence $\{C_t\}_{t=1}^N \subset \{A_\gamma : A \in E\}$ and each $x \in K$, the following inequality holds:

$$d(C_N \cdot \dots \cdot C_1x, F) \le (1 - \gamma)^N d(x, F) < 8^{-1}\varepsilon. \tag{6.23}$$

Applying Lemma 6.9 with the uniformly continuous set $\{A_\gamma : A \in E\}$, we obtain that there exists a number $\delta > 0$ such that for each sequence $\{C_t\}_{t=1}^N \subset \{A_\gamma : A \in E\}$

and each sequence of operators $B_t : K \to K$, $t = 1, \dots, N$, satisfying (6.22), the following inequality holds:

$$\|B_N \cdot \dots \cdot B_1 x - C_N \cdot \dots \cdot C_1 x\| \leq 8^{-1}\varepsilon, \quad x \in K. \tag{6.24}$$

Assume that $\{C_t\}_{t=1}^N \subset \{A_\gamma : A \in E\}$, $B_t : K \to K, t = 1, \dots, N$, and that (6.22) holds. Then (6.23) and (6.24) are valid for each $x \in K$. This implies that

$$d(B_N \cdot \dots \cdot B_1 x, F) \leq \varepsilon, \quad x \in K.$$

The lemma is proved. □

6.5 Proofs of Theorems 6.1 and 6.2

Fix $x_* \in K$. Let $\{A_t\}_{t=1}^\infty \in \mathcal{A}_{ne}$ and $\gamma \in (0, 1)$. For $t = 1, 2, \dots$, define $A_{t\gamma} : K \to K$ by

$$A_{t\gamma} x = (1 - \gamma) A_t x + \gamma x_*, \quad x \in K. \tag{6.25}$$

Clearly, $\{A_{t\gamma}\}_{t=1}^\infty \in \mathcal{A}_{ne}$ and

$$\|A_{t\gamma} x - A_{t\gamma} y\| \leq (1 - \gamma)\|x - y\|, \quad x, y \in K, t = 1, 2, \dots. \tag{6.26}$$

It is easy to see that the set

$$\left\{\{A_{t\gamma}\}_{t=1}^\infty : \{A_t\}_{t=1}^\infty \in \mathcal{A}_{ne}, \gamma \in (0, 1)\right\}$$

is an everywhere dense subset of $\mathcal{A}_{ne}$ in the strong topology.

Proof of Theorem 6.1 Let $\{A_t\}_{t=1}^\infty \in \mathcal{A}_{ne}$, $\gamma \in (0, 1)$ and let $i \geq 1$ be an integer. Choose a natural number $N(\gamma, i)$ such that

$$(1 - \gamma)^{N(\gamma,i)} \operatorname{rad}(K) \leq 2^{-i-4}. \tag{6.27}$$

Inequalities (6.26) and (6.27) imply that

$$\|A_{T\gamma} \cdot \dots \cdot A_{1\gamma} x - A_{T\gamma} \cdot \dots \cdot A_{1\gamma} y\| \leq 2^{-i-3}$$
$$\text{for all } x, y \in K \text{ and all integers } T \geq N(\gamma, i). \tag{6.28}$$

By Lemma 6.9, there exists an open neighborhood $U(\{A_t\}_{t=1}^\infty, \gamma, i)$ of $\{A_{t\gamma}\}_{t=1}^\infty$ in the space $\mathcal{A}_{ne}$ with the weak topology such that for each $\{B_t\}_{t=1}^\infty \in U(\{A_t\}_{t=1}^\infty, \gamma, i)$ and each $x \in K$,

$$\|A_{N(\gamma,i)\gamma} \cdot \dots \cdot A_{1\gamma} x - B_{N(\gamma,i)} \cdot \dots \cdot B_1 x\| \leq 2^{-i-3}.$$

Together with (6.28) this implies that for each $\{B_t\}_{t=1}^{\infty} \in U(\{A_t\}_{t=1}^{\infty}, \gamma, i)$, each $x, y \in K$ and each integer $T \geq N(\gamma, i)$,

$$\|B_T \cdot \cdots \cdot B_1 x - B_T \cdot \cdots \cdot B_1 y\| \leq 2^{-i-1}. \tag{6.29}$$

Define

$$\mathcal{F} = \bigcap_{q=1}^{\infty} \bigcup \{U(\{A_t\}_{t=1}^{\infty}, \gamma, i) : \{A_t\}_{t=1}^{\infty} \in \mathcal{A}_{ne}, \gamma \in (0,1), i = q, q+1, \ldots\}.$$

Evidently $\mathcal{F}$ is a countable intersection of open (in the weak topology) everywhere dense (in the strong topology) subsets of $\mathcal{A}_{ne}$.

Assume that $\{B_t\}_{t=1}^{\infty} \in \mathcal{F}$ and $\varepsilon > 0$. Choose a natural number q such that

$$2^{4-q} < \varepsilon. \tag{6.30}$$

There exist $\{A_t\}_{t=1}^{\infty} \in \mathcal{A}_{ne}$, $\gamma \in (0,1)$ and an integer $i \geq q$ such that

$$\{B_t\}_{t=1}^{\infty} \in U(\{A_t\}_{t=1}^{\infty}, \gamma, i).$$

It follows from (5.29) and (6.30) that for each $\{C_t\}_{t=1}^{\infty} \in U(\{A_t\}_{t=1}^{\infty}, \gamma, i)$, each pair of points $x, y \in K$ and each integer $T \geq N(\gamma, i)$,

$$\|C_T \cdot \cdots \cdot C_1 x - C_T \cdot \cdots \cdot C_1 y\| \leq 2^{-i-1} \leq 2^{-q-1} < \varepsilon.$$

This completes the proof of the theorem. □

Proof of Theorem 6.2 Let $\{A_t\}_{t=1}^{\infty} \in \mathcal{A}_{ne}$, $\gamma \in (0,1)$ and let $i \geq 1$ be an integer. Choose a natural number $N(\gamma, i)$ such that (6.27) is valid. Inequalities (6.26) and (6.27) imply that for each integer $T \geq N(\gamma, i)$, each $r : \{1, \ldots, T\} \to \{1, 2, \ldots\}$ and each $x, y \in K$,

$$\|A_{r(T)\gamma} \cdot \cdots \cdot A_{r(1)\gamma} x - A_{r(T)\gamma} \cdot \cdots \cdot A_{r(1)\gamma} y\| \leq 2^{-i-3}. \tag{6.31}$$

By Lemma 6.9, there is an open neighborhood $U(\{A_t\}_{t=1}^{\infty}, \gamma, i)$ of $\{A_{t\gamma}\}_{t=1}^{\infty}$ in the space $\mathcal{A}_{ne}$ with the strong topology such that for each $\{C_t\}_{t=1}^{\infty} \in U(\{A_t\}_{t=1}^{\infty}, \gamma, i)$, each $x \in K$ and each $r : \{1, \ldots, N(\gamma, i)\} \to \{1, 2, \ldots\}$, the following inequality holds:

$$\|A_{r(N(\gamma,i))\gamma} \cdot \cdots \cdot A_{r(1)\gamma} x - C_{r(N(\gamma,i))} \cdot \cdots \cdot C_{r(1)} x\| \leq 2^{-i-3}.$$

Together with (6.31) this implies that the following property holds:

(a) For each $\{C_t\}_{t=1}^{\infty} \in U(\{A_t\}_{t=1}^{\infty}, \gamma, i)$, each integer $T \geq N(\gamma, i)$, each $x, y \in K$ and each $r : \{1, \ldots, T\} \to \{1, 2, \ldots\}$, the following inequality is valid:

$$\|C_{r(T)} \cdot \cdots \cdot C_{r(1)} x - C_{r(T)} \cdot \cdots \cdot C_{r(1)} y\| \leq 2^{-i-1}. \tag{6.32}$$

Define

$$\mathcal{F} = \bigcap_{q=1}^{\infty} \bigcup \{U(\{A_t\}_{t=1}^{\infty}, \gamma, i) : \{A_t\}_{t=1}^{\infty} \in \mathcal{A}_{ne}, \gamma \in (0,1), i = q, q+1, \ldots\}.$$

Evidently, $\mathcal{F}$ is a countable intersection of open everywhere dense (in the strong topology) subsets of $\mathcal{A}_{ne}$.

Assume that $\{B_t\}_{t=1}^{\infty} \in \mathcal{F}$ and $\varepsilon > 0$ are given. Choose a natural number q which satisfies (5.30).

There exist $\{A_t\}_{t=1}^{\infty} \in \mathcal{A}_{ne}$, $\gamma \in (0,1)$ and an integer $i \geq q$ such that

$$\{B_t\}_{t=1}^{\infty} \in U(\{A_t\}_{t=1}^{\infty}, \gamma, i).$$

The validity of Theorem 6.2 now follows from property (a) and (6.30). □

6.6 Proofs of Theorems 6.3 and 6.4

Here we prove Theorem 6.4. Theorem 6.3 is proved analogously.

Proof of Theorem 6.4 For each $\mathbf{A} = \{A_t\}_{t=1}^{\infty} \in \mathcal{A}_{ue}^*$, there exists $x(\mathbf{A}) \in K$ such that

$$\begin{aligned} &A_t x(\mathbf{A}) = x(\mathbf{A}), \quad t = 1, 2, \ldots, \\ &\|A_t y - x(\mathbf{A})\| \leq \|y - x(\mathbf{A})\|, \quad y \in K, t = 1, 2, \ldots. \end{aligned} \tag{6.33}$$

Let $\{A_t\}_{t=1}^{\infty} \in \mathcal{A}_{ue}^*$ and $\gamma \in (0,1)$. For $t = 1, 2, \ldots$, define $A_{t\gamma} : K \to K$ by

$$A_{t\gamma} x = (1-\gamma) A_t x + \gamma x(\mathbf{A}), \quad x \in K. \tag{6.34}$$

It is easy to see that $\{A_{t\gamma}\}_{t=1}^{\infty} \in \mathcal{A}_{ue}$,

$$A_{t\gamma} x(\mathbf{A}) = x(\mathbf{A}), \quad t = 1, 2, \ldots$$

and

$$\|A_{t\gamma}(y) - x(\mathbf{A})\| \leq (1-\gamma) \|y - x(\mathbf{A})\|, \quad y \in K, t = 1, 2, \ldots. \tag{6.35}$$

Therefore $\{A_{t\gamma}\}_{t=1}^{\infty} \in \mathcal{A}_{ue}^*$. It is easy to see that the set

$$\{\{A_{t\gamma}\}_{t=1}^{\infty} : \{A_t\}_{t=1}^{\infty} \in \mathcal{A}_{ue}^*, \gamma \in (0,1)\}$$

is an everywhere dense subset of $\mathcal{A}_{ue}^*$ in the strong topology.

Let $\{A_t\}_{t=1}^{\infty} \in \mathcal{A}_{ue}^*$, $\gamma \in (0,1)$, and let $i \geq 1$ be an integer. Choose a natural number $N(\gamma, i)$ such that

$$(1-\gamma)^{N(\gamma,i)} \operatorname{rad}(K) \leq 4^{-i-2}. \tag{6.36}$$

Inequalities (6.35) and (6.36) imply that the following property holds:

(a) For each integer $T \geq N(\gamma, i)$, each $r : \{1, \ldots, T\} \to \{1, 2, \ldots\}$ and each $x \in K$,

$$\left\| A_{r(T)\gamma} \cdot \cdots \cdot A_{r(1)\gamma} x - x(\mathbf{A}) \right\| \leq 2^{-2i-3}.$$

By Lemma 6.9, there exists an open neighborhood $U(\{A_t\}_{t=1}^{\infty}, \gamma, i)$ of $\{A_{t\gamma}\}_{t=1}^{\infty}$ in $\bar{\mathcal{A}}_{ue}^*$ with the strong topology such that:

For each $\{C_t\}_{t=1}^{\infty} \in U(\{A_t\}_{t=1}^{\infty}, \gamma, i)$,

$$\rho_s\left(\{C_t\}_{t=1}^{\infty}, \{A_{t\gamma}\}_{t=1}^{\infty}\right) < 4^{-i-2}. \tag{6.37}$$

For each $\{C_t\}_{t=1}^{\infty} \in U(\{A_t\}_{t=1}^{\infty}, \gamma, i)$, each $x \in K$ and each mapping

$$r : \left\{1, \ldots, N(\gamma, i)\right\} \to \{1, 2, \ldots\},$$

the following inequality holds:

$$\|A_{r(N(\gamma,i))\gamma} \cdot \cdots \cdot A_{r(1)\gamma} x - C_{r(N(\gamma,i))} \cdot \cdots \cdot C_{r(1)} x\| \leq 2^{-2i-3}.$$

Together with property (a) this implies that the following property holds:

(b) For each $\{C_t\}_{t=1}^{\infty} \in U(\{A_t\}_{t=1}^{\infty}, \gamma, i)$, each $r : \{1, \ldots, N(\gamma, i)\} \to \{1, 2, \ldots\}$ and each $x \in K$, the following inequality holds:

$$\left\| C_{r(N(\gamma,i))} \cdot \cdots \cdot C_{r(1)} x - x(\mathbf{A}) \right\| \leq 4^{-i-1}.$$

Define

$$\mathcal{F} := \bigcap_{q=1}^{\infty} \bigcup \left\{ U\left(\{A_t\}_{t=1}^{\infty}, \gamma, i\right) : \{A_t\}_{t=1}^{\infty} \in \mathcal{A}_{ue}^*, \gamma \in (0, 1), i = q, q+1, \ldots \right\}.$$

Evidently, $\mathcal{F}$ is a countable intersection of open everywhere dense (in the strong topology) subsets of $\bar{\mathcal{A}}_{ue}^*$.

Assume that $\{B_t\}_{t=1}^{\infty} \in \mathcal{F}$ and $\varepsilon > 0$ are given. Choose a natural number q such that

$$2^{6-q} < \varepsilon. \tag{6.38}$$

There exist $\{A_t\}_{t=1}^{\infty} \in \mathcal{A}_{ue}^*$, $\gamma \in (0, 1)$, and an integer $i \geq q$ such that

$$\{B_t\}_{t=1}^{\infty} \in U\left(\{A_t\}_{t=1}^{\infty}, \gamma, i\right).$$

By (6.38) and property (b), for each pair of integers $t \geq 1$, $p \geq N(\gamma, i)$, and each $x \in K$,

$$\left\| (B_t)^p x - x(\mathbf{A}) \right\| \leq 4^{-i-1} < \varepsilon. \tag{6.39}$$

Since ε is an arbitrary positive number, this implies that for each integer $t \geq 1$ and each $x \in K$, there exists $\lim_{p\to\infty}(B_t)^p x$. Together with (6.39) this implies that for each integer $t \geq 1$ and each $x \in K$,

$$\left\| \lim_{p\to\infty} (B_t)^p x - x(\mathbf{A}) \right\| \leq 4^{-i-1} < \varepsilon.$$

Since ε is an arbitrary positive number, this implies, in turn, that there exists $x_* \in K$ such that

$$B_t x_* = x_*, \quad t = 1, 2, \ldots, \qquad \left\| x(\mathbf{A}) - x_* \right\| \leq 4^{-i-1} < \varepsilon. \tag{6.40}$$

It follows from (6.37), (6.40), (6.35) and (6.38) that for each integer $t \geq 1$ and each $y \in K$,

$$\begin{aligned}\|B_t y - x_*\| &\leq \left\| A_{t\gamma} y - x(\mathbf{A}) \right\| + \|B_t y - A_{t\gamma} y\| + \left\| x(\mathbf{A}) - x_* \right\| \\ &\leq 4^{-i-1} + 4^{-i-2} + (1-\gamma)\left\| y - x(\mathbf{A}) \right\| \\ &\leq \|y - x_*\| + 2^{1-2i} < \|y - x_*\| + \varepsilon.\end{aligned}$$

Since ε is an arbitrary positive number, we conclude that for each integer $t \geq 1$ and each $y \in K$,

$$\|B_t y - x_*\| \leq \|y - x_*\|.$$

Therefore (6.38), (6.40) and property (b) now imply that for each $\{C_t\}_{t=1}^{\infty} \in U(\{A_t\}_{t=1}^{\infty}, \gamma, i)$, each integer $T \geq N(\gamma, i)$, each $r : \{1, \ldots, T\} \to \{1, 2, \ldots\}$ and each $x \in K$,

$$\begin{aligned}\|C_{r(T)} \cdot \cdots \cdot C_{r(1)} x - x_*\| &\leq \left\| x_* - x(\mathbf{A}) \right\| + \left\| C_{r(T)} \cdot \cdots \cdot C_{r(1)} x - x(\mathbf{A}) \right\| \\ &\leq 2^{-2i-1} < \varepsilon.\end{aligned}$$

This completes the proof of Theorem 6.4. □

6.7 Proofs of Theorems 6.5, 6.6 and 6.7

In this section we prove Theorems 6.5 and 6.6. The proof of Theorem 6.7 is analogous to that of Theorem 6.6.

Let $\{A_t\}_{t=1}^{\infty} \in \mathcal{A}_{ne}^{(F)}$ and $\gamma \in (0, 1)$. For $t = 1, 2, \ldots$, define $A_{t\gamma} : K \to K$ by

$$A_{t\gamma} x = (1-\gamma) A_t x + \gamma Q x, \quad x \in K. \tag{6.41}$$

It is easy to see that

$$\{A_{t\gamma}\}_{t=1}^{\infty} \in \mathcal{A}_{ne}^{(F)} \quad \text{and} \quad \rho_s\left(\{A_t\}_{t=1}^{\infty}, \{A_{t\gamma}\}_{t=1}^{\infty}\right) \leq 2\gamma \operatorname{rad}(K). \tag{6.42}$$

Let $i \geq 1$ be an integer. Choose a natural number $N(\gamma, i)$ such that

$$(1-\gamma)^{N(\gamma,i)} \operatorname{rad}(K) \leq 4^{-i-2}. \tag{6.43}$$

By Lemmata 6.9 and 6.10, there exists a number $\delta(\{A_t\}_{t=1}^{\infty}, \gamma, i) > 0$ such that the following property holds:

(a) For each $r : \{1, \ldots, N(\gamma, i)\} \to \{1, 2, \ldots\}$ and each sequence of (not necessarily continuous) operators $C_t : K \to K, t = 1, \ldots, N(\gamma, i)$, satisfying

$$\|C_t - A_{r(t)\gamma}\| \le \delta\big(\{A_t\}_{t=1}^{\infty}, \gamma, i\big), \quad t = 1, \ldots, N(\gamma, i),$$

the following relations hold:

$$d(C_{N(\gamma,i)} \cdot \cdots \cdot C_1 x, F) \le 4^{-i}, \quad x \in K,$$

$$\|C_{N(\gamma,i)} \cdot \cdots \cdot C_1 x - A_{r(N(\gamma,i))\gamma} \cdot \cdots \cdot A_{r(1)\gamma} x\| \le 4^{-i-2}, \quad x \in K.$$

Proof of Theorem 6.5 Let $\{A_t\}_{t=1}^{\infty} \in \mathcal{A}_{ne}^{(F)}$, $\gamma \in (0, 1)$ and let $i \ge 1$ be an integer. There exists an open neighborhood $U(\{A_t\}_{t=1}^{\infty}, \gamma, i)$ of $\{A_{t\gamma}\}_{t=1}^{\infty}$ in the space $\mathcal{A}_{ne}^{(F)}$ with the weak topology such that for each $\{C_t\}_{t=1}^{\infty} \in U(\{A_t\}_{t=1}^{\infty}, \gamma, i)$,

$$\|C_t - A_{t\gamma}\| < 4^{-1}\delta\big(\{A_t\}_{t=1}^{\infty}, \gamma, i\big), \quad t = 1, \ldots, N(\gamma, i). \tag{6.44}$$

Define

$$\mathcal{F} := \bigcap_{q=1}^{\infty} \bigcup \big\{U\big(\{A_t\}_{t=1}^{\infty}, \gamma, i\big) : \{A_t\}_{t=1}^{\infty} \in \mathcal{A}_{ne}^{(F)}, \gamma \in (0, 1), i = q, q+1, \ldots\big\}.$$

Clearly, $\mathcal{F}$ is a countable intersection of open (in the weak topology) everywhere dense (in the strong topology) subsets of $\mathcal{A}_{ne}^{(F)}$.

Assume that $\{B_t\}_{t=1}^{\infty} \in \mathcal{F}$ and $\varepsilon > 0$ are given. Choose a natural number q such that

$$2^{6-q} < \varepsilon. \tag{6.45}$$

There exist $\{A_t\}_{t=1}^{\infty} \in \mathcal{A}_{ne}^{(F)}$, $\gamma \in (0, 1)$ and an integer $i \ge q$ such that

$$\{B_t\}_{t=1}^{\infty} \in U\big(\{A_t\}_{t=1}^{\infty}, \gamma, i\big). \tag{6.46}$$

It follows from (6.44) and property (a) that the following property holds:

(b) For each $\{C_t\}_{t=1}^{\infty} \in U(\{A_t\}_{t=1}^{\infty}, \gamma, i)$,

$$d(C_{N(\gamma,i)} \cdot \cdots \cdot C_1 x, F) \le 4^{-i}, \quad x \in K.$$

When combined with (6.46) and (6.45), this implies that for each $x \in K$, there is $f(x) \in F$ such that

$$\big\|B_{N(\gamma,i)} \cdot \cdots \cdot B_1 x - f(x)\big\| \le 2 \cdot 4^{-i} < \varepsilon. \tag{6.47}$$

Since $\{B_t\}_{t=1}^{\infty} \in \mathcal{A}_{ne}^{(F)}$, (6.47) and (6.45) imply that for each pair of integers $T, S \geq N(\gamma, i)$, and each $x \in K$,

$$\begin{aligned} \|B_T \cdot \cdots \cdot B_1 x - f(x)\|, \|B_S \cdot \cdots \cdot B_1 x - f(x)\| &< 2 \cdot 4^{-i}, \\ \|B_T \cdot \cdots \cdot B_1 x - B_S \cdot \cdots \cdot B_1 x\| &< 4^{1-i} < \varepsilon. \end{aligned} \tag{6.48}$$

Since ε is an arbitrary positive number, we conclude that there exists an operator $P_* : K \to K$ such that

$$P_* x = \lim_{t \to \infty} B_t \cdot \cdots \cdot B_1 x, \quad x \in K, \tag{6.49}$$

and

$$\|P_* x - f(x)\| \leq 2 \cdot 4^{-i}, \quad x \in K.$$

It is clear that the operator $P_* : K \to K$ is nonexpansive and $P_* x = x$, $x \in F$. Since ε is an arbitrary positive number, (6.49) and (6.45) imply that $P_*(K) \subset F$.

By (6.49) and (6.47), for each $x \in K$,

$$\|B_{N(\gamma,i)} \cdot \cdots \cdot B_1 x - P_*(x)\| \leq 4^{1-i}. \tag{6.50}$$

Property (a), (6.50), (6.44) and (6.45) imply that for each

$$\{C_t\}_{t=1}^{\infty} \in U(\{A_t\}_{t=1}^{\infty}, \gamma, i),$$

each $x \in K$, and each integer $T \geq N(\gamma, i)$,

$$\begin{aligned} &\|C_T \cdot \cdots \cdot C_1 x - P_*(x)\| \\ &\quad \leq \|C_{N(\gamma,i)} \cdot \cdots \cdot C_1 x - P_*(x)\| \\ &\quad \leq \|B_{N(\gamma,i)} \cdot \cdots \cdot B_1 x - P_*(x)\| + \|C_{N(\gamma,i)} \cdot \cdots \cdot C_1 x - A_{N(\gamma,i)\gamma} \cdot \cdots \cdot A_{1\gamma} x\| \\ &\qquad + \|A_{N(\gamma,i)\gamma} \cdot \cdots \cdot A_{1\gamma} x - B_{N(\gamma,i)} \cdot \cdots \cdot B_1 x\| \leq 4^{1-i} + 2 \cdot 4^{-2-i} < \varepsilon. \end{aligned}$$

This completes the proof of Theorem 6.5. □

Proof of Theorem 6.6 Let $\{A_t\}_{t=1}^{\infty} \in \mathcal{A}_{ne}^{(F)}$, $\gamma \in (0, 1)$, and let $i \geq 1$ be an integer. There exists an open neighborhood $U(\{A_t\}_{t=1}^{\infty}, \gamma, i)$ of $\{A_{t\gamma}\}_{t=1}^{\infty}$ in the space $\mathcal{A}_{ne}^{(F)}$ with the strong topology such that for each $\{C_t\}_{t=1}^{\infty} \in U(\{A_t\}_{t=1}^{\infty}, \gamma, i)$,

$$\|C_t - A_{t\gamma}\| < 4^{-1} \delta(\{A_t\}_{t=1}^{\infty}, \gamma, i), \quad t = 1, 2, \ldots. \tag{6.51}$$

Define

$$\mathcal{F} := \bigcap_{q=1}^{\infty} \bigcup \{U(\{A_t\}_{t=1}^{\infty}, \gamma, i) : \{A_t\}_{t=1}^{\infty} \in \mathcal{A}_{ne}^{(F)}, \gamma \in (0, 1), i = q, q+1, \ldots\}.$$

It is easy to see that $\mathcal{F}$ is a countable intersection of open everywhere dense (in the strong topology) subsets of $\mathcal{A}_{ne}^{(F)}$.

Assume that $\{B_t\}_{t=1}^{\infty} \in \mathcal{F}$ and $\varepsilon > 0$ are given. Choose a natural number q for which (6.45) holds. There exist $\{A_t\}_{t=1}^{\infty} \in \mathcal{A}_{ne}^{(F)}$, $\gamma \in (0,1)$, and an integer $i \geq q$ such that

$$\{B_t\}_{t=1}^{\infty} \in U\big(\{A_t\}_{t=1}^{\infty}, \gamma, i\big). \tag{6.52}$$

Assume that $r : \{1, 2, \ldots\} \to \{1, 2, \ldots\}$. It follows from (6.52), (6.45), (6.51) and property (a) that for each $x \in K$, there is $f(x) \in F$ such that

$$\big\| B_{r(N(\gamma,i))} \cdot \cdots \cdot B_{r(1)}x - f(x) \big\| < 2 \cdot 4^{-i} < \varepsilon. \tag{6.53}$$

Since $\{B_t\}_{t=1}^{\infty} \in \mathcal{A}_{ne}^{(F)}$, (6.53) and (6.45) imply that for each pair of integers $T, S \geq N(\gamma, i)$ and each $x \in K$,

$$\big\| B_{r(T)} \cdot \cdots \cdot B_{r(1)}x - f(x) \big\|, \big\| B_{r(S)} \cdot \cdots \cdot B_{r(1)}x - f(x) \big\| < 2 \cdot 4^{-i},$$
$$\| B_{r(T)} \cdot \cdots \cdot B_{r(1)}x - B_{r(S)} \cdot \cdots \cdot B_{r(1)}x \| \leq 4^{1-i} < \varepsilon.$$

Since ε is an arbitrary positive number, we conclude that there exists an operator $P_r : K \to K$ such that for each $x \in K$,

$$P_r x = \lim_{t \to \infty} B_{r(t)} \cdot \cdots \cdot B_{r(1)}x, \qquad \big\| P_r x - f(x) \big\| \leq 2 \cdot 4^{-i} < \varepsilon. \tag{6.54}$$

Clearly, the operator $P_r : K \to K$ is nonexpansive and $P_r x = x$, $x \in F$. Since ε is an arbitrary positive number, (6.54) implies that $P_r(K) \subset F$.

By (6.54) and (6.53), for each $x \in K$,

$$\big\| B_{r(N(\gamma,i))} \cdot \cdots \cdot B_{r(1)}x - P_r(x) \big\| \leq 4^{1-i}. \tag{6.55}$$

Property (a), (6.55), the definition of $U(\{A_t\}_{t=1}^{\infty}, \gamma, i)$ (see (6.51)) and (6.45) now imply that for each $\{C_t\}_{t=1}^{\infty} \in U(\{A_t\}_{t=1}^{\infty}, \gamma, i)$, each $r : \{1, 2, \ldots\} \to \{1, 2, \ldots\}$, each integer $T \geq N(\gamma, i)$ and each $x \in K$,

$$\begin{aligned}
&\big\| C_{r(T)} \cdot \cdots \cdot C_{r(1)}x - P_r(x) \big\| \\
&\leq \big\| C_{r(N(\gamma,i))} \cdot \cdots \cdot C_{r(1)}x - P_r(x) \big\| \\
&\leq \| C_{r(N(\gamma,i))} \cdot \cdots \cdot C_{r(1)}x - A_{r(N(\gamma,i))\gamma} \cdot \cdots \cdot A_{r(1)\gamma}x \| \\
&\quad + \| A_{r(N(\gamma,i))\gamma} \cdot \cdots \cdot A_{r(1)\gamma}x - B_{r(N(\gamma,i))} \cdot \cdots \cdot B_{r(1)}x \| \\
&\quad + \| B_{r(N(\gamma,i))} \cdot \cdots \cdot B_{r(1)}x - P_r x \| \\
&\leq 2 \cdot 4^{-2-i} + 4^{1-i} < \varepsilon.
\end{aligned}$$

This completes the proof of Theorem 6.6. □

6.8 Hyperbolic Spaces

Let (X, ρ) be a complete hyperbolic space (see Sect. 1.1) and let K be a bounded, closed and ρ-convex subset of X.

Analogously to the case of a bounded, closed and convex subset K of a Banach space (see Sects. 6.1, 6.2 and 6.3), we may define the hyperbolic analogs of the spaces $\mathcal{A}_{ne}$, $\mathcal{A}_{ue}$, $\bar{\mathcal{A}}^*_{ne}$, $\bar{\mathcal{A}}^*_{ue}$, $\mathcal{A}^{(F)}_{ne}$ and $\mathcal{A}^{(F)}_{ue}$. One can then easily formulate extensions of Theorems 6.1–6.7 to this case and verify that these extensions can be established by arguments similar to those we have used in the present chapter. These extensions provide a partial answer to a question raised in [121].

6.9 Infinite Products of Order-Preserving Mappings

Order-preserving mappings find application in many areas of mathematics. See, for example, [3, 4, 62, 107] and the references mentioned there. We study the asymptotic behavior of (random) infinite products of generic sequences of order-preserving continuous mappings on intervals of an ordered Banach space. More precisely, we show that in appropriate spaces of sequences of operators there exists a subset which is a countable intersection of open and everywhere dense sets such that for each sequence belonging to this subset, the corresponding infinite products converge.

Let $(X, \|\cdot\|)$ be a Banach space ordered by a closed and convex cone X_+ such that $\|x\| \le \|y\|$ for each $x, y \in X_+$ satisfying $x \le y$. For each $u, v \in X$ such that $u \le v$ denote

$$\langle u, v\rangle := \{x \in X : u \le x \le v\}.$$

For each $x, y \in X_+$, we define

$$\lambda(x, y) := \sup\{r \ge 0 : rx \le y\}. \tag{6.56}$$

Let $b \in X_+ \setminus \{0\}$. We consider the space $\langle 0, b\rangle \subset X$ with the topology induced by the norm $\|\cdot\|$. Denote by $\mathcal{A}$ the set of all continuous operators $A : \langle 0, b\rangle \to \langle 0, b\rangle$ such that

$$Ax \le Ay \quad \text{for each } x, y \in \langle 0, b\rangle \text{ satisfying } x \le y$$

and

$$A(\alpha z) \ge \alpha Az \quad \text{for each } z \in \langle 0, b\rangle \text{ and each } \alpha \in [0, 1].$$

For the space $\mathcal{A}$ we define a metric $\rho : \mathcal{A} \times \mathcal{A} \to [0, \infty)$ by

$$\rho(A, B) = \sup\big\{\|Ax - Bx\| : x \in \langle 0, b\rangle\big\}, \quad A, B \in \mathcal{A}. \tag{6.57}$$

It is easy to see that the metric space $\mathcal{A}$ is complete.

We will show (see Theorem 6.11 below) that for a generic operator B in the space $\mathcal{A}$, there exists a unique fixed point x_B and the powers of B converge to x_B, uniformly for all $x \in \langle 0, b \rangle$.

Assume now that b is an interior point of the cone X_+. Define

$$\|x\|_b = \inf\{r \in [0, \infty) : -rb \le x \le rb\}, \quad x \in X. \tag{6.58}$$

Clearly, $\|\cdot\|_b$ is a norm on X which is equivalent to the norm $\|\cdot\|$.

Denote by $\mathcal{M}$ the set of all sequences $\{A_t\}_{t=1}^{\infty}$, where each $A_t \in \mathcal{A}$, $t = 1, 2, \ldots$. Such a sequence will occasionally be denoted by a boldface **A**. For the set $\mathcal{M}$ we consider the metric $\rho_s : \mathcal{M} \times \mathcal{M} \to [0, \infty)$ defined by

$$\rho_s\big(\{A_t\}_{t=1}^{\infty}, \{B_t\}_{t=1}^{\infty}\big) = \sup\big\{\|A_t x - B_t x\|_b : x \in \langle 0, b \rangle, t = 1, 2, \ldots\big\},$$
$$\{A_t\}_{t=1}^{\infty}, \{B_t\}_{t=1}^{\infty} \in \mathcal{M}. \tag{6.59}$$

It is easy to see that this metric space $(\mathcal{M}, \rho_s)$ is complete. The topology generated in $\mathcal{M}$ by the metric ρ_s will be called the strong topology.

In addition to this topology on $\mathcal{M}$, we will also consider the uniformity which is determined by the base

$$E(N, \varepsilon) = \big\{\big(\{A_t\}_{t=1}^{\infty}, \{B_t\}_{t=1}^{\infty}\big) \in \mathcal{M} \times \mathcal{M} :$$
$$\|A_t x - B_t x\|_b \le \varepsilon, t = 1, \ldots, N, x \in \langle 0, b \rangle\big\}, \tag{6.60}$$

where N is a natural number and $\varepsilon > 0$. It is easy to see that the space $\mathcal{M}$ with this uniformity is metrizable (by a metric $\rho_w : \mathcal{M} \times \mathcal{M} \to [0, \infty)$) and complete. The topology generated by ρ_w will be called the weak topology. We will show (see Theorem 6.16) that for a generic sequence $\{C_t\}_{t=1}^{\infty}$ in the space $\mathcal{M}$ with the weak topology,

$$\lambda\big(C_T \cdot \cdots \cdot C_1 x, C_T \cdot \cdots \cdot C_1(0)\big) \to 1,$$

uniformly for all $x \in \langle 0, b \rangle$. We will also establish Theorem 6.17, which shows that for a generic sequence $\{C_t\}_{t=1}^{\infty}$ in $\mathcal{M}$ with the strong topology, this type of uniform convergence holds for random products of the operators $\{C_t\}_{t=1}^{\infty}$.

Let $a \in \langle 0, b \rangle$ be an interior point of X_+. Denote by $\mathcal{M}_a$ the set of all sequences $\{A_t\}_{t=1}^{\infty} \in \mathcal{M}$ such that

$$A_t a = a, \quad t = 1, 2, \ldots.$$

Clearly, $\mathcal{M}_a$ is a closed subset of $\mathcal{M}$ with the weak topology. We consider the topological subspace $\mathcal{M}_a \subset \mathcal{M}$ with the relative weak and strong topologies.

We will show (Theorem 6.18) that for a generic sequence of operators $\{C_t\}_{t=1}^{\infty}$ in the space $\mathcal{M}_a$ with the weak topology,

$$\|C_T \cdot \cdots \cdot C_1 z - a\|_b \to 0 \quad \text{as } T \to \infty,$$

uniformly for all $x \in \langle 0, b \rangle$. We will also establish Theorem 6.19, which shows that for a generic sequence of operators $\{C_t\}_{t=1}^{\infty}$ in the space $\mathcal{M}_a$ with the strong topology, all its random products

$$C_{r(T)} \cdot \dots \cdot C_{r(1)} z \to a \quad \text{as } T \to \infty,$$

uniformly for all $x \in \langle 0, b \rangle$. Here $r : \{1, 2, \dots\} \to \{1, 2, \dots\}$ is arbitrary.

Finally, denote by $\mathcal{M}_*$ the set of all sequences $\{A_t\}_{t=1}^{\infty} \in \mathcal{M}$ such that

$$A_t a = a, \quad t = 1, 2, \dots$$

for some $a \in \langle 0, b \rangle$ such that a is an interior point of X_+. Denote by $\bar{\mathcal{M}}_*$ the closure of $\mathcal{M}_*$ in the space $\mathcal{M}$ with the strong topology and consider the topological subspace $\bar{\mathcal{M}}_* \subset \mathcal{M}$ with the relative strong topology. We will show (Theorem 6.20) that for a generic sequence $\{C_t\}_{t=1}^{\infty}$ in the space $\bar{\mathcal{M}}_*$, there exists a unique common fixed point x_*, which is an interior point of the cone X_+ and all random products of the operators $\{C_t\}_{t=1}^{\infty}$ converge to x_*, uniformly for all $x \in \langle 0, b \rangle$.

Theorems 6.11 and 6.16–6.20 appeared in [127].

6.10 Existence of a Unique Fixed Point

In this section we will prove the following result.

Theorem 6.11 *There exists a set $\mathcal{F} \subset \mathcal{A}$, which is a countable intersection of open and everywhere dense sets in $\mathcal{A}$, such that for each $B \in \mathcal{F}$, the following two assertions hold*:

1. *There exists $x_B \in \langle 0, b \rangle$ such that $Bx_B = x_B$,*

$$B^T x \to x_B \quad \text{as } T \to \infty, \text{ uniformly on } \langle 0, b \rangle.$$

2. *For each $\varepsilon > 0$, there exist a neighborhood U of B in $\mathcal{A}$ and an integer $N \geq 1$ such that for each $C \in U$, $z \in \langle 0, b \rangle$ and each integer $T \geq N$,*

$$\left\| C^T z - x_B \right\| \leq \varepsilon. \tag{6.61}$$

Before proving Theorem 6.11 we need several preliminary lemmata.

Lemma 6.12 *Let $n \geq 1$ be an integer, and let $A \in \mathcal{A}$, $\varepsilon > 0$, and $z \in \langle 0, b \rangle$ be given. Then there exists a neighborhood U of A in $\mathcal{A}$ such that for each $C \in U$,*

$$\left\| C^n z - A^n z \right\| < \varepsilon. \tag{6.62}$$

Proof We prove the assertion of the lemma by induction. For $n = 1$ the assertion of the lemma is valid. Assume that the assertion of the lemma is valid for an integer $n \geq 1$. We will show that this implies that the lemma also holds for $n + 1$.

There exists

$$\delta \in \left(0, 8^{-1} \varepsilon\right) \tag{6.63}$$

such that

$$\|Ay - A(A^n z)\| \le 8^{-1}\varepsilon \tag{6.64}$$

for each y satisfying $\|y - A^n z\| \le \delta$. There exists a neighborhood U_0 of A in $\mathcal{A}$ such that for each $C \in U_0$,

$$\|C^n z - A^n z\| < \delta.$$

Set

$$U := \{C \in U_0 : \rho(C, A) < 8^{-1}\varepsilon\}. \tag{6.65}$$

Let $C \in U$. The definition of U implies that

$$\begin{aligned}\|A^{n+1}z - C^{n+1}z\| &\le \|A^{n+1}z - AC^n z\| + \|AC^n z - C^{n+1}z\| \\ &\le \|A^{n+1}z - AC^n z\| + 8^{-1}\varepsilon.\end{aligned} \tag{6.66}$$

By the definition of U_0,

$$\|A^n z - C^n z\| < \delta.$$

It follows from this inequality and the definition of δ (see (6.63) and (6.64)) that

$$\|AC^n z - A(A^n z)\| \le 8^{-1}\varepsilon.$$

Together with (6.66) this implies that

$$\|A^{n+1}z - C^{n+1}z\| \le 4^{-1}\varepsilon.$$

This completes the proof of the lemma. □

For each $A \in \mathcal{A}$ and each $\gamma \in (0, 1)$, define

$$A_\gamma : \langle 0, b\rangle \to \langle 0, b\rangle$$

by

$$A_\gamma x := (1-\gamma)Ax + \gamma b, \quad x \in \langle 0, b\rangle. \tag{6.67}$$

It is easy to see that $A_\gamma \in \mathcal{A}$ for each $A \in \mathcal{A}$ and each $\gamma \in (0, 1)$, and that the set

$$\{A_\gamma : A \in \mathcal{A}, \gamma \in (0, 1)\}$$

is everywhere dense in $\mathcal{A}$.

Lemma 6.13 *Let $A \in \mathcal{A}$ and $\gamma \in (0, 1)$ be given. Then for each integer $t \ge 0$,*

$$A_\gamma^{t+1}(0) \ge A_\gamma^t(0), \qquad A_\gamma^{t+1}(b) \le A_\gamma^t(b), \qquad A_\gamma^t(0) \le A_\gamma^t(b) \tag{6.68}$$

and

$$\lim_{t\to\infty} \lambda(A_\gamma^t(b), A_\gamma^t(0)) = 1. \tag{6.69}$$

Proof Clearly, (6.68) is valid for each integer $t \geq 0$. We show that (6.69) holds.

To this end, let $t \geq 0$ be an integer. By (6.68),

$$\lambda\big(A_\gamma^t(b), A_\gamma^t(0)\big) \leq 1.$$

It follows from (6.67) that

$$\begin{aligned}
A_\gamma^{t+1}(0) &= A_\gamma\big(A_\gamma^t(0)\big) = (1-\gamma)A\big(A_\gamma^t(0)\big) + \gamma b \\
&= \gamma b + (1-\gamma)A\big(\lambda\big(A_\gamma^t(b), A_\gamma^t(0)\big)A_\gamma^t(b)\big) \\
&\geq \gamma b + (1-\gamma)\lambda\big(A_\gamma^t(b), A_\gamma^t(0)\big)A\big(A_\gamma^t(b)\big) \\
&= \lambda\big(A_\gamma^t(b), A_\gamma^t(0)\big)\big((1-\gamma)A\big(A_\gamma^t(b)\big) + \gamma b\big) \\
&\quad + \big(1 - \lambda\big(A_\gamma^t(b), A_\gamma^t(0)\big)\big)\gamma b \\
&\geq \lambda\big(A_\gamma^t(b), A_\gamma^t(0)\big)A_\gamma^{t+1}b + \big(1 - \lambda\big(A_\gamma^t(b), A_\gamma^t(0)\big)\big)\gamma A_\gamma^{t+1}b \\
&= \big[\lambda\big(A_\gamma^t(b), A_\gamma^t(0)\big) + \gamma\big(1 - \lambda\big(A_\gamma^t(b), A_\gamma^t(0)\big)\big)\big]A_\gamma^{t+1}b.
\end{aligned}$$

Hence

$$\lambda\big(A_\gamma^{t+1}(b), A_\gamma^{t+1}(0)\big) \geq \lambda\big(A_\gamma^t(b), A_\gamma^t(0)\big) + \gamma\big(1 - \lambda\big(A_\gamma^t(b), A_\gamma^t(0)\big)\big). \tag{6.70}$$

By (6.70), the limit

$$\Lambda = \lim_{t\to\infty} \lambda\big(A_\gamma^t(b), A_\gamma^t(0)\big)$$

exists and

$$1 \geq \Lambda \geq \Lambda + \gamma(1-\Lambda).$$

Therefore $\Lambda = 1$ and the lemma is proved. □

Lemma 6.14 *Let $A \in \mathcal{A}$ and $\gamma \in (0,1)$ be given. Then there exists $x(A,\gamma) \in \langle 0, b\rangle$ such that $A_\gamma x(A,\gamma) = x(A,\gamma)$ and*

$$\lim_{t\to\infty} A_\gamma^t(0) = \lim_{t\to\infty} A_\gamma^t(b) = x(A,\gamma). \tag{6.71}$$

Proof By Lemma 6.13,

$$\lim_{t\to\infty}\big(A_\gamma^t(b) - A_\gamma^t(0)\big) = 0$$

and $\{A_\gamma^t(0)\}_{t=1}^\infty$, $\{A_\gamma^t(b)\}_{t=1}^\infty$ are Cauchy sequences. This yields (6.71) and the lemma itself. □

Let $A \in \mathcal{A}$, $\gamma \in (0,1)$ and let $i \geq 1$ be an integer. By Lemma 6.14 there exists an integer $N(A,\gamma,i) \geq 2$ such that

$$\big\|A_\gamma^{N(A,\gamma,i)}(0) - A_\gamma^{N(A,\gamma,i)}(b)\big\| \leq 8^{-i}. \tag{6.72}$$

By Lemma 6.12, there exists an open neighborhood $U(A, \gamma, i)$ of A_γ in $\mathcal{A}$ such that for each $C \in U(A, \gamma, i)$,

$$\|C^{N(A,\gamma,i)}(0) - A_\gamma^{N(A,\gamma,i)}(0)\|, \|C^{N(A,\gamma,i)}(b) - A_\gamma^{N(A,\gamma,i)}(b)\| \le 8^{-i}. \tag{6.73}$$

Lemma 6.15 *Let $A \in \mathcal{A}$, $\gamma \in (0, 1)$ and let $i \ge 1$ be an integer. Assume that $C \in U(A, \gamma, i)$, $z \in \langle 0, b\rangle$ and that $T \ge N(A, \gamma, i)$ is an integer. Then*

$$\|C^T z - x(A, \gamma)\| \le 6 \cdot 8^{-i}.$$

Proof It is easy to see that

$$C^T z \in \langle C^T(0), C^T(b)\rangle \subset \langle C^{N(A,\gamma,i)}(0), C^{N(A,\gamma,i)}(b)\rangle. \tag{6.74}$$

By Lemma 6.14, the definition of $N(A, \gamma, i)$ (see (6.72)) and (6.73),

$$\begin{aligned}
&\|C^{N(A,\gamma,i)}(0) - x(A, \gamma)\| \\
&\quad \le \|C^{N(A,\gamma,i)}(0) - A_\gamma^{N(A,\gamma,i)}(0)\| + \|A_\gamma^{N(A,\gamma,i)}(0) - x(A, \gamma)\| \\
&\quad \le 8^{-i} + \|A_\gamma^{N(A,\gamma,i)}(b) - A_\gamma^{N(A,\gamma,i)}(0)\| \le 2 \cdot 8^{-i}, \\
&\|C^{N(A,\gamma,i)}(b) - x(A, \gamma)\| \\
&\quad \le \|C^{N(A,\gamma,i)}(b) - A_\gamma^{N(A,\gamma,i)}(b)\| + \|A_\gamma^{N(A,\gamma,i)}(b) - x(A, \gamma)\| \\
&\quad \le 8^{-i} + \|A_\gamma^{N(A,\gamma,i)}(b) - A_\gamma^{N(A,\gamma,i)}(0)\| \le 2 \cdot 8^{-i}.
\end{aligned}$$

It follows from these inequalities and (6.74) that

$$\begin{aligned}
\|C^T z - x(A, \gamma)\| &\le \|C^T z - C^{N(A,\gamma,i)}(0)\| + \|C^{N(A,\gamma,i)}(0) - x(A, \gamma)\| \\
&\le \|C^{N(A,\gamma,i)}(b) - C^{N(A,\gamma,i)}(0)\| + 2 \cdot 8^{-i} \le 6 \cdot 8^{-i}.
\end{aligned}$$

The lemma is proved. □

Proof of Theorem 6.11 Define

$$\mathcal{F} := \bigcap_{q=1}^{\infty} \bigcup \{U(A, \gamma, i) : A \in \mathcal{A}, \gamma \in (0, 1), i = q, q + 1, \dots\}.$$

Clearly, $\mathcal{F}$ is a countable intersection of open everywhere dense sets in $\mathcal{A}$.

Assume that $B \in \mathcal{F}$ and $\varepsilon > 0$ are given. Choose an integer $q \ge 1$ such that

$$6 \cdot 2^{-q} < 64^{-1}\varepsilon. \tag{6.75}$$

There exist $A \in \mathcal{A}$, $\gamma \in (0, 1)$ and an integer $i \ge q$ such that

$$B \in U(A, \gamma, i). \tag{6.76}$$

By Lemma 6.15, (6.75) and (6.76) for each $z \in \langle 0, b\rangle$ and each integer $T \geq N(A, \gamma, i)$,

$$\|B^T z - x(A, \gamma)\| \leq 6 \cdot 8^{-i} < 64^{-1}\varepsilon. \tag{6.77}$$

Since ε is an arbitrary positive number, we conclude that all the trajectories of B converge and there is $x_B \in \langle 0, b\rangle$ for which

$$Bx_B = x_B. \tag{6.78}$$

Relations (6.77) and (6.78) imply that

$$\|x(A, \gamma) - x_B\| \leq 6 \cdot 8^{-i}, \tag{6.79}$$

and that for each $z \in \langle 0, b\rangle$ and each integer $T \geq N(A, \gamma, i)$,

$$\|B^T z - x_B\| \leq 12 \cdot 8^{-i} < 32^{-1}\varepsilon.$$

Since ε is an arbitrary positive number, we conclude that

$$B^t x \to x_B \quad \text{as } T \to \infty, \text{ uniformly on } \langle 0, b\rangle.$$

Finally, assume that $C \in U(A, \gamma, i)$, $z \in \langle 0, b\rangle$ and that $T \geq N(A, \gamma, i)$ is an integer. It follows from (6.79), Lemma 6.15 and (6.75) that

$$\begin{aligned} \|C^T z - x_B\| &\leq \|C^T z - x(A, \gamma)\| + \|x(A, \gamma) - x_B\| \\ &< 6 \cdot 8^{-i} + 6 \cdot 8^{-i} < 32^{-1}\varepsilon. \end{aligned}$$

The proof of the theorem is complete. □

6.11 Asymptotic Behavior

In this section we assume that b is an interior point of the cone X_+. We will first formulate precisely our weak ergodic theorems.

Theorem 6.16 *There exists a set $\mathcal{F} \subset \mathcal{M}$, which is a countable intersection of open (in the weak topology) everywhere dense (in the strong topology) sets in $\mathcal{M}$, such that for each $\{B_t\}_{t=1}^{\infty} \in \mathcal{F}$, the following assertion holds:*

For each $\varepsilon \in (0, 1)$, there exist a neighborhood U of $\{B_t\}_{t=1}^{\infty}$ in $\mathcal{M}$ with the weak topology and an integer $N \geq 1$ such that for each $\{C_t\}_{t=1}^{\infty} \in U$, each integer $T \geq N$ and each $x \in \langle 0, b\rangle$,

$$\lambda\big(C_T \cdot \cdots \cdot C_1 x, C_T \cdot \cdots \cdot C_1(0)\big) \geq 1 - \varepsilon.$$

Theorem 6.17 *There exists a set $\mathcal{F} \subset \mathcal{M}$, which is a countable intersection of open everywhere dense sets in $\mathcal{M}$ with the strong topology, such that for each $\{B_t\}_{t=1}^{\infty} \in \mathcal{F}$, the following assertion holds*:

For each $\varepsilon \in (0,1)$, there exists a neighborhood U of $\{B_t\}_{t=1}^{\infty}$ in $\mathcal{M}$ with the strong topology and an integer $N \geq 1$ such that for each $\{C_t\}_{t=1}^{\infty} \in U$, each $r : \{1,2,\ldots\} \to \{1,2,\ldots\}$, each integer $T \geq N$ and each $x \in \langle 0,b \rangle$,

$$\lambda\big(C_{r(T)} \cdot \cdots \cdot C_{r(1)}x, C_{r(T)} \cdot \cdots \cdot C_{r(1)}(0)\big) \geq 1 - \varepsilon.$$

Let $a \in \langle 0,b \rangle$ be an interior point of X_+. Now we present the theorems which establish generic convergence to a unique fixed point in the space $\mathcal{M}_a$.

Theorem 6.18 *There exists a set $\mathcal{F} \subset \mathcal{M}_a$, which is a countable intersection of open (in the relative weak topology) everywhere dense (in the relative strong topology) sets in $\mathcal{M}_a$, such that the following assertion holds*:

For each $\{B_t\}_{t=1}^{\infty} \in \mathcal{F}$ and each $\varepsilon > 0$, there exist a neighborhood U of $\{B_t\}_{t=1}^{\infty}$ in $\mathcal{M}_a$ with the relative weak topology and a natural number N such that for each $\{C_t\}_{t=1}^{\infty} \in U$, each integer $T \geq N$ and each $z \in \langle 0,b \rangle$,

$$\|C_T \cdot \cdots \cdot C_1 z - a\|_b \leq \varepsilon.$$

Theorem 6.19 *There exists a set $\mathcal{F} \subset \mathcal{M}_a$, which is a countable intersection of open everywhere dense sets in $\mathcal{M}_a$ with the relative strong topology, such that the following assertion holds*:

For each $\{B_t\}_{t=1}^{\infty} \in \mathcal{F}$ and each $\varepsilon > 0$, there exist a neighborhood U of $\{B_t\}_{t=1}^{\infty}$ in $\mathcal{M}_a$ with the relative strong topology and a natural number N such that for each $\{C_t\}_{t=1}^{\infty} \in U$, each $r : \{1,2,\ldots\} \to \{1,2,\ldots\}$, each integer $T \geq N$ and each $z \in \langle 0,b \rangle$,

$$\|C_{r(T)} \cdot \cdots \cdot C_{r(1)} z - a\|_b \leq \varepsilon.$$

The next theorem establishes generic uniform convergence of random infinite products to a unique common fixed point in the space $\bar{\mathcal{M}}_*$.

Theorem 6.20 *There exists a set $\mathcal{F} \subset \bar{\mathcal{M}}_*$, which is a countable intersection of open everywhere dense sets in $\bar{\mathcal{M}}_*$, such that for each $\{B_t\}_{t=1}^{\infty} \in \mathcal{F}$, the following two assertions hold*:

1. *There exists an interior point $x(\mathbf{B}) \in \langle 0,b \rangle$ of the cone X_+ which satisfies*

$$B_t x(\mathbf{B}) = x(\mathbf{B}), \quad t = 1,2,\ldots.$$

2. *For each $\varepsilon > 0$, there exist a neighborhood U of $\{B_t\}_{t=1}^{\infty}$ in $\bar{\mathcal{M}}_*$ and a natural number N such that for each $\{C_t\}_{t=1}^{\infty} \in U$, each $r : \{1,2,\ldots\} \to \{1,2,\ldots\}$, each integer $T \geq N$ and each $z \in \langle 0,b \rangle$,*

$$\big\|C_{r(T)} \cdot \cdots \cdot C_{r(1)} z - x(\mathbf{B})\big\|_b \leq \varepsilon.$$

6.12 Preliminary Lemmata for Theorems 6.16–6.20

For the space $\mathcal{A}$ we define a metric $\rho : \mathcal{A} \times \mathcal{A} \to [0, \infty)$ by

$$\rho(A, B) := \sup\{\|Ax - Bx\|_b : x \in \langle 0, b\rangle\}, \quad A, B \in \mathcal{A}.$$

Fix $a \in \langle 0, b\rangle$ such that a is an interior point of X_+. Clearly,

$$0 < \lambda(b, a) \leq 1. \tag{6.80}$$

For each $\mathbf{A} = \{A_t\}_{t=1}^{\infty}$ and each $\gamma \in (0, 1)$, define $\mathbf{A}_{\gamma}^{(a)} = \{A_{t\gamma}^{(a)}\}_{t=1}^{\infty}$, where $A_{t\gamma}^{(a)} : \langle 0, b\rangle \to \langle 0, b\rangle$, $t = 1, 2, \ldots$, is defined by

$$A_{t\gamma}^{(a)} x := (1 - \gamma)A_t x + \gamma a, \quad x \in \langle 0, b\rangle, t = 1, 2, \ldots. \tag{6.81}$$

It is easy to see that $\{A_{t\gamma}^{(a)}\}_{t=1}^{\infty} \in \mathcal{M}$ for each $\{A_t\}_{t=1}^{\infty} \in \mathcal{M}$ and each $\gamma \in (0, 1)$, and that the set

$$\{\{A_{t\gamma}^{(a)}\}_{t=1}^{\infty} : \{A_t\}_{t=1}^{\infty} \in \mathcal{M}, \gamma \in (0, 1)\} \tag{6.82}$$

is everywhere dense in $\mathcal{M}$ with the strong topology.

For each ε and $\gamma \in (0, 1)$, we choose a natural number $Q(\gamma, \varepsilon)$ such that

$$Q(\gamma, \varepsilon) > 4(\varepsilon\gamma\lambda(b, a))^{-1} + 4. \tag{6.83}$$

Lemma 6.21 *Let $\{A_t\}_{t=1}^{\infty} \in \mathcal{M}$, $\varepsilon, \gamma \in (0, 1)$ and let $r : \{1, 2, \ldots\} \to \{1, 2, \ldots\}$. Then*

$$\lambda\big(A_{r(Q(\gamma,\varepsilon))\gamma}^{(a)} \cdot \cdots \cdot A_{r(1)\gamma}^{(a)}(b), A_{r(Q(\gamma,\varepsilon))\gamma}^{(a)} \cdot \cdots \cdot A_{r(1)\gamma}^{(a)}(0)\big) > 1 - \varepsilon.$$

Proof It is clear that for each integer $T \geq 1$,

$$1 \geq \lambda\big(A_{r(T)\gamma}^{(a)} \cdot \cdots \cdot A_{r(1)\gamma}^{(a)} b, A_{r(T)\gamma}^{(a)} \cdot \cdots \cdot A_{r(1)\gamma}^{(a)}\big)$$

and

$$\begin{aligned} &\lambda\big(A_{r(T+1)\gamma}^{(a)} \cdot A_{r(T)\gamma}^{(a)} \cdot \cdots \cdot A_{r(1)\gamma}^{(a)} b, A_{r(T+1)\gamma}^{(a)} \cdot A_{r(T)\gamma}^{(a)} \cdot \cdots \cdot A_{r(1)\gamma}^{(a)}(0)\big) \\ &\quad \geq \lambda\big(A_{r(T)\gamma}^{(a)} \cdot \cdots \cdot A_{r(1)\gamma}^{(a)}(b), A_{r(T)\gamma}^{(a)} \cdot \cdots \cdot A_{r(1)\gamma}^{(a)}(0)\big). \end{aligned} \tag{6.84}$$

For each integer $T \geq 1$, we have by (6.81), (6.84) and (6.80),

$$\begin{aligned} &A_{r(T+1)\gamma}^{(a)} A_{r(T)\gamma}^{(a)} \cdot \cdots \cdot A_{r(1)\gamma}^{(a)}(0) \\ &\quad = A_{r(T+1)\gamma}^{(a)}\big(A_{r(T)\gamma}^{(a)} \cdot \cdots \cdot A_{r(1)\gamma}^{(a)}(0)\big) \\ &\quad = (1 - \gamma)A_{r(T+1)}\big(A_{r(T)\gamma}^{(a)} \cdot \cdots \cdot A_{r(1)\gamma}^{(a)}(0)\big) + \gamma a \end{aligned}$$

$$
\begin{aligned}
&= \gamma a + (1-\gamma)A_{r(T+1)}\big(\lambda\big(A^{(a)}_{r(T)\gamma}\cdot\dots\cdot A^{(a)}_{r(1)\gamma}b,\\
&\quad A^{(a)}_{r(T)\gamma}\cdot\dots\cdot A^{(a)}_{r(1)\gamma}(0)\big)A^{(a)}_{r(T)\gamma}\cdot\dots\cdot A^{(a)}_{r(1)\gamma}(b)\big)\\
&\ge \gamma a + (1-\gamma)\lambda\big(A^{(a)}_{r(T)\gamma}\cdot\dots\cdot A^{(a)}_{r(1)\gamma}b,\\
&\quad A^{(a)}_{r(T)\gamma}\cdot\dots\cdot A^{(a)}_{r(1)\gamma}(0)\big)A_{r(T+1)}\cdot A^{(a)}_{r(T)\gamma}\cdot\dots\cdot A^{(a)}_{r(1)\gamma}(b)\\
&\ge \lambda\big(A^{(a)}_{r(T)\gamma}\cdot\dots\cdot A^{(a)}_{r(1)\gamma}b,\\
&\quad A^{(a)}_{r(T)\gamma}\cdot\dots\cdot A^{(a)}_{r(1)\gamma}(0)\big)\big[(1-\gamma)A_{r(T+1)}\big(A^{(a)}_{r(T)\gamma}\cdot\dots\cdot A^{(a)}_{r(1)\gamma}b\big)+\gamma a\big]\\
&\quad + \big(1-\lambda\big(A^{(a)}_{r(T)\gamma}\cdot\dots\cdot A^{(a)}_{r(1)\gamma}b, A^{(a)}_{r(T)\gamma}\cdot\dots\cdot A^{(a)}_{r(1)\gamma}(0)\big)\big)\gamma a\\
&= \lambda\big(A^{(a)}_{r(T)\gamma}\cdot\dots\cdot A^{(a)}_{r(1)\gamma}b,\\
&\quad A^{(a)}_{r(T)\gamma}\cdot\dots\cdot A^{(a)}_{r(1)\gamma}(0)\big)A^{(a)}_{r(T+1)\gamma}\cdot A^{(a)}_{r(T)\gamma}\cdot\dots\cdot A^{(a)}_{r(1)\gamma}(b)\\
&\quad + \big(1-\lambda\big(A^{(a)}_{r(T)\gamma}\cdot\dots\cdot A^{(a)}_{r(1)\gamma}(b), A^{(a)}_{r(T)\gamma}\cdot\dots\cdot A^{(a)}_{r(1)\gamma}(0)\big)\big)\gamma a\\
&\ge \big[\lambda\big(A^{(a)}_{r(T)\gamma}\cdot\dots\cdot A^{(a)}_{r(1)\gamma}b,\\
&\quad A^{(a)}_{r(T)\gamma}\cdot\dots\cdot A^{(a)}_{r(1)\gamma}(0)\big)+\gamma\lambda(b,a)\big(1-\lambda\big(A^{(a)}_{r(T)\gamma}\cdot\dots\cdot A^{(a)}_{r(1)\gamma}b,\\
&\quad A^{(a)}_{r(T)\gamma}\cdot\dots\cdot A^{(a)}_{r(1)\gamma}(0)\big)\big)\big]A^{(a)}_{r(T+1)\gamma}A^{(a)}_{r(T)\gamma}\cdot\dots\cdot A^{(a)}_{r(1)\gamma}(b)
\end{aligned}
$$

and

$$
\begin{aligned}
&\lambda\big(A^{(a)}_{r(T+1)\gamma}A^{(a)}_{r(T)\gamma}\cdot\dots\cdot A^{(a)}_{r(1)\gamma}b, A^{(a)}_{r(T+1)\gamma}A^{(a)}_{r(T)\gamma}\cdot\dots\cdot A^{(a)}_{r(1)\gamma}(0)\big)\\
&\quad\ge \lambda\big(A^{(a)}_{r(T)\gamma}\cdot\dots\cdot A^{(a)}_{r(1)\gamma}b, A^{(a)}_{r(T)\gamma}\cdot\dots\cdot A^{(a)}_{r(1)\gamma}(0)\big)\\
&\quad\quad + \gamma\big(1-\lambda\big(A^{(a)}_{r(T)\gamma}\cdot\dots\cdot A^{(a)}_{r(1)\gamma}b, A^{(a)}_{r(T)\gamma}\cdot\dots\cdot A^{(a)}_{r(1)\gamma}(0)\big)\big)\lambda(b,a). \qquad (6.85)
\end{aligned}
$$

Assume now that

$$
\lambda\big(A^{(a)}_{r(Q(\gamma,\varepsilon))\gamma}\cdot\dots\cdot A^{(a)}_{r(1)\gamma}b, A^{(a)}_{r(Q(\gamma,\varepsilon))\gamma}\cdot\dots\cdot A^{(a)}_{r(1)\gamma}(0)\big)\le 1-\varepsilon. \qquad (6.86)
$$

Inequalities (6.86), (6.85) and (6.84) imply that for each integer $T\in[1,N]$,

$$
\begin{aligned}
&\lambda\big(A^{(a)}_{r(T+1)\gamma}A^{(a)}_{r(T)\gamma}\cdot\dots\cdot A^{(a)}_{r(1)\gamma}b, A^{(a)}_{r(T+1)\gamma}A^{(a)}_{r(T)\gamma}\cdot\dots\cdot A^{(a)}_{r(1)\gamma}(0)\big)\\
&\quad\ge \lambda\big(A^{(a)}_{r(T)\gamma}\cdot\dots\cdot A^{(a)}_{r(1)\gamma}b, A^{(a)}_{r(T)\gamma}\cdot\dots\cdot A^{(a)}_{r(1)\gamma}(0)\big)+\gamma\varepsilon\lambda(b,a).
\end{aligned}
$$

Together with (6.83) this implies that

$$
\begin{aligned}
&\lambda\big(A^{(a)}_{r(Q(\gamma,\varepsilon))\gamma}\cdot\dots\cdot A^{(a)}_{r(1)\gamma}(b), A^{(a)}_{r(Q(\gamma,\varepsilon))\gamma}\cdot\dots\cdot A^{(a)}_{r(1)\gamma}(0)\big)\\
&\quad\ge \lambda(b,a)\gamma\varepsilon\big(Q(\gamma,\varepsilon)-1\big)>4.
\end{aligned}
$$

Since this contradicts (6.86), the lemma is proved. □

Lemma 6.22 *Let $\{A_t\}_{t=1}^{\infty} \in \mathcal{M}$, $\varepsilon, \gamma \in (0,1)$ be given, and let $n \geq 1$ be an integer. Then there is $\Delta > 0$ such that for each $r : \{1, 2, \ldots, n\} \to \{1, 2, \ldots\}$, each sequence $\{C_i\}_{i=1}^{n} \subset \mathcal{A}$ satisfying*

$$\rho\big(C_i, A^{(a)}_{r(i)\gamma}\big) \leq \Delta, \quad i = 1, \ldots, n,$$

and each $z \in \langle 0, b\rangle$, the inequality

$$\big\| C_n \cdot \cdots \cdot C_1 z - A^{(a)}_{r(n)\gamma} \cdot \cdots \cdot A^{(a)}_{r(1)\gamma} z \big\|_b \leq \varepsilon$$

holds.

Proof We will prove the assertion of the lemma by induction. Clearly for $n = 1$ the assertion of the lemma is valid. Assume that the assertion of the lemma is valid for an integer $n \geq 1$. To prove that the assertion also holds for $n + 1$, choose first a positive number

$$\delta < 8^{-1}\gamma^2 \varepsilon \lambda(b, a)^2. \tag{6.87}$$

Since the assertion of the lemma holds for n, there exists $\Delta_0 > 0$ such that for each $r : \{1, \ldots, n\} \to \{1, 2, \ldots\}$, each $\{C_i\}_{i=1}^{n} \subset \mathcal{A}$ satisfying

$$\rho\big(C_i, A^{(a)}_{r(i)\gamma}\big) \leq \Delta_0, \quad i = 1, \ldots, n,$$

and each $z \in \langle 0, b\rangle$, the inequality

$$\big\| C_n \cdot \cdots \cdot C_1 z - A^{(a)}_{r(n)\gamma} \cdot \cdots \cdot A^{(a)}_{r(1)\gamma} z \big\|_b \leq \delta$$

holds. Set

$$\Delta := 8^{-1} \min\big\{\Delta_0, 8^{-1}\varepsilon\big\}. \tag{6.88}$$

Assume that $\{C_i\}_{i=1}^{n+1} \subset \mathcal{A}$, $r : \{1, \ldots, n+1\} \to \{1, 2, \ldots\}$,

$$\rho\big(C_i, A^{(a)}_{r(i)\gamma}\big) \leq \Delta, \quad i = 1, \ldots, n+1, \tag{6.89}$$

and that $z \in \langle 0, b\rangle$. Relations (6.88) and (6.89) imply that

$$\begin{aligned}
&\big\| A^{(a)}_{r(n+1)\gamma} \cdot A^{(a)}_{r(n)\gamma} \cdot \cdots \cdot A^{(a)}_{r(1)\gamma} z - C_{n+1} \cdot C_n \cdot \cdots \cdot C_1 z \big\|_b \\
&\quad \leq \big\| A^{(a)}_{r(n+1)\gamma} \cdot A^{(a)}_{r(n)\gamma} \cdot \cdots \cdot A^{(a)}_{r(1)\gamma} z - A^{(a)}_{r(n+1)\gamma} \cdot C_n \cdot \cdots \cdot C_1(z) \big\|_b \\
&\qquad + \big\| A^{(a)}_{r(n+1)\gamma} \cdot C_n \cdot \cdots \cdot C_1 z - C_{n+1} \cdot C_n \cdot \cdots \cdot C_1 z \big\|_b \\
&\quad \leq \big\| A^{(a)}_{r(n+1)\gamma} \cdot A^{(a)}_{r(n)\gamma} \cdot \cdots \cdot A^{(a)}_{r(1)\gamma} z - A^{(a)}_{r(n+1)\gamma} \cdot C_n \cdot \cdots \cdot C_1 z \big\|_b + 8^{-1}\varepsilon.
\end{aligned} \tag{6.90}$$

By (6.88), (6.89) and the definition of Δ_0,

$$\big\| A^{(a)}_{r(n)\gamma} \cdot \cdots \cdot A^{(a)}_{r(1)\gamma} z - C_n \cdot \cdots \cdot C_1 z \big\|_b \leq \delta. \tag{6.91}$$

It follows from (6.91) and (6.81) that

$$\begin{aligned}C_n \cdot \dots \cdot C_1 z &\geq A^{(a)}_{r(n)\gamma} \cdot \dots \cdot A^{(a)}_{r(1)\gamma} z - \delta b \\ &\geq A^{(a)}_{r(n)\gamma} \cdot \dots \cdot A^{(a)}_{r(1)\gamma} z - \lambda(b,a)^{-1}\delta\big(\gamma^{-1} A^{(a)}_{r(n)\gamma} \cdot \dots \cdot A^{(a)}_{r(1)\gamma} z\big) \\ &= \big(1 - \delta\gamma^{-1}\lambda(b,a)^{-1}\big) A^{(a)}_{r(n)\gamma} \cdot \dots \cdot A^{(a)}_{r(1)\gamma} z,\end{aligned}$$

$$\begin{aligned}A^{(a)}_{r(n)\gamma} \cdot \dots \cdot A^{(a)}_{r(1)\gamma} z &\geq C_n \cdot \dots \cdot C_1 z - \delta b \\ &\geq C_n \cdot \dots \cdot C_1 z - \lambda(b,a)^{-1}\delta\gamma^{-1}\big(A^{(a)}_{r(n)\gamma} \cdot \dots \cdot A^{(a)}_{r(1)\gamma} z\big)\end{aligned}$$

and

$$A^{(a)}_{r(n)\gamma} \cdot \dots \cdot A^{(a)}_{r(1)\gamma} \geq \big(1 + \lambda(b,a)^{-1}\delta\gamma^{-1}\big)^{-1} C_n \cdot \dots \cdot C_1 z.$$

Together with (6.87) this implies that

$$A^{(a)}_{r(n+1)\gamma} \cdot A^{(a)}_{r(n)\gamma} \cdot \dots \cdot A^{(a)}_{r(1)\gamma} z \geq \big(1 + 8\gamma^{-1}\lambda(b,a)^{-1}\big)^{-1} A^{(a)}_{r(n+1)\gamma} \cdot C_n \cdot \dots \cdot C_1 z,$$

$$A^{(a)}_{r(n+1)\gamma} \cdot C_n \cdot \dots \cdot C_1 z \geq \big(1 - \lambda(b,a)^{-1}\delta\gamma^{-1}\big) A^{(a)}_{r(n+1)\gamma} \cdot A^{(a)}_{r(n)\gamma} \cdot \dots \cdot A^{(a)}_{r(1)\gamma} z,$$

$$\begin{aligned}&\big(\big(1 + \lambda(b,a)^{-1}\delta\gamma^{-1}\big)^{-1} - 1\big)b \\ &\quad\leq A^{(a)}_{r(n+1)\gamma} \cdot A^{(a)}_{r(n)\gamma} \cdot \dots \cdot A^{(a)}_{r(1)\gamma} z - A^{(a)}_{r(n+1)\gamma} \cdot C_n \cdot \dots \cdot C_1 z \\ &\quad\leq \delta\gamma^{-1}\lambda(b,a)^{-1} b\end{aligned}$$

and

$$\begin{aligned}&\big\| A^{(a)}_{r(n+1)\gamma} \cdot A^{(a)}_{r(n)\gamma} \cdot \dots \cdot A^{(a)}_{r(1)\gamma} z - A^{(a)}_{r(n+1)\gamma} \cdot C_n \cdot \dots \cdot C_1 z \big\|_b \\ &\quad\leq \lambda(b,a)^{-1}\delta\gamma^{-1} < 8^{-1}\varepsilon.\end{aligned}$$

It follows from this and (6.90) that

$$\big\| A^{(a)}_{r(n+1)\gamma} \cdot A^{(a)}_{r(n)\gamma} \cdot \dots \cdot A^{(a)}_{r(1)\gamma} z - C_{n+1} \cdot C_n \cdot \dots \cdot C_1 z \big\|_b \leq 4^{-1}\varepsilon.$$

This completes the proof of the lemma. □

Lemma 6.23 *Let $\{A_t\}_{t=1}^{\infty} \in \mathcal{M}$ and $\gamma, \varepsilon \in (0,1)$ be given. Then there exist an integer $Q \geq 4$ and a number $\Delta > 0$ such that for each $r : \{1, \dots, Q\} \to \{1, 2, \dots\}$ and each $\{C_i\}_{i=1}^{Q} \subset \mathcal{A}$ satisfying*

$$\rho\big(C_i, A^{(a)}_{r(i)\gamma}\big) \leq \Delta, \quad i = 1, \dots, Q, \tag{6.92}$$

the following inequality holds:

$$\lambda\big(C_Q \cdot \dots \cdot C_1(b), C_Q \cdot \dots \cdot C_1(0)\big) \geq 1 - \varepsilon. \tag{6.93}$$

Proof Choose a positive number ε_0 such that

$$\varepsilon_0 < 8^{-1}\varepsilon\gamma^2\lambda(b,a) \tag{6.94}$$

and

$$\big(1-\varepsilon_0\big(\gamma\lambda(b,a)\big)^{-1}\big)(1-\varepsilon_0)\big(1+\varepsilon_0\big(\gamma\lambda(b,a)\big)^{-1}\big) > (1-\varepsilon).$$

Set

$$Q = Q(\gamma, \varepsilon_0) \tag{6.95}$$

(see (6.83)). By Lemma 6.21, for each $r : \{1,2,\ldots\} \to \{1,2,\ldots\}$,

$$\lambda\big(A^{(a)}_{r(Q)\gamma}\cdot\dots\cdot A^{(a)}_{r(1)\gamma}(b), A^{(a)}_{r(Q)\gamma}\cdot\dots\cdot A^{(a)}_{r(1)\gamma}(0)\big) > 1-\varepsilon_0. \tag{6.96}$$

By Lemma 6.22, there is $\Delta > 0$ such that for each $r : \{1,\ldots,Q\} \to \{1,2,\ldots\}$, each sequence $\{C_i\}_{i=1}^{Q} \subset \mathcal{A}$ satisfying (6.92) and each $z \in \langle 0, b\rangle$, the following inequality holds:

$$\big\|C_Q\cdot\dots\cdot C_1 z - A^{(a)}_{r(Q)\gamma}\cdot\dots\cdot A^{(a)}_{r(1)\gamma}z\big\|_b \le \varepsilon_0. \tag{6.97}$$

Assume that $r : \{1,\ldots,Q\} \to \{1,2,\ldots\}$, $\{C_i\}_{i=1}^{Q} \subset \mathcal{A}$ and that (6.92) is valid. Then (6.96) is valid too. It follows from the definition of Δ (see (6.97)), (6.92), (6.81), (6.96) and (6.94) that

$$\begin{aligned}
C_Q\cdot\dots\cdot C_1(0) &\ge A^{(a)}_{r(Q)\gamma}\cdot\dots\cdot A^{(a)}_{r(1)\gamma}(0) - \varepsilon_0 b\\
&\ge A^{(a)}_{r(Q)\gamma}\cdot\dots\cdot A^{(a)}_{r(1)\gamma}(0) - \lambda(b,a)^{-1}\varepsilon_0\gamma^{-1}A^{(a)}_{r(Q)\gamma}\cdot\dots\cdot A^{(a)}_{r(1)\gamma}(0)\\
&= \big(1-\varepsilon_0\big(\gamma\lambda(b,a)\big)^{-1}\big)A^{(a)}_{r(Q)\gamma}\cdot\dots\cdot A^{(a)}_{r(1)\gamma}(0)\\
&\ge \big(1-\varepsilon_0\gamma^{-1}\lambda(b,a)^{-1}\big)(1-\varepsilon_0)A^{(a)}_{r(Q)\gamma}\cdot\dots\cdot A^{(a)}_{r(1)\gamma}(b),
\end{aligned}$$

$$\begin{aligned}
A^{(a)}_{r(Q)\gamma}\cdot\dots\cdot A^{(a)}_{r(1)\gamma}(b) &\ge C_Q\cdot\dots\cdot C_1(b) - \varepsilon_0 b\\
&\ge C_Q\cdot\dots\cdot C_1 b - \lambda(b,a)^{-1}\varepsilon_0\gamma^{-1}A^{(a)}_{r(Q)\gamma}\cdot\dots\cdot A^{(a)}_{r(1)\gamma}(b),\\
A^{(a)}_{r(Q)\gamma}\cdot\dots\cdot A^{(a)}_{r(1)\gamma}b &\ge \big(1+\lambda(b,a)^{-1}\varepsilon_0\gamma^{-1}\big)^{-1}C_Q\cdot\dots\cdot C_1(b)
\end{aligned}$$

and

$$\begin{aligned}
C_Q\cdot\dots\cdot C_1(0) &\ge \big(1-\lambda(b,a)^{-1}\varepsilon_0\gamma^{-1}\big)\\
&\quad\cdot(1-\varepsilon_0)\big(1+\lambda(b,a)^{-1}\varepsilon_0\gamma^{-1}\big)^{-1}C_Q\cdot\dots\cdot C_1(b)\\
&\ge (1-\varepsilon)C_Q\cdot\dots\cdot C_1(b).
\end{aligned}$$

This completes the proof of the lemma. □

6.13 Proofs of Theorems 6.16 and 6.17

Set $a = b$.

Let $\mathbf{A} = \{A_t\}_{t=1}^{\infty} \in \mathcal{M}$, $\gamma \in (0, 1)$ and let $j \geq 1$ be an integer. By Lemma 6.23, there exist an integer $Q(\mathbf{A}, \gamma, j) \geq 4$ and a number $\Delta(\mathbf{A}, \gamma, j) > 0$ such that the following property holds:

(a) for each $r : \{1, \ldots, Q(\mathbf{A}, \gamma, j)\} \to \{1, 2, \ldots\}$ and each $\{C_i\}_{i=1}^{Q(\mathbf{A},\gamma,j)} \subset \mathcal{A}$ satisfying

$$\rho\big(C_i, A_{r(i)\gamma}^{(b)}\big) \leq \Delta(\mathbf{A}, \gamma, j), \quad i = 1, \ldots, Q(\mathbf{A}, \gamma, j),$$

the inequality

$$\lambda\big(C_{Q(\mathbf{A},\gamma,j)} \cdot \cdots \cdot C_1(b), C_{Q(\mathbf{A},\gamma,j)} \cdot \cdots \cdot C_1(0)\big) \geq 1 - 16^{-j}$$

holds.

Proof of Theorem 6.16 For each $\{A_t\}_{t=1}^{\infty} \in \mathcal{M}$, each $\gamma \in (0, 1)$ and each integer $j \geq 1$, there exists an open neighborhood $U(\mathbf{A}, \gamma, j)$ of $\{A_{t\gamma}^{(b)}\}_{t=1}^{\infty}$ in the space $\mathcal{M}$ with the weak topology such that

$$\begin{aligned} U(\mathbf{A}, \gamma, j) \subset \Big\{&\{C_t\}_{t=1}^{\infty} \in \mathcal{M} : \big\|A_{t\gamma}^{(b)}x - C_t x\big\|_b \leq \Delta(\mathbf{A}, \gamma, j), \\ & x \in \langle 0, b\rangle, t = 1, \ldots, Q(\mathbf{A}, \gamma, j)\Big\}. \end{aligned} \tag{6.98}$$

Define

$$\mathcal{F} := \bigcap_{q=1}^{\infty} \bigcup \big\{U(\mathbf{A}, \gamma, j) : \mathbf{A} \in \mathcal{M}, \gamma \in (0, 1), j = q, q + 1, \ldots\big\}.$$

Clearly, $\mathcal{F}$ is a countable intersection of open (in the weak topology) everywhere dense (in the strong topology) subsets of $\mathcal{M}$.

Assume that $\{B_t\}_{t=1}^{\infty} \in \mathcal{F}$ and $\varepsilon \in (0, 1)$ are given. Choose an integer $q \geq 1$ for which

$$2^{-q} < 64^{-1}\varepsilon. \tag{6.99}$$

There exists $\{A_t\}_{t=1}^{\infty} \in \mathcal{M}$, $\gamma \in (0, 1)$ and an integer $j \geq q$ such that

$$\{B_t\}_{t=1}^{\infty} \in U\big(\{A_t\}_{t=1}^{\infty}, \gamma, j\big).$$

It follows from (6.98), (6.99) and property (a) that for each

$$\{C_t\}_{t=1}^{\infty} \in U\big(\{A_t\}_{t=1}^{\infty}, \gamma, j\big),$$

the following relation holds:

$$\lambda\big(C_{Q(\mathbf{A},\gamma,j)} \cdot \cdots \cdot C_1(b), C_{Q(\mathbf{A},\gamma,j)} \cdot \cdots \cdot C_1(0)\big) \geq 1 - 16^{-j} > 1 - 2^{-1}\varepsilon.$$

This completes the proof of the theorem. □

Proof of Theorem 6.17 For each $\{A_t\}_{t=1}^{\infty} \in \mathcal{M}$, each $\gamma \in (0,1)$ and each integer $j \geq 1$, we first define

$$U(\mathbf{A},\gamma,j) = \big\{\{B_t\}_{t=1}^{\infty} \in \mathcal{M} : \rho_s\big(\{A_{t\gamma}^{(b)}\}_{t=1}^{\infty}, \{B_t\}_{t=1}^{\infty}\big) < \Delta(\mathbf{A},\gamma,j)\big\}. \quad (6.100)$$

Next we set

$$\mathcal{F} := \bigcap_{q=1}^{\infty} \bigcup \{U(\mathbf{A},\gamma,j) : \mathbf{A} \in \mathcal{M}, \gamma \in (0,1), j = q, q+1, \ldots\}.$$

It is clear that $\mathcal{F}$ is a countable intersection of open and everywhere dense subsets of $\mathcal{M}$ with the strong topology.

Assume that $\{B_t\}_{t=1}^{\infty} \in \mathcal{F}$ and $\varepsilon \in (0,1)$ are given. Choose an integer $q \geq 1$ for which (6.99) holds. There exist $\{A_t\}_{t=1}^{\infty} \in \mathcal{M}$, $\gamma \in (0,1)$ and an integer $j \geq q$ such that

$$\{B_t\}_{t=1}^{\infty} \in U\big(\{A_t\}_{t=1}^{\infty}, \gamma, j\big).$$

It follows from (6.100), (6.99) and property (a) that for each

$$\{C_t\}_{t=1}^{\infty} \in U\big(\{A_t\}_{t=1}^{\infty}, \gamma, j\big)$$

and each $r : \{1,2,\ldots\} \to \{1,2,\ldots\}$,

$$\begin{aligned}&\lambda\big(C_{r(Q(\mathbf{A},\gamma,j))} \cdot \cdots \cdot C_{r(1)}(b), C_{r(Q(\mathbf{A},\gamma,j))} \cdot \cdots \cdot C_{r(1)}(0)\big)\\ &\geq 1 - 16^{-j} > 1 - 2^{-1}\varepsilon.\end{aligned}$$

The theorem is proved. □

6.14 Proofs of Theorems 6.18 and 6.19

It is easy to see that $\{A_{t\gamma}^{(a)}\}_{t=1}^{\infty} \in \mathcal{M}_a$ for each $\{A_t\}_{t=1}^{\infty} \in \mathcal{M}_a$ and each $\gamma \in (0,1)$, and that the set

$$\big\{\{A_{t\gamma}^{(a)}\}_{t=1}^{\infty} : \{A_t\}_{t=1}^{\infty} \in \mathcal{M}_a, \gamma \in (0,1)\big\}$$

is everywhere dense in $\mathcal{M}_a$ with the strong topology.

Let $\mathbf{A} = \{A_t\}_{t=1}^{\infty} \in \mathcal{M}$, $\gamma \in (0,1)$ and let $j \geq 1$ be an integer. By Lemma 6.23, there exist an integer $Q(\mathbf{A},\gamma,j) \geq 4$ and a number $\Delta(\mathbf{A},\gamma,j) > 0$ such that the following property holds:

(a) For each $r : \{1,\ldots,Q(\mathbf{A},\gamma,j)\} \to \{1,2,\ldots\}$ and each $\{C_i\}_{i=1}^{Q(\mathbf{A},\gamma,j)} \subset \mathcal{A}$ satisfying

$$\rho\big(C_i, A_{r(i)\gamma}^{(a)}\big) \leq \Delta(\mathbf{A},\gamma,j), \quad i = 1,\ldots,Q(\mathbf{A},\gamma,j),$$

the following inequality holds:

$$\lambda\big(C_{Q(\mathbf{A},\gamma,j)} \cdot \cdots \cdot C_1(b), C_{Q(\mathbf{A},\gamma,j)} \cdot \cdots \cdot C_1(0)\big) \geq 1 - 16^{-j}.$$

Proof of Theorem 6.18 For each $\{A_t\}_{t=1}^{\infty} \in \mathcal{M}_a$, each $\gamma \in (0,1)$ and each integer $j \geq 1$, there exists an open neighborhood $U(\mathbf{A}, \gamma, j)$ of $\{A_{t\gamma}^{(a)}\}_{t=1}^{\infty}$ in the space $\mathcal{M}_a$ with the relative weak topology such that

$$U(\mathbf{A}, \gamma, j) \subset \big\{\{C_t\}_{t=1}^{\infty} \in \mathcal{M}_a : \|A_{t\gamma}^{(a)} x - C_t x\|_b \leq \Delta(\mathbf{A}, \gamma, j),\\ x \in \langle 0, b \rangle, t = 1, \ldots, Q(\mathbf{A}, \gamma, j)\big\}. \tag{6.101}$$

Define

$$\mathcal{F} := \bigcap_{q=1}^{\infty} \bigcup \{U(\mathbf{A}, \gamma, j) : A \in \mathcal{M}_a, \gamma \in (0,1), j = q, q+1, \ldots\}.$$

Clearly, $\mathcal{F}$ is a countable intersection of open (in the relative weak topology) everywhere dense (in the relative strong topology) sets in $\mathcal{M}_a$.

Assume that $\{B_t\}_{t=1}^{\infty} \in \mathcal{F}$ and $\varepsilon > 0$ are given. Choose a natural number q for which

$$2^{-q} < 64^{-1}\varepsilon. \tag{6.102}$$

There exist $\mathbf{A} = \{A_t\}_{t=1}^{\infty} \in \mathcal{M}_a$, $\gamma \in (0,1)$ and an integer $j \geq q$ such that

$$\{B_t\}_{t=1}^{\infty} \in U(\mathbf{A}, \gamma, j). \tag{6.103}$$

Assume that $\{C_t\}_{t=1}^{\infty} \in U(\mathbf{A}, \gamma, j)$. It follows from property (a) and (6.101) that

$$\lambda\big(C_{Q(\mathbf{A},\gamma,j)} \cdot \cdots \cdot C_1(b), C_{Q(\mathbf{A},\gamma,j)} \cdot \cdots \cdot C_1(0)\big) \geq 1 - 16^{-j}. \tag{6.104}$$

Since $C_t a = a$, $t = 1, 2, \ldots$, it follows from (6.104) that

$$C_{Q(\mathbf{A},\gamma,j)} \cdot \cdots \cdot C_1(0) \leq a \leq C_{Q(\mathbf{A},\gamma,j)} \cdot \cdots \cdot C_1(b),\\ C_{Q(\mathbf{A},\gamma,j)} \cdot \cdots \cdot C_1(0) \geq (1 - 16^{-j})a$$

and

$$a \geq (1 - 16^{-j}) C_{Q(\mathbf{A},\gamma,j)} \cdot \cdots \cdot C_1(b).$$

By these relations and (6.102), for each integer $T \geq Q(\mathbf{A}, \gamma, j)$ and each $z \in \langle 0, b \rangle$, we have

$$C_T \cdot \cdots \cdot C_1(0) \geq (1 - 16^{-j})a, \qquad a \geq (1 - 16^{-j}) C_T \cdot \cdots \cdot C_1(b),\\ C_T \cdot \cdots \cdot C_1 z - a \in \langle C_T \cdot \cdots \cdot C_1(0) - a, C_T \cdot \cdots \cdot C_1(b) - a \rangle$$

and finally,

$$\|C_T \cdot \cdots \cdot C_1 z - a\|_b \leq \|C_T \cdot \cdots \cdot C_1 b - a\|_b + \|C_T \cdot \cdots \cdot C_1(0) - a\|_b\\ \leq 2 \cdot 16^{-j} < \varepsilon,$$

as claimed. □

Proof of Theorem 6.19 For each $\{A_t\}_{t=1}^{\infty} \in \mathcal{M}_a$, each $\gamma \in (0, 1)$ and each integer $j \geq 1$, define

$$U(\mathbf{A}, \gamma, j) = \{\{B_t\}_{t=1}^{\infty} \in \mathcal{M}_a : \rho_s(\{A_{t\gamma}^{(a)}\}_{t=1}^{\infty}, \{B_t\}_{t=1}^{\infty}) < \Delta(\mathbf{A}, \gamma, j)\}, \quad (6.105)$$

and set

$$\mathcal{F} = \bigcap_{q=1}^{\infty} \bigcup \{U(\mathbf{A}, \gamma, j) : \mathbf{A} \in \mathcal{M}_a, \gamma \in (0, 1), j = q, q+1, \ldots\}.$$

Clearly, $\mathcal{F}$ is a countable intersection of open and everywhere dense subsets of $\mathcal{M}_a$ with the relative strong topology.

Assume that $\{B_t\}_{t=1}^{\infty} \in \mathcal{F}$ and $\varepsilon > 0$ are given. Choose an integer $q \geq 1$ for which (6.102) is valid. There exist $\{A_t\}_{t=1}^{\infty} \in \mathcal{M}_a$, $\gamma \in (0, 1)$ and an integer $j \geq q$ such that

$$\{B_t\}_{t=1}^{\infty} \in U(\{A_t\}_{t=1}^{\infty}, \gamma, j).$$

Assume that $\{C_t\}_{t=1}^{\infty} \in U(\{A_t\}_{t=1}^{\infty}, \gamma, j)$ and $r : \{1, 2, \ldots\} \to \{1, 2, \ldots\}$. It follows from (6.105) and property (a) that

$$\lambda(C_{r(Q(\mathbf{A},\gamma,j))} \cdot \cdots \cdot C_{r(1)}(b), C_{r(Q(\mathbf{A},\gamma,j))} \cdot \cdots \cdot C_{r(1)}(0)) \geq 1 - 16^{-j}. \quad (6.106)$$

Since $C_t a = a$, $t = 1, 2, \ldots$, it follows from (6.106) that

$$C_{r(Q(\mathbf{A},\gamma,j))} \cdot \cdots \cdot C_{r(1)}(0) \leq a \leq C_{r(Q(\mathbf{A},\gamma,j))} \cdot \cdots \cdot C_{r(1)}(b),$$
$$C_{r(Q(\mathbf{A},\gamma,j))} \cdot \cdots \cdot C_{r(1)}(0) \geq (1 - 16^{-j})a,$$

and

$$a \geq (1 - 16^{-j}) C_{r(Q(\mathbf{A},\gamma,j))} \cdot \cdots \cdot C_{r(1)}(b).$$

By these relations and (6.102), for each integer $T \geq Q(\mathbf{A}, \gamma, j)$ and each $z \in \langle 0, b \rangle$,

$$C_{r(T)} \cdot \cdots \cdot C_{r(1)}(0) \geq (1 - 16^{-j})a, \qquad a \geq (1 - 16^{-j}) C_{r(T)} \cdot \cdots \cdot C_{r(1)}(b),$$
$$C_{r(T)} \cdot \cdots \cdot C_{r(1)}(z) - a \in \langle C_{r(T)} \cdot \cdots \cdot C_{r(1)}(0) - a, C_{r(T)} \cdot \cdots \cdot C_{r(1)}(b) - a \rangle,$$

and finally,

$$\begin{aligned} \|C_{r(T)} \cdot \cdots \cdot C_{r(1)}(z) - a\|_b &\leq \|C_{r(T)} \cdot \cdots \cdot C_{r(1)}(b) - a\|_b \\ &\quad + \|C_{r(T)} \cdot \cdots \cdot C_{r(1)}(0) - a\|_b \\ &\leq 2 \cdot 16^{-j} < \varepsilon, \end{aligned}$$

as required. This completes the proof of Theorem 6.19. □

6.15 Proof of Theorem 6.20

Let $\{A_t\}_{t=1}^{\infty} \in \mathcal{M}_*$. There exists $x(\mathbf{A}) \in \langle 0, b \rangle$ which is an interior point of the cone X_+ such that

$$A_t\big(x(\mathbf{A})\big) = x(\mathbf{A}), \quad t = 1, 2, \ldots. \tag{6.107}$$

For each $\mathbf{A} = \{A_t\}_{t=1}^{\infty} \in \mathcal{M}_*$ and each $\gamma \in (0, 1)$, we set

$$A_{t\gamma} = A_{t\gamma}^{(x(\mathbf{A}))}, \quad t = 1, 2, \ldots. \tag{6.108}$$

It is easy to see that $\{A_{t\gamma}\}_{t=1}^{\infty} \in \mathcal{M}_*$ for each $\{A_t\}_{t=1}^{\infty} \in \mathcal{M}_*$ and each $\gamma \in (0, 1)$, and that the set

$$\big\{\{A_{t\gamma}\}_{t=1}^{\infty} : \{A_t\}_{t=1}^{\infty} \in \mathcal{M}_*, \gamma \in (0, 1)\big\}$$

is everywhere dense in $\bar{\mathcal{M}}_*$.

Let $\mathbf{A} = \{A_t\}_{t=1}^{\infty} \in \mathcal{M}_*$ and $\gamma \in (0, 1)$, and let $j \geq 1$ be an integer. By Lemma 6.23, there exist an integer $Q(\mathbf{A}, \gamma, j) \geq 4$ and a number $\Delta_1(\mathbf{A}, \gamma, j) > 0$ such that the following property holds:

(a) For each $r : \{1, \ldots, Q(\mathbf{A}, \gamma, j)\} \to \{1, 2, \ldots\}$ and each $\{C_i\}_{i=1}^{Q(\mathbf{A},\gamma,j)} \subset \mathcal{A}$ satisfying

$$\rho(C_i, A_{r(i)\gamma}) \leq \Delta_1(\mathbf{A}, \gamma, j), \quad i = 1, \ldots, Q(\mathbf{A}, \gamma, j),$$

the following inequality holds:

$$\lambda\big(C_{Q(\mathbf{A},\gamma,j)} \cdot \cdots \cdot C_1(b), C_{Q(\mathbf{A},\gamma,j)} \cdot \cdots \cdot C_1(0)\big) \geq 1 - 16^{-j}.$$

Choose now a positive number

$$\delta(\mathbf{A}, \gamma, j) < 16^{-j}\gamma^2\lambda\big(b, x(\mathbf{A})\big). \tag{6.109}$$

By Lemma 6.22, there is a number $\Delta_2(\mathbf{A}, \gamma, j) > 0$ such that the following property holds:

(b) For each $r : \{1, \ldots, Q(\mathbf{A}, \gamma, j)\} \to \{1, 2, \ldots\}$, each sequence

$$\{C_i\}_{i=1}^{Q(\mathbf{A},\gamma,j)} \subset \mathcal{A}$$

satisfying

$$\rho(C_i, A_{r(i)\gamma}) \leq \Delta_2(\mathbf{A}, \gamma, j), \quad i = 1, \ldots, Q(\mathbf{A}, \gamma, j),$$

and each $z \in \langle 0, b \rangle$, the following inequality holds:

$$\|C_{Q(\mathbf{A},\gamma,j)} \cdot \cdots \cdot C_1 z - A_{r(Q(\mathbf{A}\gamma,j))\gamma} \cdot \cdots \cdot A_{r(1)\gamma} z\|_b \leq \delta(\mathbf{A}, \gamma, j).$$

Define

$$U(\mathbf{A}, \gamma, j) = \big\{\{B_t\}_{t=1}^{\infty} \in \bar{\mathcal{M}}_* :$$
$$\rho_s\big(\{A_{t\gamma}\}_{t=1}^{\infty}, \{B_t\}_{t=1}^{\infty}\big) < \inf\big\{\Delta_1(\mathbf{A}, \gamma, j), \Delta_2(\mathbf{A}, \gamma, j)\big\}\big\} \tag{6.110}$$

and

$$\mathcal{F} = \bigcap_{q=1}^{\infty} \bigcup \{U(\mathbf{A}, \gamma, j) : \mathbf{A} \in \mathcal{M}_*, \gamma \in (0, 1), j = q, q+1, \dots\}.$$

Clearly, $\mathcal{F}$ is a countable intersection of open and everywhere dense sets in $\bar{\mathcal{M}}_*$. Assume that $\{B_t\}_{t=1}^{\infty} \in \mathcal{F}$ and $\varepsilon > 0$ are given. Choose a natural number q such that

$$2^{-q} < 64^{-1}\varepsilon. \tag{6.111}$$

There exist $\mathbf{A} = \{A_t\}_{t=1}^{\infty} \in \mathcal{M}_*$, $\gamma \in (0, 1)$ and an integer $j \geq q$ such that

$$\{B_t\}_{t=1}^{\infty} \in U(\mathbf{A}, \gamma, j). \tag{6.112}$$

It follows from property (a) that for each $r : \{1, \dots, Q(\mathbf{A}, \gamma, j)\} \to \{1, 2, \dots\}$ and each $z \in \langle 0, b \rangle$,

$$\left\| A_{r(Q(\mathbf{A},\gamma,j))\gamma} \cdot \dots \cdot A_{r(1)\gamma} z - x(\mathbf{A}) \right\|_b \leq 2 \cdot 16^{-j}. \tag{6.113}$$

Together with property (b), (6.110) and (6.109) this implies that the following property holds:

(c) For each $\{C_t\}_{t=1}^{\infty} \in U(\mathbf{A}, \gamma, j)$, each $r : \{1, \dots, Q(\mathbf{A}, \gamma, j)\} \to \{1, 2, \dots\}$ and each $z \in \langle 0, b \rangle$,

$$\left\| C_{r(Q(\mathbf{A},\gamma,j))} \cdot \dots \cdot C_{r(1)} z - x(\mathbf{A}) \right\|_b \leq 3 \cdot 16^{-j}. \tag{6.114}$$

Property (c), when combined with (6.112) and (6.111), implies that for each integer $\tau \geq 1$, each integer $T \geq r(Q(\mathbf{A}, \gamma, j))$ and each $z \in \langle 0, b \rangle$,

$$\left\| B_\tau^T z - x(\mathbf{A}) \right\|_b \leq 3 \cdot 16^{-j} \leq \varepsilon. \tag{6.115}$$

Since ε is an arbitrary positive number, we conclude that there exists $x(\mathbf{B}) \in \langle 0, b \rangle$ such that

$$\lim_{T\to\infty} B_\tau^T z = x(\mathbf{B}) \quad \text{for each integer } \tau \geq 1 \text{ and each } z \in \langle 0, b \rangle. \tag{6.116}$$

Clearly,

$$B_t\big(x(\mathbf{B})\big) = x(\mathbf{B}), \quad t = 1, 2, \dots. \tag{6.117}$$

By (6.115) and (6.116),

$$\left\| x(\mathbf{B}) - x(\mathbf{A}) \right\|_b \leq 3 \cdot 16^{-j}. \tag{6.118}$$

We will show that $x(\mathbf{B})$ is an interior point of X_+. To this end, note that property (b), (6.112), (6.110), (6.107), (6.109) and (6.117) yield

$$\left\| x(\mathbf{B}) - A_{Q(\mathbf{A},\gamma,j)\gamma} \cdot \dots \cdot A_{1\gamma}\big(x(\mathbf{B})\big) \right\|_b \leq \delta(\mathbf{A}, \gamma, j),$$

$$x(\mathbf{B}) \geq -\delta(\mathbf{A}, \gamma, j)b + A_{Q(\mathbf{A},\gamma,j)} \cdot \dots \cdot A_{1\gamma}\big(x(\mathbf{B})\big)$$
$$\geq \gamma x(\mathbf{A}) - \delta(\mathbf{A}, \gamma, j)b \geq \gamma x(\mathbf{A}) - 16^{-j}\gamma^2 x(\mathbf{A}).$$

This implies that $x(\mathbf{B})$ is indeed an interior point of X_+.

Assume that $\{C_t\}_{t=1}^{\infty} \in U(\mathbf{A}, \gamma, j)$ and $r : \{1, 2, \dots\} \to \{1, 2, \dots\}$ are given. It follows from property (c), (6.118) and (6.111) that for each $z \in \langle 0, b\rangle$ and each integer $T \geq Q(\mathbf{A}, \gamma, j)$,

$$\big\| C_{r(T)} \cdot \dots \cdot C_{r(1)}z - x(\mathbf{A}) \big\|_b \leq 3 \cdot 16^{-j}$$

and

$$\big\| C_{r(T)} \cdot \dots \cdot C_{r(1)}z - x(\mathbf{B}) \big\|_b \leq 6 \cdot 16^{-j} < \varepsilon.$$

This completes the proof of Theorem 6.20.

6.16 Infinite Products of Positive Linear Operators

Infinite products of linear operators are of interest in many areas of mathematics and its applications. See, for instance, [5, 22, 55–58, 71, 72, 91, 95, 110, 175] and the references mentioned there. Since many linear operators between Banach spaces arising in classical and modern analysis are, in fact, positive operators, the theory of positive linear operators and its applications have drawn the attention of more and more mathematicians. See, for example, [3, 86, 96, 170] and the references cited therein.

In this section we study (random) infinite products of generic sequences of positive linear operators on an ordered Banach space. In addition to a weak ergodic theorem (Theorem 6.27), we also obtain generic convergence to an operator of the form $f(\cdot)\eta$, where f is a bounded linear functional and η is a common fixed point. More precisely, having chosen an appropriate space of sequences of positive linear operators, we construct a subset which is a countable intersection of open and everywhere dense sets such that for each sequence belonging to this subset, the corresponding infinite products converge.

Let $(X, \|\cdot\|)$ be a real Banach space with norm $\|\cdot\|$, which is ordered by a closed and convex cone X_+. For each $u, v \in X$ such that $u \leq v$, we define

$$\langle u, v\rangle = \{z \in X : u \leq z \leq v\}.$$

For each set $E \subset X$, we denote by $\mathrm{int}(E)$ the interior of E. We assume that the cone X_+ has a nonempty interior $\mathrm{int}(X_+)$ and that for each $x, y \in X_+$ satisfying $x \leq y$, the inequality $\|x\| \leq \|y\|$ holds.

Fix an interior point η of the cone X_+ and define

$$\|x\|_\eta := \inf\{r \in [0, \infty) : -r\eta \leq x \leq r\eta\}, \quad x \in X. \tag{6.119}$$

It is clear that $\|\cdot\|_\eta$ is a norm on X which is equivalent to the original norm $\|\cdot\|$.

Let X' be the space of all linear continuous functionals $f : X \to R^1$. Define

$$X'_+ := \{f \in X' : f(x) \geq 0 \text{ for all } x \in X_+\}.$$

Denote by $\mathcal{A}$ the set of all linear operators $A : X \to X$ such that $A(X_+) \subset X_+$. Let $\mathcal{M}$ be the set of all sequences $\{A_t\}_{t=1}^{\infty}$, where $A_t \in \mathcal{A}$, $t = 1, 2, \ldots$. Such a sequence will occasionally be denoted by a boldface $\mathbf{A}$. Define

$$\mathcal{M}_\eta := \{\{A_t\}_{t=1}^{\infty} \in \mathcal{M} : A_t \eta = \eta, t = 1, 2, \ldots\}. \tag{6.120}$$

For the set $\mathcal{M}_\eta$ we consider the metric $\rho_s : \mathcal{M}_\eta \times \mathcal{M}_\eta \to [0, \infty)$ defined by

$$\rho_s(\{A_t\}_{t=1}^{\infty}, \{B_t\}_{t=1}^{\infty}) = \sup\{\|A_t x - B_t x\|_\eta : x \in \langle 0, \eta\rangle, t = 1, 2, \ldots\},$$
$$\{A_t\}_{t=1}^{\infty}, \{B_t\}_{t=1}^{\infty} \in \mathcal{M}_\eta. \tag{6.121}$$

It is easy to see that the metric space $(\mathcal{M}_\eta, \rho_s)$ is complete. We shall refer to the topology generated by the metric ρ_s as the strong topology. For the set $\mathcal{M}_\eta$ we also consider the uniformity which is determined by the base

$$E(N, \varepsilon) = \{(\{A_t\}_{t=1}^{\infty}, \{B_t\}_{t=1}^{\infty}) \in \mathcal{M}_\eta \times \mathcal{M}_\eta :$$
$$\|A_t x - B_t x\|_\eta \leq \varepsilon, t = 1, \ldots, N, x \in \langle 0, \eta\rangle\},$$

where N is a natural number and $\varepsilon > 0$. The topology generated by this uniformity on $\mathcal{M}_\eta$ will be called the weak topology. It is easy to see that the space $\mathcal{M}_\eta$ with this uniformity is metrizable (by a metric $\rho_w : \mathcal{M}_\eta \times \mathcal{M}_\eta \to [0, \infty)$) and complete ([80]).

We now state our first two results [125]. The second one deals with random products.

Theorem 6.24 *There exists a set $\mathcal{F} \subset \mathcal{M}_\eta$, which is a countable intersection of open (in the weak topology) everywhere dense (in the strong topology) sets in $\mathcal{M}_\eta$, such that for each $\mathbf{B} = \{B_t\}_{t=1}^{\infty} \in \mathcal{F}$, the following two assertions hold:*

1. *There exists a continuous linear functional $f_{\mathbf{B}} : X \to R^1$ such that*

$$\lim_{T\to\infty} B_T \cdot \cdots \cdot B_1 x = f_{\mathbf{B}}(x)\eta \quad \text{for each } x \in X.$$

2. *For each $\varepsilon > 0$, there exists a neighborhood U of $\mathbf{B} = \{B_t\}_{t=1}^{\infty}$ in $\mathcal{M}_\eta$ with the weak topology and a natural number N such that for each $\{C_t\}_{t=1}^{\infty} \in U$, each integer $T \geq N$ and each $x \in \langle -\eta, \eta\rangle$,*

$$\|C_T \cdot \cdots \cdot C_1 x - f_{\mathbf{B}}(x)\eta\|_\eta \leq \varepsilon.$$

Theorem 6.25 *There exists a set $\mathcal{F} \subset \mathcal{M}_\eta$, which is a countable intersection of open everywhere dense in the strong topology sets in $\mathcal{M}_\eta$, such that for each $\mathbf{B} = \{B_t\}_{t=1}^{\infty} \in \mathcal{F}$, the following two assertions hold:*

1. *For each* $r : \{1, 2, \ldots\} \to \{1, 2, \ldots\}$, *there exists a linear functional* $f_r \in X'_+$ *such that*

$$\lim_{T\to\infty} B_{r(T)} \cdot \cdots \cdot B_{r(1)}x = f_r(x)\eta \quad \text{for each } x \in X.$$

2. *For each* $\varepsilon > 0$, *there exists a neighborhood* U *of* $\{B_t\}_{t=1}^{\infty}$ *in* $\mathcal{M}_\eta$ *with the strong topology and a natural number* N *such that for each* $\{C_t\}_{t=1}^{\infty} \in U$, *each integer* $T \geq N$, *each* $r : \{1, 2, \ldots\} \to \{1, 2, \ldots\}$ *and each* $x \in \langle -\eta, \eta \rangle$,

$$\left\| C_{r(T)} \cdot \cdots \cdot C_{r(1)}x - f_r(x)\eta \right\|_\eta \leq \varepsilon.$$

We now turn our attention to another metric space of sequences.

Define

$$\mathcal{M}_b := \left\{ \{A_t\}_{t=1}^{\infty} \in \mathcal{M} : \sup\left\{ \|A_t\eta\|_\eta : t = 1, 2, \ldots \right\} < \infty \right\}. \tag{6.122}$$

For the set $\mathcal{M}_b$ we consider the metric $\rho_s : \mathcal{M}_b \times \mathcal{M}_b \to [0, \infty)$ defined by

$$\rho_s\left(\{A_t\}_{t=1}^{\infty}, \{B_t\}_{t=1}^{\infty}\right) = \sup\left\{ \|A_t x - B_t x\|_\eta : x \in \langle 0, \eta \rangle, t = 1, 2, \ldots \right\},$$
$$\{A_t\}_{t=1}^{\infty}, \{B_t\}_{t=1}^{\infty} \in \mathcal{M}_b. \tag{6.123}$$

It is easy to see that the metric space $(\mathcal{M}_b, \rho_s)$ is complete.

Denote by $\mathcal{M}_b^*$ the set of all $\{A_t\}_{t=1}^{\infty} \in \mathcal{M}_b$ such that there exists an interior point $\xi_{\mathbf{A}}$ of X_+ for which

$$A_t \xi_{\mathbf{A}} = \xi_{\mathbf{A}}, \quad t = 1, 2, \ldots.$$

Finally, denote by $\bar{\mathcal{M}}_b^*$ the closure of $\mathcal{M}_b^*$ in $\mathcal{M}_b$.

Theorem 6.26 *There exists a set* $\mathcal{F} \subset \bar{\mathcal{M}}_b^*$, *which is a countable intersection of open and everywhere dense sets in* $\bar{\mathcal{M}}_b^*$, *such that for each* $\mathbf{B} = \{B_t\}_{t=1}^{\infty} \in \mathcal{F}$, *there exists an interior point* $\xi_{\mathbf{B}}$ *of* X_+ *satisfying*

$$B_t \xi_{\mathbf{B}} = \xi_{\mathbf{B}}, \quad t = 1, 2, \ldots, \qquad \|\xi_{\mathbf{B}}\|_\eta = 1,$$

and the following two assertions hold:

1. *For each* $r : \{1, 2, \ldots\} \to \{1, 2, \ldots\}$, *there exists a linear functional* $f_r \in X'_+$ *such that*

$$\lim_{T\to\infty} B_{r(T)} \cdot \cdots \cdot B_{r(1)}x = f_r(x)\xi_{\mathbf{B}}, \quad x \in X.$$

2. *For each* $\varepsilon > 0$, *there exist a neighborhood* U *of* $\{B_t\}_{t=1}^{\infty}$ *in* $\bar{\mathcal{M}}_b^*$ *and a natural number* N *such that for each* $\{C_t\}_{t=1}^{\infty} \in U \cap \mathcal{M}_b^*$, *each integer* $T \geq N$, *each* $r : \{1, 2, \ldots\} \to \{1, 2, \ldots\}$ *and each* $x \in \langle -\eta, \eta \rangle$,

$$\left\| C_{r(T)} \cdot \cdots \cdot C_{r(1)}x - f_r(x)\xi_{\mathbf{B}} \right\|_\eta \leq \varepsilon.$$

For each $x, y \in X_+$, define

$$\begin{aligned} \lambda(x, y) &= \sup\{r \in [0, \infty) : rx \le y\}, \\ r(x, y) &= \inf\{\lambda \in [0, \infty) : y \le \lambda x\}. \end{aligned} \tag{6.124}$$

Here we use the usual convention that the infimum of the empty set is ∞.

Denote by $\mathcal{M}_{reg}$ the set of all sequences $\mathbf{A} = \{A_t\}_{t=1}^{\infty} \in \mathcal{M}$ such that there exist positive constants $c_1 < c_2$ satisfying

$$c_2\eta \ge A_T \cdot \cdots \cdot A_1\eta \ge c_1\eta, \quad T = 1, 2, \ldots. \tag{6.125}$$

For the set $\mathcal{M}$ we consider the uniformity which is determined by the base

$$\begin{aligned} E(N, \varepsilon) = \big\{ &(\{A_t\}_{t=1}^{\infty}, \{B_t\}_{t=1}^{\infty}) \in \mathcal{M} \times \mathcal{M} : \\ &\|A_t x - B_t x\|_\eta \le \varepsilon, t = 1, \ldots, N, x \in \langle 0, \eta \rangle \big\}, \end{aligned}$$

where N is a natural number and $\varepsilon > 0$. It is easy to see that the space $\mathcal{M}$ with this uniformity is metrizable (by a metric $\rho_w : \mathcal{M} \times \mathcal{M} \to [0, \infty)$) and complete. The topology generated by this uniformity on $\mathcal{M}$ will be called the weak topology.

For the set $\mathcal{M}$ we also consider the uniformity which is determined by the following base:

$$\begin{aligned} E(\varepsilon) = \big\{ &(\{A_t\}_{t=1}^{\infty}, \{B_t\}_{t=1}^{\infty}) \in \mathcal{M} \times \mathcal{M} : \\ &\|A_t x - B_t x\|_\eta \le \varepsilon, t = 1, 2, \ldots, x \in \langle 0, \eta \rangle \big\}, \end{aligned}$$

where $\varepsilon > 0$. It is easy to see that the space $\mathcal{M}$ with this uniformity is metrizable (by a metric $\rho_s : \mathcal{M} \times \mathcal{M} \to [0, \infty)$) and complete. The topology generated by this uniformity on $\mathcal{M}$ is obviously stronger than the weak topology defined above. Therefore we will refer to it as the strong topology.

Denote by $\bar{\mathcal{M}}_{reg}$ the closure of $\mathcal{M}_{reg}$ in the space $\mathcal{M}$ with the weak topology generated by the metric ρ_w. We consider the topological subspace $\bar{\mathcal{M}}_{reg} \subset \mathcal{M}$ with the relative weak and strong topologies. Our next result is a weak ergodic theorem in the sense of [43].

Theorem 6.27 *There exists a set $\mathcal{F} \subset \bar{\mathcal{M}}_{reg}$, which is a countable intersection of open (in the weak topology) and everywhere dense (in the strong topology) subsets of $\bar{\mathcal{M}}_{reg}$, such that for each $\mathbf{B} = \{B_t\}_{t=1}^{\infty} \in \mathcal{F}$, the following two assertions hold:*

1. *$B_T \cdot \cdots \cdot B_1\eta$ is an interior point of X_+ for each integer $T \ge 1$.*

2. *For each $\varepsilon \in (0, 1)$, there exist a neighborhood U of $\{B_t\}_{t=1}^{\infty}$ in $\bar{\mathcal{M}}_{reg}$ with the relative weak topology and a natural number N such that for each $\{C_t\}_{t=1}^{\infty} \in U$, the point $C_T \cdot \cdots \cdot C_1\eta \in \operatorname{int}(X_+)$ for all $T \in \{1, \ldots, N\}$, and*

$$r(C_N \cdot \cdots \cdot C_1\eta, C_N \cdot \cdots \cdot C_1 x) - \lambda(C_N \cdot \cdots \cdot C_1\eta, C_N \cdot \cdots \cdot C_1 x) \le \varepsilon, \quad x \in \langle \varepsilon\eta, \eta \rangle.$$

For the set $\mathcal{A}$ itself we can also define a metric $\rho(\cdot,\cdot)$ by

$$\rho(A,B) := \sup\{\|Ax - Bx\|_\eta : x \in \langle 0,\eta\rangle\}, \quad A,B \in \mathcal{A}.$$

It is clear that the metric space $(\mathcal{A},\rho)$ is complete.

For each interior point ξ of X_+, define

$$\mathcal{A}_\xi := \{A \in \mathcal{A} : A\xi = \xi\}.$$

Clearly, $\mathcal{A}_\xi$ is a closed subset of $\mathcal{A}$ for each $\xi \in \text{int}(X_+)$. For such ξ, we equip the topological subspace $\mathcal{A}_\xi \subset \mathcal{A}$ with the relative topology.

Denote by $\mathcal{A}_*$ the set of all $A \in \mathcal{A}$ such that

$$A\xi = \xi \quad \text{for some } \xi \in \text{int}(X_+).$$

Let $\bar{\mathcal{A}}_*$ be the closure of $\mathcal{A}_*$ in $\mathcal{A}$. The topological subspace $\bar{\mathcal{A}}_* \subset \mathcal{A}$ is also equipped with the relative topology.

We can now formulate our last two results. The second one deals with powers of a single operator.

Theorem 6.28 *Let ξ be an interior point of X_+. Then there exists a set $\mathcal{F} \subset \mathcal{A}_\xi$, which is a countable intersection of open and everywhere dense sets in $\mathcal{A}_\xi$, such that for each $B \in \mathcal{F}$, there exists a continuous linear functional $f_B : X \to R^1$ satisfying*

$$f_B(Bx) = f_B(x), \quad x \in X, \qquad f_B(x) \ge 0, \quad x \in X_+, \qquad f_B(\xi) = 1,$$

$$\lim_{T\to\infty} B^T x = f_B(x)\xi, \quad x \in X,$$

and the following assertion holds:

For each $\varepsilon > 0$, there exist a neighborhood U of B in $\mathcal{A}_\xi$ and a natural number N such that for each $\{C_t\}_{t=1}^\infty \in U$, each integer $T \ge N$ and each $x \in \langle -\eta,\eta\rangle$,

$$\|C_T \cdot \dots \cdot C_1 x - f_B(x)\xi\|_\eta \le \varepsilon.$$

Theorem 6.29 *There exists a set $\mathcal{F} \subset \bar{\mathcal{A}}_*$, which is a countable intersection of open and everywhere dense sets in $\bar{\mathcal{A}}_*$, such that for each $B \in \mathcal{F}$, there exists an interior point ξ_B of X_+ satisfying $B\xi_B = \xi_B$, $\|\xi_B\|_\eta = 1$, and the following two assertions hold:*

1. *There exists $f_B \in X'_+$ such that*

$$\lim_{T\to\infty} B^T x = f_B(x)\xi_B, \quad x \in X.$$

2. *For each $\varepsilon > 0$, there exist a neighborhood U of B in $\bar{\mathcal{A}}_*$ and a natural number N such that for each $C \in U \cap \mathcal{A}_*$, each integer $T \ge N$ and each $x \in \langle -\eta,\eta\rangle$,*

$$\|C^T x - f_B(x)\xi_B\| \le \varepsilon.$$

Note that Theorems 6.24–6.29 were obtained in [125]. Theorems 6.24, 6.26 and 6.27 will be proved in the next three sections. The proof of Theorem 6.28 is analogous to that of Theorem 6.24, while the proofs of Theorems 6.25 and 6.29 are analogous to the proof of Theorem 6.26. Therefore these proofs will be omitted.

6.17 Proof of Theorem 6.24

Recall that X' is the space of all continuous linear functionals $f : X \to R^1$ and that X'_+ is the cone of all $f \in X'$ such that $f(x) \geq 0$, $x \in X_+$.

Lemma 6.30 *Let $A \in \mathcal{A}$ satisfy $A\eta = \eta$. Then there is $f_A \in X'$ such that*

$$f_A(x) \geq 0, \quad x \in X_+, \qquad f_A(\eta) = 1 \quad \text{and} \quad f_A \circ A = f_A. \tag{6.126}$$

Proof Define $S = \{f \in X' : f(x) \geq 0, x \in X_+, f(\eta) = 1\}$. Clearly, the nonempty set S is convex and compact in the weak topology. The operator A' defined by $A'(f) = f \circ A$, $f \in X'$, is weakly continuous and $A'(S) \subset S$. By Tychonoff's fixed point theorem, there exists $f_A \in S$ for which $f_A \circ A = f_A$. This completes the proof of the lemma. □

By Lemma 6.30, for each $A \in \mathcal{A}$ satisfying $A\eta = \eta$, there exists $f_A \in X'$ which satisfies (6.126). For each $\mathbf{A} = \{A_t\}_{t=1}^{\infty} \in \mathcal{M}_\eta$ and each $\gamma \in (0, 1)$, we define $\mathbf{A}_\gamma = \{A_{t\gamma}\}_{t=1}^{\infty} \in \mathcal{M}_\eta$ by

$$A_{t\gamma}x := \gamma f_{A_t}(x)\eta + (1-\gamma)A_t x, \quad x \in X, t = 1, 2, \ldots. \tag{6.127}$$

It is clear that the set

$$\left\{\{A_{t\gamma}\}_{t=1}^{\infty} : \{A_t\}_{t=1}^{\infty} \in \mathcal{M}_\eta, \gamma \in (0, 1)\right\}$$

is everywhere dense in the space $\mathcal{M}_\eta$ with the strong topology.

Lemma 6.31 *Let $\{A_t\}_{t=1}^{\infty} \in \mathcal{M}_\eta$ and let $\gamma \in (0, 1)$. Then for each integer $T \geq 1$, there is $\gamma_T \in X'_+$ such that for each $x \in X$,*

$$A_{T\gamma} \cdot \cdots \cdot A_{1\gamma}x = (1-\gamma)^T A_T \cdot \cdots \cdot A_1 x + \gamma_T(x)\eta. \tag{6.128}$$

Proof We will show by induction that for each integer $T \geq 1$, there is $\gamma_T \in X'_+$ such that (6.128) holds for all $x \in X$. It is clear that for $T = 1$ equality (6.128) is valid with $\gamma_1 = \gamma f_{A_1}$.

Assume that for some integer $T \geq 1$ and $\gamma_T \in X'_+$ equality (6.128) holds for all $x \in X$. Then by (6.128) and (6.127), we have, for every $x \in X$,

$$\begin{aligned} &A_{(T+1)\gamma} \cdot A_{T\gamma} \cdot \cdots \cdot A_{1\gamma}x \\ &= A_{(T+1)\gamma}\left((1-\gamma)^T A_T \cdot \cdots \cdot A_1 x + \gamma_T(x)\eta\right) \end{aligned}$$

$$= \gamma_T(x)\eta + A_{(T+1)\gamma}\big((1-\gamma)^T A_T \cdot \dots \cdot A_1 x\big)$$
$$= \gamma_T(x)\eta + (1-\gamma)^T \gamma f_{A_{T+1}}(A_T \cdot \dots \cdot A_1 x)\eta$$
$$+ (1-\gamma)^{T+1} A_{T+1} A_T \cdot \dots \cdot A_1 x$$
$$= (1-\gamma)^{T+1} A_{T+1} \cdot A_T \cdot \dots \cdot A_1 x$$
$$+ \gamma_T(x)\eta + (1-\gamma)^T f_{A_{T+1}}(A_T \cdot \dots \cdot A_1 x)\eta.$$

This implies that (6.128) holds for $T+1$ too. The lemma follows. □

Lemma 6.32 *Let $\{A_t\}_{t=1}^{\infty} \in \mathcal{M}_\eta$ and let $\gamma, \varepsilon \in (0,1)$. Then there exist a neighborhood U of $\{A_{t\gamma}\}_{t=1}^{\infty}$ in the space $\mathcal{M}_\eta$ with the weak topology, a functional $l \in X'_+$ and an integer $N \ge 1$ such that for each $\{C_t\}_{t=1}^{\infty} \in U$, each $x \in \langle -\eta, \eta\rangle$ and each integer $T \ge N$,*

$$-\varepsilon\eta \le C_T \cdot \dots \cdot C_1 x - l(x)\eta \le \varepsilon\eta. \tag{6.129}$$

Proof Choose a natural number $N \ge 2$ for which

$$(1-\gamma)^N < 64^{-1}\varepsilon. \tag{6.130}$$

By Lemma 6.31, there exists $l \in X'_+$ such that for each $x \in X$,

$$A_{N\gamma} \cdot \dots \cdot A_{1\gamma} x = (1-\gamma)^N A_N \cdot \dots \cdot A_1 x + l(x)\eta. \tag{6.131}$$

Choose

$$\varepsilon_0 \in \big(0, (64N)^{-1}\varepsilon\big), \tag{6.132}$$

and define

$$U := \big\{\{B_t\}_{t=1}^{\infty} \in \mathcal{M}_\eta : \|B_t x - A_{t\gamma} x\|_\eta \le \varepsilon_0,$$
$$t = 1, \dots, N, x \in \langle 0, \eta\rangle\big\}. \tag{6.133}$$

Assume that $\{C_t\}_{t=1}^{\infty} \in U$ and $x \in \langle 0, \eta\rangle$. To prove the lemma it is sufficient to show that

$$-8^{-1}\varepsilon\eta \le C_N \cdot \dots \cdot C_1 x - l(x)\eta \le 8^{-1}\varepsilon\eta. \tag{6.134}$$

By induction we will show that for $s = 1, \dots, N$,

$$-s\varepsilon_0\eta \le C_s \cdot \dots \cdot C_1 x - A_{s\gamma} \cdot \dots \cdot A_{1\gamma} x \le s\varepsilon_0\eta. \tag{6.135}$$

It is clear that (6.135) is valid for $s = 1$.

Assume now that (6.135) is valid for some natural number $s < N$. Then it follows from (6.135) and (6.133) that

$$C_{s+1} \cdot C_s \cdot \dots \cdot C_1 x - A_{(s+1)\gamma} A_{s\gamma} \cdot \dots \cdot A_{1\gamma} x$$
$$= (C_{s+1} - A_{(s+1)\gamma}) C_s \cdot \dots \cdot C_1 x + A_{(s+1)\gamma}(C_s \cdot \dots \cdot C_1 x - A_{s\gamma} \cdot \dots \cdot A_{1\gamma} x)$$

$$\in \langle -\varepsilon_0\eta, \varepsilon_0\eta\rangle + A_{(s+1)\gamma}\big(\langle -s\varepsilon_0\eta, s\varepsilon_0\eta\rangle\big)$$
$$\subset \big\langle -(s+1)\varepsilon_0\eta, (s+1)\varepsilon_0\eta\big\rangle.$$

Therefore (6.135) holds for all $s = 1, \dots, N$. Together with (6.130), (6.131) and (6.132) this implies that

$$\begin{aligned} &C_N \cdot \dots \cdot C_1 x - l(x)\eta \\ &\quad = C_N \cdot \dots \cdot C_1 x - A_{N\gamma} \cdot \dots \cdot A_{1\gamma} x + A_{N\gamma} \cdot \dots \cdot A_{1\gamma} x - l(x)\eta \\ &\quad \in \big\langle -64^{-1}\varepsilon\eta, 64^{-1}\varepsilon\eta\big\rangle + \big\langle -(1-\gamma)^N\eta, (1-\gamma)^N\eta\big\rangle \subset 32^{-1}\langle -\varepsilon\eta, \varepsilon\eta\rangle. \end{aligned}$$

This implies (6.134). The lemma is proved. □

Construction of the set $\mathcal{F}$: Let $\{A_t\}_{t=1}^\infty \in \mathcal{M}_\eta$, $\gamma \in (0,1)$ and let $i \ge 1$ be an integer. By Lemma 6.32, there exist an open neighborhood $U(\{A_t\}_{t=1}^\infty, \gamma, i)$ of $\{A_{t\gamma}\}_{t=1}^\infty$ in the space $\mathcal{M}_\eta$ with the weak topology, a functional $l_{\gamma i}^{(\mathbf{A})} \in X'_+$ and a natural number $N(\{A_t\}_{t=1}^\infty, \gamma, i)$ such that the following property holds:

(a) for each $\{C_t\}_{t=1}^\infty \in U(\{A_t\}_{t=1}^\infty, \gamma, i)$, each $x \in \langle -\eta, \eta\rangle$ and each integer $T \ge N(\{A_t\}_{t=1}^\infty, \gamma, i)$,

$$\big\| C_T \cdot \dots \cdot C_1 x - l_{\gamma,i}^{(\mathbf{A})}(x)\eta \big\|_\eta \le 4^{-i}. \tag{6.136}$$

Define

$$\mathcal{F} = \bigcap_{q=1}^\infty \bigcup \big\{ U\big(\{A_t\}_{t=1}^\infty, \gamma, i\big) : \{A_t\}_{t=1}^\infty \in \mathcal{M}_\eta,$$
$$\gamma \in (0,1), i = q, q+1, \dots \big\}. \tag{6.137}$$

Clearly $\mathcal{F}$ is a countable intersection of open (in the weak topology) everywhere dense (in the strong topology) sets in $\mathcal{M}_\eta$.

Assume that $\{B_t\}_{t=1}^\infty \in \mathcal{F}$ and $\varepsilon \in (0,1)$. Choose an integer $q \ge 1$ such that

$$64 \cdot 2^{-q} < \varepsilon. \tag{6.138}$$

There exist $\{A_t\}_{t=1}^\infty \in \mathcal{M}_\eta$, $\gamma \in (0,1)$ and an integer $i \ge q$ such that

$$\{B_t\}_{t=1}^\infty \in U\big(\{A_t\}_{t=1}^\infty, \gamma, i\big). \tag{6.139}$$

It follows from property (a), (6.139) and (6.138) that for each $x \in \langle -\eta, \eta\rangle$ and each integer $T \ge N(\{A_t\}_{t=1}^\infty, \gamma, i)$,

$$\big\| B_T \cdot \dots \cdot B_1 x - l_{\gamma,i}^{(\mathbf{A})}(x)\eta \big\|_\eta \le 4^{-i} < 64^{-1}\varepsilon. \tag{6.140}$$

Since ε is an arbitrary positive number, we conclude that there exists a linear operator $P : X \to X$ such that

$$\lim_{T\to\infty} B_T \cdot \dots \cdot B_1 x = Px, \quad x \in X. \tag{6.141}$$

By (6.140) and (6.141), for each $x \in \langle -\eta, \eta \rangle$,

$$\left\| Px - l^{(\mathbf{A})}_{\gamma,i}(x)\eta \right\|_\eta \leq 4^{-i} < 64^{-1}\varepsilon. \tag{6.142}$$

Once again, since ε is an arbitrary positive number, we conclude that there is a linear functional $f_B : X \to R^1$ such that

$$Px = f_B(x)\eta, \quad x \in X. \tag{6.143}$$

It is clear that $f_B \in X_+$. It follows from (6.143), (6.142), (6.138) and property (a) that for each $\{C_t\}_{t=1}^\infty \in U(\{A_t\}_{t=1}^\infty, \gamma, i)$, each $x \in \langle -\eta, \eta \rangle$ and each integer $T \geq N(\{A_t\}_{t=1}^\infty, \gamma, i)$,

$$\left\| C_T \cdot \dots \cdot C_1 x - f_B(x)\eta \right\|_\eta \leq 2 \cdot 4^{-i} < 32^{-1}\varepsilon.$$

This completes the proof of Theorem 6.24.

6.18 Proof of Theorem 6.26

Assume that $A \in \mathcal{A}$ and $A\xi = \xi$ for some $\xi \in \mathrm{int}(X_+)$. Then by Lemma 6.30 there exists $f_A \in X'_+$ such that

$$f_A \circ A = f_A, \qquad f_A(\xi) = 1. \tag{6.144}$$

For each $\mathbf{A} = \{A_t\}_{t=1}^\infty \in \mathcal{M}_b^*$, there exist $\xi_A \in \mathrm{int}(X_+)$ and a real number $M_\mathbf{A} \geq 2$ such that

$$A_t\xi_\mathbf{A} = \xi_\mathbf{A}, \quad t = 1, 2, \dots, \qquad \|\xi_\mathbf{A}\|_\eta = 1, \quad \text{and} \quad M_\mathbf{A}\xi_\mathbf{A} \geq \eta. \tag{6.145}$$

For each $\mathbf{A} = \{A_t\}_{t=1}^\infty \in \mathcal{M}_b^*$ and each $\gamma \in (0, 1)$ we define $\{A_{t\gamma}\}_{t=1}^\infty \in \mathcal{M}_b^*$ by

$$A_{t\gamma}x = \gamma f_{A_t}(x)\xi_\mathbf{A} + (1-\gamma)A_t x, \quad x \in X, t = 1, 2, \dots. \tag{6.146}$$

Clearly, the set $\{\{A_{t\gamma}\}_{t=1}^\infty : \{A_t\}_{t=1}^\infty \in \mathcal{M}_b^*, \gamma \in (0, 1)\}$ is everywhere dense in the space $\bar{\mathcal{M}}_b^*$.

Let $\mathbf{A} = \{A_t\}_{t=1}^\infty \in \mathcal{M}_b^*$, $\gamma \in (0, 1)$ and let $i \geq 1$ be an integer. Choose a natural number $N(\mathbf{A}, \gamma, i) \geq 4$ for which

$$(1-\gamma)^{N(\mathbf{A},\gamma,i)} < 64^{-1}8^{-i}(M_\mathbf{A} + 1)^{-6}, \tag{6.147}$$

and then choose a real number $\delta(\mathbf{A}, \gamma, i)$ such that

$$\delta(\mathbf{A}, \gamma, i) \in \left(0, 64^{-1}8^{-i}(M_\mathbf{A} + 1)^{-6}N(\mathbf{A}, \gamma, i)^{-1}4^{-N(\mathbf{A},\gamma,i)}\right) \tag{6.148}$$

and

$$\delta(\mathbf{A}, \gamma, i) < M_\mathbf{A}^{-2}.$$

Now define

$$U(\mathbf{A},\gamma,i)=\big\{\{B_t\}_{t=1}^{\infty}\in\bar{\mathcal{M}}_b^*:\rho_s\big(\{A_{t\gamma}\}_{t=1}^{\infty},\{B_t\}_{t=1}^{\infty}\big)<\delta(\mathbf{A},\gamma,i)\big\} \tag{6.149}$$

and

$$\mathcal{F}=\bigcap_{q=1}^{\infty}\bigcup\big\{U\big(\{A_t\}_{t=1}^{\infty},\gamma,i\big):\{A_t\}_{t=1}^{\infty}\in\mathcal{M}_b^*,\ \gamma\in(0,1),i=q,q+1,\dots\big\}. \tag{6.150}$$

It is clear that $\mathcal{F}$ is a countable intersection of open everywhere dense subsets of $\bar{\mathcal{M}}_b^*$.

Lemma 6.33 *Let $\{A_t\}_{t=1}^{\infty}\in\mathcal{M}_b^*$, $\gamma\in(0,1)$ and let $i\geq 1$ be an integer. Assume that $r:\{1,2,\dots\}\to\{1,2,\dots\}$. Then there exists $l\in X'_+$ such that*

$$l(\xi_{\mathbf{A}})<1 \tag{6.151}$$

and for each $\{C_t\}_{t=1}^{\infty}\in U(\{A_t\}_{t=1}^{\infty},\gamma,i)$ and each $x\in\langle 0,\eta\rangle$,

$$C_{r(N(\mathbf{A},\gamma,i))}\cdot\dots\cdot C_{r(1)}x-l(x)\xi_{\mathbf{A}}\in 32^{-1}8^{-i}(M_{\mathbf{A}}+1)^{-4}\langle-\eta,\eta\rangle. \tag{6.152}$$

Proof Set

$$N=N\big(\{A_t\}_{t=1}^{\infty},\gamma,i\big)\quad\text{and}\quad\delta=\delta\big(\{A_t\}_{t=1}^{\infty},\gamma,i\big). \tag{6.153}$$

By Lemma 6.31, there exists $l\in X'_+$ such that for each $x\in X$,

$$A_{r(N)\gamma}\cdot\dots\cdot A_{r(1)\gamma}=(1-\gamma)^N A_{r(N)}\cdot\dots\cdot A_{r(1)}x+l(x)\xi_{\mathbf{A}}. \tag{6.154}$$

Let

$$x\in\langle 0,\eta\rangle\quad\text{and}\quad\{C_t\}_{t=1}^{\infty}\in U\big(\{A_t\}_{t=1}^{\infty},\gamma,i\big). \tag{6.155}$$

We will show by induction that for $s=1,\dots,N$,

$$-4^s\delta M_{\mathbf{A}}^2\xi_{\mathbf{A}}\leq C_{r(s)}\cdot\dots\cdot C_{r(1)}x-A_{r(s)\gamma}\cdot\dots\cdot A_{r(1)\gamma}x\leq M_{\mathbf{A}}^2 4^s\delta\xi_{\mathbf{A}}. \tag{6.156}$$

Clearly for $s=1$ the induction assumption is valid. Assume now that (6.156) is valid for a natural number $s<N$. Then it follows from (6.155), (6.156), (6.145), (6.149), (6.148) and (6.153) that

$$\begin{aligned}
&C_{r(s+1)}C_{r(s)}\cdot\dots\cdot C_{r(1)}x-A_{r(s+1)\gamma}A_{r(s)\gamma}\cdot\dots\cdot A_{r(1)\gamma}\\
&=(C_{r(s+1)}-A_{r(s+1)\gamma})C_{r(s)}\cdot\dots\cdot C_{r(1)}x\\
&\quad+A_{r(s+1)\gamma}(C_{r(s)}\cdot\dots\cdot C_{r(1)}x-A_{r(s)\gamma}\cdot\dots\cdot A_{r(1)\gamma}x)\\
&\in(C_{r(s+1)}-A_{r(s+1)\gamma})\big\langle 0,M_{\mathbf{A}}^2 4^s\delta\xi_{\mathbf{A}}+M_{\mathbf{A}}\xi_{\mathbf{A}}\big\rangle
\end{aligned}$$

$$+ A_{r(s+1)\gamma}\langle 4^{-s}\delta M_{\mathbf{A}}^2\xi_{\mathbf{A}}, M_{\mathbf{A}}^2 4^s\delta\xi_{\mathbf{A}}\rangle$$
$$\subset M_{\mathbf{A}}(1+4^s\delta M_{\mathbf{A}})\delta M_{\mathbf{A}}\langle -\xi_{\mathbf{A}}, \xi_{\mathbf{A}}\rangle + \langle 4^{-s}\delta M_{\mathbf{A}}^2\xi_{\mathbf{A}}, M_{\mathbf{A}}^2 4^s\delta\xi_{\mathbf{A}}\rangle$$
$$\subset (4^{s+1}\delta M_{\mathbf{A}}^2)\langle -\xi_{\mathbf{A}}, \xi_{\mathbf{A}}\rangle.$$

Therefore (6.156) is valid for all $s = 1, \dots, N$. When combined with (6.154), (6.145), (6.153), (6.148) and (6.147), this implies that for each $x \in \langle 0, \eta\rangle$,

$$C_{r(N)} \cdot \dots \cdot C_{r(1)}x - l(x)\xi_{\mathbf{A}}$$
$$\in M_{\mathbf{A}}^2 4^N\delta\langle \xi_{\mathbf{A}}, \xi_{\mathbf{A}}\rangle + (1-\gamma)^N M_{\mathbf{A}}\langle 0, \xi_{\mathbf{A}}\rangle$$
$$\subset 64^{-1}\cdot 8^{-i}(M_{\mathbf{A}}+1)^{-4}\langle -\xi_{\mathbf{A}}, \xi_{\mathbf{A}}\rangle + 64^{-1}\cdot 8^{-i}(M_{\mathbf{A}}+1)^{-2}\langle 0, \xi_{\mathbf{A}}\rangle$$
$$\subset 32^{-1}\cdot 8^{-i}(M_{\mathbf{A}}+1)^{-4}\langle -\xi_{\mathbf{A}}, \xi_{\mathbf{A}}\rangle$$
$$\subset 32^{-1}\cdot 8^{-i}(M_{\mathbf{A}}+1)^{-4}\langle -\eta, \eta\rangle.$$

The lemma is proved. □

Lemma 6.34 *Let $\{A_t\}_{t=1}^\infty \in \mathcal{M}_b^*$ and $\gamma \in (0,1)$, let $i \geq 1$ be an integer and let $r: \{1,2,\dots\} \to \{1,2,\dots\}$. Let $l \in X'_+$ be as guaranteed by Lemma 6.33. Assume that*

$$\{C_t\}_{t=1}^\infty \in U(\{A_t\}_{t=1}^\infty, \gamma, i), \qquad y \in X_+,$$
$$\|y\|_\eta = 1, \quad \text{and} \quad C_t y = y, \quad t = 1, 2, \dots. \tag{6.157}$$

Then

$$\|y - \xi_{\mathbf{A}}\|_\eta \leq 16^{-1}8^{-i}, \quad y \in \operatorname{int}(X_+), \tag{6.158}$$

and for each $x \in \langle 0, \eta\rangle$ and each integer $T \geq N(\mathbf{A}, \gamma, i)$,

$$\left\|C_{r(T)} \cdot \dots \cdot C_{r(1)}x - l(x)y\right\|_\eta \leq 8^{-i} \tag{6.159}$$

and

$$\left\|C_{r(T)} \cdot \dots \cdot C_{r(1)}x - l(x)\xi_{\mathbf{A}}\right\|_\eta \leq 2\cdot 8^{-i}. \tag{6.160}$$

Proof By Lemma 6.33 and the definition of l, for each $x \in \langle 0, \eta\rangle$,

$$l(\xi_{\mathbf{A}}) \leq 1 \tag{6.161}$$

and

$$C_{r(N(\mathbf{A},\gamma,i))} \cdot \dots \cdot C_{r(1)}x - l(x)\xi_{\mathbf{A}} \in 32^{-1}\cdot 8^{-i}(M_{\mathbf{A}}+1)^{-4}\langle -\eta, \eta\rangle. \tag{6.162}$$

Together with (6.157) and (6.145) this implies that

$$\left\|y - l(y)\xi_{\mathbf{A}}\right\|_\eta \leq 32^{-1}\cdot 8^{-i}(M_{\mathbf{A}}+1)^{-4}, \tag{6.163}$$

$$\left|l(y)-1\right| \le \left\|l(y)\xi_{\mathbf{A}}-y\right\|_{\eta} \le 32^{-1}\cdot 8^{-i}(M_{\mathbf{A}}+1)^{-4}, \tag{6.164}$$

$$\|y-\xi_{\mathbf{A}}\|_{\eta} \le 16^{-1}\cdot 8^{-i}(M_{\mathbf{A}}+1)^{-4}, \tag{6.165}$$

and

$$y \ge \xi_{\mathbf{A}} - 16^{-1}\cdot 8^{-i}(M_{\mathbf{A}}+1)^{-4}\eta \ge \left(M_{\mathbf{A}}^{-1} - 16^{-1}\cdot 8^{-i}(M_{\mathbf{A}}+1)^{-4}\right)\eta. \tag{6.166}$$

It follows from (6.145), (6.161), (6.162), (6.165) and (6.157) that for each $x \in \langle 0, \eta\rangle$ and each integer $T \ge N(\mathbf{A}, \gamma, i)$,

$$l(x) \le l(\eta) \le M_{\mathbf{A}} l(\xi_{\mathbf{A}}) \le M_{\mathbf{A}} \tag{6.167}$$

and

$$\begin{aligned}
&\left\|C_{r(N(A,\gamma,i))}\cdot\dots\cdot C_{r(1)}x - l(x)y\right\|_{\eta}\\
&\le \left\|C_{r(N(A,\gamma,i))}\cdot\dots\cdot C_{r(1)}x - l(x)\xi_{\mathbf{A}}\right\|_{\eta} + \left\|l(x)(\xi_{\mathbf{A}}-y)\right\|_{\eta}\\
&\le 32^{-1}\cdot 8^{-i}(M_{\mathbf{A}}+1)^{-4} + 16^{-1}\cdot 8^{-i}M_{\mathbf{A}}(M_{\mathbf{A}}+1)^{-4}\\
&\le 8^{-i-1}(M_{\mathbf{A}}+1)^{-3}.
\end{aligned} \tag{6.168}$$

By (6.157), (6.168) and (6.166),

$$\begin{aligned}
&C_{r(T)}\cdot\dots\cdot C_{r(1)}x - l(x)y\\
&= C_{r(T)}\cdot\dots\cdot C_{r(N(\mathbf{A},\gamma,i)+1)}\left(C_{r(N(\mathbf{A},\gamma,i))}\cdot\dots\cdot C_{r(1)}x - l(x)y\right)\\
&\in C_{r(T)}\cdot\dots\cdot C_{r(N(\mathbf{A},\gamma,i)+1)}\left(\langle -8^{-i-1}(M_{\mathbf{A}}+1)^{-3}\eta, 8^{-i-1}(M_{\mathbf{A}}+1)^{-3}\eta\rangle\right)\\
&\subset 8^{-i-1}(M_{\mathbf{A}}+1)^{-3}\left(M_{\mathbf{A}}^{-1} - 16^{-1}8^{-i}(M_{\mathbf{A}}+1)^{-4}\right)^{-1}\langle -y, y\rangle\\
&\subset (M_{\mathbf{A}}+1)^{-2}8^{-i}\langle -y, y\rangle \subset 8^{-i}(M_{\mathbf{A}}+1)^{-2}\langle -\eta, \eta\rangle.
\end{aligned} \tag{6.169}$$

Now by using (6.169), (6.167) and (6.165), we deduce that

$$\begin{aligned}
&\left\|C_{r(T)}\cdot\dots\cdot C_{r(1)}x - l(x)\xi_{\mathbf{A}}\right\|_{\eta}\\
&\le 8^{-i}(M_{\mathbf{A}}+1)^{-2} + \left|l(x)\right|\|y-\xi_{\mathbf{A}}\|_{\eta}\\
&\le 8^{-i}(M_{\mathbf{A}}+1)^{-2} + M_{\mathbf{A}}16^{-1}8^{-i}(M_{\mathbf{A}}+1)^{-4} < 2\cdot 8^{-i}.
\end{aligned}$$

The lemma is proved. □

Completion of the proof of Theorem 6.26: Assume that $\{B_t\}_{t=1}^{\infty} \in \mathcal{F}$. There exist $\mathbf{A}^{(k)} = \{A_t^{(k)}\}_{t=1}^{\infty} \in \mathcal{M}_b^*$, $k = 1, 2, \dots$, $\{\gamma_k\}_{k=1}^{\infty} \subset (0,1)$, and a strictly increasing sequence of natural numbers $\{i_k\}_{k=1}^{\infty}$ such that

$$\{B_t\}_{t=1}^{\infty} \in U\left(\{A_t^{(k)}\}_{t=1}^{\infty}, \gamma_k, i_k\right), \quad k = 1, 2, \dots, \tag{6.170}$$

and

$$U(\{A_t^{(k+1)}\}_{t=1}^{\infty}, \gamma_{k+1}, i_{k+1}) \subset U(\{A_t^{(k)}\}_{t=1}^{\infty}, \gamma_k, i_k), \quad k = 1, 2, \ldots.$$

By Lemma 6.34, $\{\xi_{\mathbf{A}^{(k)}}\}_{k=1}^{\infty}$ is a Cauchy sequence and there exists

$$\xi_{\mathbf{B}} = \lim_{k \to \infty} \xi_{\mathbf{A}^{(k)}}. \tag{6.171}$$

It follows from (6.170), (6.149), (6.148), (6.145) and (6.171) that for $t = 1, 2, \ldots$,

$$A_t^{(k)} \xi_{\mathbf{A}^{(k)}} - B_t \xi_{\mathbf{B}} = (A_t^{(k)} - B_t)(\xi_{\mathbf{A}^{(k)}}) + B_t(\xi_{\mathbf{A}^{(k)}} - \xi_{\mathbf{B}}) \to 0 \quad \text{as } k \to \infty.$$

Together with (6.145) and (6.171) this implies that

$$B_t \xi_{\mathbf{B}} = \xi_{\mathbf{B}}, \quad t = 1, 2, \ldots, \quad \text{and} \quad \|\xi_{\mathbf{B}}\|_\eta = 1. \tag{6.172}$$

Lemma 6.34, (6.170) and (6.172) imply that $\xi_{\mathbf{B}}$ is an interior point of X_+.

Let $\varepsilon > 0$ be given. There is an integer $k \geq 1$ such that

$$2^{-i_k} < 64^{-1}\varepsilon. \tag{6.173}$$

Assume that $r : \{1, 2, \ldots\} \to \{1, 2, \ldots\}$. By Lemma 6.34, there exists $l \in X'_+$ such that the following property holds:

(a) Assume that $\{C_t\}_{t=1}^{\infty} \in U(\{A_t^{(k)}\}_{t=1}^{\infty}, \gamma_k, i_k)$, $y \in X'_+$,

$$\|y\|_\eta = 1, \quad \text{and} \quad C_t y = y, \quad t = 1, 2, \ldots.$$

Then y is an interior point of X_+, $\|y - \xi_{\mathbf{A}^{(k)}}\|_\eta \leq 16^{-1} 8^{-i_k}$ and for each $x \in \langle 0, \eta \rangle$ and each integer $T \geq N(\mathbf{A}^{(k)}, \gamma_k, i_k)$,

$$\left\| C_{r(T)} \cdot \cdots \cdot C_{r(1)} x - l(x) y \right\|_\eta \leq 8^{-i_k}$$

and

$$\left\| C_{r(T)} \cdot \cdots \cdot C_{r(1)} x - l(x) \xi_{\mathbf{A}^{(k)}} \right\|_\eta \leq 2 \cdot 8^{-i_k}.$$

It follows from property (a), (6.170), (6.172) and (6.173) that

$$\|\xi_{\mathbf{B}} - \xi_{\mathbf{A}^{(k)}}\| \leq 16^{-1} 8^{-i_k}, \tag{6.174}$$

and for each $x \in \langle 0, \eta \rangle$ and each integer $T \geq N(\mathbf{A}^{(k)}, \gamma_k, i_k)$,

$$\left\| B_{r(T)} \cdot \cdots \cdot B_{r(1)} x - l(x) \xi_{\mathbf{B}} \right\|_\eta \leq 8^{-i_k} < 64^{-1}\varepsilon. \tag{6.175}$$

Since ε is any positive number, we conclude that

$$\lim_{T \to \infty} B_{r(T)} \cdot \cdots \cdot B_{r(1)} x = f_r(x) \xi_{\mathbf{B}}, \quad x \in X, \tag{6.176}$$

where $f_r \in X'_+$.

By (6.176), (6.175), (6.172) and (6.173),

$$\left|l(x) - f_r(x)\right| \le 8^{-i_k} < 64^{-1}\varepsilon, \quad x \in \langle 0, \eta \rangle. \tag{6.177}$$

There is $M_0 > 1$ such that

$$\eta \le M_0 \xi_{\mathbf{B}}. \tag{6.178}$$

We may assume that

$$2^{-i_k} < 64^{-1}\varepsilon(4 + M_0)^{-1}. \tag{6.179}$$

Assume that $\{C_t\}_{t=1}^{\infty} \in U(\{A_t^{(k)}\}_{t=1}^{\infty}, \gamma_k, i_k) \cap \mathcal{M}_b^*$, $T \ge N(\mathbf{A}^{(k)}, \gamma, i)$ is an integer and $x \in \langle 0, \eta \rangle$. To complete the proof of the theorem it is sufficient to show that

$$\left\| C_{r(T)} \cdot \cdots \cdot C_{r(1)} x - f_r(x)\xi_{\mathbf{B}} \right\|_\eta \le 4^{-1}\varepsilon. \tag{6.180}$$

Indeed, it follows from property (a), (6.145), (6.177), (6.174), the definition of M_0 (see (6.173)), (6.176) and (6.179) that

$$\begin{aligned}
&\left\| C_{r(T)} \cdot \cdots \cdot C_{r(1)} x - f_r(x)\xi_{\mathbf{B}} \right\|_\eta \\
&\le \left\| C_{r(T)} \cdot \cdots \cdot C_{r(1)} x - l(x)\xi_{\mathbf{A}^{(k)}} \right\|_\eta \\
&\quad + \left\| l(x)\xi_{\mathbf{A}^{(k)}} - f_r(x)\xi_{\mathbf{A}^{(k)}} \right\|_\eta + \left|f_r(x)\right| \|\xi_{\mathbf{A}^{(k)}} - \xi_{\mathbf{B}}\|_\eta \\
&\le 2 \cdot 8^{-i_k} + 8^{-i_k} + 16^{-1} 8^{-i_k} f_r(\eta) \\
&\le 2 \cdot 8^{-i_k} + 8^{-i_k} + 16^{-1} \cdot 8^{-i_k} M_0 < 64^{-1}\varepsilon.
\end{aligned}$$

This completes the proof of Theorem 6.26.

6.19 Proof of Theorem 6.27

Fix $f \in X'_+$ such that $f(\eta) = 1$. Assume that $\{A_t\}_{t=1}^{\infty} \in \mathfrak{M}_{reg}$, $0 < c_1 < 1 < c_2$, $\gamma \in (0, 1)$ and

$$c_2\eta \ge A_T \cdot \cdots \cdot A_1\eta \ge c_1\eta, \quad T = 1, 2, \ldots. \tag{6.181}$$

Define a sequence of operators $A_t^\gamma : X \to X$, $t = 1, 2, \ldots$, by

$$\begin{aligned}
A_1^\gamma x &:= (1-\gamma)A_1 x + \gamma\big(f(\eta)\big)^{-1} f(x) A_1\eta, \quad x \in X, \\
A_{t+1}^\gamma x &:= (1-\gamma)A_{t+1} x \\
&\quad + \gamma\big(f(A_t \cdot \cdots \cdot A_1\eta)\big)^{-1} f(x) A_{t+1} A_t \cdot \cdots \cdot A_1\eta, \\
&x \in X, t = 1, 2, \ldots.
\end{aligned} \tag{6.182}$$

Clearly,

$$\{A_t^\gamma\}_{t=1}^{\infty} \in \mathfrak{M}. \tag{6.183}$$

Lemma 6.35 *For each integer $T \geq 1$,*

$$A_T^{\gamma} \cdot \dots \cdot A_1^{\gamma} \eta = A_T \cdot \dots \cdot A_1 \eta. \tag{6.184}$$

Proof We will prove the lemma by induction. Clearly for $T = 1$, (6.181) is valid. Assume that $T \geq 1$ is an integer and that (6.184) holds. It follows from (6.184) and (6.182) that

$$\begin{aligned} A_{T+1}^{\gamma} A_T^{\gamma} \cdot \dots \cdot A_1^{\gamma} \eta &= A_{T+1}^{\gamma} A_T \cdot \dots \cdot A_1 \eta \\ &= (1-\gamma) A_{T+1} A_T \cdot \dots \cdot A_1 \eta + \gamma A_{T+1} A_T \cdot \dots \cdot A_1 \eta. \end{aligned}$$

This completes the proof of the lemma. □

We omit the easy proof of our next lemma.

Lemma 6.36 *For each integer $T \geq 1$,*

$$\sup\{\|A_T x - A_T^{\gamma} x\|_{\eta} : x \in \langle 0, \eta \rangle\} \leq 2\gamma c_1^{-1} c_2.$$

Lemma 6.37 *For each integer $T \geq 1$, there exists $f_T \in X'_+$ such that*

$$A_T^{\gamma} \cdot \dots \cdot A_1^{\gamma} x = (1-\gamma)^T A_T \cdot \dots \cdot A_1 x + f_T(x) A_T \cdot \dots \cdot A_1 \eta, \quad x \in X. \tag{6.185}$$

Proof We will prove the lemma by induction. Clearly for $T = 1$ the assertion of the lemma is valid. Assume that there is $f_T \in X'_+$ such that (6.185) holds. It follows from (6.185), (6.182) and Lemma 6.35 that for each $x \in X$,

$$\begin{aligned} &A_{T+1}^{\gamma} A_T^{\gamma} \cdot \dots \cdot A_1^{\gamma} x \\ &= A_{T+1}^{\gamma} (A_T^{\gamma} \cdot \dots \cdot A_1^{\gamma} x) \\ &= (1-\gamma)^T A_{T+1}^{\gamma} (A_T \cdot \dots \cdot A_1 x) + f_T(x) A_{T+1}^{\gamma} A_T \cdot \dots \cdot A_1 \eta \\ &= (1-\gamma)^T \big((1-\gamma) A_{T+1} A_T \cdot \dots \cdot A_1 x \\ &\quad + \gamma f(A_T \cdot \dots \cdot A_1 \eta)^{-1} f(A_T \cdot \dots \cdot A_1 x) A_{T+1} A_T \cdot \dots \cdot A_1 \eta\big) \\ &\quad + f_T(x) A_{T+1} A_T \cdot \dots \cdot A_1 \eta \\ &= (1-\gamma)^{T+1} A_{T+1} A_T \cdot \dots \cdot A_1 x \\ &\quad + \big[(1-\gamma)^T \gamma f(A_T \cdot \dots \cdot A_1 \eta)^{-1} f(A_T \cdot \dots \cdot A_1 x) \\ &\quad + f_T(x)\big] A_{T+1} A_T \cdot \dots \cdot A_1 \eta. \end{aligned}$$

This completes the proof of the lemma. □

Lemma 6.38 *Let $\varepsilon \in (0, 2^{-1})$ and let N be a natural number for which*

$$(1-\gamma)^N < 2^{-1} \varepsilon. \tag{6.186}$$

Then for each $x \in \langle 0, \eta \rangle$,

$$r\big(A_N \cdot \dots \cdot A_1\eta, A_N^\gamma \cdot \dots \cdot A_1^\gamma x\big) - \lambda\big(A_N \cdot \dots \cdot A_1\eta, A_N^\gamma \cdot \dots \cdot A_1^\gamma x\big) \le 2^{-1}\varepsilon. \quad (6.187)$$

Proof By Lemma 6.37 there is $f_N \in X'_+$ such that (6.185) holds with $T = N$. Together with Lemma 6.35 this implies that

$$f_N(\eta)A_N \cdot \dots \cdot A_1\eta = \big(1 - (1-\gamma)^N\big)A_N \cdot \dots \cdot A_1, \qquad f_N(\eta) = -(1-\gamma)^N + 1, \quad (6.188)$$

and for each $x \in X_+$,

$$\begin{aligned} & f_N(x)A_N \cdot \dots \cdot A_1\eta \\ & \le A_N^\gamma \cdot \dots \cdot A_1^\gamma x \\ & \le f_N(x)A_N \cdot \dots \cdot A_1\eta + (1-\gamma)^T r(x,\eta)A_N \cdot \dots \cdot A_1\eta. \end{aligned} \quad (6.189)$$

If $x \in \langle 0, \eta \rangle$, then (6.187) follows from (6.189) and (6.186). The lemma is proved. □

Lemma 6.39 *Let* $0 < \Delta_1 < 1 < \Delta_2$, $\Gamma > 1$, *and let* $n \ge 1$ *be an integer. Then there is a number* $\delta > 0$ *such that for each sequence* $\{B_i\}_{i=1}^n \subset \mathcal{A}$ *satisfying*

$$\sup\big\{\big\|B_i x - A_i^\gamma x\big\|_\eta : x \in \langle 0, \eta \rangle, i = 1, \dots, n\big\} \le \delta \quad (6.190)$$

and for each $z \in X_+$ *satisfying*

$$z \in \langle 0, \Delta_2\eta \rangle \quad \text{and} \quad z \ge \Delta_1\eta, \quad (6.191)$$

the following relation holds:

$$B_n \cdot \dots \cdot B_1 z \in \big\langle \Gamma^{-1}A_n^\gamma \cdot \dots \cdot A_1^\gamma z, \Gamma A_n^\gamma \cdot \dots \cdot A_1^\gamma z\big\rangle. \quad (6.192)$$

Proof We will prove this lemma by induction. Let $n = 1$. Choose a positive number δ such that

$$c_1^{-1}\Delta_1^{-1}\delta\Delta_2 < \gamma(\Gamma - 1)\Gamma^{-1}. \quad (6.193)$$

Assume that

$$\begin{aligned} & B_1 \in \mathcal{A}, \quad \sup\big\{\big\|B_1 x - A_1^\gamma x\big\|_\eta : x \in \langle 0, \eta \rangle\big\} \le \delta, \\ & z \in \langle 0, \Delta_2\eta \rangle \quad \text{and} \quad z \ge \Delta_1\eta. \end{aligned} \quad (6.194)$$

It follows from (6.194) that

$$\big\|B_1 z - A_1^\gamma z\big\|_\eta \le \delta\Delta_2 \quad \text{and} \quad B_1 z \in \big\langle A_1^\gamma z - \delta\Delta_2\eta, A_1^\gamma z + \delta\Delta_2\eta\big\rangle. \quad (6.195)$$

By (6.182), (6.194), (6.181) and (6.193),

$$A_1^\gamma z \geq \gamma\big(f(\eta)\big)^{-1} f(z) A_1 \eta \geq \gamma \Delta_1 A_1 \eta \geq \gamma \Delta_1 c_1 \eta$$

and

$$\delta \Delta_2 \eta \leq \gamma(\Gamma - 1)\Gamma^{-1} c_1 \Delta_1 \eta \leq (\Gamma - 1)\Gamma^{-1} A_1^\gamma z.$$

Together with (6.195) this implies that

$$B_1 z \in \big\langle \Gamma^{-1} A_1^\gamma z, \Gamma A_1^\gamma z\big\rangle.$$

Thus for $n = 1$ the assertion of the lemma is true.

Assume now that the assertion of the lemma holds for $n = 1, \ldots, k$. Choose a positive number Γ_0 satisfying

$$1 < \Gamma_0 < \Gamma^{1/2}. \tag{6.196}$$

Since the assertion of the lemma holds for $n = k$, there is a number $\delta_0 > 0$ such that for each sequence $\{B_i\}_{i=1}^k \subset \mathfrak{A}$ satisfying

$$\sup\big\{\big\|B_i x - A_i^\gamma x\big\|_\eta : x \in \langle 0, \eta\rangle, i = 1, \ldots, k\big\} \leq \delta_0, \tag{6.197}$$

and each $z \in X_+$ satisfying (6.191), the following relation holds:

$$B_k \cdot \cdots \cdot B_1 z \in \big\langle \Gamma_0^{-1} A_k^\gamma \cdot \cdots \cdot A_1^\gamma z, \Gamma_0 A_k^\gamma \cdot \cdots \cdot A_1^\gamma z\big\rangle. \tag{6.198}$$

Choose a number $\delta \in (0, \delta_0)$ such that

$$\delta \Delta_2 c_2 c_1^{-1} \Delta_1^{-1} \leq \gamma(\Gamma_0 - 1)\Gamma_0^{-1}. \tag{6.199}$$

Assume that $\{B_i\}_{i=1}^{k+1} \subset \mathcal{A}$,

$$\sup\big\{\big\|B_i x - A_i^\gamma x\big\|_\eta : x \in \langle 0, \eta\rangle, i = 1, \ldots, k+1\big\} \leq \delta, \tag{6.200}$$

and that $z \in X_+$ satisfies (6.191). Then (6.198) is valid. This implies that

$$B_{k+1} B_k \cdot \cdots \cdot B_1 z \in \big\langle \Gamma_0^{-1} B_{k+1} A_k^\gamma \cdot \cdots \cdot A_1^\gamma z, \Gamma_0 B_{k+1} A_k^\gamma \cdot \cdots \cdot A_1^\gamma z\big\rangle. \tag{6.201}$$

It follows from (6.191), (6.182), Lemma 6.35 and (6.200) that

$$A_{k+1}^\gamma A_k^\gamma \cdot \cdots \cdot A_1^\gamma z, A_k^\gamma \cdot \cdots \cdot A_1^\gamma z \in \langle \Delta_1 c_1 \eta, c_2 \Delta_2 \eta\rangle, \tag{6.202}$$

$$\big\|B_{k+1} A_k^\gamma \cdot \cdots \cdot A_1^\gamma z - A_{k+1}^\gamma A_k^\gamma \cdot \cdots \cdot A_1^\gamma z\big\|_\eta \leq \delta c_2 \Delta_2 \tag{6.203}$$

and

$$B_{k+1} A_k^\gamma \cdot \cdots \cdot A_1^\gamma z \in \big\langle A_{k+1}^\gamma A_k^\gamma \cdot \cdots \cdot A_1^\gamma z - \delta c_2 \Delta_2 \eta, A_{k+1}^\gamma A_k^\gamma \cdot \cdots \cdot A_1^\gamma z + \delta c_2 \Delta_2 \eta\big\rangle.$$

By (6.199), (6.202) and (6.203),

$$\delta c_2 \Delta_2 c_1 \eta \le c_1 \Delta_1 \gamma (\Gamma_0 - 1)\Gamma_0^{-1}\eta \le (\Gamma_0 - 1)\Gamma_0^{-1} A_{k+1}^{\gamma} A_k^{\gamma} \cdot \cdots \cdot A_1^{\gamma} z$$

and

$$B_{k+1} A_k^{\gamma} \cdot \cdots \cdot A_1^{\gamma} z \in \langle \Gamma_0^{-1} A_{k+1}^{\gamma} A_k^{\gamma} \cdot \cdots \cdot A_1^{\gamma} z, \Gamma_0 A_{k+1}^{\gamma} A_k^{\gamma} \cdot \cdots \cdot A_1^{\gamma} z \rangle.$$

It follows from this last relation, (6.201) and (6.196) that

$$B_{k+1} B_k \cdot \cdots \cdot B_1 z \in \langle \Gamma^{-1} A_{k+1}^{\gamma} A_k^{\gamma} \cdot \cdots \cdot A_1^{\gamma} z, \Gamma A_{k+1}^{\gamma} A_k^{\gamma} \cdot \cdots \cdot A_1^{\gamma} z \rangle.$$

This completes the proof of Lemma 6.39. □

Lemma 6.40 *Let $0 < \Delta < 1$, $0 < \varepsilon < \Delta/2$ and let N be a natural number for which*

$$(1-\gamma)^N < 2^{-1}\varepsilon. \tag{6.204}$$

Then there exist a neighborhood U of $\{A_t^{\gamma}\}_{t=1}^{\infty}$ in $\mathcal{M}$ and a number $\kappa > 0$ such that for each $\{B_t\}_{t=1}^{\infty} \in U$, the following two assertions hold:

1. $B_T \cdot \cdots \cdot B_1 \eta \ge \kappa \eta$, $T = 1, \dots, N$.
2. *For each $x \in \langle \Delta\eta, \eta \rangle$,*

$$r(B_N \cdot \cdots \cdot B_1 \eta, B_N \cdot \cdots \cdot B_1 x) - \lambda(B_N \cdot \cdots \cdot B_1 \eta, B_N \cdot \cdots \cdot B_1 x) \le (3/4)\varepsilon. \tag{6.205}$$

Proof By Lemma 6.38, for each $x \in \langle 0, \eta \rangle$,

$$r\big(A_N \cdot \cdots \cdot A_1 \eta, A_N^{\gamma} \cdot \cdots \cdot A_1^{\gamma} x\big) - \lambda\big(A_N \cdot \cdots \cdot A_1 \eta, A_N^{\gamma} \cdot \cdots \cdot A_1^{\gamma} x\big) \le 2^{-1}\varepsilon. \tag{6.206}$$

Choose a real number Γ for which

$$\Gamma > 1 \quad \text{and} \quad \big(\Gamma^2 - 1\big) < 8^{-1}\varepsilon. \tag{6.207}$$

By Lemma 6.39, there exists a neighborhood U of $\{A_t^{\gamma}\}_{t=1}^{\infty}$ in $\mathcal{M}$ such that for each $\{B_t\}_{t=1}^{\infty} \in U$, each $z \in \langle \Delta\eta, \eta \rangle$ and each integer $T \in [1, N]$,

$$B_T \cdot \cdots \cdot B_1 z \in \langle \Gamma^{-1} A_T^{\gamma} \cdot \cdots \cdot A_1^{\gamma} z, \Gamma A_T^{\gamma} \cdot \cdots \cdot A_1^{\gamma} z \rangle. \tag{6.208}$$

Assume that $\{B_t\}_{t=1}^{\infty} \in U$. It follows from the definition of U (see (6.208)), Lemma 6.35 and (6.181) that for $T = 1, \dots, N$,

$$B_T \cdot \cdots \cdot B_1 \eta \ge \Gamma^{-1} A_T^{\gamma} \cdot \cdots \cdot A_1^{\gamma} \eta = \Gamma^{-1} A_T \cdot \cdots \cdot A_1 \eta \ge \Gamma^{-1} c_1 \eta. \tag{6.209}$$

Therefore assertion 1 holds with $\kappa = \Gamma^{-1} c_1$. Now we will show that assertion 2 holds too.

Assume that

$$\{B_t\}_{t=1}^{\infty} \in U \quad \text{and} \quad x \in \langle \Delta\eta, \eta \rangle. \tag{6.210}$$

Then (6.206) is valid. By (6.210) and the definition of U (see (6.208)),

$$B_N \cdot \cdots \cdot B_1 x \in \langle \Gamma^{-1} A_N^\gamma \cdot \cdots \cdot A_1^\gamma x, \Gamma A_N^\gamma \cdot \cdots \cdot A_1^\gamma x \rangle \tag{6.211}$$

and

$$B_N \cdot \cdots \cdot B_1 \eta \in \langle \Gamma^{-1} A_N^\gamma \cdot \cdots \cdot A_1^\gamma \eta, \Gamma A_N^\gamma \cdot \cdots \cdot A_1^\gamma \eta \rangle.$$

It follows from (6.211), Lemma 6.35 and (6.144) that

$$\begin{aligned} r(B_N \cdot \cdots \cdot B_1 \eta, B_N \cdot \cdots \cdot B_1 x) &\le r(\Gamma^{-1} A_N \cdot \cdots \cdot A_1 \eta, \Gamma A_N^\gamma \cdot \cdots \cdot A_1^\gamma x) \\ &= \Gamma^2 r(A_N \cdot \cdots \cdot A_1 \eta, A_N^\gamma \cdot \cdots \cdot A_1^\gamma x) \end{aligned}$$

and

$$\begin{aligned} \lambda(B_N \cdot \cdots \cdot B_1 \eta, B_N \cdot \cdots \cdot B_1 x) &\ge \lambda(\Gamma A_N \cdot \cdots \cdot A_1 \eta, \Gamma^{-1} A_N^\gamma \cdot \cdots \cdot A_1^\gamma x) \\ &= \lambda(A_N \cdot \cdots \cdot A_1 \eta, A_N^\gamma \cdot \cdots \cdot A_1^\gamma x) \Gamma^{-2}. \end{aligned}$$

By these inequalities, (6.206), (6.210) and (6.207),

$$\begin{aligned} & r(B_N \cdot \cdots \cdot B_1 \eta, B_N \cdot \cdots \cdot B_1 x) - \lambda(B_N \cdot \cdots \cdot B_1 \eta, B_N \cdot \cdots \cdot B_1 x) \\ & \le \Gamma^2 r(A_N \cdot \cdots \cdot A_1 \eta, A_N^\gamma \cdot \cdots \cdot A_1^\gamma x) - \Gamma^{-2} \lambda(A_N \cdot \cdots \cdot A_1 \eta, A_N^\gamma \cdot \cdots \cdot A_1^\gamma x) \\ & \le r(A_N \cdot \cdots \cdot A_1 \eta, A_N^\gamma \cdot \cdots \cdot A_1^\gamma x) - \lambda(A_N \cdot \cdots \cdot A_1 \eta, A_N^\gamma \cdot \cdots \cdot A_1^\gamma x) \\ & \quad + (\Gamma^2 - 1) r(\eta, x) + (1 - \Gamma^{-2}) r(A_N \cdot \cdots \cdot A_1 \eta, A_N^\gamma \cdot \cdots \cdot A_1^\gamma x) \\ & \le 2^{-1} \varepsilon + 2(\Gamma^2 - 1) < 3\varepsilon/4. \end{aligned}$$

This completes the proof of Lemma 6.40. □

Completion of the proof of Theorem 6.27: By Lemmas 6.36 and 6.35 and by (6.181) and (6.183), the set $\{\{A_t^\gamma\}_{t=1}^\infty : \{A_t\}_{t=1}^\infty \in \mathcal{M}_{reg}, \gamma \in (0,1)\}$ is an everywhere dense subset of $\bar{\mathcal{M}}_{reg}$ with the strong topology.

Let $\{A_t\}_{t=1}^\infty \in \mathcal{M}_{reg}, \gamma \in (0,1)$ and let $i \ge 1$ be an integer. By Lemma 6.40, there exist an open neighborhood $U(\mathbf{A}, \gamma, i)$ of $\{A_t^\gamma\}_{t=1}^\infty$ in the space $\bar{\mathcal{M}}_{reg}$ with the weak topology and an integer $N(\mathbf{A}, \gamma, i) \ge 2i + 2$ such that for each $\{C_t\}_{t=1}^\infty \in U(\mathbf{A}, \gamma, i)$, the following two properties hold:

(a) $C_T \cdot \cdots \cdot C_1 \eta$ is an interior point of X_+ for $T = 1, \dots, N(\mathbf{A}, \gamma, i)$;
(b) for each $x \in \langle 4^{-i}\eta, \eta \rangle$,

$$\begin{aligned} & r(C_{N(\mathbf{A},\gamma,i)} \cdot \cdots \cdot C_1 \eta, C_{N(\mathbf{A},\gamma,i)} \cdot \cdots \cdot C_1 x) \\ & \quad - \lambda(C_{N(\mathbf{A},\gamma,i)} \cdot \cdots \cdot C_1 \eta, C_{N(\mathbf{A},\gamma,i)} \cdot \cdots \cdot C_1 x) \le 8^{-i}. \end{aligned} \tag{6.212}$$

Define

$$\mathcal{F} := \bigcap_{q=1}^{\infty} \bigcup \{U(\mathbf{A}, \gamma, i) : A \in \mathcal{M}_{reg}, \gamma \in (0,1), i = q, q+1, \dots\}.$$

Clearly, $\mathcal{F}$ is a countable intersection of open (in the weak topology) everywhere dense (in the strong topology) sets in $\bar{\mathcal{M}}_{reg}$. Let $\{B_t\}_{t=1}^{\infty}$ belong to $\mathcal{F}$. It is easy to verify that assertion 1 of Theorem 6.27 holds. We will show that assertion 2 of the theorem is valid too.

Let $\varepsilon \in (0,1)$. Choose an integer $q \geq 1$ such that

$$2^{-q} < \varepsilon/64. \tag{6.213}$$

There are $\{A_t\}_{t=1}^{\infty} \in \mathcal{M}_{reg}$, $\gamma \in (0,1)$ and an integer $i \geq q$ such that

$$\{B_t\}_{t=1}^{\infty} \in U(\mathbf{A}, \gamma, i). \tag{6.214}$$

Assume that $\{C_t\}_{t=1}^{\infty} \in U(\mathbf{A}, \gamma, i)$ and that

$$x \in \langle \varepsilon\eta, \eta \rangle. \tag{6.215}$$

Then property (a) holds. It follows from property (b), (2.215) and (6.213) that (6.212) is also valid. This completes the proof of Theorem 6.27.

6.20 Homogeneous Order-Preserving Mappings

In this section we study the asymptotic behavior of (random) infinite products of generic sequences of homogeneous order-preserving mappings on a cone in an ordered Banach space. Infinite products of such mappings have been studied by Fujimoto and Krause [62] and by Nussbaum [106, 107]. The interest in their asymptotic behavior stems, for instance, from population biology (see [43]). We show that in appropriate spaces of sequences of mappings there exists a subset which is a countable intersection of open and everywhere dense sets such that for each sequence belonging to this subset the corresponding infinite products converge.

Let $(X, \|\cdot\|)$ be a Banach space ordered by a closed cone X_+ with a nonempty interior such that $\|x\| \leq \|y\|$ for each $x, y \in X_+$ satisfying $x \leq y$. When $u, v \in X$ and $u \leq v$ we set

$$\langle u, v \rangle = \{x \in X : u \leq x \leq v\}.$$

For each $x, y \in X_+$ we define

$$\begin{aligned} \lambda(x, y) &= \sup\{r \in [0, \infty) : rx \leq y\}, \\ r(x, y) &= \inf\{\lambda \in [0, \infty) : y \leq \lambda x\}. \end{aligned} \tag{6.216}$$

(We assume that the infimum of the empty set is ∞.) Note that other authors use the notations $m(y/x)$ and $M(y/x)$ instead of $\lambda(x, y)$ and $r(x, y)$, respectively.

For an interior point η of the cone X_+ we define

$$\|x\|_\eta = \inf\{r \in [0, \infty) : -r\eta \leq x \leq r\eta\}. \tag{6.217}$$

Clearly, $\|\cdot\|_\eta$ is a norm on X which is equivalent to the norm $\|\cdot\|$.

Denote by $\mathcal{A}$ the set of all mappings $A : X_+ \to X_+$ such that

$$
\begin{aligned}
Ax &\le Ay \quad \text{for each } x \in X_+ \text{ and each } y \ge x, \\
A(\alpha z) &= \alpha Az \quad \text{for each } \alpha \in [0, \infty) \text{ and each } x \in X_+.
\end{aligned}
\tag{6.218}
$$

Fix an interior point η of the cone X_+.

For the space $\mathcal{A}$ we define a metric $\rho : \mathcal{A} \times \mathcal{A} \to [0, \infty)$ by

$$\rho(A, B) := \sup\{\|Ax - Bx\|_\eta : x \in \langle 0, \eta\rangle\}, \quad A, B \in \mathcal{A}. \tag{6.219}$$

It is easy to see that the metric space $(\mathcal{A}, \rho)$ is complete.

Denote by $\mathcal{M}$ the set of all sequences $\{A_t\}_{t=1}^{\infty} \subset \mathcal{A}$. A member of $\mathcal{M}$ will occasionally be denoted by a boldface $\mathbf{A}$. For the set $\mathcal{M}$ we consider the uniformity which is determined by the following base:

$$E(N, \varepsilon) = \{(\{A_t\}_{t=1}^{\infty}, \{B_t\}_{t=1}^{\infty}) \in \mathcal{M} \times \mathcal{M} : \rho(A_t, B_t) \le \varepsilon, t = 1, \ldots, N\}, \tag{6.220}$$

where N is a natural number and $\varepsilon > 0$. It is easy to see that the uniform space $\mathcal{M}$ is metrizable (by a metric $\rho_w : \mathcal{M} \times \mathcal{M} \to [0, \infty)$) and complete. This uniformity generates a topology which we call the weak topology in $\mathcal{M}$.

For the set $\mathcal{M}$ we also consider the uniformity which is determined by the following base:

$$E(N, \varepsilon) = \{(\{A_t\}_{t=1}^{\infty}, \{B_t\}_{t=1}^{\infty}) \in \mathcal{M} \times \mathcal{M} : \rho(A_t, B_t) \le \varepsilon, t = 1, 2, \ldots\}, \tag{6.221}$$

where $\varepsilon > 0$. It is easy to see that the space $\mathcal{M}$ with this uniformity is metrizable (by a metric $\rho_s : \mathcal{M} \times \mathcal{M} \to [0, \infty)$) and complete. This uniformity generates a topology which we call the strong topology in $\mathcal{M}$. We do not write down the explicit expressions for the metrics ρ_w and ρ_s because we are not going to use them in the sequel.

Denote by $\mathcal{M}_{reg}$ the set of all sequences $\{A_t\}_{t=1}^{\infty} \in \mathcal{M}$ for which there exist positive constants $c_1 < c_2$ such that for each integer $T \ge 1$,

$$c_2\eta \ge A_T \cdot \cdots \cdot A_1\eta \ge c_1\eta.$$

Denote by $\bar{\mathcal{M}}_{reg}$ the closure of $\mathcal{M}_{reg}$ in $\mathcal{M}$ with the weak topology. We consider the topological subspace $\bar{\mathcal{M}}_{reg} \subset \mathcal{M}$ with the relative weak and strong topologies.

We now list the results which were obtained in [130].

Theorem 6.41 *There exists a set $\mathcal{F} \subset \bar{\mathcal{M}}_{reg}$ which is a countable intersection of open (in the weak topology) everywhere dense (in the strong topology) sets in $\bar{\mathcal{M}}_{reg}$ such that for each $\{B_t\}_{t=1}^{\infty} \in \mathcal{F}$ the following two assertions hold:*

1. *$B_T \cdot \cdots \cdot B_1\eta$ is an interior point of X_+ for each integer $T \ge 1$.*

2. *For each* $\Delta > 1$ *and each* $\varepsilon \in (0,1)$, *there exist an integer* $N \geq 1$ *and an open neighborhood* U *of* $\{B_t\}_{t=1}^{\infty}$ *in* $\bar{\mathcal{M}}_{reg}$ *with the weak topology such that for each* $\{C_t\}_{t=1}^{\infty} \in U$,

$$C_T \cdot \dots \cdot C_1 \eta$$

is an interior point of X_+ *for all* $T \in \{1, \dots, N\}$ *and*

$$r(C_N \cdot \dots \cdot C_1 \eta, C_N \cdot \dots \cdot C_1 x) - \lambda(C_N \cdot \dots \cdot C_1 \eta, C_N \cdot \dots \cdot C_1 x) \leq \varepsilon$$

for all $x \in \langle 0, \Delta\eta \rangle$.

Such results are usually called weak ergodic theorems in the population biology literature [43, 107]. This result shows that a weak ergodic theorem holds for most of the elements in $\bar{\mathcal{M}}_{reg}$. Clearly, if such a theorem holds for a sequence $\{A_t\}_{t=1}^{\infty}$ it also holds for all sequences of the form $\{a_t A_t\}_{t=1}^{\infty}$, where $\{a_t\}_{t=1}^{\infty} \subset R^1$ is a positive sequence. Therefore Theorem 6.41 shows that a weak ergodic theorem actually holds for most of those elements $\{A_t\}_{t=1}^{\infty} \in \mathcal{M}$ for which there exists a positive constant c such that for each integer $T \geq 1$,

$$\|A_T \cdot \dots \cdot A_1 \eta\|_{\eta}^{-1} A_T \cdot \dots \cdot A_1 \eta \geq c\eta.$$

Let $0 < c_1 < c_2$. Denote by $\mathcal{M}(c_1, c_2)$ the set of all sequences $\{A_t\}_{t=1}^{\infty} \in \mathcal{M}$ such that

$$A_T \cdot \dots \cdot A_1 \eta \in \langle c_1\eta, c_2\eta \rangle \quad \text{for all integers } T \geq 1.$$

It is easy to verify that $\mathcal{M}(c_1, c_2)$ is a closed subset of $\mathcal{M}$ with the weak topology. We first consider the topological subspace $\mathcal{M}(c_1, c_2) \subset \mathcal{M}$ with the relative weak and strong topologies.

Theorem 6.42 *There exists a set* $\mathcal{F}_0 \subset \mathcal{M}(c_1, c_2)$ *which is a countable intersection of open (in the weak topology) and everywhere dense (in the strong topology) sets in* $\mathcal{M}(c_1, c_2)$ *such that for each* $\{B_t\}_{t=1}^{\infty} \in \mathcal{F}_0$ *assertion 2 of Theorem 6.41 is valid.*

Denote by $\mathrm{int}(X_+)$ the set of interior points of the cone X_+. Let $\mathcal{M}_*$ be the set of all $\{A_t\}_{t=1}^{\infty} \in \mathcal{M}$ for which there exists a point $\xi \in \mathrm{int}(X_+)$ such that

$$A_t \xi = \xi, \quad t = 1, 2, \dots.$$

Denote by $\bar{\mathcal{M}}_*$ the closure of $\mathcal{M}_*$ in the strong topology. Next we consider the topological subspace $\bar{\mathcal{M}}_* \subset \mathcal{M}$ with the relative strong topology.

Theorem 6.43 *There exists a set* $\mathcal{F} \subset \bar{\mathcal{M}}_*$ *which is a countable intersection of open everywhere dense sets in* $\bar{\mathcal{M}}_*$ *such that for each* $\{B_t\}_{t=1}^{\infty} \in \mathcal{F}$ *there exists an interior point* ξ_B *of* X_+ *satisfying*

$$B_t \xi_B = \xi_B, \quad t = 1, 2, \dots, \qquad \|\xi_B\|_{\eta} = 1,$$

and the following two assertions hold:

1. *For each* $s : \{1, 2, \dots\} \to \{1, 2, \dots\}$ *there exists a function* $g_s : X_+ \to [0, \infty)$ *such that*

$$\lim_{T\to\infty} B_{s(T)} \cdot \dots \cdot B_{s(1)}x = g_s(x)\xi_B, \quad x \in X_+.$$

2. *For each* $\varepsilon > 0$, *there exist a neighborhood* U *of* $\{B_t\}_{t=1}^{\infty}$ *in* $\bar{\mathcal{M}}_*$ *and an integer* $N \geq 1$ *such that for each* $\{C_t\}_{t=1}^{\infty} \in U \cap \mathcal{M}_*$, *each integer* $T \geq N$, *each* $s : \{1, 2, \dots\} \to \{1, 2, \dots\}$ *and each* $x \in \langle 0, \eta\rangle$,

$$\left\| C_{s(T)} \cdot \dots \cdot C_{s(1)}x - g_s(x)\xi_B \right\|_\eta \leq \varepsilon.$$

Denote by $\mathcal{M}_\eta$ the set of all sequences $\{A_t\}_{t=1}^{\infty} \in \mathcal{M}$ such that $A_t\eta = \eta$, $t = 1, 2, \dots$. Clearly $\mathcal{M}_\eta$ is a closed subset of $\mathcal{M}$ with the weak topology. We now consider the topological subspace $\mathcal{M}_\eta \subset \mathcal{M}$ with the relative weak and strong topologies and state the following two results.

Theorem 6.44 *There exists a set* $\mathcal{F} \subset \mathcal{M}_\eta$ *which is a countable intersection of open (in the weak topology) everywhere dense (in the strong topology) sets in* $\mathcal{M}_\eta$ *such that for each* $\{B_t\}_{t=1}^{\infty} \in \mathcal{F}$ *the following two assertions holds:*

1. *There exists* $f : X_+ \to R^1$ *such that*

$$\lim_{T\to\infty} B_T \cdot \dots \cdot B_1x = f(x)\eta, \quad x \in X_+.$$

2. *For each* $\varepsilon > 0$, *there exist a neighborhood* U *of* $\{B_t\}_{t=1}^{\infty}$ *in* $\mathcal{M}_\eta$ *with the weak topology and an integer* $N \geq 1$ *such that for each* $\{C_t\}_{t=1}^{\infty} \in U$, *each integer* $T \geq N$ *and each* $x \in \langle 0, \eta\rangle$,

$$\left\| C_T \cdot \dots \cdot C_1x - f(x)\eta \right\|_\eta \leq \varepsilon.$$

Theorem 6.45 *There exists a set* $\mathcal{F} \subset \mathcal{M}_\eta$ *which is a countable intersection of open everywhere dense sets in* $\mathcal{M}_\eta$ *with the strong topology such that for each* $\{B_t\}_{t=1}^{\infty} \in \mathcal{F}$ *the following two assertions hold:*

1. *For each* $s : \{1, 2, \dots\} \to \{1, 2, \dots\}$, *there exists a function* $g_s : X_+ \to R^1$ *such that*

$$\lim_{T\to\infty} B_{s(T)} \cdot \dots \cdot B_{s(1)}x = g_s(x)\eta, \quad x \in X_+.$$

2. *For each* $\varepsilon > 0$, *there exist a neighborhood* U *of* $\{B_t\}_{t=1}^{\infty}$ *in* $\mathcal{M}_\eta$ *with the strong topology and an integer* $N \geq 1$ *such that for each* $\{C_t\}_{t=1}^{\infty} \in U$, *each integer* $T \geq N$, *each* $s : \{1, 2, \dots\} \to \{1, 2, \dots\}$ *and each* $x \in \langle 0, \eta\rangle$,

$$\left\| C_{s(T)} \cdot \dots \cdot C_{s(1)}x - g_s(x)\eta \right\|_\eta \leq \varepsilon.$$

Denote by $\mathcal{A}_*$ the set of all $A \in \mathcal{A}$ such that there is an interior point ξ_A of X_+ satisfying $A\xi_A = \xi_A$. Denote by $\bar{\mathcal{A}}_*$ the closure of $\mathcal{A}_*$ in $\mathcal{A}$. We equip the topological subspace $\bar{\mathcal{A}}_* \subset \mathcal{A}$ with the relative topology.

Theorem 6.46 *There exists a set $\mathcal{F} \subset \bar{\mathcal{A}}_*$ which is a countable intersection of open everywhere dense sets in $\bar{\mathcal{A}}_*$ such that for each $B \in \mathcal{F}$ there exists an interior point ξ_B of X_+ satisfying*

$$B\xi_B = \xi_B, \qquad \|\xi_B\|_\eta = 1,$$

and the following two assertions hold:

1. *There exists a function $g_B : X_+ \to R^1$ such that*

$$\lim_{T\to\infty} B^T x = g_B(x)\xi_B, \quad x \in X_+.$$

2. *For each $\varepsilon > 0$, there exist a neighborhood U of B in $\bar{\mathcal{A}}_*$ and an integer $N \geq 1$ such that for each $C \in U \cap \mathcal{A}_*$, each integer $T \geq N$ and each point $x \in \langle 0, \eta\rangle$,*

$$\left\|C^T x - g_B(x)\xi_B\right\|_\eta \leq \varepsilon.$$

Finally, denote by $\mathcal{A}_\eta$ the set of all $A \in \mathcal{A}$ satisfying $A\eta = \eta$. Clearly $\mathcal{A}_\eta$ is a closed subset of $\mathcal{A}$. We endow the topological subspace $\mathcal{A}_\eta \subset \mathcal{A}$ with the relative topology.

Theorem 6.47 *There exists a set $\mathcal{F} \subset \mathcal{A}_\eta$ which is a countable intersection of open everywhere dense sets in $\mathcal{A}_\eta$ such that for each $B \in \mathcal{F}$ the following two assertions hold*:

1. *There exists a functional $g_B : X_+ \to R^1$ such that*

$$\lim_{T\to\infty} B^T x = g_B(x)\eta, \quad x \in X_+.$$

2. *For each $\varepsilon > 0$, there exist a neighborhood U of B in $\mathcal{A}_\eta$ and an integer $N \geq 1$ such that for each $C \in U$, each integer $T \geq N$ and each $x \in \langle 0, \eta\rangle$,*

$$\left\|C^T x - g_B(x)\eta\right\|_\eta \leq \varepsilon.$$

In the next sections we prove Theorems 6.41–6.43. Theorem 6.44 is proved by a simple modification of the proof of Theorem 6.41 while Theorems 6.45–6.47 can be proved by slightly modifying the proof of Theorem 6.43.

6.21 Preliminary Lemmata for Theorems 6.41–6.43

We begin with the following simple observation.

Lemma 6.48 *Assume that $\{A_t\}_{t=1}^\infty \in \mathcal{M}$ and that for each integer $T \geq 1$,*

$$c_2\eta \geq A_T \cdot \cdots \cdot A_1\eta \geq c_1\eta$$

with some constants $c_2 > c_1 > 0$ which do not depend on T. Let ξ be an interior point of X_+. Then there exist constants $c_{2\xi}, c_{1\xi} > 0$ such that for each integer $T \geq 1$,

$$c_{2\xi}\xi \geq A_T \cdot \dots \cdot A_1\xi \geq c_{1\xi}\xi.$$

Clearly, for each interior point ξ of X_+ (see (6.216), (6.217)), we have

$$r(\xi, y) = \|y\|_\xi, \quad y \in X_+. \tag{6.222}$$

Assume that ξ is an interior point of X_+, $\{A_t\}_{t=1}^\infty \in \mathcal{M}$ and that there are numbers $c_1 \in (0, 1)$, $c_2 > 1$ such that for each integer $T \geq 1$,

$$c_2\xi \geq A_T \cdot \dots \cdot A_1\xi \geq c_1\xi. \tag{6.223}$$

Let $\gamma \in (0, 1)$. Clearly,

$$0 < \lambda(\eta, \xi). \tag{6.224}$$

Define a sequence of operators $A_t^\gamma : X_+ \to X_+$, $t = 1, 2, \dots$, by

$$\begin{aligned} A_1^\gamma x &= (1-\gamma)A_1x + \gamma r(\xi, x)A_1\xi, \quad x \in X_+, \\ A_{t+1}^\gamma x &= (1-\gamma)A_{t+1}x + \gamma r(A_t \cdot \dots \cdot A_1\xi, x)A_{t+1} \cdot A_t \cdot \dots \cdot A_1\xi, \\ &x \in X_+, t = 1, 2, \dots. \end{aligned} \tag{6.225}$$

Clearly,

$$\{A_t^\gamma\}_{t=1}^\infty \in \mathcal{M}. \tag{6.226}$$

Lemma 6.49 *For each integer $T \geq 1$,*

$$A_T^\gamma \cdot \dots \cdot A_1^\gamma\xi = A_T \cdot \dots \cdot A_1\xi. \tag{6.227}$$

Proof We will prove the lemma by induction. Clearly, for $T = 1$ (6.227) is valid. Assume that $T \geq 1$ is an integer and (6.227) holds. It follows from (6.227), (6.225) and (6.222) that

$$\begin{aligned} A_{T+1}^\gamma \cdot A_T^\gamma \cdot \dots \cdot A_1^\gamma\xi &= A_{T+1}^\gamma \cdot A_T \cdot \dots \cdot A_1\xi \\ &= (1-\gamma)A_{T+1} \cdot A_T \cdot \dots \cdot A_1\xi + \gamma A_{T+1} \cdot A_T \cdot \dots \cdot A_1\xi \\ &= A_{T+1} \cdot A_T \cdot \dots \cdot A_1\xi. \end{aligned}$$

This completes the proof of the lemma. □

Lemma 6.50 *For each integer $T \geq 1$,*

$$\rho(A_T, A_T^\gamma) \leq \gamma c_1^{-1} c_2 \lambda(\eta, \xi)^{-1} r(\eta, \xi).$$

Proof Let $x \in \langle 0, \eta \rangle$. Then by (6.216), (6.217) and (6.219),

$$0 \le x \le \eta \le \lambda(\eta,\xi)^{-1}\xi \le c_1^{-1}\lambda(\eta,\xi)^{-1}A_T \cdot \cdots \cdot A_1\xi, \quad T = 1, 2, \ldots, \tag{6.228}$$

$$\begin{aligned} A_1^\gamma x - A_1 x &= \gamma r(\xi, x)A_1\xi - \gamma A_1 x \\ &\in \langle -\gamma\lambda(\eta,\xi)^{-1}c_2 r(\eta,\xi)\eta, \gamma\lambda(\eta,\xi)^{-1}c_2 r(\eta,\xi)\eta \rangle \end{aligned} \tag{6.229}$$

and

$$\|A_1^\gamma x - A_1 x\|_\eta \le \gamma\lambda(\eta,\xi)^{-1}c_2 r(\eta,\xi).$$

For each $T \ge 1$, it now follows from (6.225), (6.228) and (6.223) that

$$\begin{aligned} A_{T+1}^\gamma x - A_{T+1}x &= \gamma r(A_T \cdot \cdots \cdot A_1\xi, x)A_{T+1} \cdot A_T \cdot \cdots \cdot A_1\xi - \gamma A_{T+1}x \\ &\in \langle -\gamma\lambda(\eta,\xi)^{-1}c_1^{-1}c_2\xi, \gamma c_1^{-1}\lambda(\eta,\xi)^{-1}c_2\xi \rangle \\ &\subset \langle -\gamma\lambda(\eta,\xi)^{-1}c_1^{-1}c_2 r(\eta,\xi)\eta, \gamma c_1^{-1}\lambda(\eta,\xi)^{-1}c_2 r(\eta,\xi)\eta \rangle \end{aligned}$$

and

$$\|A_{T+1}^\gamma x - A_{T+1}x\|_\eta \le \gamma c_1^{-1}c_2\lambda(\eta,\xi)^{-1}r(\eta,\xi).$$

This completes the proof of the lemma. □

Lemma 6.51 *For each* $x \in X_+$,

$$\lambda(A_1\xi, A_1^\gamma x) \ge (1-\gamma)\lambda(\xi, x) + \gamma r(\xi, x), \qquad r(A_1\xi, A_1^\gamma x) \le r(\xi, x), \tag{6.230}$$

and for each integer $T \ge 1$,

$$\begin{aligned} &\lambda(A_{T+1} \cdot A_T \cdot \cdots \cdot A_1\xi, A_{T+1}^\gamma \cdot A_T^\gamma \cdot \cdots \cdot A_1^\gamma x) \\ &\quad \ge (1-\gamma)\lambda(A_T \cdot \cdots \cdot A_1\xi, A_T^\gamma \cdot \cdots \cdot A_1^\gamma x) \\ &\qquad + \gamma r(A_T \cdot \cdots \cdot A_1\xi, A_T^\gamma \cdot \cdots \cdot A_1^\gamma x) \end{aligned} \tag{6.231}$$

and

$$\begin{aligned} &r(A_{T+1} \cdot A_T \cdot \cdots \cdot A_1\xi, A_{T+1}^\gamma \cdot A_T^\gamma \cdot \cdots \cdot A_1^\gamma x) \\ &\quad \le r(A_T \cdot \cdots \cdot A_1\xi, A_T^\gamma \cdot \cdots \cdot A_1^\gamma x). \end{aligned} \tag{6.232}$$

Proof By (6.225), we have for each $x \in X_+$,

$$\begin{aligned} A_1^\gamma x &= (1-\gamma)A_1 x + \gamma r(\xi, x)A_1\xi \ge (1-\gamma)A_1(\lambda(\xi, x)\xi) + \gamma r(\xi, x)A_1\xi \\ &= [(1-\gamma)\lambda(\xi, x) + \gamma r(\xi, x)]A_1\xi, \\ &\quad \lambda(A_1\xi, A_1^\gamma x) \ge (1-\gamma)\lambda(\xi, x) + \gamma r(\xi, x); \end{aligned}$$

$$A_1^\gamma x \le (1-\gamma)A_1\bigl(r(\xi,x)\xi\bigr) + \gamma r(\xi,x)A_1\xi = r(\xi,x)A_1\xi,$$
$$r\bigl(A_1\xi, A_1^\gamma x\bigr) \le r(\xi,x).$$

Again by (6.225), we also have for each integer $T \ge 1$ and each $x \in \mathrm{int}(X_+)$,

$$\begin{aligned}
&A_{T+1}^\gamma \cdot A_T^\gamma \cdot \cdots \cdot A_1^\gamma x \\
&\quad = (1-\gamma)A_{T+1} \cdot A_T^\gamma \cdot \cdots \cdot A_1^\gamma x \\
&\qquad + \gamma r\bigl(A_T \cdot \cdots \cdot A_1\xi, A_T^\gamma \cdot \cdots \cdot A_1^\gamma x\bigr)A_{T+1} \cdot A_T \cdot \cdots \cdot A_1\xi \\
&\quad \ge (1-\gamma)A_{T+1}\bigl(\lambda\bigl(A_T \cdot \cdots \cdot A_1\xi, A_T^\gamma \cdot \cdots \cdot A_1^\gamma x\bigr)A_T \cdot \cdots \cdot A_1\xi\bigr) \\
&\qquad + \gamma r\bigl(A_T \cdot \cdots \cdot A_1\xi, A_T^\gamma \cdot \cdots \cdot A_1^\gamma x\bigr)A_{T+1} \cdot A_T \cdot \cdots \cdot A_1\xi \\
&\quad = \bigl[(1-\gamma)\lambda\bigl(A_T \cdot \cdots \cdot A_1\xi, A_T^\gamma \cdot \cdots \cdot A_1^\gamma x\bigr) \\
&\qquad + \gamma r\bigl(A_T \cdot \cdots \cdot A_1\xi, A_T^\gamma \cdot \cdots \cdot A_1^\gamma x\bigr)\bigr]A_{T+1} \cdot A_T \cdot \cdots \cdot A_1\xi, \\
&\qquad \lambda\bigl(A_{T+1} \cdot A_T \cdot \cdots \cdot A_1\xi, A_{T+1}^\gamma \cdot A_T^\gamma \cdot \cdots \cdot A_1^\gamma x\bigr) \\
&\qquad\quad \ge (1-\gamma)\lambda\bigl(A_T \cdot \cdots \cdot A_1\xi, A_T^\gamma \cdot \cdots \cdot A_1^\gamma x\bigr) \\
&\qquad\qquad + \gamma r\bigl(A_T \cdot \cdots \cdot A_1\xi, A_T^\gamma \cdot \cdots \cdot A_1^\gamma x\bigr)
\end{aligned}$$

and

$$\begin{aligned}
&A_{T+1}^\gamma \cdot A_T^\gamma \cdot \cdots \cdot A_1^\gamma x \\
&\quad \le (1-\gamma)A_{T+1}\bigl(r\bigl(A_T \cdot \cdots \cdot A_1\xi, A_T^\gamma \cdot \cdots \cdot A_1^\gamma x\bigr)A_T \cdot \cdots \cdot A_1\xi\bigr) \\
&\qquad + \gamma r\bigl(A_T \cdot \cdots \cdot A_1\xi, A_T^\gamma \cdot \cdots \cdot A_1^\gamma x\bigr)A_{T+1} \cdot A_T \cdot \cdots \cdot A_1\xi \\
&\quad = r\bigl(A_T \cdot \cdots \cdot A_1\xi, A_T^\gamma \cdot \cdots \cdot A_1^\gamma x\bigr)A_{T+1} \cdot A_T \cdot \cdots \cdot A_1\xi.
\end{aligned}$$

Thus

$$r\bigl(A_{T+1} \cdot A_T \cdot \cdots \cdot A_1\xi, A_{T+1}^\gamma \cdot A_T^\gamma \cdot \cdots \cdot A_1^\gamma x\bigr) \le r\bigl(A_T \cdot \cdots \cdot A_1\xi, A_T^\gamma \cdot \cdots \cdot A_1^\gamma x\bigr)$$

and the lemma is proved. □

Lemma 6.52 *Let $0 < \varepsilon < \Delta$ and let N be a natural number for which*

$$\Delta(1-\gamma)^N < 2^{-1}\varepsilon. \tag{6.233}$$

Then for each $x \in \langle 0, \Delta\xi\rangle$,

$$r\bigl(A_N \cdot \cdots \cdot A_1\xi, A_N^\gamma \cdot \cdots \cdot A_1^\gamma x\bigr) - \lambda\bigl(A_N \cdot \cdots \cdot A_1\xi, A_N^\gamma \cdot \cdots \cdot A_1^\gamma x\bigr) \le 2^{-1}\varepsilon.$$

Proof Let $x \in X_+$ satisfy

$$0 \le x \le \Delta\xi. \tag{6.234}$$

Set

$$\begin{aligned} r_0 &= r(\xi, x), \qquad \lambda_0 = \lambda(\xi, x), \\ r_t &= r\big(A_t \cdot \dots \cdot A_1\xi, A_t^\gamma \cdot \dots \cdot A_1^\gamma x\big), \\ \lambda_t &= \lambda\big(A_t \cdot \dots \cdot A_1\xi, A_t^\gamma \cdot \dots \cdot A_1^\gamma x\big), \quad t = 1, 2, \dots. \end{aligned} \tag{6.235}$$

By (6.216), (6.234) and Lemma 6.51,

$$r_0 \le \Delta, \qquad r_t \ge \lambda_t, \quad t = 0, 1, \dots,$$

$$r_{t+1} \le r_t, \qquad \lambda_{t+1} \ge \lambda_t, \quad t = 0, 1, \dots$$

and

$$\lambda_{t+1} \ge (1-\gamma)\lambda_t + \gamma r_t, \quad t = 0, 1, \dots.$$

Together with (6.233) this implies that for all $t = 0, 1, \dots,$

$$r_{t+1} - \lambda_{t+1} \le (1-\gamma)(r_t - \lambda_t)$$

and

$$r_N - \lambda_N \le (1-\gamma)^N(r_0 - \lambda_0) = (1-\gamma)^N \Delta < 2^{-1}\varepsilon.$$

This completes the proof of the lemma. □

Lemma 6.53 *Let* $0 < \Delta_1 < 1 < \Delta_2$, $\Gamma > 1$ *and let* $n \ge 1$ *be an integer. Then there is a number* $\delta > 0$ *such that for each sequence* $\{B_i\}_{i=1}^n \subset \mathcal{A}$ *satisfying*

$$\rho\big(B_i, A_i^\gamma\big) \le \delta, \quad i = 1, \dots, n, \tag{6.236}$$

and each $z \in \langle 0, \Delta_2\xi\rangle$ *satisfying* $r(\xi, z) \ge \Delta_1$, *the following relation holds*:

$$B_n \cdot \dots \cdot B_1 z \in \big\langle \Gamma^{-1} A_n^\gamma \cdot \dots \cdot A_1^\gamma z, \Gamma A_n^\gamma \cdot \dots \cdot A_1^\gamma z\big\rangle.$$

Proof We prove this lemma by induction. Let $n = 1$. Choose a positive number δ such that

$$c_1^{-1}\Delta_1^{-1}\delta\Delta_2 r(\eta, \xi)\lambda(\eta, \xi)^{-1} < \gamma(\Gamma - 1)\Gamma^{-1}. \tag{6.237}$$

Assume that $B_1 \in \mathcal{A}$,

$$z \in \langle 0, \Delta_2\xi\rangle, \qquad \rho\big(B_1, A_1^\gamma\big) \le \delta \quad \text{and} \quad r(\xi, z) \ge \Delta_1. \tag{6.238}$$

It follows from (6.238), (6.216) and (2.219) that

$$z \le \Delta_2\xi \le \Delta_2 r(\eta, \xi)\eta, \qquad \big\|B_1 z - A_1^\gamma z\big\|_\eta \le \delta\Delta_2 r(\eta, \xi) \tag{6.239}$$

and

$$B_1 z \in \langle A_1^\gamma z - \delta \Delta_2 r(\eta, \xi)\eta, A_1^\gamma z + \delta \Delta_2 r(\eta, \xi)\eta \rangle$$
$$\subset \langle A_1^\gamma z - \delta \Delta_2 r(\eta, \xi)\lambda(\eta, \xi)^{-1}\xi, A_1^\gamma z + \delta \Delta_2 r(\eta, \xi)\lambda(\eta, \xi)^{-1}\xi \rangle. \quad (6.240)$$

By (6.238), Lemma 6.49, (6.223), (6.237) and (6.225),

$$A_1^\gamma z \geq \gamma r(\xi, z) A_1 \xi \geq \gamma \Delta_1 c_1 \xi$$

and

$$\delta \Delta_2 r(\eta, \xi)\lambda(\eta, \xi)^{-1}\xi \leq \gamma (1 - \Gamma^{-1}) \Delta_1 c_1 \xi_1.$$

Together with (6.240) this implies that

$$B_1 z \in \langle \Gamma^{-1} A_1^\gamma z, \Gamma A_1^\gamma z \rangle.$$

Thus for $n = 1$ the assertion of the lemma is valid.

Assume now that the assertion of the lemma holds for $n = 1, \dots, k$. We now show that the assertion of the lemma also holds for $n = k + 1$. To this end, choose a positive number $\Gamma_0 > 1$ such that

$$1 < \Gamma_0 < \Gamma^{1/2}. \quad (6.241)$$

Since the assertion of the lemma holds for $n = k$, there is a number $\delta_0 > 0$ such that for each sequence $\{B_i\}_{i=1}^k \subset \mathcal{A}$ satisfying

$$\rho(B_i, A_i^\gamma) \leq \delta_0, \quad i = 1, \dots, k,$$

and each $z \in \langle 0, \Delta_2 \xi \rangle$ satisfying $r(\xi, z) \geq \Delta_1$ the following relation holds:

$$B_k \cdot \cdots \cdot B_1 z \in \langle \Gamma_0^{-1} A_k^\gamma \cdot \cdots \cdot A_1^\gamma z, \Gamma_0 A_k^\gamma \cdot \cdots \cdot A_1^\gamma z \rangle. \quad (6.242)$$

Choose a number $\delta \in (0, \delta_0)$ such that

$$\delta \Delta_2 c_2 r(\eta, \xi)\lambda(\xi, \eta)^{-1} c_1^{-1} \Delta_1^{-1} \leq \gamma (\Gamma_0 - 1)\Gamma_0^{-1}. \quad (6.243)$$

Assume that $\{B_i\}_{i=1}^{k+1} \subset \mathcal{A}$,

$$\rho(B_i, A_i^\gamma) \leq \delta, \quad i = 1, \dots, k+1,$$
$$z \in \langle 0, \Delta_2 \xi \rangle \quad \text{and} \quad r(\xi, z) \geq \Delta_1. \quad (6.244)$$

Then relation (6.242) is valid. This implies that

$$B_{k+1} \cdot B_k \cdot \cdots \cdot B_1 z$$
$$\in \langle \Gamma_0^{-1} B_{k+1} \cdot A_k^\gamma \cdot \cdots \cdot A_1^\gamma z, \Gamma_0 B_{k+1} \cdot A_k^\gamma \cdot \cdots \cdot A_1^\gamma z \rangle. \quad (6.245)$$

It follows from (6.244), (6.223), (6.225), (6.219) and Lemma 6.49 that

$$\begin{aligned} A_1^{\gamma} z \ge \gamma \Delta_1 A_1 \xi, \qquad A_{k+1}^{\gamma} \cdot A_k^{\gamma} \cdot \dots \cdot A_1^{\gamma} z \ge \Delta_1 c_1 \gamma \xi, \\ A_k^{\gamma} \cdot \dots \cdot A_1^{\gamma} z \in \langle \Delta_1 c_1 \gamma \xi, \Delta_2 c_2 \xi \rangle, \\ \| A_k^{\gamma} \cdot \dots \cdot A_1^{\gamma} z \|_{\eta} \le \Delta_2 c_2 r(\eta, \xi), \\ \| B_{k+1} \cdot A_k^{\gamma} \cdot \dots \cdot A_1^{\gamma} z - A_{k+1}^{\gamma} \cdot A_k^{\gamma} \cdot \dots \cdot A_1^{\gamma} z \|_{\eta} \le \delta \Delta_2 c_2 r(\eta, \xi) \end{aligned} \tag{6.246}$$

and

$$\begin{aligned} & B_{k+1} \cdot A_k^{\gamma} \cdot \dots \cdot A_1^{\gamma} z \\ & \in \big\langle A_{k+1}^{\gamma} \cdot A_k^{\gamma} \cdot \dots \cdot A_1^{\gamma} z - \delta \Delta_2 c_2 r(\eta, \xi) \eta, \\ & \quad A_{k+1}^{\gamma} \cdot A_k^{\gamma} \cdot \dots \cdot A_1^{\gamma} z + \delta \Delta_2 c_2 r(\eta, \xi) \eta \big\rangle. \end{aligned} \tag{6.247}$$

By (6.216), (6.217), (6.244), Lemma 6.49, (6.243) and (6.246),

$$\begin{aligned} \delta \Delta_2 c_2 r(\eta, \xi) \eta & \le \delta \Delta_2 c_2 r(\eta, \xi) \lambda(\eta, \xi)^{-1} \xi \\ & \le \delta \Delta_2 c_2 r(\eta, \xi) \lambda(\xi, \eta)^{-1} c_1^{-1} \Delta_1^{-1} \gamma A_{k+1}^{\gamma} \cdot A_k^{\gamma} \cdot \dots \cdot A_1^{\gamma} z \\ & \le (\Gamma_0 - 1) \Gamma_0^{-1} A_{k+1}^{\gamma} \cdot A_k^{\gamma} \cdot \dots \cdot A_1^{\gamma} z. \end{aligned}$$

Together with (6.247) this implies that

$$B_{k+1} \cdot A_k^{\gamma} \cdot \dots \cdot A_1^{\gamma} z \in \big\langle \Gamma_0^{-1} A_{k+1}^{\gamma} \cdot A_k^{\gamma} \cdot \dots \cdot A_1^{\gamma} z, \Gamma_0 A_{k+1}^{\gamma} \cdot A_k^{\gamma} \cdot \dots \cdot A_1^{\gamma} z \big\rangle.$$

It follows from this relation, (6.245) and (6.241) that

$$B_{k+1} \cdot B_k \cdot \dots \cdot B_1 z \in \big\langle \Gamma^{-1} A_{k+1}^{\gamma} \cdot A_k^{\gamma} \cdot \dots \cdot A_1^{\gamma} z, \Gamma A_{k+1}^{\gamma} \cdot A_k^{\gamma} \cdot \dots \cdot A_1^{\gamma} z \big\rangle.$$

This completes the proof of the lemma. □

Lemma 6.54 *Let* $1 < \Delta$, $0 < \varepsilon < 1$ *and let* N *be a natural number for which*

$$\Delta (1 - \gamma)^N < 2^{-1} \varepsilon. \tag{6.248}$$

Then there exist a neighborhood U *of* $\{A_t^{\gamma}\}_{t=1}^{\infty}$ *in* $\mathcal{M}$ *with the weak topology and a number* $\kappa > 0$ *such that for each* $\{B_t\}_{t=1}^{\infty} \in U$, *the following two assertions hold*:

1. $B_T \cdot \dots \cdot B_1 \xi \ge \kappa \xi$, $T = 1, \dots, N$.
2. *For each* $x \in \langle 0, \Delta \xi \rangle$,

$$\begin{aligned} & r(B_N \cdot \dots \cdot B_1 \xi, B_N \cdot \dots \cdot B_1 x) - \lambda(B_N \cdot \dots \cdot B_1 \xi, B_N \cdot \dots \cdot B_1 x) \\ & \le (3/4) \varepsilon. \end{aligned} \tag{6.249}$$

Proof By Lemma 6.52 for each $x \in \langle 0, \Delta\xi \rangle$,

$$r\big(A_N \cdot \dots \cdot A_1\xi, A_N^\gamma \cdot \dots \cdot A_1^\gamma x\big) - \lambda\big(A_N \cdot \dots \cdot A_1\xi, A_N^\gamma \cdot \dots \cdot A_1^\gamma x\big) \\ \le \varepsilon/2. \tag{6.250}$$

Choose a number Γ for which

$$\Gamma > 1 \quad \text{and} \quad \big(\Gamma^2 - 1\big)\Delta < 8^{-1}\varepsilon. \tag{6.251}$$

By Lemma 6.53, there exists a neighborhood U of $\{A_t^\gamma\}_{t=1}^\infty$ in $\mathcal{M}$ such that for each $\{B_t\}_{t=1}^\infty \in U$, each $z \in \langle 0, \Delta\xi \rangle$ satisfying $r(\xi, z) \ge \varepsilon/2$ and each integer $T \in [1, N]$,

$$B_T \cdot \dots \cdot B_1 z \in \big\langle \Gamma^{-1} A_T^\gamma \cdot \dots \cdot A_1^\gamma z, \Gamma A_T^\gamma \cdot \dots \cdot A_1^\gamma z \big\rangle. \tag{6.252}$$

Assume that $\{B_t\}_{t=1}^\infty \in U$. It follows from the definition of U (see (6.252)), Lemma 6.49 and (6.223) that for $T = 1, \dots, N$,

$$B_T \cdot \dots \cdot B_1\xi \ge \Gamma^{-1} A_T^\gamma \cdot \dots \cdot A_1^\gamma \xi = \Gamma^{-1} A_T \cdot \dots \cdot A_1\xi \ge \Gamma^{-1} c_1\xi. \tag{6.253}$$

Therefore assertion 1 holds with $\kappa = \Gamma^{-1}c_1$. Now we will show that assertion 2 also holds.

Assume that

$$\{B_t\}_{t=1}^\infty \in U \quad \text{and} \quad x \in \langle 0, \Delta\xi \rangle. \tag{6.254}$$

Then (6.250) is valid. We will show that (6.249) holds. To this end, we may assume without loss of generality that $r(\xi, \eta) \ge \varepsilon/2$. By (6.254) and the definition of U (see (6.252)),

$$B_N \cdot \dots \cdot B_1 x \in \big\langle \Gamma^{-1} A_N^\gamma \cdot \dots \cdot A_1^\gamma x, \Gamma A_N^\gamma \cdot \dots \cdot A_1^\gamma x \big\rangle \tag{6.255}$$

and

$$B_N \cdot \dots \cdot B_1\xi \in \big\langle \Gamma^{-1} A_N^\gamma \cdot \dots \cdot A_1^\gamma \xi, \Gamma A_N^\gamma \cdot \dots \cdot A_1^\gamma \xi \big\rangle.$$

It follows from (6.255), Lemma 6.49 and (6.216) that

$$\begin{aligned} r(B_N \cdot \dots \cdot B_1\xi, B_N \cdot \dots \cdot B_1 x) &\le r\big(\Gamma^{-1} A_N \cdot \dots \cdot A_1\xi, \Gamma A_N^\gamma \cdot \dots \cdot A_1^\gamma x\big) \\ &\le \Gamma^2 r\big(A_N \cdot \dots \cdot A_1\xi, A_N^\gamma \cdot \dots \cdot A_1^\gamma x\big), \\ \lambda(B_N \cdot \dots \cdot B_1\xi, B_N \cdot \dots \cdot B_1 x) &\ge \lambda\big(\Gamma A_N \cdot \dots \cdot A_1\xi, \Gamma^{-1} A_N^\gamma \cdot \dots \cdot A_1^\gamma x\big) \\ &= \Gamma^{-2}\lambda\big(A_N \cdot \dots \cdot A_1\xi, A_N^\gamma \cdot \dots \cdot A_1^\gamma x\big). \end{aligned}$$

By these relations, (6.250), (6.254) and (6.251),

$$r(B_N \cdot \dots \cdot B_1\xi, B_N \cdot \dots \cdot B_1 x) - \lambda(B_N \cdot \dots \cdot B_1\xi, B_N \cdot \dots \cdot B_1 x) \\ \le \Gamma^2 r\big(A_N \cdot \dots \cdot A_1\xi, A_N^\gamma \cdot \dots \cdot A_1^\gamma x\big) - \Gamma^{-2}\lambda\big(A_N \cdot \dots \cdot A_1\xi, A_N^\gamma \cdot \dots \cdot A_1^\gamma x\big)$$

$$\begin{aligned}&\leq r\big(A_N\cdot\dots\cdot A_1\xi, A_N^\gamma\cdot\dots\cdot A_1^\gamma x\big)-\lambda\big(A_N\cdot\dots\cdot A_1\xi, A_N^\gamma\cdot\dots\cdot A_1^\gamma x\big)\\&\quad+\big(\Gamma^2-1\big)r(\xi,x)+\big(1-\Gamma^{-2}\big)r\big(A_N\cdot\dots\cdot A_1\xi, A_N^\gamma\cdot\dots\cdot A_1^\gamma x\big)\\&\leq 2^{-1}\varepsilon+2\big(\Gamma^2-1\big)r(\xi,x)\leq 2^{-1}\varepsilon+\big(\Gamma^2-1\big)\Delta^2\leq(3/4)\varepsilon.\end{aligned}$$

Thus (6.249) is indeed valid and this completes the proof of the lemma. □

Our next claim is a direct consequence of (6.225).

Lemma 6.55 *Assume that $A_t\xi=\xi$, $t=1,2,\dots$. Then*

$$A_t^\gamma x=(1-\gamma)A_tx+\gamma r(\xi,x)\xi,\quad x\in X_+, t=1,2,\dots.$$

Lemma 6.51 implies the following fact.

Lemma 6.56 *Assume that $A_t\xi=\xi$, $t=1,2,\dots$, $s:\{1,2,\dots\}\to\{1,2,\dots\}$ and $x\in X_+$. Then*

$$\lambda\big(\xi, A_{s(1)}^\gamma x\big)\geq(1-\gamma)\lambda(\xi,x)+\gamma r(\xi,x),$$

and for each integer $T\geq 1$,

$$\begin{aligned}\lambda\big(\xi, A_{s(T+1)}^\gamma\cdot A_{s(T)}^\gamma\cdot\dots\cdot A_{s(1)}^\gamma x\big)&\geq(1-\gamma)\lambda\big(\xi, A_{s(T)}^\gamma\cdot\dots\cdot A_{s(1)}^\gamma x\big)\\&\quad+\gamma r\big(\xi, A_{s(T)}^\gamma\cdot\dots\cdot A_{s(1)}^\gamma x\big).\end{aligned}$$

By using Lemma 6.56 and an analogue of the proof of Lemma 6.52, we can prove the following lemma.

Lemma 6.57 *Assume that $A_t\xi=\xi$, $t=1,2,\dots$. Let $0<\varepsilon<\Delta$ and let N be a natural number for which*

$$\Delta(1-\gamma)^N<2^{-1}\varepsilon.$$

Then for each $x\in\langle 0,\Delta\xi\rangle$ and each $s:\{1,2,\dots\}\to\{1,2,\dots\}$,

$$r\big(\xi, A_{s(N)}^\gamma\cdot\dots\cdot A_{s(1)}^\gamma x\big)-\lambda\big(\xi, A_{s(N)}^\gamma\cdot\dots\cdot A_{s(1)}^\gamma x\big)\leq 2^{-1}\varepsilon.$$

Analogously to the proof of Lemma 6.53 we can also establish our last preliminary result.

Lemma 6.58 *Assume that $A_t\xi=\xi$, $t=1,2,\dots$. Let $0<\Delta_1<1<\Delta_2$, $\Gamma>1$ and let n be a natural number. Then there exists a number $\delta>0$ such that for each $s:\{1,2,\dots\}\to\{1,2,\dots\}$, each sequence $\{B_i\}_{i=1}^n\subset\mathcal{A}$ satisfying*

$$\rho\big(B_i, A_{s(i)}^\gamma\big)\leq\delta,\quad i=1,\dots,n,$$

and each $z\in\langle 0,\Delta_2\xi\rangle$ satisfying $r(\xi,z)\geq\Delta_1$, the following relation holds:

$$B_n\cdot\dots\cdot B_1z\in\big\langle\Gamma^{-1}A_{s(n)}^\gamma\cdot\dots\cdot A_{s(1)}^\gamma z, \Gamma A_{s(n)}^\gamma\cdot\dots\cdot A_{s(1)}^\gamma z\big\rangle.$$

6.22 Proofs of Theorems 6.41 and 6.42

In this section we use the notations of Sects. 6.20 and 6.21 with $\xi = \eta$. For each $\{A_t\}_{t=1}^{\infty} \in \mathcal{M}_{reg}$ and each $\gamma \in (0,1)$, define a sequence of operators $\{A_t^{\gamma}\}_{t=1}^{\infty}$ by (6.225) with $\xi = \eta$. By (6.226), Lemmas 6.49 and 6.50, the set

$$\left\{\{A_t^{\gamma}\}_{t=1}^{\infty} : \{A_t\}_{t=1}^{\infty} \in \mathcal{M}, \gamma \in (0,1)\right\} \subset \mathcal{M}$$

is an everywhere dense subset of $\bar{\mathcal{M}}_{reg}$ with the strong topology.

Let $\{A_t\}_{t=1}^{\infty} \in \mathcal{M}_{reg}$, $\gamma \in (0,1)$ and $i \geq 1$ be an integer. By Lemma 6.54, there exist an open neighborhood $U(\mathbf{A},\gamma,i)$ of $\{A_t^{\gamma}\}_{t=1}^{\infty}$ in the space $\bar{\mathcal{M}}_{reg}$ with the weak topology and an integer $N(A,\gamma,i) \geq 2i+2$ such that for each $\{C_t\}_{t=1}^{\infty} \in U(\mathbf{A},\gamma,i)$, the following two properties hold:

(a) $C_T \cdot \cdots \cdot C_1\eta$ is an interior point of X_+ for $T = 1,\dots,N(\mathbf{A},\gamma,i)$;

(b) for each $x \in \langle 0, 4^i\eta\rangle$,

$$\begin{aligned} &r(C_{N(\mathbf{A},\gamma,i)} \cdot \cdots \cdot C_1\eta, C_{N(\mathbf{A},\gamma,i)} \cdot \cdots \cdot C_1 x) \\ &\quad - \lambda(C_{N(\mathbf{A},\gamma,i)} \cdot \cdots \cdot C_1\eta, C_{N(\mathbf{A},\gamma,i)} \cdot \cdots \cdot C_1 x) \leq 8^{-i}. \end{aligned} \tag{6.256}$$

Proof of Theorem 6.41 Define

$$\mathcal{F} := \bigcap_{q=1}^{\infty} \bigcup \left\{U(\mathbf{A},\gamma,i) : \mathbf{A} \in \mathcal{M}_{reg}, \gamma \in (0,1), i = q, q+1, \dots\right\}.$$

Clearly $\mathcal{F}$ is a countable intersection of open (in the weak topology) everywhere dense (in the strong topology) sets in $\bar{\mathcal{M}}_{reg}$.

Assume that $\{B_t\}_{t=1}^{\infty} \in \mathcal{F}$. It is easy to verify that assertion 1 of Theorem 6.41 holds. We now show that assertion 2 of Theorem 6.41 is also valid.

Let $\Delta > 1$ and $\varepsilon \in (0,1)$ be given. Choose an integer $q \geq 1$ such that

$$2^q > \max\left\{\Delta, \varepsilon^{-1}\right\}. \tag{6.257}$$

There are $\{A_t\}_{t=1}^{\infty} \in \mathcal{M}_{reg}$, $\gamma \in (0,1)$ and an integer $i \geq q$ such that

$$\{B_t\}_{t=1}^{\infty} \in U(\mathbf{A},\gamma,i). \tag{6.258}$$

Assume that $\{C_t\}_{t=1}^{\infty} \in U(\mathbf{A},\gamma,i)$ and

$$x \in \langle 0, \Delta\eta\rangle. \tag{6.259}$$

Then property (a) holds. It follows from property (b), (6.259) and (6.257) that relation (6.256) is valid. Since $8^{-i} < \varepsilon$, this completes the proof of the theorem. □

Proof of Theorem 6.42 Let $\mathcal{F}$ be defined as in the proof of Theorem 6.41. By (6.226), Lemmas 6.49 and 6.50,

$$\left\{\{A_t^{\gamma}\}_{t=1}^{\infty} : \{A_t\}_{t=1}^{\infty} \in \mathcal{M}(c_1,c_2), \gamma \in (0,1)\right\}$$

is an everywhere dense subset of $\mathcal{M}(c_1, c_2)$ with the strong topology. Define

$$\mathcal{F}_0 := \bigcap_{q=1}^{\infty} \bigcup \{U(\mathbf{A}, \gamma, i) \cap \mathcal{M}(c_1, c_2) :$$
$$\mathbf{A} \in \mathcal{M}(c_1, c_2), \gamma \in (0,1), i = q, q+1, \dots\}.$$

Clearly, $\mathcal{F}_0$ is a countable intersection of open (in the weak topology) everywhere dense (in the strong topology) sets in $\mathcal{M}(c_1, c_2)$. Since $\mathcal{F}_0 \subset \mathcal{F}$ we conclude that assertion 2 of Theorem 6.41 is valid for each $\{B_t\}_{t=1}^{\infty} \in \mathcal{F}_0$. The proof is complete. □

6.23 Proof of Theorem 6.43

Assume that $\mathbf{A} = \{A_t\}_{t=1}^{\infty} \in \mathcal{M}_*$. There exists an interior point $\xi_{\mathbf{A}}$ of X_+ such that

$$A_t \xi_{\mathbf{A}} = \xi_{\mathbf{A}}, \quad t = 1, 2, \dots, \quad \text{and} \quad \|\xi_{\mathbf{A}}\|_\eta = 1. \tag{6.260}$$

For each $\gamma \in (0,1)$, define $\{A_t^\gamma\}_{t=1}^{\infty}$ by (6.225) with $\eta = \xi_{\mathbf{A}}$. The sequence $\{A_t^\gamma\}_{t=1}^{\infty} \in \mathcal{M}$ by (6.226). Lemma 6.55 implies that

$$A_t^\gamma x = (1-\gamma) A_t x + \gamma r(\xi_{\mathbf{A}}, x) \xi_{\mathbf{A}}, \quad x \in X_+, t = 1, 2, \dots. \tag{6.261}$$

Together with (6.260) this implies that

$$A_t^\gamma \xi_{\mathbf{A}} = \xi_{\mathbf{A}}, \quad t = 1, 2, \dots. \tag{6.262}$$

By (6.262) and Lemma 6.30, the set

$$\{\{A_t^\gamma\}_{t=1}^{\infty} : \{A_t\}_{t=1}^{\infty} \in \mathcal{M}_*, \gamma \in (0,1)\}$$

is everywhere dense in $\bar{\mathcal{M}}_*$.

Let $\mathbf{A} = \{A_t\}_{t=1}^{\infty} \in \mathcal{M}_*$, $\gamma \in (0,1)$ and let $i \geq 1$ be an integer. Choose a natural number $N(\mathbf{A}, \gamma, i) \geq 4$ for which

$$2 \cdot 16^i \lambda(\eta, \xi_{\mathbf{A}})^{-1} (1-\gamma)^{N(\mathbf{A},\gamma,i)}$$
$$< 64^{-1} \cdot 16^{-i} \left(\lambda(\eta, \xi_{\mathbf{A}})^{-1} + 1\right)^{-4} \lambda(\eta, \xi_{\mathbf{A}}). \tag{6.263}$$

Fix a number $\Gamma(\mathbf{A}, \gamma, i)$ such that

$$\Gamma(\mathbf{A}, \gamma, i) > 1 \quad \text{and}$$
$$2\left(\Gamma(\mathbf{A}, \gamma, i) - \Gamma(\mathbf{A}, \gamma, i)^{-1}\right) 16^i \lambda(\eta, \xi_{\mathbf{A}})^{-1} \tag{6.264}$$
$$< 64^{-1} \cdot 16^{-i} \left(\lambda(\eta, \xi_{\mathbf{A}})^{-1} + 1\right)^{-4} \lambda(\eta, \xi_{\mathbf{A}}).$$

By Lemma 6.58, there exists a number $\delta(\mathbf{A}, \gamma, i) \in (0, 8^{-i})$ such that for each $s : \{1, 2, \ldots\} \to \{1, 2, \ldots\}$, each sequence $\{B_i\}_{i=1}^{N(\mathbf{A},\gamma,i)} \subset \mathcal{A}$ satisfying

$$\rho(B_i, A^{\gamma}_{s(i)}) \le \delta(\mathbf{A}, \gamma, i), \quad i = 1, \ldots, N(\mathbf{A}, \gamma, i), \tag{6.265}$$

and each $z \in \langle 0, 16^i \lambda(\eta, \xi_{\mathbf{A}})^{-1}\eta\rangle$ satisfying $r(\xi_{\mathbf{A}}, z) \ge 16^{-i}\lambda(\eta, \xi_{\mathbf{A}})$, the following relations hold:

$$\begin{aligned} &B_{N(\mathbf{A},\gamma,i)} \cdot \cdots \cdot B_1 z \\ &\quad \in \big\langle \Gamma^{-1}(\mathbf{A}, \gamma, i) A^{\gamma}_{s(N(\mathbf{A},\gamma,i))} \cdot \cdots \cdot A^{\gamma}_{s(1)} z, \\ &\quad \Gamma(\mathbf{A}, \gamma, i) A^{\gamma}_{s(N(\mathbf{A},\gamma,i))} \cdot \cdots \cdot A^{\gamma}_{s(1)} z \big\rangle. \end{aligned} \tag{6.266}$$

Set

$$U(\mathbf{A}, \gamma, i) = \big\{\{B_t\}_{t=1}^{\infty} \in \bar{\mathcal{M}}_* : \rho_s\big(\{A^{\gamma}_t\}_{t=1}^{\infty}, \{B_t\}_{t=1}^{\infty}\big) < \delta(\mathbf{A}, \gamma, i)\big\}. \tag{6.267}$$

Define

$$\mathcal{F} := \bigcap_{q=1}^{\infty} \bigcup \{U(\mathbf{A}, \gamma, i) : \mathbf{A} \in \mathcal{M}_*, \gamma \in (0, 1), i = q, q+1, \ldots\}. \tag{6.268}$$

It is clear that $\mathcal{F}$ is a countable intersection of open everywhere dense subsets in $\bar{\mathcal{M}}_*$.

Lemma 6.59 *Let* $\{A_t\}_{t=1}^{\infty} \in \mathcal{M}_*$, $\gamma \in (0, 1)$, $s : \{1, 2, \ldots\} \to \{1, 2, \ldots\}$, *and let* $i \ge 1$ *be an integer. Define*

$$f_s(x) := r(\xi_{\mathbf{A}}, A_{s(N(\mathbf{A},\gamma,i))} \cdot \cdots \cdot A_{s(1)} x), \quad x \in X_+. \tag{6.269}$$

Then for each $\{C_t\}_{t=1}^{\infty} \in U(\{A_t\}_{t=1}^{\infty}, \gamma, i)$ *and each*

$$x \in \big\langle 0, 16^i \lambda(\eta, \xi_{\mathbf{A}})^{-1} \xi_{\mathbf{A}} \big\rangle$$

satisfying

$$r(\xi_{\mathbf{A}}, x) \ge 16^{-i} \lambda(\eta, \xi_{\mathbf{A}}),$$

the following inequality holds:

$$\begin{aligned} &\big\| C_{s(N(\mathbf{A},\gamma,i))} \cdot \cdots \cdot C_{s(1)} x - f_s(x) \xi_{\mathbf{A}} \big\|_{\eta} \\ &\quad \le \lambda(\eta, \xi_{\mathbf{A}}) 32^{-1} 16^{-i} \big(\lambda(\eta, \xi_{\mathbf{A}})^{-1} + 1\big)^{-4}. \end{aligned} \tag{6.270}$$

Proof Set

$$N = N(\mathbf{A}, \gamma, i), \qquad \delta = \delta(\mathbf{A}, \gamma, i) \quad \text{and} \quad \Gamma = \Gamma(\mathbf{A}, \gamma, i). \tag{6.271}$$

It follows from (6.271), (6.260), the definition of $N(\mathbf{A}, \gamma, i)$ (see (6.263)), and Lemma 6.57 that for each $x \in \langle 0, 16^i \lambda(\eta, \xi_{\mathbf{A}})^{-1} \xi_{\mathbf{A}} \rangle$, the following inequality holds:

$$r\bigl(\xi_{\mathbf{A}}, A^{\gamma}_{s(N)} \cdot \dots \cdot A^{\gamma}_{s(1)} x\bigr) - \lambda\bigl(\xi_{\mathbf{A}}, A^{\gamma}_{s(N)} \cdot \dots \cdot A^{\gamma}_{s(1)} x\bigr)$$
$$\le \lambda(\eta, \xi_{\mathbf{A}}) 64^{-1} 16^{-i} \bigl(\lambda(\eta, \xi_{\mathbf{A}})^{-1} + 1\bigr)^{-4}. \tag{6.272}$$

Assume that $\{C_t\}_{t=1}^{\infty} \in U(\{A_t\}_{t=1}^{\infty}, \gamma, i)$ and that

$$x \in \bigl\langle 0, 16^i \lambda(\eta, \xi_{\mathbf{A}})^{-1} \xi_{\mathbf{A}} \bigr\rangle, \quad r(\xi_{\mathbf{A}}, x) \ge 16^{-i} \lambda(\eta, \xi_{\mathbf{A}}). \tag{6.273}$$

We will show that (6.270) holds. Clearly (6.272) holds. By (6.269), (6.272), (6.271) and (6.260),

$$\bigl\| A^{\gamma}_{s(N)} \cdot \dots \cdot A^{\gamma}_{s(1)}(x) - f_s(x) \xi_{\mathbf{A}} \bigr\|_{\eta}$$
$$\le r\bigl(\xi_{\mathbf{A}}, A^{\gamma}_{s(N)} \cdot \dots \cdot A^{\gamma}_{s(1)} x\bigr) - \lambda\bigl(\xi_{\mathbf{A}}, A^{\gamma}_{s(N)} \cdot \dots \cdot A^{\gamma}_{s(1)} x\bigr)$$
$$\le 64^{-1} 16^{-i} \bigl(\lambda(\eta, \xi_{\mathbf{A}})^{-1} + 1\bigr)^{-4} \lambda(\eta, \xi_{\mathbf{A}}). \tag{6.274}$$

We now estimate

$$\bigl\| C_{s(N)} \cdot \dots \cdot C_{s(1)} x - A^{\gamma}_{s(N)} \cdot \dots \cdot A^{\gamma}_{s(1)} x \bigr\|_{\eta}.$$

It follows from (6.267), the definition of $\delta(\mathbf{A}, \gamma, i)$ (see (6.265), (6.266)), (6.271), (6.264), (6.273) and (6.262) that

$$C_{s(N)} \cdot \dots \cdot C_{s(1)} x \in \bigl\langle \Gamma^{-1} A^{\gamma}_{s(N)} \cdot \dots \cdot A^{\gamma}_{s(1)} x, \Gamma A^{\gamma}_{s(N)} \cdot \dots \cdot A^{\gamma}_{s(1)} x \bigr\rangle,$$
$$\bigl\| C_{s(N)} \cdot \dots \cdot C_{s(1)} x - A^{\gamma}_{s(N)} \cdot \dots \cdot A^{\gamma}_{s(1)} x \bigr\|_{\eta}$$
$$\le 2\bigl(\Gamma - \Gamma^{-1}\bigr) \bigl\| A^{\gamma}_{s(N)} \cdot \dots \cdot A^{\gamma}_{s(1)} x \bigr\|_{\eta}$$
$$\le 2\bigl(\Gamma - \Gamma^{-1}\bigr) 16^i \lambda(\eta, \xi_{\mathbf{A}})^{-1} < 64^{-1} \cdot 16^{-i} \bigl(\lambda(\eta, \xi_{\mathbf{A}})^{-1} + 1\bigr)^{-4} \lambda(\eta, \xi_{\mathbf{A}}).$$

Together with (6.274) this implies inequality (6.270). The proof of the lemma is complete. □

Lemma 6.60 *Let $\{A_t\}_{t=1}^{\infty} \in \mathcal{M}_*$, $\gamma \in (0, 1)$, $s : \{1, 2, \dots\} \to \{1, 2, \dots\}$ and let $i \ge 1$ be an integer. Let $f_s : X_+ \to R^1$ be defined by* (6.269).

Assume that

$$\{C_t\}_{t=1}^{\infty} \in U\bigl(\{A_t\}_{t=1}^{\infty}, \gamma, i\bigr), \qquad y \in X_+,$$
$$\|y\|_{\eta} = 1, \qquad C_t y = y, \quad t = 1, 2, \dots. \tag{6.275}$$

Then

$$\|y - \xi_{\mathbf{A}}\|_{\eta} \le 16^{-i-1}, \qquad y \text{ is an interior point of } X_+,$$
$$y \ge 2^{-1} \lambda(\eta, \xi_{\mathbf{A}}) \eta \tag{6.276}$$

and for each $x \in \langle 0, \eta \rangle$ *and each integer* $T \geq N(\mathbf{A}, \gamma, i)$,

$$\left\| C_{s(T)} \cdot \cdots \cdot C_{s(1)} x - f_s(x) y \right\|_\eta \leq 4 \cdot 16^{-i-1} \tag{6.277}$$

and

$$\left\| C_{s(T)} \cdot \cdots \cdot C_{s(1)} x - f_s(x) \xi_{\mathbf{A}} \right\|_\eta \leq 5 \cdot 16^{-i-1}. \tag{6.278}$$

Proof It follows from (6.275) that

$$y \leq \eta \leq \lambda(\eta, \xi_{\mathbf{A}})^{-1} \xi_{\mathbf{A}}, \qquad r(\xi_{\mathbf{A}}, y) \geq 1 \geq \lambda(\eta, \xi_{\mathbf{A}}). \tag{6.279}$$

By Lemma 6.59, for each $x \in X_+$ which satisfies

$$x \in \left\langle 0, 16^i \lambda(\eta, \xi_{\mathbf{A}})^{-1} \xi_{\mathbf{A}} \right\rangle \quad \text{and} \quad r(\xi_{\mathbf{A}}, x) \geq 16^{-i} \lambda(\eta, \xi_{\mathbf{A}}) \tag{6.280}$$

relation (6.270) holds. Together with (6.279), (6.275) and (6.260) this implies that

$$\begin{gathered} \left\| y - f_s(y) \xi_{\mathbf{A}} \right\|_\eta \leq 32^{-1} \lambda(\eta, \xi_{\mathbf{A}}) 16^{-i} \left(\lambda(\eta, \xi_{\mathbf{A}})^{-1} + 1 \right)^{-4}, \\ \left| f_s(y) - 1 \right| \leq \left\| f_s(y) \xi_{\mathbf{A}} - y \right\|_\eta, \end{gathered} \tag{6.281}$$

$$\|y - \xi_{\mathbf{A}}\|_\eta \leq \left\| y - f_s(y) \xi_{\mathbf{A}} \right\|_\eta + \left| f_s(y) - 1 \right| \leq 2 \left\| f_s(y) \xi_{\mathbf{A}} - y \right\|_\eta,$$

$$y - \xi_{\mathbf{A}}, \xi_{\mathbf{A}} - y \leq 16^{-i-1} \lambda(\eta, \xi_{\mathbf{A}}) \left(\lambda(\eta, \xi_{\mathbf{A}})^{-1} + 1 \right)^{-4} \eta \tag{6.282}$$

and

$$y \geq \lambda(\eta, \xi_{\mathbf{A}}) \left(1 - 16^{-i-1} \right) \left(\lambda(\eta, \xi_{\mathbf{A}})^{-1} + 1 \right)^{-4} \eta. \tag{6.283}$$

It follows from the definition of f_s (see (6.269)), (6.270) and (6.282) that for each $x \in \langle 0, \eta \rangle$ and each integer $T \geq N(\mathbf{A}, \gamma, i)$,

$$f_s(x) \leq f_s(\eta) \leq f_s\left(\lambda(\eta, \xi_{\mathbf{A}})^{-1} \xi_{\mathbf{A}} \right) = \lambda(\eta, \xi_{\mathbf{A}})^{-1} \tag{6.284}$$

and

$$\begin{aligned} & \left\| C_{s(N(\mathbf{A},\gamma,i))} \cdot \cdots \cdot C_{s(1)} x - f_s(x) y \right\|_\eta \\ & \leq \left\| C_{s(N(\mathbf{A},\gamma,i))} \cdot \cdots \cdot C_{s(1)} x - f_s(x) \xi_{\mathbf{A}} \right\|_\eta + f_s(x) \|\xi_{\mathbf{A}} - y\|_\eta \\ & \leq 32^{-1} \lambda(\eta, \xi_{\mathbf{A}}) 16^{-i} \left(\lambda(\eta, \xi_{\mathbf{A}})^{-1} + 1 \right)^{-4} + 16^{-i-1} \left(\lambda(\eta, \xi_{\mathbf{A}})^{-1} + 1 \right)^{-4} \\ & \leq 2 \cdot 16^{-i-1} \left(\lambda(\eta, \xi_{\mathbf{A}})^{-1} + 1 \right)^{-3}. \end{aligned} \tag{6.285}$$

We also have by (6.275) and (6.283),

$$\begin{aligned} & C_{s(T)} \cdot \cdots \cdot C_{s(1)} x - f_s(x) y \\ & = C_{s(T)} \cdot \cdots \cdot C_{s(N(\mathbf{A},\gamma,i)+1)} \left(C_{s(N(\mathbf{A},\gamma,i))} \cdot \cdots \cdot C_{s(1)} x - f_s(x) y \right) \end{aligned}$$

$$\begin{aligned}
&\in C_{s(T)} \cdot \cdots \cdot C_{s(N(\mathbf{A},\gamma,i)+1)}\big(2 \cdot 16^{-i-1}\big(\lambda(\eta, \xi_{\mathbf{A}})^{-1}+1\big)^{-3}\langle -\eta, \eta\rangle\big) \\
&\subset 2 \cdot 16^{-i-1}\big(\lambda(\eta, \xi_{\mathbf{A}})^{-1}+1\big)^{-3} \cdot 2\lambda(\eta, \xi_{\mathbf{A}})^{-1}\langle -y, y\rangle \\
&\subset 4 \cdot 16^{-i-1}\big[\lambda(\eta, \xi_{\mathbf{A}})^{-1}+1\big]^{-2}\langle -\eta, \eta\rangle.
\end{aligned}$$

Hence

$$\big\|C_{s(T)} \cdot \cdots \cdot C_{s(1)}x - f_s(x)y\big\|_\eta \le 4 \cdot 16^{-i-1}\big[\lambda(\eta, \xi_{\mathbf{A}})^{-1}+1\big]^{-2}. \tag{6.286}$$

Let $x \in \langle 0, \eta\rangle$ and let $T \ge N(\mathbf{A}, \gamma, i)$ be an integer. By (6.284), (6.285), (6.286) and (6.282),

$$\begin{aligned}
&\big\|C_{s(T)} \cdot \cdots \cdot C_{s(1)}x - f_s(x)\xi_{\mathbf{A}}\big\|_\eta \\
&\quad\le \big\|C_{s(T)} \cdot \cdots \cdot C_{s(1)}x - f_s(x)y\big\|_\eta + f_s(x)\|y - \xi_{\mathbf{A}}\|_\eta \\
&\quad\le 4 \cdot 16^{-i-1}\big[\lambda(\eta, \xi_{\mathbf{A}})^{-1}+1\big]^{-2} + 16^{-i-1}\big[\lambda(\eta, \xi_{\mathbf{A}})^{-1}+1\big]^{-4} \\
&\quad\le 5 \cdot 16^{-i-1}\big[\lambda(\eta, \xi_{\mathbf{A}})^{-1}+1\big]^{-2}.
\end{aligned}$$

The lemma is proved. □

Assume that $\{B_t\}_{t=1}^\infty \in \mathcal{F}$. There exist $\mathbf{A}^{(k)} = \{A_t^{(k)}\}_{t=1}^\infty \in \mathcal{M}_*$, $k = 1, 2, \ldots$, $\{\gamma_k\}_{k=1}^\infty \subset (0, 1)$, and a strictly increasing sequence of natural numbers $\{i_k\}_{k=1}^\infty$ such that for all integers $k \ge 1$,

$$\begin{aligned}
&\{B_t\}_{t=1}^\infty \in U\big(\mathbf{A}^{(k)}, \gamma_k, i_k\big) \quad \text{and} \\
&U\big(\mathbf{A}^{(k+1)}, \gamma_{k+1}, i_{k+1}\big) \subset U\big(\mathbf{A}^{(k)}, \gamma_k, i_k\big).
\end{aligned} \tag{6.287}$$

By Lemma 6.60, $\{\xi_{\mathbf{A}^{(k)}}\}_{k=1}^\infty$ is a Cauchy sequence and there exists

$$\xi_{\mathbf{B}} = \lim_{k\to\infty} \xi_{\mathbf{A}^{(k)}}, \quad \text{where } \xi_{\mathbf{B}} \in \operatorname{int}(X_+). \tag{6.288}$$

It follows from (6.288) and (6.287) that for $t = 1, 2, \ldots$,

$$A_t^{(k)}\xi_{\mathbf{A}^{(k)}} - B_t\xi_{\mathbf{B}} = \big(A_t^{(k)} - B_t\big)(\xi_{\mathbf{A}^{(k)}}) + B_t(\xi_{\mathbf{A}^{(k)}} - \xi_{\mathbf{B}}) \to 0 \tag{6.289}$$

as $k \to \infty$. Together with (6.260) and (6.288) this implies that

$$B_t\xi_{\mathbf{B}} = \xi_{\mathbf{B}}, \quad t = 1, 2, \ldots, \qquad \|\xi_{\mathbf{B}}\|_\eta = 1. \tag{6.290}$$

Let $\varepsilon > 0$ be given. There is an integer $k \ge 1$ such that

$$2^{-i_k} < 64^{-1}\varepsilon\big(4 + \lambda(\eta, \xi_{\mathbf{B}})^{-1}\big)^{-1} \cdot 4^{-1}. \tag{6.291}$$

Assume that $s : \{1, 2, \dots\} \to \{1, 2, \dots\}$. Define $f_s : X_+ \to R^1$ by

$$f_s(x) = r\big(\xi_{\mathbf{A}^{(k)}}, A^{(k)}_{s(N(\mathbf{A}^{(k)}, \gamma_k, i_k))} \cdot \dots \cdot A^{(k)}_{s(1)} x\big), \quad x \in X_+. \tag{6.292}$$

By Lemma 6.60 the following property holds:

(a) Assume that $\{C_t\}_{t=1}^{\infty} \in U(\{A_t^{(k)}\}_{t=1}^{\infty}, \gamma_k, i_k)$, $y \in X_+$, $\|y\|_\eta = 1$ and $C_t y = y$, $t = 1, 2, \dots$. Then y is an interior point of X_+, $\|y - \xi_{\mathbf{A}^{(k)}}\|_\eta \le 16^{-1-i_k}$, and for each $x \in \langle 0, \eta \rangle$ and each integer $T \ge N(\mathbf{A}^{(k)}, \gamma_k, i_k)$,

$$\big\| C_{s(T)} \cdot \dots \cdot C_{s(1)} x - f_s(x) y \big\|_\eta \le 4 \cdot 16^{-i_k},$$

and

$$\big\| C_{s(T)} \cdot \dots \cdot C_{s(1)} x - f_s(x) \xi_{\mathbf{A}^{(k)}} \big\|_\eta \le 5 \cdot 16^{-i_k}.$$

It follows from property (a), (6.287), (6.290) and (6.291) that

$$\|\xi_{\mathbf{B}} - \xi_{\mathbf{A}^{(k)}}\| = 16^{-i_k - 1}, \tag{6.293}$$

and for each $x \in \langle 0, \eta \rangle$ and each integer $T \ge N(\mathbf{A}^{(k)}, \gamma_k, i_k)$,

$$\big\| B_{s(T)} \cdot \dots \cdot B_{s(1)} x - f_s(x) \xi_{\mathbf{B}} \big\|_\eta \le 4 \cdot 16^{-i_k} < \varepsilon \cdot 64^{-1}. \tag{6.294}$$

Since ε is any positive number, we conclude that there exists $g_s : X_+ \to R^1$ such that

$$\lim_{T \to \infty} B_{s(T)} \cdot \dots \cdot B_{s(1)} x = g_s(x) \xi_{\mathbf{B}}, \quad x \in X_+. \tag{6.295}$$

By (6.294) and (6.295),

$$\big| g_s(x) - f_s(x) \big| \le 4 \cdot 16^{-i_k}, \quad x \in \langle 0, \eta \rangle. \tag{6.296}$$

Assume that $\{C_t\}_{t=1}^{\infty} \in U(\{A_t^{(k)}\}_{t=1}^{\infty}, \gamma_k, i_k) \cap \mathcal{M}_*$, $T \ge N(\mathbf{A}^{(k)}, \gamma_k, i_k)$ is a natural number, and $x \in \langle 0, \eta \rangle$. To complete the proof of the theorem it is sufficient to show that

$$\big\| C_{s(T)} \cdot \dots \cdot C_{s(1)} x - g_s(x) \xi_{\mathbf{B}} \big\|_\eta \le 4^{-1} \varepsilon. \tag{6.297}$$

Indeed it follows from property (a), (6.296), (6.293), (6.295), (6.290) and (6.291) that

$$\begin{aligned}
&\big\| C_{s(T)} \cdot \dots \cdot C_{s(1)} x - g_s(x) \xi_{\mathbf{B}} \big\|_\eta \\
&\le \big\| C_{s(T)} \cdot \dots \cdot C_{s(1)} x - f_s(x) \xi_{\mathbf{A}^{(k)}} \big\|_\eta \\
&\quad + \big\| f_s(x) \xi_{\mathbf{A}^{(k)}} - g_s(x) \xi_{\mathbf{A}^{(k)}} \big\|_\eta + g_s(x) \|\xi_{\mathbf{A}^{(k)}} - \xi_{\mathbf{B}}\|_\eta \\
&\le 5 \cdot 16^{-i_k} + 4 \cdot 16^{-i_k} + 16^{-i_k - 1} g_s(\eta) \\
&\le 9 \cdot 16^{-i_k} + 16^{-i_k - 1} \lambda(\eta, \xi_{\mathbf{B}})^{-1} < 4^{-1} \varepsilon.
\end{aligned}$$

This completes the proof of Theorem 6.43.

6.24 Infinite Products of Affine Operators

In this section we study the asymptotic behavior of random infinite products of generic sequences of affine uniformly continuous operators on bounded, closed and convex subsets of a Banach space. More precisely, we show that in appropriate spaces of sequences of operators there exists a subset which is a countable intersection of open and everywhere dense sets such that for each sequence belonging to this subset, the corresponding random infinite products converge. We remark in passing that common fixed point theorems for families of affine mappings (e.g., those of Markov-Kakutani and Ryll-Nardzewski) have applications in various mathematical areas. See, for example, [48] and the references mentioned there.

Let $(X, \|\cdot\|)$ be a Banach space and let K be a nonempty, bounded, closed and convex subset of X with the topology induced by the norm $\|\cdot\|$.

Denote by $\mathcal{A}$ the set of all sequences $\{A_t\}_{t=1}^{\infty}$, where each $A_t : K \to K$ is a continuous operator, $t = 1, 2, \ldots$. Such a sequence will occasionally be denoted by a boldface $\mathbf{A}$.

We equip the set $\mathcal{A}$ with the metric $\rho_s : \mathcal{A} \times \mathcal{A} \to [0, \infty)$ defined by

$$\begin{aligned} &\rho_s\big(\{A_t\}_{t=1}^{\infty}, \{B_t\}_{t=1}^{\infty}\big) = \sup\big\{\|A_t x - B_t x\| : x \in K, t = 1, 2, \ldots\big\}, \\ &\{A_t\}_{t=1}^{\infty}, \{B_t\}_{t=1}^{\infty} \in \mathcal{A}. \end{aligned} \tag{6.298}$$

It is easy to see that the metric space $(\mathcal{A}, \rho_s)$ is complete. We will always consider the set $\mathcal{A}$ with the topology generated by the metric ρ_s.

We say that a set E of operators $A : K \to K$ is uniformly equicontinuous (ue) if for any $\varepsilon > 0$, there exists $\delta > 0$ such that $\|Ax - Ay\| \le \varepsilon$ for all $A \in E$ and all $x, y \in K$ satisfying $\|x - y\| \le \delta$.

An operator $A : K \to K$ is called uniformly continuous if the singleton $\{A\}$ is a (ue) set.

Define

$$\mathcal{A}_{ue} := \big\{\{A_t\}_{t=1}^{\infty} \in \mathcal{A} : \{A_t\}_{t=1}^{\infty} \text{ is a (ue) set}\big\}. \tag{6.299}$$

It is clear that $\mathcal{A}_{ue}$ is a closed subset of $\mathcal{A}$.

We endow the topological subspace $\mathcal{A}_{ue} \subset \mathcal{A}$ with the relative topology.

We say that an operator $A : K \to K$ is affine if

$$A\big(\alpha x + (1 - \alpha)y\big) = \alpha Ax + (1 - \alpha)Ay$$

for each $x, y \in K$ and all $\alpha \in [0, 1]$.

Denote by $\mathcal{M}$ the set of all uniformly continuous affine mappings $A : K \to K$. For the space $\mathcal{M}$ we consider the metric

$$\rho(A, B) = \sup\big\{\|Ax - Bx\| : x \in K\big\}, \quad A, B \in \mathcal{M}.$$

It is easy to see that the metric space $(\mathcal{M}, \rho)$ is complete.

In the next sections we analyze the convergence of infinite products of operators in $\mathcal{M}$ and other mappings of affine type and prove several convergence results which were obtained in [126].

We begin by showing (Theorem 6.61) that for a generic operator B in the space $\mathcal{M}$ there exists a unique fixed point x_B and the powers of B converge to x_B for all $x \in K$. We continue with a study of the asymptotic behavior of infinite products of this kind of operators and prove a weak ergodic theorem. Finally, we present several theorems on the generic convergence of infinite product trajectories to a common fixed point and to a common fixed point set, respectively.

Denote by $\mathcal{A}_{ue}^{af}$ the set of all $\{A_t\}_{t=1}^{\infty} \in \mathcal{A}_{ue}$ such that for each integer $t \geq 1$, each $x, y \in K$ and all $\alpha \in [0, 1]$,

$$A_t\big(\alpha x + (1-\alpha)y\big) = \alpha A_t x + (1-\alpha)A_t y.$$

Clearly, $\mathcal{A}_{ue}^{af}$ is a closed subset of $\mathcal{A}_{ue}$. We consider the topological subspace $\mathcal{A}_{ue}^{af} \subset \mathcal{A}_{ue}$ with the relative topology.

We will show (Theorem 6.63) that for a generic sequence $\{C_t\}_{t=1}^{\infty}$ in the space $\mathcal{A}_{ue}^{af}$,

$$\|C_{r(T)} \cdot \cdots \cdot C_{r(1)}x - C_{r(T)} \cdot \cdots \cdot C_{r(1)}y\| \to 0,$$

uniformly for all $x, y \in K$ and all mappings $r : \{1, 2, \ldots\} \to \{1, 2, \ldots\}$. Such results are usually called weak ergodic theorems in the population biology literature [43].

Denote by $\mathcal{A}_{ue}^{0}$ the set of all $\mathbf{A} = \{A_t\}_{t=1}^{\infty} \in \mathcal{A}_{ue}$ for which there exists $x_{\mathbf{A}} \in K$ such that

$$A_t x_{\mathbf{A}} = x_{\mathbf{A}}, \quad t = 1, 2, \ldots, \tag{6.300}$$

and for each $\gamma \in (0, 1)$, $x \in K$ and each integer $t \geq 1$,

$$A_t\big(\gamma x_{\mathbf{A}} + (1-\gamma)x\big) = \lambda_t(\gamma, x)x_{\mathbf{A}} + \big(1 - \lambda_t(\gamma, x)\big)A_t x \tag{6.301}$$

with some constant $\lambda_t(\gamma, x) \in [\gamma, 1]$.

Denote by $\bar{\mathcal{A}}_{ue}^{0}$ the closure of $\mathcal{A}_{ue}^{0}$ in the space $\mathcal{A}_{ue}$. We will consider the topological subspace $\bar{\mathcal{A}}_{ue}^{0}$ with the relative topology and show (Theorem 6.64) that for a generic sequence $\{C_t\}_{t=1}^{\infty}$ in the space $\bar{\mathcal{A}}_{ue}^{0}$, there exists a unique common fixed point x_* and all random products of the operators $\{C_t\}_{t=1}^{\infty}$ converge to x_*, uniformly for all $x \in K$. We will also show that this convergence of random infinite products to a unique common fixed point holds for a generic sequence from certain subspaces of the space $\bar{\mathcal{A}}_{ue}^{0}$.

Assume now that $F \subset K$ is a nonempty, closed and convex set, $Q : K \to F$ is a uniformly continuous operator such that

$$Qx = x, \quad x \in F, \tag{6.302}$$

and for each $y \in K$, $x \in F$ and $\alpha \in [0, 1]$,

$$Q\big(\alpha x + (1-\alpha)y\big) = \alpha x + (1-\alpha)Qy. \tag{6.303}$$

Denote by $\mathcal{A}_{ue}^{(F,0)}$ the set of all $\{A_t\}_{t=1}^{\infty} \in \mathcal{A}_{ue}$ such that

$$A_t x = x, \quad t = 1, 2, \ldots, x \in F,$$

and for each integer $t \geq 1$, each $y \in K$, $x \in F$ and $\alpha \in (0, 1]$,

$$A_t\big(\alpha x + (1-\alpha)y\big) = \alpha x + (1-\alpha)A_t y.$$

Clearly, $\mathcal{A}_{ue}^{(F,0)}$ is a closed subset of $\mathcal{A}_{ue}$.

The topological subspace $\mathcal{A}_{ue}^{(F,0)} \subset \mathcal{A}_{ue}$ will be equipped with the relative topology.

We will show (Theorem 6.67) that for a generic sequence of operators $\{C_t\}_{t=1}^{\infty}$ in the space $\mathcal{A}_{ue}^{(F,0)}$, all its random infinite products

$$C_{r(t)} \cdot \cdots \cdot C_{r(1)} x$$

tend to the set F, uniformly for all $x \in K$. Moreover, under a certain additional assumption on F, these random products converge to a uniformly continuous retraction $P_r : K \to F$, uniformly for all $x \in K$ (Theorem 6.69).

For each bounded operator $A : K \to X$, we set

$$\|A\| = \sup\big\{\|Ax\| : x \in K\big\}. \tag{6.304}$$

For each $x \in K$ and each $E \subset X$, we set

$$d(x, E) = \inf\big\{\|x - y\| : y \in E\big\}, \qquad \text{rad}(E) = \sup\big\{\|y\| : y \in E\big\}. \tag{6.305}$$

6.25 A Generic Fixed Point Theorem for Affine Mappings

This section is devoted to the proof of the following result.

Theorem 6.61 *There exists a set $\mathcal{F} \subset \mathcal{M}$, which is a countable intersection of open and everywhere dense subsets of $\mathcal{M}$, such that for each $A \in \mathcal{F}$, the following two assertions hold*:

1. *There exists a unique $x_A \in K$ such that $Ax_A = x_A$*;

2. *For each $\varepsilon > 0$, there exist a neighborhood U of A in $\mathcal{M}$ and a natural number N such that for each $\{B_t\}_{t=1}^{\infty} \subset U$ and each $x \in K$,*

$$\|B_T \cdot \cdots \cdot B_1 x - x_A\| \leq \varepsilon \quad \textit{for all integers } T \geq N.$$

In the proof of Theorem 6.61 we will need the following lemma.

Lemma 6.62 *Let $B \in \mathcal{M}$ and $\varepsilon \in (0, 1)$ be given. Then there exist $B_\varepsilon \in \mathcal{M}$, an integer $q \geq 1$ and $y_\varepsilon \in K$ such that*

$$\rho(B, B_\varepsilon) \leq \varepsilon, \qquad \big\|B_\varepsilon^t y_\varepsilon - y_\varepsilon\big\| \leq \varepsilon, \quad t = 1, \ldots, q,$$

and for each $z \in K$ the following inequality holds:

$$\| B_\varepsilon^q z - y_\varepsilon \| \le \varepsilon.$$

Proof Choose a number $\gamma \in (0, 1)$ for which

$$8\gamma\bigl(\mathrm{rad}(K) + 1\bigr) \le \varepsilon, \tag{6.306}$$

and then an integer $q \ge 1$ such that

$$(1-\gamma)^q\bigl(\mathrm{rad}(K) + 1\bigr) \le 16^{-1}\varepsilon \tag{6.307}$$

and a natural number N such that

$$16qN^{-1}\bigl(\mathrm{rad}(K) + 1\bigr) \le 8^{-1}\varepsilon. \tag{6.308}$$

Fix $x_0 \in K$ and define a sequence $\{x_t\}_{t=0}^{\infty} \subset K$ by

$$x_{t+1} = Bx_t, \quad t = 0, 1, \ldots. \tag{6.309}$$

For each integer $k \ge 0$, define

$$y_k = N^{-1} \sum_{i=k}^{k+N-1} x_i. \tag{6.310}$$

It is easy to see that

$$By_k = y_{k+1}, \quad k = 0, 1, \ldots, \tag{6.311}$$

and for each $k \in \{0, \ldots, q\}$,

$$\|y_0 - y_k\| \le 2kN^{-1}\,\mathrm{rad}(K) \le 2qN^{-1}\,\mathrm{rad}(K). \tag{6.312}$$

Define $B_\varepsilon : K \to K$ by

$$B_\varepsilon z := (1-\gamma)Bz + \gamma y_0, \quad z \in K. \tag{6.313}$$

It is easy to see that

$$B_\varepsilon \in \mathcal{M} \quad \text{and} \quad \rho(B, B_\varepsilon) < 2^{-1}\varepsilon. \tag{6.314}$$

Now let z be an arbitrary point in K. We will show by induction that for each integer $n \ge 1$,

$$B_\varepsilon^n z = (1-\gamma)^n B^n z + \sum_{i=0}^{n-1} c_{ni} y_i, \tag{6.315}$$

where

$$c_{ni} > 0, \quad i = 0, \ldots, n-1, \quad \text{and} \quad \sum_{i=0}^{n-1} c_{ni} + (1-\gamma)^n = 1. \tag{6.316}$$

It is easy to see that for $n = 1$ our assertion holds.

Assume that it is also valid for an integer $n \geq 1$. It follows from (6.313), (6.315), (6.316), (6.314) and (6.311) that

$$\begin{aligned} B_\varepsilon^{n+1} z &= \gamma y_0 + (1-\gamma) B\big(B_\varepsilon^n z\big) \\ &= \gamma y_0 + (1-\gamma) \left[(1-\gamma)^n B^{n+1} z + \sum_{i=0}^{n-1} c_{ni} B y_i \right] \\ &= (1-\gamma)^{n+1} B^{n+1} z + \gamma y_0 + (1-\gamma) \sum_{i=0}^{n-1} c_{ni} y_{i+1}. \end{aligned}$$

This implies that our assertion is also valid for $n + 1$. Therefore for each integer $n \geq 1$, equality (6.315) holds with some constants c_{ni}, $i = 0, \ldots, n-1$, satisfying (6.316).

Now we will show that

$$\big\| B_\varepsilon^q z - y_0 \big\| \leq \varepsilon.$$

We have already shown that there exist positive numbers $c_{qi} > 0$, $i = 0, \ldots, q-1$, such that

$$\sum_{i=0}^{q-1} c_{qi} + (1-\gamma)^q = 1 \quad \text{and} \quad B_\varepsilon^q z = (1-\gamma)^q B^q z + \sum_{i=0}^{q-1} c_{qi} y_i. \tag{6.317}$$

By (6.317), (6.312), (6.307) and (6.308),

$$\begin{aligned} \big\| B_\varepsilon^q z - y_0 \big\| &\leq (1-\gamma)^q \big\| B^q z - y_0 \big\| + \sum_{i=0}^{q-1} c_{qi} \| y_0 - y_i \| \\ &\leq 2(1-\gamma)^q \operatorname{rad}(K) + 2qN^{-1} \operatorname{rad}(K) \\ &< 16^{-1}\varepsilon + 8^{-1}\varepsilon < 2^{-1}\varepsilon. \end{aligned}$$

Thus we have shown that

$$\big\| B_\varepsilon^q z - y_0 \big\| \leq 2^{-1}\varepsilon \quad \text{for each } z \in K.$$

Let $t \in \{1, \ldots, q\}$. To finish the proof we will show that

$$\big\| B_\varepsilon^t y_0 - y_0 \big\| \leq \varepsilon.$$

By (6.315) and (6.316), there exist positive numbers c_{ti}, $i = 0, \ldots, t-1$, such that

$$\sum_{i=0}^{t-1} c_{ti} + (1-\gamma)^t = 1 \quad \text{and} \quad B_\varepsilon^t y_0 = (1-\gamma)^t B^t y_0 + \sum_{i=0}^{t-1} c_{ti} y_i.$$

Together with (6.311), (6.312) and (6.308) this implies that

$$\begin{aligned} \|y_0 - B_\varepsilon^t y_0\| &= \left\| y_0 - \sum_{i=0}^{t-1} c_{ti} y_i - (1-\gamma)^t y_t \right\| \\ &\le 4qN^{-1} \operatorname{rad}(K) < 8^{-1}\varepsilon. \end{aligned}$$

This completes the proof of Lemma 6.62 (with $y_\varepsilon = y_0$). □

Proof of Theorem 6.61 To begin the construction of the set $\mathcal{F}$, let $B \in \mathcal{M}$ and let $i \ge 1$ be an integer. By Lemma 6.62, there exist $C^{(B,i)} \in \mathcal{M}$, $y(B,i) \in K$ and an integer $q(B,i) \ge 1$ such that

$$\begin{aligned} &\rho\big(B, C^{(B,i)}\big) \le 8^{-i}, \\ &\big\|\big(C^{(B,i)}\big)^t y(B,i) - y(B,i)\big\| \le 8^{-i}, \quad t = 0, \ldots, q(B,i), \end{aligned} \tag{6.318}$$

and

$$\big\|\big(C^{(B,i)}\big)^{q(B,i)} z - y(B,i)\big\| \le 8^{-i} \quad \text{for each } z \in K. \tag{6.319}$$

By Lemma 6.9, there exists an open neighborhood $U(B,i)$ of $C^{(B,i)}$ in $\mathcal{M}$ such that for each $\{A_j\}_{j=1}^{q(B,i)} \subset U(B,i)$ and each $z \in K$,

$$\big\|A_{q(B,i)} \cdot \cdots \cdot A_1 z - \big(C^{(B,i)}\big)^{q(B,i)} z\big\| \le 64^{-i}. \tag{6.320}$$

It follows from (6.319) and (6.320) that for each $\{A_i\}_{j=1}^{q(B,i)} \subset U(B,i)$ and each $z \in K$,

$$\big\|A_{q(B,i)} \cdot \cdots \cdot A_1 z - y(B,i)\big\| \le 8^{-i} + 64^{-i}. \tag{6.321}$$

Define

$$\mathcal{F} := \bigcap_{k=1}^{\infty} \bigcup \{U(B,i) : B \in \mathcal{M}, i = k, k+1, \ldots\}.$$

It is easy to see that $\mathcal{F}$ is a countable intersection of open and everywhere dense subsets of $\mathcal{M}$.

Assume that $A \in \mathcal{F}$ and $\varepsilon > 0$. Choose a natural number k for which

$$64 \cdot 2^{-k} < \varepsilon. \tag{6.322}$$

There exist $B \in \mathcal{M}$ and an integer $i \geq k$ such that

$$A \in U(B, i). \tag{6.323}$$

When combined with (6.321) and (6.322), this implies that for each $z \in K$,

$$\| A^{q(B,i)} z - y(B, i) \| \leq 8^{-i} + 64^{-i} < \varepsilon.$$

Since ε is an arbitrary positive number, we conclude that there exists a unique $x_A \in K$ such that $Ax_A = x_A$. Clearly,

$$\| x_A - y(B, i) \| \leq 8^{-i} + 64^{-i}.$$

Together with (6.321) and (6.322) this last inequality implies that for each $\{A_j\}_{j=1}^{\infty} \subset U(B, i)$, each $z \in K$ and each integer $T \geq q(B, i)$,

$$\| A_T \cdot \dots \cdot A_1 z - x_A \| \leq 2 \big(8^{-i} + 64^{-i} \big) < \varepsilon.$$

This completes the proof of Theorem 6.61. □

6.26 A Weak Ergodic Theorem for Affine Mappings

In this section we will prove the following result.

Theorem 6.63 *There exists a set $\mathcal{F} \subset \mathcal{A}_{ue}^{af}$, which is a countable intersection of open and everywhere dense subsets of $\mathcal{A}_{ue}^{af}$, such that for each $\{B_t\}_{t=1}^{\infty} \in \mathcal{F}$ and each $\varepsilon > 0$, there exist a neighborhood U of $\{B_t\}_{t=1}^{\infty}$ in $\mathcal{A}_{ue}^{af}$ and a natural number N such that for each $\{C_t\}_{t=1}^{\infty} \in U$, each integer $T \geq N$, each $r : \{1, \dots, T\} \to \{1, 2, \dots\}$ and each $x, y \in K$,*

$$\| C_{r(T)} \cdot \dots \cdot C_{r(1)} x - C_{r(T)} \cdot \dots \cdot C_{r(1)} y \| \leq \varepsilon.$$

Proof Fix $y_* \in K$. Let $\{A_t\}_{t=1}^{\infty} \in \mathcal{A}_{ue}^{af}$ and $\gamma \in (0, 1)$. For $t = 1, 2, \dots$, define $A_{t\gamma} : K \to K$ by

$$A_{t\gamma} x = (1 - \gamma) A_t x + \gamma y_*, \quad x \in K.$$

Clearly,

$$\{A_{t\gamma}\}_{t=1}^{\infty} \in \mathcal{A}_{ue}^{af}, \quad \rho_s \big(\{A_t\}_{t=1}^{\infty}, \{A_{t\gamma}\}_{t=1}^{\infty} \big) \leq 2\gamma \operatorname{rad}(K). \tag{6.324}$$

Let $i \geq 1$ be an integer. Choose a natural number $N(\gamma, i) \geq 4$ such that

$$(1 - \gamma)^{N(\gamma,i)} \big(\operatorname{rad}(K) + 1 \big) < 16^{-1} 4^{-i}. \tag{6.325}$$

We will show by induction that for each integer $T \geq 1$, the following assertion holds:

For each $r : \{1, \dots, T\} \to \{1, 2, \dots\}$, there exists $y_{r,T} \in K$ such that

$$\begin{aligned} &A_{r(T)\gamma} \cdot \dots \cdot A_{r(1)\gamma} x \\ &= (1-\gamma)^T A_{r(T)} \cdot \dots \cdot A_{r(1)} x + \big(1-(1-\gamma)^T\big) y_{r,T} \end{aligned} \tag{6.326}$$

for each $x \in K$.

It is clear that for $T = 1$ this assertion is true. Assume that it is also true for an integer $T \ge 1$. It follows from (6.327) that for each $r : \{1, \dots, T+1\} \to \{1, 2, \dots\}$ and each $x \in K$,

$$\begin{aligned} &A_{r(T+1)\gamma} \cdot \dots \cdot A_{r(1)\gamma} x \\ &= A_{r(T+1)\gamma}[A_{r(T)\gamma} \cdot \dots \cdot A_{r(1)\gamma} x] \\ &= A_{r(T+1)\gamma}\big[(1-\gamma)^T A_{r(T)} \cdot \dots \cdot A_{r(1)} x + \big(1-(1-\gamma)^T\big) y_{r,T}\big] \\ &= \gamma y_* + (1-\gamma) A_{r(T+1)}\big[(1-\gamma)^T A_{r(T)} \cdot \dots \cdot A_{r(1)} x + \big(1-(1-\gamma)^T\big) y_{r,T}\big] \\ &= (1-\gamma)^{T+1} A_{r(T+1)} \cdot \dots \cdot A_{r(1)} x + (1-\gamma)\big(1-(1-\gamma)^T\big) A_{r(T+1)} y_{r,T} \\ &\quad + \gamma y_*. \end{aligned}$$

This implies that the assertion is also valid for $T + 1$. Therefore we have shown that our assertion is true for any integer $T \ge 1$. Together with (6.325) this implies that the following property holds:

(a) For each integer $T \ge N(\gamma, i)$, each $r : \{1, \dots, T\} \to \{1, 2, \dots\}$ and each $x, y \in K$,

$$\begin{aligned} &\|A_{r(T)\gamma} \cdot \dots \cdot A_{r(1)\gamma} x - A_{r(T)\gamma} \cdot \dots \cdot A_{r(1)\gamma} y\| \\ &\le 2(1-\gamma)^T \operatorname{rad}(K) \le 8^{-1} \cdot 4^{-i}. \end{aligned}$$

By Lemma 6.9, there is an open neighborhood $U(\{A_t\}_{t=1}^\infty, \gamma, i)$ of $\{A_{t\gamma}\}_{t=1}^\infty$ in $\mathcal{A}_{ue}^{af}$ such that for each $\{C_t\}_{t=1}^\infty \in U(\{A_t\}_{t=1}^\infty, \gamma, i)$, each

$$r : \{1, \dots, N(\gamma, i)\} \to \{1, 2, \dots\}$$

and each $x \in K$,

$$\|C_{r(N(\gamma,i))} \cdot \dots \cdot C_{r(1)} x - A_{r(N(\gamma,i))\gamma} \cdot \dots \cdot A_{r(1)\gamma} x\| \le 64^{-1} \cdot 4^{-i}.$$

When combined with property (a) this implies that the following property also holds:

(b) For each integer $T \ge N(\gamma, i)$, each $r : \{1, \dots, T\} \to \{1, 2, \dots\}$, each $x, y \in K$ and each $\{C_t\}_{t=1}^\infty \in U(\{A_t\}_{t=1}^\infty, \gamma, i)$,

$$\|C_{r(T)} \cdot \dots \cdot C_{r(1)} x - C_{r(T)} \cdot \dots \cdot C_{r(1)} y\| \le 4^{-i-1}. \tag{6.327}$$

Define

$$\mathcal{F} := \bigcap_{q=1}^{\infty} \bigcup \{ U(\{A_t\}_{t=1}^{\infty}, \gamma, i) : \{A_t\}_{t=1}^{\infty} \in \mathcal{A}_{ue}^{af}, \gamma \in (0,1), i = q, q+1, \dots \}.$$

Clearly, $\mathcal{F}$ is a countable intersection of open and everywhere dense subsets of $\mathcal{A}_{ue}^{af}$. Let $\{B_t\}_{t=1}^{\infty} \in \mathcal{F}$ and $\varepsilon > 0$ be given. Choose a natural number q for which

$$64 \cdot 2^{-q} < \varepsilon. \tag{6.328}$$

There exist $\{A_t\}_{t=1}^{\infty} \in \mathcal{A}_{ue}^{af}$, $\gamma \in (0,1)$ and an integer $i \geq q$ such that

$$\{B_t\}_{t=1}^{\infty} \in U(\{A_t\}_{t=1}^{\infty}, \gamma, i).$$

By property (b) and (6.328), for each $\{C_t\}_{t=1}^{\infty} \in U(\{A_t\}_{t=1}^{\infty}, \gamma, i)$, each $T \geq N(\gamma, i)$, each $r : \{1, \dots, T\} \to \{1, 2, \dots\}$ and each $x, y \in K$,

$$\|C_{r(T)} \cdot \dots \cdot C_{r(1)} x - C_{r(T)} \cdot \dots \cdot C_{r(1)} y\| \leq 4^{-i-1} < \varepsilon.$$

This completes the proof of Theorem 6.63. □

6.27 Affine Mappings with a Common Fixed Point

In this section we will state three theorems which will be proved in the next section.

Theorem 6.64 *There exists a set $\mathcal{F} \subset \bar{\mathcal{A}}_{ue}^{0}$, which is a countable intersection of open and everywhere dense subsets of $\bar{\mathcal{A}}_{ue}^{0}$, such that $\mathcal{F} \subset \mathcal{A}_{ue}^{0}$ and for each $\mathbf{B} = \{B_t\}_{t=1}^{\infty} \in \mathcal{F}$, the following assertion holds:*

Let $x_{\mathbf{B}} \in K$, $B_t x_{\mathbf{B}} = x_{\mathbf{B}}$, $t = 1, 2, \dots$, and let $\varepsilon > 0$. Then there exist a neighborhood U of $\mathbf{B} = \{B_t\}_{t=1}^{\infty}$ in $\bar{\mathcal{A}}_{ue}^{0}$ and a natural number N such that for each $\{C_t\}_{t=1}^{\infty} \in U$, each integer $T \geq N$, each $r : \{1, \dots, T\} \to \{1, 2, \dots\}$ and each $x \in K$,

$$\|C_{r(T)} \cdot \dots \cdot C_{r(1)} x - x_{\mathbf{B}}\| \leq \varepsilon.$$

Denote by $\mathcal{A}_{ue}^{(1)}$ the set of all $\mathbf{A} = \{A_t\}_{t=1}^{\infty} \in \mathcal{A}_{ue}$ for which there exists $x_{\mathbf{A}} \in K$ such that

$$A_t x_{\mathbf{A}} = x_{\mathbf{A}}, \quad t = 1, 2, \dots, \tag{6.329}$$

and for each $\alpha \in (0,1)$, $x \in K$ and an integer $t \geq 1$,

$$A_t(\alpha x_{\mathbf{A}} + (1-\alpha)x) = \alpha x_{\mathbf{A}} + (1-\alpha) A_t x.$$

Denote by $\bar{\mathcal{A}}_{ue}^{(1)}$ the closure of $\mathcal{A}_{ue}^{(1)}$ in the space $\mathcal{A}_{ue}$. We equip the topological subspace $\bar{\mathcal{A}}_{ue}^{(1)} \subset \mathcal{A}_{ue}$ with the relative topology.

Theorem 6.65 *Let a set $\mathcal{F} \subset \bar{\mathcal{A}}^0_{ue}$ be as guaranteed in Theorem 6.64. There exists a set $\mathcal{F}^{(1)} \subset \mathcal{F} \cap \mathcal{A}^{(1)}_{ue}$ which is a countable intersection of open and everywhere dense subsets of $\bar{\mathcal{A}}^{(1)}_{ue}$.*

Denote by $\mathcal{A}^{af}_{ue,0}$ the set of all $\mathbf{A} = \{A_t\}_{t=1}^{\infty} \in \mathcal{A}^{af}_{ue}$ for which there exists $x_{\mathbf{A}} \in K$ such that (6.329) holds.

Denote by $\bar{\mathcal{A}}^{af}_{ue,0}$ the closure of $\mathcal{A}^{af}_{ue,0}$ in the space $\mathcal{A}_{ue}$. We also consider the topological subspace $\bar{\mathcal{A}}^{af}_{ue,0} \subset \mathcal{A}_{ue}$ with the relative topology.

Theorem 6.66 *Let a set $\mathcal{F}^{(1)}$ be as guaranteed in Theorem 6.65. There exists a set $\mathcal{F}_* \subset \mathcal{F}^{(1)} \cap \mathcal{A}^{af}_{ue,0}$ which is a countable intersection of open and everywhere dense subsets of $\bar{\mathcal{A}}^{af}_{ue,0}$.*

Theorems 6.65 and 6.66 show that the generic convergence established in Theorem 6.64 is also valid for certain subspaces of $\bar{\mathcal{A}}^0_{ue}$.

6.28 Proofs of Theorems 6.64, 6.65 and 6.66

Proof of Theorem 6.64 Let $\mathbf{A} = \{A_t\}_{t=1}^{\infty} \in \mathcal{A}^0_{ue}$ and $\gamma \in (0, 1)$. There exists $x_{\mathbf{A}} \in K$ such that

$$A_t x_{\mathbf{A}} = x_{\mathbf{A}}, \quad t = 1, 2, \ldots, \tag{6.330}$$

and for each integer $t \geq 1$, $x \in K$ and $\alpha \in (0, 1)$,

$$A_t\big(\alpha x_{\mathbf{A}} + (1-\alpha)x\big) = \lambda_t(\alpha, x)x_{\mathbf{A}} + \big(1 - \lambda_t(\alpha, x)\big)A_t x \tag{6.331}$$

with some constant $\lambda_t(\alpha, x) \in [\alpha, 1]$.

For $t = 1, 2, \ldots$, define $A_{t\gamma} : K \to K$ by

$$A_{t\gamma}x = (1-\gamma)A_t x + \gamma x_{\mathbf{A}}, \quad x \in K. \tag{6.332}$$

Clearly,

$$\{A_{t\gamma}\}_{t=1}^{\infty} \in \mathcal{A}_{ue}, \quad A_{t\gamma}x_{\mathbf{A}} = x_{\mathbf{A}}, \quad t = 1, 2, \ldots. \tag{6.333}$$

Let $x \in K$, $\alpha \in [0, 1)$ and let $t \geq 1$ be an integer. Then there exists $\lambda_t(\alpha, x) \in [\alpha, 1]$ such that (6.331) holds. Also, by (6.331) and (6.332),

$$\begin{aligned}
&A_{t\gamma}\big(\alpha x_{\mathbf{A}} + (1-\alpha)x\big) \\
&\quad= (1-\gamma)A_t\big(\alpha x_{\mathbf{A}} + (1-\alpha)x\big) + \gamma x_{\mathbf{A}} \\
&\quad= \gamma x_{\mathbf{A}} + (1-\gamma)\big[\lambda_t(\alpha, x)x_{\mathbf{A}} + \big(1 - \lambda_t(\alpha, x)\big)A_t x\big]
\end{aligned}$$

$$= (1-\gamma)\big(1-\lambda_t(\alpha,x)\big)A_t x + \big[\gamma + (1-\gamma)\lambda_t(\alpha,x)\big]x_{\mathbf{A}}$$
$$= \big(1-\lambda_t(\alpha,x)\big)A_{t\gamma}x + \big[\gamma + (1-\gamma)\lambda_t(\alpha,x) - \gamma\big(1-\lambda_t(\alpha,x)\big)\big]x_{\mathbf{A}}$$
$$= \big(1-\lambda_t(\alpha,x)\big)A_{t\gamma}x + \lambda_t(\alpha,x)x_{\mathbf{A}}. \tag{6.334}$$

Thus property (6.301) is satisfied and therefore

$$\{A_{t\gamma}\}_{t=1}^{\infty} \in \mathcal{A}_{ue}^{0}. \tag{6.335}$$

Let $z \in K$. We will show by induction that for each integer $T \geq 1$ and each $r : \{1, \ldots, T\} \to \{1, 2, \ldots\}$, there exists $\lambda(z, T, r) \in [0, (1-\gamma)^T]$ such that

$$A_{r(T)\gamma} \cdot \cdots \cdot A_{r(1)\gamma} z$$
$$= \lambda(z, T, r) A_{r(T)} \cdot \cdots \cdot A_{r(1)} z + \big(1 - \lambda(z, T, r)\big)x_{\mathbf{A}}. \tag{6.336}$$

It is clear that for $T = 1$ our assertion is valid.

Assume that it is also valid for an integer $T \geq 1$. Let $r : \{1, \ldots, T+1\} \to \{1, 2, \ldots\}$. There exists $\lambda(z, T, r) \in [0, (1-\gamma)^T]$ such that (6.336) is valid. It follows from (6.336) and (6.334) that

$$A_{r(T+1)\gamma} \cdot \cdots \cdot A_{r(1)\gamma} z$$
$$= A_{r(T+1)\gamma}\big[\lambda(z, T, r) A_{r(T)} \cdot \cdots \cdot A_{r(1)} z + \big(1 - \lambda(z, T, r)\big)x_{\mathbf{A}}\big]$$
$$= (1-\gamma)(1-\kappa) A_{r(T+1)} A_{r(T)} \cdot \cdots \cdot A_{r(1)} z + \big[\gamma + (1-\gamma)\kappa\big]x_{\mathbf{A}}$$

with $\kappa \in [1 - \lambda(z, T, r), 1]$. Set

$$\lambda(z, T+1, r) = (1-\gamma)(1-\kappa).$$

It is easy to see that

$$0 \leq \lambda(z, T+1, r) \leq (1-\gamma)\lambda(z, T, r) \leq (1-\gamma)^{T+1}$$

and

$$A_{r(T+1)\gamma} \cdot \cdots \cdot A_{r(1)\gamma} z$$
$$= \lambda(z, T+1, r) A_{r(T+1)} \cdot \cdots \cdot A_{r(1)} z + \big(1 - \lambda(z, T+1, r)\big)x_{\mathbf{A}}.$$

Therefore the assertion is valid for $T + 1$. Thus we have shown that for each integer $T \geq 1$ and each $r : \{1, \ldots, T\} \to \{1, 2, \ldots\}$, there exists $\lambda(z, T, r) \in [0, (1-\gamma)^T]$ such that (6.336) holds.

Let $i \geq 1$ be an integer. Choose a natural number $N(\gamma, i)$ for which

$$64(1-\gamma)^{N(\gamma,i)}\big(\mathrm{rad}(K) + 1\big) < 8^{-i}. \tag{6.337}$$

We will show that for each $z \in K$, each integer $T \geq N(\gamma, i)$ and each $r : \{1, \ldots, T\} \to \{1, 2, \ldots\}$,

$$\|A_{r(T)\gamma} \cdot \cdots \cdot A_{r(1)\gamma} z - x_{\mathbf{A}}\| \leq 8^{-i-1}. \tag{6.338}$$

Let $T \ge N(\gamma, i)$ be an integer, $z \in K$ and $r : \{1, \dots, T\} \to \{1, 2, \dots\}$. There exists $\lambda(z, T, r) \in [0, (1-\gamma)^T]$ such that (6.336) holds. It is easy to see that (6.336) and (6.337) imply (6.338).

By Lemma 6.9, there exists a number

$$\delta\big(\{A_t\}_{t=1}^{\infty}, \gamma, i\big) \in \big(0, 16^{-1}8^{-i}\big) \tag{6.339}$$

such that for each $\{C_t\}_{t=1}^{\infty} \in \bar{\mathcal{A}}_{ue}^0$ satisfying

$$\rho_s\big(\{C_t\}_{t=1}^{\infty}, \{A_{t\gamma}\}_{t=1}^{\infty}\big) \le \delta\big(\{A_t\}_{t=1}^{\infty}, \gamma, i\big),$$

each $r : \{1, \dots, N(\gamma, i)\} \to \{1, 2, \dots\}$ and each $x \in K$,

$$\|C_{r(N(\gamma,i))} \cdot \dots \cdot C_{r(1)}x - A_{r(N(\gamma,i))\gamma} \cdot \dots \cdot A_{r(1)\gamma}x\| \le 16^{-1} \cdot 8^{-i}. \tag{6.340}$$

Set

$$U\big(\{A_t\}_{t=1}^{\infty}, \gamma, i\big) = \big\{\{C_t\}_{t=1}^{\infty} \in \bar{\mathcal{A}}_{ue}^0 : \rho_s\big(\{C_t\}_{t=1}^{\infty}, \{A_{t\gamma}\}_{t=1}^{\infty}\big) < \delta\big(\{A_t\}_{t=1}^{\infty}, \gamma, i\big)\big\}. \tag{6.341}$$

It follows from (6.341), the choice of $\delta(\{A_t\}_{t=1}^{\infty}, \gamma, i)$ (see (6.339), (6.340)) and (6.338) that the following property holds:

(a) For each $\{C_t\}_{t=1}^{\infty} \in U(\{A_t\}_{t=1}^{\infty}, \gamma, i)$, each integer $T \ge N(\gamma, i)$, each $r : \{1, \dots, T\} \to \{1, 2, \dots\}$ and each $x \in K$,

$$\|C_{r(T)} \cdot \dots \cdot C_{r(1)}x - x_{\mathbf{A}}\| \le 8^{-i}.$$

Define

$$\mathcal{F} := \bigcap_{q=1}^{\infty} \bigcup \{U\big(\{A_t\}_{t=1}^{\infty}, \gamma, i\big) : \{A_t\}_{t=1}^{\infty} \in \mathcal{A}_{ue}^0, \gamma \in (0, 1), i = q, q+1, \dots\}.$$

It is easy to see that $\mathcal{F}$ is a countable intersection of open and everywhere dense subsets of $\bar{\mathcal{A}}_{ue}^0$.

Assume now that $\mathbf{B} = \{B_t\}_{t=1}^{\infty} \in \mathcal{F}$ and $\varepsilon > 0$. Choose a natural number q such that

$$64 \cdot 2^{-q} < \varepsilon. \tag{6.342}$$

There exist $\{A_t\}_{t=1}^{\infty} \in \mathcal{A}_{ue}^0$, $\gamma \in (0, 1)$ and an integer $i \ge q$ such that

$$\{B_t\}_{t=1}^{\infty} \in U\big(\{A_t\}_{t=1}^{\infty}, \gamma, i\big). \tag{6.343}$$

By property (a), (6.343) and (6.342), for each $x \in K$, each integer $T \ge N(\gamma, i)$ and each integer $\tau \ge 1$,

$$\big\|B_\tau^T x - x_{\mathbf{A}}\big\| \le 8^{-i} < \varepsilon. \tag{6.344}$$

Since ε is an arbitrary positive number, we conclude that there exists $x_{\mathbf{B}} \in K$ such that

$$\lim_{T\to\infty} B_\tau^T x = x_{\mathbf{B}}$$

for each $x \in K$ and each integer $\tau \geq 1$. It is easy to see that

$$B_t x_{\mathbf{B}} = x_{\mathbf{B}}, \quad t = 1, 2, \ldots, \|x_{\mathbf{B}} - x_{\mathbf{A}}\| \leq 8^{-i} < \varepsilon. \tag{6.345}$$

It follows from property (a), (6.345) and (6.342) that for each sequence $\{C_t\}_{t=1}^\infty \in U(\{A_t\}_{t=1}^\infty, \gamma, i)$, each integer $T \geq N(\gamma, i)$, each $r : \{1, \ldots, T\} \to \{1, 2, \ldots\}$ and each $x \in K$,

$$\|C_{r(T)} \cdot \cdots \cdot C_{r(1)} x - x_{\mathbf{B}}\| < \varepsilon. \tag{6.346}$$

We will show that for each integer $t \geq 1$, $x \in K$ and $\alpha \in (0, 1)$, there exists $\lambda \in [\alpha, 1]$ such that

$$B_t\big(\alpha x_{\mathbf{B}} + (1-\alpha)x\big) = \lambda x_{\mathbf{B}} + (1-\lambda)B_t x. \tag{6.347}$$

Let $t \geq 1$ be an integer, $x \in K$ and let $\alpha \in (0, 1)$. By (6.331) and (6.334), there exists $\lambda_\varepsilon \in [\alpha, 1]$ such that

$$A_{t\gamma}\big(\alpha x_{\mathbf{A}} + (1-\alpha)x\big) = \lambda_\varepsilon x_{\mathbf{A}} + (1-\lambda_\varepsilon)A_{t\gamma}x. \tag{6.348}$$

Since ε is an arbitrary positive number, it follows from (6.348), (6.345), (6.343), (6.341), (6.339) and (6.342) that for each $\varepsilon > 0$, there exist $\lambda_\varepsilon \in [\alpha, 1]$ and $z_\varepsilon \in K$ such that

$$\|z_\varepsilon - x_{\mathbf{B}}\| \leq \varepsilon, \qquad \big\|B_t\big(\alpha z_\varepsilon + (1-\alpha)x\big) - \big(\lambda_\varepsilon x_{\mathbf{B}} + (1-\lambda_\varepsilon)B_t x\big)\big\| \leq \varepsilon.$$

This implies that (6.347) holds with some $\lambda \in [\alpha, 1]$ and completes the proof of Theorem 6.64. □

Proof of Theorem 6.65 Let $\mathcal{F}$ be as constructed in the proof of Theorem 6.64. Let $\mathbf{A} = \{A_t\}_{t=1}^\infty \in \mathcal{A}_{ue}^{(1)}$, $\gamma \in (0, 1)$ and let $i \geq 1$ be an integer. There exists $x_{\mathbf{A}} \in K$ such that (6.340) holds, and for each $x \in K$, each integer $t \geq 1$ and each $\alpha \in [0, 1]$, equality (6.331) holds with $\lambda_t(\alpha, x) = \alpha$. For $t = 1, 2, \ldots$, define $A_{t\gamma} : K \to K$ by (6.332). It is easy to see that $\{A_{t\gamma}\}_{t=1}^\infty \in \mathcal{A}_{ue}^{(1)}$. Choose a natural number $N(\gamma, i)$ for which (6.337) holds. Let $\delta(\{A_t\}_{t=1}^\infty, \gamma, i)$ and $U(\{A_t\}_{t=1}^\infty, \gamma, i)$ be defined as in the proof of Theorem 6.64. Set

$$\mathcal{F}^{(1)} := \Bigg[\bigcap_{q=1}^\infty \bigcup \{U(\{A_t\}_{t=1}^\infty, \gamma, i) : \{A_t\}_{t=1}^\infty \in \mathcal{A}_{ue}^{(1)}, \gamma \in (0, 1), i = q, q+1, \ldots\}\Bigg] \cap \bar{\mathcal{A}}_{ue}^{(1)}.$$

Clearly, $\mathcal{F}^{(1)}$ is a countable intersection of open and everywhere dense subsets of $\bar{\mathcal{A}}_{ue}^{(1)}$ and $\mathcal{F}^{(1)} \subset \mathcal{F}$. Arguing as in the proof of Theorem 6.64, we can show that $\mathcal{F}^{(1)} \subset \mathcal{A}_{ue}^{(1)}$. This completes the proof of Theorem 6.65. □

The proof of Theorem 6.66 is analogous to that of Theorem 6.65.

6.29 Weak Convergence

In this section we present two theorems concerning the space $\mathcal{A}_{ue}^{(F,0)}$ defined in Sect. 6.24. Recall that F is a nonempty, closed and convex subset of K for which there exists a uniformly continuous operator $Q : K \to F$ such that

$$Qx = x, \quad x \in F, \tag{6.349}$$

and for each $y \in K$, $x \in F$ and $\alpha \in [0, 1]$,

$$Q\big(\alpha x + (1 - \alpha)y\big) = \alpha x + (1 - \alpha)Qy. \tag{6.350}$$

We now state our first theorem.

Theorem 6.67 *There exists a set $\mathcal{F} \subset \mathcal{A}_{ue}^{(F,0)}$ which is a countable intersection of open everywhere dense sets in $\mathcal{A}_{ue}^{(F,0)}$ and such that for each $\{B_t\}_{t=1}^{\infty} \in \mathcal{F}$, the following assertion holds*:

For each $\varepsilon > 0$, there exist a neighborhood U of $\{B_t\}_{t=1}^{\infty}$ in the space $\mathcal{A}_{ue}^{(F,0)}$ and a natural number N such that for each $\{C_t\}_{t=1}^{\infty} \in U$, each integer $T \geq N$, each $r : \{1, 2, \ldots, T\} \to \{1, 2, \ldots\}$ and each $x \in K$,

$$d(C_{r(T)} \cdot \cdots \cdot C_{r(1)}x, F) \leq \varepsilon.$$

Assume now that for each $x, y \in K$ and $\alpha \in [0, 1]$,

$$Q\big(\alpha x + (1 - \alpha)y\big) = \alpha Qx + (1 - \alpha)Qy. \tag{6.351}$$

Denote by $\mathcal{A}_{ue}^{(F,1)}$ the set of all $\{A_t\}_{t=1}^{\infty} \in \mathcal{A}_{ue}$ such that

$$A_t x = x, \quad t = 1, 2, \ldots, x \in F,$$

and for each $t \in \{1, 2, \ldots\}$, each $x, y \in K$ and each $\alpha \in [0, 1]$,

$$A_t\big(\alpha x + (1 - \alpha)y\big) = \alpha A_t x + (1 - \alpha)A_t y.$$

It is clear that $\mathcal{A}_{ue}^{(F,1)}$ is a closed subset of $\mathcal{A}_{ue}^{(F,0)}$. We consider the topological subspace $\mathcal{A}_{ue}^{(F,1)} \subset \mathcal{A}_{ue}^{(F,0)}$ with the relative topology.

Here is our second theorem.

Theorem 6.68 *Let the set $\mathcal{F}$ be as guaranteed in Theorem 6.67. Then there exists a set $\mathcal{F}_1 \subset \mathcal{F} \cap \mathcal{A}_{ue}^{(F,1)}$ which is a countable intersection of open everywhere dense subsets of $\mathcal{A}_{ue}^{(F,1)}$.*

6.30 Proofs of Theorems 6.67 and 6.68

Proof of Theorem 6.67 Let $\{A_t\}_{t=1}^{\infty} \in \mathcal{A}_{ue}^{(F,0)}$ and $\gamma \in (0,1)$ be given. For $t = 1, 2, \ldots$ we define $A_{t\gamma} : K \to K$ by

$$A_{t\gamma}x = (1-\gamma)A_t x + \gamma Qx, \quad x \in K. \tag{6.352}$$

It is easy to see that

$$\{A_{t\gamma}\}_{t=1}^{\infty} \in \mathcal{A}_{ue}^{(F,0)}. \tag{6.353}$$

Let $z \in K$. By induction we will show that for each integer $T \geq 1$, the following assertion holds:

For each $r : \{1, \ldots, T\} \to \{1, 2, \ldots\}$,

$$A_{r(T)\gamma} \cdot \cdots \cdot A_{r(1)\gamma} z = (1-\gamma)^T A_{r(T)} \cdot \cdots \cdot A_{r(1)} z + \big(1 - (1-\gamma)^T\big) y_T \tag{6.354}$$

for some $y_T \in F$.

Clearly, for $T = 1$ our assertion is valid. Assume that it is also valid for $T \geq 1$ and that $r : \{1, \ldots, T+1\} \to \{1, 2, \ldots\}$. Evidently, (6.354) holds with some $y_T \in F$. By (6.354), (6.353) and (6.352),

$$\begin{aligned}
&A_{r(T+1)\gamma} \cdot \cdots \cdot A_{r(1)\gamma} z \\
&\quad = A_{r(T+1)\gamma}\big[(1-\gamma)^T A_{r(T)} \cdot \cdots \cdot A_{r(1)} z + \big(1-(1-\gamma)^T\big) y_T\big] \\
&\quad = (1-\gamma)^T A_{r(T+1)\gamma}[A_{r(T)} \cdot \cdots \cdot A_{r(1)} z] + \big(1-(1-\gamma)^T\big) y_T \\
&\quad = (1-\gamma)^{T+1} A_{r(T+1)} \cdot \cdots \cdot A_{r(1)} z \\
&\qquad + \gamma(1-\gamma)^T Q[A_{r(T)} \cdot \cdots \cdot A_{r(1)} z] + \big(1-(1-\gamma)^T\big) y_T \\
&\quad = (1-\gamma)^{T+1} A_{r(T+1)} \cdot \cdots \cdot A_{r(1)} z \\
&\qquad + \big(1-(1-\gamma)^{T+1}\big)\big[\big(1-(1-\gamma)^{T+1}\big)^{-1}\gamma(1-\gamma)^T \\
&\qquad \times Q[A_{r(T)} \cdot \cdots \cdot A_{r(1)} z] + \big(1-(1-\gamma)^{T+1}\big)^{-1}\big(1-(1-\gamma)^T\big) y_T\big].
\end{aligned}$$

This implies that our assertion also holds for $T+1$.

Therefore we have shown that it is valid for all integers $T \geq 1$.

Let $i \geq 1$ be an integer. Choose a natural number $N(\gamma, i)$ for which

$$64(1-\gamma)^{N(\gamma,i)}\big(\mathrm{rad}(K) + 1\big) < 8^{-i}. \tag{6.355}$$

It follows from (6.354) that for each $z \in K$, each integer $T \geq N(\gamma, i)$ and each $r : \{1, \dots, T\} \to \{1, 2, \dots\}$,

$$d(A_{r(T)\gamma} \cdot \dots \cdot A_{r(1)\gamma} z, F) \leq 8^{-i-1}. \tag{6.356}$$

By Lemma 6.9, there exists an open neighborhood $U(\{A_t\}_{t=1}^{\infty}, \gamma, i)$ of $\{A_{t\gamma}\}_{t=1}^{\infty}$ in $\mathcal{A}_{ue}^{(F,0)}$ such that the following property holds:

(a) for each $\{C_t\}_{t=1}^{\infty} \in U(\{A_t\}_{t=1}^{\infty}, \gamma, i)$, each $r : \{1, \dots, N(\gamma, i)\} \to \{1, 2, \dots\}$ and each $x \in K$,

$$\|C_{r(N(\gamma,i))} \cdot \dots \cdot C_{r(1)} x - A_{r(N(\gamma,i))\gamma} \cdot \dots \cdot A_{r(1)\gamma} x\| \leq 16^{-1} 8^{-i}.$$

It follows from the definition of $U(\{A_t\}_{t=1}^{\infty}, \gamma, i)$ and (6.356) that the following property is also true:

(b) For each $\{C_t\}_{t=1}^{\infty} \in U(\{A_t\}_{t=1}^{\infty}, \gamma, i)$, each integer $T \geq N(\gamma, i)$, each $r : \{1, \dots, T\} \to \{1, 2, \dots\}$ and each $x \in K$,

$$d(C_{r(T)} \cdot \dots \cdot C_{r(1)} x, F) \leq 8^{-i}.$$

Define

$$\mathcal{F} := \bigcap_{q=1}^{\infty} \bigcup \{U(\{A_t\}_{t=1}^{\infty}, \gamma, i) : \{A_t\}_{t=1}^{\infty} \in \mathcal{A}_{ue}^{(F,0)}, \gamma \in (0,1), i = q, q+1, \dots\}.$$

It is easy to see that $\mathcal{F}$ is a countable intersection of open and everywhere dense subsets of $\mathcal{A}_{ue}^{(F,0)}$.

Assume that $\{B_t\}_{t=1}^{\infty} \in \mathcal{F}$ and $\varepsilon > 0$. Choose a natural number q such that

$$64 \cdot 2^{-q} < \varepsilon. \tag{6.357}$$

There exist $\{A_t\}_{t=1}^{\infty} \in \mathcal{A}_{ue}^{(F,0)}$, $\gamma \in (0,1)$ and an integer $i \geq q$ such that $\{B_t\}_{t=1}^{\infty} \in U(\{A_t\}_{t=1}^{\infty}, \gamma, i)$. By (6.357) and property (b), for each $\{C_t\}_{t=1}^{\infty} \in U(\{A_t\}_{t=1}^{\infty}, \gamma, i)$, each integer $T \geq N(\gamma, i)$, each $r : \{1, \dots, T\} \to \{1, 2, \dots\}$ and each $x \in K$,

$$d(C_{r(T)} \cdot \dots \cdot C_{r(1)} x, F) \leq \varepsilon.$$

This completes the proof of Theorem 6.67. □

Analogously to the proof of Theorem 6.65 we can prove Theorem 6.68 by modifying the proof of Theorem 6.67.

6.31 Affine Mappings with a Common Set of Fixed Points

In this section we assume that F is a nonempty, closed and convex subset of K, and $Q : K \to F$ is a uniformly continuous retraction satisfying (6.350).

We assume, in addition, that there exists a number $\Delta > 0$ such that

$$\{x \in X : d(x, F) < \Delta\} \subset K.$$

In this setting we can strengthen Theorem 6.67.

Theorem 6.69 *Let the set $\mathcal{F} \subset \mathcal{A}_{ue}^{(F,0)}$ be as constructed in the proof of Theorem 6.67. Then for each $\{B_t\}_{t=1}^{\infty} \in \mathcal{F}$, the following two assertions hold:*

1. *For each $r : \{1, 2, \ldots\} \to \{1, 2, \ldots\}$, there exists a uniformly continuous operator $P_r : K \to F$ such that*

$$\lim_{T\to\infty} B_{r(T)} \cdot \cdots \cdot B_{r(1)}x = P_r x \quad \textit{for each } x \in K.$$

2. *For each $\varepsilon > 0$, there exist a neighborhood U of $\{B_t\}_{t=1}^{\infty}$ in the space $\mathcal{A}_{ue}^{(F,0)}$ and a natural number N such that for each $\{C_t\}_{t=1}^{\infty} \in U$, each $r : \{1, 2, \ldots\} \to \{1, 2, \ldots\}$ and each integer $T \ge N$,*

$$\|C_{r(T)} \cdot \cdots \cdot C_{r(1)}x - P_r x\| \le \varepsilon \quad \textit{for all } x \in K.$$

Proof As in the previous section, given $\{A_t\}_{t=1}^{\infty} \in \mathcal{A}_{ue}^{(F,0)}$, $\gamma \in (0, 1)$ and an integer $i \ge 1$, we define $\{A_{t\gamma}\}_{t=1}^{\infty} \in \mathcal{A}_{ue}^{(F,0)}$ (see (6.352)), a natural number $N(\gamma, i)$ (see (6.355)) and an open neighborhood $U(\{A_t\}_{t=1}^{\infty}, \gamma, i)$ of $\{A_{t\gamma}\}_{t=1}^{\infty}$ in $\mathcal{A}_{ue}^{(F,0)}$ (see property (a)). Again, as in the previous section, we define a set $\mathcal{F}$ which is a countable intersection of open and everywhere dense sets in $\mathcal{A}_{ue}^{(F,0)}$ by

$$\mathcal{F} := \bigcap_{q=1}^{\infty} \bigcup \{U(\{A_t\}_{t=1}^{\infty}, \gamma, i) : \{A_t\}_{t=1}^{\infty} \in \mathcal{A}_{ue}^{(F,0)}, \gamma \in (0, 1), i = q, q+1, \ldots\}.$$

Assume that $\{B_t\}_{t=1}^{\infty} \in \mathcal{F}$ and $\varepsilon \in (0, 1)$. Choose a number ε_0 such that

$$\varepsilon_0 < 64^{-1}(\min\{\varepsilon, \Delta\}), \qquad 8\varepsilon_0 \Delta^{-1}(\mathrm{rad}(K) + 1) < 8^{-1}\varepsilon. \tag{6.358}$$

Choose a natural number q such that

$$64 \cdot 2^{-q} < \varepsilon_0. \tag{6.359}$$

There exist $\{A_t\}_{t=1}^{\infty} \in \mathcal{A}_{ue}^{(F,0)}$, $\gamma \in (0, 1)$ and an integer $i \ge q$ such that

$$\{B_t\}_{t=1}^{\infty} \in U(\{A_t\}_{t=1}^{\infty}, \gamma, i). \tag{6.360}$$

It was shown in the previous section (see (6.356)) that the following property holds:

(c) For each $z \in K$, each integer $T \ge N(\gamma, i)$ and each $r : \{1, \ldots, T\} \to \{1, 2, \ldots\}$,

$$d(A_{r(T)\gamma} \cdot \cdots \cdot A_{r(1)\gamma}z, F) \le 8^{-i-1}.$$

By the definition of $U(\{A_t\}_{t=1}^{\infty}, \gamma, i)$ (see Sect. 6.30 and property (a)), the following property holds:

(d) for each $\{C_t\}_{t=1}^{\infty} \in U(\{A_t\}_{t=1}^{\infty}, \gamma, i)$, each $r : \{1, \ldots, N(\gamma, i)\} \to \{1, 2, \ldots\}$ and each $x \in K$,

$$\|C_{r(N(\gamma,i))} \cdot \cdots \cdot C_{r(1)}x - A_{r(N(\gamma,i))\gamma} \cdot \cdots \cdot A_{r(1)\gamma}x\| \leq 16^{-1} \cdot 8^{-i}.$$

Assume that $r : \{1, 2, \ldots\} \to \{1, 2, \ldots\}$. Then by property (c), for each $x \in K$ there exists $f_r(x) \in K$ such that

$$\left\| A_{r(N(\gamma,i))\gamma} \cdot \cdots \cdot A_{r(1)\gamma}x - f_r(x) \right\| \leq 2 \cdot 8^{-i-1}. \tag{6.361}$$

We will show that for each $\{C_t\}_{t=1}^{\infty} \in U(\{A_t\}_{t=1}^{\infty}, \gamma, i)$, each integer $T \geq N(\gamma, i)$ and each $x \in K$,

$$\left\| C_{r(T)} \cdot \cdots \cdot C_{r(1)}x - f_r(x) \right\| \leq 8^{-1}\varepsilon. \tag{6.362}$$

Let $\{C_t\}_{t=1}^{\infty} \in U(\{A_t\}_{t=1}^{\infty}, \gamma, i)$ and let $x \in K$. By (6.361) and property (d),

$$\left\| C_{r(N(\gamma,i))} \cdot \cdots \cdot C_{r(1)}x - f_r(x) \right\| \leq 8^{-i}\left(16^{-1} + 4^{-1}\right). \tag{6.363}$$

Set

$$z = f_r(x) + 8^i \Delta\left[C_{r(N(\gamma,i))} \cdot \cdots \cdot C_{r(1)}x - f_r(x)\right]. \tag{6.364}$$

It follows from (6.363), (6.364) and the definition of Δ that $z \in K$ and

$$C_{r(N(\gamma,i))} \cdot \cdots \cdot C_{r(1)}x = 8^{-i}\Delta^{-1}z + \left(1 - 8^{-i}\Delta^{-1}\right)f_r(x). \tag{6.365}$$

It follows from (6.365), (6.358) and (6.359) that for each integer $T > N(\gamma, i)$,

$$C_{r(T)} \cdot \cdots \cdot C_{r(1)}x = 8^{-i}\Delta^{-1}C_{r(T)} \cdot \cdots \cdot C_{r(N(\gamma,i)+1)}z + \left(1 - 8^{-i}\Delta^{-1}\right)f_r(x).$$

Together with (6.366) and (6.358) this implies that for each integer $T \geq N(\gamma, i)$,

$$\left\| C_{r(T)} \cdot \cdots \cdot C_{r(1)}x - f_r(x) \right\| \leq 2\,\mathrm{rad}(K)8^{-i}\Delta^{-1} < 8^{-1}\varepsilon.$$

Therefore we have shown that for each $r : \{1, 2, \ldots\} \to \{1, 2, \ldots\}$ and each $x \in K$, there exists $f_r(x) \in F$ such that the following property holds:

(e) For each $\{C_t\}_{t=1}^{\infty} \in U(\{A_t\}_{t=1}^{\infty}, \gamma, i)$, each integer $T \geq N(\gamma, i)$ and each $x \in K$, inequality (6.362) is valid.

Since ε is an arbitrary positive number, this implies that for each $r : \{1, 2, \ldots\} \to \{1, 2, \ldots\}$, there exists an operator $P_r : K \to K$ such that

$$\lim_{T \to \infty} B_{r(T)} \cdot \cdots \cdot B_{r(1)}x = P_r x, \quad x \in K. \tag{6.366}$$

Let $r : \{1, 2, \ldots\} \to \{1, 2, \ldots\}$ be given. By (6.366), property (e) and (6.362),

$$\left\| P_r x - f_r(x) \right\| \leq 8^{-1}\varepsilon, \quad x \in K,$$

and for each $\{C_t\}_{t=1}^{\infty} \in U(\{A_t\}_{t=1}^{\infty}, \gamma, i)$, each integer $T \geq N(\gamma, i)$ and each $x \in K$,

$$\left\| C_{r(T)} \cdot \cdots \cdot C_{r(1)}x - P_r(x) \right\| \leq 4^{-1}\varepsilon.$$

This completes the proof of Theorem 6.69. □

6.32 Infinite Products of Resolvents of Accretive Operators

Accretive operators and their resolvents play an important role in nonlinear functional analysis [16, 20, 24, 46]. Infinite products of resolvents of accretive operators and their applications were investigated, for example, in [21, 25, 70, 104, 120, 167, 172].

We use Baire's category to study the asymptotic behavior of infinite products of resolvents of a generic m-accretive operator on a general Banach space X. We prove a weak ergodic theorem (Theorem 6.71) and Theorem 6.72, which provides strong convergence of infinite products to the unique zero of such an operator. These results were obtained in [134]. More precisely, we consider two spaces of m-accretive operators on X. The first space is the space of all m-accretive operators endowed with an appropriate complete metrizable uniformity. The second space is the closure in the first space of all those operators which have a zero. For the first space we construct a subset which is a countable intersection of open and everywhere dense sets such that for each operator belonging to this subset, all infinite products of resolvents have the same asymptotics. For the second space we again construct a subset which is a countable intersection of open and everywhere dense sets such that for each operator belonging to this subset, all infinite products of resolvents converge uniformly on bounded subsets of X to the unique zero of the operator.

Let $(X, \|\cdot\|)$ be a Banach space. We denote by $I : X \to X$ the identity operator on X (that is, $Ix = x$, $x \in X$). Recall that a set-valued operator $A : X \to 2^X$ with a nonempty domain

$$D(A) = \{x \in X : Ax \neq \emptyset\}$$

and range

$$R(A) = \left\{ y \in X : y \in Ax \text{ for some } x \in D(A) \right\}$$

is said to be accretive if

$$\|x - y\| \leq \left\| x - y + r(u - v) \right\| \tag{6.367}$$

for all $x, y \in D(A)$, $u \in Ax$, $v \in Ay$ and $r > 0$. When the operator A is accretive, then it follows from (6.367) that its resolvents

$$J_r^A = (I + rA)^{-1} : R(I + rA) \to D(A) \tag{6.368}$$

are single-valued nonexpansive operators for all positive r. In other words,

$$\left\| J_r^A x - J_r^A y \right\| \leq \|x - y\| \tag{6.369}$$

for all x and y in $D(J_r^A) = R(I + rA)$. As usual, the graph of the operator A is defined by

$$\text{graph}(A) = \{(x, y) \in X \times X : y \in Ax\}.$$

Note that if A is accretive, then the operator $\bar{A} : X \to 2^X$, the graph of which is the closure of graph(A) in the norm topology of $X \times X$, is also accretive. We will say that the operator A is closed if its graph is closed in $X \times X$.

An accretive operator $A : X \to X$ is said to be m-accretive if

$$R(I + rA) = X \quad \text{for all } r > 0.$$

Note that if X is a Hilbert space $(H, \langle \cdot, \cdot \rangle)$, then an operator A is accretive if and only if it is monotone; that is, if and only if

$$\langle u - v, x - y \rangle \geq 0 \quad \text{for all } (x, u), (y, v) \in \text{graph}(A).$$

It is well known that in a Hilbert space an operator A is m-accretive if and only if it is maximal monotone. It is not difficult to see that in any Banach space an m-accretive operator is maximal accretive; that is, if $\tilde{A} : X \to X$ is accretive and graph$(A) \subset$ graph$(\tilde{A})$, then $\tilde{A} = A$. However, the converse is not true in general.

In the sequel we are going to use a certain topology on the space of nonempty closed subsets of $Y = X \times X$. We will now define this topology in a more general setting (cf. [11]). Let (Y, ρ) be a complete metric space. Fix $\theta \in Y$. For each $r > 0$, define

$$Y_r = \{y \in Y : \rho(y, \theta) \leq r\}.$$

For each $y \in Y$ and each $E \subset Y$, define

$$\rho(y, E) = \inf\{\rho(y, z) : z \in E\}.$$

Denote by $S(Y)$ the set of all nonempty and closed subsets of Y. For $F, G \in S(Y)$ and an integer $n \geq 1$, define

$$h_n(F, G) = \sup_{y \in Y_n} |\rho(y, F) - \rho(y, G)|.$$

Clearly, $h_n(F, G) < \infty$ for each integer $n \geq 1$ and each pair of sets $F, G \in S(Y)$. For the set $S(Y)$ we consider the uniformity generated by the following base:

$$\tilde{E}(n) = \{(F, G) \in S(Y) \times S(Y) : h_n(F, G) < n^{-1}\}, \quad n = 1, 2, \ldots. \tag{6.370}$$

This uniform space is metrizable by the metric

$$h(F, G) = \sum_{n=1}^{\infty} 2^{-n} [h_n(F, G)/(1 + h_n(F, G))] \tag{6.371}$$

and the metric space $(S(Y), h)$ is complete.

From now on we apply the above to the space $Y = X \times X$ with the metric

$$\rho\big((x_1, x_2), (z_1, z_2)\big) = \|x_1 - z_1\| + \|x_2 - z_2\|, \quad x_i, z_i \in X, i = 1, 2,$$

and with $\theta = (0, 0)$.

Denote by $\mathcal{M}_a$ the set of all closed accretive operators $A : X \to 2^X$. For each $A, B \in \mathcal{M}_a$, define

$$h_a(A, B) = h\big(\mathrm{graph}(A), \mathrm{graph}(B)\big). \tag{6.372}$$

Clearly, $(\mathcal{M}_a, h_a)$ is a metric space and the set $\{\mathrm{graph}(A) : A \in \mathcal{M}_a\}$ is a closed subset of $S(X \times X)$. Therefore $(\mathcal{M}_a, h_a)$ is a complete metric space. Denote by $\mathcal{M}_m$ the set of all m-accretive operators $A \in \mathcal{M}_a$.

Proposition 6.70 *$\mathcal{M}_m$ is a closed subset of $\mathcal{M}_a$.*

Proof Suppose that $\{A_i\}_{i=1}^{\infty} \subset \mathcal{M}_m$, $A \in \mathcal{M}_a$, and that $A_i \to A$ as $i \to \infty$ in $\mathcal{M}_a$. Assume that r is a positive number. We have to show that $R(I + rA) = X$. To this end, let $z \in X$. For each integer $n \geq 1$, there exists $y_n \in X$ for which

$$z \in (I + rA_n)y_n \quad \text{or, equivalently,} \quad y_n = (I + rA_n)^{-1}z. \tag{6.373}$$

We will show that the sequence $\{y_n\}_{n=1}^{\infty}$ is bounded. To this end, fix $(x, u) \in \mathrm{graph}(A)$. There is a sequence $\{(x_n, u_n)\}_{n=1}^{\infty} \subset X \times X$ such that

$$(x_n, u_n) \in \mathrm{graph}(A_n), \quad n = 1, 2, \ldots, \quad \text{and} \quad \lim_{n\to\infty} (x_n, u_n) = (x, u). \tag{6.374}$$

For each integer $n \geq 1$,

$$x_n = (I + rA_n)^{-1}(x_n + ru_n) \quad \text{and} \quad \|x_n - y_n\| \leq \|x_n + ru_n - z\|. \tag{6.375}$$

By (6.374) and (6.375), the sequence $\{y_n\}_{n=1}^{\infty}$ is indeed bounded. By (6.373), for each integer $n \geq 1$, there exists v_n for which

$$v_n \in A_n(y_n) \quad \text{and} \quad z = y_n + rv_n. \tag{6.376}$$

Clearly, the sequence $\{(y_n, v_n)\}_{n=1}^{\infty}$ is also bounded. There exists a sequence

$$\big\{(\tilde{y}_n, \tilde{v}_n)\big\}_{n=1}^{\infty} \subset \mathrm{graph}(A)$$

such that

$$\|\tilde{y}_n - y_n\| + \|\tilde{v}_n - v_n\| \to 0 \quad \text{as } n \to \infty. \tag{6.377}$$

Set, for all integers $n \geq 1$,

$$z_n = \tilde{y}_n + r\tilde{v}_n \in (I + rA)\tilde{y}_n. \tag{6.378}$$

By (6.376)–(6.378),

$$\lim_{n\to\infty} z_n = z \quad \text{and} \quad \|z_n - z_k\| \geq \|\tilde{y}_n - \tilde{y}_k\| \quad \text{for all integers } n, k.$$

Therefore the sequence $\{(\tilde{y}_n, \tilde{v}_n)\}_{n=1}^{\infty}$ converges to $(y, v) \in \text{graph}(A)$. Clearly, $z = y + rv$. Proposition 6.70 is proved. □

Denote by $\mathcal{M}_m^*$ the set of all $A \in \mathcal{M}_m$ such that there exists x_A for which $0 \in A(x_A)$ and denote by $\bar{\mathcal{M}}_m^*$ the closure of $\mathcal{M}_m^*$ in $\mathcal{M}_m$. The two complete metric spaces $(\mathcal{M}_m, h_a)$ and $(\bar{\mathcal{M}}_m^*, h_a)$ are the focal points of our investigations. Finally, we denote by $\mathcal{M}_0^*$ the set of all $A \in \mathcal{M}_m^*$ for which there exists $x_A \in X$ such that

$$0 \in A(x_A) \quad \text{and} \quad \left(J_1^A\right)^n(x) \to x_A \quad \text{as } n \to \infty \text{ for all } x \in X.$$

Let $\{\bar{r}_n\}_{n=1}^{\infty}$ be a sequence of positive numbers such that

$$\bar{r}_n < 1, \quad n = 1, 2, \ldots, \qquad \lim_{n\to\infty} \bar{r}_n = 0 \quad \text{and} \quad \sum_{n=1}^{\infty} \bar{r}_n = \infty \tag{6.379}$$

and let $\tilde{r} > 1$.

Theorem 6.71 *There exists a set $\mathcal{F} \subset \mathcal{M}_m$, which is a countable intersection of open and everywhere dense sets in $\mathcal{M}_m$ such that for each $A \in \mathcal{F}$, each $\delta > 0$ and each $K > 0$ the following assertion holds*:

There exist a neighborhood U of A in $\mathcal{M}_m$ and an integer $n_0 \geq 1$ such that for each sequence of positive numbers $\{r_n\}_{n=1}^{\infty}$ satisfying $\tilde{r} > r_n \geq \bar{r}_n$, $n = 1, 2, \ldots$, each $B \in U$ and each $x, y \in X$ satisfying $\|x\|, \|y\| \leq K$, we have

$$\left\| J_{r_n}^B \cdot J_{r_{n-1}}^B \cdot \cdots \cdot J_{r_1}^B x - J_{r_n}^B \cdot J_{r_{n-1}}^B \cdot \cdots \cdot J_{r_1}^B y \right\| \leq \delta$$

for all integers $n \geq n_0$.

We remark in passing that such a result is called a weak ergodic theorem in population biology [43]. It means that for a generic operator in $\mathcal{M}_m$ all infinite products of its resolvents become eventually close to each other.

Theorem 6.72 *There exists a set $\mathcal{F} \subset \mathcal{M}_0^* \cap \bar{\mathcal{M}}_m^*$, which is a countable intersection of open and everywhere dense sets in $\bar{\mathcal{M}}_m^*$ such that for each $A \in \mathcal{F}$, the following two assertions hold*:

(i) *There exists a unique $x_A \in X$ such that $0 \in A(x_A)$.*
(ii) *For each $\delta > 0$ and each $K > 0$, there exist a neighborhood U of A in $\mathcal{M}_m$ and an integer $n_0 \geq 1$ such that for each sequence of positive numbers $\{r_n\}_{n=1}^{\infty}$ satisfying $\tilde{r} > r_n \geq \bar{r}_n$, $n = 1, 2, \ldots$, each $B \in U \cap \mathcal{M}_0^*$ and each $x \in X$ satisfying $\|x\| \leq K$, we have*

$$\left\| J_{r_n}^B \cdot J_{r_{n-1}}^B \cdot \cdots \cdot J_{r_1}^B x - x_A \right\| \leq \delta$$

for all integers $n \geq n_0$.

This result means that a generic operator in $\bar{\mathcal{M}}_m^*$ has a unique zero and all the infinite products of its resolvents converge uniformly on bounded subsets of X to this zero.

6.33 Auxiliary Results

Let $\{\bar{r}_n\}_{n=1}^{\infty} \subset (0, 1)$ satisfy (6.379) and let $\tilde{r} > 1$.

Lemma 6.73 *Let $A \in \mathcal{M}_m$, $K_0 > 0$ and let $n_0 \geq 2$ be an integer. Then there exist a neighborhood U of A in $\mathcal{M}_m$ and a number $c_0 > 0$ such that for each $B \in U$, each sequence $\{r_i\}_{i=1}^{n_0-1} \subset (0, \tilde{r})$ and each sequence $\{x_i\}_{i=1}^{n_0} \subset X$ satisfying $\|x_1\| \leq K_0$, $x_{i+1} = J_{r_i}^B(x_i)$, $i = 1, \ldots, n_0 - 1$, we have $\|x_i\| \leq c_0$ for all $i = 1, \ldots, n_0$.*

Proof Choose $(x_A, u_A) \in \text{graph}(A)$. There exists a neighborhood U of A in $\mathcal{M}_m$ such that for each $B \in U$ there exists $(x_B, u_B) \in \text{graph}(B)$ satisfying

$$\|x_B - x_A\| + \|u_A - u_B\| < 1. \tag{6.380}$$

Assume that $B \in U$,

$$\begin{gathered} \{r_i\}_{i=1}^{n_0-1} \subset (0, \tilde{r}), \qquad x_1 \in X, \qquad \|x_1\| \leq K_0 \quad \text{and} \\ x_{i+1} = J_{r_i}^B(x_i), \quad i = 1, \ldots, n_0 - 1. \end{gathered} \tag{6.381}$$

We will estimate $\|x_i\|$ for $i = 1, \ldots, n_0$. To this end, set

$$z_i = x_B + r_i u_B, \quad i = 1, \ldots, n_0 - 1. \tag{6.382}$$

For such i we clearly have by (6.380)–(6.382),

$$x_B = J_{r_i}^B(z_i), \qquad \|x_B - x_{i+1}\| \leq \|z_i - x_i\|$$

and

$$\begin{aligned} \|x_{i+1}\| &\leq \|x_B\| + \|x_i\| + \|z_i\| \leq \|x_i\| + \|x_A\| + 1 + \|x_B + r_i u_B\| \\ &\leq \|x_i\| + 1 + \|x_A\| + \|x_B\| + \tilde{r}\|u_B\| \\ &\leq \|x_i\| + 1 + 2\|x_A\| + 1 + \tilde{r}\big(\|u_A\| + 1\big). \end{aligned}$$

This implies that for $i = 1, \ldots, n_0 - 1$,

$$\|x_{i+1}\| \leq i\big(2\|x_A\| + 2 + \tilde{r}\big(\|u_A\| + 1\big)\big) + K_0.$$

The proof of Lemma 6.73 is complete. □

Assumption (6.379) and Lemma 6.73 imply the following result.

Lemma 6.74 *Let $A \in \mathcal{M}_m$, $K_0 > 0$ and let $n_0 \geq 2$ be an integer. Then there exist a neighborhood U of A in $\mathcal{M}_m$ and a number $c_1 > 0$ such that for each $B \in U$, each sequence $r_i \in [\bar{r}_i, \tilde{r})$, $i = 1, \dots, n_0 - 1$, and each two sequences $\{x_i\}_{i=1}^{n_0} \subset X$, $\{y_i\}_{i=2}^{n_0} \subset X$ satisfying*

$$\|x_1\| \leq K_0, \qquad x_{i+1} = J_{r_i}^B(x_i),$$

$$x_i = x_{i+1} + r_i y_{i+1}, \qquad y_{i+1} \in B(x_{i+1}), \quad i = 1, \dots, n_0 - 1,$$

the following two estimates hold:

$$\|x_i\| \leq c_1, \quad i = 1, \dots, n_0, \quad \text{and} \quad \|y_i\| \leq c_1, \quad i = 2, \dots, n_0.$$

Lemma 6.75 *Let $A \in \mathcal{M}_m$, $x_* \in X$, $0 \in A(x_*)$, $\varepsilon > 0$ and let $n_0 \geq 2$ be an integer. Then there exists a neighborhood U of A in $\mathcal{M}_m$ such that for each $B \in U$ and each sequence $r_i \in (0, \tilde{r})$, $i = 1, \dots, n_0 - 1$, there exists a sequence $\{x_i\}_{i=1}^{n_0} \subset X$ such that*

$$x_{i+1} = J_{r_i}^B(x_i), \quad i = 1, \dots, n_0 - 1, \quad \text{and} \quad \|x_i - x_*\| \leq \varepsilon, \quad i = 1, \dots, n_0.$$

Proof Choose a natural number p such that

$$p > 4 + n_0 + \|x_*\| \quad \text{and} \quad p > \tilde{r}(n_0 + 1)\big(\inf\{1, \varepsilon\}\big)^{-1} \tag{6.383}$$

and define

$$U = \big\{B \in \mathcal{M}_m : h_p\big(\text{graph}(A), \text{graph}(B)\big) < p^{-1}\big\}. \tag{6.384}$$

Assume that $B \in U$ and $r_i \in (0, \tilde{r})$, $i = 1, \dots, n_0 - 1$. By (6.383) and (6.384), there exists $(x_1, y_1) \in \text{graph}(B)$ such that

$$\|x_1 - x_*\| + \|y_1\| < p^{-1}. \tag{6.385}$$

Set

$$\xi_i = x_1 + r_i y_1, \quad i = 1, \dots, n_0 - 1. \tag{6.386}$$

Then

$$x_1 = J_{r_i}^B(\xi_i) \quad \text{and} \quad \|x_1 - \xi_i\| < \tilde{r}/p, \quad i = 1, \dots, n_0 - 1. \tag{6.387}$$

Set

$$x_{i+1} = J_{r_i}^B(x_i), \quad i = 1, \dots, n_0 - 1. \tag{6.388}$$

Since for $i = 1, \dots, n_0 - 1$, $J_{r_i}^B$ is a nonexpansive operator, it follows from (6.385)–(6.388) that for each integer $k \in [2, n_0]$, we have

$$\begin{aligned}\|x_k - x_1\| &\leq \|x_{k-1} - \xi_{k-1}\| \leq \|x_{k-1} - x_1\| + \tilde{r}\|y_1\| \\ &< \|x_{k-1} - x_1\| + \tilde{r}/p, \qquad \|x_k - x_1\| \leq k\tilde{r}/p\end{aligned}$$

and

$$\|x_k - x_*\| < \|x_k - x_1\| + \|x_1 - x_*\| < (k+1)\tilde{r}/p \le (n_0+1)\tilde{r}/p < \varepsilon.$$

This completes the proof of Lemma 6.75. □

6.34 Proof of Theorem 6.71

For each $A \in \mathcal{M}_m$, $\xi \in X$ and each positive number γ, let the operator $A_{\gamma,\xi}$ be defined by

$$A_{\gamma,\xi}x = Ax + \gamma(x - \xi), \quad x \in X.$$

We begin the proof with the following three observations.

Lemma 6.76 *If $A \in \mathcal{M}_m$, $\xi \in X$ and $\gamma > 0$, then $A_{\gamma,\xi} \in \mathcal{M}_m$.*

Lemma 6.77 *Let $A \in \mathcal{M}_m$, $\xi \in X$, $\gamma, r > 0$ and let $x, y \in X$. Then*

$$\left\| J_r^{A_{\gamma,\xi}}(x) - J_r^{A_{\gamma,\xi}}(y)\right\| \le (1+\gamma r)^{-1}\|x - y\|.$$

Lemma 6.78 *For each fixed $\xi \in X$, the set $\{A_{\gamma,\xi} : A \in \mathcal{M}_m, \gamma \in (0,1)\}$ is everywhere dense in $\mathcal{M}_m$.*

In the rest of the proof we assume that (cf. (6.379))

$$\tilde{r} > 1, \qquad \{\bar{r}_n\}_{n=1}^{\infty} \subset (0,1), \qquad \lim_{n\to\infty} \bar{r}_n = 0 \quad \text{and} \quad \sum_{n=1}^{\infty} \bar{r}_n = \infty. \tag{6.389}$$

Lemma 6.79 *Let $A \in \mathcal{M}_m$, $\xi \in X$, $\gamma \in (0,1)$ and $\delta, K > 0$. Then there exist a neighborhood U of $A_{\gamma,\xi}$ in $\mathcal{M}_m$ and an integer $n_0 \ge 4$ such that for each $B \in U$, each sequence of numbers $r_i \in [\bar{r}_i, \tilde{r})$, $i = 1, \dots, n_0 - 1$, and each $x, y \in X$ satisfying $\|x\|, \|y\| \le K$, the following estimate holds:*

$$\left\| J^B_{r_{n_0-1}} \cdot J^B_{r_{n_0-2}} \cdot \dots \cdot J^B_{r_1} x - J^B_{r_{n_0-1}} \cdot J^B_{r_{n_0-2}} \cdot \dots \cdot J^B_{r_1} y \right\| \le \delta. \tag{6.390}$$

Proof Choose a number γ_0 such that

$$\gamma_0 \in (0, \gamma). \tag{6.391}$$

Clearly

$$\prod_{i=1}^{n} (1 + \gamma_0 \bar{r}_i) \to \infty \quad \text{as } n \to \infty. \tag{6.392}$$

Therefore there exists an integer $n_0 \geq 4$ such that

$$(2K+2)\prod_{i=1}^{n_0-1}(1+\gamma_0\bar{r}_i)^{-1} < \delta/2. \tag{6.393}$$

By Lemma 6.74, there exist a neighborhood U_1 of $A_{\gamma,\xi}$ in $\mathcal{M}_m$ and a number $c_1 > 0$ such that for each $B \in U_1$, each sequence $r_i \in [\bar{r}_i, \tilde{r}_i)$, $i = 1, \dots, n_0 - 1$, and each pair of sequences $\{x_i\}_{i=1}^{n_0} \subset X$ and $\{u_i\}_{i=2}^{n_0} \subset X$ satisfying

$$\begin{aligned} &\|x_1\| \leq K, \qquad x_{i+1} = J_{r_i}^B(x_i), \qquad x_i = x_{i+1} + r_i u_{i+1}, \\ &u_{i+1} \in B(x_{i+1}), \quad i = 1, \dots, n_0 - 1, \end{aligned} \tag{6.394}$$

the following estimates hold:

$$\|x_i\| \leq c_1, \quad i = 1, \dots, n_0, \quad \text{and} \quad \|u_i\| \leq c_1, \quad i = 2, \dots, n_0. \tag{6.395}$$

Choose a natural number m_1 such that

$$\begin{aligned} &m_1 > 4\big(n_0 + 8(c_1 + 1)\big), \\ &\big[(1+\gamma_0\bar{r}_i)^{-1} - (1+\gamma\bar{r}_i)^{-1}\big]\delta > 2(2+\tilde{r})m_1^{-1}, \quad i = 1, \dots, n_0, \end{aligned} \tag{6.396}$$

and set

$$U = \big\{B \in U_1 : h_{m_1}\big(\mathrm{graph}(A_{\gamma,\xi}), \mathrm{graph}(B)\big) < m_1^{-1}\big\}. \tag{6.397}$$

Assume that $B \in U$, $r_i \in [\bar{r}_i, \tilde{r})$, $i = 1, \dots, n_0 - 1$, and

$$x, y \in X \quad \text{and} \quad \|x\|, \|y\| \leq K. \tag{6.398}$$

Set

$$\begin{aligned} &x_1 = x, \qquad y_1 = y, \\ &x_{i+1} = J_{r_i}^B(x_i) \quad \text{and} \quad y_{i+1} = J_{r_i}^B(y_i), \quad i = 1, \dots, n_0 - 1. \end{aligned} \tag{6.399}$$

For each $i = 1, \dots, n_0 - 1$, there exist u_{i+1} and $v_{i+1} \in X$ such that

$$\begin{aligned} &u_{i+1} \in B(x_{i+1}), \qquad v_{i+1} \in B(y_{i+1}), \\ &x_i = x_{i+1} + r_i u_{i+1} \quad \text{and} \quad y_i = y_{i+1} + r_i v_{i+1}. \end{aligned} \tag{6.400}$$

It follows from the definition of U_1 (see (6.394)) and (6.400) that

$$\|x_i\|, \|y_i\| \leq c_1, \quad i = 1, \dots, n_0 \quad \text{and} \quad \|u_i\| \|v_i\| \leq c_1, \quad i = 2, \dots, n_0. \tag{6.401}$$

To prove the lemma it is sufficient to show that

$$\|x_{n_0} - y_{n_0}\| \leq \delta. \tag{6.402}$$

Assume the contrary. Then

$$\|x_i - y_i\| > \delta, \quad i = 1, \dots, n_0. \tag{6.403}$$

Let $i \in \{1, \dots, n_0 - 1\}$. It follows from (6.400), (6.401), (6.397) and (6.396) that there exist

$$(\bar{x}_{i+1}, \bar{u}_{i+1}) \in \text{graph}(A_{\gamma,\xi}) \quad \text{and} \quad (\bar{y}_{i+1}, \bar{v}_{i+1}) \in \text{graph}(A_{\gamma,\xi}) \tag{6.404}$$

such that

$$\begin{aligned} &\|\bar{x}_{i+1} - x_{i+1}\| + \|\bar{u}_{i+1} - u_{i+1}\| < m_1^{-1} \quad \text{and} \\ &\|\bar{y}_{i+1} - y_{i+1}\| + \|\bar{v}_{i+1} - v_{i+1}\| < m_1^{-1}. \end{aligned} \tag{6.405}$$

Set

$$\bar{x}_i = \bar{x}_{i+1} + r_i \bar{u}_{i+1} \quad \text{and} \quad \bar{y}_i = \bar{y}_{i+1} + r_i \bar{v}_{i+1}. \tag{6.406}$$

By Lemma 6.77, (6.404) and (6.406),

$$\begin{aligned} \|\bar{x}_{i+1} - \bar{y}_{i+1}\| &= \left\| J_{r_i}^{A_{\gamma,\xi}} \bar{x}_i - J_{r_i}^{A_{\gamma,\xi}} \bar{y}_i \right\| \le (1 + \gamma r_i)^{-1} \|\bar{x}_i - \bar{y}_i\| \\ &\le (1 + \gamma \bar{r}_i)^{-1} \|\bar{x}_i - \bar{y}_i\|. \end{aligned} \tag{6.407}$$

It follow from (6.406), (6.400) and (6.405) that

$$\|\bar{x}_i - x_i\| \le \|\bar{x}_{i+1} - x_{i+1}\| + r_i \|\bar{u}_{i+1} - u_{i+1}\| \le m_1^{-1}(1 + \tilde{r}) \tag{6.408}$$

and

$$\|\bar{y}_i - y_i\| \le \|\bar{y}_{i+1} - y_{i+1}\| + r_i \|\bar{v}_{i+1} - v_{i+1}\| \le m_1^{-1}(1 + \tilde{r}).$$

By (6.405), (6.407) and (6.408),

$$\begin{aligned} \|x_{i+1} - y_{i+1}\| &\le \|\bar{x}_{i+1} - \bar{y}_{i+1}\| + 2m_1^{-1} \le 2m_1^{-1} + (1 + \gamma \bar{r}_i)^{-1} \|\bar{x}_i - \bar{y}_i\| \\ &\le 2m_1^{-1} + (1 + \gamma \bar{r}_i)^{-1} \big(\|x_i - y_i\| + 2m_1^{-1}(1 + \tilde{r})\big) \\ &\le (1 + \gamma \bar{r}_i)^{-1} \|x_i - y_i\| + 2m_1^{-1} \big(1 + (1 + \gamma \bar{r}_i)^{-1}(1 + \tilde{r})\big) \\ &\le (1 + \gamma \bar{r}_i)^{-1} \|x_i - y_i\| + 2m_1^{-1}(2 + \tilde{r}). \end{aligned} \tag{6.409}$$

Now (6.409), (6.396) and (6.403) imply that

$$\|x_{i+1} - y_{i+1}\| \le (1 + \gamma_0 \bar{r}_i)^{-1} \|x_i - y_i\|$$

and since these inequalities are valid for all $i \in \{1, \dots, n_0 - 1\}$, it follows from (6.398), (6.399) and (6.393) that

$$\|x_{n_0} - y_{n_0}\| \le 2K \prod_{i=1}^{n_0 - 1} (1 + \gamma_0 \bar{r}_i)^{-1} < \delta/2.$$

This contradicts (6.403). Therefore (6.402) is true and Lemma 6.79 is proved. □

Completion of the proof of Theorem 6.71 Let $A \in \mathcal{M}_m$, $\xi = 0$, $\gamma \in (0,1)$ and let $i \geq 1$ be an integer. By Lemma 6.79, there exist an open neighborhood $U(A, \gamma, i)$ of $A_{\gamma,0}$ in $\mathcal{M}_m$ and an integer $q(A, \gamma, i) \geq 4$ such that for each $B \in U(A, \gamma, i)$, each sequence of numbers $r_i \in [\bar{r}_i, \tilde{r})$, $i = 1, \dots, q(A, \gamma, i) - 1$, and each $x, y \in X$ satisfying $\|x\|, \|y\| \leq 2^{i+1}$, the following estimate holds:

$$\left\| J^B_{r_{q(A,\gamma,i)-1}} \cdot \dots \cdot J^B_{r_1} x - J^B_{r_{q(A,\gamma,i)-1}} \cdot \dots \cdot J^B_{r_1} y \right\| \leq 2^{-i-1}.$$

Define

$$\mathcal{F} := \bigcap_{n=1}^{\infty} \bigcup \left\{ U(A, \gamma, i) : A \in \mathcal{M}_m, \gamma \in (0,1), i \geq n \right\}.$$

Clearly (see Lemma 6.78), $\mathcal{F}$ is a countable intersection of open and everywhere dense sets in $\mathcal{M}_m$. Let $A \in \mathcal{F}$, $\delta > 0$ and $K > 0$ be given. Choose an integer $n > 2K + 2 + 8\delta^{-1}$. There exist $C \in \mathcal{M}_m$, $\gamma \in (0,1)$ and $i \geq n$ such that $A \in U(C, \gamma, i)$. The validity of Theorem 6.71 now follows from the definitions of $U(C, \gamma, i)$ and $q(C, \gamma, i)$.

6.35 Proof of Theorem 6.72

Let

$$\tilde{r} > 1, \qquad \{\bar{r}_n\}_{n=1}^{\infty} \subset (0,1), \qquad \lim_{n \to \infty} \bar{r}_n = 0 \quad \text{and} \quad \sum_{n=1}^{\infty} \bar{r}_n = \infty. \tag{6.410}$$

By definition, for each $A \in \mathcal{M}_m^*$ there exists $x_A \in X$ such that

$$0 \in A(x_A). \tag{6.411}$$

Recalling the definition of $A_{\gamma,\xi}$ at the beginning of Sect. 6.34, we will use in this section the operator A_{γ,x_A}. In other words,

$$A_{\gamma,x_A} x = Ax + \gamma(x - x_A), \quad x \in X. \tag{6.412}$$

By Lemma 6.76 and (6.411), for each $A \in \mathcal{M}_m^*$ and each $\gamma \in (0,1)$,

$$A_{\gamma,x_A} \in \mathcal{M}_m^* \quad \text{and} \quad 0 \in A_{\gamma,x_A}(x_A). \tag{6.413}$$

The following observation is also clear.

Lemma 6.80 *The set* $\{A_{\gamma,x_A} : A \in \mathcal{M}_m^*, \gamma \in (0,1)\}$ *is everywhere dense in* $\bar{\mathcal{M}}_m^*$.

Let $A \in \mathcal{M}_m^*$, $\gamma \in (0,1)$ and let $i \geq 1$ be an integer. By Lemma 6.79 with $\xi = x_A$, there exist an open neighborhood $U_1(A, \gamma, i)$ of A_{γ,x_A} in $\mathcal{M}_m$ and an integer $n(A, \gamma, i) \geq 4$ such that the following property holds:

(a) For each $B \in U_1(A, \gamma, i)$, each sequence

$$r_j \in [\bar{r}_j, \tilde{r}), \quad j = 1, \ldots, n(A, \gamma, i) - 1$$

and each $x, y \in X$ satisfying

$$\|x\|, \|y\| \leq 8^{i+1}\left(4 + 4\|x_A\|\right), \tag{6.414}$$

the following estimate holds:

$$\left\| J^B_{r_{n(A,\gamma,i)-1}} \cdot \cdots \cdot J^B_{r_1} x - J^B_{r_{n(A,\gamma,i)-1}} \cdot \cdots \cdot J^B_{r_1} y \right\| \leq 8^{-i-1}. \tag{6.415}$$

By Lemma 6.75, there exists an open neighborhood $U(A, \gamma, i)$ of A_{γ, x_A} in $\mathcal{M}_m$ such that

$$U(A, \gamma, i) \subset U_1(A, \gamma, i) \tag{6.416}$$

and the following property holds:
(b) For each $B \in U(A, \gamma, i)$ and each sequence

$$r_j \in (0, \tilde{r}), \quad j = 1, \ldots, 8n(A, \gamma, i) - 1,$$

there exists a sequence $\{x_j : j = 1, \ldots, 8n(A, \gamma, j)\} \subset X$ such that

$$x_{j+1} = J^B_{r_j}(x_j), \quad j = 1, \ldots, 8n(A, \gamma, i) - 1, \tag{6.417}$$

and

$$\|x_j - x_A\| \leq 8^{-i-1}, \quad j = 1, \ldots, 8n(A, \gamma, i).$$

We will now show that the following property also holds:
(c) For each $B \in U(A, \gamma, i)$, each $x \in X$ satisfying $\|x\| \leq 8^{i+1}(2 + 2\|x_A\|)$ and each integer $m \geq n(A, \gamma, i) - 1$,

$$\left\| \left(J^B_1\right)^m (x) - x_A \right\| \leq 2 \cdot 8^{-i-1}. \tag{6.418}$$

Indeed, let $B \in U(A, \gamma, i)$. By property (b), there exists a sequence

$$\left\{\bar{x}_j : j = 1, \ldots, 8n(A, \gamma, i)\right\} \subset X \tag{6.419}$$

such that

$$\bar{x}_{j+1} = J^B_1(\bar{x}_j), \quad j = 1, \ldots, 8n(A, \gamma, i) - 1, \tag{6.420}$$

and

$$\|\bar{x}_j - x_A\| < 8^{-i-1}, \quad j = 1, \ldots, 8n(A, \gamma, i).$$

Let $x \in X$ with

$$\|x\| \leq 8^{i+1}\left(2 + 2\|x_A\|\right) \tag{6.421}$$

and consider the sequence $\{(J_1^B)^j(x)\}_{j=1}^{\infty}$. Since the operator J_1^B is nonexpansive, it follows from (6.420) and (6.421) that for $j = 1, \ldots, 8n(A, \gamma, i) - 1$,

$$\begin{aligned}\|(J_1^B)^j x\| &\le \|\bar{x}_{j+1}\| + \|(J_1^B)^j x - \bar{x}_{j+1}\| \\ &\le \|x_A\| + \|\bar{x}_{j+1} - x_A\| + \|(J_1^B)^j x - (J_1^B)^j(\bar{x}_1)\| \\ &\le \|x_A\| + 8^{-i-1} + \|x - \bar{x}_1\| \\ &\le 2(\|x_A\| + 8^{-i-1}) + \|x\| \\ &\le 8^{i+1}(2 + 2\|x_A\|) + 2(\|x_A\| + 2^{-1}) \\ &< 8^{i+1}(4 + 4\|x_A\|). \end{aligned} \tag{6.422}$$

We now show by induction that (6.418) is valid for all integers $m \ge n(A, \gamma, i) - 1$. Let $m = n(A, \gamma, i) - 1$. Then by property (a) and (6.420),

$$\begin{aligned}\|(J_1^B)^m(x) - x_A\| &\le \|(J_1^B)(x) - (J_1^B)^m(\bar{x}_1)\| + \|(J_1^B)^m(\bar{x}_1) - x_A\| \\ &\le 8^{-i-1} + \|\bar{x}_{m+1} - x_A\| \le 2 \cdot 8^{-i-1}.\end{aligned}$$

Therefore for $m = n(A, \gamma, i) - 1$ (6.418) is valid. Assume that $q \ge n(A, \gamma, i) - 1$ and that (6.418) is valid for all integers $m \in [n(A, \gamma, i) - 1, q]$. Consider

$$y = (J_1^B)^p(x) \quad \text{with } p = q - (n(A, \gamma, i) - 1) + 1. \tag{6.423}$$

It follows from (6.418), which is valid by our inductive assumption for all integers $m \in [n(A, \gamma, i) - 1, q]$, and (6.422), which holds for all $j = 1, \ldots, 8n(A, \gamma, i) - 1$, that

$$\|y\| \le 8^{i+1}(4 + 4\|x_A\|).$$

By this estimate, (6.423), (6.420) and property (a),

$$\begin{aligned}&\|(J_1^B)^{q+1}(x) - x_A\| \\ &= \|(J_1^B)^{n(A,\gamma,i)-1}(y) - x_A\| \\ &\le \|(J_1^B)^{n(A,\gamma,i)-1}y - (J_1^B)^{n(A,\gamma,i)-1}(\bar{x}_1)\| + \|\bar{x}_{n(A,\gamma,i)} - x_A\| \\ &\le 2 \cdot 8^{-i-1}.\end{aligned}$$

Therefore (6.418) is valid for all integers $m \ge n(A, \gamma, i) - 1$ and property (c) holds. Next we define

$$\mathcal{F} := \left[\bigcap_{k=1}^{\infty} \bigcup \{U(A, \gamma, i) : A \in \mathcal{M}_m^*, \gamma \in (0, 1), i \ge k\}\right] \cap \bar{\mathcal{M}}_m^*.$$

Clearly, $\mathcal{F}$ is a countable intersection of open and everywhere dense sets in $\bar{\mathcal{M}}_m^*$. We will show that $\mathcal{F} \subset \mathcal{M}_0^*$.

Let $A \in \mathcal{F}$. Then there exist sequences $\{A_k\}_{k=1}^{\infty} \subset \mathcal{M}_m^*$, $\{\gamma_k\}_{k=1}^{\infty} \subset (0,1)$ and a strictly increasing sequence of natural numbers $\{i_k\}_{k=1}^{\infty}$ such that $A \in U(A_k, \gamma, i_k)$ for all natural numbers k. Property (c) implies that there exists $x_A \in X$ such that

$$\lim_{j\to\infty} \left(J_1^A\right)^j(x) = x_A \quad \text{for all } x \in X.$$

Clearly, $0 \in A(x_A)$ and if $y \in X$ satisfies $0 \in A(y)$, then $y = x_A$. Therefore $\mathcal{F} \subset \mathcal{M}_0^*$.

Let $\delta, K > 0$ be given. Choose a natural number q such that

$$4^q > 4K + 4 \quad \text{and} \quad 4^q > \delta^{-1}, \tag{6.424}$$

and consider the open set $U(A_q, \gamma_q, i_q)$.

Let $r_i \in [\bar{r}_i, \tilde{r})$, $i = 1, 2, \dots$, and let

$$B \in \mathcal{M}_0^* \cap U(A_g, \gamma_q, i_q). \tag{6.425}$$

There exists a unique $x_B \in X$ such that

$$0 \in B(x_B) \tag{6.426}$$

and

$$\left(J_1^B\right)^n y \to x_B \quad \text{as } n \to \infty \text{ for all } y \in X. \tag{6.427}$$

It follows from (6.427) and property (c) that

$$\|x_A - x_{A_q}\|, \|x_B - x_{A_q}\| \le 2 \cdot 8^{-i_q - 1}. \tag{6.428}$$

Let $x \in X$ with

$$\|x\| \le K. \tag{6.429}$$

Set $\bar{n} = n(A_q, \gamma_q, i_q)$. It follows from (6.425), (6.428), (6.429), (6.424) and property (a) that

$$\left\| J_{r_{\bar{n}-1}}^B \cdot \dots \cdot J_{r_1}^B x - J_{r_{\bar{n}-1}}^B \cdot \dots \cdot J_{r_1}^B x_B \right\| \le 8^{-i_q - 1}. \tag{6.430}$$

By (6.426), (6.430) and (6.428), we now have, for each integer $n \ge \bar{n}$,

$$\left\| J_{r_{n-1}}^B \cdot \dots \cdot J_{r_1}^B x - x_B \right\| \le \left\| J_{r_{\bar{n}-1}}^B \cdot \dots \cdot J_{r_1}^B x - x_B \right\| \le 8^{-i_q - 1}$$

and

$$\left\| J_{r_{n-1}}^B \cdot \dots \cdot J_{r_1}^B x - x_A \right\| \le 5 \cdot 8^{-i_q - 1} < \delta.$$

This completes the proof of Theorem 6.72.

Let $x \in Z$. Then there exist sequences [illegible] $\in (0, 1)$ and a strictly increasing sequence of natural numbers [illegible] such that [illegible] for all natural numbers k. Property (e) implies that there exists [illegible] such that

[illegible]

Lemma [illegible]

Let [illegible] Choose a natural number q such that

[illegible] (6.324)

[illegible]

[illegible] (6.325)

There exists a number [illegible] such that

[illegible] (6.326)

and

[illegible] (6.327)

It follows from [illegible] that

[illegible] (6.328)

[illegible] (6.329)

[illegible] (6.324), (6.325), (6.329), (6.331) and property [illegible] that

[illegible] (6.330)

[illegible] we have [illegible]

[illegible]

and

[illegible]

This completes the proof of Theorem 6.72.

Chapter 7
Best Approximation

7.1 Well-Posedness and Porosity

Given a nonempty closed subset A of a Banach space $(X, \|\cdot\|)$ and a point $x \in X$, we consider the minimization problem

$$\min\{\|x - y\| : y \in A\}. \tag{P}$$

It is well known that if A is convex and X is reflexive, then problem (P) always has at least one solution. This solution is unique when X is strictly convex.

If A is merely closed but X is uniformly convex, then according to classical results of Stechkin [173] and Edelstein [59], the set of all points in X having a unique nearest point in A is G_δ and dense in X. Since then there has been a lot of activity in this direction. In particular, it is known [84, 88] that the following properties are equivalent for any Banach space X:

(A) X is reflexive and has a Kadec-Klee norm.
(B) For each nonempty closed subset A of X, the set of points in $X \setminus A$ with nearest points in A is dense in $X \setminus A$.
(C) For each nonempty closed subset A of X, the set of points in $X \setminus A$ with nearest points in A is generic (that is, a dense G_δ subset) in $X \setminus A$.

A more recent result of De Blasi, Myjak and Papini [52] establishes well-posedness of problem (P) for a uniformly convex X, closed A and a generic $x \in X$.

In this connection we recall that the minimization problem (P) is said to be well posed if it has a unique solution, say a_0, and every minimizing sequence of (P) converges to a_0.

A more precise formulation of the De Blasi-Myjak-Papini result mentioned above involves the notion of porosity.

Using this terminology and denoting by F the set of all points such that the minimization problem (P) is well posed, we note that De Blasi, Myjak and Papini [52] proved, in fact, that the complement $X \setminus F$ is σ-porous in X.

However, the fundamental restriction in all these results is that they hold only under certain assumptions on the space X. In view of the Lau-Konjagin result

S. Reich, A.J. Zaslavski, *Genericity in Nonlinear Analysis*,
Developments in Mathematics 34, DOI 10.1007/978-1-4614-9533-8_7,

mentioned above these assumptions cannot be removed. On the other hand, many generic results in nonlinear functional analysis hold in any Banach space. Therefore the following natural question arises: can generic results for best approximation problems be obtained in general Banach spaces? In [138] we answer this question in the affirmative. In this chapter we present the results obtained in [138].

To this end, we change our point of view and consider a new framework. The main feature of this new framework is that the set A in problem (P) may also vary. In our first result (Theorem 7.3) we fix x and consider the space $S(X)$ of all nonempty closed subsets of X equipped with an appropriate complete metric, say h. We then show that the collection of all sets $A \in S(X)$ for which problem (P) is well posed has a σ-porous complement.

In the second result (Theorem 7.4) we consider the space of pairs $S(X) \times X$ with the metric $h(A, B) + \|x - y\|$, where $A, B \in S(X)$ and $x, y \in X$. Once again we show that the family of all pairs $(A, x) \in S(X) \times X$ for which problem (P) is well-posed has a σ-porous complement.

In our third result (Theorem 7.5) we show that for any nonempty, separable and closed subset X_0 of X, there exists a subset $\mathcal{F}$ of $(S(X), h)$ with a σ-porous complement such that any $A \in \mathcal{F}$ has the following property:

There exists a dense G_δ subset F of X_0 such that for any $x \in F$, the minimization problem (P) is well posed.

In order to prove these results we now provide more information on porous sets.

Let (Y, ρ) be a metric space. We denote by $B_\rho(y, r)$ the closed ball of center $y \in Y$ and radius $r > 0$.

The following simple observation was made in [180].

Proposition 7.1 *Let E be a subset of the metric space (Y, ρ). Assume that there exist $r_0 > 0$ and $\beta \in (0, 1)$ such that the following property holds*:

(P1) *For each $x \in Y$ and each $r \in (0, r_0]$, there exists $z \in Y \setminus E$ such that $\rho(x, z) \leq r$ and $B_\rho(z, \beta r) \cap E = \emptyset$.*

Then E is porous with respect to ρ.

Proof Let $x \in Y$ and $r \in (0, r_0]$. By property (P1), there exists $z \in Y \setminus E$ such that

$$\rho(x, z) \leq r/2 \quad \text{and} \quad B_\rho(z, \beta r/2) \cap E = \emptyset.$$

Hence $B_\rho(z, \beta r/2) \subset B_\rho(x, r) \setminus E$ and Proposition 7.1 is proved. □

As a matter of fact, property (P1) can be weakened.

Proposition 7.2 *Let E be a subset of the metric space (Y, ρ). Assume that there exist $r_0 > 0$ and $\beta \in (0, 1)$ such that the following property holds*:

(P2) *For each $x \in E$ and each $r \in (0, r_0]$, there exists $z \in Y \setminus E$ such that $\rho(x, z) \leq r$ and $B_\rho(z, \beta r) \cap E = \emptyset$.*

Then E is porous with respect to ρ.

Proof We may assume that $\beta < 1/2$. Let $x \in Y$ and $r \in (0, r_0]$. We will show that there exists $z \in Y \setminus E$ such that

$$\rho(x, z) \le r \quad \text{and} \quad B_\rho(z, \beta r/2) \cap E = \emptyset. \tag{7.1}$$

If $B_\rho(x, r/4) \cap E = \emptyset$, then (7.1) holds with $z = x$. Assume now that $B_\rho(x, r/4) \cap E \neq \emptyset$. Then there exists

$$x_1 \in B_\rho(x, r/4) \cap E. \tag{7.2}$$

By property (P2), there exists $z \in Y \setminus E$ such that

$$\rho(x_1, z) \le r/2 \quad \text{and} \quad B_\rho(z, \beta r/2) \cap E = \emptyset. \tag{7.3}$$

The relations (7.2) and (7.3) imply that

$$\rho(x, z) \le \rho(x, x_1) + \rho(x_1, z) \le 3r/4.$$

Thus there indeed exists $z \in Y \setminus E$ satisfying (7.1). Proposition 7.2 is now seen to follow from Proposition 7.1. □

The following definition was introduced in [180].

Assume that a set Y is equipped with two metrics ρ_1 and ρ_2 such that $\rho_1(x, y) \le \rho_2(x, y)$ for all $x, y \in Y$ and that the metric spaces (Y, ρ_1) and (Y, ρ_2) are complete.

We say that a set $E \subset Y$ is porous with respect to the pair (ρ_1, ρ_2) if there exist $r_0 > 0$ and $\alpha \in (0, 1)$ such that for each $x \in E$ and each $r \in (0, r_0]$, there exists $z \in Y \setminus E$ such that $\rho_2(z, x) \le r$ and $B_{\rho_1}(z, \alpha r) \cap E = \emptyset$.

Proposition 7.2 implies that if E is porous with respect to (ρ_1, ρ_2), then it is porous with respect to both ρ_1 and ρ_2.

A set $E \subset Y$ is called σ-porous with respect to (ρ_1, ρ_2) if it is a countable union of sets which are porous with respect to (ρ_1, ρ_2).

As a matter of fact, it turns out that our results are true not only for Banach spaces, but also for all complete hyperbolic spaces.

Let (X, ρ, M) be a complete hyperbolic space. For each $x \in X$ and each $A \subset X$, set

$$\rho(x, A) = \inf\{\rho(x, y) : y \in A\}.$$

Denote by $S(X)$ the family of all nonempty closed subsets of X. For each $A, B \in S(X)$, define

$$H(A, B) := \max\{\sup\{\rho(x, B) : x \in A\}, \sup\{\rho(y, A) : y \in B\}\} \tag{7.4}$$

and

$$\tilde{H}(A, B) := H(A, B)\big(1 + H(A, B)\big)^{-1}.$$

It is easy to see that $\tilde{H}$ is a metric on $S(X)$ and that the space $(S(X), \tilde{H})$ is complete.

Fix $\theta \in X$. For each natural number n and each $A, B \in S(X)$, we set

$$h_n(A, B) = \sup\{|\rho(x, A) - \rho(x, B)| : x \in X \text{ and } \rho(x, \theta) \leq n\} \tag{7.5}$$

and

$$h(A, B) = \sum_{n=1}^{\infty} [2^{-n} h_n(A, B)(1 + h_n(A, B))^{-1}].$$

Once again it is not difficult to see that h is a metric on $S(X)$ and that the metric space $(S(X), h)$ is complete. Clearly,

$$\tilde{H}(A, B) \geq h(A, B) \quad \text{for all } A, B \in S(X).$$

We equip the set $S(X)$ with the pair of metrics $\tilde{H}$ and h.

We now state the following three results which were obtained in [138]. Their proofs are given later in this chapter.

Theorem 7.3 *Let (X, ρ, M) be a complete hyperbolic space and let $\tilde{x} \in X$. Then there exists a set $\Omega \subset S(X)$ such that its complement $S(X) \setminus \Omega$ is σ-porous with respect to the pair $(h, \tilde{H})$ and such that for each $A \in \Omega$, the following property holds:*

(C1) *There exists a unique $\tilde{y} \in A$ such that $\rho(\tilde{x}, \tilde{y}) = \rho(\tilde{x}, A)$. Moreover, for each $\varepsilon > 0$, there exists $\delta > 0$ such that if $x \in A$ satisfies $\rho(\tilde{x}, x) \leq \rho(\tilde{x}, A) + \delta$, then $\rho(x, \tilde{y}) \leq \varepsilon$.*

To state the following result we endow the Cartesian product $S(X) \times X$ with the pair of metrics d_1 and d_2 defined by

$$d_1((A, x), (B, y)) = h(A, B) + \rho(x, y),$$

$$d_2((A, x), (B, y)) = \tilde{H}(A, B) + \rho(x, y), \quad x, y \in X, A, B \in S(X).$$

Theorem 7.4 *Let (X, ρ, M) be a complete hyperbolic space. There exists a set $\Omega \subset S(X) \times X$ such that its complement $[S(X) \times X] \setminus \Omega$ is σ-porous with respect to the pair (d_1, d_2) and such that for each $(A, \tilde{x}) \in \Omega$, the following property holds:*

(C2) *There exists a unique $\tilde{y} \in A$ such that $\rho(\tilde{x}, \tilde{y}) = \rho(\tilde{x}, A)$. Moreover, for each $\varepsilon > 0$, there exists $\delta > 0$ such that if $z \in X$ satisfies $\rho(\tilde{x}, z) \leq \delta$, $B \in S(X)$ satisfies $h(A, B) \leq \delta$, and $y \in B$ satisfies $\rho(y, z) \leq \rho(z, B) + \delta$, then $\rho(y, \tilde{y}) \leq \varepsilon$.*

In classical generic results the set A was fixed and x varied in a dense G_δ subset of X. In our first two results the set A is also variable. However, in our third result we show that if X_0 is a nonempty, separable and closed subset of X, then for every fixed A in a dense G_δ subset of $S(X)$ with a σ-porous complement, the set of all $x \in X_0$ for which problem (P) is well posed contains a dense G_δ subset of X_0.

Theorem 7.5 *Let (X, ρ, M) be a complete hyperbolic space. Assume that X_0 is a nonempty, separable and closed subset of X. Then there exists a set $\mathcal{F} \subset S(X)$ such that $S(X) \setminus \mathcal{F}$ is σ-porous with respect to the pair $(h, \tilde{H})$ and such that for each $A \in \mathcal{F}$, the following property holds*:

(C3) *There exists a set $F \subset X_0$ which is a countable intersection of open and everywhere dense subsets of X_0 with the relative topology such that for each $\tilde{x} \in F$, there exists a unique $\tilde{y} \in A$ for which $\rho(\tilde{x}, \tilde{y}) = \rho(\tilde{x}, A)$. Moreover, if $\{y_i\}_{i=1}^{\infty} \subset A$ satisfies $\lim_{i\to\infty} \rho(\tilde{x}, y_i) = \rho(\tilde{x}, A)$, then $y_i \to \tilde{y}$ as $i \to \infty$.*

7.2 Auxiliary Results

Let (X, ρ, M) be a complete hyperbolic space and let $S(X)$ be the family of all nonempty closed subsets of X.

Lemma 7.6 *Let $A \in S(X)$, $\tilde{x} \in X$ and let $r, \varepsilon \in (0, 1)$. Then there exists $\bar{x} \in X$ such that $\rho(\bar{x}, A) \le r$ and for the set $\tilde{A} = A \cup \{\bar{x}\}$ the following properties hold*:

$$\rho(\tilde{x}, \bar{x}) = \rho(\tilde{x}, \tilde{A});$$

$$\text{if } x \in \tilde{A} \text{ and } \rho(\tilde{x}, x) \le \rho(\tilde{x}, \tilde{A}) + \varepsilon r/4, \text{ then } \rho(\bar{x}, x) \le \varepsilon.$$

Proof If $\rho(\tilde{x}, A) \le r$, then the lemma holds with $\bar{x} = \tilde{x}$ and $\tilde{A} = A \cup \{\tilde{x}\}$. Therefore we may restrict ourselves to the case where

$$\rho(\tilde{x}, A) > r. \tag{7.6}$$

Choose $x_0 \in A$ such that

$$\rho(\tilde{x}, x_0) \le \rho(\tilde{x}, A) + r/2. \tag{7.7}$$

There exists

$$\bar{x} \in \{\gamma\tilde{x} \oplus (1 - \gamma)x_0 : \gamma \in (0, 1)\} \tag{7.8}$$

such that

$$\rho(\bar{x}, x_0) = r \quad \text{and} \quad \rho(\tilde{x}, \bar{x}) = \rho(\tilde{x}, x_0) - r. \tag{7.9}$$

Set $\tilde{A} = A \cup \{\bar{x}\}$. We have by (7.9) and (7.7),

$$\rho(\tilde{x}, \bar{x}) = \rho(\tilde{x}, x_0) - r \le \rho(\tilde{x}, A) + r/2 - r = \rho(\tilde{x}, A) - r/2.$$

Therefore $\rho(\tilde{x}, \bar{x}) = \rho(\tilde{x}, \tilde{A})$, and if $x \in \tilde{A}$ and $\rho(\tilde{x}, x) < \rho(\tilde{x}, \tilde{A}) + r/2$, then $x = \bar{x}$. This completes the proof of Lemma 7.6. □

Before stating our next lemma we choose, for each $\varepsilon \in (0, 1)$ and each natural number n, a number

$$\alpha(\varepsilon, n) \in (0, 16^{-n-2}\varepsilon). \tag{7.10}$$

Lemma 7.7 *Let $A \in S(X)$, $\tilde{x} \in X$ and let $r, \varepsilon \in (0,1)$. Suppose that n is a natural number, let*

$$\alpha = \alpha(\varepsilon, n) \tag{7.11}$$

and assume that

$$\rho(\tilde{x}, \theta) \le n \quad \textit{and} \quad \{x \in X : \rho(x, \theta) \le n\} \cap A \ne \emptyset. \tag{7.12}$$

Then there exists $\bar{x} \in X$ such that $\rho(\bar{x}, A) \le r$ and such that the set $\tilde{A} = A \cup \{\bar{x}\}$ has the following two properties:

$$\rho(\tilde{x}, \bar{x}) = \rho(\tilde{x}, \tilde{A}); \tag{7.13}$$

if

$$\tilde{y} \in X, \quad \rho(\tilde{y}, \tilde{x}) \le \alpha r, \tag{7.14}$$

$$B \in S(X), \quad h(\tilde{A}, B) \le \alpha r, \tag{7.15}$$

and

$$z \in B, \quad \rho(\tilde{y}, z) \le \rho(\tilde{y}, B) + \varepsilon r/16, \tag{7.16}$$

then

$$\rho(z, \bar{x}) \le \varepsilon. \tag{7.17}$$

Proof By Lemma 7.6, there exists $\bar{x} \in X$ such that

$$\rho(\bar{x}, A) \le r \tag{7.18}$$

and such that for the set $\tilde{A} = A \cup \{\bar{x}\}$, equality (7.13) is true and the following property holds:

$$\text{If } x \in \tilde{A} \text{ and } \rho(\tilde{x}, x) \le \rho(\tilde{x}, \tilde{A}) + \varepsilon r/8, \text{ then } \rho(\bar{x}, x) \le \varepsilon/2. \tag{7.19}$$

Assume that $\tilde{y} \in X$ satisfies (7.14) and $B \in S(X)$ satisfies (7.15). We will show that

$$\rho(\tilde{y}, B) < \rho(\tilde{x}, \tilde{A}) + 4\alpha r 16^n. \tag{7.20}$$

By (7.14),

$$\left|\rho(\tilde{y}, \tilde{A}) - \rho(\tilde{x}, \tilde{A})\right| \le \alpha r.$$

When combined with (7.13), this implies that

$$\left|\rho(\tilde{y}, \tilde{A}) - \rho(\tilde{x}, \bar{x})\right| \le \alpha r. \tag{7.21}$$

Relations (7.13) and (7.12) imply that

$$\rho(\tilde{x}, \bar{x}) \le \rho(\tilde{x}, A) \le 2n \quad \text{and} \quad \rho(\bar{x}, \theta) \le 3n. \tag{7.22}$$

It follows from (7.5) and (7.15) that

$$h_{4n}(\tilde{A}, B)\big(1 + h_{4n}(\tilde{A}, B)\big)^{-1} \leq 2^{4n} h(\tilde{A}, B) \leq 2^{4n} \alpha r.$$

When combined with (7.10) and (7.11), this inequality implies that

$$h_{4n}(\tilde{A}, B) \leq 2^{4n} \alpha r \big(1 - 2^{4n} \alpha r\big)^{-1} < 2^{4n+1} \alpha r. \tag{7.23}$$

Since $\bar{x} \in \tilde{A}$, it now follows from (7.23), (7.22) and (7.5) that $\rho(\bar{x}, B) < 2^{4n+1}\alpha r$ and there exists $\bar{y} \in X$ such that

$$\bar{y} \in B \quad \text{and} \quad \rho(\bar{x}, \bar{y}) < 2\alpha r 16^n. \tag{7.24}$$

By (7.24), (7.14) and (7.13),

$$\begin{aligned} \rho(\tilde{y}, B) &\leq \rho(\tilde{y}, \bar{y}) \leq \rho(\tilde{y}, \bar{x}) + \rho(\bar{x}, \bar{y}) \\ &< \rho(\tilde{y}, \tilde{x}) + \rho(\tilde{x}, \bar{x}) + 2\alpha r 16^n \\ &\leq 2\alpha r 16^n + \alpha r + \rho(\tilde{x}, \tilde{A}). \end{aligned}$$

This certainly implies (7.20), as claimed.

Assume now that $z \in B$ satisfies (7.16). It follows from (7.16), (7.20), (7.11) and (7.10) that

$$\begin{aligned} \rho(\tilde{y}, z) &\leq \rho(\tilde{y}, B) + \epsilon r/16 \leq \rho(\tilde{x}, \tilde{A}) + 4\alpha r 16^n + \epsilon r/16 \\ &\leq \rho(\tilde{x}, \tilde{A}) + \epsilon r/8. \end{aligned} \tag{7.25}$$

Relations (7.25), (7.22) and (7.14) imply that

$$\rho(\tilde{y}, z) \leq \rho(\tilde{x}, \tilde{A}) + \epsilon r/8 \leq 2n + r/8. \tag{7.26}$$

By (7.26), (7.14), (7.11) and (7.12),

$$\begin{aligned} \rho(z, \theta) &\leq \rho(z, \tilde{y}) + \rho(\tilde{y}, \theta) \leq 2n + r/8 + \rho(\tilde{y}, \theta) \\ &\leq 2n + r/8 + \rho(\tilde{y}, \tilde{x}) + \rho(\tilde{x}, \theta) \\ &\leq 2n + r/8 + \alpha r + n \leq 4n. \end{aligned} \tag{7.27}$$

It follows from (7.23), (7.5), (7.16) and (7.27) that

$$\rho(z, \tilde{A}) = \big|\rho(z, \tilde{A}) - \rho(z, B)\big| \leq h_{4n}(\tilde{A}, B) < 2\alpha r 16^n.$$

Hence there exists $\tilde{z} \in X$ such that

$$\tilde{z} \in \tilde{A} \quad \text{and} \quad \rho(z, \tilde{z}) < 2\alpha r 16^n. \tag{7.28}$$

By (7.14), (7.28) and (7.16) we have

$$\begin{aligned}\rho(\tilde{x}, \tilde{z}) &\le \rho(\tilde{x}, \tilde{y}) + \rho(\tilde{y}, z) + \rho(z, \tilde{z}) \\ &\le \alpha r + \rho(\tilde{y}, z) + 2\alpha r 16^n \\ &\le \alpha r + 2\alpha r 16^n + \rho(\tilde{y}, B) + \varepsilon r/16.\end{aligned}$$

It follows from this inequality, (7.20), (7.11) and (7.10) that

$$\begin{aligned}\rho(\tilde{x}, \tilde{z}) &\le \alpha r + 2\alpha r 16^n + \varepsilon r/16 + \rho(\tilde{x}, \tilde{A}) + 4\alpha r 16^n \\ &\le \rho(\tilde{x}, \tilde{A}) + 8\alpha r 16^n + \varepsilon r/16 \le \rho(\tilde{x}, \tilde{A}) + \varepsilon r/8.\end{aligned}$$

Thus

$$\rho(\tilde{x}, \tilde{z}) \le \rho(\tilde{x}, \tilde{A}) + \varepsilon r/8.$$

Using this inequality, (7.28) and (7.19), we see that $\rho(\bar{x}, \tilde{z}) \le \varepsilon/2$. Combining this fact with (7.28), (7.11) and (7.10), we conclude that

$$\rho(z, \bar{x}) \le \rho(z, \tilde{z}) + \rho(\tilde{z}, \bar{x}) \le 2\alpha r 16^n + \varepsilon/2 \le \varepsilon.$$

Thus (7.17) holds and Lemma 7.7 is proved. □

7.3 Proofs of Theorems 7.3–7.5

Proof of Theorem 7.3 For each integer $k \ge 1$, denote by Ω_k the set of all $A \in S(X)$ which have the following property:

(P3) There exist $x_A \in X$ and $\delta_A > 0$ such that if $x \in A$ satisfies $\rho(x, \tilde{x}) \le \rho(\tilde{x}, A) + \delta_A$, then $\rho(x, x_A) \le 1/k$.

Clearly, $\Omega_{k+1} \subset \Omega_k$, $k = 1, 2, \dots$. Set

$$\Omega = \bigcap_{k=1}^{\infty} \Omega_k.$$

First we will show that $S(X) \setminus \Omega$ is σ-porous with respect to the pair $(h, \tilde{H})$. To meet this goal it is sufficient to show that $S(X) \setminus \Omega_k$ is σ-porous with respect to $(h, \tilde{H})$ for all sufficiently large integers k.

There exists a natural number k_0 such that $\rho(\theta, \tilde{x}) \le k_0$. Let $k \ge k_0$ be an integer. We will show that the set $S(X) \setminus \Omega_k$ is σ-porous with respect to $(h, \tilde{H})$. For each integer $n \ge k_0$, set

$$E_{nk} = \{A \in S(X) \setminus \Omega_k : \{z \in X : \rho(z, \theta) \le n\} \cap A \ne \emptyset\}.$$

By Lemma 7.7, the set E_{nk} is porous with respect to $(h, \tilde{H})$ for all integers $n \ge k_0$. Since $S(X) \setminus \Omega_k = \bigcup_{n=k_0}^{\infty} E_{nk}$, we conclude that $S(X) \setminus \Omega_k$ is σ-porous with respect to $(h, \tilde{H})$. Therefore $S(X) \setminus \Omega$ is also σ-porous with respect to $(h, \tilde{H})$.

Let $A \in \Omega$ be given. We will show that A has property (C1). By the definition of Ω_k and property (P3), for each integer $k \geq 1$, there exist $x_k \in X$ and $\delta_k > 0$ such that the following property holds:

(P4) If $x \in A$ satisfies $\rho(x, \tilde{x}) \leq \rho(\tilde{x}, A) + \delta_k$, then $\rho(x, x_k) \leq 1/k$.

Let $\{z_i\}_{i=1}^{\infty} \subset A$ be such that

$$\lim_{i \to \infty} \rho(\tilde{x}, z_i) = \rho(\tilde{x}, A). \tag{7.29}$$

Fix an integer $k \geq 1$. It follows from property (P4) that for all large enough natural numbers i,

$$\rho(\tilde{x}, z_i) \leq \rho(\tilde{x}, A) + \delta_k$$

and

$$\rho(z_i, x_k) \leq 1/k.$$

Since k is an arbitrary natural number, we conclude that $\{z_i\}_{i=1}^{\infty}$ is a Cauchy sequence which converges to some $\tilde{y} \in A$. It is clear that $\rho(\tilde{x}, \tilde{y}) = \rho(\tilde{x}, A)$. If the minimizer $\tilde{y}$ were not unique, we would be able to construct a nonconvergent minimizing sequence $\{z_i\}_{i=1}^{\infty}$. Thus $\tilde{y}$ is the unique solution to problem (P) (with $x = \tilde{x}$) and any sequence $\{z_i\}_{i=1}^{\infty} \subset A$ satisfying (7.29) converges to $\tilde{y}$. This completes the proof of Theorem 7.3. □

Proof of Theorem 7.4 For each integer $k \geq 1$, denote by Ω_k the set of all $(A, \tilde{x}) \in S(X) \times X$ which have the following property:

(P5) There exist $\bar{x} \in X$ and $\bar{\delta} > 0$ such that if $x \in X$ satisfies $\rho(x, \tilde{x}) \leq \bar{\delta}$, $B \in S(X)$ satisfies $h(A, B) \leq \bar{\delta}$, and $y \in B$ satisfies $\rho(y, x) \leq \rho(x, B) + \bar{\delta}$, then $\rho(y, \bar{x}) \leq 1/k$.

Clearly $\Omega_{k+1} \subset \Omega_k$, $k = 1, 2, \ldots$. Set

$$\Omega = \bigcap_{k=1}^{\infty} \Omega_k.$$

First we will show that $[S(X) \times X] \setminus \Omega$ is σ-porous with respect to the pair (d_1, d_2). For each pair of natural numbers n and k, set

$$E_{nk} = \big\{(A, x) \in \big[S(X) \times X\big] \setminus \Omega_k : \rho(x, \theta) \leq n, B_\rho(\theta, n) \cap A \neq \emptyset\big\}.$$

By Lemma 7.7, the set E_{nk} is porous with respect to (d_1, d_2) for all natural numbers n and k. Since

$$\big[S(X) \times X\big] \setminus \Omega = \bigcup_{k=1}^{\infty}\big(\big[S(X) \times X\big] \setminus \Omega_k\big) = \bigcup_{k=1}^{\infty} \bigcup_{n=1}^{\infty} E_{nk},$$

the set $[S(X) \times X] \setminus \Omega$ is σ-porous with respect to (d_1, d_2), by definition.

Let $(A, \tilde{x}) \in \Omega$. We will show that $(A, \tilde{x})$ has property (C2).

By the definition of Ω_k and property (P5), for each integer $k \geq 1$, there exist $x_k \in X$ and $\delta_k > 0$ with the following property:

(P6) If $x \in X$ satisfies $\rho(x, \tilde{x}) \leq \delta_k$, $B \in S(X)$ satisfies $h(A, B) \leq \delta_k$, and $y \in B$ satisfies $\rho(y, x) \leq \rho(x, B) + \delta_k$, then $\rho(y, x_k) \leq 1/k$.

Let $\{z_i\}_{i=1}^{\infty} \subset A$ be such that

$$\lim_{i \to \infty} \rho(\tilde{x}, z_i) = \rho(\tilde{x}, A). \tag{7.30}$$

Fix an integer $k \geq 1$. It follows from property (P6) that for all large enough natural numbers i,

$$\rho(\tilde{x}, z_i) \leq \rho(\tilde{x}, A) + \delta_k$$

and

$$\rho(z_i, x_k) \leq 1/k.$$

Since k is an arbitrary natural number, we conclude that $\{z_i\}_{i=1}^{\infty}$ is a Cauchy sequence which converges to some $\tilde{y} \in A$. Clearly, $\rho(\tilde{x}, \tilde{y}) = \rho(\tilde{x}, A)$. It is not difficult to see that $\tilde{y}$ is the unique solution to the minimization problem (P) with $x = \tilde{x}$.

Let $\varepsilon > 0$ be given. Choose an integer $k > 4/\min\{1, \varepsilon\}$. By property (P6),

$$\rho(\tilde{y}, x_k) \leq 1/k. \tag{7.31}$$

Assume that $z \in X$ satisfies $\rho(z, \tilde{x}) \leq \delta_k$, $B \in S(X)$ satisfies $h(A, B) \leq \delta_k$ and $y \in B$ satisfies $\rho(y, z) \leq \rho(z, B) + \delta_k$. Then it follows from property (P6) that $\rho(y, x_k) \leq 1/k$. When combined with (7.31), this implies that $\rho(y, \tilde{y}) \leq 2/k < \varepsilon$. This completes the proof of Theorem 7.4. □

Proof of Theorem 7.5 Let $\{x_i\}_{i=1}^{\infty} \subset X_0$ be an everywhere dense subset of X_0. For each natural number p, there exists a set $\mathcal{F}_p \subset S(X)$ such that Theorem 7.3 holds with $\tilde{x} = x_p$ and $\Omega = \mathcal{F}_p$. Set $\mathcal{F} = \bigcap_{p=1}^{\infty} \mathcal{F}_p$. Clearly, $S(X) \setminus \mathcal{F}$ is σ-porous with respect to the pair $(h, \tilde{H})$.

Let $A \in \mathcal{F}$ and let $p \geq 1$ be an integer. By Theorem 7.3, which holds with $\tilde{x} = x_p$ and $\Omega = \mathcal{F}_p$, there exists a unique $\bar{x}_p \in A$ such that

$$\rho(x_p, \bar{x}_p) = \rho(x_p, A) \tag{7.32}$$

and the following property holds:

(P7) For each integer $k \geq 1$, there exists $\delta(p, k) > 0$ such that if $x \in A$ satisfies $\rho(x, x_p) \leq \rho(x_p, A) + 4\delta(p, k)$, then $\rho(x, \bar{x}_p) \leq 1/k$.

For each pair of natural numbers p and k, set

$$V(p, k) = \big\{z \in X_0 : \rho(z, x_p) < \delta(p, k)\big\}.$$

It follows from property (P7) that for each pair of integers $p, k \geq 1$, the following property holds:

(P8) If $x \in A$, $z \in X_0$, $\rho(z, x_p) \leq \delta(p, k)$ and $\rho(z, x) \leq \rho(z, A) + \delta(p, k)$, then $\rho(x, \bar{x}_p) \leq 1/k$.

Set

$$F := \bigcap_{k=1}^{\infty} \Big[\bigcup \{V(p,k) : p = 1, 2, \ldots\} \Big].$$

Clearly, F is a countable intersection of open and everywhere dense subsets of X_0.

Let $x \in F$ be given. Consider a sequence $\{x_i\}_{i=1}^{\infty} \subset A$ such that

$$\lim_{i \to \infty} \rho(x, x_i) = \rho(x, A). \tag{7.33}$$

Let $\varepsilon > 0$. Choose a natural number $k > 8^{-1}/\min\{1, \varepsilon\}$. There exists an integer $p \geq 1$ such that $x \in V(p, k)$. By the definition of $V(p, k)$, $\rho(x, x_p) < \delta(p, k)$. It follows from this inequality and property (P8) that for all sufficiently large integers i, $\rho(x, x_i) \leq \rho(x, A) + \delta(p, k)$ and $\rho(x_i, \bar{x}_p) \leq 1/k < \varepsilon/2$. Since ε is an arbitrary positive number, we conclude that $\{x_i\}_{i=1}^{\infty}$ is a Cauchy sequence which converges to $\tilde{y} \in A$. Clearly, $\tilde{y}$ is the unique minimizer of the minimization problem $z \to \rho(x, z)$, $z \in A$. Note that we have shown that any sequence $\{x_i\}_{i=1}^{\infty} \subset A$ satisfying (7.33) converges to $\tilde{y}$. This completes the proof of Theorem 7.5. □

7.4 Generalized Best Approximation Problems

Given a closed subset A of a Banach space X, a point $x \in X$ and a continuous function $f : X \to R^1$, we consider the problem of finding a solution to the minimization problem $\min\{f(x - y) : y \in A\}$. For a fixed function f, we define an appropriate complete metric space $\mathcal{M}$ of all pairs (A, x) and construct a subset Ω of $\mathcal{M}$, which is a countable intersection of open and everywhere dense sets such that for each pair in Ω, our minimization problem is well posed.

Let $(X, \|\cdot\|)$ be a Banach space and let $f : X \to R^1$ be a continuous function. Assume that

$$\inf\big\{f(x) : x \in X\big\} \text{ is attained at a unique point } x_* \in X, \tag{7.34}$$

$$\lim_{\|u\| \to \infty} f(u) = \infty, \tag{7.35}$$

$$\text{if } \{x_i\}_{i=1}^{\infty} \subset X \text{ and } \lim_{i \to \infty} f(x_i) = f(x_*), \text{ then } \lim_{i \to \infty} x_i = x_*, \tag{7.36}$$

and that for each integer $n \geq 1$, there exists an increasing function $\phi_n : (0, 1) \to (0, 1)$ such that

$$f\big(\alpha x + (1 - \alpha)x_*\big) \leq \phi_n(\alpha) f(x) + \big(1 - \phi_n(\alpha)\big) f(x_*) \tag{7.37}$$

for all $x \in X$ satisfying $\|x\| \le n$ and all $\alpha \in (0, 1)$. It is clear that (7.37) holds if f is convex.

Given a closed subset A of X and a point $x \in X$, we consider the minimization problem

$$\min\{f(x-y) : y \in A\}. \tag{P}$$

This problem was studied by many mathematicians mostly in the case where $f(x) = \|x\|$. We recall that the minimization problem (P) is said to be well posed if it has a unique solution, say a_0, and every minimizing sequence of (P) converges to a_0. In other words, if $\{y_i\}_{i=1}^{\infty} \subset A$ and $\lim_{i\to\infty} f(x - y_i) = f(x - a_0)$, then $\lim_{i\to\infty} y_i = a_0$.

Note that in the studies of problem (P) [52, 59, 84, 88, 173], the function f is the norm of the space X. There are some additional results in the literature where either f is a Minkowski functional [51, 93] or the function $\|x - y\|$, $y \in A$, is perturbed by some convex function [42].

However, the fundamental restriction in all these results is that they only hold under certain assumptions on either the space X or the set A. In view of the Lau-Konjagin result mentioned above, these assumptions cannot be removed. On the other hand, many generic results in nonlinear functional analysis hold in any Banach space. Therefore a natural question is whether generic existence results for best approximation problems can be obtained for general Banach spaces. Positive answers to this question in the special case where $f = \|\cdot\|$ can be found in Sects. 7.1–7.3. In the next sections, which are based on [143], we answer this question in the affirmative for a general function f satisfying (7.34)–(7.37).

To this end, we change our point of view and consider another framework, the main feature of which is that the set A in problem (P) can also vary. We prove four theorems which were established in [143]. In our first result (Theorem 7.8), we fix x and consider the space $S(X)$ of all nonempty closed subsets of X equipped with an appropriate complete metric, say h. We then show that the collection of all sets $A \in S(X)$ for which problem (P) is well posed contains an everywhere dense G_δ set. In the second result (Theorem 7.9), we consider the space of pairs $S(X) \times X$ with the metric $h(A, B) + \|x - y\|$, $A, B \in S(X)$, $x, y \in X$. Once again, we show that the family of all pairs $(A, x) \in S(X) \times X$ for which problem (P) is well posed contains an everywhere dense G_δ set. In our third result (Theorem 7.10), we show that for any separable closed subset X_0 of X, there exists an everywhere dense G_δ subset $\mathcal{F}$ of $(S(X), h)$ such that any $A \in \mathcal{F}$ has the following property: there exists a G_δ dense subset F of X_0 such that for any $x \in F$, problem (P) is well posed.

In our fourth result (Theorem 7.11), we show that a continuous coercive convex $f : X \to R^1$ which has a unique minimizer and a certain well-posedness property (on the whole space X) has a unique minimizer and the same well-posedness property on a generic closed subset of X.

7.5 Theorems 7.8–7.11

We recall that $(X, \|\cdot\|)$ is a Banach space, $f : X \to R^1$ is a continuous function satisfying (7.34)–(7.36) and that for each integer $n \geq 1$, there exists an increasing function $\phi_n : (0,1) \to (0,1)$ such that (7.37) is true.

For each $x \in X$ and each $A \subset X$, set

$$\rho(x, A) = \inf\{\rho(x, y) : y \in A\} \tag{7.38}$$

and

$$\rho_f(x, A) = \inf\{f(x - y) : y \in A\}. \tag{7.39}$$

Denote by $S(X)$ the collection of all nonempty closed subsets of X. For each $A, B \in S(X)$, define

$$H(A, B) := \max\{\sup\{\rho(x, B) : x \in A\}, \sup\{\rho(y, A) : y \in B\}\} \tag{7.40}$$

and

$$\tilde{H}(A, B) := H(A, B)\big(1 + H(A, B)\big)^{-1}.$$

Here we use the convention that $\infty/\infty = 1$.

It is not difficult to see that the metric space $(S(X), \tilde{H})$ is complete.

For each natural number n and each $A, B \in S(X)$, we set

$$h_n(A, B) := \sup\{|\rho(x, A) - \rho(x, B)| : x \in X \text{ and } \|x\| \leq n\} \tag{7.41}$$

and

$$h(A, B) := \sum_{n=1}^{\infty} \big[2^{-n} h_n(A, B)\big(1 + h_n(A, B)\big)^{-1}\big].$$

Once again, it is not difficult to see that h is a metric on $S(X)$ and that the metric space $(S(X), h)$ is complete. Clearly, $\tilde{H}(A, B) \geq h(A, B)$ for all $A, B \in S(X)$.

We equip the set $S(X)$ with the pair of metrics $\tilde{H}$ and h. The topologies induced by the metrics $\tilde{H}$ and h on $S(X)$ will be called the strong topology and the weak topology, respectively.

We now state Theorems 7.8–7.11.

Theorem 7.8 *Let $\tilde{x} \in X$. Then there exists a set $\Omega \subset S(X)$, which is a countable intersection of open (in the weak topology) everywhere dense (in the strong topology) subsets of $S(X)$, such that for each $A \in \Omega$, the following property holds:*

(C1) *There exists a unique $\tilde{y} \in A$ such that $f(\tilde{x} - \tilde{y}) = \rho_f(\tilde{x}, A)$. Moreover, for each $\varepsilon > 0$, there exists $\delta > 0$ such that if $x \in A$ satisfies $f(\tilde{x} - x) \leq \rho_f(\tilde{x}, A) + \delta$, then $\|x - \tilde{y}\| \leq \varepsilon$.*

To state our second result we endow the Cartesian product $S(X) \times X$ with the pair of metrics d_1 and d_2 defined by

$$d_1\big((A,x),(B,y)\big) = h(A,B) + \rho(x,y),$$
$$d_2\big((A,x),(B,y)\big) = \tilde{H}(A,B) + \rho(x,y), \quad x,y \in X,\ A,B \in S(X).$$

We will refer to the topologies induced on $S(X) \times X$ by d_2 and d_1 as the strong and weak topologies, respectively.

Theorem 7.9 *There exists a set $\Omega \subset S(X) \times X$, which is a countable intersection of open (in the weak topology) everywhere dense (in the strong topology) subsets of $S(X) \times X$, such that for each $(A,\tilde{x}) \in \Omega$, the following property holds:*

(C2) *There exists a unique $\tilde{y} \in A$ such that $f(\tilde{x} - \tilde{y}) = \rho_f(\tilde{x}, A)$. Moreover, for each $\varepsilon > 0$, there exists $\delta > 0$ such that if $z \in X$ satisfies $\|z - \tilde{x}\| \le \delta$, $B \in S(X)$ satisfies $h(A,B) \le \delta$, and $y \in B$ satisfies $f(z-y) \le \rho_f(z,B) + \delta$, then $\|y - \tilde{y}\| \le \varepsilon$.*

In most classical generic results the set A was fixed and x varied in a dense G_δ subset of X. In our first two results the set A is also variable. However, our third result shows that for every fixed A in a dense G_δ subset of $S(X)$, the set of all $x \in X$ for which problem (P) is well posed contains a dense G_δ subset of X.

Theorem 7.10 *Assume that X_0 is a closed separable subset of X. Then there exists a set $\mathcal{F} \subset S(X)$, which is a countable intersection of open (in the weak topology) everywhere dense (in the strong topology) subsets of $S(X)$, such that for each $A \in \mathcal{F}$, the following property holds:*

(C3) *There exists a set $F \subset X_0$, which is a countable intersection of open and everywhere dense subsets of X_0 with the relative topology, such that for each $\tilde{x} \in F$, there exists a unique $\tilde{y} \in A$ for which $f(\tilde{x} - \tilde{y}) = \rho_f(\tilde{x}, A)$. Moreover, if $\{y_i\}_{i=1}^{\infty} \subset A$ satisfies $\lim_{i\to\infty} f(\tilde{x} - y_i) = \rho_f(\tilde{x}, A)$, then $y_i \to \tilde{y}$ as $i \to \infty$.*

Now we will show that Theorem 7.8 implies the following result.

Theorem 7.11 *Assume that $g : X \to R^1$ is a continuous convex function such that $\inf\{g(x) : x \in X\}$ is attained at a unique point $y_* \in X$, $\lim_{\|u\|\to\infty} g(u) = \infty$, and if $\{y_i\}_{i=1}^{\infty} \subset X$ and $\lim_{i\to\infty} g(y_i) = g(y_*)$, then $y_i \to y_*$ as $i \to \infty$. Then there exists a set $\Omega \subset S(X)$, which is a countable intersection of open (in the weak topology) everywhere dense (in the strong topology) subsets of $S(X)$, such that for each $A \in \Omega$, the following property holds:*

(C4) *There is a unique $y_A \in A$ such that $g(y_A) = \inf\{g(y) : y \in A\}$. Moreover, for each $\varepsilon > 0$, there exists $\delta > 0$ such that if $y \in A$ satisfies $g(y) \le g(y_A) + \delta$, then $\|y - y_A\| \le \varepsilon$.*

Proof Define $f(x) = g(-x)$, $x \in X$. Clearly, f is convex and satisfies (7.34)–(7.36). Therefore Theorem 7.8 is valid with $\tilde{x} = 0$ and there exists a set $\Omega \subset S(X)$, which is a countable intersection of open (in the weak topology) everywhere dense (in the strong topology) subsets of $S(X)$, such that for each $A \in \Omega$, the following property holds:

There is a unique $\tilde{y} \in A$ such that

$$g(\tilde{y}) = f(-\tilde{y}) = \inf\{f(-y) : y \in A\} = \inf\{g(y) : y \in A\}.$$

Moreover, for each $\varepsilon > 0$, there exists $\delta > 0$ such that if $x \in A$ satisfies

$$g(x) = f(-x) \le \rho_f(0, A) + \delta = \inf\{f(-y) : y \in A\} + \delta = \inf\{g(y) : y \in A\} + \delta,$$

then $\|x - \tilde{y}\| \le \varepsilon$. Theorem 7.11 is proved. □

It is easy to see that in the proofs of Theorems 7.8–7.10 we may assume without loss of generality that $\inf\{f(x) : x \in X\} = 0$. It is also not difficult to see that we may assume without loss of generality that $x_* = 0$. Indeed, instead of the function $f(\cdot)$ we can consider $f(\cdot + x_*)$. This new function also satisfies (7.34)–(7.37). Once Theorems 7.8–7.10 are proved for this new function, they will also hold for the original function f because the mapping $(A, x) \to (A, x + x_*)$, $(A, x) \in S(X) \times A$, is an isometry with respect to both metrics d_1 and d_2.

7.6 A Basic Lemma

Lemma 7.12 *Let $A \in S(X)$, $\tilde{x} \in X$, and let $r, \varepsilon \in (0, 1)$. Then there exists $\tilde{A} \in S(X)$, $\bar{x} \in \tilde{A}$, and $\delta > 0$ such that*

$$\tilde{H}(A, \tilde{A}) \le r, \qquad f(\tilde{x} - \bar{x}) = \rho_f(\tilde{x}, \tilde{A}), \tag{7.42}$$

and such that the following property holds:

For each $\tilde{y} \in X$ satisfying $\|\tilde{y} - \tilde{x}\| \le \delta$, each $B \in S(X)$ satisfying $h(B, \tilde{A}) \le \delta$, and each $z \in B$ satisfying

$$f(\tilde{y} - z) \le \rho_f(\tilde{y}, B) + \delta, \tag{7.43}$$

the inequality $\|z - \bar{x}\| \le \varepsilon$ holds.

Proof There are two cases: either $\rho(\tilde{x}, A) \le r$ or $\rho(\tilde{x}, A) > r$. Consider the first case where

$$\rho(\tilde{x}, A) \le r. \tag{7.44}$$

Set

$$\bar{x} = \tilde{x} \quad \text{and} \quad \tilde{A} = A \cup \{\tilde{x}\}. \tag{7.45}$$

Clearly, (7.42) is true. Fix an integer $n > \|\tilde{x}\|$. By (7.36), there is $\xi \in (0, 1)$ such that

$$\text{if } z \in X \text{ and } f(z) \leq 4\xi, \text{ then } \|z\| \leq \varepsilon/2. \tag{7.46}$$

Using (7.34), we choose a number $\delta \in (0, 1)$ such that

$$\delta < 2^{-n-4} \min\{\varepsilon, \xi\} \tag{7.47}$$

and

$$\text{if } z \in X \text{ and } \|z\| \leq 2^{n+4}\delta, \text{ then } f(z) \leq \xi. \tag{7.48}$$

Let

$$\tilde{y} \in X, \quad \|\tilde{y} - \tilde{x}\| \leq \delta, \qquad B \in S(X), \quad h(B, \tilde{A}) \leq \delta \tag{7.49}$$

and let $z \in B$ satisfy (7.43). By (7.49) and (7.41), $h_n(\tilde{A}, B)(1 + h_n(\tilde{A}, B))^{-1} \leq 2^n\delta$. This implies that $h_n(\tilde{A}, B)(1 - 2^n\delta) \leq 2^n\delta$. When combined with (7.47), this inequality shows that $h_n(\tilde{A}, B) \leq 2^{n+1}\delta$. Since $n > \|\tilde{x}\|$, the last inequality, when combined with (7.44) and (7.41), implies that $\rho(\tilde{x}, B) \leq 2^{n+1}\delta$. Hence there is $x_0 \in B$ such that $\|\tilde{x} - x_0\| \leq 2^{n+2}\delta$. This inequality and (7.49) imply in turn that $\|\tilde{y} - x_0\| \leq 2^{n+3}\delta$. The definition of δ (see (7.48)) now shows that $f(\tilde{y} - x_0) \leq \xi$. Combining this inequality with (7.43), (7.47) and the inclusion $x_0 \in B$, we see that

$$f(\tilde{y} - z) \leq \delta + f(\tilde{y} - x_0) \leq \xi + \delta \leq 2\xi. \tag{7.50}$$

It now follows from (7.46) that $\|z - \tilde{y}\| \leq \varepsilon/2$. Hence (7.47), (7.49) and (7.45) imply that $\|\bar{x} - z\| \leq \varepsilon$. This concludes the proof of the lemma in the first case.

Now we turn our attention to the second case where

$$\rho(\tilde{x}, A) > r. \tag{7.51}$$

For each $t \in [0, r]$, set

$$A_t = \{v \in X : \rho(v, A) \leq t\} \in S(X) \tag{7.52}$$

and

$$\mu(t) = \rho_f(\tilde{x}, A_t). \tag{7.53}$$

By (7.51) and (7.36),

$$\mu(t) > 0, \quad t \in [0, r]. \tag{7.54}$$

It is clear that $\mu(t)$, $t \in [0, r]$, is a decreasing function. Choose a number

$$t_0 \in (0, r/4) \tag{7.55}$$

such that μ is continuous at t_0. By (7.35), there exists a natural number n which satisfies the following conditions:

$$n > 4\|\tilde{x}\| + 8 \tag{7.56}$$

and

$$\text{if } z \in X,\ f(x) \leq \mu(0) + 1, \text{ then } \|z\| \leq n/4. \tag{7.57}$$

Let $\phi_n : (0, 1) \to (0, 1)$ be an increasing function for which (7.37) is true. Choose a positive number $\gamma \in (0, 1)$ such that

$$\gamma < \mu(t_0)\big(1 - \phi(1 - 2r/n)\big)/8. \tag{7.58}$$

Next, choose a positive number $\delta_0 < 1/4$ such that

$$2^{n+3}\delta_0 < \min\{\varepsilon, \gamma\}, \tag{7.59}$$

$$[t_0 - 4\delta_0, t_0 + 4\delta_0] \subset (0, r/4), \tag{7.60}$$

and

$$\big|\mu(t) - \mu(t_0)\big| \leq \gamma, \quad t \in [t_0 - 4\delta_0, t_0 + 4\delta_0]. \tag{7.61}$$

Finally, choose a vector x_0 such that

$$x_0 \in A_{t_0} \quad \text{and} \quad f(\tilde{x} - x_0) \leq \mu(t_0) + \gamma. \tag{7.62}$$

It follows from (7.62), (7.52) and (7.55) that

$$\|x_0 - \tilde{x}\| \geq \rho(\tilde{x}, A) - \rho(x_0, A) \geq \rho(\tilde{x}, A) - t_0 \geq \rho(\tilde{x}, A) - r/2, \tag{7.63}$$

and hence by (6.51),

$$\|x_0 - \tilde{x}\| > r/2. \tag{7.64}$$

It follows from (7.62) and (7.57) that

$$\|x_0 - \tilde{x}\| \leq n/4. \tag{7.65}$$

There exist $\bar{x} \in \{\alpha x_0 + (1 - \alpha)\tilde{x} : \alpha \in (0, 1)\}$ and $\alpha_0 \in (0, 1)$ such that

$$\|\bar{x} - x_0\| = r/2 \tag{7.66}$$

and

$$\bar{x} = \alpha_0 x_0 + (1 - \alpha_0)\tilde{x}. \tag{7.67}$$

By (7.67) and (7.66), $r/2 = \|\bar{x} - x_0\| = \|\alpha_0 x_0 + (1 - \alpha_0)\tilde{x} - x_0\| = (1 - \alpha_0)\|\tilde{x} - x_0\|$ and

$$\alpha_0 = 1 - r\big(2\|\tilde{x} - x_0\|\big)^{-1}. \tag{7.68}$$

Relations (7.68) and (7.65) imply that

$$\alpha_0 \leq 1 - r/(2n/4) = 1 - 2r/n. \tag{7.69}$$

Set

$$\tilde{A} = A_{t_0} \cup \{\bar{x}\}. \tag{7.70}$$

Now we will estimate $f(\tilde{x} - \bar{x})$. By (7.67), (7.65), (7.37), (7.62) and (7.69),

$$\begin{aligned} f(\tilde{x} - \bar{x}) &= f\big(\tilde{x} - (\alpha_0 x_0 + (1 - \alpha_0)\tilde{x})\big) = f\big(\alpha_0(\tilde{x} - x_0)\big) \\ &\le \phi_n(\alpha_0) f(\tilde{x} - x_0) \le \phi_n(\alpha_0)\big(\mu(t_0) + \gamma\big) \\ &\le \phi_n(1 - 2r/n)\big(\mu(t_0) + \gamma\big). \end{aligned}$$

Thus

$$f(\tilde{x} - \bar{x}) \le \phi_n(1 - 2r/n)\big(\mu(t_0) + \gamma\big) \le \mu(t_0)\phi_n(1 - 2r/n) + \gamma. \tag{7.71}$$

By (7.70), (7.53), (7.58) and (7.71), for each $x \in \tilde{A} \setminus \{\bar{x}\} \subset A_{t_0}$,

$$f(\tilde{x} - x) \ge \mu(t_0) > f(\tilde{x} - \bar{x}) \tag{7.72}$$

and therefore

$$f(\tilde{x} - \bar{x}) = \rho_f(\tilde{x}, \tilde{A}). \tag{7.73}$$

There exists $\delta \in (0, \delta_0)$ such that

$$2^{n+4}\delta < \delta_0 \tag{7.74}$$

and

$$\begin{aligned} &\big|f(z) - f(\tilde{x} - \bar{x})\big| \le \gamma/4 \\ &\quad \text{for all } z \in X \text{ satisfying } \big\|z - (\tilde{x} - \bar{x})\big\| \le 2^{n+3}\delta. \end{aligned} \tag{7.75}$$

By (7.70), (7.40), (7.66), (7.62), (7.55) and (7.52),

$$\tilde{H}(\tilde{A}, A) \le H(\tilde{A}, A) \le r. \tag{7.76}$$

Relations (7.76) and (7.73) imply (7.42). Assume now that

$$\tilde{y} \in X, \quad \|\tilde{y} - \tilde{x}\| \le \delta \tag{7.77}$$

and

$$B \in S(X) \quad \text{and} \quad h(\tilde{A}, B) \le \delta. \tag{7.78}$$

First we will show that

$$\rho_f(\tilde{y}, B) \le \mu(t_0)\phi_n(1 - 2r/n) + 2\gamma. \tag{7.79}$$

By (7.78) and the definition of h (see (7.41)), $h_n(\tilde{A}, B)(1 + h_n(\tilde{A}, B))^{-1} \le 2^n\delta$. When combined with (7.74), this inequality implies that

$$h_n(\tilde{A}, B) \le 2^n\delta\big(1 - 2^n\delta\big)^{-1} \le 2^{n+1}\delta. \tag{7.80}$$

It follows from (7.41) and the definition of n (see (7.57), (7.56)) that $\|\tilde{x} - \bar{x}\| \le n/2$ and $\|\bar{x}\| \le n$. When combined with (7.70) and (7.80), this implies that $\rho(\bar{x}, B) \le 2^{n+1}\delta$. Therefore there exists $\bar{y} \in B$ such that $\|\bar{x} - \bar{y}\| \le 2^{n+2}\delta$. Combining this inequality with (7.77), we see that $\|(\bar{y}-\tilde{y})-(\bar{x}-\tilde{x})\| \le \|\bar{x}-\bar{y}\|+\|\tilde{y}-\tilde{x}\| \le 2^{n+3}\delta$. It follows from this inequality and (7.75) that $f(\tilde{y} - \bar{y}) \le f(\tilde{x} - \bar{x}) + \gamma/4$. By the last inequality and (7.71), $f(\tilde{y}-\bar{y}) \le \mu(t_0)\phi_n(1-2r/n)+2\gamma$. This implies (7.79).

Assume now that $z \in B$ satisfies (7.43). To complete the proof of the lemma it is sufficient to show that $\|\tilde{x} - z\| \le \varepsilon$. Assume the contrary. Then

$$\|\tilde{x} - z\| > \varepsilon. \tag{7.81}$$

We will show that there exists $\bar{z} \in \tilde{A}$ such that

$$\|z - \bar{z}\| \le 2^{n+2}\delta. \tag{7.82}$$

We have already shown that (7.80) holds. By (7.43), (7.79), (7.58) and (7.74),

$$f(\tilde{y} - z) \le \rho_f(\tilde{y}, B) + \delta \le \phi_n(1 - 2r/n)\mu(t_0) + 2\gamma + \delta \le \mu(0) + 1/2.$$

Hence $\|z - \tilde{y}\| \le n/4$ by (7.57), and by (7.77) and (7.56),

$$\|z\| \le n/4 + \|\tilde{y}\| \le n/4 + \|\tilde{x}\| + \|\tilde{y} - \tilde{x}\| \le n.$$

Thus $\|z\| \le n$. The inclusion $z \in B$ and (7.80) now imply that $\rho(z, \tilde{A}) \le h_n(B, \tilde{A}) \le 2^{n+1}\delta$. Therefore there exists $\bar{z} \in \tilde{A}$ such that (7.82) holds. It follows from (7.82), (7.81), (7.70), (7.74) and (7.59) that

$$\bar{z} \in A_{t_0}. \tag{7.83}$$

By (7.82) and (7.77), $\|z + \tilde{x} - \tilde{y} - \bar{z}\| \le \|\tilde{x} - \tilde{y}\| + \|z - \bar{z}\| \le 2^{n+2}\delta + \delta \le 2^{n+3}\delta$. It follows from this inequality, (7.83), (7.52) and (7.74) that

$$\rho(z + \tilde{x} - \tilde{y}, A) \le \|z + \tilde{x} - \tilde{y} - \bar{z}\| + \rho(\bar{z}, A) \le 2^{n+3}\delta + t_0 \le t_0 + \delta_0.$$

Thus $z + \tilde{x} - \tilde{y} \in A_{t_0+\delta_0}$. By this inclusion, (7.52), (7.53) and (7.61),

$$f(\tilde{y} - z) = f\big(\tilde{x} - (z + \tilde{x} - \tilde{y})\big) \ge \rho_f(\tilde{x}, A_{t_0+\delta_0}) = \mu(t_0 + \delta_0) \ge \mu(t_0) - \gamma.$$

Hence, by (7.43), (7.79), (7.59) and (7.74),

$$\begin{aligned}\mu(t_0) - \gamma \le f(\tilde{y} - z) &\le \rho_f(\tilde{y}, B) + \delta \le \phi_n(1 - 2r/n)\mu(t_0) + 2\gamma + \delta \\ &\le \phi_n(1 - 2r/n)\mu(t_0) + 3\gamma.\end{aligned}$$

Thus $\mu(t_0) - \gamma \le \phi_n(1-2r/n)\mu(t_0)+3\gamma$, which contradicts (7.58). This completes the proof of Lemma 7.12. □

7.7 Proofs of Theorems 7.8–7.11

The cornerstone of our proofs is the property established in Lemma 7.12.

By Lemma 7.12, for each $(A, x) \in S(X) \times X$ and each integer $k \geq 1$, there exist $A(x, k) \in S(X)$, $\bar{x}(A, k) \in A(x, k)$, and $\delta(x, A, k) > 0$ such that

$$\tilde{H}\big(A, A(x, k)\big) \leq 2^{-k}, \qquad f\big(x - \bar{x}(A, k)\big) = \rho_f\big(x, A(x, k)\big), \tag{7.84}$$

and the following property holds:

(P1) For each $y \in X$ satisfying $\|y - x\| \leq 2\delta(x, A, k)$, each $B \in S(X)$ satisfying $h(B, A(x, k)) \leq 2\delta(x, A, k)$ and each $z \in B$ satisfying $f(y - z) \leq \rho_f(y, B) + 2\delta(x, A, k)$, the inequality $\|z - \bar{x}(A, k)\| \leq 2^{-k}$ holds.

For each $(A, x) \in S(X) \times X$ and each integer $k \geq 1$, define

$$V(A, x, k) = \big\{(B, y) \in S(X) \times X : \\ h\big(B, A(x, k)\big) < \delta(x, A, k) \text{ and } \|y - x\| < \delta(x, A, k)\big\} \tag{7.85}$$

and

$$U(A, x, k) = \big\{B \in S(X) : h\big(B, A(x, k)\big) < \delta(x, A, k)\big\}. \tag{7.86}$$

Now set

$$\Omega = \bigcap_{n=1}^{\infty} \bigcup \big\{V(A, x, k) : (A, x) \in S(X) \times X, k \geq n\big\}, \tag{7.87}$$

and for each $x \in X$ let

$$\Omega_x = \bigcap_{n=1}^{\infty} \bigcup \big\{U(A, x, k) : A \in S(X), k \geq n\big\}. \tag{7.88}$$

It is easy to see that $\Omega_x \times \{x\} \subset \Omega$ for all $x \in X$, Ω_x is a countable intersection of open (in the weak topology) everywhere dense (in the strong topology) subsets of $S(X)$ for all $x \in X$, and Ω is a countable intersection of open (in the weak topology) everywhere dense (in the strong topology) subsets of $S(X) \times X$.

Completion of the proof of Theorem 7.9 Let $(A, \tilde{x}) \in \Omega$. We will show that $(A, \tilde{x})$ has property (C2). By the definition of Ω (see (7.87)), for each integer $n \geq 1$, there exist an integer $k_n \geq n$ and a pair $(A_n, x_n) \in S(X) \times X$ such that

$$(A, \tilde{x}) \in V(A_n, x_n, k_n). \tag{7.89}$$

Let $\{z_i\}_{i=1}^{\infty} \subset A$ be such that

$$\lim_{i \to \infty} f(\tilde{x} - z_i) = \rho_f(\tilde{x}, A). \tag{7.90}$$

Fix an integer $n \geq 1$. It follows from (7.89), (7.85) and property (P1) that for all large enough integers i,

$$f(\tilde{x} - z_i) < \rho_f(\tilde{x}, A) + \delta(x_n, A_n, k_n)$$

and

$$\left\| z_i - \bar{x}_n(A_n, k_n) \right\| \leq 2^{-n}.$$

Since $n \geq 1$ is arbitrary, we conclude that $\{z_i\}_{i=1}^{\infty}$ is a Cauchy sequence which converges to some $\tilde{y} \in A$. Clearly $f(\tilde{x} - \tilde{y}) = \rho_f(\tilde{x}, A)$. If the minimizer $\tilde{y}$ were not unique we would be able to construct a nonconvergent minimizing sequence $\{z_i\}_{i=1}^{\infty}$. Thus $\tilde{y}$ is the unique solution to problem (P) (with $x = \tilde{x}$).

Let $\varepsilon > 0$ be given. Choose an integer $n > 4/\min\{1, \varepsilon\}$. By property (P1), (7.89) and (7.85),

$$\left\| \tilde{y} - \bar{x}_n(A_n, k_n) \right\| \leq 2^{-n}. \tag{7.91}$$

Assume that $z \in X$ satisfies $\|z - \tilde{x}\| \leq \delta(x_n, A_n, k_n)$, $B \in S(X)$ satisfies $h(A, B) \leq \delta(x_n, A_n, k_n)$, and $y \in B$ satisfies $f(z - y) \leq \rho_f(z, B) + \delta(x_n, A_n, k_n)$. Then

$$h\big(B, A_n(x_n, k_n)\big) \leq 2\delta(x_n, A_n, k_n) \quad \text{and} \quad \left\| z - \bar{x}_n(A_n, k_n) \right\| \leq 2\delta(x_n, A_n, k_n)$$

by (7.89) and (7.85). Now it follows from property (P1) that

$$\left\| y - \bar{x}_n(A_n, k_n) \right\| \leq 2^{-n}.$$

When combined with (7.91), this implies that

$$\|y - \tilde{y}\| \leq 2^{1-n} < \varepsilon.$$

The proof of Theorem 7.9 is complete. □

Theorem 7.8 follows from Theorem 7.9 and the inclusion $\Omega_{\tilde{x}} \times \{\tilde{x}\} \subset \Omega$.

Although a variant of Theorem 7.10 also follows from Theorem 7.9 by a classical result of Kuratowski and Ulam [87], the following direct proof may also be of interest.

Proof of Theorem 7.10 Let the sequence $\{x_i\}_{i=1}^{\infty} \subset X_0$ be everywhere dense in X_0. Set $\mathcal{F} = \bigcap_{p=1}^{\infty} \Omega_{x_p}$. Clearly, $\mathcal{F}$ is a countable intersection of open (in the weak topology) everywhere dense (in the strong topology) subsets of $S(X)$.

Let $A \in \mathcal{F}$ and let $p, n \geq 1$ be integers. Clearly, $A \in \Omega_{x_p}$ and by (7.88) and (7.86), there exist $A_n \in S(X)$ and an integer $k_n \geq n$ such that

$$h\big(A, A_n(x_p, k_n)\big) < \delta(x_p, A_n, k_n) \quad \text{with } A \in S(X). \tag{7.92}$$

It follows from this inequality and property (P1) that the following property holds:

(P2) For each $y \in X$ satisfying $\|y - x_p\| \le \delta(x_p, A_n, k_n)$ and each $z \in A$ satisfying $f(y - z) \le \rho_f(y, A) + 2\delta(x_p, A_n, k_n)$, the inequality $\|z - \bar{x}_p(A_n, k_n)\| \le 2^{-n}$ holds.

Set $W(p, n) = \{z \in X_0 : \|z - x_p\| < \delta(x_p, A_n, k_n)\}$ and

$$F = \bigcap_{n=1}^{\infty} \bigcup \{W(p, n) : p = 1, 2, \dots\}.$$

It is clear that F is a countable intersection of open and everywhere dense subsets of X_0.

Let $x \in F$ be given. Consider a sequence $\{z_i\}_{i=1}^{\infty} \subset A$ such that

$$\lim_{i \to \infty} f(x - z_i) = \rho_f(x, A). \tag{7.93}$$

Let $\varepsilon > 0$. Choose an integer $n > 8/\min\{1, \varepsilon\}$. There exists an integer $p \ge 1$ such that $x \in W(p, n)$. By the definition of $W(p, n)$, $\|x - x_p\| < \delta(x_p, A_n, k_n)$. It follows from this inequality, (7.93) and property (P2) that for all sufficiently large integers i, $f(x - z_i) \le \rho_f(x, A) + \delta(x_p, A_n, k_n)$ and $\|z_i - \bar{x}_p(A_n, k_n)\| \le 2^{-n} < \varepsilon$. Since $\varepsilon > 0$ is arbitrary, we conclude that $\{z_i\}_{i=1}^{\infty}$ is a Cauchy sequence which converges to $\tilde{y} \in A$. Clearly, $\tilde{y}$ is the unique minimizer of the minimization problem $z \to f(x - z)$, $z \in A$. Note that we have shown that any sequence $\{z_i\}_{i=1}^{\infty} \subset A$ satisfying (7.93) converges to $\tilde{y}$. This completes the proof of Theorem 7.10. □

7.8 A Porosity Result in Best Approximation Theory

Let D be a nonempty compact subset of a complete hyperbolic space (X, ρ, M) and denote by $S(X)$ the family of all nonempty closed subsets of X. We endow $S(X)$ with a pair of natural complete metrics and show that there exists a set $\Omega \subset S(X)$ such that its complement $S(X) \setminus \Omega$ is σ-porous with respect to this pair of metrics and such that for each $A \in \Omega$ and each $\tilde{x} \in D$, the following property holds: the set $\{y \in A : \rho(\tilde{x}, y) = \rho(\tilde{x}, A)\}$ is nonempty and compact, and each sequence $\{y_i\}_{i=1}^{\infty} \subset A$ which satisfies $\lim_{i \to \infty} \rho(\tilde{x}, y_i) = \rho(\tilde{x}, A)$ has a convergent subsequence. This result was obtained in [147].

Let (X, ρ, M) be a complete hyperbolic space. For each $x \in X$ and each $A \subset X$, set

$$\rho(x, A) = \inf\{\rho(x, y) : y \in A\}.$$

Denote by $S(X)$ the family of all nonempty closed subsets of X. For each $A, B \in S(X)$, define

$$H(A, B) := \max\{\sup\{\rho(x, B) : x \in A\}, \sup\{\rho(y, A) : y \in B\}\} \tag{7.94}$$

and

$$\tilde{H}(A, B) := H(A, B)(1 + H(A, B))^{-1}.$$

Here we use the convention that $\infty/\infty = 1$. It is easy to see that $\tilde{H}$ is a metric on $S(X)$ and that the metric space $(S(X), \tilde{H})$ is complete.

Fix $\theta \in X$. For each natural number n and each $A, B \in S(X)$, we set

$$h_n(A, B) = \sup\{|\rho(x, A) - \rho(x, B)| : x \in X \text{ and } \rho(x, \theta) \le n\} \tag{7.95}$$

and

$$h(A, B) = \sum_{n=1}^{\infty} [2^{-n} h_n(A, B)(1 + h_n(A, B))^{-1}].$$

Once again, it is not difficult to see that h is a metric on $S(X)$ and that the metric space $(S(X), h)$ is complete. Clearly,

$$\tilde{H}(A, B) \ge h(A, B) \quad \text{for all } A, B \in S(X).$$

We equip the set $S(X)$ with the pair of metrics $\tilde{H}$ and h and prove the following theorem which is the main result of [147].

Theorem 7.13 *Given a nonempty compact subset D of a complete hyperbolic space (X, ρ, M), there exists a set $\Omega \subset S(X)$ such that its complement $S(X) \setminus \Omega$ is σ-porous with respect to the pair of metrics $(h, \tilde{H})$, and such that for each $A \in \Omega$ and each $\tilde{x} \in D$, the following property holds:*

The set $\{y \in A : \rho(\tilde{x}, y) = \rho(\tilde{x}, A)\}$ is nonempty and compact and each sequence $\{y_i\}_{i=1}^{\infty} \subset A$ which satisfies $\lim_{i\to\infty} \rho(\tilde{x}, y_i) = \rho(\tilde{x}, A)$ has a convergent subsequence.

7.9 Two Lemmata

Let (X, ρ, M) be a complete hyperbolic space and let D be a nonempty compact subset of X. In the proof of Theorem 7.13 we will use the following two lemmata.

Lemma 7.14 *Let q be a natural number, $A \in S(X)$, $\varepsilon \in (0, 1)$, $r \in (0, 1]$, and let $Q = \{\xi_1, \dots, \xi_q\}$ be a finite subset of D. Then there exists a finite set $\{\tilde{\xi}_1, \dots, \tilde{\xi}_q\} \subset X$ such that*

$$\rho(\tilde{\xi}_i, A) \le r, \quad i = 1, \dots, q, \tag{7.96}$$

and such that the set $\tilde{A} := A \cup \{\tilde{\xi}_1, \dots, \tilde{\xi}_q\}$ has the following properties:

$$\rho(\xi_i, \{\tilde{\xi}_1, \dots, \tilde{\xi}_q\}) = \rho(\xi_i, \tilde{A}), \quad i = 1, \dots, q; \tag{7.97}$$

(P3) *if $i \in \{1, \dots, q\}$, $x \in \tilde{A}$, and $\rho(\xi_i, x) \le \rho(\xi_i, \tilde{A}) + \varepsilon r/4$, then*

$$\rho(x, \{\tilde{\xi}_1, \dots, \tilde{\xi}_q\}) \le \varepsilon.$$

Proof Let $i \in \{1, \dots, q\}$. There are two cases: (1) $\rho(\xi_i, A) \le r$; (2) $\rho(\xi_i, A) > r$. In the first case we set

$$\tilde{\xi}_i = \xi_i. \tag{7.98}$$

In the second case, we first choose $x_i \in A$ for which

$$\rho(\xi_i, x_i) \le \rho(\xi_i, A) + r/4, \tag{7.99}$$

and then choose

$$\tilde{\xi}_i \in \{\gamma x_i \oplus (1-\gamma)\xi_i : \gamma \in (0,1)\} \tag{7.100}$$

such that

$$\rho(\tilde{\xi}_i, x_i) = r \quad \text{and} \quad \rho(\tilde{\xi}_i, \xi_i) = \rho(x_i, \xi_i) - r. \tag{7.101}$$

Clearly, (7.96) holds. Consider now the set $\tilde{A} = A \cup \{\tilde{\xi}_1, \dots, \tilde{\xi}_q\}$.

Let $i \in \{1, \dots, q\}$. It is not difficult to see that if $\rho(\xi_i, A) \le r$, then the assertion of the lemma is true. Consider the case where $\rho(\xi_i, A) > r$. It follows from (7.99) and (7.101) that

$$\begin{aligned}\rho(\xi_i, \{\tilde{\xi}_1, \dots, \tilde{\xi}_q\}) &\le \rho(\xi_i, \tilde{\xi}_i) = \rho(x_i, \xi_i) - r \\ &\le \rho(\xi_i, A) + r/4 - r = \rho(\xi_i, A) - 3r/4.\end{aligned}$$

Therefore

$$\rho(\xi_i, \{\tilde{\xi}_1, \dots, \tilde{\xi}_q\}) = \rho(\xi_i, \tilde{A}),$$

and if $x \in \tilde{A}$ and $\rho(\xi_i, x) \le \rho(\xi_i, \tilde{A}) + r/2$, then $x \in \{\tilde{\xi}_1, \dots, \tilde{\xi}_q\}$. This completes the proof of Lemma 7.14. □

For each $\varepsilon \in (0,1)$ and each natural number n, choose a number

$$\alpha(\varepsilon, n) \in (0, 16^{-n-2}\varepsilon) \tag{7.102}$$

and a natural number n_0 such that

$$\rho(x, \theta) \le n_0, \quad x \in D. \tag{7.103}$$

Lemma 7.15 *Let $n \ge n_0$ be a natural number, $A \in S(X)$, $\varepsilon \in (0,1)$, $r \in (0,1]$, and*

$$\alpha = \alpha(\varepsilon, n). \tag{7.104}$$

Assume that

$$\{z \in A : \rho(z, \theta) \le n\} \ne \emptyset. \tag{7.105}$$

Then there exist a natural number q and a finite set $\{\tilde{\xi}_1, \dots, \tilde{\xi}_q\} \subset X$ such that

$$\rho(\tilde{\xi}_i, A) \le r, \quad i = 1, \dots, q, \tag{7.106}$$

and if $\tilde{A} := A \cup \{\tilde{\xi}_1, \ldots, \tilde{\xi}_q\}$, $u \in D$, $B \in S(X)$,

$$h(\tilde{A}, B) \le \alpha r, \tag{7.107}$$

and

$$z \in B, \quad \rho(u, z) \le \rho(u, B) + \varepsilon r/16, \tag{7.108}$$

then

$$\rho\big(z, \{\tilde{\xi}_1, \ldots, \tilde{\xi}_q\}\big) \le \varepsilon. \tag{7.109}$$

Proof Since D is compact, there are a natural number q and a finite subset $\{\xi_1, \ldots, \xi_q\}$ of D such that

$$D \subset \bigcup_{i=1}^{q} \{z \in X : \rho(z, \xi_i) < \alpha r\}. \tag{7.110}$$

By Lemma 7.14, there exists a finite set $\{\tilde{\xi}_1, \ldots \tilde{\xi}_q\} \subset X$ such that (7.106) holds, and the set $\tilde{A} := A \cup \{\tilde{\xi}_1, \ldots, \tilde{\xi}_q\}$ satisfies (7.97) and has the following property:

(P4) If $i \in \{1, \ldots, q\}$, $x \in \tilde{A}$, and $\rho(\xi_i, x) \le \rho(\xi_i, \tilde{A}) + \varepsilon r/8$, then

$$\rho\big(x, \{\tilde{\xi}_1, \ldots, \tilde{\xi}_q\}\big) \le \varepsilon/2.$$

Assume that $u \in D$, $B \in S(X)$, and that (7.107) holds. By (7.110), there is $j \in \{1, \ldots, q\}$ such that

$$\rho(\xi_j, u) < \alpha r. \tag{7.111}$$

We will show that

$$\rho(u, B) < \rho(\xi_j, \tilde{A}) + 4 \cdot 16^n \alpha r. \tag{7.112}$$

Indeed, there exists $p \in \{1, \ldots, q\}$ such that

$$\rho(\xi_j, \tilde{\xi}_p) = \rho\big(\xi_j, \{\tilde{\xi}_1, \ldots, \tilde{\xi}_q\}\big).$$

By (7.97),

$$\rho(\xi_j, \tilde{\xi}_p) = \rho(\xi_j, \tilde{A}). \tag{7.113}$$

By (7.111),

$$\big|\rho(u, \tilde{A}) - \rho(\xi_j, \tilde{A})\big| \le \alpha r. \tag{7.114}$$

When combined with (7.113), this inequality implies that

$$\big|\rho(u, \tilde{A}) - \rho(\xi_j, \tilde{\xi}_p)\big| \le \alpha r. \tag{7.115}$$

Now (7.113), (7.105) and (7.103) imply that

$$\rho(\xi_j, \tilde{\xi}_p) \le \rho(\xi_j, A) \le 2n \quad \text{and} \quad \rho(\tilde{\xi}_p, \theta) \le 3n. \tag{7.116}$$

It follows from (7.95) and (7.107) that

$$h_{4n}(\tilde{A}, B)\big(1 + h_{4n}(\tilde{A}, B)\big)^{-1} \le 2^{4n} h(\tilde{A}, B) \le 2^{4n} \alpha r,$$

and when combined with (7.104) and (7.102), this inequality yields

$$h_{4n}(\tilde{A}, B) \le 2^{4n} \alpha r \big(1 - 2^{4n} \alpha r\big)^{-1} < 2^{4n+1} \alpha r. \tag{7.117}$$

Since $\tilde{\xi}_p \in \tilde{A}$, it follows from (7.117), (7.116) and (7.97) that $\rho(\tilde{\xi}_p, B) < 2^{4n+1}\alpha r$ and there exists $v \in X$ such that

$$v \in B \quad \text{and} \quad \rho(\tilde{\xi}_p, v) < 2\alpha r 16^n. \tag{7.118}$$

By (7.118), (7.111), (7.113) and (7.118),

$$\begin{aligned} \rho(u, B) &\le \rho(u, v) \le \rho(u, \tilde{\xi}_p) + \rho(\tilde{\xi}_p, v) \le \rho(u, \xi_j) + \rho(\xi_j, \tilde{\xi}_p) + \rho(\tilde{\xi}_p, v) \\ &< \alpha r + \rho(\xi_j, \tilde{A}) + 2 \cdot 16^n \alpha r. \end{aligned}$$

Hence (7.112) is valid.

Now let (7.108) hold. Then by (7.108), (7.112) and (7.102),

$$\begin{aligned} \rho(z, u) &\le \rho(u, B) + \varepsilon r/16 < \rho(\xi_j, \tilde{A}) + 4 \cdot 16^n \alpha r + \varepsilon r/16 \\ &< \rho(\xi_j, \tilde{A}) + \varepsilon r/8. \end{aligned} \tag{7.119}$$

Therefore (7.119) and (7.116) imply that

$$\rho(z, u) \le \rho(\xi_j, \tilde{A}) + \varepsilon r/8 \le 2n + r/8.$$

It follows from this inequality, (7.111) and (7.103) that

$$\begin{aligned} \rho(z, \theta) &\le \rho(z, u) + \rho(u, \theta) \le 2n + r/8 + \rho(u, \theta) \\ &\le 2n + r/8 + \rho(u, \xi_j) + \rho(\xi_j, \theta) \le 2n + r/8 + \alpha r + n \le 4n. \end{aligned}$$

Since $z \in B$, it follows from (7.97) and (7.117) that

$$\rho(z, \tilde{A}) = \big|\rho(z, \tilde{A}) - \rho(z, B)\big| \le h_{4n}(\tilde{A}, B) < 2 \cdot 16^n \alpha r.$$

Therefore there exists $\tilde{z} \in \tilde{A}$ such that

$$\rho(z, \tilde{z}) < 2 \cdot 16^n \alpha r. \tag{7.120}$$

By (7.111), (7.120), (7.108), (7.112) and (7.102),

$$\begin{aligned} \rho(\tilde{z}, \xi_j) &\le \rho(\xi_j, u) + \rho(u, z) + \rho(z, \tilde{z}) < \alpha r + \rho(u, z) + 2 \cdot 16^n \alpha r \\ &\le \alpha r + 2 \cdot 16^n \alpha r + \rho(u, B) + \varepsilon r/16 \end{aligned}$$

$$< \varepsilon r/16 + \alpha r + 2 \cdot 16^n \alpha r + \rho(\xi_j, \tilde{A}) + 4 \cdot 16^n \alpha r$$
$$\leq \rho(\xi_j, \tilde{A}) + 8 \cdot 16^n \alpha r + \varepsilon r/16 \leq \rho(\xi_j, \tilde{A}) + \varepsilon r/8$$

and

$$\rho(\tilde{z}, \xi_j) < \rho(\xi_j, \tilde{A}) + \varepsilon r/8. \tag{7.121}$$

Since $\tilde{z} \in \tilde{A}$, it follows from (7.121) and property (P4) that $\rho(\tilde{z}, \{\tilde{\xi}_1, \ldots, \tilde{\xi}_q\}) \leq \varepsilon/2$. When combined with (7.120) and (7.102), this inequality implies that

$$\rho\big(z, \{\tilde{\xi}_1, \ldots, \tilde{\xi}_q\}\big) \leq \varepsilon.$$

This completes the proof of Lemma 7.15. □

7.10 Proof of Theorem 7.13

For each integer $k \geq 1$, denote by Ω_k the set of all $A \in S(X)$ which have the following property:

(P5) There exist a nonempty finite set $Q \subset X$ and a number $\delta > 0$ such that if $u \in D$, $x \in A$ and $\rho(u, x) \leq \rho(u, A) + \delta$, then $\rho(x, Q) \leq 1/k$.

It is clear that $\Omega_{k+1} \subset \Omega_k$, $k = 1, 2, \ldots$. Set $\Omega = \bigcap_{k=1}^{\infty} \Omega_k$.

Let $k \geq n_0$ (see (7.103)) be an integer. We will show that $S(X) \setminus \Omega_k$ is σ-porous with respect to the pair $(h, \tilde{H})$. For any integer $n \geq k$, define

$$E_{nk} = \big\{A \in S(X) \setminus \Omega_k : \{z \in A : \rho(z, \theta) \leq n\} \neq \emptyset\big\}.$$

By Lemma 7.15, E_{nk} is porous with respect to the pair $(h, \tilde{H})$ for all integers $n \geq k$. Thus $S(X) \setminus \Omega_k = \bigcup_{n=k}^{\infty} E_{nk}$ is σ-porous with respect to $(h, \tilde{H})$. Hence $S(X) \setminus \Omega = \bigcup_{k=n_0}^{\infty} (S(X) \setminus \Omega_k)$ is also σ-porous with respect to the pair of metrics $(h, \tilde{H})$.

Let $A \in \Omega$. Since $A \in \Omega_k$ for each integer $k \geq 1$, it follows from property (P5) that for any integer $k \geq 1$, there exist a nonempty finite set $Q_k \subset X$ and a number $\delta_k > 0$ such that the following property also holds:

(P6) If $u \in D$, $x \in A$, and $\rho(u, x) \leq \rho(x, A) + \delta_k$, then $\rho(x, Q_k) \leq 1/k$.

Let $u \in D$. Consider a sequence $\{x_i\}_{i=1}^{\infty} \subset A$ such that $\lim_{i \to \infty} \rho(u, x_i) = \rho(u, D)$. By property (P6), for each integer $k \geq 1$, there exists a subsequence $\{x_i^{(k)}\}_{i=1}^{\infty}$ of $\{x_i\}_{i=1}^{\infty}$ such that the following two properties hold:

(i) $\{x_i^{(k+1)}\}_{i=1}^{\infty}$ is a subsequence of $\{x_i^{(k)}\}_{i=1}^{\infty}$ for all integers $k \geq 1$;
(ii) for any integer $k \geq 1$, $\rho(x_j^{(k)}, x_s^{(k)}) \leq 2/k$ for all integers $j, s \geq 1$.

These properties imply that there exists a subsequence $\{x_i^*\}_{i=1}^{\infty}$ of $\{x_i\}_{i=1}^{\infty}$ which is a Cauchy sequence. Therefore $\{x_i^*\}_{i=1}^{\infty}$ converges to a point $\tilde{x} \in A$ which satisfies $\rho(\tilde{x}, u) = \lim_{i \to \infty} \rho(x_i, u) = \rho(u, D)$. This completes the proof of Theorem 7.13.

7.11 Porous Sets and Generalized Best Approximation Problems

Given a closed subset A of a Banach space X, a point $x \in X$ and a Lipschitzian (on bounded sets) function $f : X \to R^1$, we consider the problem of finding a solution to the minimization problem $\min\{f(x-y) : y \in A\}$. For a fixed function f, we define an appropriate complete metric space $\mathcal{M}$ of all pairs (A, x) and construct a subset Ω of $\mathcal{M}$, with a σ-porous complement $\mathcal{M} \setminus \Omega$, such that for each pair in Ω, our minimization problem is well posed.

Let $(X, \|\cdot\|)$ be a Banach space and let $f : X \to R^1$ be a Lipschitzian (on bounded sets) function. Assume that

$$\inf\{f(x) : x \in X\} \text{ is attained at a unique point } x_* \in X, \tag{7.122}$$

$$\lim_{\|u\|\to\infty} f(u) = \infty, \tag{7.123}$$

$$\text{if } \{x_i\}_{i=1}^{\infty} \subset X \text{ and } \lim_{i\to\infty} f(x_i) = f(x_*), \text{ then } \lim_{i\to\infty} x_i = x_*, \tag{7.124}$$

$$f\big(\alpha x + (1-\alpha)x_*\big) \le \alpha f(x) + (1-\alpha) f(x_*)$$
$$\text{for all } x \in X \text{ and all } \alpha \in (0, 1), \tag{7.125}$$

and that for each natural number n, there exists $k_n > 0$ such that

$$\big|f(x) - f(y)\big| \le k_n \|x-y\| \quad \text{for each } x, y \in X \text{ satisfying } \|x\|, \|y\| \le n. \tag{7.126}$$

Clearly, (7.125) holds if f is convex.

Given a closed subset A of X and a point $x \in X$, we consider the minimization problem

$$\min\{f(x-y) : y \in A\}. \tag{P}$$

For each $x \in X$ and each $A \subset X$, set

$$\rho(x, A) = \inf\{\|x-y\| : y \in A\}$$

and

$$\rho_f(x, A) = \inf\{f(x-y) : y \in A\}.$$

Denote by $S(X)$ the collection of all nonempty closed subsets of X. For each $A, B \in S(X)$, define

$$H(A, B) := \max\{\sup\{\rho(x, B) : x \in A\}, \sup\{\rho(y, A) : y \in B\}\} \tag{7.127}$$

and

$$\tilde{H}(A, B) := H(A, B)\big(1 + H(A, B)\big)^{-1}.$$

Here we use the convention that $\infty/\infty = 1$.

It is not difficult to see that the metric space $(S(X), \tilde{H})$ is complete.

For each natural number n and each $A, B \in S(X)$, we set

$$h_n(A, B) := \sup\{|\rho(x, A) - \rho(x, B)| : x \in X \text{ and } \|x\| \le n\} \tag{7.128}$$

and

$$h(A, B) := \sum_{n=1}^{\infty} [2^{-n} h_n(A, B)(1 + h_n(A, B))^{-1}].$$

Once again, it is not difficult to see that h is a metric on $S(X)$ and that the metric space $(S(X), h)$ is complete. Clearly, $\tilde{H}(A, B) \ge h(A, B)$ for all $A, B \in S(X)$.

We equip the set $S(X)$ with the pair of metrics $\tilde{H}$ and h. The topologies induced by the metrics $\tilde{H}$ and h on $S(X)$ will be called the strong topology and the weak topology, respectively.

Let $A \in S(X)$ and $\tilde{x} \in X$ be given. We say that the best approximation problem

$$f(\tilde{x} - y) \to \min, \quad y \in A,$$

is strongly well posed if there exists a unique $\bar{x} \in A$ such that

$$f(\tilde{x} - \bar{x}) = \inf\{f(\tilde{x} - y) : y \in A\}$$

and the following property holds:

For each $\varepsilon > 0$, there exists $\delta > 0$ such that if $z \in X$ satisfies $\|z - \tilde{x}\| \le \delta$, $B \in S(X)$ satisfies $h(A, B) \le \delta$, and $y \in B$ satisfies $f(z - y) \le \rho_f(z, B) + \delta$, then $\|y - \bar{x}\| \le \varepsilon$.

We now state four results obtained in [151]. Their proofs will be given in the next sections.

Theorem 7.16 *Let $\tilde{x} \in X$ be given. Then there exists a set $\Omega \subset S(X)$ such that its complement $S(X) \setminus \Omega$ is σ-porous with respect to $(h, \tilde{H})$ and for each $A \in \Omega$, the problem $f(\tilde{x} - y) \to$ min, $y \in A$, is strongly well posed.*

To state our second result, we endow the Cartesian product $S(X) \times X$ with the pair of metrics d_1 and d_2 defined by

$$d_1((A, x), (B, y)) = h(A, B) + \|x - y\|,$$
$$d_2((A, x), (B, y)) = \tilde{H}(A, B) + \|x - y\|, \quad x, y \in X, A, B \in S(X).$$

We will refer to the metrics induced on $S(X) \times X$ by d_2 and d_1 as the strong and weak metrics, respectively.

Theorem 7.17 *There exists a set $\Omega \subset S(X) \times X$ such that its complement $(S(X) \times X) \setminus \Omega$ is σ-porous with respect to (d_1, d_2) and for each $(A, \tilde{x}) \in \Omega$, the minimization problem*

$$f(\tilde{x} - y) \to \min, \quad y \in A,$$

is strongly well posed.

In most classical generic results the set A was fixed and x varied in a dense G_δ subset of X. In our first two results the set A is also variable. However, our third result shows that for every fixed A in a subset of $S(X)$ which has a σ-porous complement, the set of all $x \in X$ for which problem (P) is strongly well posed contains a dense G_δ subset of X.

Theorem 7.18 *Assume that X_0 is a closed separable subset of X. Then there exists a set $\mathcal{F} \subset S(X)$ such that its complement $S(X) \setminus \mathcal{F}$ is σ-porous with respect to $(h, \tilde{H})$ and for each $A \in \mathcal{F}$, the following property holds:*

There exists a set $F \subset X_0$, which is a countable intersection of open and everywhere dense subsets of X_0 with the relative topology, such that for each $\tilde{x} \in F$, the minimization problem

$$f(\tilde{x} - y) \to \min, \quad y \in A,$$

is strongly well posed.

Now we will show that Theorem 7.16 implies the following result.

Theorem 7.19 *Assume that $g : X \to R^1$ is a convex function which is Lipschitzian on bounded subsets of X and that $\inf\{g(x) : x \in X\}$ is attained at a unique point $y_* \in X$, $\lim_{\|u\| \to \infty} g(u) = \infty$, and if $\{y_i\}_{i=1}^{\infty} \subset X$ and $\lim_{i \to \infty} g(y_i) = g(y_*)$, then $y_i \to y_*$ as $i \to \infty$. Then there exists a set $\Omega \subset S(X)$ such that its complement $S(X) \setminus \Omega$ is σ-porous with respect to $(h, \tilde{H})$ and for each $A \in \Omega$, the following property holds:*

There is a unique $y_A \in A$ such that $g(y_A) = \inf\{g(y) : y \in A\}$. Moreover, for each $\varepsilon > 0$, there exists $\delta > 0$ such that if $y \in A$ satisfies $g(y) \le g(y_A) + \delta$, then $\|y - y_A\| \le \varepsilon$.

Proof Define $f(x) = g(-x)$, $x \in X$. It is clear that f is convex and satisfies (7.122)–(7.126). Therefore Theorem 7.16 is valid with $\tilde{x} = 0$ and there exists a set $\Omega \subset S(X)$ such that its complement $S(X) \setminus \Omega$ is σ-porous with respect to $(h, \tilde{H})$ and for each $A \in \Omega$, the following property holds:

There is a unique $\tilde{y} \in A$ such that

$$g(\tilde{y}) = f(-\tilde{y}) = \inf\{f(-y) : y \in A\} = \inf\{g(y) : y \in A\}.$$

Moreover, for each $\varepsilon > 0$, there exists $\delta > 0$ such that if $B \in S(X)$ satisfies $h(A, B) \le \delta$ and $x \in B$ satisfies

$$g(x) = f(-x) \le \rho_f(0, B) + \delta = \inf\{f(-y) : y \in B\} + \delta = \inf\{g(y) : y \in B\} + \delta,$$

then $\|x - \tilde{y}\| \le \varepsilon$. Theorem 7.19 is proved. □

It is easy to see that in the proofs of Theorems 7.16–7.18 we may assume without any loss of generality that $\inf\{f(x) : x \in X\} = 0$. It is also not difficult to see that we may assume without loss of generality that $x_* = 0$. Indeed, instead of the function

$f(\cdot)$ we can consider $f(\cdot + x_*)$. This new function also satisfies (7.122)–(7.126). Once Theorems 7.16–7.18 are proved for this new function, they will also hold for the original function f because the mapping $(A, x) \to (A, x + x_*)$, $(A, x) \in S(X) \times X$, is an isometry with respect to both metrics d_1 and d_2.

7.12 A Basic Lemma

Let m and n be two natural numbers. Choose a number

$$c_m > \sup\{f(u) : u \in X \text{ and } \|u\| \le 2m + 4\} + 2 \tag{7.129}$$

(see (7.126)). By (7.123), there exists a natural number

$$a_m > m + 2$$

such that

$$\text{if } u \in X \text{ and } f(u) \le c_m, \text{ then } \|u\| \le a_m. \tag{7.130}$$

By (7.126), there is $k_m > 1$ such that

$$|f(x) - f(y)| \le k_m \|x - y\|$$
$$\text{for each } x, y \in X \text{ satisfying } \|x\|, \|y\| \le 4a_m + 4. \tag{7.131}$$

By (7.131), there exists a positive number

$$\alpha(m, n) < 2^{-4a_m - 4} 16^{-1} n^{-1} \tag{7.132}$$

such that

$$\text{if } u \in X \text{ satisfies } f(u) \le 320 a_m \alpha(m, n), \text{ then } \|u\| \le (4n)^{-1}. \tag{7.133}$$

Finally, we choose a positive number

$$\bar{\alpha}(m, n) < \alpha(m, n)\left[(k_m + 1)^{-1} 2^{-4a_m - 16}\right]. \tag{7.134}$$

Lemma 7.20 *Let*

$$\alpha = \alpha(m, n), \qquad \bar{\alpha} = \bar{\alpha}(m, n), \tag{7.135}$$

$A \in S(X)$, $\tilde{x} \in X$, $r \in (0, 1]$, *and assume that*

$$\|\tilde{x}\| \le m \quad \textit{and} \quad \{z \in X : \|z\| \le m\} \cap A \ne \emptyset. \tag{7.136}$$

Then there exists $\bar{x} \in X$ *such that*

$$\rho(\bar{x}, A) \le r/8 \tag{7.137}$$

and for the set $\tilde{A} := A \cup \{\bar{x}\}$, *the following property holds:*
If

$$B \in S(X), \quad h(\tilde{A}, B) \le \bar{\alpha} r, \tag{7.138}$$

$$\tilde{y} \in X, \quad \|\tilde{y} - \bar{x}\| \le \bar{\alpha} r, \tag{7.139}$$

and

$$z \in B, \quad f(\tilde{y} - z) \le \rho_f(\tilde{y}, B) + \alpha r, \tag{7.140}$$

then

$$h(A, B) \le r \tag{7.141}$$

and

$$\|z - \bar{x}\| \le n^{-1}. \tag{7.142}$$

Proof First we choose $\tilde{x} \in X$. There are two cases: (1) $\rho(\tilde{x}, A) \le r/8$; (2) $\rho(\tilde{x}, A) > r/8$. If

$$\rho(\tilde{x}, A) \le r/8, \tag{7.143}$$

then we set

$$\bar{x} = \tilde{x} \quad \text{and} \quad \tilde{A} = A \cup \{\tilde{x}\}. \tag{7.144}$$

Now consider the second case where

$$\rho(\tilde{x}, A) > r/8. \tag{7.145}$$

First, choose $x_0 \in A$ such that

$$f(\tilde{x} - x_0) \le \rho_f(\tilde{x}, A) + \alpha(m, n) r \tag{7.146}$$

and then choose

$$\bar{x} \in \{\gamma \tilde{x} + (1 - \gamma) x_0 : \gamma \in (0, 1)\} \tag{7.147}$$

such that

$$\|\bar{x} - x_0\| = r/8 \quad \text{and} \quad \|\tilde{x} - \bar{x}\| = \|\tilde{x} - x_0\| - r/8. \tag{7.148}$$

Finally, set

$$\tilde{A} = A \cup \{\bar{x}\}. \tag{7.149}$$

Clearly, there is $\gamma \in (0, 1)$ such that

$$\bar{x} = \gamma \tilde{x} + (1 - \gamma) x_0. \tag{7.150}$$

It is easy to see that in both cases (7.137) holds and

$$\tilde{H}(A, \tilde{A}) \le H(A, \tilde{A}) \le r/8. \tag{7.151}$$

Now assume that $z \in X$ satisfies

$$z \in \tilde{A} \quad \text{and} \quad f(\tilde{x} - z) \le \rho_f(\tilde{x}, \tilde{A}) + 8\alpha(m,n)r. \tag{7.152}$$

We will show that $\|\bar{x} - z\| \le (2n)^{-1}$. First consider case (1). Then by (7.152), (7.144) and (7.149),

$$f(\bar{x} - z) = f(\tilde{x} - z) \le 8\alpha(m,n)r.$$

When combined with (7.133), this inequality implies that

$$\|\bar{x} - z\| \le (4n)^{-1}.$$

Now consider case (2). We first estimate $f(\tilde{x} - \bar{x})$. By (7.150) and (7.125) (with $x_* = 0$ and $f(x_*) = 0$),

$$\begin{aligned} f(\tilde{x} - \bar{x}) &= f\big(\tilde{x} - \gamma\tilde{x} - (1-\gamma)x_0\big) \\ &= f\big((1-\gamma)(\tilde{x} - x_0)\big) \le (1-\gamma)f(\tilde{x} - x_0). \end{aligned} \tag{7.153}$$

By (7.136), there is $z_0 \in X$ such that

$$z_0 \in A \quad \text{and} \quad \|z_0\| \le m. \tag{7.154}$$

Thus (7.146), (7.132), (7.154) and (7.136) imply that

$$\begin{aligned} f(\tilde{x} - x_0) &\le \rho_f(\tilde{x}, \tilde{A}) + 1 \le f(\tilde{x} - z_0) + 1 \\ &\le \sup\big\{f(u) : u \in X,\ \|u\| \le 2m + 1\big\} + 1 < c_m. \end{aligned} \tag{7.155}$$

Relations (7.155) and (7.130) imply that

$$\|x_0 - \tilde{x}\| \le a_m. \tag{7.156}$$

It follows from (7.148), (7.150) and (7.156) that

$$\begin{aligned} \|\tilde{x} - x_0\| - r/8 = \|\tilde{x} - \bar{x}\| &= \big\|\tilde{x} - \gamma\tilde{x} - (1-\gamma)x_0\big\| \\ &= (1-\gamma)\|\tilde{x} - x_0\|, \end{aligned}$$

$$\begin{aligned} 1 - \gamma &= \big(\|\tilde{x} - x_0\| - r/8\big)\|\tilde{x} - x_0\|^{-1} = 1 - r\big(8\|\tilde{x} - x_0\|\big)^{-1} \\ &\le 1 - r(8a_m)^{-1} \end{aligned}$$

and that

$$1 - \gamma \le 1 - r(8a_m)^{-1}. \tag{7.157}$$

By (7.153) and (7.157),

$$f(\tilde{x} - \bar{x}) = (1-\gamma)f(\tilde{x} - x_0) \le \big(1 - r(8a_m)^{-1}\big)f(\tilde{x} - x_0). \tag{7.158}$$

Relations (7.152) and (7.158) now imply that

$$f(\tilde{x}-z) \le f(\tilde{x}-\bar{x}) + 8\alpha r \le 8\alpha r + \bigl(1 - r(8a_m)^{-1}\bigr) f(\tilde{x}-x_0). \tag{7.159}$$

There are two cases:

$$f(\tilde{x}-x_0) \ge 8 \cdot 18\alpha a_m \tag{7.160}$$

and

$$f(\tilde{x}-x_0) \le 8 \cdot 18\alpha a_m. \tag{7.161}$$

Assume that (7.160) holds. Then it follows from (7.159), (7.146) and (7.160) that

$$\begin{aligned} f(\tilde{x}-z) &\le 8\alpha r + f(\tilde{x}-x_0) - r(8a_m)^{-1} f(\tilde{x}-x_0) \\ &\le 8\alpha r + \rho_f(\tilde{x}, A) + \alpha r - 8^{-1} \cdot 18\alpha r < \rho_f(\tilde{x}, A). \end{aligned}$$

Thus $z \notin A$ and by (7.152) and (7.149),

$$z = \bar{x}. \tag{7.162}$$

Now assume that (7.161) is true. By (7.161) and (7.152),

$$f(\tilde{x}-z) \le f(\tilde{x}-x_0) + 8\alpha r \le 8 \cdot 18\alpha a_m + 8\alpha \le 160\alpha a_m.$$

When combined with (7.133), (7.148) and (7.161), this estimate implies that

$$\|\tilde{x}-z\| \le (4n)^{-1}, \qquad \|\tilde{x}-x_0\| \le (4n)^{-1},$$
$$\|\tilde{x}-\bar{x}\| < \|\tilde{x}-x_0\| < (4n)^{-1},$$

and

$$\|\bar{x}-z\| < (2n)^{-1}.$$

Thus in both cases,

$$\|\bar{x}-z\| < (2n)^{-1}.$$

In other words, we have shown that the following property holds:

(P1) If $z \in X$ satisfies (7.152), then $\|\bar{x}-z\| \le (2n)^{-1}$.

Now assume that (7.138)–(7.140) hold. By (7.136) and (7.139), we have

$$\|\tilde{x}\| \le m \quad \text{and} \quad \|\tilde{y}\| \le m+1. \tag{7.163}$$

Relation (7.136) implies that there is $z_0 \in X$ such that

$$z_0 \in A \quad \text{and} \quad \|z_0\| \le m. \tag{7.164}$$

It follows from (7.128), (7.138), (7.164), (7.134) and (7.128) that

$$\begin{aligned} h_{4a_m+4}(\tilde{A}, B)\big(1 + h_{4a_m+4}(\tilde{A}, B)\big)^{-1} &\le 2^{4a_m+4} h(\tilde{A}, B) \le 2^{4a_m+4}\bar{\alpha} r, \\ h_{4a_m+4}(\tilde{A}, B) &\le 2^{4a_m+4}\bar{\alpha} r\big(1 - 2^{4a_m+4}\bar{\alpha} r\big) \le 2^{4a_m+5}\bar{\alpha} r \end{aligned} \tag{7.165}$$

and

$$\begin{aligned} \rho(z_0, B) &\le \rho(z_0, \tilde{A}) + \big|\rho(z_0, B) - \rho(z_0, \tilde{A})\big| \\ &\le h_{4a_m+4}(\tilde{A}, B) \le 2^{4a_m+5}\bar{\alpha} r. \end{aligned} \tag{7.166}$$

Inequalities (7.166), (7.134) and (7.132) imply that $\rho(z_0, B) < 1$, and that there is $\tilde{z}_0 \in X$ such that

$$\tilde{z}_0 \in B \quad \text{and} \quad \|\tilde{z}_0 - z_0\| < 1. \tag{7.167}$$

Clearly, by (7.164) and (7.167),

$$\|\tilde{z}_0\| < m + 1. \tag{7.168}$$

Let

$$\big\{(L, l)\big\} \in \big\{(\tilde{A}, \tilde{x}), (B, \tilde{y})\big\}. \tag{7.169}$$

By (7.136), (7.163), (7.164), (7.168) and (7.167),

$$\|l\| \le m + 1 \tag{7.170}$$

and there is $\bar{u} \in X$ such that

$$\bar{u} \in L \quad \text{and} \quad \|\bar{u}\| \le m + 1. \tag{7.171}$$

Relations (7.171), (7.170) and (7.129) imply that

$$\rho_f(l, L) \le f(l - \bar{u}) \le \sup\big\{f(u) : u \in X,\ \|u\| \le 2m + 2\big\} \le c_m - 2. \tag{7.172}$$

Also, relations (7.172), (7.130) and (7.170) imply the following property:

(P2) If $u \in L$ and $f(l - u) \le \rho_f(l, L) + 2$, then $\|l - u\| \le a_m$ and $\|u\| \le \|l\| + a_m \le 2a_m$.

Now assume that $L_i \in S(X)$ and $l_i \in X$, $i = 1, 2$, satisfy

$$\big\{(L_1, l_1), (L_2, l_2)\big\} = \big\{(\tilde{A}, \tilde{x}), (B, \tilde{y})\big\}. \tag{7.173}$$

Let

$$u \in L_1 \text{ be such that } f(l_1 - u) \le \rho_f(l_1, L_1) + 2. \tag{7.174}$$

By (7.174), (7.173) and property (P2),

$$\|u\| \le 2a_m. \tag{7.175}$$

Relations (7.174), (7.173), (7.175), (7.165) and (7.128) imply that

$$\rho(u, L_2) = \left|\rho(u, L_1) - \rho(u, L_2)\right| \le h_{2a_m}(L_1, L_2)$$
$$\le h_{4a_m+4}(\tilde{A}, B) \le 2^{4a_m+5}\bar{\alpha}r.$$

When combined with (7.132) and (7.134), this inequality implies that there is $v \in X$ such that

$$v \in L_2 \quad \text{and} \quad \|u - v\| \in 2^{4a_m+6}\bar{\alpha}r \le 1. \tag{7.176}$$

Inequalities (7.175) and (7.176) imply that

$$\|v\| \le 1 + 2a_m. \tag{7.177}$$

By (7.177), (7.175), (7.173) and (7.163),

$$\|l_1 - u\|, \|l_2 - v\| \le 1 + 2a_m + m + 1 < 3a_m. \tag{7.178}$$

It follows from (7.176), (7.139) and (7.173) that

$$\left\|(l_1 - u) - (l_2 - v)\right\| \le \bar{\alpha}r + 2^{4a_m+6}\bar{\alpha}r. \tag{7.179}$$

By (7.179), (7.178), (7.134) and the definition of k_m (see (7.131)),

$$\left|f(l_1 - u) - f(l_2 - v)\right| \le k_m\left\|(l_1 - u) - (l_2 - v)\right\|$$
$$\le k_m\bar{\alpha}r\left(1 + 2^{4a_m+6}\right) \le r\alpha 2^{-9}. \tag{7.180}$$

Inequalities (7.180) and (7.176) imply that

$$\rho_f(l_2, L_2) \le f(l_2 - v) \le f(l_1 - u) + 2^{-9}\alpha r$$

and

$$\rho_f(l_2, L_2) \le 2^{-9}\alpha r + f(l_1 - u). \tag{7.181}$$

Since (7.181) holds for any u satisfying (7.174), we conclude that

$$\rho_f(l_2, L_2) \le 2^{-9}\alpha r + \rho_f(l_1, L_1).$$

This fact implies, in turn, that

$$\left|\rho_f(l_1, L_1) - \rho_f(l_2, L_2)\right| = \left|\rho_f(\tilde{x}, \tilde{A}) - \rho_f(\tilde{y}, B)\right| \le 2^{-9}\alpha r. \tag{7.182}$$

By property (P2), (7.169) and (7.140),

$$\|\tilde{y} - z\| \le a_m \quad \text{and} \quad \|z\| \le 2a_m. \tag{7.183}$$

It follows from (7.140), (7.183), (7.165) and (7.128) that

$$\rho(z, \tilde{A}) \le \rho(z, B) + |\rho(z, B) - \rho(z, \tilde{A})|$$
$$= |\rho(z, B) - \rho(z, \tilde{A})| \le h_{4a_m+4}(\tilde{A}, B) \le 2^{4a_m+5}\bar{\alpha}r.$$

Thus there exists $\tilde{z} \in X$ such that

$$\tilde{z} \in \tilde{A} \quad \text{and} \quad \|z - \tilde{z}\| \le 2^{4a_m+6}\bar{\alpha}r. \tag{7.184}$$

By (7.136), (7.183), (7.184), (7.134) and (7.132), we have

$$\|\tilde{x} - \tilde{z}\| \le \|\tilde{x}\| + \|\tilde{z}\| \le m + \|z\| + \|\tilde{z} - z\|$$
$$\le m + 2a_m + 2^{4a_m+6}\bar{\alpha}r \le 3a_m + 1.$$

When combined with (7.134), (7.184), (7.139), (7.140) and (7.182), this inequality implies that

$$f(\tilde{x} - \tilde{z}) \le f(\tilde{y} - z) + |f(\tilde{x} - \tilde{z}) - f(\tilde{y} - z)|$$
$$\le f(\tilde{y} - z) + k_m \|\tilde{x} - \tilde{z} - (\tilde{y} - z)\| \le f(\tilde{y} - z)$$
$$\le k_m \|\tilde{x} - \tilde{y}\| + k_m \|\tilde{z} - z\| \le f(\tilde{y} - z) + k_m\bar{\alpha}r + k_m 2^{4a_m+6}\bar{\alpha}r$$
$$\le \rho_f(\tilde{y}, B) + \alpha r + k_m\bar{\alpha}r(1 + 2^{4a_m+6})$$
$$\le \alpha r + k_m\bar{\alpha}r(1 + 2^{4a_m+6}) + \rho_f(\tilde{x}, \tilde{A}) + 2^{-9}\alpha r \le \alpha r + \alpha r + \rho_f(\tilde{x}, \tilde{A}).$$

Thus we see that

$$f(\tilde{x} - \tilde{z}) \le \rho_f(\tilde{x}, \tilde{A}) + 2\alpha r. \tag{7.185}$$

It follows from property (P1), (7.152), (7.185) and (7.184) that

$$\|\tilde{z} - \tilde{x}\| \le (2n)^{-1}.$$

When combined with (7.184), (7.134) and (7.132), this inequality implies that

$$\|z - \tilde{x}\| \le \|z - \tilde{z}\| + \|\tilde{z} - \tilde{x}\| \le 2^{4a_m+6}\bar{\alpha}r + (2n)^{-1} \le n^{-1}.$$

Thus (7.142) is proved. Inequality (7.141) follows from (7.138), (7.151), (7.134) and (7.132). Thus we have shown that (7.138)–(7.140) imply (7.141) and (7.142). Lemma 7.20 is proved. □

7.13 Proofs of Theorems 7.16–7.18

We use the notations and the definitions from the previous section.

For each natural number n, denote by $\mathcal{F}_n$ the set of all $(x, A) \in X \times S(X)$ such that the following property holds:

(P3) There exist $y \in A$ and $\delta > 0$ such that for each $\tilde{x} \in X$ satisfying $\|\tilde{x} - x\| \leq \delta$, each $B \in S(X)$ satisfying $h(A, B) \leq \delta$, and each $z \in B$ satisfying $f(\tilde{x} - z) \leq \rho_f(\tilde{x}, B) + \delta$, the inequality $\|z - y\| \leq n^{-1}$ holds.

Set

$$\mathcal{F} = \bigcap_{n=1}^{\infty} \mathcal{F}_n. \tag{7.186}$$

Lemma 7.21 *If*

$$(x, A) \in \mathcal{F}, \tag{7.187}$$

then the problem $f(x - y) \to \min$, $y \in A$, *is strongly well posed.*

Proof Let $(x, A) \in \mathcal{F}$ and let n be a natural number. Since $(x, A) \in \mathcal{F} \subset \mathcal{F}_n$, there exist $x_n \in A$ and $\delta_n > 0$ such that the following property holds:

(P4) For each $\tilde{x} \in X$ satisfying $\|\tilde{x} - x\| \leq \delta_n$, each $B \in S(X)$ satisfying $h(A, B) \leq \delta_n$, and each $z \in B$ satisfying $f(\tilde{x} - z) \leq \rho_f(\tilde{x}, B) + \delta_n$, the inequality $\|z - x_n\| \leq n^{-1}$ holds.

Suppose that

$$\{z_i\}_{i=1}^{\infty} \subset A \quad \text{and} \quad \lim_{i \to \infty} f(x - z_i) = \rho_f(x, A). \tag{7.188}$$

Let n be any natural number. By (7.188) and property (P4), for all sufficiently large i we have

$$f(x - z_i) \leq \rho_f(x, A) + \delta_n \quad \text{and} \quad \|z_i - x_n\| \leq n^{-1}. \tag{7.189}$$

The second inequality of (7.189) implies that $\{z_i\}_{i=1}^{\infty}$ is a Cauchy sequence and there exists

$$\bar{x} = \lim_{i \to \infty} z_i. \tag{7.190}$$

Limits (7.190) and (7.188) imply that

$$f(x - \bar{x}) = \rho_f(x, A).$$

Clearly, $\bar{x}$ is the unique solution of the problem $f(x - z) \to \min$, $z \in A$. Otherwise we would be able to construct a nonconvergent sequence $\{z_i\}_{i=1}^{\infty}$ satisfying (7.188). By (7.190) and (7.189),

$$\|\bar{x} - x_n\| \leq n^{-1}, \quad n = 1, 2, \ldots. \tag{7.191}$$

Let $\varepsilon > 0$ be given. Choose a natural number

$$n > 8\varepsilon^{-1}. \tag{7.192}$$

Assume that

$$\tilde{x} \in X, \quad \|\tilde{x} - x\| \le \delta_n, \qquad B \in S(X), \quad h(A, B) \le \delta_n,$$
$$z \in B, \quad \text{and} \quad f(\tilde{x} - z) \le \rho_f(\tilde{x}, B) + \delta_n.$$

By Property (P4), $\|z - x_n\| \le 1/n$. When combined with (7.192) and (7.191), this inequality implies that

$$\|z - \bar{x}\| \le \|z - x_n\| + \|x_n - \bar{x}\| \le (2n)^{-1} < \varepsilon.$$

Thus the problem $f(x - z) \to \min$, $z \in A$, is strongly well posed. Lemma 7.21 is proved. □

Proof of Theorem 7.16 For each integer $n \ge 1$, set

$$\Omega_n := \{A \in S(X) : (\tilde{x}, A) \in \mathcal{F}_n\} \tag{7.193}$$

and let

$$\Omega := \bigcap_{n=1}^{\infty} \Omega_n. \tag{7.194}$$

By Lemma 7.21, (7.193) and (7.194), for each $A \in \Omega$, the problem $f(\tilde{x} - z) \to \min$, $z \in A$, is strongly well posed. In order to prove the theorem, it is sufficient to show that for each natural number n, the set $S(X) \setminus \Omega_n$ is σ-porous with respect to $(h, \tilde{H})$. To this end, let n be any natural number.

Fix a natural number

$$m_0 > \|\tilde{x}\|. \tag{7.195}$$

For each integer $m \ge m_0$, define

$$E_m := \{A \in S(X) : A \cap \{z \in X : \|z\| \le m\} \ne \emptyset\}. \tag{7.196}$$

Since

$$S(X) \setminus \Omega_n = \bigcup_{m=m_0}^{\infty} (E_m \setminus \Omega_n),$$

in order to prove the theorem, it is sufficient to show that for any natural number $m \ge m_0$, the set $E_m \setminus \Omega_n$ is porous with respect to $(h, \tilde{H})$. Let $m \ge m_0$ be a natural number. Define

$$\alpha_* = \bar{\alpha}(m + 1, n)/2 \tag{7.197}$$

(see (7.132) and (7.134)). Let $A \in S(X)$ and $r \in (0, 1]$. There are two cases:

case (1), where

$$A \cap \{z \in X : \|z\| \le m + 1\} = \emptyset \tag{7.198}$$

and case (2), where

$$A \cap \{z \in X : \|z\| \le m+1\} \ne \emptyset. \tag{7.199}$$

Consider the first case.

Let

$$B \in S(X) \text{ be such that } h(A,B) \le 2^{-m-2}. \tag{7.200}$$

We claim that $B \notin E_m$. Assume the contrary. Then there is $u \in X$ such that

$$u \in B \quad \text{and} \quad \|u\| \le m. \tag{7.201}$$

By (7.201) and (7.128),

$$\rho(u,A) \le \rho(u,B) + |\rho(u,B) - \rho(u,A)| \le h_m(A,B). \tag{7.202}$$

The definition of h_m (see (7.128)) and (7.200) imply that

$$h_m(A,B)\big(1+h_m(A,B)\big)^{-1} \le h(A,B)2^m \le 2^{-2},$$
$$h_m(A,B) \le h_m(A,B)2^{-2} + 2^{-2}$$

and

$$h_m(A,B) \le 1/3.$$

When combined with (7.202), this implies that there is $v \in A$ such that $\|u-v\| \le 1/2$. Together with (7.201) this inequality implies that $\|v\| \le m + 1/2$, a contradiction (see (7.198)). Therefore $B \notin E_m$, as claimed. Thus we have shown that

$$\{B \in S(X) : h(A,B) \le 2^{-m-2}\} \cap E_m = \emptyset. \tag{7.203}$$

Now consider the second case. Then by Lemma 7.20, (7.195) and (7.199), there exists $\bar{x} \in X$ such that

$$\rho(\bar{x}, A) \le r/8$$

and such that for the set $\tilde{A} = A \cup \{\bar{x}\}$, the following property holds:

(P5) if $B \in S(X)$, $h(\tilde{A}, B) \le \bar{\alpha}(m+1,n)r$, $\tilde{y} \in X$, $\|\tilde{y} - \bar{x}\| \le \bar{\alpha}(m+1,n)r$, and $z \in B$ satisfies

$$f(\tilde{y} - z) \le \rho_f(\tilde{y}, B) + \bar{\alpha}(m+1,n),$$

then

$$\|z - \bar{x}\| \le n^{-1} \quad \text{and} \quad h(A,B) \le r.$$

Clearly,

$$\tilde{H}(A, \tilde{A}) \le r/8.$$

Property (P5), (7.193) and the definition of $\mathcal{F}_n$ (see (P3)) imply that

$$\{B \in S(X) : h(\tilde{A}, B) \le \bar{\alpha}(m+1, n)r/2\} \subset \Omega_n.$$

Thus in both cases we have

$$\{B \in S(X) : h(\tilde{A}, B) \le \alpha_* r/2\} \cap (E_m \setminus \Omega_n) = \emptyset. \tag{7.204}$$

(Note that in the first case (7.204) is true with $\tilde{A} = A$.)

Therefore we have shown that the set $E_m \setminus \Omega_n$ is porous with respect to $(h, \tilde{H})$. Theorem 7.16 is proved. □

Proof of Theorem 7.17 By Lemma 7.21, in order to prove the theorem, it is sufficient to show that for any natural number n, the set $(X \times S(X)) \setminus \mathcal{F}_n$ is σ-porous in $X \times S(X)$ with respect to $(h, \tilde{H})$. To this end, let n be a natural number. For each natural number m, define

$$E_m = \{(x, A) \in X \times S(X) : \|x\| \le m \text{ and } A \cap \{z \in X : \|z\| \le m\} \ne \emptyset\}. \tag{7.205}$$

Since

$$(X \times S(X)) \setminus \mathcal{F}_n = \bigcup_{m=1}^{\infty} E_m \setminus \mathcal{F}_n,$$

in order to prove the theorem it is sufficient to show that for each natural number m, the set $E_m \setminus \mathcal{F}_n$ is porous in $X \times S(X)$ with respect to $(h, \tilde{H})$.

Let m be a natural number. Define α_* by (7.197). Assume that $(\tilde{x} \times A) \in X \times S(X)$ and $r \in (0, 1]$.

There are three cases:

case (1), where

$$\|\tilde{x}\| > m + 1,$$

case (2), where

$$\|\tilde{x}\| \le m + 1 \quad \text{and} \quad \{z \in A : \|z\| \le m + 1\} = \emptyset, \tag{7.206}$$

and case (3), where

$$\|\tilde{x}\| \le m + 1 \quad \text{and} \quad \{z \in A : \|z\| \le m + 1\} \ne \emptyset. \tag{7.207}$$

In the first case,

$$\{(y, B) \in X \times S(X) : d_1((\tilde{x}, A), (y, B)) \le 2^{-1}\} \cap E_m = \emptyset. \tag{7.208}$$

Next, consider the second case. In the proof of Theorem 7.16 we have shown that

$$\text{if } B \in S(X) \text{ satisfies } h(A, B) \le 2^{-m-2}, \text{ then}$$
$$B \cap \{z \in X : \|z\| \le m\} = \emptyset$$

and

$$\{(y, B) \in X \times S(X) : d_1((y, B), (\tilde{x}, A)) \le 2^{-m-2}\} \cap E_m = \emptyset. \tag{7.209}$$

Finally, consider the third case. Then by Lemma 7.20, there exists $\bar{x} \in X$ such that $\rho(\bar{x}, A) \le r/8$ and such that for the set $\tilde{A} = A \cup \{\bar{x}\}$, property (P5) holds. Clearly,

$$d_2((\tilde{x}, A), (\tilde{x}, \tilde{A})) = \tilde{H}(A, \tilde{A}) \le r/8.$$

Property (P5) implies that

$$\{(\tilde{y}, B) \in X \times S(X) : d_1((\tilde{y}, B), (\tilde{x}, \tilde{A})) \le \bar{\alpha}(m+1, n)r/2\} \subset \mathcal{F}_n.$$

Hence in all three cases we have

$$\{(\tilde{y}, B) \in X \times S(X) : d_1((\tilde{y}, B), (\tilde{x}, \tilde{A})) \le \alpha_* r\} \cap (E_m \setminus \mathcal{F}_n) = \emptyset. \tag{7.210}$$

Note that in the first and second cases, (7.210) is true with $A = \tilde{A}$. Therefore we have shown that the set $E_m \setminus \mathcal{F}_n$ is porous with respect to (d_1, d_2). Theorem 7.17 is proved. □

Proof of Theorem 7.18 Let $\{x_i\}_{i=1}^{\infty}$ be a countable dense subset of X_0. By countable dense subset of X_0. By Theorem 7.16, for each $\mathcal{F}_i \subset S(X)$ such that $S(X) \setminus \mathcal{F}_i$ is σ-porous in $S(X)$ with respect to $(h, \tilde{H})$ and such that for each $A \in S(X)$, the problem $f(x_i - z) \to \min$, $z \in X$, is strongly well posed. Set

$$\mathcal{F} := \bigcap_{i=1}^{\infty} \mathcal{F}_i. \tag{7.211}$$

Clearly, $S(X) \setminus \mathcal{F}$ is a σ-porous subset of $S(X)$ with respect to $(h, \tilde{H})$.

Let $A \in \mathcal{F}$. Assume that n and i are natural numbers. Since the problem $f(x_i - z) \to \min$, $z \in A$, is strongly well posed, there exists a number $\delta_{in} > 0$ and a unique $\bar{x}_i \in A$ such that

$$f(x_i - \bar{x}_i) = \rho_f(x_i, A) \tag{7.212}$$

and the following property holds:

(P6) if $y \in X$ satisfies $\|y - x_i\| \le \delta_{in}$, $B \in S(X)$ satisfies $h(A, B) \le \delta_{in}$, and $z \in B$ satisfies

$$f(y - z) \le \rho_f(y, B) + \delta_{in}, \tag{7.213}$$

then $\|z - \bar{x}_i\| \le (2n)^{-1}$.

Define

$$F = \bigcap_{q=1}^{\infty} \bigcup \{\{z \in X : \|z - x_i\| < \delta_{in}\} : i = 1, 2, \dots, n = q, q+1, \dots\} \cap X_0. \tag{7.214}$$

Clearly, F is a countable intersection of open everywhere dense subsets of X_0. Let

$$\tilde{x} \in F. \tag{7.215}$$

For each natural number q, there exist natural numbers $n_q \geq q$ and i_q such that

$$\|\tilde{x} - x_{iq}\| < \delta_{i_q n_q}. \tag{7.216}$$

Assume that

$$\{y_k\}_{k=1}^{\infty} \subset A \quad \text{and} \quad \lim_{k \to \infty} f(\tilde{x} - y_k) = \rho_f(\tilde{x}, A). \tag{7.217}$$

Let q be a natural number. Then for all sufficiently large natural numbers k,

$$f(\tilde{x} - y_k) \leq \rho_f(\tilde{x}, A) + \delta_{i_q n_q},$$

and by property (P6) and (7.216),

$$\|y_k - \bar{x}_{i_q}\| \leq (2n_q)^{-1} \leq (2q)^{-1}. \tag{7.218}$$

This implies that $\{y_k\}_{k=1}^{\infty}$ is a Cauchy sequence and there exists $\bar{x} = \lim_{k \to \infty} y_k$. By (7.217), $f(\tilde{x} - \bar{x}) = \rho_f(\tilde{x}, A)$. Clearly, $\bar{x}$ is the unique minimizer for the problem $f(\tilde{x} - z) \to \min$, $z \in A$. Otherwise, we would be able to construct a nonconvergent sequence $\{y_k\}_{k=1}^{\infty}$. By (7.218),

$$\|\bar{x} - x_{i_q}\| \leq (2q)^{-1}, \quad q = 1, 2, \ldots. \tag{7.219}$$

Let $\varepsilon > 0$ be given. Choose a natural number

$$q > 8\varepsilon^{-1}.$$

Set

$$\delta = \delta_{i_q n_q} - \|\tilde{x} - x_{i_q}\|. \tag{7.220}$$

By (7.216), $\delta > 0$. Assume that

$$y \in X, \quad \|y - \tilde{x}\| \leq \delta, \qquad B \in S(X), \quad h(A, B) \leq \delta, \tag{7.221}$$

and

$$z \in B, \quad f(y - z) \leq \rho_f(y, B) + \delta.$$

By (7.220) and (7.221),

$$\|y - x_{i_q}\| \leq \|y - \tilde{x}\| + \|\tilde{x} - x_{i_q}\| \leq \delta_{i_q n_q}. \tag{7.222}$$

By (7.222), (7.220) and property (P6), $\|z - \bar{x}_{i_q}\| \leq (2q)^{-1}$. When combined with (7.219), this inequality implies that $\|z - \bar{x}\| \leq q^{-1} < \varepsilon$. This completes the proof of Theorem 7.18. □

Chapter 8
Descent Methods

8.1 Discrete Descent Methods for a Convex Objective Function

Given a Lipschitzian convex function f on a Banach space X, we consider a complete metric space $\mathcal{A}$ of vector fields V on X with the topology of uniform convergence on bounded subsets. With each such vector field we associate two iterative processes. We introduce the class of regular vector fields $V \in \mathcal{A}$ and prove (under two mild assumptions on f) that the complement of the set of regular vector fields is not only of the first category, but also σ-porous. We then show that for a locally uniformly continuous regular vector field V and a coercive function f, the values of f tend to its infimum for both processes. These results were obtained in [136].

Assume that $(X, \|\cdot\|)$ is a Banach space with norm $\|\cdot\|$, $(X^*, \|\cdot\|_*)$ is its dual space with the norm $\|\cdot\|_*$, and $f : X \to R^1$ is a convex continuous function which is bounded from below. Recall that for each pair of sets $A, B \subset X^*$,

$$H(A, B) = \max\Big\{\sup_{x \in A} \inf_{y \in B} \|x - y\|_*, \sup_{y \in B} \inf_{x \in A} \|x - y\|_*\Big\}$$

is the Hausdorff distance between A and B.

For each $x \in X$, let

$$\partial f(x) := \big\{l \in X^* : f(y) - f(x) \geq l(y - x) \text{ for all } y \in X\big\}$$

be the subdifferential of f at x. It is well known that the set $\partial f(x)$ is nonempty and bounded (in the norm topology). Set

$$\inf(f) := \inf\big\{f(x) : x \in X\big\}.$$

Denote by $\mathcal{A}$ the set of all mappings $V : X \to X$ such that V is bounded on every bounded subset of X (i.e., for each $K_0 > 0$ there is $K_1 > 0$ such that $\|Vx\| \leq K_1$ if $\|x\| \leq K_0$), and for each $x \in X$ and each $l \in \partial f(x)$, $l(Vx) \leq 0$. We denote by $\mathcal{A}_c$ the set of all continuous $V \in \mathcal{A}$, by $\mathcal{A}_u$ the set of all $V \in \mathcal{A}$ which are uniformly

S. Reich, A.J. Zaslavski, *Genericity in Nonlinear Analysis*,
Developments in Mathematics 34, DOI 10.1007/978-1-4614-9533-8_8,

continuous on each bounded subset of X, and by $\mathcal{A}_{au}$ the set of all $V \in \mathcal{A}$ which are uniformly continuous on the subsets

$$\{x \in X : \|x\| \le n \text{ and } f(x) \ge \inf(f) + 1/n\}$$

for each integer $n \ge 1$. Finally, let $\mathcal{A}_{auc} = \mathcal{A}_{au} \cap \mathcal{A}_c$.

Next we endow the set $\mathcal{A}$ with a metric ρ: For each $V_1, V_2 \in \mathcal{A}$ and each integer $i \ge 1$, we first set

$$\rho_i(V_1, V_2) := \sup\{\|V_1 x - V_2 x\| : x \in X \text{ and } \|x\| \le i\} \tag{8.1}$$

and then define

$$\rho(V_1, V_2) := \sum_{i=1}^{\infty} 2^{-i}\left[\rho_i(V_1, V_2)\left(1 + \rho_i(V_1, V_2)\right)^{-1}\right]. \tag{8.2}$$

Clearly $(\mathcal{A}, \rho)$ is a complete metric space. It is also not difficult to see that the collection of the sets

$$E(N, \varepsilon) = \{(V_1, V_2) \in \mathcal{A} \times \mathcal{A} : \|V_1 x - V_2 x\| \le \varepsilon, x \in X, \|x\| \le N\}, \tag{8.3}$$

where $N, \varepsilon > 0$, is a base for the uniformity generated by the metric ρ. Evidently $\mathcal{A}_c$, $\mathcal{A}_u$, $\mathcal{A}_{au}$ and $\mathcal{A}_{auc}$ are closed subsets of the metric space $(\mathcal{A}, \rho)$. In the sequel we assign to all these spaces the same metric ρ.

To compute $\inf(f)$, we are going to associate with each vector field $W \in \mathcal{A}$ two gradient-like iterative processes (see (8.5) and (8.7) below).

The study of steepest descent and other minimization methods is a central topic in optimization theory. See, for example, [2, 19, 44, 47, 69, 73, 103] and the references mentioned therein. Note, in particular, that the counterexample studied in Sect. 2.2 of Chap. VIII of [73] shows that, even for two-dimensional problems, the simplest choice for a descent direction, namely the normalized steepest descent direction,

$$V(x) = \operatorname{argmin}\left\{\max_{l \in \partial f(x)} \langle l, d\rangle : \|d\| = 1\right\},$$

may produce sequences the functional values of which fail to converge to the infimum of f. This vector field V belongs to $\mathcal{A}$ and the Lipschitzian function f attains its infimum. The steepest descent scheme (Algorithm 1.1.7) presented in Sect. 1.1 of Chap. VIII of [73] corresponds to any of the two iterative processes we consider below.

In infinite dimensions the problem is even more difficult and less understood. Moreover, positive results usually require special assumptions on the space and the functions. However, as shown in our paper [135] (under certain assumptions on the function f), for an arbitrary Banach space X and a generic vector field $V \in \mathcal{A}$, the values of f tend to its infimum for both processes. In that paper, instead of considering a certain convergence property for a method generated by a single vector field

V, we investigated it for the whole space $\mathcal{A}$ and showed that this property held for most of the vector fields in $\mathcal{A}$.

Here we introduce the class of regular vector fields $V \in \mathcal{A}$. Our first result, Theorem 8.2, shows (under the two mild assumptions A(i) and A(ii) on f stated below) that the complement of the set of regular vector fields is not only of the first category, but also σ-porous in each of the spaces $\mathcal{A}$, $\mathcal{A}_c$, $\mathcal{A}_u$, $\mathcal{A}_{au}$ and $\mathcal{A}_{auc}$. We then show (Theorem 8.3) that for any regular vector field $V \in \mathcal{A}_{au}$, if the constructed sequence $\{x_i\}_{i=0}^{\infty} \subset X$ has a bounded subsequence (in the case of the first process) or is bounded (in the case of the second one), then the values of the function f tend to its infimum for both processes. If, in addition to A(i) and A(ii), f also satisfies the assumption A(iii), then this convergence result is valid for any regular $V \in \mathcal{A}$. Note that if the function f is coercive, then the constructed sequences will always stay bounded. Thus we see, by Theorem 8.2, that for a coercive f the set of divergent descent methods is σ-porous. Our last result, Theorem 8.4, shows that in this case we obtain not only convergence, but also stability.

Our results are established in any Banach space and for those convex functions which satisfy the following two assumptions.

A(i) There exists a bounded (in the norm topology) set $X_0 \subset X$ such that

$$\inf(f) = \inf\{f(x) : x \in X\} = \inf\{f(x) : x \in X_0\};$$

A(ii) for each $r > 0$, the function f is Lipschitzian on the ball $\{x \in X : \|x\| \leq r\}$.

Note that we may assume that the set X_0 in A(i) is closed and convex. It is clear that assumption A(i) holds if $\lim_{\|x\|\to\infty} f(x) = \infty$.

We say that a mapping $V \in \mathcal{A}$ is regular if for any natural number n, there exists a positive number $\delta(n)$ such that for each $x \in X$ satisfying

$$\|x\| \leq n \quad \text{and} \quad f(x) \geq \inf(f) + 1/n,$$

and each $l \in \partial f(x)$, we have

$$l(Vx) \leq -\delta(n).$$

Denote by $\mathcal{F}$ the set of all regular vector fields $V \in \mathcal{A}$.

It is not difficult to verify the following property of regular vector fields. It means, in particular, that $\mathcal{G} = \mathcal{A} \setminus \mathcal{F}$ is a face of the convex cone $\mathcal{A}$ in the sense that if a non-trivial convex combination of two vector fields in $\mathcal{A}$ belongs to $\mathcal{G}$, then both of them must belong to $\mathcal{G}$.

Proposition 8.1 *Assume that $V_1, V_2 \in \mathcal{A}$, V_1 is regular, $\phi : X \to [0, 1]$, and that for each integer $n \geq 1$,*

$$\inf\{\phi(x) : x \in X \text{ and } \|x\| \leq n\} > 0.$$

Then the mapping $x \to \phi(x)V_1x + (1 - \phi(x))V_2x$, $x \in X$, also belongs to $\mathcal{F}$.

Our first result shows that in a very strong sense most of the vector fields in $\mathcal{A}$ are regular.

Theorem 8.2 *Assume that both* A(i) *and* A(ii) *hold. Then* $\mathcal{A} \setminus \mathcal{F}$ *(respectively,* $\mathcal{A}_c \setminus \mathcal{F}$*,* $\mathcal{A}_{au} \setminus \mathcal{F}$ *and* $\mathcal{A}_{auc} \setminus \mathcal{F}$*) is a* σ*-porous subset of the space* $\mathcal{A}$ *(respectively,* $\mathcal{A}_c$*,* $\mathcal{A}_{au}$ *and* $\mathcal{A}_{auc}$*). Moreover, if* f *attains its infimum, then the set* $\mathcal{A}_u \setminus \mathcal{F}$ *is also a* σ*-porous subset of the space* $\mathcal{A}_u$*.*

Now let $W \in \mathcal{A}$. We associate with W two iterative processes.

For $x \in X$ we denote by $P_W(x)$ the set of all

$$y \in \{x + \alpha Wx : \alpha \in [0, 1]\}$$

such that

$$f(y) = \inf\{f(x + \beta Wx) : \beta \in [0, 1]\}. \tag{8.4}$$

Given any initial point $x_0 \in X$, one can construct a sequence $\{x_i\}_{i=0}^{\infty} \subset X$ such that for all $i = 0, 1, \ldots,$

$$x_{i+1} \in P_W(x_i). \tag{8.5}$$

This is our first iterative process.

Next we describe the second iterative process.

Given a sequence $\mathbf{a} = \{a_i\}_{i=0}^{\infty} \subset (0, 1]$ such that

$$\lim_{i\to\infty} a_i = 0 \quad \text{and} \quad \sum_{i=0}^{\infty} a_i = \infty, \tag{8.6}$$

we construct for each initial point $x_0 \in X$, a sequence $\{x_i\}_{i=0}^{\infty} \subset X$ according to the following rule:

$$\begin{aligned} x_{i+1} &= x_i + a_i W(x_i) \quad \text{if } f\big(x_i + a_i W(x_i)\big) < f(x_i), \\ x_{i+1} &= x_i \quad \text{otherwise}, \end{aligned} \tag{8.7}$$

where $i = 0, 1, \ldots.$

We will also make use of the following assumption:

A(iii) For each integer $n \geq 1$, there exists $\delta > 0$ such that for each $x_1, x_2 \in X$ satisfying

$$\|x_1\|, \|x_2\| \leq n, \qquad f(x_i) \geq \inf(f) + 1/n, \quad i = 1, 2, \quad \text{and}$$
$$\|x_1 - x_2\| \leq \delta,$$

the following inequality holds:

$$H\big(\partial f(x_1), \partial f(x_2)\big) \leq 1/n.$$

This assumption is certainly satisfied if f is differentiable and its derivative is uniformly continuous on those bounded subsets of X over which the infimum of f is larger than $\inf(f)$.

Our next result is a convergence theorem for those iterative processes associated with regular vector fields. It is of interest to note that we obtain convergence when either the regular vector field W or the subdifferential ∂f enjoy a certain uniform continuity property.

Theorem 8.3 *Assume that $W \in \mathcal{A}$ is regular,* A(i), A(ii) *are valid and that at least one of the following conditions holds*: 1. $W \in \mathcal{A}_{au}$; 2. A(iii) *is valid. Then the following two assertions are true*:

(i) *Let the sequence $\{x_i\}_{i=0}^{\infty} \subset X$ satisfy* (8.5) *for all $i = 0, 1, \ldots$. If*

$$\liminf_{i \to \infty} \|x_i\| < \infty,$$

then $\lim_{i \to \infty} f(x_i) = \inf(f)$.

(ii) *Let a sequence $\mathbf{a} = \{a_i\}_{i=0}^{\infty} \subset (0, 1]$ satisfy* (8.6) *and let the sequence $\{x_i\}_{i=0}^{\infty} \subset X$ satisfy* (8.7) *for all $i = 0, 1, \ldots$. If $\{x_i\}_{i=0}^{\infty}$ is bounded, then*

$$\lim_{i \to \infty} f(x_i) = \inf(f).$$

Finally, we impose an additional coercivity condition on f and establish the following stability theorem. Note that this coercivity condition implies A(i).

Theorem 8.4 *Assume that $f(x) \to \infty$ as $\|x\| \to \infty$, $V \in \mathcal{A}$ is regular,* A(ii) *is valid and that at least one of the following conditions holds*: 1. $V \in \mathcal{A}_{au}$; 2. A(iii) *is valid.*

Let $K, \varepsilon > 0$ be given. Then there exist a neighborhood $\mathcal{U}$ of V in $\mathcal{A}$ and a natural number N_0 such that the following two assertions are true:

(i) *For each $W \in \mathcal{U}$ and each sequence $\{x_i\}_{i=0}^{N_0} \subset X$ which satisfies $\|x_0\| \leq K$ and* (8.5) *for all $i = 0, \ldots, N_0 - 1$, the inequality $f(x_{N_0}) \leq \inf(f) + \varepsilon$ holds.*
(ii) *For each sequence of numbers $\mathbf{a} = \{a_i\}_{i=0}^{\infty} \subset (0, 1]$ satisfying* (8.6), *there exists a natural number N such that for each $W \in \mathcal{U}$ and each sequence $\{x_i\}_{i=0}^{N} \subset X$ which satisfies $\|x_0\| \leq K$ and* (8.7) *for all $i = 0, \ldots, N - 1$, the inequality $f(x_N) \leq \inf(f) + \varepsilon$ holds.*

8.2 An Auxiliary Result

Assume that $\mathcal{K}$ is a nonempty, closed and convex subset of X. We consider the topological subspace $\mathcal{K} \subset X$ with the relative topology. For each function $h : \mathcal{K} \to R^1$ define $\inf(h) := \inf\{h(x) : x \in \mathcal{K}\}$.

Proposition 8.5 *Let $g : \mathcal{K} \to R^1$ be a convex, bounded from below, function which is uniformly continuous on bounded subsets of $\mathcal{K}$. Assume that there exists a bounded and convex set $\mathcal{K}_0 \subset \mathcal{K}$ such that for each $x \in \mathcal{K}$, there exists $y \in \mathcal{K}_0$ for which $g(y) \le g(x)$.*

Then there exists a continuous mapping $A_g : \mathcal{K} \to \mathcal{K}_0$ which satisfies $g(A_g x) \le g(x)$ for all $x \in \mathcal{K}$ and has the following two properties:

B(i) *For each integer $n \ge 1$, the mapping A_g is uniformly continuous on the set*

$$\{x \in \mathcal{K} : \|x\| \le n \text{ and } g(x) \ge \inf(g) + 1/n\};$$

B(ii) *if $g(x) \ge \inf(g) + \varepsilon$ for some $\varepsilon > 0$ and $x \in \mathcal{K}$, then*

$$g(A_g x) \le g(x) - \varepsilon/2.$$

Proof If there exists $x \in \mathcal{K}$ for which $g(x) = \inf(g)$, then there exists $x^* \in \mathcal{K}_0$ for which $g(x^*) = \inf(g)$ and we can set $A_g(y) = x^*$ for all $y \in \mathcal{K}$. Therefore we may assume that

$$\{x \in \mathcal{K} : g(x) = \inf(g)\} = \emptyset.$$

For each integer $i \ge 0$, there exists $y_i \in \mathcal{K}_0$ such that

$$g(y_i) \le \left(4(i+1)\right)^{-1} + \inf(g). \tag{8.8}$$

Consider now the linear segments which join $y_0, y_1, \dots, y_n, \dots$ (all contained in $\mathcal{K}_0$ by the convexity of $\mathcal{K}_0$), represented as a continuous curve $\gamma : [0, \infty) \to \mathcal{K}_0$ and parametrized so that

$$\gamma(t) = y_i + (t - i)(y_{i+1} - y_i) \quad \text{if } i \le t < i + 1 \ (i = 0, 1, 2, \dots). \tag{8.9}$$

The curve γ is Lipschitzian because the set $\mathcal{K}_0$ is bounded. Define

$$A_g x = \gamma\left(g(x) - \left(\inf(g)\right)^{-1}\right), \quad x \in \mathcal{K}. \tag{8.10}$$

It is easy to see that $A_g x \in \mathcal{K}_0$ for all $x \in \mathcal{K}$, the mapping A_g is continuous on $\mathcal{K}$ and that it is uniformly continuous on the subsets

$$\{x \in \mathcal{K} : \|x\| \le n \text{ and } g(x) \ge \inf(g) + 1/n\}$$

for each integer $n \ge 1$.

Assume that

$$x \in \mathcal{K}, \qquad \varepsilon > 0 \quad \text{and} \quad g(x) \ge \inf(g) + \varepsilon. \tag{8.11}$$

There is an integer $i \ge 0$ such that

$$g(x) - \inf(g) \in \left((i+1)^{-1}, i^{-1}\right] \tag{8.12}$$

(here $0^{-1} = \infty$). Then

$$\left(g(x) - \inf(g)\right)^{-1} \in [i, i+1) \tag{8.13}$$

and by (8.10), (8.9) and (8.13),

$$A_g x = \gamma\left(g(x) - \left(\inf(g)\right)^{-1}\right) = y_i + \left(\left(g(x) - \inf(g)\right)^{-1} - i\right)(y_{i+1} - y_i).$$

It follows from this relation, (8.8), (8.11), (8.12) and the convexity of g that

$$\begin{aligned} g(A_g x) &\le \max\{g(y_i), g(y_{i+1})\} \le \inf(g) + \left(4(i+1)\right)^{-1} \\ &\le \inf(g) + 4^{-1}\left(g(x) - \inf(g)\right) = g(x) - 3 \cdot 4^{-1}\left(g(x) - \inf(g)\right) \\ &\le g(x) - 3 \cdot 4^{-1}\varepsilon. \end{aligned}$$

This completes the proof of Proposition 8.5. □

8.3 Proof of Theorem 8.2

We first note the following simple lemma.

Lemma 8.6 *Assume that $V_1, V_2 \in \mathcal{A}$, $\phi : X \to [0, 1]$, and that*

$$Vx = \left(1 - \phi(x)\right)V_1 x + \phi(x)V_2 x, \quad x \in X.$$

Then $V \in \mathcal{A}$. If $V_1, V_2 \in \mathcal{A}_c$ and ϕ is continuous on X, then $V \in \mathcal{A}_c$. If $V_1, V_2 \in \mathcal{A}_u$ (respectively, $\mathcal{A}_{au}$, $\mathcal{A}_{auc}$) and ϕ is uniformly continuous on bounded subsets of X, then $V \in \mathcal{A}_u$ (respectively, $\mathcal{A}_{au}$, $\mathcal{A}_{auc}$).

For each pair of integers $m, n \ge 1$, denote by Ω_{mn} the set of all $V \in \mathcal{A}$ such that

$$\|Vx\| \le m \quad \text{for all } x \in X \text{ satisfying } \|x\| \le n+1 \tag{8.14}$$

and

$$\sup\{l(Vx) : x \in X, \|x\| \le n, f(x) \ge \inf(f) + 1/n, l \in \partial f(x)\} = 0. \tag{8.15}$$

Clearly,

$$\bigcup_{m=1}^{\infty}\bigcup_{n=1}^{\infty} \Omega_{mn} = \mathcal{A} \setminus \mathcal{F}. \tag{8.16}$$

Therefore in order to prove Theorem 8.2 it is sufficient to show that for each pair of integers $m, n \ge 1$, the set Ω_{mn} (respectively, $\Omega_{mn} \cap \mathcal{A}_c$, $\Omega_{mn} \cap \mathcal{A}_{au}$, $\Omega_{mn} \cap \mathcal{A}_{auc}$) is a porous subset of $\mathcal{A}$ (respectively, $\mathcal{A}_c$, $\mathcal{A}_{au}$, $\mathcal{A}_{auc}$), and if f attains its minimum, then $\Omega_{mn} \cap \mathcal{A}_u$ is a porous subset of $\mathcal{A}_u$.

By assumption A(i), there is a bounded and convex set $X_0 \subset X$ with the following property:

C(i) For each $x \in X$, there is $x_0 \in X_0$ such that $f(x_0) \le f(x)$. If f attains its minimum, then X_0 is a singleton.

By Proposition 8.5, there is a continuous mapping $A_f : X \to X$ such that

$$A_f(X) \subset X_0, \qquad f(A_f x) \le f(x) \quad \text{for all } x \in X, \tag{8.17}$$

and which has the following two properties:

C(ii) If $x \in X$, $\varepsilon > 0$ and $f(x) \ge \inf(f) + \varepsilon$, then $f(A_f x) \le f(x) - \varepsilon/2$;
C(iii) for any natural number n, the mapping A_f is uniformly continuous on the set

$$\{x \in X : \|x\| \le n \text{ and } f(x) \ge \inf(f) + 1/n\}.$$

Let $m, n \ge 1$ be integers. In the sequel we will use the piecewise linear function $\phi : R^1 \to R^1$ defined by

$$\phi(x) = 1, \quad x \in [-n, n], \qquad \phi(x) = 0, \quad |x| \ge n + 1 \tag{8.18}$$

and

$$\phi(-n - 1 + t) = t, \quad t \in [0, 1], \qquad \phi(n + t) = 1 - t, \quad t \in [0, 1].$$

By assumption A(ii), there is $c_0 > 1$ such that

$$|f(x) - f(y)| \le c_0 \|x - y\| \tag{8.19}$$

for all $x, y \in X$ satisfying $\|x\|, \|y\| \le n + 2$. Choose $\alpha \in (0, 1)$ such that

$$\alpha c_0 2^{n+2} < (2n)^{-1} 2^{-1} (1 - \alpha)\big(m + n + 2 + \sup\{\|x\| : x \in X_0\}\big)^{-1}. \tag{8.20}$$

Assume that $V \in \Omega_{mn}$ and $r \in (0, 1]$. Let

$$\gamma = 2^{-1}(1 - \alpha) r \big(m + n + 2 + \sup\{\|x\| : x \in X_0\}\big)^{-1} \tag{8.21}$$

and define $V_\gamma : X \to X$ by

$$V_\gamma x = \big(1 - \gamma\phi(\|x\|)\big) V x + \gamma\phi(\|x\|)(A_f x - x), \quad x \in X. \tag{8.22}$$

By Lemma 8.6, $V_\gamma \in \mathcal{A}$ and moreover, if $V \in \mathcal{A}_c$ (respectively, $\mathcal{A}_{au}$, $\mathcal{A}_{auc}$), then $V_\gamma \in \mathcal{A}_c$ (respectively, $\mathcal{A}_{au}$, $\mathcal{A}_{auc}$), and if $V \in \mathcal{A}_u$ and f attains its minimum, then A_f is constant (see C(i)) and $V_\gamma \in \mathcal{A}_u$.

Next we estimate the distance $\rho(V_\gamma, V)$. It follows from (8.22) and the definition of ϕ (see (8.18)) that $V_\gamma x = V x$ for all $x \in X$ satisfying $\|x\| \ge n + 1$ and

$$\rho_i(V_\gamma, V) = \rho_{n+1}(V_\gamma, V) \quad \text{for all integers } i \ge n + 1.$$

Since $V \in \Omega_{mn}$, the above equality, when combined with (8.2), (8.1), (8.22), (8.18) and (8.17), yields

$$\begin{aligned} \rho(V_\gamma, V) &\le \sum_{i=1}^{\infty} 2^{-i} \rho_i(V, V_\gamma) \le \rho_{n+1}(V, V_\gamma) \\ &= \sup\{\|Vx - V_\gamma x\| : x \in X, \|x\| \le n+1\} \\ &\le \sup\{\gamma\phi(\|x\|)(\|Vx\| + \|A_f x - x\|) : x \in X, \|x\| \le n+1\} \\ &\le \gamma(m+1) + \gamma(n+1) + \gamma \sup\{\|x\| : x \in X_0\}. \end{aligned} \tag{8.23}$$

Assume that $W \in \mathcal{A}$ with

$$\rho(W, V_\gamma) \le \alpha r. \tag{8.24}$$

By (8.24), (8.23) and (8.21),

$$\rho(W, V) \le \alpha r + \gamma(m + n + 2 + \sup\{\|x\| : x \in X_0\}) \le 2^{-1}(1+\alpha)r < r. \tag{8.25}$$

Assume now that

$$x \in X, \quad \|x\| \le n, \quad f(x) \ge \inf(f) + 1/n \quad \text{and} \quad l \in \partial f(x). \tag{8.26}$$

Inequality (8.19) implies that

$$\|l\|_* \le c_0.$$

By (8.22), (8.26), the definition of ϕ (see (8.18)) and C(ii),

$$\begin{aligned} l(V_\gamma x) &= l((1 - \gamma\phi(\|x\|))Vx + \gamma\phi(\|x\|)(A_f x - x)) \le \gamma\phi(\|x\|) l(A_f x - x) \\ &= \gamma l(A_f x - x) \le \gamma(f(A_f x) - f(x)) \le -\gamma(2n)^{-1}. \end{aligned} \tag{8.27}$$

It follows from (8.26) and (8.1) that

$$\|Wx - V_\gamma x\| \le \rho_n(W, V_\gamma). \tag{8.28}$$

By (8.24), (8.28) and the inequality $\|l\|_* \le c_0$, we have

$$\begin{gathered} 2^{-n}\rho_n(W, V_\gamma)(1 + \rho_n(W, V_\gamma))^{-1} \le \rho(W, V_\gamma) \le \alpha r, \\ \rho_n(W, V_\gamma)(1 + \rho_n(W, V_\gamma))^{-1} \le 2^n \alpha r, \\ \rho_n(W, V_\gamma)(1 - 2^n\alpha r) \le 2^n \alpha r, \qquad \|Wx - V_\gamma x\| \le 2^n \alpha r(1 - 2^n \alpha r)^{-1}, \end{gathered} \tag{8.29}$$

and

$$|l(Wx) - l(V_\gamma x)| \le c_0 2^n \alpha r(1 - 2^n \alpha r)^{-1}. \tag{8.30}$$

By (8.30), (8.27), (8.21) and (8.20),

$$
\begin{aligned}
l(Wx) &\le l(V_\gamma x) + c_0 2^n \alpha r\big(1 - 2^n \alpha r\big)^{-1} \\
&\le -\gamma (2n)^{-1} + c_0 2^n \alpha r\big(1 - 2^n \alpha r\big)^{-1} \\
&= c_0 2^n \alpha r\big(1 - 2^n \alpha r\big)^{-1} \\
&\quad - (2n)^{-1} 2^{-1}(1-\alpha) r\big(m+n+2+\sup\{\|x\| : x \in X_0\}\big)^{-1} \\
&\le -r\big[-c_0 2^n \alpha \cdot 2 + (2n)^{-1}2^{-1}(1-\alpha)\big(m+n+2+\sup\{\|x\| : x \in X_0\}\big)^{-1}\big] \\
&\le -2r c_0 2^n \alpha.
\end{aligned}
$$

Thus

$$\{W \in \mathcal{A} : \rho(W, V_\gamma) \le \alpha r\} \cap \Omega_{mn} = \emptyset.$$

In view of (8.25), we can conclude that Ω_{mn} is porous in $\mathcal{A}$, $\Omega_{mn} \cap \mathcal{A}_c$ is porous in $\mathcal{A}_c$, $\Omega_{mn} \cap \mathcal{A}_{au}$ is porous in $\mathcal{A}_{au}$, $\Omega_{mn} \cap \mathcal{A}_{auc}$ is porous in $\mathcal{A}_{auc}$, and if f attains its minimum, then $\Omega_{mn} \cap \mathcal{A}_u$ is porous in $\mathcal{A}_u$. This completes the proof of Theorem 8.2.

8.4 A Basic Lemma

The following result is our key lemma.

Lemma 8.7 *Assume that $V \in \mathcal{A}$ is regular,* A(i), A(ii) *are valid and that at least one of the following conditions holds*: 1. $V \in \mathcal{A}_{au}$; 2. A(iii) *is valid.*

Let $\bar{K}$ and $\bar{\varepsilon}$ be positive. Then there exist a neighborhood $\mathcal{U}$ of V in $\mathcal{A}$ and positive numbers $\bar{\alpha}$ and γ such that for each $W \in \mathcal{U}$, each $x \in X$ satisfying

$$\|x\| \le \bar{K}, \qquad f(x) \ge \inf(f) + \bar{\varepsilon}, \tag{8.31}$$

and each $\beta \in (0, \bar{\alpha}]$,

$$f(x) - f(x + \beta W x) \ge \beta\gamma. \tag{8.32}$$

Proof There exists $K_0 > \bar{K} + 1$ such that

$$\|Vx\| \le K_0 \quad \text{if } x \in X \text{ and } \|x\| \le \bar{K} + 2. \tag{8.33}$$

By Assumption A(ii), there exists a constant $L_0 > 4$ such that

$$\big|f(x_1) - f(x_2)\big| \le L_0\|x_1 - x_2\| \tag{8.34}$$

for all $x_1, x_2 \in X$ satisfying $\|x_1\|, \|x_2\| \le 2K_0 + 4$. Since V is regular, there exists a positive number $\delta_0 \in (0, 1)$ such that

$$\xi(Vy) \le -\delta_0 \tag{8.35}$$

for each $y \in X$ satisfying $\|y\| \le K_0 + 4$, $f(y) \ge \inf(f) + \bar{\varepsilon}/4$, and each $\xi \in \partial f(y)$. Choose $\delta_1 \in (0, 1)$ such that

$$4\delta_1(K_0 + L_0) < \delta_0. \tag{8.36}$$

There exists a positive number $\bar{\alpha}$ such that the following conditions hold:

$$8\bar{\alpha}(L_0 + 1)(K_0 + 1) < \min\{1, \bar{\varepsilon}\}; \tag{8.37}$$

(a) if $V \in \mathcal{A}_{au}$, then for each $x_1, x_2 \in X$ satisfying

$$\begin{aligned} \|x_1\|, \|x_2\| \le \bar{K} + 4, \qquad \min\{f(x_1), f(x_2)\} \ge \inf(f) + \bar{\varepsilon}/4, \\ \text{and} \quad \|x_1 - x_2\| \le \bar{\alpha}(K_0 + 1), \end{aligned} \tag{8.38}$$

the following inequality is true:

$$\|Vx_1 - Vx_2\| \le \delta_1; \tag{8.39}$$

(b) if A(iii) is valid, then for each $x_1, x_2 \in X$ satisfying (8.38), the following inequality is true:

$$H(\partial f(x_1), \partial f(x_2)) < \delta_1. \tag{8.40}$$

Next choose a positive number δ_2 such that

$$8\delta_2(L_0 + 1) < \delta_1\delta_0. \tag{8.41}$$

Now choose a positive number γ such that

$$\gamma < \delta_0/8 \tag{8.42}$$

and define

$$\mathcal{U} := \{W \in \mathcal{A} : \|Wx - Vx\| \le \delta_2, x \in X \text{ and } \|x\| \le \bar{K}\}. \tag{8.43}$$

Assume that $W \in \mathcal{U}$, $x \in X$ satisfies (8.31), and that $\beta \in (0, \bar{\alpha}]$. We intend to show that (8.32) holds. To this end, we first note that (8.31), (8.33), (8.37), (8.43) and (8.41) yield

$$\|x + \beta Vx\| \le \bar{K} + \beta K_0 \le \bar{K} + \bar{\alpha} K_0 \le \bar{K} + 1$$

and

$$\|x + \beta Wx\| \le \delta_2\beta + \|x + \beta Vx\| \le \bar{K} + 1 + \bar{\alpha}\delta_2 \le \bar{K} + 2.$$

By these inequalities, the definition of L_0 (see (8.34)) and (8.43),

$$|f(x + \beta Vx) - f(x + \beta Wx)| \le L_0\beta\|Wx - Vx\| \le L_0\beta\delta_2. \tag{8.44}$$

Next we will estimate $f(x) - f(x + \beta Vx)$. There exist $\theta \in [0, \beta]$ and $l \in \partial f(x + \theta Vx)$ such that

$$f(x + \beta Vx) - f(x) = l(Vx)\beta. \tag{8.45}$$

By (8.31), (8.33) and (8.37),

$$\|x\| \le \bar{K}, \qquad \|Vx\| \le K_0, \qquad \|\theta Vx\| \le \bar{\alpha} K_0, \quad \text{and} \\ \|x + \theta Vx\| \le \bar{K} + 1. \tag{8.46}$$

It follows from (8.46) and the definition of L_0 (see (8.34)) that

$$\|l\|_* \le L_0. \tag{8.47}$$

It follows from (8.46), the definition of L_0 (see (8.34)), (8.37) and (8.31) that

$$\begin{aligned} f(x + \theta Vx) &\ge f(x) - L_0 \|\theta Vx\| \\ &\ge f(x) - L_0 \bar{\alpha} K_0 \ge f(x) - 8^{-1}\bar{\varepsilon} \ge \inf(f) + \bar{\varepsilon}/2. \end{aligned} \tag{8.48}$$

Consider the case where $V \in \mathcal{A}_{au}$. By (8.47), condition (a), (8.46), (8.31) and (8.48),

$$\begin{aligned} \beta l(Vx) &\le \beta l\big(V(x + \theta Vx)\big) + \beta \|l\|_* \big(\|V(x + \theta Vx) - Vx\|\big) \\ &\le \beta l\big(V(x + \theta Vx)\big) + \beta L_0 \|V(x + \theta Vx) - Vx\| \\ &\le \beta l\big(V(x + \theta Vx)\big) + \beta L_0 \delta_1. \end{aligned} \tag{8.49}$$

By (8.46), (8.48) and the definition of δ_0 (see (8.35)),

$$l\big(V(x + \theta Vx)\big) \le -\delta_0.$$

When combined with (8.49) and (8.36), this inequality implies that

$$\beta l(Vx) \le -\beta\delta_0 + \beta L_0 \delta_1 \le -\beta\delta_0/2.$$

By these inequalities and (8.45),

$$f(x + \beta Vx) - f(x) \le -\beta\delta_0/2. \tag{8.50}$$

Assume now that A(iii) is valid. It then follows from condition (b), (8.46), (8.31) and (8.48) that

$$H\big(\partial f(x), \partial f(x + \theta Vx)\big) < \delta_1.$$

Therefore there exists $\bar{l} \in \partial f(x)$ such that $\|\bar{l} - l\|_* \le \delta_1$. When combined with (8.45) and (8.46), this fact implies that

$$\begin{aligned} f(x + \beta Vx) - f(x) = \beta l(Vx) &\le \beta \bar{l}(Vx) + \beta \|\bar{l} - l\|_* \|Vx\| \\ &\le \beta \bar{l}(Vx) + \beta \delta_1 K_0. \end{aligned} \tag{8.51}$$

It follows from the definition of δ_0 (see (8.35)) and (8.31) that $\beta\bar{l}(Vx) \le -\beta\delta_0$. Combining this inequality with (8.51) and (8.36), we see that

$$f(x+\beta Vx) - f(x) \le -\beta\delta_0 + \beta\delta_1 K_0 \le -\beta\delta_0/2.$$

Thus in both cases (8.50) is true. It now follows from (8.50), (8.44), (8.41) and (8.42) that

$$\begin{aligned} f(x+\beta Wx) - f(x) &\le f(x+\beta Vx) - f(x) + f(x+\beta Wx) - f(x+\beta Vx) \\ &\le -\beta\delta_0/2 + L_0\beta\delta_2 \le -\beta\delta_0/4 \le -\gamma\beta. \end{aligned}$$

Thus (8.32) holds. Lemma 8.7 is proved. □

8.5 Proofs of Theorems 8.3 and 8.4

Proof of Theorem 8.3 To show that assertion (i) holds, suppose that

$$\{x_i\}_{i=0}^{\infty} \subset X, \qquad x_{i+1} \in P_W x_i, \quad i = 0, 1, \ldots, \quad \text{and} \quad \liminf_{i\to\infty} \|x_i\| < \infty. \tag{8.52}$$

We will show that

$$\lim_{i\to\infty} f(x_i) = \inf(f). \tag{8.53}$$

Assume the contrary. Then there exists $\varepsilon > 0$ such that

$$f(x_i) \ge \inf(f) + \varepsilon, \quad i = 0, 1, \ldots. \tag{8.54}$$

There exists a number $S > 0$ and a strictly increasing sequence of natural numbers $\{i_k\}_{k=1}^{\infty}$ such that

$$\|x_{i_k}\| \le S, \quad k = 1, 2, \ldots. \tag{8.55}$$

By Lemma 8.7, there exist numbers $\alpha, \gamma \in (0, 1)$ such that for each $x \in X$ satisfying

$$\|x\| \le S, \qquad f(x) \ge \inf(f) + \varepsilon, \tag{8.56}$$

and each $\beta \in (0, \alpha]$,

$$f(x) - f(x+\beta Wx) \ge \gamma\beta. \tag{8.57}$$

It follows from (8.52), (8.4), (8.5), the definitions of α and γ, (8.55) and (8.54) that for each integer $k \ge 1$,

$$f(x_{i_k}) - f(x_{i_k+1}) \ge f(x_{i_k}) - f(x_{i_k} + \alpha W x_{i_k}) \ge \gamma\alpha.$$

Since this inequality holds for all integers $k \geq 1$, we conclude that

$$\lim_{n\to\infty}\big(f(x_0) - f(x_n)\big) = \infty.$$

This contradicts our assumption that f is bounded from below. Therefore (8.53) and assertion (i) are indeed true, as claimed.

We turn now to assertion (ii). Let $\mathbf{a} = \{a_i\}_{i=0}^{\infty} \subset (0, 1]$ satisfy (8.6) and let a bounded $\{x_i\}_{i=0}^{\infty} \subset X$ satisfy (8.7) for all integers $i \geq 0$. We will show that (8.53) holds. Indeed, assume that (8.53) is not true. Then there exists $\varepsilon > 0$ such that (8.54) holds. Since the sequence $\{x_i\}_{1=0}^{\infty}$ is bounded, there exists a number $S > 0$ such that

$$S > \|x_i\|, \quad i = 0, 1, \ldots. \tag{8.58}$$

By Lemma 8.7, there exist numbers $\alpha, \gamma \in (0, 1)$ such that for each $x \in X$ satisfying (8.56) and each $\beta \in (0, \alpha]$, inequality (8.57) holds. Since $a_i \to 0$ as $i \to \infty$, there exists a natural number i_0 such that

$$a_i < \alpha \quad \text{for all integers } i \geq i_0. \tag{8.59}$$

Let $i \geq i_0$ be an integer. Then it follows from (8.58), (8.54), the definitions of α and γ, and (8.59) that

$$f(x_i) - f(x_i + a_i W x_i) \geq \gamma a_i, \quad x_{i+1} = x_i + a_i W x_i,$$

and

$$f(x_i) - f(x_{i+1}) \geq \gamma a_i.$$

Since $\sum_{i=0}^{\infty} a_i = \infty$, we conclude that

$$\lim_{n\to\infty}\big(f(x_0) - f(x_n)\big) = \infty.$$

The contradiction we have reached shows that (8.53), assertion (ii) and Theorem 8.3 itself are all true. □

Proof of Theorem 8.4 Let

$$K_0 > \sup\big\{f(x) : x \in X,\ \|x\| \leq K + 1\big\} \tag{8.60}$$

and set

$$E_0 = \big\{x \in X : f(x) \leq K_0 + 1\big\}. \tag{8.61}$$

Clearly, E_0 is bounded and closed. Choose

$$K_1 > \sup\big\{\|x\| : x \in E_0\big\} + 1 + K. \tag{8.62}$$

By Lemma 8.7, there exist a neighborhood $\mathcal{U}$ of V in $\mathcal{A}$ and numbers $\alpha, \gamma \in (0, 1)$ such that for each $W \in \mathcal{U}$, each $x \in X$ satisfying

$$\|x\| \leq K_1, \qquad f(x) \geq \inf(f) + \varepsilon, \tag{8.63}$$

and each $\beta \in (0, \alpha]$,

$$f(x) - f(x + \beta Wx) \geq \gamma\beta. \tag{8.64}$$

Now choose a natural number N_0 which satisfies

$$N_0 > (\alpha\gamma)^{-1}\big(K_0 + 4 + |\inf(f)|\big). \tag{8.65}$$

First we will show that assertion (i) is true. Assume that $W \in \mathcal{U}$, $\{x_i\}_{i=0}^{N_0} \subset X$,

$$\|x_0\| \leq K, \quad \text{and} \quad x_{i+1} \in P_W x_i, \quad i = 0, \dots, N_0 - 1. \tag{8.66}$$

Our aim is to show that

$$f(x_{N_0}) \leq \inf(f) + \varepsilon. \tag{8.67}$$

Assume that (8.67) is not true. Then

$$f(x_i) > \inf(f) + \varepsilon, \quad i = 0, \dots, N_0. \tag{8.68}$$

By (8.66) and (8.60)–(8.62), we also have

$$\|x_i\| \leq K_1, \quad i = 0, \dots, N_0. \tag{8.69}$$

Let $i \in \{0, \dots, N_0 - 1\}$. It follows from (8.69), (8.68) and the definitions of $\mathcal{U}$, α and γ (see (8.63) and (8.64)) that

$$f(x_i) - f(x_{i+1}) \geq f(x_i) - f(x_i + \alpha W x_i) \geq \gamma\alpha.$$

Summing up from $i = 0$ to $N_0 - 1$, we conclude that

$$f(x_0) - f(x_{N_0}) \geq N_0\gamma\alpha.$$

It follows from this inequality, (8.60), (8.65) and (8.66) that

$$\inf(f) \leq f(x_{N_0}) \leq f(x_0) - N_0\gamma\alpha \leq K_0 - N_0\gamma\alpha \leq -4 - |\inf(f)|.$$

Since we have reached a contradiction, we see that (8.67) must be true and assertion (i) is proved.

Now we will show that assertion (ii) is also valid. To this end, let a sequence $\mathbf{a} = \{a_i\}_{i=0}^{\infty} \subset (0, 1]$ satisfy

$$\lim_{i\to\infty} a_i = 0 \quad \text{and} \quad \sum_{i=0}^{\infty} a_i = \infty. \tag{8.70}$$

Evidently, there exists a natural number N_1 such that

$$a_i \leq \alpha \quad \text{for all } i \geq N_1. \tag{8.71}$$

Choose a natural number $N > N_1 + 4$ such that

$$\gamma \sum_{i=N_1}^{N-1} a_i > K_0 + 4 + \left|\inf(f)\right|. \tag{8.72}$$

Now assume that $W \in \mathcal{U}$, $\{x_i\}_{i=0}^N \subset X$, $\|x_0\| \le K$, and that (8.7) holds for all $i = 0, \dots, N-1$. We claim that

$$f(x_N) \le \inf(f) + \varepsilon. \tag{8.73}$$

Assume the contrary. Then

$$f(x_i) > \inf(f) + \varepsilon, \quad i = 0, \dots, N. \tag{8.74}$$

Since $\|x_0\| \le K$, we see by (8.7) and (8.60)–(8.62) that

$$\|x_i\| \le K_1, \quad i = 0, \dots, N. \tag{8.75}$$

Let $i \in \{N_1, \dots, N-1\}$. It follows from (8.75), (8.74), (8.71) and the definitions of α and γ (see (8.63) and (8.64)) that

$$f(x_i) - f(x_i + a_i W x_i) \ge \gamma a_i.$$

This implies that

$$f(x_{N_1}) - f(x_N) \ge \gamma \sum_{i=N_1}^{N-1} a_i.$$

By this inequality, (8.7), the inequality $\|x_0\| \le K$, (8.60) and (8.72), we obtain

$$\inf(f) \le f(x_N) \le f(x_{N_1}) - \gamma \sum_{i=N_1}^{N-1} a_i$$

$$\le K_0 - \gamma \sum_{i=N_1}^{N-1} a_i < -4 - |\inf(f)|.$$

The contradiction we have reached proves (8.73) and assertion (ii). This completes the proof of Theorem 8.4. □

8.6 Methods for a Nonconvex Objective Function

Assume that $(X, \|\cdot\|)$ is a Banach space, $(X^*, \|\cdot\|_*)$ is its dual space, and $f : X \to R^1$ is a function which is bounded from below and Lipschitzian on bounded subsets of X. Recall that for each pair of sets $A, B \subset X^*$,

$$H(A, B) = \max\left\{\sup_{x\in A}\inf_{y\in B}\|x - y\|_*, \sup_{y\in B}\inf_{x\in A}\|x - y\|_*\right\}$$

is the Hausdorff distance between A and B. For each $x \in X$, let

$$f^0(x, h) = \limsup_{t\to 0^+, y\to x} \big[f(y + th) - f(y)\big]/t, \quad h \in X, \tag{8.76}$$

be the Clarke derivative of f at the point x [41],

$$\partial f(x) = \big\{l \in X^* : f^0(x, h) \geq l(h) \text{ for all } h \in X\big\} \tag{8.77}$$

the Clarke subdifferential of f at x, and

$$\Xi(x) := \inf\big\{f^0(x, h) : h \in X \text{ and } \|h\| = 1\big\}. \tag{8.78}$$

It is well known that the set $\partial f(x)$ is nonempty and bounded. It should be mentioned that the functional Ξ was introduced in [176] and used in [182] in order to study penalty methods in constrained optimization.

Set $\inf(f) = \inf\{f(x) : x \in X\}$. Denote by $\mathcal{A}$ the set of all mappings $V : X \to X$ such that V is bounded on every bounded subset of X, and for each $x \in X$, $f^0(x, Vx) \leq 0$. We denote by $\mathcal{A}_c$ the set of all continuous $V \in \mathcal{A}$ and by $\mathcal{A}_b$ the set of all $V \in \mathcal{A}$ which are bounded on X. Finally, let $\mathcal{A}_{bc} = \mathcal{A}_b \cap \mathcal{A}_c$. Next we endow the set $\mathcal{A}$ with two metrics, ρ_s and ρ_w. To define ρ_s, we set, for each $V_1, V_2 \in \mathcal{A}$, $\tilde{\rho}_s(V_1, V_2) = \sup\{\|V_1x - V_2x\| : x \in X\}$ and

$$\rho_s(V_1, V_2) = \tilde{\rho}_s(V_1, V_2)\big(1 + \tilde{\rho}_s(V_1, V_2)\big)^{-1}. \tag{8.79}$$

(Here we use the convention that $\infty/\infty = 1$.) It is clear that $(\mathcal{A}, \rho_s)$ is a complete metric space. To define ρ_w, we set, for each $V_1, V_2 \in \mathcal{A}$ and each integer $i \geq 1$,

$$\rho_i(V_1, V_2) := \sup\big\{\|V_1x - V_2x\| : x \in X \text{ and } \|x\| \leq i\big\}, \tag{8.80}$$

$$\rho_w(V_1, V_2) := \sum_{i=1}^{\infty} 2^{-i}\big[\rho_i(V_1, V_2)\big(1 + \rho_i(V_1, V_2)\big)^{-1}\big]. \tag{8.81}$$

Clearly, $(\mathcal{A}, \rho_w)$ is a complete metric space. It is also not difficult to see that the collection of the sets

$$E(N, \varepsilon) = \big\{(V_1, V_2) \in \mathcal{A} \times \mathcal{A} : \|V_1x - V_2x\| \leq \varepsilon, x \in X, \|x\| \leq N\big\},$$

where $N, \varepsilon > 0$, is a base for the uniformity generated by the metric ρ_w. It is easy to see that $\rho_w(V_1, V_2) \leq \rho_s(V_1, V_2)$ for all $V_1, V_2 \in \mathcal{A}$. The metric ρ_w induces on $\mathcal{A}$ a topology which is called the weak topology and the metric ρ_s induces a topology which is called the strong topology. Clearly, $\mathcal{A}_c$ is a closed subset of $\mathcal{A}$ with the weak topology while $\mathcal{A}_b$ and $\mathcal{A}_{bc}$ are closed subsets of $\mathcal{A}$ with the strong topology.

We consider the subspaces $\mathcal{A}_c$, $\mathcal{A}_b$ and $\mathcal{A}_{bc}$ with the metrics ρ_s and ρ_w which induce the strong and the weak topologies, respectively.

When the function f is convex, one usually looks for a sequence $\{x_i\}_{i=1}^{\infty}$ which tends to a minimum point of f (if such a point exists) or at least such that $\lim_{i\to\infty} f(x_i) = \inf(f)$. If f is not necessarily convex, but X is finite-dimensional, then we expect to construct a sequence which tends to a critical point z of f, namely a point z for which $0 \in \partial f(z)$. If f is not necessarily convex and X is infinite-dimensional, then the problem is more difficult and less understood because we cannot guarantee, in general, the existence of a critical point and a convergent subsequence. To partially overcome this difficulty, we have introduced the function $\Xi : X \to R^1$. Evidently, a point z is a critical point of f if and only if $\Xi(z) \geq 0$. Therefore we say that z is ε-critical for a given $\varepsilon > 0$ if $\Xi(z) \geq -\varepsilon$. We look for sequences $\{x_i\}_{i=1}^{\infty}$ such that either $\liminf_{i\to\infty} \Xi(x_i) \geq 0$ or at least $\limsup_{i\to\infty} \Xi(x_i) \geq 0$. In the first case, given $\varepsilon > 0$, all the points x_i, except possibly a finite number of them, are ε-critical, while in the second case this holds for a subsequence of $\{x_i\}_{i=1}^{\infty}$.

We show, under certain assumptions on f, that for most (in the sense of Baire's categories) vector fields $W \in \mathcal{A}$, the iterative processes defined below (see (8.84) and (8.85)) yield sequences with the desirable properties. Moreover, we show that the complement of the set of "good" vector fields is not only of the first category, but also σ-porous. These results, which were obtained in [141], are stated in this section. Their proofs are relegated to subsequent sections.

For each set $E \subset X$, we denote by $\mathrm{cl}(E)$ the closure of E in the norm topology. Our results hold for any Banach space and for those functions which satisfy the following two assumptions.

A(i) For each $\varepsilon > 0$, there exists $\delta \in (0, \varepsilon)$ such that

$$\mathrm{cl}(\{x \in X : \Xi(x) < -\varepsilon\}) \subset \{x \in X : \Xi(x) < -\delta\};$$

A(ii) for each $r > 0$, the function f is Lipschitzian on the ball $\{x \in X : \|x\| \leq r\}$.

We say that a mapping $V \in \mathcal{A}$ is regular if for any natural number n, there exists a positive number $\delta(n)$ such that for each $x \in X$ satisfying $\|x\| \leq n$ and $\Xi(x) < -1/n$, we have $f^0(x, Vx) \leq -\delta(n)$.

This concept of regularity is a non-convex analog of the regular vector fields introduced in [136]. We denote by $\mathcal{F}$ the set of all regular vector fields $V \in \mathcal{A}$.

Theorem 8.8 *Assume that both* A(i) *and* A(ii) *hold. Then* $\mathcal{A} \setminus \mathcal{F}$ *(respectively,* $\mathcal{A}_c \setminus \mathcal{F}$, $\mathcal{A}_b \setminus \mathcal{F}$ *and* $\mathcal{A}_{bc} \setminus \mathcal{F}$*) is a* σ*-porous subset of the space* $\mathcal{A}$ *(respectively,* $\mathcal{A}_c$, $\mathcal{A}_b$ *and* $\mathcal{A}_{bc}$*) with respect to the pair* (ρ_w, ρ_s).

Now let $W \in \mathcal{A}$. We associate with W two iterative processes. For $x \in X$ we denote by $P_W(x)$ the set of all $y \in \{x + \alpha Wx : \alpha \in [0, 1]\}$ such that

$$f(y) = \inf\{f(x + \beta Wx) : \beta \in [0, 1]\}. \tag{8.82}$$

Given any initial point $x_0 \in X$, one can construct a sequence $\{x_i\}_{i=0}^{\infty} \subset X$ such that for all $i = 0, 1, \ldots,$

$$x_{i+1} \in P_W(x_i). \tag{8.83}$$

This is our first iterative process. Next we describe the second iterative process. Given a sequence $\mathbf{a} = \{a_i\}_{i=0}^{\infty} \subset (0, 1)$ such that

$$\lim_{i\to\infty} a_i = 0 \quad \text{and} \quad \sum_{i=0}^{\infty} a_i = \infty, \tag{8.84}$$

we construct for each initial point $x_0 \in X$, a sequence $\{x_i\}_{i=0}^{\infty} \subset X$ according to the following rule:

$$\begin{aligned} x_{i+1} &= x_i + a_i W(x_i) \quad \text{if } f\big(x_i + a_i W(x_i)\big) < f(x_i), \\ x_{i+1} &= x_i \quad \text{otherwise, where } i = 0, 1, \ldots. \end{aligned} \tag{8.85}$$

In the sequel we will also make use of the following assumption:

A(iii) For each integer $n \geq 1$, there exists $\delta > 0$ such that for each $x_1, x_2 \in X$ satisfying $\|x_1\|, \|x_2\| \leq n$, $\min\{\Xi(x_i) : i = 1, 2\} \leq -1/n$, and $\|x_1 - x_2\| \leq \delta$, the following inequality holds: $H(\partial f(x_1), \partial f(x_2)) \leq 1/n$.

We denote by $\text{Card}(B)$ the cardinality of a set B.

Theorem 8.9 *Assume that $W \in \mathcal{A}$ is regular, and that* A(i), A(ii) *and* A(iii) *are all valid. Then the following two assertions are true*:

(i) *Let the sequence $\{x_i\}_{i=0}^{\infty} \subset X$ satisfy* (8.83) *for all $i = 0, 1, \ldots$. If $\{x_i\}_{i=0}^{\infty}$ is bounded, then* $\liminf_{i\to\infty} \Xi(x_i) \geq 0$.
(ii) *Let a sequence $\mathbf{a} = \{a_i\}_{i=0}^{\infty} \subset (0, 1)$ satisfy* (8.84) *and let the sequence $\{x_i\}_{i=0}^{\infty} \subset X$ satisfy* (8.85) *for all $i = 0, 1, \ldots$. If $\{x_i\}_{i=0}^{\infty}$ is bounded, then*

$$\limsup_{i\to\infty} \Xi(x_i) \geq 0.$$

Theorem 8.10 *Assume that $f(x) \to \infty$ as $\|x\| \to \infty$, $V \in \mathcal{A}$ is regular, and that* A(i), A(ii) *and* A(iii) *are all valid. Let $K, \varepsilon > 0$ be given. Then there exist a neighborhood $\mathcal{U}$ of V in $\mathcal{A}$ with the weak topology and a natural number N_0 such that the following two assertions are true*:

(i) *For each $W \in \mathcal{U}$, each integer $n \geq N_0$ and each sequence $\{x_i\}_{i=0}^{n} \subset X$ which satisfies $\|x_0\| \leq K$ and* (8.83) *for all $i = 0, \ldots, n-1$, we have*

$$\text{Card}\big\{i \in \{0, \ldots, N-1\} : \Xi(x_i) \leq -\varepsilon\big\} \leq N_0.$$

(ii) *For each sequence of numbers $\mathbf{a} = \{a_i\}_{i=0}^{\infty} \subset (0, 1)$ satisfying* (8.84), *there exists a natural number N such that for each $W \in \mathcal{U}$ and each sequence*

$\{x_i\}_{i=0}^N \subset X$ which satisfies $\|x_0\| \le K$ and (8.85) *for all $i = 0, \dots, N-1$, we have*

$$\max\{\Xi(x_i) : i = 0, \dots, N\} \ge -\varepsilon.$$

8.7 An Auxiliary Result

For each positive number λ, set

$$E_\lambda := \{x \in X : \Xi(x) < -\lambda\}. \tag{8.86}$$

Proposition 8.11 *Let $\varepsilon > 0$ be given. Suppose that*

$$\mathrm{cl}(E_\varepsilon) \subset E_{\delta(\varepsilon)} \tag{8.87}$$

for some $\delta(\varepsilon) \in (0, \varepsilon)$. Then there exists a locally Lipschitzian vector field $V \in \mathcal{A}_b$ such that $f^0(y, Vy) < -\delta(\varepsilon)$ for all $y \in X$ satisfying $\Xi(y) < -\varepsilon$.

Proof It easily follows from definitions (8.76) and (8.78) that E_λ is an open set for all $\lambda > 0$. Let $x \in E_{\delta(\varepsilon)}$. Then there exist $h_x \in X$ such that $\|h_x\| = 1$ and $f^0(x, h_x) < -\delta(\varepsilon)$, and (see (8.76)) an open neighborhood U_x of x in X such that

$$f^0(y, h_x) < -\delta(\varepsilon) \quad \text{for all } y \in U_x. \tag{8.88}$$

For $x \in X \setminus E_{\delta(\varepsilon)}$, set

$$h_x = 0 \quad \text{and} \quad U_x = X \setminus \mathrm{cl}(E_\varepsilon). \tag{8.89}$$

Clearly, $\{U_x\}_{x \in X}$ is an open covering of X. Since any metric space is paracompact, there is a locally finite refinement $\{Q_\alpha : \alpha \in A\}$ of $\{U_x : x \in X\}$, i.e., an open covering of X such that each $x \in X$ has a neighborhood $Q(x)$ with $Q(x) \cap Q_\alpha \neq \emptyset$ only for finitely many $\alpha \in A$, and such that for each $\alpha \in A$, there exists $x_\alpha \in X$ with $Q_\alpha \subset U(x_\alpha)$. Let $\alpha \in A$. Define $\mu_\alpha : X \to [0, \infty)$ by $\mu_\alpha(x) = 0$ if $x \notin Q_\alpha$ and by $\mu_\alpha(x) = \inf\{\|x - y\| : y \in \partial Q_\alpha\}$ otherwise. (Here ∂B is the boundary of a set $B \subset X$.) The function μ_α is clearly Lipschitzian on all of X with Lipschitz constant 1. Let $\omega_\alpha(x) = \mu_\alpha(x)(\sum_{\beta \in A} \mu_\beta(x))^{-1}$, $x \in X$. Since $\{Q_\alpha : \alpha \in A\}$ is locally finite, each ω_α is well defined and locally Lipschitzian on X. Define a locally Lipschitzian, bounded mapping $V : X \to X$ by

$$V(y) := \sum_{\alpha \in A} \omega_\alpha(y) h_{x_\alpha}, \quad y \in X. \tag{8.90}$$

Let $y \in X$. There are a neighborhood Q of y in X and $\alpha_1, \dots, \alpha_n \in A$ such that

$$\{\alpha \in A : Q_\alpha \cap Q \neq \emptyset\} = \{\alpha_1, \dots, \alpha_n\}. \tag{8.91}$$

We have

$$V(y)=\sum_{i=1}^{n}\omega_{\alpha_i}(y)h_{x_{\alpha_i}},\qquad \sum_{i=1}^{n}\omega_{\alpha_i}(y)=1, \tag{8.92}$$

$$f^0(y,Vy)=f^0\left(y,\sum_{i=1}^{n}\omega_{\alpha_i}(y)h_{x_{\alpha_i}}\right)\le\sum_{i=1}^{n}\omega_{\alpha_i}(y)f^0(y,h_{x_{\alpha_i}}). \tag{8.93}$$

Let $i\in\{1,\dots,n\}$ with $\omega_{\alpha_i}(y)>0$. Then

$$y\in\operatorname{supp}\{\omega_{\alpha_i}\}\subset Q_{\alpha_i}\subset U_{x_{\alpha_i}}. \tag{8.94}$$

If $x_{\alpha_i}\in X\setminus E_{\delta(\varepsilon)}$, then by (8.89), $h_{x_{\alpha_i}}=0$ and $f^0(y,h_{x_{\alpha_i}})=0$. If $x_{\alpha_i}\in E_{\delta(\varepsilon)}$, then by (8.88) and (8.94), $f^0(y,h_{x_{\alpha_i}})<0$. Therefore $f^0(y,h_{x_{\alpha_i}})\le 0$ in both cases and $f^0(y,Vy)\le 0$. Thus $V\in\mathcal{A}$. Assume that $y\in E_\varepsilon$, $i\in\{1,\dots,n\}$ and $\omega_{\alpha_i}(y)>0$. Then (8.94) holds. We assert that $x_{\alpha_i}\in E_{\delta(\varepsilon)}$. Assume the contrary. Then $x_{\alpha_i}\in X\setminus E_{\delta(\varepsilon)}$ and by (8.89), $U_{x_{\alpha_i}}=X\setminus\mathrm{cl}(E_\varepsilon)$. When combined with (8.94), this implies that $y\in E_\varepsilon\cap U_{x_{\alpha_i}}=E_\varepsilon\cap(X\setminus\mathrm{cl}(E_\varepsilon))$, a contradiction. Thus $x_{\alpha_i}\in E_{\delta(\varepsilon)}$, as asserted. By the definition of $U_{x_{\alpha_i}}$ (see (8.88)) and (8.94), $f^0(y,h_{x_{\alpha_i}})<-\delta(\varepsilon)$. When combined with (8.93), this implies that $f^0(y,Vy)<-\delta(\varepsilon)$. □

8.8 Proof of Theorem 8.8

For each pair of integers $m,n\ge 1$, denote by Ω_{mn} the set of all $V\in\mathcal{A}$ such that

$$\|Vx\|\le m\quad\text{for all } x\in X \text{ satisfying } \|x\|\le n+1\quad\text{and} \tag{8.95}$$

$$\sup\{f^0(x,Vx):x\in X,\ \|x\|\le n,\ \Xi(x)<-1/n\}=0. \tag{8.96}$$

Clearly,

$$\bigcup_{m=1}^{\infty}\bigcup_{n=1}^{\infty}\Omega_{mn}=\mathcal{A}\setminus\mathcal{F}. \tag{8.97}$$

Therefore in order to prove Theorem 8.8 it is sufficient to show that for each pair of integers $m,n\ge 1$, the set Ω_{mn} (respectively, $\Omega_{mn}\cap\mathcal{A}_c$, $\Omega_{mn}\cap\mathcal{A}_b$, $\Omega_{mn}\cap\mathcal{A}_{bc}$) is a porous subset of $\mathcal{A}$ (respectively, $\mathcal{A}_c$, $\mathcal{A}_b$, $\mathcal{A}_{bc}$) with respect to the pair (ρ_w,ρ_s). Let $m,n\ge 1$ be integers. By Proposition 8.11, there exists a vector field $V_*\in\mathcal{A}$ such that (i) V_* is bounded on X and V_* is locally Lipschitzian on X; (ii) there exists $\delta_*\in(0,1)$ such that

$$f^0(y,V_*y)<-\delta_*\quad\text{for all } y\in X \text{ satisfying } \Xi(y)<-(4n)^{-1}. \tag{8.98}$$

By assumption A(ii), there is $c_0>1$ such that

$$|f(x)-f(y)|\le c_0\|x-y\| \tag{8.99}$$

for all $x, y \in X$ satisfying $\|x\|, \|y\| \le n+2$. Choose $\alpha \in (0,1)$ such that

$$\alpha c_0 2^{n+2} < (2n)^{-1} 2^{-1}(1-\alpha)\delta_*\bigl(m+1+\sup\{\|V_*x\| : x \in X\}\bigr)^{-1}. \tag{8.100}$$

Assume that $V \in \mathcal{A}$ and $r \in (0,1]$. There are two cases: (a) $\sup\{\|Vx\| : x \in X, \|x\| \le n+1\} \le m+1$; (b) $\sup\{\|Vx\| : x \in X, \|x\| \le n+1\} > m+1$. We first assume that (b) holds. Let $W \in \mathcal{A}$ with $\rho_w(W, V) \le 2^{-n-4}$. Then $\rho_{n+1}(W, V)(1+\rho_{n+1}(V, W))^{-1} \le 8^{-1}$, $\rho_{n+1}(W, V) \le 1/7$, and $\sup\{\|Wx\| : x \in X, \|x\| \le n+1\} > m$. Thus $\{W \in \mathcal{A} : \rho_w(W, V) \le 2^{-n-4}\} \cap \Omega_{mn} = \emptyset$. Assume now that (a) holds. Let

$$\gamma = 2^{-1}(1-\alpha)r\bigl(m+1+\sup\{\|V_*x\| : x \in X\}\bigr)^{-1} \tag{8.101}$$

and define $V_\gamma \in \mathcal{A}$ by

$$V_\gamma x = Vx + \gamma V_* x, \quad x \in X. \tag{8.102}$$

If $V \in \mathcal{A}_c$ (respectively, $\mathcal{A}_b$, $\mathcal{A}_{bc}$), then $V_\gamma \in \mathcal{A}_c$ (respectively, $\mathcal{A}_b$, $\mathcal{A}_{bc}$). Next we estimate the distance $\rho_s(V_\gamma, V)$. It follows from (8.102), (8.101) and (8.76) that

$$\rho_s(V_\gamma, V) \le \tilde{\rho}_s(V_\gamma, V) \le \gamma \sup\{\|V_*(x)\| : x \in X\} \le 2^{-1}(1-\alpha)r. \tag{8.103}$$

Assume that $W \in \mathcal{A}$ with

$$\rho_w(W, V_\gamma) \le \alpha r. \tag{8.104}$$

By (8.104) and (8.103),

$$\begin{aligned}\rho_w(W, V) &\le \rho_w(W, V_\gamma) + \rho_w(V_\gamma, V) \le \alpha r + 2^{-1}(1-\alpha)r \\ &\le 2^{-1}(1+\alpha)r < r.\end{aligned} \tag{8.105}$$

Assume now that

$$x \in X, \qquad \|x\| \le n, \qquad \Xi(x) < -1/n \quad \text{and} \quad l \in \partial f(x). \tag{8.106}$$

Inequality (8.99) implies that

$$\|l\|_* \le c_0. \tag{8.107}$$

By (8.102), (8.98) and (8.106),

$$l(V_\gamma x) = l(Vx) + \gamma l\bigl(V_*(x)\bigr) \le \gamma l(V_* x) \le \gamma f^0(x, V_* x) \le \gamma(-\delta_*). \tag{8.108}$$

It follows from (8.106) and (8.80) that

$$\|Wx - V_\gamma x\| \le \rho_n(W, V_\gamma). \tag{8.109}$$

By (8.104) and (8.81), we have $2^{-n}\rho_n(W, V_\gamma)(1+\rho_n(W, V_\gamma))^{-1} \le \rho_w(W, V_\gamma) \le \alpha r$, $\rho_n(W, V_\gamma)(1+\rho_n(W, V_\gamma))^{-1} \le 2^n \alpha r$, and $\rho_n(W, V_\gamma)(1-2^n\alpha r) \le 2^n \alpha r$.

When combined with (8.109), the last inequality implies that $\|Wx - V_\gamma x\| \le 2^n \alpha r(1 - 2^n \alpha r)^{-1}$, and when combined with (8.107), this implies that

$$\left|l(Wx) - l(V_\gamma x)\right| \le c_0 2^n \alpha r\left(1 - 2^n \alpha r\right)^{-1}. \tag{8.110}$$

By (8.110), (8.108), (8.101) and (8.100),

$$\begin{aligned} l(Wx) &\le l(V_\gamma x) + c_0 2^n \alpha r\left(1 - 2^n \alpha r\right)^{-1} \le -\gamma \delta_* + c_0 2^n \alpha r\left(1 - 2^n \alpha r\right)^{-1} \\ &= c_0 2^n \alpha r\left(1 - 2^n \alpha r\right)^{-1} \\ &\quad - \delta_*\left[2^{-1}(1 - \alpha) r\left(m + 1 + \sup\left\{\|V_* x\| : x \in X\right\}\right)\right]^{-1} \\ &= -r\left[-c_0 2^n \alpha\left(1 - 2^n \alpha r\right)^{-1}\right. \\ &\quad \left. + \delta_* 2^{-1}(1 - \alpha)\left(m + 1 + \sup\left\{\|V_* x\| : x \in X\right\}\right)^{-1}\right] \\ &\le -2 r c_0 2^n \alpha. \end{aligned}$$

Since l is an arbitrary element of $\partial f(x)$, we conclude that $f^0(x, Wx) \le -2 r c_0 2^n \alpha$. Thus $\{W \in \mathcal{A} : \rho_w(W, V_\gamma) \le \alpha r\} \cap \Omega_{mn} = \emptyset$. Recall that in case (b), $\{W \in \mathcal{A} : \rho_w(W, V) \le 2^{-n-4}\} \cap \Omega_{mn} = \emptyset$. Therefore Ω_{mn} is porous in $\mathcal{A}$, $\Omega_{mn} \cap \mathcal{A}_c$ is porous in $\mathcal{A}_c$, $\Omega_{mn} \cap \mathcal{A}_b$ is porous in $\mathcal{A}_b$, and $\Omega_{mn} \cap \mathcal{A}_{bc}$ is porous in $\mathcal{A}_{bc}$, as asserted.

8.9 A Basic Lemma for Theorems 8.9 and 8.10

Lemma 8.12 *Assume that $V \in \mathcal{A}$ is regular, and that* A(i), A(ii) *and* A(iii) *are all valid. Let $\bar{K}$ and $\bar{\varepsilon}$ be positive. Then there exist a neighborhood $\mathcal{U}$ of V in $\mathcal{A}$ with the weak topology and positive numbers $\bar{\alpha}$ and γ such that for each $W \in \mathcal{U}$, each $x \in X$ satisfying*

$$\|x\| \le \bar{K} \quad \textit{and} \quad \Xi(x) \le -\bar{\varepsilon}, \tag{8.111}$$

and each $\beta \in (0, \bar{\alpha}]$, we have

$$f(x) - f(x + \beta W x) \ge \beta \gamma. \tag{8.112}$$

Proof There exists $K_0 > \bar{K} + 1$ such that

$$\|Vx\| \le K_0 \quad \text{if } x \in X \text{ and } \|x\| \le \bar{K} + 2. \tag{8.113}$$

By Assumption A(ii), there exists a constant $L_0 > 4$ such that

$$\left|f(x_1) - f(x_2)\right| \le L_0 \|x_1 - x_2\| \tag{8.114}$$

for all $x_1, x_2 \in X$ satisfying $\|x_1\|, \|x_2\| \le 2K_0 + 4$. There is $\delta_0 \in (0, 1)$ such that

$$f^0(y, Vy) \le -\delta_0 \tag{8.115}$$

for each $y \in X$ satisfying $\|y\| \le K_0 + 4$ and $\Xi(y) \le -\bar{\varepsilon}/4$. Choose $\delta_1 \in (0, 1)$ such that

$$4\delta_1(K_0 + L_0) < \delta_0. \tag{8.116}$$

By A(iii), there is a positive $\bar{\alpha}$ such that the following conditions hold:

$$8\bar{\alpha}(L_0 + 1)(K_0 + 1) < \min\{1, \bar{\varepsilon}\}; \tag{8.117}$$

for each $x_1, x_2 \in X$ satisfying

$$\begin{gathered} \|x_1\|, \|x_2\| \le \bar{K} + 4, \qquad \min\{\Xi(x_1), \Xi(x_2)\} \le -\bar{\varepsilon}/4, \\ \|x_1 - x_2\| \le \bar{\alpha}(K_0 + 1), \end{gathered} \tag{8.118}$$

the following inequality is true:

$$H\big(\partial f(x_1), \partial f(x_2)\big) < \delta_1/2. \tag{8.119}$$

Next, choose a positive number δ_2 such that

$$8\delta_2(L_0 + 1) < \delta_1\delta_0. \tag{8.120}$$

Finally, choose a positive number γ and define a neighborhood $\mathcal{U}$ such that

$$\gamma < \delta_0/4, \tag{8.121}$$

$$\mathcal{U} = \big\{W \in \mathcal{A} : \|Wx - Vx\| \le \delta_2, x \in X \text{ and } \|x\| \le \bar{K}\big\}. \tag{8.122}$$

Assume that $W \in \mathcal{U}$, $x \in X$ satisfies (8.111), and that $\beta \in (0, \bar{\alpha}]$. We intend to show that (8.112)) holds. To this end, we first note that (8.111), (8.113), (8.117) and (8.122) yield

$$\begin{gathered} \|x + \beta Vx\| \le \bar{K} + \beta K_0 \le \bar{K} + \bar{\alpha}K_0 \le \bar{K} + 1, \\ \|x + \beta Wx\| \le \delta_2\beta + \|x + \beta Vx\| \le \bar{K} + 1 + \bar{\alpha}\delta_2 \le \bar{K} + 2. \end{gathered} \tag{8.123}$$

By these inequalities, the definition of L_0 (see (8.114)) and (8.122),

$$\big|f(x + \beta Vx) - f(x + \beta Wx)\big| \le L_0\beta\|Wx - Vx\| \le L_0\beta\delta_2. \tag{8.124}$$

Next we estimate $f(x) - f(x + \beta Vx)$. By [89], there exist $\theta \in [0, \beta]$ and $l \in \partial f(x + \theta Vx)$ such that

$$f(x + \beta Vx) - f(x) = l(Vx)\beta. \tag{8.125}$$

By (8.111), (8.114) and (8.117),

$$\begin{gathered} \|x\| \le \bar{K}, \qquad \|Vx\| \le K_0, \qquad \|\theta Vx\| \le \bar{\alpha}K_0, \quad \text{and} \\ \|x + \theta Vx\| \le \bar{K} + 1. \end{gathered} \tag{8.126}$$

Note that (8.126) and the definition of L_0 (see (8.114)) imply that

$$\|l\|_* \le L_0. \tag{8.127}$$

It also follows from (8.111), (8.126) and the definition of $\bar{\alpha}$ (see (8.118) and (8.119)) that $H(\partial f(x), \partial f(x + \theta V x)) < \delta_1$. Therefore there exists $\bar{l} \in \partial f(x)$ such that $\|\bar{l} - l\|_* \le \delta_1$. When combined with (8.125) and (8.126), this fact implies that

$$\begin{aligned} f(x + \beta V x) - f(x) &= \beta l(Vx) \le \beta \bar{l}(Vx) + \beta \|\bar{l} - l\|_* \|Vx\| \\ &\le \beta \bar{l}(Vx) + \beta \delta_1 K_0. \end{aligned} \tag{8.128}$$

It follows from the definition of δ_0 (see (8.115)) and (8.111) that $\beta \bar{l}(Vx) \le -\beta \delta_0$. Combining this inequality with (8.128) and (8.116), we see that $f(x + \beta V x) - f(x) \le -\beta \delta_0 + \beta \delta_1 K_0 \le -\beta \delta_0 / 2$. It now follows from this inequality, (8.120), (8.124) and (8.121) that $f(x + \beta W x) - f(x) \le f(x + \beta V x) - f(x) + f(x + \beta W x) - f(x + \beta V x) \le -\beta \delta_0 / 2 + L_0 \beta \delta_2 \le -\beta \delta_0 / 4 \le -\gamma \beta$. Thus (8.112) holds and Lemma 8.12 is proved. □

8.10 Proofs of Theorems 8.9 and 8.10

Proof of Theorem 8.9 To show that assertion (i) holds, suppose that

$$\begin{aligned} \{x_i\}_{i=0}^{\infty} \subset X, \qquad x_{i+1} \in P_W x_i, \quad i = 0, 1, \ldots, \\ \sup\{\|x_i\| : i = 0, 1, \ldots\} < \infty. \end{aligned} \tag{8.129}$$

We claim that

$$\liminf_{i \to \infty} \Xi(x_i) \ge 0. \tag{8.130}$$

Assume the contrary. Then there exist $\varepsilon > 0$ and a strictly increasing sequence of natural numbers $\{i_k\}_{k=1}^{\infty}$ such that

$$\Xi(x_{i_k}) \le -\varepsilon, \quad k = 1, 2, \ldots. \tag{8.131}$$

Choose a number $S > 0$ such that

$$\|x_i\| \le S, \quad i = 1, 2, \ldots. \tag{8.132}$$

By Lemma 8.12, there exist numbers $\alpha, \gamma \in (0, 1)$ such that for each $x \in X$ satisfying

$$\|x\| \le S \quad \text{and} \quad \Xi(x) \le -\varepsilon, \tag{8.133}$$

and each $\beta \in (0, \alpha]$, we have

$$f(x) - f(x + \beta W x) \ge \gamma \beta. \tag{8.134}$$

It follows from (8.129), (8.82), (8.83), the definitions of α and γ, (8.132) and (8.131) that for each integer $k \geq 1$, $f(x_{i_k}) - f(x_{i_k+1}) \geq f(x_{i_k}) - f(x_{i_k} + \alpha W x_{i_k}) \geq \gamma\alpha$. Since this inequality holds for all integers $k \geq 1$, we conclude that $\lim_{n\to\infty}(f(x_0) - f(x_n)) = \infty$. This contradicts our assumption that f is bounded from below. Therefore (8.130) and assertion (i) are indeed true, as claimed.

We turn now to assertion (ii). Let $\mathbf{a} = \{a_i\}_{i=0}^{\infty} \subset (0, 1)$ satisfy (8.84) and let a bounded $\{x_i\}_{i=0}^{\infty} \subset X$ satisfy (8.85) for all integers $i \geq 0$. We will show that

$$\limsup_{i\to\infty} \Xi(x_i) \geq 0. \tag{8.135}$$

Indeed, assume that (8.135) is not true. Then there exist $\varepsilon > 0$ and an integer $i_1 \geq 0$ such that

$$\Xi(x_i) \leq -\varepsilon, \quad i \geq i_1. \tag{8.136}$$

Since the sequence $\{x_i\}_{1=0}^{\infty}$ is bounded, there exists a number $S > 0$ such that

$$S > \|x_i\|, \quad i = 0, 1, \ldots. \tag{8.137}$$

By Lemma 8.12, there exist numbers $\alpha, \gamma \in (0, 1)$ such that for each $x \in X$ satisfying (8.133) and each $\beta \in (0, \alpha]$, inequality (8.134) holds. Since $a_i \to 0$ as $i \to \infty$, there exists a natural number $i_0 \geq i_1$ such that

$$a_i < \alpha \quad \text{for all integers } i \geq i_0. \tag{8.138}$$

Let $i \geq i_0$ be an integer. Then it follows from (8.137), (8.136), the definitions of α and γ, and (8.138) that $f(x_i) - f(x_i + a_i W x_i) \geq \gamma a_i$, $x_{i+1} = x_i + a_i W x_i$, and $f(x_i) - f(x_{i+1}) \geq \gamma a_i$. Since $\sum_{i=0}^{\infty} a_i = \infty$, we conclude that $\lim_{n\to\infty}(f(x_0) - f(x_n)) = \infty$. The contradiction we have reached shows that (8.135), assertion (ii) and Theorem 8.9 itself are all true. □

Proof of Theorem 8.10 Let

$$K_0 > \sup\{f(x) : x \in X, \|x\| \leq K + 1\}, \tag{8.139}$$

$$E_0 = \{x \in X : f(x) \leq K_0 + 1\}. \tag{8.140}$$

It is clear that E_0 is bounded and closed. Choose

$$K_1 > \sup\{\|x\| : x \in E_0\} + 1 + K. \tag{8.141}$$

By Lemma 8.12, there exist a neighborhood $\mathcal{U}$ of V in $\mathcal{A}$ and numbers $\alpha, \gamma \in (0, 1)$ such that for each $W \in \mathcal{U}$, each $x \in X$ satisfying

$$\|x\| \leq K_1 \quad \text{and} \quad \Xi(x) \leq -\varepsilon, \tag{8.142}$$

and each $\beta \in (0, \alpha]$,

$$f(x) - f(x + \beta W x) \geq \gamma\beta. \tag{8.143}$$

Now choose a natural number N_0 which satisfies

$$N_0 > (\alpha\gamma)^{-1}\big(K_0 + 4 + |\inf(f)|\big). \tag{8.144}$$

Let $W \in \mathcal{U}$, $\{x_i\}_{i=0}^{n} \subset X$, where the integer $n \geq N_0$,

$$\|x_0\| \leq K, \quad \text{and} \quad x_{i+1} \in P_W x_i, \quad i = 0, \ldots, n-1, \tag{8.145}$$

$$B = \big\{i \in \{0, \ldots, n-1\} : \Xi(x_i) \leq -\varepsilon\big\} \quad \text{and} \quad m = \mathrm{Card}(B). \tag{8.146}$$

By (8.145) and (8.139)–(8.141), we have

$$\|x_i\| \leq K_1, \quad i = 0, \ldots, n. \tag{8.147}$$

Let $i \in B$. It follows from (8.147), (8.146) and the definitions of $\mathcal{U}$, α and γ (see (8.142) and (8.143)) that $f(x_i) - f(x_{i+1}) \geq f(x_i) - f(x_i + \alpha W x_i) \geq \gamma\alpha$. Summing up from $i = 0$ to $n-1$, we conclude that

$$f(x_0) - f(x_n) \geq \gamma\alpha\,\mathrm{Card}(B) = m\gamma\alpha.$$

It follows from this inequality, (8.139), (8.145) and (8.144) that

$$m \leq \big[|\inf(f)| + K_0\big](\alpha\gamma)^{-1} < N_0.$$

Thus we see that assertion (i) is proved.

To prove assertion (ii), let a sequence $\mathbf{a} = \{a_i\}_{i=0}^{\infty} \subset (0, 1)$ satisfy

$$\lim_{i\to\infty} a_i = 0 \quad \text{and} \quad \sum_{i=0}^{\infty} a_i = \infty. \tag{8.148}$$

Clearly, there exists a natural number N_1 such that

$$a_i \leq \alpha \quad \text{for all } i \geq N_1. \tag{8.149}$$

Choose a natural number $N > N_1 + 4$ such that

$$\gamma \sum_{i=N_1}^{N-1} a_i > K_0 + 4 + |\inf(f)|. \tag{8.150}$$

Now assume that $W \in \mathcal{U}$, $\{x_i\}_{i=0}^{N} \subset X$, $\|x_0\| \leq K$, and that (8.85) holds for all $i = 0, \ldots, N-1$. We will show that

$$\max\big\{\Xi(x_i) : i = 0, \ldots, N\big\} \geq -\varepsilon. \tag{8.151}$$

Assume the contrary. Then

$$\Xi(x_i) \leq -\varepsilon, \quad i = 0, \ldots, N. \tag{8.152}$$

Since $\|x_0\| \le K$, we see by (8.85) and (8.139)–(8.141) that

$$\|x_i\| \le K_1, \quad i = 0, \dots, N. \tag{8.153}$$

Let $i \in \{N_1, \dots, N-1\}$. It follows from (8.153), (8.152), (8.149) and the definitions of α and γ (see (8.142)) and (8.143)) that

$$f(x_i) - f(x_i + a_i W x_i) \ge \gamma a_i.$$

This implies that

$$f(x_{N_1}) - f(x_N) \ge \gamma \sum_{i=N_1}^{N-1} a_i.$$

By this inequality, (8.85), the inequality $\|x_0\| \le K$, (8.139) and (8.150), we obtain that

$$\inf(f) \le f(x_N) \le f(x_{N_1}) - \gamma \sum_{i=N_1}^{N-1} a_i \le K_0 - \gamma \sum_{i=N_1}^{N-1} a_i < -4 - |\inf(f)|.$$

The contradiction we have reached proves (8.151) and assertion (ii). □

8.11 Continuous Descent Methods

Let $(X^*, \|\cdot\|_*)$ be the dual space of the Banach space $(X, \|\cdot\|)$, and let $f : X \to R^1$ be a convex continuous function which is bounded from below. Recall that for each pair of sets $A, B \subset X^*$,

$$H(A, B) = \max\Big\{\sup_{x \in A} \inf_{y \in B} \|x - y\|_*, \sup_{y \in B} \inf_{x \in A} \|x - y\|_*\Big\}$$

is the Hausdorff distance between A and B.

For each $x \in X$, let

$$\partial f(x) := \big\{l \in X^* : f(y) - f(x) \ge l(y - x) \text{ for all } y \in X\big\}$$

be the subdifferential of f at x. It is well known that the set $\partial f(x)$ is nonempty and norm-bounded. Set

$$\inf(f) := \inf\big\{f(x) : x \in X\big\}.$$

Denote by $\mathcal{A}$ the set of all mappings $V : X \to X$ such that V is bounded on every bounded subset of X (that is, for each $K_0 > 0$, there is $K_1 > 0$ such that $\|Vx\| \le K_1$ if $\|x\| \le K_0$), and for each $x \in X$ and each $l \in \partial f(x)$, $l(Vx) \le 0$. We denote by $\mathcal{A}_c$ the set of all continuous $V \in \mathcal{A}$, by $\mathcal{A}_u$ the set of all $V \in \mathcal{A}$ which are uniformly continuous on each bounded subset of X, and by $\mathcal{A}_{au}$ the set of all $V \in \mathcal{A}$ which are uniformly continuous on the subsets

$$\{x \in X : \|x\| \le n \text{ and } f(x) \ge \inf(f) + 1/n\}$$

for each integer $n \ge 1$. Finally, let $\mathcal{A}_{auc} = \mathcal{A}_{au} \cap \mathcal{A}_c$.

Our results are valid in any Banach space and for those convex functions which satisfy the following two assumptions.

A(i) There exists a bounded set $X_0 \subset X$ such that

$$\inf(f) = \inf\{f(x) : x \in X\} = \inf\{f(x) : x \in X_0\};$$

A(ii) for each $r > 0$, the function f is Lipschitzian on the ball $\{x \in X : \|x\| \le r\}$.

Note that assumption A(i) clearly holds if $\lim_{\|x\| \to \infty} f(x) = \infty$.

We recall that a mapping $V \in \mathcal{A}$ is regular if for any natural number n, there exists a positive number $\delta(n)$ such that for each $x \in X$ satisfying

$$\|x\| \le n \quad \text{and} \quad f(x) \ge \inf(f) + 1/n,$$

and for each $l \in \partial f(x)$, we have

$$l(Vx) \le -\delta(n).$$

Denote by $\mathcal{F}$ the set of all regular vector fields $V \in \mathcal{A}$.

Let $T > 0$, $x_0 \in X$ and let $u : [0, T] \to X$ be a Bochner integrable function. Set

$$x(t) = x_0 + \int_0^t u(s)\,ds, \quad t \in [0, T].$$

Then $x : [0, T] \to X$ is differentiable and $x'(t) = u(t)$ for almost every $t \in [0, T]$. Recall that the function $f : X \to R^1$ is assumed to be convex and continuous, and therefore it is, in fact, locally Lipschitzian. It follows that its restriction to the set $\{x(t) : t \in [0, T]\}$ is Lipschitzian. Indeed, since the set $\{x(t) : t \in [0, T]\}$ is compact, the closure of its convex hull C is both compact and convex, and so the restriction of f to C is Lipschitzian. Hence the function $(f \cdot x)(t) := f(x(t))$, $t \in [0, T]$, is absolutely continuous. It follows that for almost every $t \in [0, T]$, both the derivatives $x'(t)$ and $(f \cdot x)'(t)$ exist:

$$x'(t) = \lim_{h \to 0} h^{-1}[x(t+h) - x(t)],$$

$$(f \cdot x)'(t) = \lim_{h \to 0} h^{-1}[f(x(t+h)) - f(x(t))].$$

We continue with the following fact.

Proposition 8.13 *Assume that $t \in [0, T]$ and that both the derivatives $x'(t)$ and $(f \cdot x)'(t)$ exist. Then*

$$(f \cdot x)'(t) = \lim_{h \to 0} h^{-1}[f(x(t) + hx'(t)) - f(x(t))]. \tag{8.154}$$

Proof There exist a neighborhood $\mathcal{U}$ of $x(t)$ in X and a constant $L > 0$ such that

$$|f(z_1) - f(z_2)| \le L\|z_1 - z_2\| \quad \text{for all } z_1, z_2 \in \mathcal{U}. \tag{8.155}$$

Let $\varepsilon > 0$ be given. There exists $\delta > 0$ such that

$$x(t+h), x(t) + hx'(t) \in \mathcal{U} \quad \text{for each } h \in [-\delta, \delta] \cap [-t, T-t], \tag{8.156}$$

and such that for each $h \in [(-\delta, \delta) \setminus \{0\}] \cap [-t, T-t]$,

$$\|x(t+h) - x(t) - hx'(t)\| < \varepsilon|h|. \tag{8.157}$$

Let

$$h \in \big[(-\delta, \delta) \setminus \{0\}\big] \cap [-t, T-t]. \tag{8.158}$$

It follows from (8.156), (8.155) and (8.157) that

$$\big|f\big(x(t+h)\big) - f\big(x(t) + hx'(t)\big)\big| \le L\big\|x(t+h) - x(t) - hx'(t)\big\| < L\varepsilon|h|. \tag{8.159}$$

Clearly,

$$\begin{aligned}\big[f\big(x(t+h)\big) - f\big(x(t)\big)\big]h^{-1} &= \big[f\big(x(t+h)\big) - f\big(x(t) + hx'(t)\big)\big]h^{-1} \\ &\quad + \big[f\big(x(t) + hx'(t)\big) - f\big(x(t)\big)\big]h^{-1}.\end{aligned} \tag{8.160}$$

Relations (8.159) and (8.160) imply that

$$\begin{aligned}&\big|\big[f\big(x(t+h)\big) - f\big(x(t)\big)\big]h^{-1} - \big[f\big(x(t) + hx'(t)\big) - f\big(x(t)\big)\big]h^{-1}\big| \\ &\quad \le \big|f\big(x(t+h)\big) - f\big(x(t) + hx'(t)\big)\big|\big|h^{-1}\big| \le L\varepsilon.\end{aligned}$$

Since ε is an arbitrary positive number, we conclude that (8.154) holds. □

Assume now that $V \in \mathcal{A}$ and that the differentiable function $x : [0, T] \to X$ satisfies

$$x'(t) = V\big(x(t)\big) \quad \text{for a.e. } t \in [0, T]. \tag{8.161}$$

Then by Proposition 8.13, $(f \cdot x)'(t) \le 0$ for a.e. $t \in [0, T]$, and $f(x(t))$ is decreasing on $[0, T]$.

In the sequel we denote by $\mu(E)$ the Lebesgue measure of $E \subset R^1$.

In the next two sections, we prove the following two results which were obtained in [148].

Theorem 8.14 *Let $V \in \mathcal{A}$ be regular, let $x : [0, \infty) \to X$ be differentiable and suppose that*

$$x'(t) = V\big(x(t)\big) \quad \textit{for a.e. } t \in [0, \infty). \tag{8.162}$$

Assume that there exists a positive number r such that

$$\mu\big(\{t \in [0, T] : \|x(t)\| \le r\}\big) \to \infty \quad \text{as } T \to \infty. \tag{8.163}$$

Then $\lim_{t\to\infty} f(x(t)) = \inf(f)$.

Theorem 8.15 *Let* $V \in \mathcal{A}$ *be regular, let* f *be Lipschitzian on bounded subsets of* X, *and assume that* $\lim_{\|x\|\to\infty} f(x) = \infty$. *Let* K_0 *and* $\varepsilon > 0$ *be positive. Then there exist* $N_0 > 0$ *and* $\delta > 0$ *such that for each* $T \ge N_0$ *and each differentiable mapping* $x : [0, T] \to X$ *satisfying*

$$\|x(0)\| \le K_0 \quad \text{and} \quad \|x'(t) - V(x(t))\| \le \delta \quad \text{for a.e. } t \in [0, T],$$

the following inequality holds for all $t \in [N_0, T]$:

$$f(x(t)) \le \inf(f) + \varepsilon.$$

8.12 Proof of Theorem 8.14

Assume the contrary. Since $f(x(t))$ is decreasing on $[0, \infty)$, this means that there exists $\varepsilon > 0$ such that

$$\lim_{t\to\infty} f(x(t)) > \inf(f) + \varepsilon. \tag{8.164}$$

Then by Proposition 8.13 and (8.162), we have for each $T > 0$,

$$\begin{aligned} f(x(T)) - f(x(0)) &= \int_0^T (f \cdot x)'(t)\,dt \\ &= \int_0^T f^0(x(t), x'(t))\,dt = \int_0^T f^0(x(t), V(x(t)))\,dt \\ &\le \int_{\Omega_T} f^0(x(t), V(x(t)))\,dt, \end{aligned} \tag{8.165}$$

where

$$\Omega_T = \{t \in [0, T] : \|x(t)\| \le r\}. \tag{8.166}$$

Since V is regular, there exists $\delta > 0$ such that for each $x \in X$ satisfying

$$\|x\| \le r + 1 \quad \text{and} \quad f(x) \ge \inf(f) + \varepsilon/2, \tag{8.167}$$

and each $l \in \partial f(x)$, we have

$$l(Vx) \le -\delta. \tag{8.168}$$

It follows from (8.165), (8.166), (8.164), the definition of δ (see (8.167) and (8.168)) and (8.163) that for each $T > 0$,

$$f(x(T)) - f(x(0)) \leq \int_{\Omega_T} f^0(x(t), V(x(t)))\,dt \leq -\delta\mu(\Omega_T) \to -\infty$$

as $T \to \infty$, a contradiction. The contradiction we have reached proves Theorem 8.14.

8.13 Proof of Theorem 8.15

We may assume without loss of generality that $\varepsilon < 1/2$. Choose

$$K_1 > \sup\{f(x) : x \in X \text{ and } \|x\| \leq K_0 + 1\}. \tag{8.169}$$

The set

$$\{x \in X : f(x) \leq K_1 + |\inf(f)| + 4\} \tag{8.170}$$

is bounded. Therefore there exists

$$K_2 > K_0 + K_1$$

such that

$$\text{if } f(x) \leq K_1 + |\inf(f)| + 4, \text{ then } \|x\| \leq K_2. \tag{8.171}$$

There exists a number $K_3 > K_2 + 1$ such that

$$\sup\{f(x) : x \in X \text{ and } \|x\| \leq K_2 + 1\} + 2 < \inf\{f(x) : x \in X \text{ and } \|x\| \geq K_3\}. \tag{8.172}$$

There exists a number $L_0 > 0$ such that

$$|f(x_1) - f(x_2)| \leq L_0\|x_1 - x_2\| \tag{8.173}$$

for each $x_1, x_2 \in X$ satisfying

$$\|x_1\|, \|x_2\| \leq K_3 + 1. \tag{8.174}$$

Fix an integer

$$n > K_3 + 8/\varepsilon. \tag{8.175}$$

There exists a positive number $\delta(n) < 1$ such that:

(P1) for each $x \in X$ satisfying

$$\|x\| \le n \quad \text{and} \quad f(x) \ge \inf(f) + 1/n,$$

and each $l \in \partial f(x)$, we have

$$l(Vx) \le -\delta(n).$$

Choose a natural number $N_0 > 8$ such that

$$8^{-1}\delta(n)N_0 > \left|\inf(f)\right| + \sup\left\{\left|f(z)\right| : z \in X \text{ and } \|z\| \le K_2\right\} + 4 \tag{8.176}$$

and a positive number δ which satisfies

$$8\delta(N_0+1)(L_0+1) < \varepsilon \quad \text{and} \quad (1+L_0)\delta < \delta(n)/2. \tag{8.177}$$

Let $T \ge N_0$ and let $x : [0, T] \to X$ be a differentiable function such that

$$\left\|x(0)\right\| \le K_2 \tag{8.178}$$

and

$$\left\|x'(t) - V\left(x(t)\right)\right\| \le \delta \quad \text{for a.e. } t \in [0, T]. \tag{8.179}$$

We claim that

$$\left\|x(t)\right\| \le K_3, \quad t \in \left[0, \min\{2N_0, T\}\right]. \tag{8.180}$$

Assume the contrary. Then there exists $t_0 \in (0, \min\{2N_0, T\}]$ such that

$$\left\|x(t)\right\| \le K_3, \quad t \in [0, t_0) \quad \text{and} \quad \left\|x(t_0)\right\| = K_3. \tag{8.181}$$

It follows from Proposition 8.13, the convexity of directional derivatives, the inequality $f^0(x(t), Vx(t)) \le 0$, which holds for all $t \in [0, T]$, (8.181), the definition of L_0 (see (8.173), (8.174) and (8.179)) that

$$\begin{aligned}
&f\left(x(t_0)\right) - f\left(x(0)\right) \\
&\quad = \int_0^{t_0} (f \cdot x)'(t)\,dt = \int_0^{t_0} f^0\left(x(t), x'(t)\right) dt \\
&\quad \le \int_0^{t_0} f^0\left(x(t), V\left(x(t)\right)\right) dt + \int_0^{t_0} f^0\left(x(t), x'(t) - V\left(x(t)\right)\right) dt \\
&\quad \le \int_0^{t_0} f^0\left(x(t), x'(t) - V\left(x(t)\right)\right) dt \le \int_0^{t_0} L_0\left\|x'(t) - V\left(x(t)\right)\right\| dt \le t_0 L_0 \delta.
\end{aligned}$$

Thus by (8.177),

$$f\left(x(t_0)\right) \le f\left(x(0)\right) + 2N_0 L_0 \delta < f\left(x(0)\right) + 1.$$

Since $\|x(0)\| \le K_2$ (see (8.178)) and $\|x(t_0)\| = K_3$, the inequality just obtained contradicts (8.172). The contradiction we have reached proves (8.180).

We now claim that there exists a number

$$t_0 \in [1, N_0] \tag{8.182}$$

such that

$$f\big(x(t_0)\big) \le \inf(f) + \varepsilon/8. \tag{8.183}$$

Assume the contrary. Then

$$f\big(x(t)\big) > \inf(f) + \varepsilon/8 \quad \text{and} \quad \big\|x(t)\big\| \le K_3, \quad t \in [1, N_0]. \tag{8.184}$$

It follows from (8.184), Property (P1) and (8.175) that

$$f^0\big(x(t), V\big(x(t)\big)\big) \le -\delta(n), \quad t \in [1, N_0]. \tag{8.185}$$

By (8.185), (8.184), (8.179), (8.177), the convexity of the directional derivatives of f, and the definition of L_0 (see (8.173) and (8.174)), we have, for almost every $t \in [1, N_0]$,

$$\begin{aligned} f^0\big(x(t), x'(t)\big) &\le f^0\big(x(t), V\big(x(t)\big)\big) + f^0\big(x(t), x'(t) - V\big(x(t)\big)\big) \\ &\le -\delta(n) + L_0\big\|x'(t) - V\big(x(t)\big)\big\| \le -\delta(n) + L_0\delta \\ &\le -\delta(n)/2. \end{aligned} \tag{8.186}$$

It follows from the convexity of the directional derivatives of f, the inclusion $V \in \mathcal{A}$, (8.179), (8.180) and the definition of L_0 (see (8.173) and (8.174)), that for almost every $t \in [0, 1]$,

$$\begin{aligned} f^0\big(x(t), x'(t)\big) &\le f^0\big(x(t), V\big(x(t)\big)\big) + f^0\big(x(t), x'(t) - V\big(x(t)\big)\big) \\ &\le f^0\big(x(t), x'(t) - V\big(x(t)\big)\big) \le L_0\big\|x'(t) - V\big(x(t)\big)\big\| \\ &\le L_0\delta. \end{aligned} \tag{8.187}$$

Inequalities (8.178), (8.186) and (8.187) imply that

$$\begin{aligned} &\inf(f) - \sup\big\{f(z) : z \in X, \|z\| \le K_2\big\} \\ &\quad \le f\big(x(N_0)\big) - f\big(x(0)\big) \\ &\quad = \int_0^{N_0} f^0\big(x(t), x'(t)\big)\,dt = \int_0^1 f^0\big(x(t), x'(t)\big)\,dt + \int_1^{N_0} f^0\big(x(t), x'(t)\big)\,dt \\ &\quad \le -2^{-1}\delta(n)N_0/2 + 1. \end{aligned}$$

This contradicts (8.176). The contradiction we have reached yields the existence of a point t_0 which satisfies both (8.182) and (8.183). Clearly, $\|x(t_0)| \le K_2$. Having established (8.180) and the existence of such a point t_0 for an arbitrary mapping x satisfying both (8.178) and (8.179), we now consider the mapping $x_0(t) = x(t + t_0)$, $t \in [0, T - t_0]$. Evidently, (8.178) and (8.179) hold true with x replaced by x_0 and T replaced by $T - t_0$. Hence, if $T - t_0 \geq N_0$, then we have

$$\|x(t)\| = \|x_0(t - t_0)\| \leq K_3, \quad t \in \left[t_0, t_0 + \min\{2N_0, T\}\right],$$

and there exists

$$t_1 \in [t_0 + 1, t_0 + N_0]$$

for which

$$f\big(x(t_1)\big) \leq \inf(f) + \varepsilon/8.$$

Repeating this procedure, we obtain by induction a finite sequence of points $\{t_i\}_{i=0}^q$ such that

$$t_0 \in [1, N_0], \qquad t_{i+1} - t_i \in [1, N_0], \quad i = 0, \dots, q - 1, \qquad T - t_q < N_0,$$
$$f\big(x(t_i)\big) \leq \inf(f) + \varepsilon/8, \quad i = 0, \dots, q,$$
$$\|x(t)\| \leq K_3, \quad t \in [t_0, T].$$

Let $i \in \{0, \dots, q\}$, $t \leq T$, and $0 < t - t_i \leq N_0$. Then by Proposition 8.13, the convexity of the directional derivative of f, the inclusion $V \in \mathcal{A}$, the definition of L_0 (see (8.173) and (8.174)), (7.177) and (8.179), we have

$$\begin{aligned}
f\big(x(t)\big) - f\big(x(t_i)\big) &= \int_{t_i}^{t} f^0\big(x(t), x'(t)\big)\,dt \\
&\leq \int_{t_i}^{t} f^0\big(x(t), V\big(x(t)\big)\big)\,dt + \int_{t_i}^{t} f^0\big(x(t), x'(t) - V\big(x(t)\big)\big)\,dt \\
&\leq \int_{t_i}^{t} f^0\big(x(t), x'(t) - V\big(x(t)\big)\big)\,dt \\
&\leq \int_{t_i}^{t} L_0\big\|x'(t) - V\big(x(t)\big)\big\|\,dt \\
&\leq L_0\delta(t - t_i) \leq 2N_0L_0\delta < \varepsilon/4
\end{aligned}$$

and hence

$$f\big(x(t)\big) \leq f\big(x(t_i)\big) + \varepsilon/4 \leq \inf(f) + \varepsilon/2.$$

This completes the proof of Theorem 8.15.

8.14 Regular Vector-Fields

In the previous sections of this chapter, given a continuous convex function f on a Banach space X, we associate with f a complete metric space $\mathcal{A}$ of mappings $V : X \to X$ such that $f^0(x, Vx) \le 0$ for all $x \in X$. Here $f^0(x, u)$ is the right-hand derivative of f at x in the direction of $u \in X$. We call such mappings descent vector-fields (with respect to f). We identified a regularity property of such vector-fields and showed that regular vector-fields generate convergent discrete descent methods. This has turned out to be true for continuous descent methods as well. Such results are significant because most of the elements in $\mathcal{A}$ are, in fact, regular. Here by "most" we mean an everywhere dense G_δ subset of $\mathcal{A}$. Thus it is important to know when a given descent vector-field $V : X \to X$ is regular. In [163] we established necessary and sufficient conditions for regularity: see Theorems 8.18–8.21 below.

More precisely, let $(X, \|\cdot\|)$ be a Banach space and let $(X^*, \|\cdot\|_*)$ be its dual.

For each $h : X \to R^1$, set $\inf(h) = \{h(z) : z \in X\}$.

Let U be a nonempty, open subset of X and let $f : U \to R^1$ be a locally Lipschitzian function.

For each $x \in U$, let

$$f^0(x,h) = \limsup_{t \to 0^+, y \to x} \big[f(y+th) - f(y)\big]/t, \quad h \in X, \tag{8.188}$$

be the Clarke derivative of f at the point x, and let

$$\partial f(x) = \big\{l \in X^* : f^0(x,h) \ge l(h) \text{ for all } h \in X\big\} \tag{8.189}$$

be the Clarke subdifferential of f at x.

For each $x \in U$, set

$$\Xi_f(x) := \inf\big\{f^0(x,u) : u \in X, \|u\| \le 1\big\}. \tag{8.190}$$

Clearly, $\Xi_f(x) \le 0$ for all $x \in X$ and $\Xi_f(x) = 0$ if and only if $0 \in \partial f(x)$.

For each $x \in U$, set

$$\tilde{\Xi}_f(x) = \inf\big\{f^0(x,h) : h \in X, \|h\| = 1\big\}. \tag{8.191}$$

Let $x \in U$. Clearly, $\tilde{\Xi}_f(x) \ge \Xi_f(x)$ and $0 \in \partial f(x)$ if and only if $\tilde{\Xi}_f(x) \ge 0$.

In the next section we prove the following two propositions.

Proposition 8.16 *Let $x \in U$. If $\tilde{\Xi}_f(x) \ge 0$, then $\Xi_f(x) = 0$. If $\tilde{\Xi}_f(x) < 0$, then $\Xi_f(x) = \tilde{\Xi}_f(x)$.*

Proposition 8.17 *For each $x \in U$,*

$$\Xi_f(x) = -\inf\big\{\|l\|_* : l \in \partial f(x)\big\}. \tag{8.192}$$

Assume now that $f : X \to R^1$ is a continuous and convex function which is bounded from below. It is known that f is locally Lipschitzian. It is also known (see Chap. 2, Sect. 2 of [41]) that in this case

$$f^0(x,h) = \lim_{t\to 0^+} \big[f(x+th) - f(x)\big]/t, \quad x, h \in X.$$

Recall that a mapping $V : X \to X$ is called regular if V is bounded on every bounded subset of X, $f^0(x, Vx) \le 0$ for all $x \in X$, and if for any natural number n, there exists a positive number $\delta(n)$ such that for each $x \in X$ satisfying $\|x\| \le n$ and $f(x) \ge \inf(f) + 1/n$, we have

$$f^0(x, Vx) \le -\delta(n).$$

We now present four results which were established in [163]. Their proofs are given in subsequent sections.

Theorem 8.18 *Let $f : X \to R^1$ be a convex and continuous function which is bounded from below, let $\bar{x} \in X$ satisfy*

$$f(\bar{x}) = \inf\big\{f(z) : z \in X\big\}, \tag{8.193}$$

and let the following property hold:

(P1) *for every sequence $\{y_i\}_{i=1}^{\infty} \subset X$ satisfying $\lim_{i\to\infty} f(y_i) = f(\bar{x})$, $\lim_{i\to\infty} y_i = \bar{x}$ in the norm topology.*

For each natural number n, let $\phi_n : [0,\infty) \to [0,\infty)$ be an increasing function such that $\phi_n(0) = 0$ and the following property holds:

(P2) *for each $\varepsilon > 0$, there exists $\delta := \delta(\varepsilon, n) > 0$ such that for each $t \ge 0$ satisfying $\phi_n(t) \le \delta$, the inequality $t \le \varepsilon$ holds.*

If $V : X \to X$ is bounded on bounded subsets of X,

$$f^0(x, Vx) \le 0 \quad \textit{for all } x \in X, \tag{8.194}$$

and if for each natural number n and each $x \in X$ satisfying $\|x\| \le n$, we have

$$f^0(x, Vx) \le -\phi_n\big(-\Xi_f(x)\big), \tag{8.195}$$

then V is regular.

Theorem 8.19 *Assume that $f : X \to R^1$ is a convex and continuous function, $\bar{x} \in X$,*

$$f(\bar{x}) = \inf(f),$$

property (P1) *holds and the following property also holds:*

(P3) *if* $\{x_i\}_{i=1}^{\infty} \subset X$ *converges to* $\bar{x}$ *in the norm topology, then*

$$\lim_{i\to\infty} \Xi_f(x_i) = 0.$$

Assume that $V : X \to X$ *is regular and let* $n \geq 1$ *be an integer. Then there exists an increasing function* $\phi_n : [0, \infty) \to [0, \infty)$ *such that* $\phi_n(0) = 0$, *property* (P2) *holds, and for each* $x \in X$ *satisfying* $\|x\| \leq n$, *we have*

$$f^0(x, Vx) \leq -\phi_n\big(-\Xi_f(x)\big).$$

Assume now that $f : X \to R^1$ is merely locally Lipschitzian. Recall that in this case a mapping $V : X \to X$ is called regular if V is bounded on every bounded subset of X,

$$f^0(x, Vx) \leq 0 \quad \text{for all } x \in X, \tag{8.196}$$

and for any natural number n, there exists $\delta(n) > 0$ such that for each $x \in X$ satisfying $\|x\| \leq n$ and $\Xi_f(x) \leq -1/n$, we have $f^0(x, Vx) \leq -\delta(n)$.

Theorem 8.20 *Let* $f : X \to R^1$ *be a locally Lipschitzian function. For each natural number* n, *let* $\phi_n : [0, \infty) \to [0, \infty)$ *be an increasing function such that* $\phi_n(0) = 0$ *and property* (P2) *holds.*

Assume that $V : X \to X$ *is bounded on every bounded subset of* X,

$$f^0(x, Vx) \leq 0 \quad \textit{for all } x \in X,$$

and for each natural number n *and each* $x \in X$ *satisfying* $\|x\| \leq n$, *we have*

$$f^0(x, Vx) \leq -\phi_n\big(-\Xi_f(x)\big). \tag{8.197}$$

Then V *is regular.*

Theorem 8.21 *Assume that the function* $f : X \to R^1$ *is locally Lipschitzian and that* $V : X \to X$ *is regular.*

Then for each natural number n, *there exists an increasing function* ϕ_n : $[0, \infty) \to [0, \infty)$ *such that* (P2) *holds and for each natural number* n *and each* $x \in X$ *satisfying* $\|x\| \leq n$, (8.197) *holds.*

8.15 Proofs of Propositions 8.16 and 8.17

Proof of Proposition 8.16 Assume that $\tilde{\Xi}_f(x) \geq 0$. Then $0 \in \partial f(x)$ and $\Xi_f(x) = 0$. Assume that $\tilde{\Xi}_f(x) < 0$. Then by definition (see (8.191)),

$$\inf\big\{f^0(x, h) : h \in X, \|h\| = 1\big\} = \tilde{\Xi}_f(x) < 0. \tag{8.198}$$

By (8.198) and the homogeneity of $f^0(x, \cdot)$,

$$f^0(x,h) \geq \tilde{\Xi}_f(x)\|h\| \quad \text{for all } h \in X. \tag{8.199}$$

By (8.198), (8.191), (8.190) and (8.199),

$$\begin{aligned} 0 > \tilde{\Xi}_f(x) \geq \Xi_f(x) &= \inf\{f^0(x,h) : h \in X, \|h\| \leq 1\} \\ &\geq \inf\{\tilde{\Xi}_f(x)\|h\| : h \in X, \|h\| \leq 1\} = \tilde{\Xi}_f(x). \end{aligned}$$

This implies that

$$\tilde{\Xi}_f(x) = \Xi_f(x),$$

as claimed. Proposition 8.16 is proved. □

We precede the proof of Proposition 8.17 with the following lemma.

Lemma 8.22 *Let $x \in U$ and $c > 0$ be given. Then the following statements are equivalent*:

(i) $\Xi_f(x) \geq -c$;
(ii) $\tilde{\Xi}_f(x) \geq -c$;
(iii) *there is $l \in \partial f(x)$ such that $\|l\|_* \leq c$.*

Proof By Proposition 8.16,

$$\Xi_f(x) \geq -c \quad \text{if and only if} \quad \tilde{\Xi}_f(x) \geq -c.$$

It follows from (8.191) that $\tilde{\Xi}_f(x) \geq -c$ if and only if

$$f^0(x,h) \geq -c \quad \text{for all } h \in X \text{ satisfying } \|h\| = 1,$$

which is, in its turn, equivalent to the following relation:

$$f^0(x,h) \geq -c\|h\| \quad \text{for all } h \in X.$$

Rewriting this last inequality as

$$f^0(x,h) + c\|h\| \geq 0 \quad \text{for all } h \in X,$$

we see that it is equivalent to the inclusion

$$0 \in \partial f(x) + c\{l \in X^* : \|l\|_* \leq 1\}.$$

Thus we have proved that (ii) is equivalent to (iii). This completes the proof of Lemma 8.22. □

Proof of Proposition 8.17 Clearly, equality (8.192) holds if either one of its sides equals zero. Therefore we only need to prove (8.192) in the case where

$$\Xi_f(x) < 0 \quad \text{and} \quad \inf\{\|l\|_* : l \in \partial f(x)\} > 0. \tag{8.200}$$

Assume that (8.200) holds. By Lemma 8.22, there is $\bar{l}$ such that

$$\bar{l} \in \partial f(x) \quad \text{and} \quad \|\bar{l}\|_* \le -\Xi_f(x). \tag{8.201}$$

Hence

$$-\inf\{\|l\|_* : l \in \partial f(x)\} \ge -\|\bar{l}\|_* \ge \Xi_f(x). \tag{8.202}$$

Let ε be any positive number. There is $l_\varepsilon \in \partial f(x)$ such that

$$\|l_\varepsilon\|_* \le \inf\{\|l\|_* : l \in \partial f(x)\} + \varepsilon. \tag{8.203}$$

By (8.203) and Lemma 8.22,

$$\Xi_f(x) \ge -\varepsilon - \inf\{\|l\|_* : l \in \partial f(x)\}.$$

Since ε is any positive number, we conclude that

$$\Xi_f(x) \ge -\inf\{\|l\|_* : l \in \partial f(x)\}.$$

When combined with (8.202), this inequality completes the proof of Proposition 8.17. □

8.16 An Auxiliary Result

Proposition 8.23 *Let $g : X \to R^1$ be a convex and continuous function, $\bar{x} \in X$,*

$$g(\bar{x}) = \inf\{g(z) : z \in X\}, \tag{8.204}$$

and let the following property hold:

(P4) *for any sequence $\{y_i\}_{i=1}^\infty \subset X$ satisfying $\lim_{i\to\infty} g(y_i) = g(\bar{x})$, we have $\lim_{i\to\infty} \|y_i - \bar{x}\| = 0$.*

Assume that $\{x_i\}_{i=1}^\infty \subset X$,

$$\sup\{\|x_i\| : i = 1, 2, \dots\} < \infty \quad \textit{and} \quad \lim_{i\to\infty} \Xi_g(x_i) = 0. \tag{8.205}$$

Then $\lim_{i\to\infty} \|x_i - \bar{x}\| = 0$.

Proof By (8.205) and Proposition 8.17, there exists a sequence $\{l_i\}_{i=1}^{\infty} \subset X^*$ such that

$$\lim_{i\to\infty} \|l_i\|_* = 0 \quad \text{and} \quad l_i \in \partial g(x_i) \quad \text{for all integers } i \ge 1. \tag{8.206}$$

Choose a number $M > 0$ such that

$$\|x_i\| \le M \quad \text{for all integers } i \ge 1 \tag{8.207}$$

and let $i \ge 1$ be an integer. By (8.206),

$$g(z) - l_i(z) \ge g(x_i) - l_i(x_i) \quad \text{for all } z \in X. \tag{8.208}$$

It follows from (8.208), (8.207) and (8.206) that

$$\begin{aligned} g(\bar{x}) - g(x_i) &= g(\bar{x}) - l_i(\bar{x}) - \big(g(x_i) - l_i(x_i)\big) + l_i(\bar{x} - x_i) \\ &\ge l_i(\bar{x} - x_i) \ge -\|l_i\|\|\bar{x} - x_i\| \ge -\|l_i\|\big(M + \|\bar{x}\|\big) \to 0 \quad \text{as } i \to \infty \end{aligned}$$

and therefore

$$\liminf_{i\to\infty}\big(g(\bar{x}) - g(x_i)\big) \ge 0.$$

Together with (P4) this implies that $\lim_{i\to\infty} \|x_i - \bar{x}\| = 0$. Proposition 8.23 is proved. □

8.17 Proof of Theorem 8.18

To show that V is regular, let n be a natural number. We have to find a positive number $\delta = \delta(n)$ such that for each $x \in X$ satisfying $\|x\| \le n$ and $f(x) \ge \inf(f) + 1/n$,

$$f^0(x, Vx) \le -\delta.$$

Assume the contrary. Then for each natural number k, there exists $x_k \in X$ satisfying

$$\|x_k\| \le n, \qquad f(x_k) \ge \inf(f) + 1/n, \tag{8.209}$$

and

$$f^0(x_k, Vx_k) > -1/k. \tag{8.210}$$

It follows from (8.210), (8.209) and (8.195) that for each natural number k,

$$-k^{-1} < f^0(x_k, Vx_k) \le -\phi_n\big(-\Xi_f(x_k)\big)$$

and hence $\phi_n(-\Xi_f(x_k)) < k^{-1}$.

Together with (P2) this inequality implies that $\lim_{k\to\infty} \Xi_f(x_k) = 0$. When combined with Proposition 8.23 and (8.209), this implies $\lim_{k\to\infty} \|x_k - \bar{x}\| = 0$. Since f is continuous,

$$\lim_{k\to\infty} f(x_k) = f(\bar{x}) = \inf(f).$$

This, however, contradicts (8.209). The contradiction we have reached proves that V is indeed regular, as asserted.

8.18 Proof of Theorem 8.19

In what follows we make the convention that the infimum over the empty set is infinity. Set $\phi_n(0) = 0$ and let $t > 0$. Put

$$\phi_n(t) = \min\big\{\inf\big\{-f^0(x, Vx) : x \in X, \|x\| \le n \text{ and } \Xi_f(x) \le -t\big\}, 1\big\}. \quad (8.211)$$

Clearly, $\phi_n : [0, \infty) \to [0, 1]$ is well defined and increasing.

We show that for each $x \in X$ satisfying $\|x\| \le n$,

$$f^0(x, Vx) \le -\phi_n\big(-\Xi_f(x)\big). \quad (8.212)$$

Let $x \in X$ with $\|x\| \le n$. If $\Xi_f(x) = 0$, then it is obvious that (8.212) holds. Assume now that

$$\Xi_f(x) < 0. \quad (8.213)$$

Then by (8.211)), (8.213) and the inequality $\|x\| \le n$,

$$\begin{aligned} \phi_n\big(-\Xi_f(x)\big) &= \min\big\{\inf\big\{-f^0(y, Vy) : y \in X, \|y\| \le n \text{ and } \Xi_f(y) \le \Xi_f(x)\big\}, 1\big\} \\ &\le \min\big\{1, -f^0(x, Vx)\big\} \le -f^0(x, Vx) \end{aligned}$$

and hence

$$f^0(x, Vx) \le -\phi_n\big(-\Xi_f(x)\big).$$

Thus (8.212) holds for each $x \in X$ satisfying $\|x\| \le n$.

Next we show that (P2) holds. To this end, let $\varepsilon > 0$ be given. We claim that there is $\delta > 0$ such that for each $t \ge 0$ satisfying $\phi_n(t) \le \delta$, the inequality $t \le \varepsilon$ holds.

Assume the contrary. Then for each natural number i, there exists $t_i \ge 0$ such that

$$\phi_n(t_i) \le (4i)^{-1}, \quad t_i > \varepsilon. \quad (8.214)$$

By (8.214) and (8.211), for each natural number i, there exists a point $x_i \in X$ such that

$$\|x_i\| \le n, \qquad \Xi_f(x_i) \le -t_i < -\varepsilon, \quad (8.215)$$

and

$$f^0(x_i, Vx_i) \geq -(2i)^{-1}. \tag{8.216}$$

Now it follows from (8.215), (8.216) and the definition of regularity that

$$\lim_{i \to \infty} f(x_i) = f(\bar{x}).$$

Together with (P1) this implies that $\lim_{i \to \infty} \|x_i - \bar{x}\| = 0$. When combined with (P3), this inequality implies that $\lim_{i \to \infty} \Xi_f(x_i) = 0$. This, however, contradicts (8.215). The contradiction we have reached proves Theorem 8.19.

8.19 Proof of Theorem 8.20

Let n be a given natural number. We need to show that there exists $\delta > 0$ such that for each $x \in X$ satisfying

$$\|x\| \leq n \quad \text{and} \quad \Xi_f(x) < -1/n, \tag{8.217}$$

we have

$$f^0(x, Vx) \leq -\delta.$$

Assume the contrary. Then for each natural number k, there exists $x_k \in X$ such that

$$\|x_k\| \leq n, \qquad \Xi_f(x_k) \leq -1/n, \tag{8.218}$$

and

$$f^0(x_k, Vx_k) > -1/k.$$

By (8.218) and (8.197),

$$-1/k < f^0(x_k, Vx_k) \leq -\phi_n\big(-\Xi_f(x_k)\big)$$

and

$$\phi\big(-\Xi_f(x_k)\big) \leq 1/k. \tag{8.219}$$

It now follows from (8.219) and property (P2) that

$$\limsup_{k \to \infty} \big(-\Xi_f(x_k)\big) = 0$$

and

$$\lim_{k \to \infty} \Xi_f(x_k) = 0.$$

The last equality contradicts (8.218) and this contradiction proves Theorem 8.20.

8.20 Proof of Theorem 8.21

Set $\phi_n(0) = 0$ and let $t > 0$. Define

$$\phi_n(t) = \min\{\inf\{-f^0(x, Vx) : x \in X, \|x\| \le n, \Xi_f(x) \le -t\}, 1\}. \tag{8.220}$$

Clearly, $\phi : [0, \infty) \to [0, 1]$ is well defined and increasing.

We show that for each $x \in X$ satisfying $\|x\| \le n$,

$$f^0(x, Vx) \le -\phi_n(-\Xi_f(x)). \tag{8.221}$$

Consider $x \in X$ with

$$\|x\| \le n. \tag{8.222}$$

If $\Xi_f(x) = 0$, then (8.221) clearly holds. Assume that

$$\Xi_f(x) < 0. \tag{8.223}$$

Then by (8.220), (8.221), (8.222) and (8.223),

$$\begin{aligned}\phi_n(-\Xi_f(x)) &= \min\{\inf\{-f^0(y, Vy) : y \in X, \|y\| \le n, \Xi_f(y) \le \Xi_f(x)\}, 1\} \\ &\le \min\{1, -f^0(x, Vx)\} \le -f^0(x, Vx)\end{aligned}$$

and hence (8.221) holds for all $x \in X$ satisfying $\|x\| \le n$, as claimed.

Now we show that property (P2) also holds. To this end, let ε be positive.

We claim that there is $\delta > 0$ such that for each $t \ge 0$ satisfying $\phi_n(t) \le \delta$, the inequality $t \le \varepsilon$ holds.

Assume the contrary. Then for each natural number i, there exists $t_i \ge 0$ such that

$$\phi(t_i) \le (4i)^{-1}, \qquad t_i > \varepsilon. \tag{8.224}$$

Let i be a natural number. By (8.224) and (8.220), there exists $x_i \in X$ such that

$$\|x_i\| \le n, \qquad \Xi_f(x_i) \le -t_i < -\varepsilon, \tag{8.225}$$

and

$$-f^0(x_i, Vx_i) \le (2i)^{-1}.$$

Clearly,

$$f^0(x_i, Vx_i) \ge -(2i)^{-1}. \tag{8.226}$$

Choose a natural number p such that

$$p > n \quad \text{and} \quad 1/p < \varepsilon. \tag{8.227}$$

Since V is regular, there is $\delta > 0$ such that

$$\text{if } x \in X,\ \|x\| \le p \text{ and } \Xi_f(x) < -1/p, \text{ then } f^0(x, Vx) < -\delta. \tag{8.228}$$

Choose a natural number j such that

$$1/j < \delta. \tag{8.229}$$

Then for all integers $i \geq j$, it follows from (8.225) and (8.227) that

$$\Xi_f(x_i) < -\varepsilon < -1/p \quad \text{and} \quad \|x_i\| \leq p.$$

Together with (8.228) and (8.229), this implies that for all integers $i \geq j$,

$$f^0(x_i, Vx_i) < -\delta < -j^{-1} < -(i)^{-1}.$$

Since this contradicts (8.226), the proof of Theorem 8.21 is complete.

8.21 Most Continuous Descent Methods Converge

Let $(X, \|\cdot\|)$ be a Banach space and let $f : X \to R^1$ be a convex continuous function which satisfies the following conditions:

C(i) $\lim_{\|x\|\to\infty} f(x) = \infty$;
C(ii) there is $\bar{x} \in X$ such that $f(\bar{x}) \leq f(x)$ for all $x \in X$;
C(iii) if $\{x_n\}_{n=1}^{\infty} \subset X$ and $\lim_{n\to\infty} f(x_n) = f(\bar{x})$, then

$$\lim_{n\to\infty} \|x_n - \bar{x}\| = 0.$$

By C(iii), the point $\bar{x}$, where the minimum of f is attained, is unique.
For each $x \in X$, let

$$f^0(x,u) = \lim_{t\to 0^+} \big[f(x+tu) - f(x)\big]/t, \quad u \in X. \tag{8.230}$$

Let $(X^*, \|\cdot\|_*)$ be the dual space of $(X, \|\cdot\|)$.
For each $x \in X$, let

$$\partial f(x) = \big\{l \in X^* : f(y) - f(x) \geq l(y-x) \text{ for all } y \in X\big\}$$

be the subdifferential of f at x. It is well known that the set $\partial f(x)$ is nonempty and norm-bounded.
For each $x \in X$ and $r > 0$, set

$$B(x,r) = \big\{z \in X : \|z - x\| \leq r\big\} \quad \text{and} \quad B(r) = B(0,r). \tag{8.231}$$

For each mapping $A : X \to X$ and each $r > 0$, put

$$\mathrm{Lip}(A,r) := \sup\big\{\|Ax - Ay\|/\|x - y\| : x, y \in B(r) \text{ and } x \neq y\big\}. \tag{8.232}$$

Denote by $\mathcal{A}_l$ the set of all mappings $V : X \to X$ such that $\mathrm{Lip}(V, r) < \infty$ for each positive r (this means that the restriction of V to any bounded subset of X is Lipschitzian) and $f^0(x, Vx) \le 0$ for all $x \in X$.

For the set $\mathcal{A}_l$ we consider the uniformity determined by the base

$$E_s(n, \varepsilon) = \big\{(V_1, V_2) \in \mathcal{A}_l \times \mathcal{A}_l : \mathrm{Lip}(V_1 - V_2, n) \le \varepsilon \text{ and } \|V_1 x - V_2 x\| \le \varepsilon \text{ for all } x \in B(n)\big\}. \tag{8.233}$$

Clearly, this uniform space $\mathcal{A}_l$ is metrizable and complete. The topology induced by this uniformity in $\mathcal{A}_l$ will be called the strong topology.

We also equip the space $\mathcal{A}_l$ with the uniformity determined by the base

$$E_w(n, \varepsilon) = \big\{(V_1, V_2) \in \mathcal{A}_l \times \mathcal{A}_l : \|V_1 x - V_2 x\| \le \varepsilon \text{ for all } x \in B(n)\big\} \tag{8.234}$$

where $n, \varepsilon > 0$. The topology induced by this uniformity will be called the weak topology.

The following existence result is proved in the next section.

Proposition 8.24 *Let $x_0 \in X$ and $V \in \mathcal{A}_l$. Then there exists a unique continuously differentiable mapping $x : [0, \infty) \to X$ such that*

$$x'(t) = Vx(t), \quad t \in [0, \infty),$$
$$x(0) = x_0.$$

In the subsequent sections we prove the following result which was obtained in [1].

Theorem 8.25 *There exists a set $\mathcal{F} \subset \mathcal{A}_l$ which is a countable intersection of open (in the weak topology) everywhere dense (in the strong topology) subsets of $\mathcal{A}_l$ such that for each $V \in \mathcal{F}$, the following property holds:*

For each $\varepsilon > 0$ and each $n > 0$, there exist $T_{\varepsilon n} > 0$ and a neighborhood $\mathcal{U}$ of V in $\mathcal{A}_l$ with the weak topology such that for each $W \in \mathcal{U}$ and each differentiable mapping $y : [0, \infty) \to X$ satisfying

$$\big|f\big(y(0)\big)\big| \le n \quad \text{and} \quad y'(t) = Wy(t) \quad \text{for all } t \ge 0,$$

the inequality $\|y(t) - \bar{x}\| \le \varepsilon$ holds for all $t \ge T_{\varepsilon n}$.

8.22 Proof of Proposition 8.24

Since V is locally Lipschitzian, there exists a unique differentiable function $x : I \to X$, where I is an interval of the form $[0, b)$, $b > 0$, such that

$$x(0) = x_0, \qquad x'(t) = Vx(t), \quad t \in I. \tag{8.235}$$

We may and will assume that I is the maximal interval of this form on which the solution exists.

We need to show that $b = \infty$. Suppose, by contradiction, that $b < \infty$.

By Proposition 8.13 and the relation $V \in \mathcal{A}_l$, the function $f(x(t))$ is decreasing on I. By C(i), the set $\{x(t) : t \in [0, b)\}$ is bounded. Thus there is $K_0 > 0$ such that

$$\|x(t)\| \le K_0 \quad \text{for all } t \in [0, b). \tag{8.236}$$

Since V is Lipschitzian on bounded subsets of X, there is $K_1 > 0$ such that

$$\text{if } z \in X,\ \|z\| \le K_0, \text{ then } \|Vz\| \le K_1. \tag{8.237}$$

Let $\varepsilon > 0$ be given. Then it follows from (8.235), (8.236) and (8.237) that for each $t_1, t_2 \in [0, b)$ such that $0 < t_2 - t_1 < \varepsilon/K_1$,

$$\begin{aligned}\|x(t_2) - x(t_1)\| &= \left\| \int_{t_1}^{t_2} x'(t)\, dt \right\| = \left\| \int_{t_1}^{t_2} Vx(t)\, dt \right\| \\ &\le \int_{t_1}^{t_2} \|Vx(t)\|\, dt \le \int_{t_1}^{t_2} K_1\, dt = K_1(t_2 - t_1) < \varepsilon.\end{aligned}$$

Hence there exists $z_0 = \lim_{t \to b^-} x(t)$ in the norm topology. It follows that there exists a unique solution of the initial value problem

$$z'(t) = Vz(t), \qquad z(b) = z_0,$$

defined on a neighborhood of b, and this implies that our solution $x(\cdot)$ can be extended to an open interval larger than I. The contradiction we have reached completes the proof of Proposition 8.24.

8.23 Proof of Theorem 8.25

For each $V \in \mathcal{A}_l$ and each $\gamma \in (0, 1)$, set

$$V_\gamma x = Vx + \gamma(\bar{x} - x), \quad x \in X. \tag{8.238}$$

We first prove several lemmata.

Lemma 8.26 *Let $V \in \mathcal{A}_l$ and $\gamma \in (0, 1)$. Then $V_\gamma \in \mathcal{A}_l$.*

Proof Clearly, V_γ is Lipschitzian on any bounded subset of X. Let $x \in X$. Then by (8.238), the subadditivity and positive homogeneity of the directional derivative of a convex function, the relation $V \in \mathcal{A}_l$, and C(ii),

$$\begin{aligned}f^0(x, V_\gamma x) &= f^0\big(x, Vx + \gamma(\bar{x} - x)\big) \le f^0(x, Vx) + \gamma f^0(x, \bar{x} - x) \\ &\le \gamma f^0(x, \bar{x} - x) \le \gamma\big(f(\bar{x}) - f(x)\big) \le 0.\end{aligned}$$

This completes the proof of Lemma 8.26. □

It is easy to see that the following lemma also holds.

Lemma 8.27 *Let* $V \in \mathcal{A}_l$. *Then* $\lim_{\gamma \to 0^+} V_\gamma = V$ *in the strong topology.*

Lemma 8.28 *Let* $V \in \mathcal{A}_l$, $\gamma \in (0, 1)$, $\varepsilon > 0$, *and let* $x \in X$ *satisfy* $f(x) \geq f(\bar{x}) + \varepsilon$. *Then* $f^0(x, V_\gamma x) \leq -\gamma\varepsilon$.

Proof It follows from (8.238), the properties of the directional derivative of a convex function, and the relation $V \in \mathcal{A}_l$ that

$$\begin{aligned} f^0(x, V_\gamma x) &= f^0\big(x, Vx + \gamma(\bar{x} - x)\big) \leq f^0(x, Vx) + \gamma f^0(x, \bar{x} - x) \\ &\leq \gamma f^0(x, \bar{x} - x) \leq \gamma\big(f(\bar{x}) - f(x)\big) \leq -\varepsilon\gamma. \end{aligned}$$

The lemma is proved. □

Lemma 8.29 *Let* $V \in \mathcal{A}_l$, $\gamma \in (0, 1)$, *and let* $x \in C^1([0, \infty); X)$ *satisfy*

$$x'(t) = V_\gamma x(t), \quad t \in [0, \infty). \tag{8.239}$$

Assume that $T_0, \varepsilon > 0$ *are such that*

$$T_0 > \big(f\big(x(0)\big) - f(\bar{x})\big)(\gamma\varepsilon)^{-1}. \tag{8.240}$$

Then for each $t \geq T_0$, $f(x(t)) \leq f(\bar{x}) + \varepsilon$.

Proof Since the function $f(x(\cdot))$ is decreasing on $[0, \infty)$ (see Proposition 8.13, Lemma 8.26 and (8.239)), it is sufficient to show that

$$f\big(x(T_0)\big) \leq f(\bar{x}) + \varepsilon. \tag{8.241}$$

Assume the contrary. Then $f(x(T_0)) > f(\bar{x}) + \varepsilon$, and since $f(x(\cdot))$ is decreasing on $[0, \infty)$, we have

$$f\big(x(t)\big) > f(\bar{x}) + \varepsilon \quad \text{for all } t \in [0, T_0]. \tag{8.242}$$

When combined with Lemma 8.28, inequality (8.242) implies that

$$f^0(x(t), V_\gamma\big(x(t)\big) \leq -\gamma\varepsilon \quad \text{for all } t \in [0, T_0]. \tag{8.243}$$

It now follows from Proposition 8.13, (8.239) and (8.243) that

$$\begin{aligned} f\big(x(T_0)\big) - f\big(x(0)\big) &= \int_0^{T_0} (f \circ x)'(t)\,dt = \int_0^{T_0} f^0\big(x(t), x'(t)\big)\,dt \\ &= \int_0^{T_0} f^0\big(x(t), V_\gamma x(t)\big)\,dt \leq T_0(-\gamma\varepsilon), \end{aligned}$$

whence

$$T_0\gamma\varepsilon \le f\big(x(0)\big) - f\big(x(T_0)\big) < f\big(x(0)\big) - f(\bar{x}).$$

This contradicts (8.240). The contradiction we have reached proves the lemma. □

Lemma 8.30 *Let $V \in \mathcal{A}_l$, $\gamma \in (0,1)$, $\varepsilon > 0$ and $n > 0$. Then there exist a neighborhood $\mathcal{U}$ of V_γ in $\mathcal{A}_l$ with the weak topology and $\tau > 0$ such that for each $W \in \mathcal{U}$ and each continuously differentiable mapping $x : [0,\infty) \to X$ satisfying*

$$x'(t) = Wx(t), \quad t \in [0,\infty), \tag{8.244}$$

and

$$\big|f\big(x(0)\big)\big| \le n, \tag{8.245}$$

the following inequality holds:

$$\big\|x(t) - \bar{x}\big\| \le \varepsilon \quad \textit{for all } t \ge \tau. \tag{8.246}$$

Proof By C(i), there is $n_1 > n$ such that

$$\text{if } z \in X,\ f(z) \le n, \text{ then } \|z\| \le n_1. \tag{8.247}$$

By C(iii), there is $\delta_1 > 0$ such that

$$\text{if } z \in X \text{ and } f(z) \le f(\bar{x}) + \delta_1, \text{ then } \|z - \bar{x}\| \le \varepsilon. \tag{8.248}$$

Since f is continuous, there is $\varepsilon_1 > 0$ such that

$$\big|f(\bar{x}) - f(z)\big| \le \delta_1 \quad \text{for each } z \in X \text{ satisfying } \|z - \bar{x}\| \le \varepsilon_1. \tag{8.249}$$

In view of C(iii), there exists $\delta_0 \in (0,1)$ such that

$$\text{if } z \in X \text{ and } f(z) \le f(\bar{x}) + \delta_0, \text{ then } \|z - \bar{x}\| \le \varepsilon_1/4. \tag{8.250}$$

Since $V_\gamma \in \mathcal{A}_l$, there is $L > 0$ such that

$$\|V_\gamma z_1 - V_\gamma z_2\| \le L\|z_1 - z_2\| \quad \text{for all } z_1, z_2 \in B(n_1). \tag{8.251}$$

Fix

$$\tau > \big(n - f(\bar{x}) + 1\big)(\gamma\delta_0)^{-1} + 1 \tag{8.252}$$

and choose a positive number Δ such that

$$\Delta\tau e^{L\tau} \le \varepsilon_1/4. \tag{8.253}$$

Set

$$\mathcal{U} = \big\{W \in \mathcal{A}_l : \|Wz - V_\gamma z\| \le \Delta \text{ for all } z \in B(n_1)\big\}. \tag{8.254}$$

Assume that

$$W \in \mathcal{U} \tag{8.255}$$

and that $x \in C^1([0, \infty); X)$ satisfies (8.244) and (8.245). We have to prove (8.246). In view of (8.248), it is sufficient to show that

$$f\big(x(t)\big) \le f(\bar{x}) + \delta_1 \quad \text{for all } t \ge \tau.$$

Since the function $f(x(\cdot))$ is decreasing on $[0, \infty)$, in order to prove the lemma we only need to show that

$$f\big(x(\tau)\big) \le f(\bar{x}) + \delta_1.$$

By (8.249), this inequality will follow from the inequality

$$\big\|x(\tau) - \bar{x}\big\| \le \varepsilon_1. \tag{8.256}$$

We now prove (8.256).

To this end, consider a continuously differentiable mapping $y : [0, \infty) \to X$ which satisfies

$$y'(t) = V_\gamma y(t), \quad t \in [0, \infty), \tag{8.257}$$

and

$$y(0) = x(0). \tag{8.258}$$

Since the functions $f(x(\cdot))$ and $f(y(\cdot))$ are decreasing on $[0, \infty)$, we obtain by (8.258) and (8.245) that for each $s \ge 0$,

$$f\big(x(s)\big), f\big(y(s)\big) \le f\big(x(0)\big) \le n.$$

When combined with (8.247), this inequality implies that

$$\big\|x(s)\big\|, \big\|y(s)\big\| \le n_1 \quad \text{for all } s \ge 0. \tag{8.259}$$

It follows from Lemma 8.29 (with $x = y$, $\varepsilon = \delta_0$), (8.258), (8.257), (8.252) and (8.245) that

$$f\big(y(\tau)\big) \le f(\bar{x}) + \delta_0.$$

This inequality and (8.250) imply that

$$\big\|y(\tau) - \bar{x}\big\| \le \varepsilon_1/4. \tag{8.260}$$

Now we estimate $\|x(\tau) - y(\tau)\|$. It follows from (8.257), (8.244) and (8.258) that for each $s \in [0, \tau]$,

$$\big\|y(s) - x(s)\big\| = \left\|y(0) + \int_0^s V_\gamma y(t)\,dt - \left(x(0) + \int_0^s Wx(t)\,dt\right)\right\|$$

$$= \left\| \int_0^s \left(V_\gamma y(t) - Wx(t)\right) dt \right\| \le \int_0^s \left\| V_\gamma y(t) - Wx(t) \right\| dt$$

$$\le \int_0^s \left\| V_\gamma y(t) - V_\gamma x(t) \right\| dt + \int_0^s \left\| V_\gamma x(t) - Wx(t) \right\| dt. \quad (8.261)$$

By (8.259) and (8.254), for each $s \in (0, \tau]$, we have

$$\int_0^s \left\| V_\gamma x(t) - Wx(t) \right\| dt \le \int_0^s \Delta \, dt \le \Delta s \le \Delta \tau. \quad (8.262)$$

By (8.259) and (8.251), for each $s \in [0, \tau]$,

$$\int_0^s \left\| V_\gamma y(t) - V_\gamma x(t) \right\| dt \le \int_0^s L \left\| y(t) - x(t) \right\| dt. \quad (8.263)$$

It follows from (8.261), (8.262) and (8.263) that for each $s \in [0, \tau]$,

$$\left\| y(s) - x(s) \right\| \le \Delta \tau + \int_0^s L \left\| y(t) - x(t) \right\| dt. \quad (8.264)$$

Applying Gronwall's inequality, we obtain that

$$\left\| y(\tau) - x(\tau) \right\| \le \Delta \tau e^{\int_0^\tau L \, dt} = \Delta \tau e^{L\tau}.$$

When combined with (8.253), this inequality implies that

$$\left\| y(\tau) - x(\tau) \right\| \le \varepsilon_1 / 4.$$

Together with (8.260), this implies that $\|x(\tau) - \bar{x}\| \le \varepsilon_1/2$. Lemma 8.30 is proved. □

Completion of the proof of Theorem 8.25 Let $V \in \mathcal{A}_\gamma$, $\gamma \in (0, 1)$, and let i be a natural number. By Lemma 8.30, there exist an open neighborhood $\mathcal{U}(V, \gamma, i)$ of V_γ in $\mathcal{A}_l$ with the weak topology and a positive number $\tau(V, \gamma, i)$ such that the following property holds:

(P) For each $W \in \mathcal{U}(V, \gamma, i)$ and each continuously differentiable mapping $x : [0, \infty) \to X$ satisfying

$$x'(t) = Wx(t), \quad t \in [0, \infty),$$
$$\left| f\left(x(0)\right) \right| \le i,$$

the following inequality holds:

$$\left\| x(t) - \bar{x} \right\| \le i^{-1} \quad \text{for all } t \ge \tau(V, \gamma, i).$$

Set

$$\mathcal{F} := \bigcap_{i=1}^{\infty} \bigcup \{\mathcal{U}(V, \gamma, i) : V \in \mathcal{A}_l, \gamma \in (0, 1)\}. \tag{8.265}$$

By Lemma 8.27, $\mathcal{F}$ is a countable intersection of open (in the weak topology) everywhere dense (in the strong topology) subsets of $\mathcal{A}_l$.

Let $\tilde{V} \in \mathcal{F}$ and let $n, \varepsilon > 0$ be given. Choose a natural number i such that

$$i > n, \qquad i > \varepsilon^{-1}. \tag{8.266}$$

By (8.265), there are $V \in \mathcal{A}_l$ and $\gamma \in (0, 1)$ such that

$$\tilde{V} \in \mathcal{U}(V, \gamma, i). \tag{8.267}$$

We claim show that the assertion of Theorem 8.15 holds with $\mathcal{U} = \mathcal{U}(V, \gamma, i)$ and $T_{\varepsilon n} = \tau(V, \gamma, i)$.

Assume that $W \in \mathcal{U}(V, \gamma, i)$ and that the continuously differentiable mapping $y : [0, \infty) \to X$ satisfies

$$\left| f\big(y(0)\big) \right| \le n, \qquad y'(t) = Wy(t) \quad \text{for all } t \ge 0. \tag{8.268}$$

Then by (8.268), (8.266) and property (P), it follows that

$$\left\| y(t) - \bar{x} \right\| \le i^{-1} \quad \text{for all } t \ge \tau(V, \gamma, i).$$

When combined with (8.266), this inequality implies that $\|y(t) - \bar{x}\| \le \varepsilon$ for all $t \ge \tau(V, \gamma, i)$. Theorem 8.25 is established. □

Chapter 9
Set-Valued Mappings

9.1 Contractive Mappings

We begin this chapter with a few results on single-valued contractive mappings, which will be used in subsequent sections.

Let (X, ρ) be a complete metric space. Recall that an operator $A : X \to X$ is said to be nonexpansive if

$$\rho(Ax, Ay) \le \rho(x, y) \quad \text{for all } x, y \in X.$$

We denote by $\mathfrak{A}$ the set of all nonexpansive operators $A : X \to X$. We assume that X is bounded and set

$$d(X) = \sup\{\rho(x, y) : x, y \in X\} < \infty.$$

We equip the set $\mathfrak{A}$ with the metric $\rho_{\mathfrak{A}}$ defined by

$$\rho_{\mathfrak{A}}(A, B) := \sup\{\rho(Ax, Bx) : x \in X\}, \quad A, B \in \mathfrak{A}. \tag{9.1}$$

It is clear that the metric space $(\mathfrak{A}, \rho_{\mathfrak{A}})$ is complete.

Denote by $\mathcal{A}$ the set of all sequences $\{A_t\}_{t=1}^{\infty}$, where $A_t \in \mathfrak{A}$, $t = 1, 2, \ldots$. A member of $\mathcal{A}$ will occasionally be denoted by boldface $\mathbf{A}$.

For the set $\mathcal{A}$ we define a metric $\rho_{\mathcal{A}}$ by

$$\rho_{\mathcal{A}}(\{A_t\}_{t=1}^{\infty}, \{B_t\}_{t=1}^{\infty}) = \sup\{\rho(A_t x, B_t x) : t = 1, 2, \ldots \text{ and } x \in X\}. \tag{9.2}$$

Clearly, the metric space $(\mathcal{A}, \rho_{\mathcal{A}})$ is also complete.

A sequence $\{A_t\}_{t=1}^{\infty} \in \mathcal{A}$ is called contractive if there exists a decreasing function $\phi : [0, d(X)] \to [0, 1]$ such that

$$\phi(t) < 1 \quad \text{for all } t \in (0, d(X)] \tag{9.3}$$

and

$$\rho(A_t x, A_t y) \le \phi(\rho(x, y))\rho(x, y) \quad \text{for all } x, y \in X \text{ and all integers } t \ge 1. \tag{9.4}$$

S. Reich, A.J. Zaslavski, *Genericity in Nonlinear Analysis*,
Developments in Mathematics 34, DOI 10.1007/978-1-4614-9533-8_9,

An operator $A \in \mathfrak{A}$ is called contractive if the sequence $\{A_t\}_{t=1}^{\infty}$ with $A_t = A$, $t = 1, 2, \ldots$, is contractive.

It is known that the iterates of any contractive mapping converge to its unique fixed point (see Chap. 3). The following theorem, which was obtained in [144], extends this result to infinite products.

Theorem 9.1 *Assume that the sequence $\{A_t\}_{t=1}^{\infty}$ is contractive and that $\varepsilon > 0$. Then there exists a natural number N such that for each integer $T \geq N$, each mapping $h : \{1, \ldots, T\} \to \{1, 2, \ldots\}$ and each $x, y \in X$,*

$$\rho(A_{h(T)} \cdot \cdots \cdot A_{h(1)}x, A_{h(T)} \cdot \cdots \cdot A_{h(1)}y) \leq \varepsilon. \tag{9.5}$$

Proof There exists a decreasing function $\phi : [0, d(X)] \to [0, 1]$ such that inequalities (9.3) and (9.4) hold. Choose a natural number $N > 4$ such that

$$d(X)\phi(\varepsilon)^N < \varepsilon. \tag{9.6}$$

Assume that $T \geq N$ is an integer, $h : \{1, \ldots, T\} \to \{1, 2, \ldots\}$ and that $x, y \in X$ are given. We intend to show that (9.5) holds. Assume it does not. Then

$$\rho(x, y) > \varepsilon \quad \text{and} \quad \rho(A_{h(n)} \cdot \cdots \cdot A_{h(1)}x, A_{h(n)} \cdot \cdots \cdot A_{h(1)}y) > \varepsilon,$$
$$n = 1, \ldots, N. \tag{9.7}$$

It follows from (9.7) and (9.4) that

$$\rho(A_{h(1)}x, A_{h(1)}y) \leq \phi(\rho(x, y))\rho(x, y) \leq \phi(\varepsilon)\rho(x, y)$$

and that for all integers $i = 1, \ldots, N - 1$,

$$\begin{aligned} &\rho(A_{h(i+1)}A_{h(i)} \cdot \cdots \cdot A_{h(1)}x, A_{h(i+1)}A_{h(i)} \cdot \cdots \cdot A_{h(1)}y) \\ &\quad \leq \phi(\varepsilon)\rho(A_{h(i)} \cdot \cdots \cdot A_{h(1)}x, A_{h(i)} \cdot \cdots \cdot A_{h(1)}y). \end{aligned}$$

When combined with (9.6), this inequality implies that

$$\rho(A_{h(N)} \cdot \cdots \cdot A_{h(1)}x, A_{h(N)} \cdot \cdots \cdot A_{h(1)}y) \leq \phi(\varepsilon)^N \rho(x, y) \leq d(X)\phi(\varepsilon)^N < \varepsilon,$$

a contradiction. This completes the proof of Theorem 9.1. □

Corollary 9.2 *Assume that the sequence $\{A_t\}_{t=1}^{\infty}$ is contractive. Then*

$$\rho(A_{h(T)} \cdot \cdots \cdot A_{h(1)}x, A_{h(T)} \cdot \cdots \cdot A_{h(1)}y) \to 0 \quad \text{as } T \to \infty,$$

uniformly in $h : \{1, 2, \ldots\} \to \{1, 2, \ldots\}$ and in $x, y \in X$.

We remark in passing that such results are called weak ergodic theorems in the population biology literature [43].

9.2 Star-Shaped Spaces

We say that a complete metric space (X, ρ) is star-shaped if it contains a point $x_* \in X$ with the following property:

For each $x \in X$, there exists a mapping

$$t \to tx \oplus (1-t)x_* \in X, \quad t \in (0,1), \tag{9.8}$$

such that for each $t \in (0,1)$ and each $x, y \in X$,

$$\rho\big(tx \oplus (1-t)x_*, ty \oplus (1-t)x_*\big) \le t\rho(x,y) \tag{9.9}$$

and

$$\rho\big(tx \oplus (1-t)x_*, x\big) \le (1-t)\rho(x, x_*). \tag{9.10}$$

For each $A \in \mathfrak{A}$ and each $\gamma \in (0,1)$, define $A_\gamma \in \mathfrak{A}$ by

$$A_\gamma x = (1-\gamma)Ax \oplus \gamma x_*, \quad x \in X. \tag{9.11}$$

For each $\mathbf{A} = \{A_t\}_{t=1}^\infty \in \mathcal{A}$, let $\mathbf{A}_\gamma = \{A_{\gamma t}\}_{t=1}^\infty$, where

$$A_{\gamma t}x = (1-\gamma)A_t x \oplus \gamma x_*, \quad x \in X, t = 1, 2, \ldots. \tag{9.12}$$

Theorem 9.3 *Assume that $\mathcal{B}$ is a closed subset of $\mathcal{A}$ such that for each $\mathbf{A} \in \mathcal{B}$ and each $\gamma \in (0,1)$, the sequence $\mathbf{A}_\gamma \in \mathcal{B}$. Then there exists a set $\mathcal{F}$ which is a countable intersection of open and everywhere dense subsets of $\mathcal{B}$ (with the relative topology) such that each $\mathbf{A} \in \mathcal{F}$ is contractive.*

Proof It follows from (9.10) that for each $\mathbf{A} = \{A_t\}_{t=1}^\infty \in \mathcal{B}$, each $\gamma \in (0,1)$ and each $x \in X$,

$$\rho(A_{\gamma t}x, A_t x) \le \gamma\rho(A_t x, x_*).$$

This implies that $\mathbf{A}_\gamma \to \mathbf{A}$ in $\mathcal{B}$ as $\gamma \to 0^+$ and that the set $\{\mathbf{A}_\gamma : \mathbf{A} \in \mathcal{B}, \gamma \in (0,1)\}$ is everywhere dense in $\mathcal{B}$.

Let $\mathbf{A} = \{A_t\}_{t=1}^\infty \in \mathcal{B}$ and $\gamma \in (0,1)$ be given. Inequality (9.9) implies that

$$\rho(A_{\gamma t}x, A_{\gamma t}y) \le (1-\gamma)\rho(x,y) \tag{9.13}$$

for all $x, y \in X$ and all integers $t \ge 1$. For each integer $i \ge 1$, choose a positive number

$$\delta(\mathbf{A}, \gamma, i) < (4i)^{-1}d(X)\gamma \tag{9.14}$$

and define

$$U(\mathbf{A}, \gamma, i) = \big\{\mathbf{B} \in \mathcal{B} : \rho_{\mathcal{A}}(\mathbf{A}_\gamma, \mathbf{B}) < \delta(\mathbf{A}, \gamma, i)\big\}. \tag{9.15}$$

Let $i \ge 1$ be an integer. We claim that the following property holds:

P(1) For each $\mathbf{B} \in U(\mathbf{A}, \gamma, i)$, each $x, y \in X$ satisfying $\rho(x, y) \geq i^{-1}d(X)$ and each integer $t \geq 1$, the inequality $\rho(B_t x, B_t y) \leq (1 - \gamma/2)\rho(x, y)$ is valid.

Indeed, assume that $\mathbf{B} \in U(\mathbf{A}, \gamma, i)$, the points $x, y \in X$ satisfy

$$\rho(x, y) \geq i^{-1}d(X), \tag{9.16}$$

and that $t \geq 1$ is an integer. It follows from the definition of $U(\mathbf{A}, \gamma, i)$ (see (9.15) and (9.14)), (9.13) and (9.16) that

$$\begin{aligned}\rho(B_t x, B_t y) &\leq \rho(A_{\gamma t}x, A_{\gamma t}y) + 2\delta(\mathbf{A}, \gamma, i) \\ &< 2\delta(\mathbf{A}, \gamma, i) + (1 - \gamma)\rho(x, y) \leq (1 - \gamma)\rho(x, y) + (2i)^{-1}\gamma d(X) \\ &\leq (1 - \gamma)\rho(x, y) + 2^{-1}\gamma\rho(x, y) \leq (1 - \gamma/2)\rho(x, y).\end{aligned}$$

Thus

$$\rho(B_t x, B_t y) \leq (1 - \gamma/2)\rho(x, y). \tag{9.17}$$

Now define

$$\mathcal{F} := \bigcap_{i=1}^{\infty} \bigcup \{U(\mathbf{A}, \gamma, i) : \mathbf{A} \in \mathcal{B}, \gamma \in (0, 1)\}. \tag{9.18}$$

It is clear that $\mathcal{F}$ is a countable intersection of open and everywhere dense subsets of $\mathcal{B}$ (equipped with the relative topology). We claim that any $\mathbf{B} \in \mathcal{F}$ is contractive. To show this, assume that i is a natural number. There exist $\mathbf{A} \in \mathcal{B}$ and $\gamma \in (0, 1)$ such that $\mathbf{B} \in U(\mathbf{A}, \gamma, i)$. By property P(1), for each $x, y \in X$ satisfying $\rho(x, y) \geq i^{-1}d(X)$ and each integer $t \geq 1$, inequality (9.17) holds. Since i is an arbitrary natural number we conclude that $\mathbf{B}$ is contractive. Theorem 9.3 is proved. □

Theorem 9.4 *Assume that $\mathfrak{B}$ is a closed subset of $\mathfrak{A}$ such that for each $A \in \mathfrak{B}$ and each $\gamma \in (0, 1)$, the mapping $A_\gamma \in \mathfrak{B}$. Then there exists a set $\mathcal{F}$ which is a countable intersection of open and everywhere dense subsets of $\mathfrak{B}$ (with the relative topology) such that each $A \in \mathcal{F}$ is contractive.*

Proof For each $A \in \mathfrak{B}$ denote by $Q(A)$ the sequence $\mathbf{A} = \{A_t\}_{t=1}^{\infty}$ with $A_t = A$, $t = 1, 2, \ldots$. Set

$$\mathcal{B} = \{Q(A) : A \in \mathfrak{B}\}.$$

It is easy to see that $\mathcal{B}$ is a closed subset of $\mathcal{A}$ and that for each $\mathbf{A} \in \mathcal{B}$ and each $\gamma \in (0, 1)$, the sequence $\mathbf{A}_\gamma \in \mathcal{B}$. Now Theorem 9.4 follows from Theorem 9.3 and the equality

$$\rho_{\mathfrak{A}}(A, B) = \rho_{\mathcal{A}}(Q(A), Q(B)).$$

□

9.3 Convergence of Iterates of Set-Valued Mappings

Assume that $(E, \|\cdot\|)$ is a Banach space, K is a nonempty, bounded and closed subset of E, and there exists $\theta \in K$ such that for each point $x \in K$,

$$tx + (1-t)\theta \in K, \quad t \in (0,1).$$

We consider the star-shaped complete metric space K with the metric $\|x-y\|$, $x, y \in K$. Denote by $S(K)$ the set of all nonempty closed subsets of K. For $x \in K$ and $A \subset K$, set

$$\rho(x, A) = \inf\{\|x-y\| : y \in A\},$$

and for each $A, B \in S(K)$, let

$$H(A,B) = \max\Big\{\sup_{x\in A} \rho(x,B), \sup_{y\in B} \rho(y,A)\Big\}. \tag{9.19}$$

We equip the set $S(K)$ with the Hausdorff metric $H(\cdot,\cdot)$. It is well known that the metric space $(S(K), H)$ is complete. Clearly, $\{\theta\} \in S(K)$.

For each subset $A \in S(K)$ and each $t \in [0,1]$, define

$$tA \oplus (1-t)\theta := \big\{tx + (1-t)\theta : x \in A\big\} \in S(K). \tag{9.20}$$

It is easy to see that the complete metric space $(S(K), H)$ is star-shaped.

Denote by $\mathfrak{A}$ the set of all nonexpansive operators $T : S(K) \to S(K)$. For the set $\mathfrak{A}$ we consider the metric $\rho_{\mathfrak{A}}$ defined by

$$\rho_{\mathfrak{A}}(T_1, T_2) := \sup\big\{H\big(T_1(A), T_2(A)\big) : A \in S(K)\big\}, \quad T_1, T_2 \in \mathfrak{A}. \tag{9.21}$$

Denote by $\mathcal{M}$ the set of all mappings $T : K \to S(K)$ such that

$$H\big(T(x), T(y)\big) \le \|x-y\|, \quad x, y \in K. \tag{9.22}$$

A mapping $T \in \mathcal{M}$ is called contractive if there exists a decreasing function ϕ : $[0, d(K)] \to [0,1]$ such that

$$\phi(t) < 1 \quad \text{for all } t \in \big(0, d(K)\big] \tag{9.23}$$

and

$$H\big(T(x), T(y)\big) \le \phi\big(\|x-y\|\big)\|x-y\| \quad \text{for all } x, y \in K. \tag{9.24}$$

Assume that $T \in \mathcal{M}$. For each $A \in S(K)$, denote by $\tilde{T}(A)$ the closure of the set $\bigcup\{T(x) : x \in A\}$ in the norm topology.

Proposition 9.5 *Assume that $T \in \mathcal{M}$. Then the mapping $\tilde{T}$ belongs to $\mathfrak{A}$.*

Proof Let $A, B \in S(K)$. We claim that

$$H(\tilde{T}(A), \tilde{T}(B)) \le H(A, B). \tag{9.25}$$

Given $\varepsilon > 0$, there exist $x_1 \in \tilde{T}(A)$ and $x_2 \in \tilde{T}(B)$ such that

$$\max\{\rho(x_1, \tilde{T}(B)), \rho(x_2, \tilde{T}(A))\} + \varepsilon/2 > H(\tilde{T}(A), \tilde{T}(B)). \tag{9.26}$$

We may assume that

$$\rho(x_1, \tilde{T}(B)) \ge \rho(x_2, \tilde{T}(A)).$$

Therefore

$$\rho(x_1, \tilde{T}(B)) + \varepsilon/2 > H(\tilde{T}(A), \tilde{T}(B)). \tag{9.27}$$

We may assume that $x_1 \in T(A)$. There exist points $x_0 \in A$ such that $x_1 \in T(x_0)$ and $y_0 \in B$ such that

$$\|x_0 - y_0\| < \rho(x_0, B) + \varepsilon/2 \le H(A, B) + \varepsilon/2.$$

Therefore inequality (9.22) implies that

$$\rho(x_1, \tilde{T}(B)) \le \rho(x_1, T(y_0)) \le H(T(x_0), T(y_0)) \le \|x_0 - y_0\| < H(A, B) + \varepsilon/2.$$

Now (9.27) yields

$$H(\tilde{T}(A), \tilde{T}(B)) < H(A, B) + \varepsilon.$$

Since ε is an arbitrary positive number, we conclude that (9.25) holds. Proposition 9.5 is proved. □

Proposition 9.6 *Assume that $T \in \mathcal{M}$. Then the mapping $\tilde{T}$ is contractive if and only if the mapping T is contractive.*

Proof It is clear that T is contractive if $\tilde{T}$ is contractive. Assume now that the mapping T is contractive. Then there exists a decreasing function $\phi : [0, d(K)] \to [0, 1]$ such that (9.23) and (9.24) hold.

Let $A, B \in S(K)$. We assert that

$$H(\tilde{T}(A), \tilde{T}(B)) \le \max\{1/2, \phi(H(A, B)/4)\}H(A, B). \tag{9.28}$$

To see this, we may assume that $H(A, B) > 0$ and that

$$H(\tilde{T}(A), \tilde{T}(B)) > H(A, B)/2. \tag{9.29}$$

Let

$$\varepsilon \in (0, H(A, B)/4). \tag{9.30}$$

By the definition of the Hausdorff metric, there exist $x_1 \in \tilde{T}(A)$ and $x_2 \in \tilde{T}(B)$ such that

$$\max\{\rho(x_1, \tilde{T}(B)), \rho(x_2, \tilde{T}(A))\} + \varepsilon/2 > H(\tilde{T}(A), \tilde{T}(B)). \tag{9.31}$$

We may assume that

$$\rho(x_1, \tilde{T}(B)) \geq \rho(x_2, \tilde{T}(A)).$$

Therefore

$$\rho(x_1, \tilde{T}(B)) + \varepsilon/2 > H(\tilde{T}(A), \tilde{T}(B)). \tag{9.32}$$

We may also assume that $x_1 \in T(A)$. There exist $x_0 \in A$ such that $x_1 \in T(x_0)$ and $y_0 \in B$ such that

$$\|x_0 - y_0\| < \rho(x_0, B) + \varepsilon/2 \leq H(A, B) + \varepsilon/2. \tag{9.33}$$

Therefore (9.24) implies that

$$\begin{aligned} \rho(x_1, \tilde{T}(B)) &\leq \rho(x_1, T(y_0)) \leq H(T(x_0), T(y_0)) \leq \phi(\|x_0 - y_0\|)\|x_0 - y_0\| \\ &\leq \phi(\|x_0 - y_0\|)(H(A, B) + \varepsilon/2). \end{aligned} \tag{9.34}$$

Combining this with (9.32), we see that

$$-\varepsilon/2 + H(\tilde{T}(A), \tilde{T}(B)) < \phi(\|x_0 - y_0\|)(H(A, B) + \varepsilon/2). \tag{9.35}$$

It follows from (9.22), (9.32), (9.29) and (9.30) that

$$\begin{aligned} \|x_0 - y_0\| &\geq H(T(x_0), T(y_0)) \geq \rho(x_1, T(y_0)) \geq \rho(x_1, \tilde{T}(B)) \\ &\geq -\varepsilon/2 + H(\tilde{T}(A), \tilde{T}(B)) > -\varepsilon/2 + H(A, B)/2 \geq H(A, B)/4. \end{aligned}$$

Thus

$$\|x_0 - y_0\| \geq H(A, B)/4.$$

Combining this last inequality with (9.35), we can deduce that

$$-\varepsilon/2 + H(\tilde{T}(A), \tilde{T}(B)) < \phi(H(A, B)/4)(H(A, B) + \varepsilon/2).$$

Since ε is an arbitrary positive number, we conclude that

$$H(\tilde{T}(A), \tilde{T}(B)) \leq \phi(H(A, B)/4)(H(A, B)).$$

This completes the proof of Proposition 9.6. □

We equip the set $\mathcal{M}$ with the metric $\rho_{\mathcal{M}}$ defined by

$$\rho_{\mathcal{M}}(T_1, T_2) := \sup\{H(T_1(x), T_2(x)) : x \in K\}, \quad T_1, T_2 \in \mathcal{M}. \tag{9.36}$$

It is not difficult to verify that the metric space $(\mathcal{M}, \rho_{\mathcal{M}})$ is complete.

For each $T \in \mathcal{M}$, set $P(T) = \tilde{T}$. It is easy to see that for each $T_1, T_2 \in \mathcal{M}$,

$$\rho_{\mathfrak{A}}\big(P(T_1), P(T_2)\big) = \rho_{\mathcal{M}}(T_1, T_2). \tag{9.37}$$

Denote

$$\mathfrak{B} = \big\{P(T) : T \in \mathcal{M}\big\}. \tag{9.38}$$

It is clear that the metric spaces $(\mathfrak{B}, \rho_{\mathfrak{A}})$ and $(\mathcal{M}, \rho_{\mathcal{M}})$ are isometric.

For each $T \in \mathfrak{A}$ and each $\gamma > 0$, define

$$T_\gamma(A) = (1 - \gamma)T(A) \oplus \gamma\theta.$$

It is easy to see that $T_\gamma \in \mathfrak{A}$ for each $T \in \mathfrak{A}$ and each $\gamma > 0$, and moreover, $T_\gamma \in \mathfrak{B}$ if $T \in \mathfrak{B}$. Now we can apply Theorem 9.4 and obtain the following result.

Theorem 9.7 *There exists a set $\mathcal{F}$ which is a countable intersection of open and everywhere dense subsets of $(\mathcal{M}, \rho_{\mathcal{M}})$ such that each $T \in \mathcal{F}$ is contractive.*

Theorem 3.1 and Proposition 9.6 imply the following result.

Theorem 9.8 *Assume that the operator $T \in \mathcal{M}$ is contractive. Then there exists a unique set $A_T \in S(K)$ such that $\tilde{T}(A_T) = A_T$ and $(\tilde{T})^n(B) \to A_T$ as $n \to \infty$, uniformly for all $B \in S(K)$.*

Let $T \in \mathcal{M}$. A sequence $\{x_n\}_{n=1}^N \subset K$ with $N \geq 1$ (respectively, $\{x_n\}_{n=1}^\infty \subset K$) is called a trajectory of T if $x_{i+1} \in T(x_i)$, $i = 1, \ldots, N-1$ (respectively, $i = 1, 2, \ldots$).

Theorem 9.8 leads to the following results.

Theorem 9.9 *Let the operator $T \in \mathcal{M}$ be contractive and let the set $A_T \in S(K)$ be as guaranteed by Theorem 9.8. Then for each $\varepsilon > 0$, there exists a natural number n such that for each trajectory $\{x_i\}_{i=1}^n \subset K$ of T, $\rho(x_n, A_T) < \varepsilon$.*

Theorem 9.10 *Let the operator $T \in \mathcal{M}$ be contractive and let the set $A_T \in S(K)$ be as guaranteed by Theorem 9.8. Then for each $\varepsilon > 0$, there exists a natural number n such that for each $z \in K$ and each $x \in A_T$, there exists a trajectory $\{x_i\}_{i=1}^n \subset K$ of T such that $x_1 = z$ and $\rho(x_n, x) < \varepsilon$.*

Corollary 9.11 *Let the operator $T \in \mathcal{M}$ be contractive and let the set $A_T \in S(K)$ be as guaranteed by Theorem 9.8. Then for each $x \in A_T$, there is a trajectory $\{x_i\}_{i=1}^\infty \subset A_T$ such that $x_1 = x$ and $\liminf_{i\to\infty} \|x_i - x\| = 0$.*

Corollary 9.12 *Let the operator $T \in \mathcal{M}$ be contractive and let the set $A_T \in S(K)$ be as guaranteed by Theorem 9.8. Assume that the set A_T is separable. Then for each $x \in A_T$, there is a trajectory $\{x_i\}_{i=1}^\infty \subset A_T$ such that $x_1 = x$ and for each $y \in A_T$, $\liminf_{i\to\infty} \|x_i - y\| = 0$.*

9.4 Existence of Fixed Points

We consider a complete metric space of nonexpansive set-valued mappings acting on a closed and convex subset of a Banach space with a nonempty interior, and show that a generic mapping in this space has a fixed point. We then prove analogous results for two complete metric spaces of set-valued mappings with convex graphs. These results were obtained in [145].

Let $(X, \|\cdot\|)$ be a Banach space and denote by $S(X)$ the set of all nonempty, closed and convex subsets of X. For $x \in X$ and $A \subset X$, set

$$\rho(x, A) = \inf\{\|x - y\| : y \in A\},$$

and for each $A, B \in S(X)$, let

$$H(A, B) = \max\Big\{\sup_{x \in A} \rho(x, B), \sup_{y \in B} \rho(y, A)\Big\}. \tag{9.39}$$

The interior of a subset $A \subset X$ will be denoted by $\mathrm{int}(A)$. For each $x \in X$ and each $r > 0$, set $B(x, r) = \{y \in X : \|y - x\| \le r\}$. For the set $S(X)$ we consider the uniformity determined by the following base:

$$\mathcal{G}(n) = \{(A, B) \in S(X) \times S(X) : H(A, B) \le n^{-1}\}, \tag{9.40}$$

$n = 1, 2, \ldots$. It is well known that the space $S(X)$ with this uniformity is metrizable and complete. We endow the set $S(X)$ with the topology induced by this uniformity.

Assume now that K is a nonempty, closed and convex subset of X and denote by $S(K)$ the set of all $A \in S(X)$ such that $A \subset K$. It is clear that $S(K)$ is a closed subset of $S(X)$. We equip the topological subspace $S(K) \subset S(X)$ with its relative topology.

Denote by $\mathcal{M}_{ne}$ the set of all mappings $T : K \to S(K)$ such that $T(x)$ is bounded for all $x \in K$ and

$$H\big(T(x), T(y)\big) \le \|x - y\|, \quad x, y \in K. \tag{9.41}$$

In other words, the set $\mathcal{M}_{ne}$ consists of those nonexpansive set-valued self-mappings of K which have nonempty, bounded, closed and convex point images.

Fix $\theta \in K$. For the set $\mathcal{M}_{ne}$ we consider the uniformity determined by the following base:

$$\mathcal{E}(n) = \big\{(T_1, T_2) \in \mathcal{M}_{ne} \times \mathcal{M}_{ne} : H\big(T_1(x), T_2(x)\big) \le n^{-1} \text{ for all } x \in K \text{ satisfying } \|x - \theta\| \le n\big\}, \quad n = 1, 2, \ldots. \tag{9.42}$$

It is not difficult to verify that the space $\mathcal{M}_{ne}$ with this uniformity is metrizable and complete.

The following result is well known [45, 102]; see also [116].

Theorem 9.13 *Assume that* $T : K \to S(K)$, $\gamma \in (0, 1)$, *and*

$$H\big(T(x), T(y)\big) \le \gamma \|x - y\|, \quad x, y \in K.$$

Then there exists $x_T \in K$ *such that* $x_T \in T(x_T)$.

The existence of fixed points for set-valued mappings which are merely nonexpansive is more delicate and was studied by several authors. See, for example, [67, 94, 119] and the references therein.

We prove the following result which shows that if $\operatorname{int}(K)$ is nonempty, then a generic nonexpansive mapping does have a fixed point.

Theorem 9.14 *Assume that* $\operatorname{int}(K) \neq \emptyset$. *Then there exists an open and everywhere dense set* $\mathcal{F} \subset \mathcal{M}_{ne}$ *with the following property: for each* $\widehat{S} \in \mathcal{F}$, *there exist* $\bar{x} \in K$ *and a neighborhood* $\mathcal{U}$ *of* $\widehat{S}$ *in* $\mathcal{M}_{ne}$ *such that* $\bar{x} \in S(\bar{x})$ *for each* $S \in \mathcal{U}$.

For our second result we assume, in addition, that the closed and convex subset $K \subset X$ is bounded. Denote by $\mathcal{M}_a$ the set of all mappings $T : K \to S(K)$ such that

$$\alpha T(x_1) + (1 - \alpha) T x_2 \subset T\big(\alpha x_1 + (1 - \alpha) x_2\big) \tag{9.43}$$

for each $x_1, x_2 \in K$ and all $\alpha \in (0, 1)$. In other words, the set $\mathcal{M}_a$ consists of all set-valued self-mappings of K with convex graphs. Note that convex-valued mappings and, in particular, mappings with convex graphs, as well as spaces of convex sets, find application in several areas of mathematics. See, for example, [54, 90, 92, 166, 168, 169, 177] and the references mentioned there. We denote by $\mathcal{M}_{ac}$ the set of all those continuous mappings $T : K \to S(K)$ which belong to $\mathcal{M}_a$.

For the set $\mathcal{M}_a$ we consider the uniformity determined by the following base:

$$\begin{aligned} \mathcal{E}_a(n) = \big\{ &(T_1, T_2) \in \mathcal{M}_a \times \mathcal{M}_a : H\big(T_1(x), T_2(x)\big) \le n^{-1} \\ &\text{for all } x \in K \big\}, \quad n = 1, 2, \ldots. \end{aligned} \tag{9.44}$$

It is easy to see that the space $\mathcal{M}_a$ with this uniformity is metrizable and complete. It is clear that $\mathcal{M}_{ac}$ is a closed subset of $\mathcal{M}_a$. We endow the topological subspace $\mathcal{M}_{ac} \subset \mathcal{M}_a$ with its relative topology and prove the following result [145].

Theorem 9.15 *Assume that* K *is bounded and* $\operatorname{int}(K) \neq \emptyset$. *Then there exists an open and everywhere dense subset* $\mathcal{F}_a$ *of* $\mathcal{M}_a$ *with the following property: for each* $\widehat{S} \in \mathcal{F}_a$, *there exist* $\bar{x} \in K$ *and a neighborhood* $\mathcal{U}$ *of* $\widehat{S}$ *in* $\mathcal{M}_a$ *such that* $\bar{x} \in S(\bar{x})$ *for each* $S \in \mathcal{U}$.

Moreover, $\mathcal{F}_a$ *contains an open and everywhere dense subset* $\mathcal{F}_{ac}$ *of* $\mathcal{M}_{ac}$.

Usually a generic result is obtained when it is shown that the set of "good" points in a complete metric space contains a dense G_δ subset. Note that our results are stronger because in each one of them we construct an open and everywhere dense subset of "good" points.

In both Theorems 9.14 and 9.15 we assume that the interior of K is nonempty. The following proposition, which will be proved in the next section, shows that this situation is typical.

Proposition 9.16 *The set of all elements of $S(X)$ (respectively, $S_b(X)$) with a nonempty interior contains an open and everywhere dense subset of $S(X)$ (respectively, $S_b(X)$).*

9.5 An Auxiliary Result and the Proof of Proposition 9.16

We need the following auxiliary result (see Proposition 5.1 of [179] for the finite dimensional case). If $(Y, \|\cdot\|)$ is a normed linear space, $x \in Y$ and $r > 0$, then we denote by $B(x,r)$ the closed ball of radius r centered at x.

Lemma 9.17 *Let $(Y, \|\cdot\|)$ be a normed linear space and let $r > 0$ be given. Assume that C is a closed and convex subset of Y such that for all $y \in B(0,r)$,*

$$\inf_{x \in C} \|y - x\| \le r. \tag{9.45}$$

Then $0 \in C$.

Proof If $0 \notin C$, then by the separation theorem there exists a bounded linear functional $l \in Y^*$ such that $\|l\| = 1$ and

$$p = \inf\{l(x) : x \in C\} > 0.$$

There is $y_0 \in B(0,r)$ such that $l(-y_0) > r - p/2$. By (9.45), there is $x_0 \in C$ such that $\|y_0 - x_0\| < r + p/2$. Now we have

$$\begin{aligned} p \le l(x_0) &= l(y_0) + l(x_0 - y_0) < -r + p/2 + \|x_0 - y_0\| \\ &< -r + p/2 + r + p/2 = p. \end{aligned}$$

Since we have reached a contradiction, we conclude that the origin does belong to C. □

Proof of Proposition 9.16 Let $A \in S(X)$ and $\varepsilon > 0$ be given. Denote by $\tilde{A}$ the closure of the set $A + \{y \in X : \|y\| \le \varepsilon\}$. Clearly, $\tilde{A} \in S(X)$ (if $A \in S_b(X)$, then $\tilde{A} \in S_b(X)$) and $H(A, \tilde{A}) \le \varepsilon$. To complete the proof, it is sufficient to show that each $B \in S(X)$ for which $H(\tilde{A}, B) \le \varepsilon/2$ has a nonempty interior.

To this end, let $B \in S(X)$ and $H(B, \tilde{A}) \le \varepsilon/2$. We claim that each point of A belongs to the interior of B. To see this, let $x \in A$ and $y \in B(x, \varepsilon/2)$.

Then $B - y$ is a closed and convex subset of X, $B(0, \varepsilon/2) \subset \tilde{A} - y$ and $H(B - y, \tilde{A} - y) \le \varepsilon/2$. By Lemma 9.17, $0 \in B - y$ and $y \in B$. Thus $B(x, \varepsilon/2) \subset B$. This completes the proof of Proposition 9.16. □

9.6 Proof of Theorem 9.14

Fix $x_* \in \text{int}(K)$. There exists $r_* \in (0, 1)$ such that

$$B(x_*, r_*) \subset K. \tag{9.46}$$

Let $T \in \mathcal{M}_{ne}$ and $\gamma \in (0, 1)$ be given. Define $T_\gamma : K \to S(K)$ by

$$T_\gamma(x) = (1 - \gamma)Tx + \gamma x_*, \quad x \in K. \tag{9.47}$$

It is clear that $T_\gamma \in \mathcal{M}_{ne}$ and $H(T_\gamma(x), T_\gamma(y)) \le \gamma \|x - y\|$ for all $x, y \in K$. By Theorem 9.13, there exists a point $x_{T,\gamma} \in K$ such that

$$T_\gamma(x_{T,\gamma}) = x_{T,\gamma}. \tag{9.48}$$

Consider the set

$$T_\gamma(K) = \bigcup \{T_\gamma(y) : y \in K\} \subset \{(1 - \gamma)y + \gamma x_* : y \in K\}.$$

It follows from this inclusion and (9.46) that for each $z \in T_\gamma(K)$,

$$B(z, \gamma r_*) \subset K. \tag{9.49}$$

For each $x \in K$, denote by $\tilde{T}_\gamma(x)$ the closure of $T_\gamma(x) + B(0, \gamma r_*)$ in the norm topology. By (9.49), $\tilde{T}_\gamma(x) \in S(K)$ for all $x \in K$. It is easy to see that $\tilde{T}_\gamma \in \mathcal{M}_{ne}$. By (9.48),

$$B(x_{T,\gamma}, \gamma r_*) \subset \tilde{T}_\gamma(x_{T,\gamma}). \tag{9.50}$$

Since the point images of the nonexpansive mapping T are bounded, the image under T of any bounded subset of K is also bounded. Therefore $\tilde{T}_\gamma \to T$ as $\gamma \to 0^+$.

Let $T \in \mathcal{M}_{ne}$ and $\gamma \in (0, 1)$. There exists an open neighborhood $U(T, \gamma)$ of $\tilde{T}_\gamma$ in $\mathcal{M}_{ne}$ such that for each $S \in U(T, \gamma)$,

$$H\big(\tilde{T}_\gamma(x_{T,\gamma}), S(x_{T,\gamma})\big) \le \gamma r_*. \tag{9.51}$$

Define

$$\mathcal{F} := \bigcup \{U(T, \gamma) : T \in \mathcal{M}_{ne}, \gamma \in (0, 1)\}.$$

It is clear that $\mathcal{F}$ is an open and everywhere dense subset of $\mathcal{M}_{ne}$.

Assume that $\widehat{S} \in \mathcal{F}$. There exist a mapping $T \in \mathcal{M}_{ne}$ and a number $\gamma \in (0, 1)$ such that $\widehat{S} \in U(T, \gamma)$. Let $S \in U(T, \gamma)$. Then (9.51) and (9.50) hold. Consider now the sets $\tilde{T}_\gamma(x_{T,\gamma}) - x_{T,\gamma}$ and $S(x_{T,\gamma}) - x_{T,\gamma}$. By (9.51),

$$H\big(\tilde{T}_\gamma(x_{T,\gamma}) - x_{T,\gamma}, S(x_{T,\gamma}) - x_{T,\gamma}\big) \le \gamma r_*. \tag{9.52}$$

By (9.50),

$$B(0, \gamma r_*) \subset \tilde{T}_\gamma(x_{T,\gamma}) - x_{T,\gamma}. \tag{9.53}$$

It follows from (9.52), (9.53) and Lemma 9.17 that $0 \in S(x_{T,\gamma}) - x_{T,\gamma}$. In other words, $x_{T,\gamma} \in S(x_{T,\gamma})$ and Theorem 9.14 is proved.

9.7 Proof of Theorem 9.15

Lemma 9.18 *Let $T \in \mathcal{M}_a$ and $\varepsilon > 0$ be given. Then there exist points $z_1 \in K$ and $z_2 \in T(z_1)$ such that $\|z_1 - z_2\| \le \varepsilon$.*

Proof Consider any sequence $\{y_i\}_{i=1}^{\infty} \subset K$ such that $y_{i+1} \in T(y_i)$, $i = 0, 1, \dots$. Choose a natural number n such that

$$n\varepsilon > 2 \sup\bigl\{\|x\| : x \in K\bigr\}.$$

Set $z_1 = n^{-1} \sum_{i=0}^{n-1} y_i$ and $z_2 = n^{-1} \sum_{i=1}^{n} y_i$. It is clear that $z_2 \in T(z_1)$. By the choice of n,

$$\|z_1 - z_2\| \le n^{-1} \|y_n - y_0\| \le 2n^{-1} \sup\bigl\{\|x\| : x \in K\bigr\} < \varepsilon,$$

as asserted. Lemma 9.18 is proved. □

Fix $x_* \in \operatorname{int}(K)$. There exists $r_* \in (0, 1)$ such that

$$B(x_*, r_*) \subset K. \tag{9.54}$$

Let $T \in \mathcal{M}_a$ and $\gamma \in (0, 1)$ be given. Define $T_\gamma : K \to S(K)$ by

$$T_\gamma(x) = (1 - \gamma)Tx + \gamma x_*, \quad x \in K. \tag{9.55}$$

It is obvious that $T_\gamma \in \mathcal{M}_a$ and $T_\gamma \in \mathcal{M}_{ac}$ if $T \in \mathcal{M}_{ac}$.

Consider now the set

$$T_\gamma(K) = \bigcup\bigl\{T_\gamma(y) : y \in K\bigr\} \subset \bigl\{(1 - \gamma)y + \gamma x_* : y \in K\bigr\}.$$

It follows from this inclusion and (9.54) that for each $z \in T_\gamma(K)$,

$$B(z, \gamma r_*) \subset K. \tag{9.56}$$

For each $x \in K$ denote by $\tilde{T}_\gamma(x)$ the closure of $T_\gamma(x) + B(0, \gamma r_*)$ in the norm topology. Clearly, $\tilde{T}_\gamma \in \mathcal{M}_a$ and $\tilde{T}_\gamma \in \mathcal{M}_{ac}$ if $T \in \mathcal{M}_{ac}$. By Lemma 9.18, there exist $x_{T,\gamma} \in K$ and $\bar{x}_{T,\gamma} \in T_\gamma(x_{T,\gamma})$ such that

$$\|\bar{x}_{T,\gamma} - x_{T,\gamma}\| \le 2^{-1} \gamma r_*.$$

It follows from this inequality and the definition of $\tilde{T}_\gamma(x_{T,\gamma})$ that

$$B\bigl(x_{T,\gamma}, 2^{-1}\gamma r_*\bigr) \subset \tilde{T}_\gamma(x_{T,\gamma}). \tag{9.57}$$

There exists an open neighborhood $U(T, \gamma)$ of $\tilde{T}_\gamma$ in $\mathcal{M}_a$ such that for each $S \in U(T, \gamma)$,

$$H\bigl(\tilde{T}_\gamma(x_{T,\gamma}), S(x_{T,\gamma})\bigr) \le 2^{-1} \gamma r_*. \tag{9.58}$$

Note that $\tilde{T}_\gamma \to T$ as $\gamma \to 0^+$.

Define

$$\mathcal{F}_a := \bigcup \{U(T,\gamma) : T \in \mathcal{M}_a, \gamma \in (0,1)\}$$

and

$$\mathcal{F}_{ac} := \Big[\bigcup \{U(T,\gamma) : T \in \mathcal{M}_{ac}, \gamma \in (0,1)\}\Big] \cap \mathcal{M}_{ac}.$$

It is clear that $\mathcal{F}_a$ is an open and everywhere dense subset of $\mathcal{M}_a$, and $\mathcal{F}_{ac}$ is an open and everywhere dense subset of $\mathcal{M}_{ac}$.

Assume that $\widehat{S} \in \mathcal{F}_a$. There exist $T \in \mathcal{M}_a$ and $\gamma \in (0,1)$ such that $\widehat{S} \in U(T,\gamma)$. Let $S \in U(T,\gamma)$. Then (9.58) and (9.57) hold. Consider the sets $\tilde{T}_\gamma(x_{T,\gamma}) - x_{T,\gamma}$ and $S(x_{T,\gamma}) - x_{T,\gamma}$. By (9.58),

$$H\big(\tilde{T}_\gamma(x_{T,\gamma}) - x_{T,\gamma}, S(x_{T,\gamma}) - x_{T,\gamma}\big) \le 2^{-1}\gamma r_*. \tag{9.59}$$

By (9.57),

$$B\big(0, 2^{-1}\gamma r_*\big) \subset \tilde{T}_\gamma(x_{T,\gamma}) - x_{T,\gamma}. \tag{9.60}$$

It follows from (9.59), (9.60) and Lemma 9.17 that $0 \in S(x_{T,\gamma}) - x_{T,\gamma}$ and $x_{T,\gamma} \in S(x_{T,\gamma})$. This completes the proof of Theorem 9.15.

9.8 An Extension of Theorem 9.15

Consider the complete uniform space $S(X)$ defined in the previous section. Assume that K is a nonempty, closed and convex (not necessarily bounded) subset of X. Denote by $\mathfrak{M}_a$ the set of all mappings $T : K \to S(X)$ such that

$$\alpha T x_1 + (1-\alpha) T x_2 \subset T\big(\alpha x_1 + (1-\alpha) x_2\big) \tag{9.61}$$

for all $x_1, x_2 \in K$ and each $\alpha \in (0,1)$. As we have already mentioned, such mappings find application in many areas. We denote by $\mathfrak{M}_{ac}$ the set of all continuous mappings $T : K \to S(X)$ which belong to $\mathfrak{M}_a$.

For the set $\mathfrak{M}_a$ we consider two uniformities, strong and weak, and the strong and weak topologies generated by them. (The weak uniformity is weaker than the strong one.) The strong uniformity is determined by the following base:

$$\begin{aligned}\mathcal{E}_s(n) = \big\{&(T_1, T_2) \in \mathfrak{M}_a \times \mathfrak{M}_a : H\big(T_1(x), T_2(x)\big) \le n^{-1} \\ &\text{for all } x \in K\big\}, \quad n = 1, 2, \ldots.\end{aligned} \tag{9.62}$$

It is not difficult to see that the space $\mathfrak{M}_a$ with this uniformity is metrizable and complete, and that $\mathfrak{M}_{ac}$ is a closed subset of $\mathfrak{M}_a$.

Fix $\theta \in K$. For the set $\mathfrak{M}_a$ we also consider the weak uniformity determined by the following base:

$$\begin{aligned}\mathcal{E}_w(n) = \big\{&(T_1, T_2) \in \mathfrak{M}_a \times \mathfrak{M}_a : H\big(T_1(x), T_2(x)\big) \le n^{-1} \\ &\text{for all } x \in K \text{ satisfying } \|x - \theta\| \le n\big\}, \quad n = 1, 2, \ldots.\end{aligned} \tag{9.63}$$

It is not difficult to verify that the space $\mathfrak{M}_a$ with this weaker uniformity is also metrizable and complete, and that $\mathfrak{M}_{ac}$ is, once again, a closed subset of $\mathfrak{M}_a$.

Denote by $\mathfrak{M}_a^*$ the set of all $T \in \mathfrak{M}_a$ such that there exists a bounded sequence $\{x_i\}_{i=0}^{\infty} \subset K$ with $x_{i+1} \in T(x_i)$, $i = 0, 1, \ldots$. Set $\mathfrak{M}_{ac}^* = \mathfrak{M}_a^* \cap \mathfrak{M}_{ac}$. Denote by $\bar{\mathfrak{M}}_a^{*s}$ the closure of $\mathfrak{M}_a^*$ in the space $\mathfrak{M}_a$ with the strong topology, by $\bar{\mathfrak{M}}_a^{*w}$ the closure of $\mathfrak{M}_a^*$ in the space $\mathfrak{M}_a$ with the weak topology, by $\bar{\mathfrak{M}}_{ac}^{*s}$ the closure of $\mathfrak{M}_{ac}^*$ in the space $\mathfrak{M}_a$ with the strong topology and by $\bar{\mathfrak{M}}_{ac}^{*w}$ the closure of $\mathfrak{M}_{ac}^*$ in the space $\mathfrak{M}_a$ with the weak topology. We equip the topological subspaces $\bar{\mathfrak{M}}_a^{*s}$, $\bar{\mathfrak{M}}_a^{*w}$, $\bar{\mathfrak{M}}_{ac}^{*s}$, $\bar{\mathfrak{M}}_{ac}^{*w} \subset \mathfrak{M}_a$ with both the weak and strong relative topologies.

In this section we prove the following result [145].

Theorem 9.19 *There exists an open everywhere dense (in the weak topology) subset $\mathcal{F}_a^w$ of $\bar{\mathfrak{M}}_a^{*w}$ with the following property: for each $A \in \mathcal{F}_a^w$, there exist $z_* \in K$ and a neighborhood $\mathcal{W}$ of A in $\mathfrak{M}_a$ with the weak topology such that $z_* \in S(z_*)$ for each $S \in \mathcal{W}$. Moreover, there exists an open (in the weak topology) and everywhere dense (in the strong topology) subset $\mathcal{F}_a^s$ of $\bar{\mathfrak{M}}_a^{*s}$, an open (in the weak topology) and everywhere dense (in the strong topology) subset $\mathcal{F}_{ac}^s$ of $\bar{\mathfrak{M}}_{ac}^{*s}$, and an open everywhere dense (in the weak topology) subset $\mathcal{F}_{ac}^w$ of $\bar{\mathfrak{M}}_{ac}^{*w}$ such that $\mathcal{F}_{ac}^s \subset \mathcal{F}_a^s \subset \mathcal{F}_a^w$ and $\mathcal{F}_{ac}^s \subset \mathcal{F}_{ac}^w \subset \mathcal{F}_a^w$.*

In the proof of Theorem 9.19 we will use the following auxiliary result (cf. Lemma 9.18).

Lemma 9.20 *Let $T \in \mathfrak{M}_a^*$ and $\varepsilon > 0$ be given. Then there exist $z_1 \in K$ and $z_2 \in T(z_1)$ such that $\|z_1 - z_2\| \le \varepsilon$.*

Proof of Theorem 9.19 Let $T \in \mathfrak{M}_a$ and $\gamma \in (0, 1)$ be given. For each $x \in K$, denote by $T_\gamma(x)$ the closure of $Tx + B(0, \gamma)$ in the norm topology. Clearly, $T_\gamma \in \mathfrak{M}_a$ and $T_\gamma \in \mathfrak{M}_{ac}$ if $T \in \mathfrak{M}_{ac}$. It is easy to see that for each $T \in \mathfrak{M}_a$, $T_\gamma \to T$ as $\gamma \to 0^+$ in the strong topology.

Let $T \in \mathfrak{M}_a^*$ and $\gamma \in (0, 1)$. By Lemma 9.20, there exists $x_{T,\gamma} \in K$ such that

$$B\big(x_{T,\gamma}, 2^{-1}\gamma\big) \subset T_\gamma(x_{T,\gamma}). \tag{9.64}$$

There also exists an open neighborhood $U(T, \gamma)$ of T_γ in $\mathfrak{M}_a$ with the weak topology such that for each $S \in U(T, \gamma)$,

$$H\big(T_\gamma(x_{T,\gamma}), S(x_{T,\gamma})\big) \le 2^{-1}\gamma. \tag{9.65}$$

Define

$$\mathcal{F}_a^s := \Big[\bigcup\{U(T, \gamma) : T \in \mathfrak{M}_a^*, \gamma \in (0, 1)\}\Big] \cap \bar{\mathfrak{M}}_a^{*s},$$

$$\mathcal{F}_a^w := \Big[\bigcup\{U(T, \gamma) : T \in \mathfrak{M}_a^*, \gamma \in (0, 1)\}\Big] \cap \bar{\mathfrak{M}}_a^{*w},$$

$$\mathcal{F}_{ac}^s := \Big[\bigcup\{U(T, \gamma) : T \in \mathfrak{M}_{ac}^*, \gamma \in (0, 1)\}\Big] \cap \bar{\mathfrak{M}}_{ac}^{*s}$$

and

$$\mathcal{F}^w_{ac} := \Big[\bigcup\{U(T,\gamma) : T \in \mathfrak{M}^*_{ac}, \gamma \in (0,1)\}\Big] \cap \bar{\mathfrak{M}}^{*w}_{ac}.$$

Clearly, $\mathcal{F}^s_{ac} \subset \mathcal{F}^s_a \subset \mathcal{F}^w_a$ and $\mathcal{F}^s_{ac} \subset \mathcal{F}^w_{ac} \subset \mathcal{F}^w_a$. It is easy to see that $\mathcal{F}^s_a$ is an open (in the weak topology) and everywhere dense (in the strong topology) subset of $\bar{\mathfrak{M}}^{s*}_a$, $\mathcal{F}^w_a$ is an open everywhere dense (in the weak topology) subset of $\bar{\mathfrak{M}}^{w*}_a$, $\mathcal{F}^s_{ac}$ is an open (in the weak topology) and everywhere dense (in the strong topology) subset of $\bar{\mathfrak{M}}^{*s}_{ac}$, and $\mathcal{F}^w_{ac}$ is an open everywhere dense (in the weak topology) subset of $\bar{\mathfrak{M}}^{*w}_{ac}$.

Assume that $A \in \mathcal{F}^w_a$. Then there exist $T \in \mathfrak{M}^*_a$ and $\gamma \in (0,1)$ such that $A \in U(T,\gamma)$. By (9.64),

$$B\big(0, 2^{-1}\gamma\big) \subset T_\gamma(x_{T,\gamma}) - x_{T,\gamma}. \tag{9.66}$$

Let $S \in U(T,\gamma)$. By (9.65),

$$H\big(T_\gamma(x_{T,\gamma}) - x_{T,\gamma}, S(x_{T,\gamma}) - x_{T,\gamma}\big) \le 2^{-1}\gamma. \tag{9.67}$$

It follows from (9.66), (9.67) and Lemma 9.17 that $0 \in S(x_{T,\gamma}) - x_{T,\gamma}$ and $x_{T,\gamma} \in S(x_{T,\gamma})$. This completes the proof of Theorem 9.19. □

9.9 Generic Existence of Fixed Points

Let (X,d) be a complete metric space. For $x \in X$ and a nonempty subset A of X, set $d(x,A) = \inf_{a\in A} d(x,a)$.

In the space X, an open ball and a closed ball of center a and radius $r > 0$ are denoted by $S_X(a,r)$ and $S_X[a,r]$, respectively.

Set

$$\mathcal{B}(X) = \{A \subset X : A \text{ is nonempty closed and bounded}\}.$$

The space $\mathcal{B}(X)$ is equipped with the Hausdorff metric

$$h(A,B) = \max\Big\{\sup_{a\in A} d(a,B), \sup_{b\in B} d(b,A)\Big\}, \quad A, B \in \mathcal{B}(X).$$

Note that $h(\cdot,\cdot)$ is, in fact, defined for all pairs of nonempty subsets of X (not necessarily bounded and closed).

A map $F : X \to \mathcal{B}(X)$ is said to be nonexpansive (respectively, strictly contractive with a constant $L_F \in [0,1)$) if it satisfies

$$h\big(F(x), F(y)\big) \le d(x,y) \qquad \big(\text{resp. } h\big(F(x), F(y)\big) \le L_F d(x,y)\big)$$

for all $x, y \in X$.

The set $\mathrm{fix}(F) = \{x \in X : x \in F(x)\}$ is called the fixed point set of F.

We say that most (or typical) elements of X have a given property P if the set $\tilde{X}$ of all $x \in X$ having P is residual in X, i.e., $X \setminus \tilde{X}$ is of the first Baire category in X.

Let E be a real Banach space with norm $\|\cdot\|$. Set

$$\mathcal{X}(E) = \{A \subset E : A \text{ is nonempty and compact}\}$$

and

$$\mathcal{E}(E) = \{A \subset E : A \text{ is nonempty, compact and convex}\}.$$

The spaces $\mathcal{X}(E)$ and $\mathcal{E}(E)$ are equipped with the Hausdorff metric h under which each one of them is complete.

For any star-shaped set $A \subset E$, $\mathrm{st}(A)$ denotes the set of all $a \in A$ such that $ta + (1-t)x \in A$ for every $x \in A$ and $t \in [0, 1]$.

In this section we prove that most compact-valued nonexpansive map from a closed bounded star-shaped subset of a Banach space E into itself have fixed points. This result was obtained in [53].

Let E be a real Banach space. For a nonempty, closed, bounded and star-shaped set $D \subset E$, define

$$\mathcal{X}_D = \{A \in \mathcal{X}(E) : A \subset D\}.$$

Under the Hausdorff metric h the space $\mathcal{X}_D$ is complete. Set

$$\mathcal{M} = \{F : D \to \mathcal{X}_D : F \text{ is nonexpansive}\},$$

$$\mathcal{N} = \{G : D \to \mathcal{X}_D : G \text{ is strictly contractive}\}.$$

The space $\mathcal{M}$ is equipped with the metric of uniform convergence

$$\rho(F_1, F_2) = \sup_{x \in D} h(F_1(x), F_2(x)), \quad F_1, F_2 \in \mathcal{M} \tag{9.68}$$

under which it is complete.

Given $F : D \to \mathcal{X}_D$ and $A \in \mathcal{X}_D$, set

$$\Phi_F(A) = \bigcup_{x \in A} F(x). \tag{9.69}$$

Lemma 9.21 *Let $F : D \to \mathcal{X}_D$ satisfy*

$$h(F(x), F(y)) \le L_F \|x - y\| \quad (L_f \ge 0) \textit{ for all } x, y \in D.$$

Then (9.69) *defines a map $\Phi_F : \mathcal{X}_D \to \mathcal{X}_D$ satisfying*

$$h(\phi_F(A), \phi_F(B)) \le L_F h(A, B) \quad \textit{for all } A, B \in \mathcal{X}_D. \tag{9.70}$$

Proof It is evident that $\Phi_F(A) \in \mathcal{X}_D$ for each $A \in \mathcal{X}_D$. To prove (9.70), let $A, B \in \mathcal{X}_D$. Let $u \in \phi_F(A)$. Then $u \in F(x)$ for some $x \in A$. Since B is compact, there is a point $y \in B$ such that $\|x - y\| = d(x, B)$. We have

$$\begin{aligned} d(u, \Phi_F(B)) &\le d(u, F(y)) \le d(u, F(x)) + h(F(x), F(y)) \\ &\le L_F \|x - y\| \le L_F d(x, B). \end{aligned}$$

Thus

$$d\big(u, \Phi_F(B)\big) \le L_F h(A, B) \quad \text{for each } u \in \Phi_F(A) \tag{9.71}$$

and similarly,

$$d\big(u, \Phi_F(A)\big) \le L_F h(A, B) \quad \text{for each } u \in \Phi_F(B). \tag{9.72}$$

Combining (9.71) and (9.72), we get (9.70), as asserted. □

Lemma 9.22 *Let $F, G \in \mathcal{M}$ be such that $\rho(F, G) < \delta$, where $\delta > 0$. Then*

$$h\big(\Phi_F(A), \Phi_G(A)\big) < \delta \quad \textit{for each } A \in \mathcal{X}_D. \tag{9.73}$$

Proof Let $A \in \mathcal{X}_D$ and $\varepsilon > 0$ be given. Since A is compact, and F and G are uniformly continuous, there exist a finite set $\{a_i\}_{i=1}^N \subset A$ and $\sigma > 0$ such that, setting $A_i = A \cap S_D[a_i, \sigma]$, one has

$$h\big(F(x), F(a_i)\big) \le \varepsilon, \qquad h\big(G(x), G(a_i)\big) \le \varepsilon \quad \text{for every } x \in A_i, i = 1, 2, \dots, N.$$

Hence

$$h\big(\Phi_F(A_i), F(a_i)\big) \le \varepsilon, \qquad h\big(\Phi_G(A_i), G(a_i)\big) \le \varepsilon, \quad i = 1, \dots, N.$$

Therefore

$$\begin{aligned} h\big(\Phi_F(A), \Phi_G(A)\big) &= h\left(\bigcup_{i=1}^N \Phi_F(A_i), \bigcup_{i=1}^N \phi_G(A_i)\right) \le \max_{1\le i\le N} h\big(\Phi_F(A_i), \Phi_G(A_i)\big) \\ &\le \max_{1\le i\le N} \big[h\big(\Phi_F(A_i), F(a_i)\big) + h\big(F(a_i), G(a_i)\big) \\ &\quad + h\big(G(a_i), \Phi_G(A_i)\big)\big] \\ &\le 2\varepsilon + \rho(F, G), \end{aligned}$$

which implies $h(\Phi_F(A), \Phi_G(A)) \le \rho(F, G) < \delta$. Since $A \in \mathcal{X}_D$ is arbitrary, inequality (9.73) indeed holds as claimed. □

Lemma 9.23 *The set $\mathcal{N}$ is dense in $\mathcal{M}$.*

Proof Let $F \in \mathcal{M}$. For a natural number n, define $G_n : D \to \mathcal{X}_D$ by

$$G_n(x) = n^{-1}a + \big(1 - n^{-1}\big)F(x), \quad x \in D,$$

where $a \in \mathrm{st}(D)$. Since $G_n \in \mathcal{N}$ and $\rho(G_n, F) \to 0$ as $n \to \infty$, the result follows. □

Lemma 9.24 *Let $G \in \mathcal{N}$ and let $\varepsilon > 0$ be given. Then there exists $0 < \delta_G(\varepsilon) < \varepsilon$ such that*

$$\text{if } F \in S_{\mathcal{M}}(G, \delta_G(\varepsilon)), \text{ then } h(\Phi_F^n(A), \Phi_G^n(A)) < \varepsilon \text{ for every } A \in \mathcal{X}_D \text{ and all natural numbers } n. \tag{9.74}$$

Proof Let $G \in \mathcal{N}$ be strictly contractive with constant $0 \le L_G < 1$ and let $\varepsilon > 0$ be given. By Lemma 9.21, $\Phi_G : \mathcal{X}_D \to \mathcal{X}_D$ is strictly contractive with the same constant L_G. We claim that (9.74) holds with $\delta_G(\varepsilon) = \delta$, where $0 < \delta < (1 - L_G)\varepsilon$.

Let $F \in S_{\mathcal{M}}(G, \delta)$. By Lemma 9.22, (9.73) is satisfied. Let $A \in \mathcal{X}_D$ be arbitrary. By (9.73),

$$h(\Phi_F(\Phi_F(A)), \Phi_G(\Phi_F(A))) \le \delta$$

and thus

$$\begin{aligned} h(\Phi_F^2(A), \Phi_G^2(A)) &\le h(\Phi_F(\Phi_F(A)), \Phi_G(\Phi_F(A))) \\ &\quad + h(\Phi_G(\Phi_F(A)), \Phi_G(\Phi_G(A))) \\ &< \delta + L_G h(\Phi_F(A), \Phi_G(A)) \le \delta(1 + L_g). \end{aligned}$$

Using induction, we obtain, for any natural number n,

$$h(\Phi_F^n(A), \Phi_G^n(A)) \le \delta(1 + L_G + \cdots + L_G^{n-1}).$$

Thus

$$h(\Phi_F^n(A), \Phi_G^n(A)) \le \delta(1 - L_G)^{-1}$$

for every $A \in \mathcal{X}_D$ and any natural number n.

Since $\delta < (1 - L_G)\varepsilon$, (9.74) holds, as claimed. □

Put

$$\mathcal{M}_0 = \{F \in \mathcal{M} : \text{fix}(F) \text{ is compact nonempty}\}.$$

Theorem 9.25 *The set $\mathcal{M}_0$ is residual in $\mathcal{M}$.*

Proof For $G \in \mathcal{N}$ and any natural number k, let $S_{\mathcal{M}}(G, \delta_G(1/k))$, where $\delta_G(1/k) < 1/k$ exists according to Lemma 9.24. Define

$$\mathcal{M}^* := \bigcap_{k=1}^{\infty} \bigcup_{G \in \mathcal{N}} S_{\mathcal{M}}(G, \delta_G(1/k)).$$

Clearly, $\mathcal{M}^*$ is residual in $\mathcal{M}$, since $\mathcal{M}^*$ is the countable intersection of sets which are open and, by Lemma 9.23, dense in $\mathcal{M}$. The theorem is an immediate consequence of the following assertion.

Claim.

$$\mathcal{M}^* \subset \mathcal{M}_0.$$

Let $F \in \mathcal{M}^*$ be given. By the definition of $\mathcal{M}^*$, there exists a sequence $\{G_k\}_{k=1}^{\infty} \subset \mathcal{N}$ such that

$$F \in S_{\mathcal{M}}\big(G_k, \delta_{G_k}(1/k)\big) \quad \text{for every natural number } k. \tag{9.75}$$

Thus by Lemma 9.24, for each natural number k,

$$h\big(\Phi_F^n(A), \Phi_{G_k}^n(A)\big) < 1/k$$
$$\text{for every } A \in \mathcal{X}_D \text{ and every natural number } n. \tag{9.76}$$

According to Lemma 9.21, $\Phi_{G_k} : \mathcal{X}_D \to \mathcal{X}_D$ is strictly contractive, for $G_k \in \mathcal{N}$, and hence for each natural number k, there exists $Z_k \in \mathcal{X}_D$ such that

$$Z_k = \Phi_{G_k}(Z_k).$$

(j) $\{Z_k\}_{k=1}^{\infty} \subset \mathcal{X}_D$ is a Cauchy sequence.

To see this, let $\varepsilon > 0$ be given. Let $k, k' > 4/\varepsilon$ be arbitrary natural numbers and let $A \in \mathcal{X}_D$. Since $\Phi_{G_k}^n(A) \to Z_k$ and $\Phi_{G_{k'}}^n(A) \to Z_{k'}$ as $n \to \infty$, there exists a natural number m such that

$$h\big(\Phi_{G_k}^n(A), Z_k\big) < \varepsilon/4, \qquad h\big(\Phi_{G_{k'}}^n(A), Z_{k'}\big) < \varepsilon$$
$$\text{for every integer } n \geq m. \tag{9.77}$$

In view of (9.77) and (9.76), one has

$$\begin{aligned} h(Z_k, Z_{k'}) &\leq h\big(Z_k, \Phi_{G_k}^m(A)\big) + h\big(\Phi_{G_k}^m(A), \Phi_F^m(A)\big) \\ &\quad + h\big(\Phi_F^m(A), \Phi_{G_{k'}}^m(A)\big) + h\big(\Phi_{G_{k'}}^m(A), Z_{k'}\big) \\ &< \varepsilon/4 + 1/k + \big(k'\big)^{-1} + \varepsilon/4 < \varepsilon, \end{aligned}$$

for $1/k + 1/k' < \varepsilon/2$. As $k, k' > 4/\varepsilon$ are arbitrary, (j) is proved.

Since $\{Z_k\}_{k=1}^{\infty} \subset \mathcal{X}_D$ is a Cauchy sequence and $\mathcal{X}_D$ is a complete metric space, there exists $Z \in \mathcal{X}_D$ such that $Z_k \to Z$ as $k \to \infty$

(jj) For each $A \in \mathcal{X}_D$, the sequence $\{\Phi_F^n(A)\}$ converges to Z as $n \to \infty$. Moreover, $Z = \Phi_F(Z)$ is the unique fixed point of Φ_F.

Let $A \in \mathcal{X}_D$. Given $\varepsilon > 0$, fix a natural number $k > 3/\varepsilon$ large enough so that $h(Z_k, Z) < \varepsilon/3$. Hence by (9.76), for every natural number n, one has

$$\begin{aligned} h\big(\Phi_F^n(A), Z\big) &\leq h\big(\Phi_F^n(A), \Phi_{G_k}^n(A)\big) + h\big(\Phi_{G_k}^n(A), Z_k\big) + h(Z_k, Z) \\ &< 1/k + h\big(\Phi_{G_k}^n(A), Z_k\big) + \varepsilon/3. \end{aligned}$$

Since $h(\Phi^n_{G_k}(A), Z_k)$ tends to zero as $n \to \infty$, there is a natural number n_0 such that $h(\Phi^n_{G_k}(A), Z_k) < \varepsilon/3$ for all $n \ge n_0$. Moreover, $1/k < \varepsilon/3$, and thus

$$h\big(\Phi^n_F(A), Z\big) < \varepsilon \quad \text{for every } n \ge n_0.$$

This shows that $\Phi^n_F(A) \to Z$ as $n \to \infty$. The second statement of (jj) is obvious.

(jjj) The fixed point set $\mathrm{fix}(F)$ is a nonempty compact subset of D.

First we show that the set $\mathrm{fix}(F)$ is nonempty. As $G_k \in \mathcal{N}$, by Nadler's theorem [102], for each natural number k, there is a point $a_k \in D$ such that

$$a_k \in G_k(a_k), \quad k = 1, 2, \ldots. \tag{9.78}$$

For each natural number k,

$$a_k \in \Phi^n_{G_k}(a_k) \quad \text{for every natural number } n. \tag{9.79}$$

This is obvious if $n = 1$ because $\Phi_{G_k}(a_k) = G_k(a_k)$. Assuming that (9.79) is valid for n, then for $n+1$ one has $a_k \in \Phi_{G_k}(a_k) \subset \Phi_{G_k}(\Phi^n_{G_k}(a_k)) = \Phi^{n+1}_{G_k}(a_k)$ and thus (9.79) holds for every natural number n. Since $\Phi^n_{G_k}(a_k) \to Z_k$ as $n \to \infty$, it follows that $a_k \in Z_k$. On the other hand, $Z_k \to Z$ implies $d(a_k, Z) \to 0$ as $k \to \infty$. Since Z is compact, there is a subsequence $\{a_{k_n}\}_{n=1}^{\infty}$ which converges to some $a \in D$.

We have $a \in F(a)$. In fact, (9.75) implies that

$$h\big(F(x), G_{k_n}(x)\big) < \delta_{G_{k_n}}(1/k_n)$$
$$\text{for every } x \in D \text{ and any natural number } n. \tag{9.80}$$

In view of (9.78) and (9.80), one has

$$\begin{aligned} d\big(a, F(a)\big) &\le \|a - a_{k_n}\| + d\big(a_{k_n}, G_{k_n}(a_{k_n})\big) \\ &\quad + h\big(G_{k_n}(a_{k_n}), F(a_{k_n})\big) + h\big(F(a_{k_n}), F(a)\big) \\ &< \|a - a_{k_n}\| + \delta_{G_{k_n}}(1/k_n) + h\big(F(a_{k_n}), F(a)\big) \le 2\|a - a_{k_n}\| + 1/k_n \end{aligned}$$

because $\delta_{G_{k_n}}(1/k_n) < 1/k_n$. As $n \to \infty$, the right-hand side tends to zero and thus $d(a, F(a)) = 0$, i.e. $a \in F(a)$. Hence $\mathrm{fix}(F) \ne \emptyset$, as claimed. It remains to show that $\mathrm{fix}(F)$ is compact. To see this, let $x \in \mathrm{fix}(F)$. Then $x \in \Phi^n_F(x)$ for every natural number n. Since by (jj), $\Phi^n_F(x) \to Z$ as $n \to \infty$, it follows that $z \in Z$. Thus $\mathrm{fix}(F) \subset Z$, which implies that $\mathrm{fix}(F)$ is compact for so is Z and $\mathrm{fix}(F)$ is closed. Hence (jjj) holds. Therefore $F \in \mathcal{M}_0$. This completes the proof of the claim and of Theorem 9.25 itself. □

For a nonempty, closed, bounded and star-shaped set $D \subset E$, let

$$\mathcal{E}_D = \big\{A \in \mathcal{E}(E) : A \subset D\big\}. \tag{9.81}$$

When endowed with the Hausdorff metric h, the space $\mathcal{E}_d$ is complete. Define

$$\mathcal{U} = \{F : D \to \mathcal{E}_D : F \text{ is nonexpansive}\}. \tag{9.82}$$

The set $\mathcal{U}$ is endowed with the metric ρ of uniform convergence (9.68) under which it is complete. Set

$$\mathcal{U}_0 = \{F \in \mathcal{U} : \text{fix}(F) \text{ is nonempty and compact}\}. \tag{9.83}$$

Using the same argument as in the proof of Theorem 9.25, one can also prove the following result.

Theorem 9.26 *The set $\mathcal{U}_0$ is residual in $\mathcal{U}$.*

9.10 Topological Structure of the Fixed Point Set

In this section, which is based on [53], we study the topological structure of the fixed point set for a typical compact-and convex-valued nonexpansive map from a closed, convex and bounded subset of a Banach space into itself.

Let E be a real Banach space. In this section D denotes a closed, convex and bounded subset of E with a nonempty interior int(D). Set

$$S = \{x \in E : \|x\| \le 1\}.$$

Let $\mathcal{E}_D, \mathcal{U}$ and $\mathcal{U}_0$ be given by (9.81), (9.82) and (9.83) with D as above. Define

$$\mathcal{U}_1 := \{F \in \mathcal{U} : \text{ there is } \alpha_F \text{ such that } F(x) + \sigma_F S \subset D, x \in D\}. \tag{9.84}$$

Lemma 9.27 *The set $\mathcal{U}_1$ is open and dense in $\mathcal{U}$.*

Proof First we show that $\mathcal{U}_1$ is open in $\mathcal{U}$. Let $F \in \mathcal{U}_1$ and let $\sigma_F > 0$ be the corresponding number in (9.84). For $0 < \varepsilon < \sigma_F/2$ we have $S_{\mathcal{U}}(F, \varepsilon) \subset \mathcal{U}_1$. In fact, each $G \in S_{\mathcal{U}}(F, \varepsilon)$ satisfies $G(x) \subset F(z) + \varepsilon S$ for each $x \in D$. Thus, taking $\sigma_G = \sigma_F/2$, one has

$$G(x) + \sigma_G S \subset F(x) + (\varepsilon + \sigma_G)S \subset F(x) + \sigma_F S \subset D \quad \text{for each } x \in D.$$

Hence $S_{\mathcal{U}}(F, \varepsilon) \subset \mathcal{U}_1$. Thus $\mathcal{U}_1$ is indeed open in $\mathcal{U}$.

Next we show that $\mathcal{U}_1$ is dense in $\mathcal{U}$. Let $F \in \mathcal{U}$, $0 < \varepsilon < 1$ and let $a \in \text{int}(D)$. Then $S_E[a, \theta] \subset D$ for some $\theta > 0$. Fix λ such that $0 < \lambda < \varepsilon/(2M)$, where $M = \sup_{x \in D} \|x\| + 1$, and define $G : D \to \mathcal{E}_D$ by

$$G(x) = \lambda a + (1 - \lambda)F(x), \quad x \in D.$$

Clearly, $G \in \mathcal{U}$, and $\rho(G, F) < \varepsilon$ since for each $x \in D$,

$$\begin{aligned} h(\lambda a + (1-\lambda)F(x), F(x)) &= h(\lambda a + (1-\lambda)F(x), \lambda F(x) + (1-\lambda)F(x)) \\ &\le \lambda h(a, F(x)) < 2\lambda M < \varepsilon. \end{aligned}$$

Furthermore, taking $0 < \sigma_G < \lambda\theta$, for each $x \in D$, one has

$$G(x) + \sigma_G S \subset \lambda\big(a + \sigma_G \lambda^{-1} S\big) + (1-\lambda)F(x) \subset \lambda S_E[a,\theta] + (1-\lambda)D \subset D$$

and thus $G \in U_1$. Since $\rho(G,F) < \varepsilon$ it follows that $\mathcal{U}_1$ is dense in $\mathcal{U}$, as asserted. □

Set

$$\mathcal{B}_D = \{A \subset D : A \text{ is nonempty, closed and convex}\}.$$

The following result is a special case of a theorem due to Ricceri [165].

Lemma 9.28 *Let $F : D \to \mathcal{B}_D$ be strictly contractive. Then the fixed point set* $\mathrm{fix}(F)$ *of F is a nonempty absolute retract.*

We call the subset of a metric space an R_δ-set if it is the intersection of a descending sequence of absolute retracts.

Theorem 9.29 *The fixed point set* $\mathrm{fix}(F)$ *of most $F \in \mathcal{U}$ is a nonempty and compact R_δ-set.*

Proof Let $\mathcal{U}_0$ and $\mathcal{U}_1$ be defined by (9.83) and (9.84), respectively. By Theorem 9.26 and Lemma 9.27, the set $\mathcal{U}^* = \mathcal{U}_0 \cap U_1$ is residual in $\mathcal{U}$. Our theorem is therefore an immediate consequence of the following assertion.

Claim. For each $F \in \mathcal{U}^*$, the set $\mathrm{fix}(F)$ is a nonempty and compact R_δ-set.

Let $F \in \mathcal{U}^*$. Since $F \in \mathcal{U}_1$, there exists $\sigma_F > 0$ such that

$$F(x) + \sigma_F S \subset D \quad \text{for each } x \in D. \tag{9.85}$$

Let $a \in \mathrm{int}(D)$. Then $S_E[a,\theta] \subset D$ for some $\theta > 0$. For a natural number n, define $G_n : D \to \mathcal{E}_D$ by

$$G_n(x) = \big(2^{-n}\big)a + \big(1 - 2^{-n}\big)F(x), \quad x \in D.$$

In addition, for a natural number n, set

$$Q_n(x) = G_n(x) + (1/n)S, \quad x \in D. \tag{9.86}$$

Let n_0 be a natural number such that $n \ge n_0$ implies $(1/n)/(1-2^{-n}) < \sigma_F$. For $n \ge n_0$ and $x \in D$, one has

$$\begin{aligned} Q_n(x) &= \big(2^{-n}\big)a + \big(1-2^{-n}\big)\big[F(x) + (1/n)\big(1-(1/2)^n\big)^{-1} S\big] \\ &\subset 2^{-n}a + \big(1-2^{-n}\big)\big(F(x) + \sigma_F S\big) \subset 2^{-n}a + \big(1-2^{-n}\big)D \end{aligned}$$

by (9.84).

Therefore $Q_n(x) \subset D$ and thus $Q_n(x) \in B_D$ for each $x \in D$. It follows that for each natural number $n \geq n_0$, (9.86) defines a map $Q_n : D \to \mathcal{B}_D$ which is a strict contraction and, moreover, $\rho(Q_n, F) \to 0$ as $n \to \infty$. Fix $n_1 \geq n_0$ so that

$$n(n+1)2^{-n}M < 1 \quad \text{for every } n \geq n_1, \text{ where } M = \sup_{x \in D} \|x\|. \tag{9.87}$$

We claim that for each integer $n \geq n_1$, one has

$$F(x) \subset Q_{n+1}(x) \subset Q_n(x) \quad \text{for every } x \in D. \tag{9.88}$$

To see this, let $n \geq n_1$ and $x \in D$ be arbitrary. Then

$$\begin{aligned} &Q_{n+1}(x) + (1/n)S \\ &\quad = \left(1 - 2^{-n-1}\right)F(x) + 2^{-n-1}a + (n+1)^{-1}S + (1/n)S \\ &\quad = \left(1 - 2^{-n}\right)F(x) + \left(2^{-n} - 2^{-n-1}\right)F(x) + 2^{-n}a - \left(2^{-n} + 2^{n+1}\right)a \\ &\qquad + (n+1)^{-1}S + n^{-1}S \end{aligned}$$

and thus

$$Q_{n+1}(x) + (1/n)S \subset Q_n(x) + 2^{-n}2^{-1}\big(F(x) - a\big) + (n+1)^{-1}S. \tag{9.89}$$

Now,

$$\big(F(x) - a\big)/2 \subset (D - a)/2 \subset MS$$

and hence by (9.87),

$$2^{-n}\big(F(x) - a\big) \subset \big(n(n+1)\big)^{-1}S. \tag{9.90}$$

Combining (9.90) with (9.89), we obtain

$$Q_{n+1}(x) + (1/n)S \subset Q_n(x) + (1/n)S.$$

It now follows from Radström's cancellation law [113] that

$$Q_{n+1}(x) \subset Q_n(x).$$

It remains to be shown that $F(x) \subset Q_{n+1}x$. Clearly,

$$\begin{aligned} F(x) + (n+1)^{-1}S &= \left(1 - 2^{-n-1}\right)F(x) + 2^{-n-1}F(x) + 2^{-n-1}a \\ &\quad - 2^{-n-1}a + (n+1)^{-1}S \\ &= Q_{n+1}(x) + 2^{-n}\big(F(x) - a\big)/2 \subset Q_{n+1}(x) + (n+1)^{-1}S, \end{aligned}$$

since by (9.90),

$$\left(1/2^n\right)\big(F(x) - a\big)/2 \subset \big(n(n+1)\big)^{-1}S \subset (n+1)^{-1}S.$$

Therefore by Radström's cancellation law $F(x) \subset Q_{n+1}(x)$ and thus (9.89) is valid.

For each integer $n \geq n_1$, $Q_n : D \to \mathcal{B}_D$ is a strict contraction and by Lemma 9.27 its fixed point set $\operatorname{fix}(Q_n)$ is a nonempty absolute retract. On the other hand, the set $\operatorname{fix}(F)$ is nonempty and compact because $F \in \mathcal{U}^* \subset \mathcal{U}_0$. By (9.88),

$$\operatorname{fix}(F) \subset \operatorname{fix}(Q_{n+1}) \subset \operatorname{fix}(Q_n) \quad \text{for every } n \geq n_1,$$

which implies that

$$\operatorname{fix}(F) \subset \bigcap_{n \geq n_1} \operatorname{fix}(Q_n).$$

On the other hand, let $x \in \operatorname{fix}(Q_n)$ for every $n \geq n_1$. Then $x \in F(x)$ because

$$d\big(x, F(x)\big) \leq d\big(x, Q_n(x)\big) + h\big(Q_n(x), F(x)\big) \leq \rho(Q_n, F)$$

and $\rho(Q_n, F) \to \infty$ as $n \to \infty$. Hence

$$\operatorname{fix}(F) = \bigcap_{n \geq n_1} \operatorname{fix}(Q_n)$$

and thus $\operatorname{fix}(F)$ is a nonempty and compact R_δ-set. Therefore our claim is valid and this completes the proof of Theorem 9.29. □

9.11 Approximation of Fixed Points

In this section, which is based on [53], we consider iterative schemes for approximating fixed points of closed-valued strict contractions in metric spaces.

Throughout this and the next section of this chapter, (X, ρ) is a complete metric space and $T : X \to 2^X \setminus \{\emptyset\}$ is a strict contraction such that $T(x)$ is a closed set for each $x \in X$. Thus T satisfies

$$h\big(T(x), T(y)\big) \leq c\rho(x, y) \quad \text{for all } x, y \in X, \tag{9.91}$$

where $0 \leq c < 1$.

For each $x \in X$ and each nonempty set $A \subset X$, let

$$\rho(x, A) = \inf\big\{\rho(x, y) : y \in A\big\}.$$

Theorem 9.30 *Let $T : X \to 2^X \setminus \{\emptyset\}$ be a strict contraction such that $T(x)$ is a closed set for each $x \in X$ and T satisfies (9.91) with $0 \leq c < 1$. Assume that $x_0 \in X$, $\{\varepsilon_i\}_{i=0}^\infty \subset (0, \infty)$, $\sum_{i=0}^\infty \varepsilon_i < \infty$, and that for each integer $i \geq 0$,*

$$x_{i+1} \in T(x_i), \qquad \rho(x_i, x_{i+1}) \leq \rho\big(x_i, T(x_i)\big) + \varepsilon_i. \tag{9.92}$$

Then $\{x_i\}_{i=0}^\infty$ converges to a fixed point of T.

Proof First, we claim that $\{x_i\}_{i=0}^{\infty}$ is a Cauchy sequence. Indeed, let $i \ge 0$ be an integer. Then by (9.92) and (9.91),

$$\rho(x_{i+1}, x_{i+2}) \le \rho\big(x_{i+1}, T(x_{i+1})\big) + \varepsilon_{i+1} \le h\big(T(x_i), T(x_{i+1})\big) + \varepsilon_{i+1}$$

and

$$d(x_{i+1}, x_{i+2}) \le c\rho(x_i, x_{i+1}) + \varepsilon_{i+1}. \tag{9.93}$$

By (9.93),

$$\rho(x_1, x_2) \le c\rho(x_0, x_1) + \varepsilon_1$$

and

$$\rho(x_2, x_3) \le c\rho(x_1, x_2) + \varepsilon_2 \le c^2\rho(x_0, x_1) + c\varepsilon_1 + \varepsilon_2. \tag{9.94}$$

Now we use induction to show that for each integer $n \ge 1$,

$$\rho(x_n, x_{n+1}) \le c^n\rho(x_0, x_1) + \sum_{i=0}^{n-1} c^i\varepsilon_{n-i}. \tag{9.95}$$

In view of (9.94), inequality (9.95) is valid for $n = 1, 2$.

Assume that $k \ge 1$ is an integer and that (9.95) holds for $n = k$. When combined with (9.93), this implies that

$$\rho(x_{k+1}, x_{k+2}) \le c\rho(x_k, x_{k+1}) + \varepsilon_{k+1} \le c^{k+1}\rho(x_0, x_1) + \sum_{i=0}^{k-1} c^{i+1}\varepsilon_{k-i} + \varepsilon_{k+1}$$

$$= c^{k+1}\rho(x_0, x_1) + \sum_{i=0}^{k} c^i\varepsilon_{k+1-i}.$$

Thus (9.95) holds with $n = k + 1$ and therefore (9.95) holds for all integers $n \ge 1$. By (9.95),

$$\sum_{n=1}^{\infty}\rho(x_n, x_{n+1}) \le \sum_{n=1}^{\infty}\left(c^n\rho(x_0, x_1) + \sum_{i=1}^{n} c^{n-i}\varepsilon_i\right)$$

$$\le \rho(x_0, x_1)\sum_{n=1}^{\infty} c^n + \sum_{i=1}^{\infty}\left(\sum_{j=0}^{\infty} c^j\right)\varepsilon_i$$

$$\le \left(\sum_{n=0}^{\infty} c^n\right)\left[\rho(x_0, x_1) + \sum_{n=1}^{\infty}\varepsilon_n\right] < \infty.$$

Thus $\{x_n\}_{n=0}^{\infty}$ is indeed a Cauchy sequence and there exists

$$x_* = \lim_{n\to\infty} x_n. \tag{9.96}$$

We claim that $x_* \in T(x_*)$. Let $\varepsilon > 0$ be given. By (9.96), there is an integer $n_0 \geq 1$ such that for each integer $n \geq n_0$,

$$\rho(x_n, x_*) \leq \varepsilon/8. \tag{9.97}$$

Let $n \geq n_0$ be an integer. By (9.91),

$$h\big(T(x_n), T(x_*)\big) \leq c\rho(x_n, x_*) \leq c\varepsilon/8. \tag{9.98}$$

By (9.92),

$$x_{n+1} \in T(x_n).$$

When combined with (9.98), this implies that

$$\rho\big(x_{n+1}, T(x_*)\big) \leq c\varepsilon/8.$$

Hence there is

$$y \in T(x_*) \tag{9.99}$$

such that $\rho(x_{n+1}, y) \leq \varepsilon c/4$. Together with (9.97) and (9.99), this implies that

$$\rho\big(x_*, T(x_*)\big) \leq \rho(x_*, y) \leq \rho(x_*, x_{n+1}) + \rho(x_{n+1}, y) \leq \varepsilon/8 + \varepsilon/4.$$

Since ε is an arbitrary positive number, we conclude that

$$x_* \in T(x_*),$$

as claimed. Theorem 9.30 is proved. □

Theorem 9.31 *Let $T : X \to 2^X \setminus \{\emptyset\}$ be a strict contraction such that $T(x)$ is a closed set for all $x \in X$ and T satisfies (9.91) with $0 \leq c < 1$. Let $\varepsilon > 0$ be given. Then there exists $\delta > 0$ such that if $x \in X$ and $\rho(x, T(x)) < \delta$, then there is $\bar{x} \in X$ such that $\bar{x} \in T(\bar{x})$ and $\rho(x, \bar{x}) \leq \varepsilon$.*

Proof Choose a positive number δ such that

$$4\delta(1-c)^{-1} < \varepsilon. \tag{9.100}$$

Consider

$$x \in X \quad \text{such that} \quad \rho\big(x, T(x)\big) < \delta. \tag{9.101}$$

Set

$$x_0 = x. \tag{9.102}$$

By (9.101), there is

$$x_1 \in T(x_0) \tag{9.103}$$

such that

$$\rho(x_0, x_1) < \delta. \tag{9.104}$$

For each integer $n \geq 1$, choose

$$x_{n+1} \in T(x_n) \tag{9.105}$$

such that

$$\rho(x_{n+1}, x_n) \leq \rho\big(x_n, T(x_n)\big)(1+c)/(2c). \tag{9.106}$$

By (9.91), (9.103), (9.105) and (9.106), for each integer $n \geq 1$,

$$\rho(x_n, x_{n+1}) \leq (1+c)h\big(T(x_{n-1}), T(x_n)\big)/(2c) \leq \big((1+c)/2\big)\rho(x_n, x_{n-1}).$$

When combined with (9.104), this implies that for each integer $n \geq 1$,

$$\rho(x_n, x_{n+1}) \leq \big[(1+c)/2\big]^n \rho(x_0, x_1) \leq \big[(1+c)/2\big]^n \delta. \tag{9.107}$$

Therefore

$$\sum_{n=0}^{\infty} \rho(x_n, x_{n+1}) < \infty,$$

$\{x_n\}_{n=0}^{\infty}$ is a Cauchy sequence and there exists $\bar{x} \in X$ such that

$$\bar{x} = \lim_{n\to\infty} x_n. \tag{9.108}$$

Since $x_{n+1} \in T(x_n)$ for all integers $n \geq 0$, (9.108) implies that

$$\bar{x} \in T(\bar{x}).$$

By (9.100), (9.107) and (9.108),

$$\rho(x_0, \bar{x}) = \lim_{n\to\infty} \rho(x_0, x_n) \leq \sum_{n=0}^{\infty} \rho(x_i, x_{i+1})$$

$$\leq \sum_{i=0}^{\infty} [(1+c)/2]^i \delta = 2\delta/(1-c) < \varepsilon/2.$$

This completes the proof of Theorem 9.31. □

The conclusions of the following two theorems hold uniformly for all those relevant sequences $\{x_i\}_{i=0}^{\infty}$ the initial point of which lies in a closed ball of center $\theta \in X$ and radius $M > 0$.

Theorem 9.32 *Let $T : X \to 2^X \setminus \{\emptyset\}$ be a strict contraction such that $T(x)$ is a closed set for all $x \in X$ and T satisfies* (9.91) *with $0 \leq c < 1$. Fix $\theta \in X$. Let $\varepsilon > 0$*

and $M > 0$ *be given. Then there exist* $\delta \in (0, \varepsilon)$ *and an integer* $n_0 \geq 1$ *with the following property*:

for each sequence $\{x_i\}_{i=0}^{\infty} \subset X$ *such that* $\rho(x_0, \theta) \leq M$ *and such that for each integer* $n \geq 0$,

$$x_{n+1} \in T(x_n) \quad \textit{and} \quad \rho(x_{n+1}, x_n) \leq \delta + \rho\big(x_n, T(x_n)\big),$$

we have

$$\rho(x_{n+1}, x_n) < \varepsilon \quad \textit{for all integers } n \geq n_0.$$

Proof Choose $\delta \in (0, 1)$ such that

$$\delta(1-c)^{-1} < \varepsilon/2 \tag{9.109}$$

and a natural number n_0 such that

$$c^{n_0}\big(2M + 1 + \rho\big(\theta, a(\theta)\big)\big) < \varepsilon/2. \tag{9.110}$$

Let $x_0 \in X$,

$$\{x_n\}_{n=0}^{\infty} \subset X, \qquad \rho(x_0, \theta) \leq M, \tag{9.111}$$

and assume that for each integer $n \geq 0$,

$$x_{n+1} \in T(x_n), \qquad \rho(x_{n+1}, x_n) \leq \rho\big(x_n, T(x_n)\big) + \delta. \tag{9.112}$$

We now estimate $\rho(x_0, T(x_0))$. By (9.91) and (9.111),

$$\begin{aligned} \rho\big(x_0, T(x_0)\big) &\leq \rho(x_0, \theta) + \rho\big(\theta, T(\theta)\big) + h\big(T(\theta), T(x_0)\big) \\ &\leq \rho(x_0, \theta) + \rho\big(\theta, T(\theta)\big) + \rho(\theta, x_0) \leq 2M + \rho\big(\theta, T(\theta)\big). \end{aligned} \tag{9.113}$$

By (9.112) and (9.113),

$$\rho(x_0, x_1) \leq \rho\big(x_0, T(x_0)\big) + \delta \leq 2M + 1 + \rho\big(\theta, T(\theta)\big). \tag{9.114}$$

By (9.112) and (9.91), for each integer $n \geq 0$,

$$\begin{aligned} \rho(x_{n+2}, x_{n+1}) &\leq \rho\big(x_{n+1}, T(x_{n+1})\big) + \delta \\ &\leq h\big(T(x_n), T(x_{n+1})\big) + \delta \leq c\rho(x_n, x_{n+1}) + \delta. \end{aligned} \tag{9.115}$$

Next, we show by induction that for each integer $n \geq 1$,

$$\rho(x_{n+1}, x_n) \leq \delta \sum_{i=0}^{n-1} c^i + c^n \rho(x_0, x_1). \tag{9.116}$$

By (9.115), inequality (9.116) holds for $n = 1$. Assume that $k \geq 1$ is an integer and that (9.116) holds with $n = k$. Then by (9.115),

$$\rho(x_{k+2}, x_{k+1}) \leq c\rho(x_k, x_{k+1}) + \delta \leq \delta \sum_{i=0}^{k} c^i + c^{k+1}\rho(x_0, x_1).$$

Thus (9.116) holds with $n = k + 1$ and therefore it holds for all integers $n \geq 1$. By (9.116) and (9.114), for all natural numbers n,

$$\rho(x_{n+1}, x_n) \leq \delta(1-c)^{-1} + c^n\big(2M + 1 + \rho\big(\theta, T(\theta)\big)\big). \tag{9.117}$$

Finally, by (9.117), (9.109) and (9.110), we obtain, for all integers $n \geq n_0$,

$$\rho(x_n, x_{n+1}) \leq \delta(1-c)^{-1} + c^{n_0}\big(2M + 1 + \rho\big(\theta, T(\theta)\big)\big) < \varepsilon.$$

Theorem 9.32 is proved. □

Theorems 9.30 and 9.31 imply the following additional result.

Theorem 9.33 *Let $T : X \to 2^X \setminus \{\emptyset\}$ be a strict contraction such that $T(x)$ is a closed set for all $x \in X$ and T satisfies (9.91) with $0 \leq c < 1$. Let positive numbers ε and M be given. Then there exist $\delta > 0$ and an integer $n_0 \geq 1$ such that if a sequence $\{x_i\}_{i=0}^{\infty} \subset X$ satisfies*

$$\rho(x_0, \theta) \leq M, \qquad x_{n+1} \in T(x_n) \quad \text{and} \quad \rho(x_n, x_{n+1}) \leq \rho\big(x_n, T(x_n)\big) + \delta$$

for all integers $n \geq 0$, then for each integer $n \geq n_0$, there is a point $y \in X$ such that $y \in T(y)$ and $\rho(y, x_n) < \varepsilon$.

The following example shows that Theorem 9.33 cannot be improved in the sense that the fixed point y, the existence of which is guaranteed by the theorem, is not, in general, the same for all integers $n \geq n_0$.

Example 9.34 Let $X = [0, 1]$, $\rho(x, y) = |x - y|$ and $T(x) = [0, 1]$ for all $x \in [0, 1]$. Let $\delta > 0$ be given. Choose a natural number k such that $1/k < \delta$. Put

$$x_0 = 0, \qquad x_i = i/k, \quad i = 0, \dots, k,$$
$$x_{i+k} = 1 - i/k, \quad i = 0, \dots, k,$$

and for all integers $p \geq 0$ and any $i \in \{0, \dots, 2k\}$, put

$$x_{2pk+i} = x_i.$$

Then $\{x_i\}_{i=0}^{\infty} \subset X$ and for any integer $i \geq 0$, we have

$$x_{i+1} \in T(x_i) \quad \text{and} \quad |x_i - x_{i+1}| \leq k^{-1} < \delta.$$

On the other hand, for all $x \in X$ and any integer $p \geq 0$,

$$\max\{|x - x_i| : i = 2kp, \ldots, 2pk + 2k\} \geq 1/2.$$

9.12 Approximating Fixed Points in Caristi's Theorem

We begin this section by recalling the following two versions of Caristi's fixed point theorem [36].

Theorem 9.35 ([82], Theorem 3.9) *Suppose that (X, ρ) is a complete metric space and $T : X \to X$ is a continuous mapping which satisfies for some $\phi : X \to [0, \infty)$,*

$$\rho(x, Tx) \leq \phi(x) - \phi(Tx), \quad x \in X.$$

Then $\{T^n x\}_{n=1}^{\infty}$ converges to a fixed point of T for each $x \in X$.

Theorem 9.36 ([82], Theorem 4.1) *Suppose that (X, ρ) is a complete metric space, $\phi : X \to R^1$ is a lower semicontinuous function which is bounded from below, and $T : X \to X$ satisfies*

$$\rho(x, Tx) \leq \phi(x) - \phi(Tx), \quad x \in X.$$

Then T has a fixed point.

We now present and prove a set-valued analog of Caristi's theorem with computational errors.

Theorem 9.37 *Assume that (X, ρ) is a complete metric space, $T : X \to 2^X \setminus \{\emptyset\}$,* graph$(T) := \{(x, y) \in X \times X : y \in T(x)\}$ *is closed, $\phi : X \to R^1 \cup \{\infty\}$ is bounded from below, and that for each $x \in X$,*

$$\inf\{\phi(y) + \rho(x, y) : y \in T(x)\} \leq \phi(x). \tag{9.118}$$

Let $\{\varepsilon_n\}_{n=0}^{\infty} \subset (0, \infty)$, $\sum_{n=0}^{\infty} \varepsilon_n < \infty$, and let $x_0 \in X$ satisfy $\phi(x_0) < \infty$. Assume that for each integer $n \geq 0$,

$$x_{n+1} \in T(x_n) \tag{9.119}$$

and

$$\phi(x_{n+1}) + \rho(x_n, x_{n+1}) \leq \inf\{\phi(y) + \rho(x, y) : y \in T(x_n)\} + \varepsilon_n. \tag{9.120}$$

Then $\{x_n\}_{n=0}^{\infty}$ converges to a fixed point of T.

Proof Clearly, $\phi(x_n) < \infty$ for all integers $n \geq 0$. By (9.120), for each integer $n \geq 0$,

$$\begin{aligned}\rho(x_n, x_{n+1}) &\leq -\phi(x_{n+1}) + \varepsilon_n + \inf\bigl\{\phi(y) + \rho(x, y) : y \in T(x_n)\bigr\} \\ &\leq -\phi(x_{n+1}) + \phi(x_n) + \varepsilon_n. \end{aligned} \tag{9.121}$$

By (9.121), for each integer $m \geq 1$,

$$\begin{aligned}\sum_{i=0}^{m} \rho(x_i, x_{i+1}) &\leq \phi(x_0) - \phi(x_m) + \sum_{i=0}^{\infty} \varepsilon_i \\ &\leq \phi(x_0) - \inf(\phi) + \sum_{i=0}^{\infty} \varepsilon_i < \infty.\end{aligned}$$

Thus $\{x_i\}_{i=0}^{\infty}$ is a Cauchy sequence and there exists $\bar{x} = \lim_{i \to \infty} x_i$. Since the graph of T is closed, it follows that

$$(\bar{x}, \bar{x}) = \lim_{i \to \infty} (x_i, x_{i+1}) \in \mathrm{graph}(T).$$

This completes the proof of Theorem 9.37. □

Chapter 10
Minimal Configurations in the Aubry-Mather Theory

10.1 Preliminaries

In this chapter, which is based on [181], we study (h)-minimal configurations in the Aubry-Mather theory, where $h : R^2 \to R^1$ belongs to a complete metric space of functions $\mathfrak{M}$. Such minimal configurations have a definite rotation number. We establish the existence of a set $\mathcal{F} \subset \mathfrak{M}$, which is a countable intersection of open and everywhere dense subsets of $\mathfrak{M}$, and such that, for each $h \in \mathcal{F}$ and each rational number $\alpha = p/q$ with p and q relatively prime, the following properties hold: (i) there exist (h)-minimal configurations $x^{(+)}$, $x^{(-)}$ and $x^{(0)}$ with rotation number α such that $x^{(+)}_{i-q} + p > x^{(+)}_i$, $x^{(-)}_{i-q} + p < x^{(-)}_i$ and $x^{(0)}_{i-q} + p = x^{(0)}_i$ for all integers i; (ii) any (h)-minimal configuration with rotation number α is a translation of one of the configurations $x^{(+)}, x^{(-)}, x^{(0)}$.

Let Z be the set of all integers. A configuration is a bi-infinite sequence $x = (x_i)_{i \in Z} \in R^Z$. The set R^Z will be endowed with the product topology and the partial order defined by $x < y$ if and only if $x_i < y_i$ for all $i \in Z$.

There is an order preserving action $T : Z^2 \times R^Z \to R^Z$ defined by

$$T(k, x) = T_k x = y \quad \text{iff} \quad k = (k_1, k_2) \in Z^2,$$
$$x, y \in R^Z \text{ and } y_i = x_{i-k_1} + k_2 \text{ for all } i \in Z. \tag{10.1}$$

Let $x, y \in R^Z$. We say that y is a translation of x if there is $n = (n_1, n_2) \in Z^2$ such that $y = T_n x$.

Let $h : R^2 \to R^1$ be a continuous function. We extend h to arbitrary finite segments $(x_j, \ldots, x_k)$, $j < k$, of configurations $x \in R^Z$ by

$$h(x_j, \ldots, x_k) := \sum_{i=j}^{k-1} h(x_i, x_{i+1}). \tag{10.2}$$

A segment $(x_j, \ldots, x_k)$ is called (h)-minimal if

S. Reich, A.J. Zaslavski, *Genericity in Nonlinear Analysis*,
Developments in Mathematics 34, DOI 10.1007/978-1-4614-9533-8_10,

$$h(x_j, \dots, x_k) \le h(y_j, \dots, y_k)$$

whenever $x_j = y_j$ and $x_k = y_k$.

We assume that h has the following properties [14, 15]:

(H1) For all $(\xi, \eta) \in R^2$, $h(\xi + 1, \eta + 1) = h(\xi, \eta)$.
(H2) $\lim_{|\eta| \to \infty} h(\xi, \xi + \eta) = \infty$, uniformly in ξ.
(H3) If $\xi_1 < \xi_2$, $\eta_1 < \eta_2$, then

$$h(\xi_1, \eta_1) + h(\xi_2, \eta_2) < h(\xi_1, \eta_2) + h(\xi_2, \eta_1).$$

(H4) If $(x_{-1}, x_0, x_1) \ne (y_{-1}, y_0, y_1)$ are (h)-minimal segments and $x_0 = y_0$, then

$$(x_{-1} - y_{-1})(x_1 - y_1) < 0.$$

A configuration $x \in R^Z$ is (h)-minimal if for each pair of integers j, k satisfying $j < k$ and each finite segment $\{y_i\}_{i=j}^k \subset R^1$ satisfying $y_j = x_j$ and $y_k = x_k$, the inequality $h(x_j, \dots, x_k) \le h(y_j, \dots, y_k)$ holds. Denote by $\mathcal{M}(h)$ the set of all (h)-minimal configurations. It is known that the set $\mathcal{M}(h)$ is closed [12, 14].

We briefly review the definitions, notions and some basic results from the Aubry-Mather theory [12, 14].

We say that $x \in R^Z$ and $x^* \in R^Z$ cross

(a) at $i \in Z$ if $x_i = x_i^*$ and $(x_{i-1} - x_{i-1}^*)(x_{i+1} - x_{i+1}^*) < 0$;
(b) between i and $i + 1$ if $(x_i - x_i^*)(x_{i+1} - x_{i+1}^*) < 0$.

We say that $x \in R^Z$ is periodic with period $(q, p) \in (Z \setminus \{0\}) \times Z$ if $T_{(q,p)}x = x$.

Remark 10.1 Assume that $h = h(\xi_1, \xi_2) \in C^2(R^2)$ and $(\partial^2 h/\partial\xi_1\, \partial\xi_2)(u, v) < 0$ for all $(u, v) \in R^2$. It is not difficult to show that (H3) and (H4) hold. Moreover, we can show that if $h \in C^2(R^2)$, then (H3) holds if and only if

$$\{(u, v) \in R^2 : (\partial^2 h/\partial\xi_1\, \partial\xi_2)(u, v) < 0\}$$

is an everywhere dense subset of R^2.

We recall the following result (see Corollary 3.16 and Theorem 3.17 of [14]).

Proposition 10.2 *There exists a continuous function $\alpha^{(h)} : \mathcal{M}(h) \to R^1$ with the following properties*:

For all $x \in \mathcal{M}(h)$ and $i \in Z$ we have

$$|x_i - x_0 - i\alpha^{(h)}(x)| < 1.$$

If $x \in \mathcal{M}(h)$ is periodic with period $(q, p) \in Z^2$, then $\alpha^{(h)}(x) = p/q$.

For all $\alpha \in R^1$, the set $\{x \in \mathcal{M}(h) : \alpha^{(h)}(x) = \alpha\} \ne \emptyset$.

Remark 10.3 We call $\alpha^{(h)}(x)$ the rotation number of $x \in \mathcal{M}(h)$.

For each $\alpha \in R^1$, define

$$\mathcal{M}(h,\alpha) := \{x \in \mathcal{M}(h) : \alpha^{(h)}(x) = \alpha\}. \tag{10.3}$$

We study $\mathcal{M}(h,\alpha)$ with rational $\alpha \in R^1$.

Let a rational number $\alpha = p/q$ be an irreducible fraction, where $q \geq 1$ and p are integers. Denote by $\mathcal{M}^{per}(h,\alpha)$ the set of all periodic (h)-minimal configurations $x \in \mathcal{M}(h,\alpha)$ which satisfy $T_{(q,p)}x = x$, equivalently, $x_{i-q} + p = x_i$ for all $i \in Z$. For the proof of the following result see [12, 14].

Proposition 10.4 *$\mathcal{M}^{per}(h,\alpha)$ is a nonempty, closed and totally ordered set. Moreover, if $x \in \mathcal{M}^{per}(h,\alpha)$, then x is a minimizer of $h_{qp} : P_{qp} \to R^1$, where*

$$h_{qp}(x) = h(x_0,\dots,x_q), \qquad P_{qp} = \{x \in R^Z : T_{(q,p)}x = x\}. \tag{10.4}$$

Two elements of $\mathcal{M}^{per}(h,\alpha)$ are called (h)-neighboring if there does not exist an element of $\mathcal{M}^{per}(h,\alpha)$ between them. The following two propositions describe the structure of the set $\mathcal{M}(h,\alpha)$. For their proofs see [14].

Proposition 10.5 *Suppose that $x^- < x^+$ are (h)-neighboring elements of the set $\mathcal{M}^{per}(h,\alpha)$. Then there exist $y^{(1)}, y^{(2)} \in \mathcal{M}(h,\alpha)$ such that*

$$x^- < y^{(1)} < x^+, \qquad x^- < y^{(2)} < x^+,$$

$$\lim_{i\to-\infty} y_i^{(1)} - x_i^- = 0, \qquad \lim_{i\to\infty} y_i^{(1)} - x_i^+ = 0,$$

$$\lim_{i\to-\infty} y_i^{(2)} - x_i^+ = 0, \qquad \lim_{i\to\infty} y_i^{(2)} - x_i^- = 0.$$

Suppose that $x^- < x^+$ are (h)-neighboring elements of $\mathcal{M}^{per}(h,\alpha)$. Define

$$\mathcal{M}^+(h,\alpha,x^-,x^+) = \left\{x \in \mathcal{M}(h,\alpha) : \lim_{i\to-\infty} x_i - x_i^- = 0, \lim_{i\to\infty} x_i - x_i^+ = 0\right\},$$

$$\mathcal{M}^-(h,\alpha,x^-,x^+) = \left\{x \in \mathcal{M}(h,\alpha) : \lim_{i\to-\infty} x_i - x_i^+ = 0, \lim_{i\to\infty} x_i - x_i^- = 0\right\}.$$

We denote by $\mathcal{M}^+(h,\alpha)$ (respectively, $\mathcal{M}^-(h,\alpha)$) the union of the sets $\mathcal{M}^+(h,\alpha,x^-,x^+)$ (respectively, $\mathcal{M}^-(h,\alpha,x^-,x^+)$) extended over all pairs of (h)-neighboring elements $x^- < x^+$ of $\mathcal{M}^{per}(h,\alpha)$.

Proposition 10.6

1. *If $x \in \mathcal{M}^-(h,\alpha,x^-,x^+) \cup \mathcal{M}^+(h,\alpha,x^-,x^+)$, where $x^-, x^+ \in \mathcal{M}^{per}(h,\alpha)$ are (h)-neighboring and $x^- < x^+$, then $x^- < x < x^+$.*
2. $\mathcal{M}(h,\alpha) = \mathcal{M}^{per}(h,\alpha) \cup \mathcal{M}^+(h,\alpha) \cup \mathcal{M}^-(h,\alpha)$.

3. *The sets* $\mathcal{M}^{per}(h,\alpha) \cup \mathcal{M}^{+}(h,\alpha)$ *and* $\mathcal{M}^{per}(h,\alpha) \cup \mathcal{M}^{-}(h,\alpha)$ *are totally ordered.*
4. $\mathcal{M}^{+}(h,\alpha) = \{x \in \mathcal{M}(h,\alpha) : x > T_{(q,p)}x\}$,

$$\mathcal{M}^{-}(h,\alpha) = \left\{x \in \mathcal{M}(h,\alpha) : x < T_{(q,p)}x\right\}.$$

Let $k \geq 2$ be an integer. In this chapter we consider a complete metric space of functions $h : R^2 \to R^1$ which belong to $C^k(R^2)$. This space is defined in the next section and is denoted by $\mathfrak{M}_k$. We prove the existence of a set $\mathcal{F} \subset \mathfrak{M}_k$, which is a countable intersection of open and everywhere dense subsets of $\mathfrak{M}_k$, and such that for each $h \in \mathcal{F}$ and each rational number $\alpha = p/q$ with p and q relatively prime, the following properties hold:

(i) there exist (h)-minimal configurations $x^{(+)}$, $x^{(-)}$ and $x^{(0)}$ with rotation number α such that $x^{(+)}_{i-q} + p > x^{(+)}_i$, $x^{(-)}_{i-q} + p < x^{(-)}_i$ and $x^{(0)}_{i-q} + p = x^{(0)}_i$ for all integers i;
(ii) any (h)-minimal configuration with rotation number α is a translation of one of the configurations $x^{(+)}$, $x^{(-)}$, $x^{(0)}$.

This result was obtained in [181].

10.2 Spaces of Functions

Let $k \geq 2$ be an integer. For $f = f(x_1, x_2) \in C^k(R^2)$ and $q = (q_1, q_2) \in \{0, \ldots, k\}^2$ satisfying $q_1 + q_2 \leq k$, we set

$$|q| = q_1 + q_2, \qquad D^q f = \partial^{|q|} f / \partial x_1^{q_1} \, \partial x_2^{q_2}.$$

Denote by $\mathfrak{M}_k$ the set of all $h \in C^k(R^2)$ which have property (H1), satisfy

$$\left(\partial^2 h / \partial x_1 \, \partial x_2\right)(\xi_1, \xi_2) \leq 0 \quad \text{for all } (\xi_1, \xi_2) \in R^2 \tag{10.5}$$

and also have the following property:

(H5) There exist $\delta_h \in (0, 1)$ and $c_h > 0$ such that

$$h(x_1, x_2) \geq \delta_h (x_1 - x_2)^2 - c_h \quad \text{for all } (x_1, x_2) \in R^2.$$

It is clear that (H5) implies (H2).

Denote by $\mathfrak{M}_{k0}$ the set of all $h \in \mathfrak{M}_k$ such that

$$\left(\partial^2 h / \partial x_1 \, \partial x_2\right)(\xi_1, \xi_2) < 0 \quad \text{for all } (\xi_1, \xi_2) \in R^2. \tag{10.6}$$

For each $N, \varepsilon > 0$, we set

$$\begin{aligned} E_k(N,\varepsilon) = \big\{&(h_1,h_2) \in \mathfrak{M}_k \times \mathfrak{M}_k : \big|D^q h_1(x_1,x_2) - D^q h_2(x_1,x_2)\big| \le \varepsilon \\ &\text{for each } q \in \{0,\dots,k\}^2 \text{ satisfying } |q| \le k \\ &\text{and each } (x_1,x_2) \in R^2 \text{ satisfying } |x_1|, |x_2| \le N\big\} \\ &\cap \big\{(h_1,h_2) \in \mathfrak{M}_k \times \mathfrak{M}_k : \big|h_1(x_1,x_2) - h_2(x_1,x_2)\big| \\ &< \varepsilon + \varepsilon \max\big\{\big|h_1(x_1,x_2)\big|, \big|h_2(x_1,x_2)\big|\big\} \text{ for all } (x_1,x_2) \in R^2\big\}. \end{aligned} \tag{10.7}$$

Using the following simple lemma, we can easily show that for the set $\mathfrak{M}_k$, there exists a uniformity which is determined by the base $E_k(N,\varepsilon)$, $N, \varepsilon > 0$.

Lemma 10.7 *Let $a, b \in R^1$, $\varepsilon \in (0,1)$ such that $|a-b| < \varepsilon + \varepsilon \max\{|a|, |b|\}$. Then*

$$|a-b| < \varepsilon + \varepsilon^2(1-\varepsilon)^{-1} + \varepsilon(1-\varepsilon)^{-1} \min\big\{|a|, |b|\big\}.$$

It is not difficult to see that the uniformity determined by the base $E_k(N,\varepsilon)$, $N, \varepsilon > 0$, is metrizable (by a metric d_k) and complete. For the set $\mathfrak{M}_k$, we consider the topology induced by the metric d_2, which is called the weak topology, and the topology induced by the metric d_k, which is called the strong topology.

The following result, which was obtained in [181], shows that a generic function in $\mathfrak{M}_k$ belongs to $\mathfrak{M}_{k0}$ and, by Remark 10.1, has properties (H1)–(H4).

Theorem 10.8 *There exists a set $\mathcal{F}_0 \subset \mathfrak{M}_{k0}$, which is a countable intersection of open (in the weak topology) and everywhere dense (in the strong topology) subsets of $\mathfrak{M}_k$.*

Proof For $h \in \mathfrak{M}_k$ and $\gamma \in (0,1)$, define $h_\gamma : R^2 \to R^1$ by

$$h_\gamma(x_1,x_2) = h(x_1,x_2) + \gamma(x_1 - x_2)^2, \quad (x_1,x_2) \in R^2.$$

It is easy to see that for $h \in \mathfrak{M}_k$ and $\gamma \in (0,1)$, $h_\gamma \in \mathfrak{M}_{k0}$ and

$$\big(\partial^2 h_\gamma / \partial x_1 \, \partial x_2\big)(\xi_1,\xi_2) \le -2\gamma, \quad (\xi_1,\xi_2) \in R^2 \tag{10.8}$$

and $h_\gamma \to h$ as $\gamma \to 0^+$ in the strong topology.

Let $f \in \mathfrak{M}_k$, $\gamma \in (0,1)$ and let $i \ge 1$ be an integer. By (10.7) and (10.8), there exists an open neighborhood $\mathcal{U}(f,\gamma,i)$ of f_γ in $\mathfrak{M}_k$ with the weak topology such that the following property holds:

(P1) For each $g \in \mathcal{U}(f,\gamma,i)$ and each $(\xi_1,\xi_2) \in R^2$ satisfying $|\xi_1|, |\xi_2| \le i$, the inequality $\partial^2 g/\partial x_1 \, \partial x_2(\xi_1,\xi_2) \le -\gamma$ holds.

Define

$$\mathcal{F}_0 := \bigcap_{n=1}^{\infty} \bigcup \big\{\mathcal{U}(f,\gamma,i) : f \in \mathfrak{M}_k, \gamma \in (0,1), i \ge n\big\}.$$

Clearly, $\mathcal{F}_0$ is a countable intersection of open (in the weak topology) and everywhere dense (in the strong topology) subsets of $\mathfrak{M}_k$. We claim that $\mathcal{F}_0 \subset \mathfrak{M}_{k0}$. Let $h \in \mathcal{F}_0$, $(\xi_1, \xi_2) \in R^2$. Choose a natural number n such that $|\xi_1| + |\xi_2| < n$. There exist $f \in \mathfrak{M}_k$, $\gamma \in (0, 1)$ and an integer $i \geq n$ such that $h \in \mathcal{U}(f, \gamma, i)$. It follows from property (P1) and the choice of n that $(\partial^2 h/\partial x_1\, \partial x_2)(\xi_1, \xi_2) \leq -\gamma$. Therefore $h \in \mathfrak{M}_{k0}$. This completes the proof of Theorem 10.8. □

10.3 The Main Results

In the subsequent sections we prove the following result [181].

Theorem 10.9 *Let $k \geq 2$ be an integer and α be a rational number. Then there exists a set $\mathcal{F}_\alpha \subset \mathfrak{M}_{k0}$, which is a countable intersection of open (in the weak topology) and everywhere dense (in the strong topology) subsets of $\mathfrak{M}_k$ such that, for each $f \in \mathcal{F}_\alpha$, the following assertions hold:*

1. *If $x, y \in \mathcal{M}^{per}(f, \alpha)$, then there exist integers m, n such that $y_i = x_{i-m} + n$ for all $i \in Z$.*
2. *If $x, y \in \mathcal{M}^{+}(f, \alpha)$, then there exist integers m, n such that $y_i = x_{i-m} + n$ for all $i \in Z$.*
3. *If $x, y \in \mathcal{M}^{-}(f, \alpha)$, then there exist integers m, n such that $y_i = x_{i-m} + n$ for all $i \in Z$.*

It is not difficult to see that Theorem 10.9 implies the following result.

Theorem 10.10 *Let $k \geq 2$ be an integer. Then there exists a set $\mathcal{F} \subset \mathfrak{M}_{k0}$, which is a countable intersection of open (in the weak topology) and everywhere dense (in the strong topology) subsets of $\mathfrak{M}_k$, such that for each rational number α and each $f \in \mathcal{F}$, assertions 1–3 of Theorem 10.9 hold.*

Theorem 10.9 follows from the next two propositions.

Proposition 10.11 *Let $k \geq 2$ be an integer and α be a rational number. Then there exists a set $\mathcal{F}_{\alpha^+} \subset \mathfrak{M}_{k0}$, which is a countable intersection of open (in the weak topology) and everywhere dense (in the strong topology) subsets of $\mathfrak{M}_k$, such that for each $f \in \mathcal{F}_{\alpha^+}$, assertions 1 and 2 of Theorem 10.9 hold.*

Proposition 10.12 *Let $k \geq 2$ be an integer and α be a rational number. Then there exists a set $\mathcal{F}_{\alpha^-} \subset \mathfrak{M}_{k0}$, which is a countable intersection of open (in the weak topology) and everywhere dense (in the strong topology) subsets of $\mathfrak{M}_k$, such that for each $f \in \mathcal{F}_{\alpha^-}$, assertions 1 and 3 of Theorem 10.9 hold.*

We prove Proposition 10.11. Proposition 10.12 can be proved analogously.

10.4 Preliminary Results for Assertion 1 of Theorem 10.9

Let $m \geq 1$ be an integer. Consider the manifold $(R^1/Z)^m$ and the canonical mapping $P_m : R^m \to (R^1/Z)^m$. We first recall the following result (see Proposition 6.2 of [178]).

Proposition 10.13 *Let Ω be a closed subset of $(R^1/Z)^2$. Then there exists a non-negative function $\phi \in C^\infty((R^1/Z)^2)$ such that $\Omega = \{x \in (R^1/Z)^2 : \phi(x) = 0\}$.*

Corollary 10.14 *Let Ω be a closed subset of R^1/Z. Then there exists a nonnegative function $\phi \in C^\infty(R^1/Z)$ such that $\Omega = \{x \in R^1/Z : \phi(x) = 0\}$.*

In this section we assume that $k \geq 2$ is an integer and $\alpha = p/q$ is an irreducible fraction where $q \geq 1$ and p are integers.

For each $f \in \mathfrak{M}_{k0}$, define

$$E_\alpha(f) = \sum_{i=0}^{q-1} f(x_i, x_{i+1}) \quad \text{where } x \in \mathcal{M}^{per}(f, \alpha) \tag{10.9}$$

(see Proposition 10.4).

Proposition 10.15 *Let $f \in \mathfrak{M}_k$, Q be a natural number and let $D, \varepsilon > 0$ be given. Then there exists a neighborhood $\mathcal{U}$ of f in $\mathfrak{M}_k$ with the weak topology such that for each $g \in \mathcal{U}$, each pair of integers $n_1, n_2 \in [n_1+1, n_1+Q]$ and each sequence $\{x_i\}_{i=n_1}^{n_2} \subset R^1$ which satisfies*

$$\min\left\{\sum_{i=n_1}^{n_2-1} f(x_i, x_{i+1}), \sum_{i=n_1}^{n_2-1} g(x_i, x_{i+1})\right\} \leq D, \tag{10.10}$$

the inequality

$$\left|\sum_{i=n_1}^{n_2-1} f(x_i, x_{i+1}) - \sum_{i=n_1}^{n_2-1} g(x_i, x_{i+1})\right| \leq \varepsilon$$

holds.

Proof By (H5), there exist $\delta_0 \in (0, 1)$ and $c_0 > 0$ such that

$$f(x_1, x_2) \geq \delta_0(x_1 - x_2)^2 - c_0 \quad \text{for all } (x_1, x_2) \in R^2. \tag{10.11}$$

Choose a positive number ε_1 for which

$$\varepsilon_1[Q + c_0 Q + D] < 4^{-1}\min\{1, \varepsilon\} \tag{10.12}$$

and a positive number $\varepsilon_0 < 1$ which satisfies

$$\varepsilon_0 + \varepsilon_0^2(1-\varepsilon_0)^{-1} + \varepsilon_0(1-\varepsilon_0)^{-1} < 4^{-1}\varepsilon_1. \tag{10.13}$$

Define

$$\mathcal{U} = \{g \in \mathfrak{M}_k : (f,g) \in E_k(1,\varepsilon_0)\} \tag{10.14}$$

(see (10.7)).

Assume that $g \in \mathcal{U}$, $n_1, n_2 \in Z$, $n_2 \in [n_1+1, n_1+Q]$, $\{x_i\}_{i=n_1}^{n_2} \subset R^1$ and that (10.10) holds. By (10.7) and (10.14), for every $(z_1, z_2) \in R^2$,

$$|f(z_1,z_2) - g(z_1,z_2)| < \varepsilon_0 + \varepsilon_0 \max\{|f(z_1,z_2)|, |g(z_1,z_2)|\}. \tag{10.15}$$

It follows from (10.15), (10.13) and Lemma 10.7 that for every $(z_1, z_2) \in R^2$,

$$\begin{aligned} |f(z_1,z_2) - g(z_1,z_2)| &< \varepsilon_0 + \varepsilon_0^2(1-\varepsilon_0)^{-1} \\ &\quad + \varepsilon_0(1-\varepsilon_0)^{-1} \min\{|f(z_1,z_2)|, |g(z_1,z_2)|\} \\ &< 4^{-1}\varepsilon_1 + 4^{-1}\varepsilon_1 \min\{|f(z_1,z_2)|, |g(z_1,z_2)|\}. \end{aligned} \tag{10.16}$$

Inequalities (10.16) and (10.11) imply that for every $(z_1, z_2) \in R^2$,

$$g(z_1,z_2) \ge f(z_1,z_2) - 4^{-1}\varepsilon_1 - 4^{-1}\varepsilon_1|f(z_1,z_2)| \ge -4^{-1}\varepsilon_1 - 2c_0. \tag{10.17}$$

Set

$$\lambda_i = \min\{f(x_i,x_{i+1}), g(x_i,x_{i+1})\}, \quad i = n_1, \dots, n_2-1. \tag{10.18}$$

It follows from (10.16), (10.11), (10.17) and (10.18) that for $i = n_1, \dots, n_2-1$,

$$\begin{aligned} &|f(x_i,x_{i+1}) - g(x_i,x_{i+1})| \\ &\quad < 4^{-1}\varepsilon_1 + 4^{-1}\varepsilon_1 \min\{f(x_i,x_{i+1}) + 2c_0, g(x_i,x_{i+1}) + 4c_0 + 2\} \\ &\quad \le 4^{-1}\varepsilon_1 + 4^{-1}\varepsilon_1\lambda_i + c_0\varepsilon_1 + \varepsilon_1/2. \end{aligned}$$

By these inequalities, (10.18), (10.10) and (10.12),

$$\begin{aligned} &\left|\sum_{i=n_1}^{n_2-1} \big(f(x_i,x_{i+1}) - g(x_i,x_{i+1})\big)\right| \\ &\quad \le (n_2-n_1)\big[4^{-1}\varepsilon_1 + 2^{-1}\varepsilon_1 + \varepsilon_1 c_0\big] + 4^{-1}\varepsilon_1 \sum_{i=n_1}^{n_2-1} \lambda_i \\ &\quad \le (n_2-n_1)[\varepsilon_1 + \varepsilon_1 c_0] + 4^{-1}\varepsilon_1 D \le Q(\varepsilon_1 + \varepsilon_1 c_0) + 4^{-1}\varepsilon_1 D < \varepsilon. \end{aligned}$$

This completes the proof of Proposition 10.15. □

Corollary 10.16 *Let $f \in \mathfrak{M}_{k0}$ and $\varepsilon > 0$ be given. Then there exists a neighborhood $\mathcal{U}$ of f in $\mathfrak{M}_k$ with the weak topology such that for each $g \in \mathcal{U} \cap \mathfrak{M}_{k0}$, $E_\alpha(g) \le E_\alpha(f) + \varepsilon$.*

Proposition 10.17 *Assume that $f \in \mathfrak{M}_{k0}$, $f_n \in \mathfrak{M}_{k0}$, $n = 1, 2, \dots$, $\lim_{n\to\infty} f_n = f$ in the weak topology,*

$$x^{(n)} \in \mathcal{M}(f_n), \quad n = 1, 2, \dots, x \in R^Z \quad \text{and} \\ \lim_{n\to\infty} x_i^{(n)} = x_i \quad \text{for all } i \in Z. \tag{10.19}$$

Then $x \in \mathcal{M}(f)$.

Proof Assume the contrary. Then there exist integers $i_1 < i_2$ and a sequence $\{y_i\}_{i=i_1}^{i_2} \subset R^1$ such that

$$y_{i_1} = x_{i_1}, \qquad y_{i_2} = x_{i_2}, \qquad \sum_{i=i_1}^{i_2-1} f(y_i, y_{i+1}) < \sum_{i=i_1}^{i_2-1} f(x_i, x_{i+1}). \tag{10.20}$$

Set

$$\Delta = \sum_{i=i_1}^{i_2-1} \big[f(x_i, x_{i+1}) - f(y_i, y_{i+1}) \big]. \tag{10.21}$$

For each integer $n \ge 1$, define a finite sequence $\{y_i^{(n)}\}_{i=i_1}^{i_2} \subset R^1$ as follows:

$$y_{i_1}^{(n)} = x_{i_1}^{(n)}, \qquad y_{i_2}^{(n)} = x_{i_2}^{(n)}, \qquad y_i^{(n)} = y_i, \quad i \in \{i_1, \dots, i_2\} \setminus \{i_1, i_2\}. \tag{10.22}$$

It follows from (10.19), (10.22), (10.20), (10.21) and the continuity of f that

$$\lim_{n\to\infty} \left[\sum_{i=i_1}^{i_2-1} f\big(x_i^{(n)}, x_{i+1}^{(n)}\big) - \sum_{i=i_1}^{i_2-1} f\big(y_i^{(n)}, y_{i+1}^{(n)}\big) \right] \\ = \sum_{i=i_1}^{i_2-1} f(x_i, x_{i+1}) - \sum_{i=i_1}^{i_2-1} f(y_i, y_{i+1}) = \Delta > 0. \tag{10.23}$$

In view of (10.19) and (10.23), the sequences

$$\left\{ \sum_{i=i_1}^{i_2-1} f\big(x_i^{(n)}, x_{i+1}^{(n)}\big) \right\}_{n=1}^{\infty}, \qquad \left\{ \sum_{i=i_1}^{i_2-1} f\big(y_i^{(n)}, y_{i+1}^{(n)}\big) \right\}_{n=1}^{\infty}$$

are bounded. It follows from this fact, Proposition 10.15 and the equality $f = \lim_{n\to\infty} f_n$ in the weak topology that

$$\lim_{n\to\infty}\left[\sum_{i=i_1}^{i_2-1} f\big(x_i^{(n)}, x_{i+1}^{(n)}\big) - \sum_{i=i_1}^{i_2-1} f_n\big(x_i^{(n)}, x_{i+1}^{(n)}\big)\right] = 0, \tag{10.24}$$

$$\lim_{n\to\infty}\left[\sum_{i=i_1}^{i_2-1} f\big(y_i^{(n)}, y_{i+1}^{(n)}\big) - \sum_{i=i_1}^{i_2-1} f_n\big(y_i^{(n)}, y_{i+1}^{(n)}\big)\right] = 0. \tag{10.25}$$

By (10.23)–(10.25),

$$\lim_{n\to\infty}\left[\sum_{i=i_1}^{i_2-1} f_n\big(x_i^{(n)}, x_{i+1}^{(n)}\big) - \sum_{i=i_1}^{i_2-1} f_n\big(y_i^{(n)}, y_{i+1}^{(n)}\big)\right] = \Delta > 0.$$

There is an integer $n_0 \geq 1$ such that for each integer $n \geq n_0$,

$$\sum_{i=i_1}^{i_2-1} f_n\big(x_i^{(n)}, x_{i+1}^{(n)}\big) - \sum_{i=i_1}^{i_2-1} f_n\big(y_i^{(n)}, y_{i+1}^{(n)}\big) > \Delta/2.$$

This fact contradicts the (f_n)-minimality of $x^{(n)}$ for all $n \geq n_0$. The contradiction we have reached proves Proposition 10.17. □

Proposition 10.18 *Let $f \in \mathfrak{M}_{k0}$, $f_n \in \mathfrak{M}_{k0}$, $n = 1, 2, \ldots$, $\lim_{n\to\infty} f_n = f$ in the weak topology, $x^{(n)} \in \mathcal{M}^{per}(f_n, \alpha)$, $n = 1, 2, \ldots$, and let the sequence $\{x_0^{(n)}\}_{n=1}^{\infty}$ be bounded. Then the following assertions hold:*

1. *There exists $x \in R^Z$ and a strictly increasing sequence of natural numbers $\{n_j\}_{j=1}^{\infty}$ such that*

$$x_{i+q} = x_i + p, \quad i \in Z, \tag{10.26}$$

$$x_i^{(n_j)} \to x_i \quad \text{as } j \to \infty \text{ for all } i \in Z. \tag{10.27}$$

2. *Assume that $x \in R^Z$ and $\{n_j\}_{j=1}^{\infty}$ is a strictly increasing sequence of natural numbers such that (10.26) and (10.27) hold. Then $x \in \mathcal{M}^{per}(f, \alpha)$ and*

$$E_\alpha(f) = \sum_{i=0}^{q-1} f(x_i, x_{i+1}) = \lim_{j\to\infty} \sum_{i=0}^{q-1} f_{n_j}\big(x_i^{(n_j)}, x_{i+1}^{(n_j)}\big)$$
$$= \lim_{j\to\infty} E_\alpha(f_{n_j}). \tag{10.28}$$

Proof By Proposition 10.2, the sequence $\{x_i^{(n)}\}_{n=1}^{\infty}$ is bounded for each $i \in Z$. This fact implies that there exist a strictly increasing sequence of natural numbers

$\{n_j\}_{j=1}^{\infty}$ and $x \in R^Z$ such that (10.26) and (10.27) are valid. Therefore assertion 1 is true.

Now we prove assertion 2. Assume that $x \in R^Z$ and $\{n_j\}_{j=1}^{\infty}$ is a strictly increasing sequence of natural numbers such that (10.26) and (10.27) hold. By Proposition 10.17 and (10.26), $x \in \mathcal{M}^{per}(f,\alpha)$. Since $\lim_{n\to\infty} f_n = f$ in the weak topology it follows from Corollary 10.16 that the sequence $\{E_\alpha(f_n)\}_{n=1}^{\infty}$ is bounded from above. Therefore the sequence $\{\sum_{i=0}^{q-1} f_n(x_i^{(n)}, x_{i+1}^{(n)})\}_{n=1}^{\infty}$ is also bounded from above. It follows from this fact, the equality $\lim_{n\to\infty} f_n = f$ in the weak topology and Proposition 10.15 that

$$\lim_{n\to\infty}\left[\sum_{i=0}^{q-1} f_n\left(x_i^{(n)}, x_{i+1}^{(n)}\right) - \sum_{i=0}^{q-1} f\left(x_i^{(n)}, x_{i+1}^{(n)}\right)\right] = 0. \tag{10.29}$$

By (10.9), (10.26), (10.27), (10.29) and Corollary 10.16,

$$\begin{aligned} E_\alpha(f) &\le \sum_{i=0}^{q-1} f(x_i, x_{i+1}) = \lim_{j\to\infty} \sum_{i=0}^{q-1} f\left(x_i^{(n_j)}, x_{i+1}^{(n_j)}\right) \\ &= \lim_{j\to\infty} \sum_{i=0}^{q-1} f_{n_j}\left(x_i^{(n_j)}, x_{i+1}^{(n_j)}\right) = \lim_{j\to\infty} E_\alpha(f_{n_j}) \le E_\alpha(f). \end{aligned}$$

These relations imply (10.28). Thus Proposition 10.18 is proved. □

Proposition 10.18 and Corollary 10.16 imply the following result.

Proposition 10.19 *The function $f \to E_\alpha(f)$ is continuous on $\mathfrak{M}_{k0}$ with the relative weak topology.*

Proposition 10.20 *Assume that $f \in \mathfrak{M}_{k0}$ and that the following property holds:*

If $x^{(1)}, x^{(2)} \in \mathcal{M}^{per}(f,\alpha)$, then there exists $n = (n_1, n_2) \in Z^2$ such that $x^{(2)} = T_n x^{(1)}$.

Then there exists $\bar{n} = (\bar{n}_1, \bar{n}_2) \in Z^2$ such that for each $x \in \mathcal{M}^{per}(f,\alpha)$,

$$T_{\bar{n}}x > x, \qquad \left\{y \in \mathcal{M}^{per}(f,\alpha) : x < y < T_{\bar{n}}x\right\} = \emptyset.$$

Proof Let $\bar{x} \in \mathcal{M}^{per}(f,\alpha)$. Then

$$\begin{aligned} \mathcal{M}^{per}(f,\alpha) &= \left\{T_n\bar{x} : n = (n_1, n_2) \in Z^2\right\} \\ &= \left\{T_n\bar{x} : n = (n_1, n_2) \in Z^2, 0 \le n_1 \le q-1\right\}. \end{aligned} \tag{10.30}$$

By (10.30), the set

$$\left\{y \in \mathcal{M}^{per}(f,\alpha) : \bar{x} < y < T_{(0,1)}\bar{x}\right\}$$

is either finite or empty. Therefore there exists $\bar{x}^+ \in \mathcal{M}^{per}(f,\alpha)$ such that

$$\bar{x} < \bar{x}^+, \qquad \{y \in \mathcal{M}^{per}(f,\alpha) : \bar{x} < y < \bar{x}^+\} = \emptyset. \tag{10.31}$$

There exists $\bar{n} = (\bar{n}_1, \bar{n}_2) \in Z^2$ such that

$$T_{\bar{n}}\bar{x} = \bar{x}^+. \tag{10.32}$$

Let $x \in \mathcal{M}^{per}(f,\alpha)$. There exists $n = (n_1, n_2) \in Z^2$ such that

$$x = T_n\bar{x}. \tag{10.33}$$

In view of (10.33), (10.32) and (10.31), we have

$$T_{\bar{n}}x = T_{\bar{n}}(T_n\bar{x}) = T_n(T_{\bar{n}}\bar{x}) = T_n\bar{x}^+ > T_n\bar{x} = x$$

and

$$T_{\bar{n}}x > x. \tag{10.34}$$

Assume that

$$y \in \mathcal{M}^{per}(f,\alpha), \qquad x < y < T_{\bar{n}}x. \tag{10.35}$$

Then

$$T_{-n}x < T_{-n}y < T_{-n}(T_{\bar{n}}x) \tag{10.36}$$

where $-n = (-n_1, -n_2)$. If follows from (10.36), (10.33) and (10.32) that

$$\bar{x} < T_{-n}y < T_{\bar{n}}(T_{-n}x) = T_{\bar{n}}\bar{x} = \bar{x}^+,$$

a contradiction (see (10.31)). Therefore

$$\{y \in \mathcal{M}^{per}(f,\alpha) : x < y < T_{\bar{n}}x\} = \emptyset.$$

This completes the proof of Proposition 10.20. □

Corollary 10.21 *Assume that $f \in \mathfrak{M}_{k0}$ and that the following property holds:*

If $x^{(1)}, x^{(2)} \in \mathcal{M}^{per}(f,\alpha)$, then there exists $n = (n_1, n_2) \in Z^2$ such that $T_n x^{(1)} = x^{(2)}$.

Then there exists a number $\kappa > 0$ such that for each $x, x^+ \in \mathcal{M}^{per}(f,\alpha)$ satisfying

$$x < x^+, \qquad \{y \in \mathcal{M}^{per}(f,\alpha) : x < y < x^+\} = \emptyset,$$

the inequality $x_i^+ - x_i > \kappa$ holds for all $i \in Z$.

Proposition 10.22 *Assume that $f \in \mathfrak{M}_{k0}$, $\bar{x} \in \mathcal{M}^{per}(f,\alpha)$,*

$$\mathcal{M}^{per}(f,\alpha) = \{T_n\bar{x} : n = (n_1, n_2) \in Z^2\} \tag{10.37}$$

and that $\varepsilon > 0$ is given. Then there exists a neighborhood $\mathcal{U}$ of f in $\mathfrak{M}_k$ with the weak topology such that for each $g \in \mathcal{U} \cap \mathfrak{M}_{k0}$ and each $x \in \mathcal{M}^{per}(g, \alpha)$, there is $m = (m_1, m_2) \in Z^2$ such that $|x_i - (T_m\bar{x})_i| \leq \varepsilon$, $i \in Z$.

Proof Assume the contrary. Then there exist a sequence $\{f_j\}_{j=1}^{\infty} \subset \mathfrak{M}_{k0}$ satisfying $\lim_{j\to\infty} f_j = f$ in the weak topology and a sequence $x^{(j)} \in \mathcal{M}^{per}(f_j, \alpha)$, $j = 1, 2, \ldots$, such that for each natural number j and each $n = (n_1, n_2) \in Z^2$,

$$\sup\{|x_i^{(j)} - (T_n\bar{x})_i| : i \in \{0, 1, \ldots, q\}\} > \varepsilon. \tag{10.38}$$

We may assume without loss of generality that the sequence $\{x_0^{(j)}\}_{j=1}^{\infty}$ is bounded. By Proposition 10.18, there exist $x \in \mathcal{M}^{per}(f, \alpha)$ and a strictly increasing sequence of natural numbers $\{j_s\}_{s=1}^{\infty}$ such that

$$x_i^{(j_s)} \to x_i \quad \text{as } s \to \infty \text{ for all } i \in Z. \tag{10.39}$$

By (10.37), there exists $m = (m_1, m_2) \in Z^2$ such that $x = T_m\bar{x}$. It follows from this equality and (10.39) that $x_i^{(j_s)} \to (T_m\bar{x})_i$ as $s \to \infty$ for all $i \in Z$. This fact contradicts (10.38). The contradiction we have reached proves Proposition 10.22. □

10.5 Preliminary Results for Assertion 2 of Theorem 10.9

In this section we assume that $k \geq 2$ is an integer and $\alpha = p/q$ is an irreducible fraction, where $q \geq 1$ and p are integers. Assume that $f \in \mathfrak{M}_{k0}$,

$$\bar{x}, \bar{x}^+ \in \mathcal{M}^{per}(f, \alpha), \qquad \bar{x} < \bar{x}^+, \tag{10.40}$$

$$\{y \in \mathcal{M}^{per}(f, \alpha) : \bar{x} < y < \bar{x}^+\} = \emptyset \tag{10.41}$$

and

$$\mathcal{M}^{per}(f, \alpha) = \{T_n\bar{x} : n = (n_1, n_2) \in Z^2\}. \tag{10.42}$$

By Corollary 10.21, there exists a number $\kappa > 0$ such that

$$x_i^+ - x_i > 2\kappa, \quad i \in Z, \tag{10.43}$$

for each $x, x^+ \in \mathcal{M}^{per}(f, \alpha)$ which satisfy

$$x < x^+, \qquad \{y \in \mathcal{M}^{per}(f, \alpha) : x < y < x^+\} = \emptyset. \tag{10.44}$$

Lemma 10.23 *Let $\varepsilon \in (0, \kappa/2)$ be given. Then there exists a neighborhood $\mathcal{U}$ of f in $\mathfrak{M}_{k0}$ with the weak topology such that the following property holds:*

For each $g \in \mathcal{U} \cap \mathfrak{M}_{k0}$ and each $y \in \mathcal{M}^{per}(g, \alpha)$, there exists a unique $x \in \mathcal{M}^{per}(f, \alpha)$ such that

$$|x_i - y_i| < \varepsilon, \quad i \in Z. \tag{10.45}$$

Proof By Proposition 10.22, there exists a neighborhood $\mathcal{U}$ of f in $\mathfrak{M}_k$ with the weak topology such that the following property holds:

For each $g \in \mathcal{U} \cap \mathfrak{M}_{k0}$ and each $y \in \mathcal{M}^{per}(g, \alpha)$, there exists $x \in \mathcal{M}^{per}(f, \alpha)$ such that (10.45) holds.

Assume that $g \in \mathcal{U} \cap \mathfrak{M}_{k0}$,

$$\begin{aligned} & y \in \mathcal{M}^{per}(g, \alpha), \qquad x^{(1)}, x^{(2)} \in \mathcal{M}^{per}(f, \alpha), \\ & \left|x_i^{(j)} - y_i\right| < \varepsilon, \quad i \in Z, j = 1, 2. \end{aligned} \tag{10.46}$$

To complete the proof of the lemma, it is sufficient to show that $x^{(1)} = x^{(2)}$. Assume the contrary. We may assume without loss of generality that $x^{(1)} < x^{(2)}$. By our choice of κ (see (10.43), (10.44)) and Proposition 10.20,

$$\inf\left\{x_i^{(2)} - x_i^{(1)} : i \in Z\right\} > 2\kappa. \tag{10.47}$$

On the other hand, it follows from (10.46) that for all $i \in Z$,

$$\left|x_i^{(2)} - x_i^{(1)}\right| \le \left|x_i^{(2)} - y_i\right| + \left|y_i - x_i^{(1)}\right| < 2\varepsilon < \kappa,$$

a contradiction. The contradiction we have reached proves Lemma 10.23. □

Lemma 10.24 *Let* $\varepsilon \in (0, \kappa/2)$ *and let a neighborhood* $\mathcal{U}$ *of* f *in* $\mathfrak{M}_k$ *with the weak topology be as guaranteed in Lemma* 10.23. *Assume that*

$$g \in \mathcal{U} \cap \mathfrak{M}_{k0}, \qquad y^{(1)}, y^{(2)} \in \mathcal{M}^{per}(g, \alpha), \qquad y^{(1)} < y^{(2)}, \tag{10.48}$$

$$\left\{z \in \mathcal{M}^{per}(g, \alpha) : y^{(1)} < z < y^{(2)}\right\} = \emptyset, \tag{10.49}$$

$$x^{(1)}, x^{(2)} \in \mathcal{M}^{per}(f, \alpha), \qquad \left|x_i^{(j)} - y_i^{(j)}\right| < \varepsilon, \quad i \in Z, j = 1, 2. \tag{10.50}$$

Then either $x^{(1)} = x^{(2)}$ *or*

$$x^{(1)} < x^{(2)}, \qquad \left\{z \in \mathcal{M}^{per}(f, \alpha) : x^{(1)} < z < x^{(2)}\right\} = \emptyset.$$

Proof Assume that $x^{(1)} \neq x^{(2)}$. By (10.50) and (10.48), for all $i \in Z$,

$$x_i^{(2)} - x_i^{(1)} = x_i^{(2)} - y_i^{(2)} + y_i^{(2)} - y_i^{(1)} + y_i^{(1)} - x_i^{(1)} > -2\varepsilon > -\kappa$$

and

$$x_i^{(2)} - x_i^{(1)} > -\kappa \quad \text{for all } i \in Z.$$

It follows from this inequality, (10.43) and Proposition 10.20 that $x^{(1)} < x^{(2)}$. To complete the proof of the lemma, we need to show that the set

$$\left\{z \in \mathcal{M}^{per}(f, \alpha) : x^{(1)} < z < x^{(2)}\right\} = \emptyset.$$

Assume the contrary. Then by Proposition 10.20, there exist $x^{(3)} \in \mathcal{M}^{per}(f,\alpha)$ such that

$$x^{(1)} < x^{(3)} < x^{(2)},$$
$$\{z \in \mathcal{M}^{per}(f,\alpha) : x^{(1)} < z < x^{(3)}\} = \emptyset. \tag{10.51}$$

It follows from Proposition 10.20, (10.51) and our choice of κ (see (10.43), (10.44)) that

$$x_i^{(2)} - x_i^{(3)} > 2\kappa, \qquad x_i^{(3)} - x_i^{(1)} > 2\kappa, \quad i \in Z. \tag{10.52}$$

By (10.42), there exists $m = (m_1, m_2) \in Z^2$ for which

$$x^{(3)} = T_m x^{(1)}. \tag{10.53}$$

Set

$$y^{(3)} = T_m y^{(1)}. \tag{10.54}$$

Clearly, $y^{(3)} \in \mathcal{M}^{per}(g,\alpha)$. It follows from (10.54), (10.53), (10.50) and (10.52) that for all $i \in Z$,

$$\begin{aligned} y_i^{(3)} - y_i^{(1)} &= y_i^{(3)} - x_i^{(3)} + x_i^{(3)} - x_i^{(1)} + x_i^{(1)} - y_i^{(1)} \\ &= y_{i-m_1}^{(1)} + m_2 - \big(x_{i-m_1}^{(1)} + m_2\big) + x_i^{(3)} - x_i^{(1)} + x_i^{(1)} - y_i^{(1)} \\ &> -2\varepsilon + 2\kappa > \kappa. \end{aligned}$$

Analogously, it follows from (10.54), (10.53), (10.50) and (10.52) that for all $i \in Z$,

$$\begin{aligned} y_i^{(2)} - y_i^{(3)} &= y_i^{(2)} - x_i^{(2)} + x_i^{(2)} - x_i^{(3)} + x_i^{(3)} - y_i^{(3)} \\ &= y_i^{(2)} - x_i^{(2)} + x_i^{(2)} - x_i^{(3)} + x_{i-m_1}^{(1)} + m_2 - \big(y_{i-m_1}^{(1)} + m_2\big) \\ &> -2\varepsilon + 2\kappa > \kappa. \end{aligned}$$

Therefore $y^{(1)} < y^{(3)} < y^{(2)}$. This fact contradicts (10.49). The contradiction we have reached proves Lemma 10.24. □

Suppose that $\varepsilon \in (0, \kappa/2)$, $g \in \mathfrak{M}_{k0}$, $y \in \mathcal{M}^+(g,\alpha)$, $y^+, y^- \in \mathcal{M}^{per}(g,\alpha)$,

$$y^- < y < y^+, \qquad \lim_{i\to\infty} y_i - y_i^+ = 0, \qquad \lim_{i\to-\infty} y_i - y_i^- = 0.$$

We say that y is regular with respect to (ε, g) if there exist $x^-, x^+ \in \mathcal{M}^{per}(f,\alpha)$ such that

$$|x_i^- - y_i^-| < \varepsilon, \qquad |x_i^+ - y_i^+| < \varepsilon, \quad i \in Z, \tag{10.55}$$

$$x^- < x^+, \qquad \{z \in \mathcal{M}^{per}(f,\alpha) : x^- < z < x^+\} = \emptyset. \tag{10.56}$$

We assume that there exists $\widehat{x} \in \mathcal{M}^+(f,\alpha)$ such that

$$\bar{x} < \widehat{x} < \bar{x}^+, \tag{10.57}$$

$$\mathcal{M}^+(f,\alpha) = \{T_n\widehat{x} : n = (n_1, n_2) \in Z^2\}. \tag{10.58}$$

Lemma 10.25 *Let a neighborhood $\mathcal{U}$ of f in $\mathfrak{M}_k$ with the weak topology be as guaranteed in Lemma 10.23 with $\varepsilon = \kappa/4$. Assume that $\{f_n\}_{n=1}^{\infty} \subset \mathcal{U} \cap \mathfrak{M}_{k0}$, $\lim_{n\to\infty} f_n = f$ in the weak topology and that $x^{(n)} \in \mathcal{M}^+(f_n,\alpha)$ is regular with respect to $(\kappa/4, f_n)$, $n = 1, 2, \ldots$. Then there exist a strictly increasing sequence of natural numbers $\{n_j\}_{j=1}^{\infty}$ and a sequence $s^{(j)} = (s_1^{(j)}, s_2^{(j)}) \in Z^2$, $j = 1, 2, \ldots$, such that*

$$T_{s^{(j)}} x_i^{(n_j)} \to \widehat{x}_i \quad \text{as } j \to \infty \text{ for all } i \in Z. \tag{10.59}$$

Proof By (10.40), (10.41) and (10.57),

$$\lim_{i\to-\infty} \widehat{x}_i - \bar{x}_i = 0, \qquad \lim_{i\to\infty} \widehat{x}_i - \bar{x}_i^+ = 0. \tag{10.60}$$

Let $n \geq 1$ be an integer. There exist

$$x^{(n^+)}, x^{(n^-)} \in \mathcal{M}^{per}(f_n,\alpha) \tag{10.61}$$

such that

$$x^{(n^-)} < x^{(n)} < x^{(n^+)}, \tag{10.62}$$

$$\lim_{i\to-\infty} x_i^{(n^-)} - x_i^{(n)} = 0, \qquad \lim_{i\to\infty} x_i^{(n^+)} - x_i^{(n)} = 0. \tag{10.63}$$

Since $f_n \in \mathcal{U}$, it follows from the definition of $\mathcal{U}$ and Lemma 10.23 that there exist unique $z^{(n^-)}, z^{(n^+)} \in \mathcal{M}^{per}(f,\alpha)$ such that

$$|z_i^{(n^-)} - x_i^{(n^-)}| \leq \kappa/4, \qquad |z_i^{(n^+)} - x_i^{(n^+)}| < \kappa/4, \quad i \in Z. \tag{10.64}$$

Since $x^{(n)}$ is regular with respect to $(\kappa/4, f_n)$, we have

$$z^{(n^-)} < z^{(n^+)}, \qquad \{z \in \mathcal{M}^{per}(f,\alpha) : z^{(n^-)} < z < z^{(n^+)}\} = \emptyset. \tag{10.65}$$

Since $\lim_{n\to\infty} f_n = f$ in the weak topology, it follows from Lemma 10.23 that

$$\lim_{n\to\infty} \sup\{|z_i^{(n^-)} - x_i^{(n^-)}|, |z_i^{(n^+)} - x_i^{(n^+)}| : i \in Z\} = 0. \tag{10.66}$$

Now it follows from (10.65), (10.40)–(10.43) and Proposition 10.20 that there is $l \in Z^2$ such that $z^{(n^-)} = T_l\bar{x}$ and $z^{(n^+)} = T_l\bar{x}^+$. We may assume without loss of generality that

$$z^{(n^-)} = \bar{x}, \qquad z^{(n^+)} = \bar{x}^+, \quad n = 1, 2, \ldots. \tag{10.67}$$

It follows from (10.64), (10.67) and the definition of κ (see (10.43), (10.44)) that for any integer $n \geq 1$ and any integer i,

$$x_i^{(n^+)} - x_i^{(n^-)} \geq x_i^{(n^+)} - z_i^{(n^+)} + z_i^{(n^+)} - z_i^{(n^-)} + z_i^{(n^-)} - x_i^{(n^-)}$$
$$> -\kappa/2 + \bar{x}_i^+ - \bar{x}_i > 3\kappa/2$$

and

$$x_i^{(n^+)} - x_i^{(n^-)} > 3\kappa/2. \tag{10.68}$$

Let $n \geq 1$ be an integer. It follows from (10.62), (10.63) and (10.68) that there exists an integer t_n such that

$$x_{t_n}^{(n)} - x_{t_n}^{(n^-)} \leq \kappa/2, \qquad x_{t_n+1}^{(n)} - x_{t_n+1}^{(n^-)} > \kappa/2. \tag{10.69}$$

Using translations, we may assume without loss of generality that

$$t_n \in [0, q]. \tag{10.70}$$

In view of (10.62), (10.64) and (10.67), for all integers $n \geq 1$ and all $i \in Z$,

$$\bar{x}_i - \kappa/4 < x_i^{(n^-)} < x_i^{(n)} < x_i^{(n^+)} < \bar{x}_i^+ + \kappa/4.$$

Therefore for any $i \in Z$, the sequence $\{x_i^{(n)}\}_{n=1}^{\infty}$ is bounded. Together with (10.70) this implies that there exist $u \in R^Z$ and a strictly increasing sequence of natural numbers $\{n_j\}_{j=1}^{\infty}$ such that

$$x_i^{(n_j)} \to u_i \quad \text{as } j \to \infty \text{ for all } i \in Z, \qquad t_{n_j} = t_{n_1}, \quad j = 1, 2, \dots. \tag{10.71}$$

It follows from (10.71), (10.62), (10.66) and (10.67) that for all $i \in Z$,

$$u_i = \lim_{j \to \infty} x_i^{(n_j)} \in \left[\lim_{j \to \infty} x_i^{(n_j^-)}, \lim_{j \to \infty} x_i^{(n_j^+)}\right] = [\bar{x}_i, \bar{x}_i^+]. \tag{10.72}$$

By Proposition 10.17, $u \in \mathcal{M}(f)$. Since $x^{(n)} \in \mathcal{M}^+(f_n, \alpha)$, $n = 1, 2, \dots$, we have $x^{(n)} > T_{(q,p)} x^{(n)}$, $n = 1, 2, \dots$. Therefore $x_i^{(n)} > x_{i-q}^{(n)} + p$ for any integer $n \geq 1$ and any integer i. When combined with (10.71), this fact implies that $u_i \geq u_{i-q} + p$ for all $i \in Z$ and that

$$u \in \mathcal{M}^{per}(f, \alpha) \cup \mathcal{M}^+(f, \alpha). \tag{10.73}$$

It follows from (10.71), (10.72) and (10.69) that

$$u_{t_1} - \bar{x}_{t_1} = \lim_{j \to \infty} x_{t_1}^{(n_j)} - \lim_{j \to \infty} x_{t_1}^{(n_j^-)} \leq \kappa/2,$$

$$u_{t_1+1} - \bar{x}_{t_1+1} = \lim_{j \to \infty} x_{t_1+1}^{(n_j)} - \lim_{j \to \infty} x_{t_1+1}^{(n_j^-)} \geq \kappa/2.$$

By these relations, (10.72), the definition of κ (see (10.43) and (10.44)), (10.40) and (10.41), $u \notin \{\bar{x}, \bar{x}^+\}$. When combined with (10.72), (10.73), (10.40) and (10.41), this fact implies that $u \in \mathcal{M}^+(f,\alpha)$. By (10.58), there exists $m = (m_1, m_2) \in Z^2$ such that $T_m x = \widehat{x}$. This completes the proof of Lemma 10.25. □

Lemma 10.26 *Let $Q \geq 1$ be an integer and let $\varepsilon \in (0, \kappa/4)$ be given. Then there exists a neighborhood $\mathcal{U}$ of f in $\mathfrak{M}_k$ with the weak topology such that for each $g \in \mathcal{U} \cap \mathfrak{M}_{k0}$ and each $y \in \mathcal{M}^+(g,\alpha)$, one of the following properties holds:*

(a) *There exists $n = (n_1, n_2) \in Z^2$ such that*

$$\left|(T_n y)_i - \bar{x}_i\right| < \varepsilon, \quad i \in Z.$$

(b) *There exists $n = (n_1, n_2) \in Z^2$ such that*

$$\left|(T_n y)_i - \widehat{x}_i\right| < \varepsilon, \quad i = -Q, \ldots, Q.$$

Proof Assume the contrary. Then there exists a sequence $\{f_s\}_{s=1}^{\infty} \subset \mathfrak{M}_{k0}$ such that $\lim_{s\to\infty} f_s = f$ in the weak topology and a sequence $y^{(s)} \in \mathcal{M}^+(f_s,\alpha)$, $s = 1, 2, \ldots,$ such that for any integer $s \geq 1$, the following properties hold:

(c) For any $n = (n_1, n_2) \in Z^2$

$$\sup\{\left|\left(T_n y^{(s)}\right)_i - \bar{x}_i\right| : i \in Z\} \geq \varepsilon.$$

(d) For any $n = (n_1, n_2) \in Z^2$

$$\sup\{\left|\left(T_n y^{(s)}\right)_i - \widehat{x}_i\right| : i = -Q, \ldots, Q\} \geq \varepsilon.$$

By Lemmata 10.23 and 10.24 and (10.42), $y^{(s)}$ is regular with respect to $(f_s, \varepsilon/2)$ for all sufficiently large integers s.

By Lemma 10.25, there exist a strictly increasing sequence of natural numbers $\{s_j\}_{j=1}^{\infty}$ and a sequence $n^{(j)} = (n_1^{(j)}, n_2^{(j)}) \in Z^2$, $j = 1, 2, \ldots,$ such that $(T_{n^{(j)}} y^{(s_j)})_i \to \widehat{x}_i$ as $j \to \infty$ for all $i \in Z$, a contradiction (see (d)). The contradiction we have reached proves Lemma 10.26. □

Lemma 10.27 *Let $\varepsilon \in (0, \kappa/4)$ be given. Then there exists a neighborhood $\mathcal{U}$ of f in $\mathfrak{M}_k$ with the weak topology such that for each $g \in \mathcal{U} \cap \mathfrak{M}_{k0}$ and each $y \in \mathcal{M}^+(g,\alpha)$, one of the following properties holds:*

(i) *There exists $m = (m_1, m_2) \in Z^2$ such that*

$$\left|(T_m y)_i - \bar{x}_i\right| < \varepsilon, \quad i \in Z.$$

(ii) *There exists $m = (m_1, m_2) \in Z^2$ such that*

$$\left|(T_m y)_i - \widehat{x}_i\right| < \varepsilon, \quad i \in Z.$$

Proof Choose a positive number

$$\epsilon_0 < \min\{\epsilon/6, \kappa/8\}. \tag{10.74}$$

By (10.57), (10.40) and (10.41), there exists a natural number $Q > 8q + 8$ such that

$$|\widehat{x}_i - \bar{x}_i^+| < \epsilon_0/4 \quad \text{for all integers } i \geq Q/2, \tag{10.75}$$

$$|\widehat{x}_i - \bar{x}_i| < \epsilon_0/4 \quad \text{for all integers } i \leq -Q/2. \tag{10.76}$$

By Lemmata 10.23 and 10.24, there exists a neighborhood $\mathcal{U}_1$ of f in $\mathfrak{M}_k$ with the weak topology such that the following properties hold:

(iii) For each $g \in \mathcal{U}_1 \cap \mathfrak{M}_{k0}$ and each $y \in \mathcal{M}^{per}(g, \alpha)$, there exists a unique $x \in \mathcal{M}^{per}(f, \alpha)$ such that $|x_i - y_i| < \epsilon_0$ for all $i \in Z$.

(iv) Let $g \in \mathcal{U}_1 \cap \mathfrak{M}_{k0}$, $y^{(1)}, y^{(2)} \in \mathcal{M}^{per}(g, \alpha)$,

$$y^{(1)} < y^{(2)}, \qquad \{z \in \mathcal{M}^{per}(g, \alpha) : y^{(1)} < z < y^{(2)}\} = \emptyset,$$

$$x^{(1)}, x^{(2)} \in \mathcal{M}^{per}(f, \alpha), \qquad |x_i^{(j)} - y_i^{(j)}| < \epsilon_0, \quad i \in Z, j = 1, 2.$$

Then either $x^{(1)} = x^{(2)}$ or

$$x^{(1)} < x^{(2)} \quad \text{and} \quad \{z \in \mathcal{M}^{per}(f, \alpha) : x^{(1)} < z < x^{(2)}\} = \emptyset.$$

By Lemma 10.26, there exists a neighborhood $\mathcal{U}$ of f in $\mathfrak{M}_k$ with the weak topology such that $\mathcal{U} \subset \mathcal{U}_1$ and for each $g \in \mathcal{U} \cap \mathfrak{M}_{k0}$ and each $y \in \mathcal{M}^+(g, \alpha)$, one of the following properties holds:

(v) There exists $m = (m_1, m_2) \in Z^2$ such that $|(T_m y)_i - \bar{x}_i| < \epsilon_0$ for all $i \in Z$.

(vi) There exists $m = (m_1, m_2) \in Z^2$ such that $|(T_m y)_i - \widehat{x}_i| < \epsilon_0$, $i = -Q, \ldots, Q$.

Let

$$g \in \mathcal{U} \cap \mathfrak{M}_{k0}, \qquad y \in \mathcal{M}^+(g, \alpha). \tag{10.77}$$

If (v) is true, then (ii) also holds. Therefore we may assume that (v) does not hold. Then by the definition of $\mathcal{U}$ and (10.77), property (vi) holds. We may assume without loss of generality that (vi) holds with $m = (0, 0)$. Thus

$$|y_i - \widehat{x}_i| < \epsilon_0, \quad i = -Q, \ldots, Q. \tag{10.78}$$

There exist

$$y^-, y^+ \in \mathcal{M}^{per}(g, \alpha) \tag{10.79}$$

such that

$$y^- < y < y^+, \qquad \lim_{i \to -\infty} y_i^- - y_i = 0, \qquad \lim_{i \to \infty} y_i^+ - y_i = 0. \tag{10.80}$$

By property (iii), (10.77) and (10.79), there exist unique

$$x^-, x^+ \in \mathcal{M}^{per}(f, \alpha) \tag{10.81}$$

such that

$$|x_i^- - y_i^-| < \epsilon_0, \qquad |x_i^+ - y_i^+| < \epsilon_0, \quad i \in Z. \tag{10.82}$$

By property (iv), (10.77), (10.79), (10.80), (10.81) and (10.82), either $x^- = x^+$ or

$$x^- < x^+ \quad \text{and} \quad \{z \in \mathcal{M}^{per}(f, \alpha) : x^- < z < x^+\} = \emptyset. \tag{10.83}$$

If $x^- = x^+$, then (10.80) and (10.82) imply that for all $i \in Z$,

$$y_i - x_i^+ = y_i - y_i^+ + y_i^+ - x_i^+ < y_i^+ - x_i^+ < \epsilon_0,$$

$$y_i - x_i^+ = y_i - x_i^- = y_i - y_i^- + y_i^- - x_i^- > y_i^- - x_i^- > -\epsilon_0,$$

$$|y_i - x_i^+| < \epsilon_0$$

and combining this with (10.42), we see that property (v) holds. The contradiction we have reached proves that (10.83) holds. It follows from (10.80) and (10.82) that for all $i \in Z$,

$$x_i^- - \epsilon_0 < y_i^- < y_i < y_i^+ < x_i^+ + \epsilon_0. \tag{10.84}$$

We claim that $x^+ = \bar{x}^+$, $x^- = \bar{x}$.

By (10.78) and (10.75), for $i = Q - 4q, \ldots, Q$,

$$|y_i - \bar{x}_i^+| \le |y_i - \widehat{x}_i| + |\widehat{x}_i - \bar{x}_i^+| < \epsilon_0 + \epsilon_0/4 \tag{10.85}$$

and for $i = -Q, \ldots, -Q + 4q$,

$$|y_i - \bar{x}_i| \le |y_i - \widehat{x}_i| + |\widehat{x}_i - \bar{x}_i| < \epsilon_0 + \epsilon_0/4. \tag{10.86}$$

It follows from (10.85), (10.84) and (10.86) that for $i = Q - 4q, \ldots, Q$,

$$\bar{x}_i^+ - \epsilon_0 - \epsilon_0/4 < y_i < x_i^+ + \epsilon_0$$

and that for $i = -Q, \ldots, -Q + 4q$,

$$x_i^- - \epsilon_0 < y_i < \bar{x}_i + \epsilon_0 + \epsilon_0/4.$$

Thus

$$\bar{x}_i^+ < x_i^+ + 2\epsilon_0 + \epsilon_0/4, \quad i = Q - 4q, \ldots, Q,$$

$$x_i^- < \bar{x}_i + 2\epsilon_0 + \epsilon_0/4, \quad i = -Q, \ldots, -Q + 4q.$$

It follows from these inequalities, the inequality $Q > 8q + 8$, (10.40), (10.81), (10.74) and the definition of κ (see (10.43), (10.44)) that

$$\text{either} \quad \bar{x}^+ < x^+ \quad \text{or} \quad \bar{x}^+ = x^+ \tag{10.87}$$

and

$$\text{either} \quad \bar{x}^- < \bar{x} \quad \text{or} \quad x^- = \bar{x}. \tag{10.88}$$

When combined with (10.40), (10.41), (10.42), (10.81) and (10.83), this fact implies that

$$\text{either} \quad \bar{x} = x^-, \qquad \bar{x}^+ = x^+ \quad \text{or} \quad \bar{x}^+ < x^+,$$
$$\bar{x} < x^- \quad \text{or} \quad x^+ < \bar{x}^+, \qquad x^- < \bar{x}. \tag{10.89}$$

By (10.87)–(10.89),

$$\bar{x} = x^-, \qquad x^+ = \bar{x}^+, \tag{10.90}$$

as claimed.

Next we show that

$$|y_i - \widehat{x}_i| < \varepsilon \tag{10.91}$$

for all $i \in Z$. By (10.78), it is sufficient to show that (10.91) is valid for all integers i satisfying $|i| > Q$.

Assume that $i > Q$ is an integer. Then there exist integers s, j such that

$$s > 1, \qquad j \in [Q - 2q, Q - q], \qquad i = j + sq. \tag{10.92}$$

By (10.78),

$$|y_j - \widehat{x}_j| < \varepsilon_0. \tag{10.93}$$

It follows from (10.75) that

$$\left|\widehat{x}_i - \bar{x}_i^+\right| < \varepsilon_0/4, \qquad \left|\widehat{x}_j - \bar{x}_j^+\right| < \varepsilon_0/4. \tag{10.94}$$

By (10.77), (10.79), (10.80), (10.82), (10.90), (10.94) and (10.93),

$$0 < y_j^+ - y_j = y_j^+ - x_j^+ + x_j^+ - \widehat{x}_j + \widehat{x}_j - y_j$$
$$< \varepsilon_0 + \bar{x}_j^+ - \widehat{x}_j + \widehat{x}_j - y_j < \varepsilon_0 + \varepsilon_0/4 + \varepsilon_0 < 3\varepsilon_0$$

and

$$0 < y_j^+ - y_j < 3\varepsilon_0. \tag{10.95}$$

Since $y \in \mathcal{M}^+(g, \alpha)$, it follows from (10.95), (10.79), (10.92) and (10.80) that

$$3\varepsilon_0 > y_j^+ - y_j > y_j^+ - (T_{(-q,-p)}y)_j > y_j^+ - \left((T_{(-q,-p)})^s y\right)_j$$
$$= y_j^+ - y_{j+sq} + sp = y_{j+sq}^+ - y_{j+sq} = y_i^+ - y_i > 0.$$

Thus we have shown that

$$0 < y_i^+ - y_i < 3\varepsilon_0 \quad \text{for all integers } i > Q. \tag{10.96}$$

By (10.94), (10.82) and (10.96), for all integers $i > Q$,

$$|\widehat{x}_i - y_i| \le |\widehat{x}_i - \bar{x}_i^+| + |\bar{x}_i^+ - y_i^+| + |y_i^+ - y_i|$$
$$< \varepsilon_0/4 + \varepsilon_0 + 3\varepsilon_0$$

and $|\widehat{x}_i - y_i| < 5\varepsilon_0 < \varepsilon$.

Analogously, we show that (10.91) holds for all integers $i < -Q$. Assume now that $i < -Q$ is an integer. Then there exist integers s, j such that

$$s > 1, \qquad j \in [-Q+q, -Q+2q], \qquad i = j - sq. \tag{10.97}$$

By (10.78), inequality (10.93) is valid. It follows from (10.76) that

$$|\widehat{x}_i - \bar{x}_i| < \varepsilon_0/4, \qquad |\widehat{x}_j - \bar{x}_j| < \varepsilon_0/4. \tag{10.98}$$

By (10.80), (10.93), (10.90), (10.82) and (10.98),

$$0 < y_j - y_j^- = y_j - \widehat{x}_j + \widehat{x}_j - \bar{x}_j + \bar{x}_j - y_j^- < \varepsilon_0 + \varepsilon_0/4 + \bar{x}_j - y_j^-$$
$$= \varepsilon_0 + \varepsilon_0/4 + x_j^- - y_j^- < \varepsilon_0 + \varepsilon_0/4 + \varepsilon_0 < 3\varepsilon_0$$

and

$$0 < y_j - y_j^- < 3\varepsilon_0. \tag{10.99}$$

Since $y \in \mathcal{M}^+(g, \alpha)$, it follows from (10.99), (10.79), (10.97) and (10.80) that

$$3\varepsilon_0 > y_j - y_j^- > (T_{(q,p)}y)_j - y_j^- > \left((T_{(q,p)})^s y\right)_j - y_j^-$$
$$= y_{j-sq} + sp - y_j^- = y_{j-sq} - y_{j-sq}^- = y_i - y_i^- > 0.$$

Thus we have shown that

$$0 < y_i - y_i^- < 3\varepsilon_0.$$

It follows from this inequality, (10.98), (10.82) and (10.90) that for all integers $i < -Q$,

$$|\widehat{x}_i - y_i| \le |\widehat{x}_i - \bar{x}_i| + |\bar{x}_i - y_i^-| + |y_i^- - y_i|$$
$$< \varepsilon_0/4 + \varepsilon_0 + 3\varepsilon_0 < 5\varepsilon_0 < \varepsilon.$$

This completes the proof of Lemma 10.27. □

10.6 Proof of Proposition 10.11

Let $k \ge 2$ be an integer and $\alpha = p/q$ be an irreducible fraction, where $q \ge 1$ and p are integers.

Let $f \in \mathfrak{M}_{k0}$. Choose $x^{(f)} \in \mathcal{M}^{(per)}(f,\alpha)$ such that $|x_0^{(f)}| \leq 1$. By Corollary 10.14, there exists a nonnegative function $\phi_f \in C^\infty((R^1/Z))$ such that

$$\{z \in R^1/Z : \phi_f(z) = 0\} = \{P_1(x_i^{(f)}) : i \in Z\}. \tag{10.100}$$

Let $\gamma \in (0, 1)$ be given. Define $f_\gamma : R^2 \to R^1$ by

$$f_\gamma(\xi_1, \xi_2) = f(\xi_1, \xi_2) + \gamma\phi_f(P_1(\xi_1)), \quad (\xi_1, \xi_2) \in R^2. \tag{10.101}$$

It is not difficult to see that $f_\gamma \in \mathfrak{M}_{k0}$. It follows from (10.9), (10.101), (10.100) that

$$\begin{aligned} E_\alpha(f) \leq E_\alpha(f_\gamma) &\leq \sum_{i=0}^{q-1} f_\gamma(x_i^{(f)}, x_{i+1}^{(f)}) \\ &= \sum_{i=0}^{q-1} f(x_i^{(f)}, x_{i+1}^{(f)}) + \gamma\sum_{i=0}^{q-1} \phi_f(P_1(x_i^{(f)})) \\ &= \sum_{i=0}^{q-1} f(x_i^{(f)}, x_{i+1}^{(f)}) = E_\alpha(f) \end{aligned}$$

and that

$$E_\alpha(f) = E_\alpha(f_\gamma) = \sum_{i=0}^{q-1} f_\gamma(x_i^{(f)}, x_{i+1}^{(f)}) = \sum_{i=0}^{q-1} f(x_i^{(f)}, x_{i+1}^{(f)}). \tag{10.102}$$

Assume that $y \in \mathcal{M}^{per}(f_\gamma, \alpha)$. Relations (10.9), (10.102), (10.101) and (10.100) imply that

$$\begin{aligned} \sum_{i=0}^{q-1} f(y_i, y_{i+1}) + \gamma\sum_{i=0}^{q-1} \phi_f(P_1(y_i)) &= \sum_{i=0}^{q-1} f_\gamma(y_i, y_{i+1}) \\ &= E_\alpha(f_\gamma) = E_\alpha(f) \leq \sum_{i=0}^{q-1} f(y_i, y_{i+1}), \end{aligned}$$

$$\sum_{i=0}^{q-1} f(y_i, y_{i+1}) = \sum_{i=0}^{q-1} f_\gamma(y_i, y_{i+1}) = E_\alpha(f_\gamma) = E_\alpha(f)$$

and

$$y \in \mathcal{M}^{per}(f_\gamma, \alpha), \qquad P_1(y_i) \in \{P_1(x_j^{(f)}) : j = 0, \ldots, q-1\}, \quad i = 0, \ldots, q-1.$$

Since the set $\mathcal{M}^{per}(f_\gamma, \alpha)$ is totally ordered, we conclude that y is a translation of $x^{(f)}$. Thus

$$\mathcal{M}^{per}(f_\gamma, \alpha) = \{T_n x^{(f)} : n = (n_1, n_2) \in Z^2\}. \tag{10.103}$$

By Proposition 10.20 and (10.103), there exist

$$x^{(f^+)} \in \mathcal{M}^{per}(f_\gamma, \alpha) \tag{10.104}$$

such that

$$x^{(f)} < x^{(f^+)}, \qquad \{z \in \mathcal{M}^{per}(f_\gamma, \alpha) : x^{(f)} < z < x^{(f^+)}\} = \emptyset. \tag{10.105}$$

Proposition 10.5 implies that there exists

$$y^{(f\gamma)} \in \mathcal{M}^+(f_\gamma, \alpha) \tag{10.106}$$

such that

$$x^{(f)} < y^{(f\gamma)} < x^{(f^+)}, \tag{10.107}$$

$$\lim_{i \to \infty} y_i^{(f\gamma)} - x_i^{(f^+)} = 0, \qquad \lim_{i \to -\infty} y_i^{(f\gamma)} - x_i^{(f)} = 0. \tag{10.108}$$

Define

$$\Omega = \{P_1(y_i^{(f\gamma)}) : i \in Z\} \cup \{P_1(x_i^{(f)}) : i \in Z\}. \tag{10.109}$$

It is easy to see that Ω is a closed subset of R^1/Z.

By Corollary 10.14, there exists a nonnegative function $\psi_{f\gamma} \in C^\infty(R^1/Z)$ such that

$$\{z \in R^1/Z : \psi_{f\gamma}(z) = 0\} = \Omega. \tag{10.110}$$

Let $\mu \in (0, 1)$. Define $f_{\gamma\mu} : R^2 \to R^1$ by

$$f_{\gamma\mu}(\xi_1, \xi_2) = f_\gamma(\xi_1, \xi_2) + \mu\psi_{f\gamma}(P_1(\xi_1)), \quad (\xi_1, \xi_2) \in R^2. \tag{10.111}$$

It is easy to see that $f_{\gamma\mu} \in \mathfrak{M}_{k0}$. Relations (10.111), (10.110), (10.109), (10.102) and (10.9) imply that

$$E_\alpha(f_\gamma) \le E_\alpha(f_{\gamma\mu}) \le \sum_{i=0}^{q-1} f_{\gamma\mu}(x_i^{(f)}, x_{i+1}^{(f)})$$

$$= \sum_{i=0}^{q-1} f_\gamma(x_i^{(f)}, x_{i+1}^{(f)}) + \mu \sum_{i=0}^{q-1} \psi_{f\gamma}(P_1(x_i^{(f)}))$$

$$= \sum_{i=0}^{q-1} f_\gamma(x_i^{(f)}, x_{i+1}^{(f)}) = E_\alpha(f) = E_\alpha(f_\gamma)$$

and

$$E_\alpha(f_{\gamma\mu}) = E_\alpha(f_\gamma) = E_\alpha(f) = \sum_{i=0}^{q-1} f_{\gamma\mu}(x_i^{(f)}, x_{i+1}^{(f)})$$

$$= \sum_{i=0}^{q-1} f_\gamma(x_i^{(f)}, x_{i+1}^{(f)}) = \sum_{i=0}^{q-1} f(x_i^{(f)}, x_{i+1}^{(f)}). \tag{10.112}$$

Assume that

$$y \in \mathcal{M}^{per}(f_{\gamma\mu}, \alpha). \tag{10.113}$$

By (10.111), (10.113), (10.112) and (10.9),

$$\sum_{i=0}^{q-1} f_\gamma(y_i, y_{i+1}) + \sum_{i=0}^{q-1} \mu\psi_{f\gamma}(P_1(y_i)) = \sum_{i=0}^{q-1} f_{\gamma\mu}(y_i, y_{i+1})$$

$$= E_\alpha(f_{\gamma\mu}) = E_\alpha(f_\gamma) \leq \sum_{i=0}^{q-1} f_\gamma(y_i, y_{i+1}),$$

$$\sum_{i=0}^{q-1} f_\gamma(y_i, y_{i+1}) = E_\alpha(f_\gamma)$$

and $y \in \mathcal{M}^{per}(f_\gamma, \alpha)$. Now (10.103) implies that y is a translation of $x^{(f)}$. Thus

$$\mathcal{M}^{per}(f_{\gamma\mu}, \alpha) = \{T_n x^{(f)} : n = (n_1, n_2) \in Z^2\}. \tag{10.114}$$

Lemma 10.28 *Let $z \in \mathcal{M}^+(f_{\gamma\mu}, \alpha)$. Then there exists $m = (m_1, m_2) \in Z^2$ such that $T_m y^{(f\gamma)} = z$.*

Proof By (10.114), (10.105), Proposition 10.20 and the definition of $\mathcal{M}^+(h, \alpha)$ with h satisfying (H1)–(H4) (see Sect. 10.1), we may assume without loss of generality that

$$x^{(f)} < z < x^{(f^+)}. \tag{10.115}$$

Then it follows from Propositions 10.5 and 10.6, the definition of $\mathcal{M}^+(h, \alpha)$ with h satisfying (H1)–(H4) and (10.5) that

$$\lim_{i\to\infty} x_i^{(f^+)} - z_i = 0, \qquad \lim_{i\to-\infty} x_i^{(f)} - z_i = 0. \tag{10.116}$$

Since the set $\mathcal{M}^+(f_{\gamma\mu}, \alpha)$ is totally ordered (see Proposition 10.6), in order to prove the lemma, it is sufficient to show that there exist $m = (m_1, m_2) \in Z^2$ and $i \in Z$ such that $z_i = (T_m y^{(f\gamma)})_i$. Assume the contrary. Then

$$\{P_1 z_i : i \in Z\} \cap \{P_1 y_i^{(f\gamma)} : i \in Z\} = \emptyset. \tag{10.117}$$

Since the set $\mathcal{M}^+(f_{\gamma\mu},\alpha)\cup\mathcal{M}^{per}(f_{\gamma\mu},\alpha)$ is totally ordered (see Proposition 10.5),

$$\{P_1 z_i : i \in Z\} \cap \{P_1 x_i^{(f)} : i \in Z\} = \emptyset. \tag{10.118}$$

Relations (10.118), (10.117) and (10.109) imply that

$$\{P_1 z_i : i \in Z\} \cap \Omega = \emptyset. \tag{10.119}$$

Relations (10.119) and (10.110) imply that

$$\psi_{f\gamma}(P_1 z_i) > 0 \quad \text{for all } i \in Z. \tag{10.120}$$

Choose a positive number

$$\Delta < 8^{-1}\mu \sum_{i=-q}^{q} \psi_{f\gamma}(P_1 z_i). \tag{10.121}$$

By Proposition 10.2,

$$|z_i - z_0 - i\alpha| < 1 \quad \text{for all } i \in Z, \tag{10.122}$$

$$\left|y_i^{(f\gamma)} - y_0^{(f\gamma)} - i\alpha\right| < 1 \quad \text{for all } i \in Z. \tag{10.123}$$

Since the functions f_γ, $f_{\gamma\mu}$ are continuous and periodic, there exists a number $\varepsilon \in (0,1)$ such that for each $\xi_1, \xi_2, \xi_3, \xi_4 \in R^1$ satisfying

$$\begin{aligned} &|\xi_1 - \xi_2|, |\xi_3 - \xi_4| \le 2|\alpha| + 8, \\ &|\xi_1 - \xi_3| \le 2\varepsilon, \qquad |\xi_2 - \xi_4| \le 2\varepsilon, \end{aligned} \tag{10.124}$$

the following inequality holds:

$$\left|h(\xi_1,\xi_2) - h(\xi_3,\xi_4)\right| \le \Delta/16, \quad h \in \{f_\gamma, f_{\gamma\mu}\}. \tag{10.125}$$

It follows from (10.116) and (10.108) that there exists an integer $m_0 > 4 + 4q$ such that

$$\left|z_i - y_i^{(f\gamma)}\right| < \varepsilon/2 \quad \text{for all integers } i \text{ satisfying } |i| \ge m_0. \tag{10.126}$$

Define $u \in R^Z$ as follows:

$$\begin{aligned} &u_i = z_i, \quad i \in \left[(-\infty, -m_0 - 1] \cup [m_0 + 1, \infty)\right] \cap Z, \\ &u_i = y_i^{(f\gamma)}, \quad i \in [-m_0, m_0] \cap Z. \end{aligned} \tag{10.127}$$

We will show that

$$\sum_{i=-m_0-1}^{m_0} f_{\gamma\mu}(z_i, z_{i+1}) - \sum_{i=-m_0-1}^{m_0} f_{\gamma\mu}(u_i, u_{i+1}) > 0.$$

It follows from (10.127) that

$$\sum_{i=-m_0-1}^{m_0} f_{\gamma\mu}(z_i, z_{i+1}) - \sum_{i=-m_0-1}^{m_0} f_{\gamma\mu}(u_i, u_{i+1})$$
$$= \sum_{i=-m_0-1}^{m_0} f_{\gamma\mu}(z_i, z_{i+1}) - f_{\gamma\mu}\big(z_{-m_0-1}, y_{-m_0}^{(f,\gamma)}\big)$$
$$- f_{\gamma\mu}\big(y_{m_0}^{(f\gamma)}, z_{m_0+1}\big) - \sum_{i=-m_0}^{m_0-1} f_{\gamma\mu}\big(y_i^{(f\gamma)}, y_{i+1}^{(f\gamma)}\big). \tag{10.128}$$

By the definition of ε (see (10.124), (10.125)), (10.126), (10.122) and (10.123),

$$\big| f_{\gamma\mu}(z_{-m_0-1}, z_{-m_0}) + f_{\gamma\mu}(z_{m_0}, z_{m_0+1})$$
$$- f_{\gamma\mu}\big(z_{-m_0-1}, y_{-m_0}^{(f\gamma)}\big) - f_{\gamma\mu}\big(y_{m_0}^{(f\gamma)}, z_{m_0+1}\big)\big| \le \Delta/8.$$

This inequality, (10.128), (10.111), (10.110), (10.109) and (10.121) imply that

$$\sum_{i=-m_0-1}^{m_0} f_{\gamma\mu}(z_i, z_{i+1}) - \sum_{i=-m_0-1}^{m_0} f_{\gamma\mu}(u_i, u_{i+1})$$
$$\ge \sum_{i=-m_0}^{m_0-1} f_{\gamma\mu}(z_i, z_{i+1}) - \sum_{i=-m_0}^{m_0-1} f_{\gamma\mu}\big(y_i^{(f\gamma)}, y_{i+1}^{(f\gamma)}\big) - 8^{-1}\Delta$$
$$= -8^{-1}\Delta + \sum_{i=-m_0}^{m_0-1} f_\gamma(z_i, z_{i+1}) + \mu \sum_{i=-m_0}^{m_0-1} \psi_{f\gamma}(P_1 z_i) - \sum_{i=-m_0}^{m_0-1} f_\gamma\big(y_i^{(f\gamma)}, y_{i+1}^{(f\gamma)}\big)$$
$$> 7\Delta + \sum_{i=-m_0}^{m_0-1} f_\gamma(z_i, z_{i+1}) - \sum_{i=-m_0}^{m_0-1} f_\gamma\big(y_i^{(f\gamma)}, y_{i+1}^{(f\gamma)}\big). \tag{10.129}$$

Define

$$v_i = z_i, \quad i = -m_0, \dots, m_0,$$
$$v_{-m_0-1} = y_{-m_0-1}^{(f\gamma)}, \qquad v_{m_0+1} = y_{m_0+1}^{(f\gamma)}. \tag{10.130}$$

Since $y^{(f\gamma)} \in \mathcal{M}(f_\gamma, \alpha)$, it follows from (10.123), the definition of ε (see (10.124), (10.125)) and (10.126) that

$$0 \le \sum_{i=-m_0-1}^{m_0} f_\gamma(v_i, v_{i+1}) - \sum_{i=-m_0-1}^{m_0} f_\gamma\big(y_i^{(f\gamma)}, y_{i+1}^{(f\gamma)}\big)$$
$$\le \sum_{i=-m_0}^{m_0-1} f_\gamma(z_i, z_{i+1}) - \sum_{i=-m_0}^{m_0-1} f_\gamma\big(y_i^{(f\gamma)}, y_{i+1}^{(f\gamma)}\big)$$

$$
\begin{aligned}
&+ f_\gamma\big(y^{(f\gamma)}_{-m_0-1}, z_{-m_0}\big) + f_\gamma\big(z_{m_0}, y^{(f\gamma)}_{m_0+1}\big)\\
&- f_\gamma\big(y^{(f\gamma)}_{-m_0-1}, y^{(f\gamma)}_{-m_0}\big) - f_\gamma\big(y^{(f\gamma)}_{m_0}, y^{(f\gamma)}_{m_0+1}\big)\\
&\le \sum_{i=-m_0}^{m_0-1} f_\gamma(z_i, z_{i+1}) - \sum_{i=-m_0}^{m_0-1} f_\gamma\big(y^{(f\gamma)}_i, y^{(f\gamma)}_{i+1}\big) + \Delta/8.
\end{aligned}
$$

By these inequalities and (10.129),

$$
\sum_{i=-m_0-1}^{m_0} f_{\gamma\mu}(z_i, z_{i+1}) - \sum_{i=-m_0-1}^{m_0} f_{\gamma\mu}(u_i, u_{i+1}) > 7\Delta + (-\Delta/8) > 6\Delta,
$$

a contradiction. The contradiction we have reached proves Lemma 10.28. □

Completion of the proof of Proposition 10.11 By Theorem 10.8, there exists a set $\mathcal{F}_0 \subset \mathfrak{M}_{k0}$, which is a countable intersection of open (in the weak topology) and everywhere dense (in the strong topology) subsets of $\mathfrak{M}_k$. It is easy to see that for each $f \in \mathfrak{M}_{k0}$, $\lim_{\gamma\to 0^+} f_\gamma = f$ in the strong topology and that for each $f \in \mathfrak{M}_{k0}$ and each $\gamma \in (0,1)$, $\lim_{\mu\to 0^+} f_{\gamma\mu} = f_\gamma$ in the strong topology. Therefore the set

$$
\mathcal{D} := \big\{f_{\gamma\mu} : f \in \mathfrak{M}_{k0}, \gamma, \mu \in (0,1)\big\} \tag{10.131}
$$

is an everywhere dense subset of $\mathfrak{M}_k$ with the strong topology.

Let $g \in \mathcal{D}$. By (10.131), (10.114), Propositions 10.5 and 10.20, and Lemma 10.28, there exist $x^{(g)}, x^{(g^+)} \in \mathcal{M}^{per}(g,\alpha)$ and $y^{(g)} \in \mathcal{M}^+(g,\alpha)$ such that

$$
\mathcal{M}^{per}(g,\alpha) = \big\{T_n x^{(g)} : n = (n_1, n_2) \in Z^2\big\}, \tag{10.132}
$$

$$
\mathcal{M}^{+}(g,\alpha) = \big\{T_n y^{(g)} : n = (n_1, n_2) \in Z^2\big\}, \tag{10.133}
$$

$$
x^{(g)} < y^{(g)} < x^{(g^+)}, \qquad \big\{z \in \mathcal{M}^{per}(g,\alpha) : x^{(g)} < z < x^{(g^+)}\big\} = \emptyset. \tag{10.134}
$$

Let $j \ge 1$ be an integer. By Proposition 10.22 and Lemma 10.27, there is an open neighborhood $\mathcal{U}(g,j)$ of g in $\mathfrak{M}_k$ with the weak topology such that the following properties hold:

(a) For each $f \in \mathcal{U}(g,j) \cap \mathfrak{M}_{k0}$ and each $x \in \mathcal{M}^{per}(f,\alpha)$, there exists $m = (m_1, m_2) \in Z^2$ such that $|x_i - (T_m x^{(g)})_i| < (2j)^{-1}$ for all $i \in Z$.

(b) For each $f \in \mathcal{U}(g,j) \cap \mathfrak{M}_{k0}$ and each $y \in \mathcal{M}^+(f,\alpha)$, there exists $m = (m_1, m_2) \in Z^2$ such that $|(T_m y)_i - x_i^{(g)}| < (2j)^{-1}$ for all $i \in Z$ or $|(T_m y)_i - y_i^{(g)}| < (2j)^{-1}$ for all $i \in Z$.

Define

$$
\mathcal{F}_{\alpha^+} := \mathcal{F}_0 \cap \left[\bigcap_{n=1}^{\infty} \bigcup \{U(g,j) : g \in \mathcal{D}, j \ge n\}\right].
$$

It is not difficult to see that $\mathcal{F}_{\alpha^+}$ is a countable intersection of open (in the weak topology) and everywhere dense (in the strong topology) subsets of $\mathfrak{M}_k$.

Let $f \in \mathcal{F}_{\alpha^+}$. For each integer $n \geq 1$, there exist an integer $s_n \geq n$ and $g_n \in \mathcal{D}$ such that

$$f \in \mathcal{U}(g_n, s_n). \tag{10.135}$$

Let $x, y \in \mathcal{M}^{per}(f, \alpha)$. We will show that y is a translation of x. It follows from the property (a) and (10.35) that for each integer $n \geq 1$, there exists $m^{(n)} = (m_1^{(n)}, m_2^{(n)}) \in Z^2$ such that

$$|y_i - (T_{m^{(n)}} x)_i| < s_n^{-1} \leq 1/n \quad \text{for all } i \in Z. \tag{10.136}$$

By the periodicity of y and x, we may assume without loss of generality that

$$m_1^{(n)} \in [0, q] \quad \text{for all integers } n \geq 1. \tag{10.137}$$

Then (10.136) implies that the sequence $\{m_2^{(n)}\}_{n=1}^{\infty}$ is bounded. By extracting a subsequence we may assume without loss of generality that

$$m^{(n)} = m^{(1)}, \quad n = 1, 2, \ldots.$$

Again (10.136) implies that for all integers $n \geq 1$,

$$|y_i - (T_{m^{(1)}} x)_i| < 1/n, \quad i \in Z.$$

Therefore $y = T_{m_1} x$. Fix $\bar{x} \in \mathcal{M}^{per}(f, \alpha)$. We have shown that

$$\mathcal{M}^{per}(f, \alpha) = \{T_n \bar{x} : n = (n_1, n_2) \in Z^2\}. \tag{10.138}$$

Proposition 10.20 implies that there exists $\bar{x}^+$ such that

$$\begin{gathered} \bar{x}^+ \in \mathcal{M}^{per}(f, \alpha), \\ \bar{x} < \bar{x}^+, \qquad \{z \in \mathcal{M}^{per}(f, \alpha) : \bar{x} < z < \bar{x}^+\} = \emptyset. \end{gathered} \tag{10.139}$$

By (10.139) and Proposition 10.5, there exists $y^{(0)} \in \mathcal{M}^+(f, \alpha)$ such that

$$\bar{x} < y^{(0)} < \bar{x}^+. \tag{10.140}$$

Assume that $y \in \mathcal{M}^+(f, \alpha)$. We will show that y is a translation of $y^{(0)}$. By the definition of $\mathcal{M}^+(f, \alpha)$, Proposition 10.20 and (10.138), we may assume without loss of generality that

$$\bar{x} < y < \bar{x}^+. \tag{10.141}$$

By (10.135) and property (b), for each integer $n \geq 1$, there exist $r^{(n)} = (r_1^{(n)}, r_2^{(n)}) \in Z^2$ and $l^{(n)} = (l_1^{(n)}, l_2^{(n)}) \in Z^2$ such that

$$\left|y_i^{(0)} - \left(T_{r^{(n)}} y^{(g_n)}\right)_i\right| < (2s_n)^{-1} \leq (2n)^{-1} \quad \text{for all } i \in Z \tag{10.142}$$

or

$$|y_i^{(0)} - (T_{r^{(n)}} x^{(g_n)})_i| < (2s_n)^{-1} \le (2n)^{-1} \quad \text{for all } i \in Z; \tag{10.143}$$

$$|y_i - (T_{l^{(n)}} y^{(g_n)})_i| < (2s_n)^{-1} \le (2n)^{-1} \quad \text{for all } i \in Z \tag{10.144}$$

or

$$|y_i - (T_{l^{(n)}} x^{(g_n)})_i| < (2s_n)^{-1} \le (2n)^{-1} \quad \text{for all } i \in Z. \tag{10.145}$$

Define

$$E = \{n \in Z : n \ge 1 \text{ and (10.145) holds}\}. \tag{10.146}$$

Assume that the set E is infinite. By the periodicity of $x^{(g_n)}$, $n \ge 1$, we may assume without loss of generality that

$$l_1^{(n)} \in [0, q], \quad n \in E. \tag{10.147}$$

Recall that $|x_0^{(g_n)}| \le 1$, $n = 1, 2, \ldots$. Together with Proposition 10.2 this implies that for each $i \in Z$,

$$|x_i^{(g_n)}| \le |x_0^{(g_n)}| + |i||\alpha| + 1, \quad n = 1, 2, \ldots. \tag{10.148}$$

It follows from (10.145), (10.146), (10.147) and (10.148) that the set $\{l_2^{(n)} : n \in E\}$ is bounded. Therefore the set $\{l^{(n)} : n \in E\}$ is bounded. There exists an infinite set $F \subset E$ such that $l^{(n_1)} = l^{(n_2)}$ for each $n^{(1)}, n^{(2)} \in F$. When combined with (10.145) and (10.146), this fact implies that $|(T_l y)_i - x_i^{(g_n)}| < (2n)^{-1}$ for all $i \in Z$ and all $n \in F$ with some $l \in Z^2$. This implies that $y \in \mathcal{M}^{per}(f, \alpha)$, a contradiction. Therefore E is finite. Since y is an arbitrary element of $\mathcal{M}^+(f, \alpha)$, the set

$$\{n \in Z : n \ge 1 \text{ and (10.143) holds}\}$$

is finite. We may assume without loss of generality that (10.142) and (10.144) hold for any integer $n \ge 1$. This fact implies that for each integer $n \ge 1$, there exists $j^{(n)} = (j_1^{(n)}, j_2^{(n)}) \in Z^2$ such that

$$|y_i - (T_{j^{(n)}} y^{(0)})_i| < 1/n \quad \text{for all } i \in Z. \tag{10.149}$$

It follows from (10.149), (10.140), (10.141), (10.139) and the definition of $\mathcal{M}^+(f, \alpha)$ that

$$\begin{aligned} &\lim_{i \to -\infty} \bar{x}_i - y_i^{(0)} = 0, \qquad \lim_{i \to -\infty} \bar{x}_i - y_i = 0, \\ &\lim_{i \to \infty} \bar{x}_i^+ - y_i^{(0)} = 0, \qquad \lim_{i \to \infty} \bar{x}_i^+ - y_i = 0. \end{aligned} \tag{10.150}$$

By (10.149) and (10.150), for each integer $n \ge 1$,

$$\lim_{i\to\infty}\left[\left(T_{j^{(n)}}\bar{x}^{+}\right)_i - \left(T_{j^{(n)}}y^{(0)}\right)_i\right] = 0,$$

$$\begin{aligned}\limsup_{i\to\infty}\left[\left|\bar{x}_i^{+} - \left(T_{j^{(n)}}\bar{x}^{+}\right)_i\right|\right] &\le \lim_{i\to\infty}\left|\bar{x}_i^{+} - y_i\right| + \limsup_{i\to\infty}\left[\left|y_i - \left(T_{j^{(n)}}y^{(0)}\right)_i\right|\right] \\ &\quad + \lim_{i\to\infty}\left[\left(T_{j^{(n)}}y^{(0)}\right)_i - \left(T_{j^{(n)}}\bar{x}^{+}\right)_i\right] \\ &\le 1/n.\end{aligned}$$

Since $\bar{x}^{+}$ is periodic, we obtain that for any integer $n \ge 1$,

$$\left|\bar{x}_i^{+} - \left(T_{j^{(n)}}\bar{x}^{+}\right)_i\right| \le 1/n, \quad i \in Z. \tag{10.151}$$

By Corollary 10.16 and (10.138), there exists $\kappa \in (0, 1)$ such that for each $z^{(1)}, z^{(2)} \in \mathcal{M}^{per}(f, \alpha)$ satisfying $z^{(1)} \neq z^{(2)}$,

$$\left|z_i^{(1)} - z_i^{(2)}\right| > 2\kappa, \quad i \in Z. \tag{10.152}$$

By (10.152) and (10.151), for any integer $n > 2\kappa^{-1}$,

$$\bar{x}^{+} = T_{j^{(n)}}\bar{x}^{+}, \qquad \bar{x}_i^{+} = \bar{x}^{+}_{i-j_1^{(n)}} + j_2^{(n)} \quad \text{for all } i \in Z$$

and that the rotation number α of $\bar{x}^{+}$ satisfies $\alpha = p/q = j_2^{(n)}/j_1^{(n)}$. Since p/q is an irreducible fraction, we obtain that for any integer $n > 2\kappa^{-1}$, there is an integer a_n such that

$$a_n(p, q) = j^{(n)}. \tag{10.153}$$

We have three cases:

(1) there exists a strictly increasing sequence of natural numbers $\{n_t\}_{t=1}^{\infty}$ such that $\lim_{t\to\infty} a_{n_t} = \infty$.
(2) there exists a strictly increasing sequence of natural numbers $\{n_t\}_{t=1}^{\infty}$ such that $\lim_{t\to\infty} a_{n_t} = -\infty$.
(3) there exists a strictly increasing sequence of natural numbers $\{n_t\}_{t=1}^{\infty}$ such that $a_{n_t} = a_{n_1}$ for all integers $t \ge 1$.

Assume that case (1) holds. Then by (10.153), (10.138) and (10.150), for any integer i,

$$\left(T_{j^{(n_t)}}y^{(0)}\right)_i = \left(T_{a_{n_t}(q,p)}y^{(0)}\right)_i = y^{(0)}_{i-a_{n_t}q} + a_{n_t}p,$$

$$\begin{aligned}\left(T_{j^{(n_t)}}y^{(0)}\right)_i - \bar{x}_i &= y^{(0)}_{i-a_{n_t}q} + a_{n_t}p - (\bar{x}_{i-a_{n_t}q} + a_{n_t}p) \\ &= y^{(0)}_{i-a_{n_t}q} - \bar{x}_{i-a_{n_t}q} \to 0 \quad \text{as } t \to \infty\end{aligned}$$

and

$$\left(T_{j^{(n_t)}}y^{(0)}\right)_i - \bar{x}_i \to 0 \quad \text{as } t \to \infty \text{ for all } i \in Z.$$

This contradicts (10.149). Therefore case (1) does not hold.

Analogously, we show that case (2) also does not hold. Indeed, assume that case (2) holds. Then by (10.153), (10.138) and (10.150), for any integer i,

$$\big(T_{j(n_t)}y^{(0)}\big)_i = \big(T_{a_{n_t}(q,p)}y^{(0)}\big)_i = y^{(0)}_{i-a_{n_t}q} + a_{n_t}p,$$

$$\begin{aligned}\big(T_{j(n_t)}y^{(0)}\big)_i - \bar{x}^+_i &= y^{(0)}_{i-a_{n_t}q} + a_{n_t}p - \big(\bar{x}^+_{i-a_{n_t}q} + a_{n_t}p\big)\\ &= y^{(0)}_{i-a_{n_t}q} - \bar{x}^+_{i-a_{n_t}q} \to 0 \quad \text{as } t \to \infty\end{aligned}$$

and

$$\big(T_{j(n_t)}y^{(0)}\big)_i - \bar{x}^+_i \to 0 \quad \text{as } t \to \infty \text{ for all } i \in Z.$$

This contradicts (10.149). Therefore case (2) indeed does not hold. Thus we have shown that case (3) is valid. Then it follows from (10.149) and (10.153) that for all $i \in Z$ and any integer $t \geq 1$,

$$1/n > \big|y_i - \big(T_{a_{n_t}(q,p)}y^{(0)}\big)_i\big| = \big|y_i - \big(T_{a_{n_1}(q,p)}y^{(0)}\big)_i\big|$$

and $y = T_{a_{n_1}(q,p)}y^{(0)}$. Proposition 10.11 is proved. □

References

1. Aizicovici, S., Reich, S., & Zaslavski, A. J. (2005). *Archiv der Mathematik, 85*, 268–277.
2. Alber, Y. I., Iusem, A. N., & Solodov, M. V. (1997). *Journal of Convex Analysis, 4*, 235–255.
3. Aliprantis, C. D., & Burkinshaw, O. (1985). *Positive operators*. Orlando: Academic Press.
4. Amann, H. (1976). *SIAM Review, 18*, 620–709.
5. Amemiya, I., & Ando, T. (1965). *Acta Scientiarum Mathematicarum, 26*, 239–244.
6. Arandelović, I. D. (2005). *Journal of Mathematical Analysis and Applications, 301*, 384–385.
7. Arav, M., Reich, S., & Zaslavski, A. J. (2006). *International Journal of Pure and Applied Mathematics, 32*, 65–70.
8. Arav, M., Castillo Santos, F. E., Reich, S., & Zaslavski, A. J. (2007). *Fixed Point Theory and Applications, 2007*, 39465.
9. Arav, M., Reich, S., & Zaslavski, A. J. (2007). *Fixed Point Theory, 8*, 3–9.
10. Arvanitakis, A. D. (2003). *Proceedings of the American Mathematical Society, 131*, 3647–3656.
11. Attouch, H., Lucchetti, R., & Wets, R. J. B. (1992). *Annali Di Matematica Pura Ed Applicata, 160*, 303–320.
12. Aubry, S., & Le Daeron, P. Y. (1983). *Physica D, 8*, 381–422.
13. Ayerbe Toledano, J. M., Dominguez Benavides, T., & López Acedo, G. (1997). *Measures of noncompactness in metric fixed point theory*. Basel: Birkhäuser.
14. Bangert, V. (1988). *Dynamics Reported, 1*, 1–56.
15. Bangert, V. (1994). *Calculus of Variations and Partial Differential Equations, 2*, 49–63.
16. Barbu, V. (1976). *Nonlinear semigroups and differential equations in Banach spaces*. Leyden: Noordhoff.
17. Bauschke, H. H., & Borwein, J. M. (1996). *SIAM Review, 38*, 367–426.
18. Bauschke, H. H., Borwein, J. M., & Lewis, A. S. (1997). In *Contemporary mathematics: Vol. 204. Recent developments in optimization theory and nonlinear analysis* (pp. 1–38).
19. Ben-Tal, A., & Zibulevsky, M. (1997). *SIAM Journal on Optimization, 7*, 347–366.
20. Brezis, H. (1973). *Opérateurs maximaux monotones*. Amsterdam: North-Holland.
21. Brezis, H., & Lions, P. L. (1978). *Israel Journal of Mathematics, 29*, 329–345.
22. Browder, F. E. (1958). *Journal of Mathematics and Mechanics, 7*, 69–80.
23. Browder, F. E. (1968). *Indagationes Mathematicae, 30*, 27–35.
24. Browder, F. E. (1976). *Proceedings of symposia in pure mathematics: Vol. 18. Nonlinear operators and nonlinear equations of evolution in Banach spaces* (Part 2). Providence: Am. Math. Soc.
25. Bruck, R. E., & Reich, S. (1977). *Houston Journal of Mathematics, 3*, 459–470.
26. Burke, J. V., & Ferris, M. C. (1993). *SIAM Journal on Control and Optimization, 31*, 1340–1359.

S. Reich, A.J. Zaslavski, *Genericity in Nonlinear Analysis*,
Developments in Mathematics 34, DOI 10.1007/978-1-4614-9533-8,

27. Butnariu, D., & Iusem, A. N. (1997). In *Contemporary mathematics: Vol. 204. Recent developments in optimization theory and nonlinear analysis* (pp. 61–91).
28. Butnariu, D., & Iusem, A. N. (2000). *Totally convex functions for fixed points computation and infinite dimensional optimization*. Dordrecht: Kluwer Academic.
29. Butnariu, D., Censor, Y., & Reich, S. (1997). *Computational Optimization and Applications*, *8*, 21–39.
30. Butnariu, D., Reich, S., & Zaslavski, A. J. (1999). *Numerical Functional Analysis and Optimization*, *20*, 629–650.
31. Butnariu, D., Iusem, A. N., & Burachik, R. S. (2000). *Computational Optimization and Applications*, *15*, 269–307.
32. Butnariu, D., Iusem, A. N., & Resmerita, E. (2000). *Journal of Convex Analysis*, *7*, 319–334.
33. Butnariu, D., Reich, S., & Zaslavski, A. J. (2001). *Journal of Applied Analysis*, *7*, 151–174.
34. Butnariu, D., Reich, S., & Zaslavski, A. J. (2006). In *Fixed point theory and its applications* (pp. 11–32). Yokohama: Yokohama Publishers.
35. Butnariu, D., Reich, S., & Zaslavski, A. J. (2007). *Journal of Applied Analysis*, *13*, 1–11.
36. Caristi, J. (1976). *Transactions of the American Mathematical Society*, *215*, 241–251.
37. Censor, Y., & Lent, A. (1981). *Journal of Optimization Theory and Applications*, *34*, 321–353.
38. Censor, Y., & Reich, S. (1996). *Optimization*, *37*, 323–339.
39. Censor, Y., & Zenios, S. A. (1997). *Parallel optimization*. New York: Oxford University Press.
40. Chen, Y. Z. (2005). *Fixed Point Theory and Applications*, *2005*, 213–217.
41. Clarke, F. H. (1983). *Optimization and nonsmooth analysis*. New York: Wiley.
42. Cobzas, S. (2000). *Journal of Mathematical Analysis and Applications*, *243*, 344–356.
43. Cohen, J. E. (1979). *Bulletin of the American Mathematical Society*, *1*, 275–295.
44. Correa, R., & Lemaréchal, C. (1993). *Mathematical Programming*, *62*, 261–275.
45. Covitz, H., & Nadler, S. B. Jr. (1970). *Israel Journal of Mathematics*, *8*, 5–11.
46. Crandall, M. G. (1986). Nonlinear semigroups and evolution governed by accretive operators. In *Proceedings of symposia in pure mathematics* (Vol. 45, Part 1, pp. 305–337). Providence: Am. Math. Soc.
47. Curry, H. B. (1944). *Quarterly of Applied Mathematics*, *2*, 258–261.
48. Day, M. M. (1973). *Normed linear spaces*. New York: Springer.
49. de Blasi, F. S., & Myjak, J. (1976). *Comptes Rendus de L'Académie Des Sciences. Paris*, *283*, 185–187.
50. de Blasi, F. S., & Myjak, J. (1989). *Comptes Rendus de L'Académie Des Sciences. Paris*, *308*, 51–54.
51. de Blasi, F. S., & Myjak, J. (1998). *Journal of Approximation Theory*, *94*, 54–72.
52. de Blasi, F. S., Myjak, J., & Papini, P. L. (1991). *Journal of the London Mathematical Society*, *44*, 135–142.
53. de Blasi, F. S., Myjak, J., Reich, S., & Zaslavski, A. J. (2009). *Set-Valued and Variational Analysis*, *17*, 97–112.
54. Diamond, P., Kloeden, P. E., Rubinov, A. M., & Vladimirov, A. (1997). *Set-Valued Analysis*, *5*, 267–289.
55. Dye, J. (1989). *Integral Equations and Operator Theory*, *12*, 12–22.
56. Dye, J. (1989). *Integral Equations and Operator Theory*, *12*, 155–162.
57. Dye, J., & Reich, S. (1992). In *Pitman research notes in mathematics series: Vol. 244. Optimization and nonlinear analysis* (pp. 106–118).
58. Dye, J., Khamsi, M. A., & Reich, S. (1991). *Transactions of the American Mathematical Society*, *325*, 87–99.
59. Edelstein, M. (1968). *Journal of the London Mathematical Society*, *43*, 375–377.
60. Frigon, M. (2007). In *Banach center publications: Vol. 77. Fixed point theory and its applications* (pp. 89–114). Warsaw: Polish Acad. Sci.
61. Frigon, M., Granas, A., & Guennoun, Z. E. A. (1995). *Annales Des Sciences Mathématiques Du Québec*, *19*, 65–68.

62. Fujimoto, T., & Krause, U. (1988). *SIAM Journal on Mathematical Analysis, 19*, 841–853.
63. Gabour, M., Reich, S., & Zaslavski, A. J. (2000). In *Canadian mathematical society conference proceedings: Vol. 27. Constructive, experimental, and nonlinear analysis* (pp. 83–91).
64. Gatica, J. A., & Kirk, W. A. (1974). *The Rocky Mountain Journal of Mathematics, 4*, 69–79.
65. Goebel, K. (2004). *Concise course on fixed point theory*. Yokohama: Yokohama Publishers.
66. Goebel, K., & Kirk, W. A. (1983). In *Contemporary mathematics: Vol. 21. Topological methods in nonlinear functional analysis* (pp. 115–123).
67. Goebel, K., & Kirk, W. A. (1990). *Topics in metric fixed point theory*. Cambridge: Cambridge University Press.
68. Goebel, K., & Reich, S. (1984). *Uniform convexity, hyperbolic geometry, and nonexpansive mappings*. New York: Dekker.
69. Gowda, M. S., & Teboulle, M. (1990). *SIAM Journal on Control and Optimization, 28*, 925–935.
70. Guler, O. (1991). *SIAM Journal on Control and Optimization, 29*, 403–419.
71. Halperin, I. (1962). *Acta Scientiarum Mathematicarum, 23*, 96–99.
72. Hartfiel, D. J., & Rothblum, U. G. (1998). *Linear Algebra and Its Applications, 277*, 1–9.
73. Hiriart-Urruty, J.-B., & Lemaréchal, C. (1993). *Convex analysis and minimization algorithms*. Berlin: Springer.
74. Ioffe, A. D., & Zaslavski, A. J. (2000). *SIAM Journal on Control and Optimization, 38*, 566–581.
75. Jachymski, J. R. (1997). *Aequationes Mathematicae, 53*, 242–253.
76. Jachymski, J. R., & Jóźwik, I. (2004). *Journal of Mathematical Analysis and Applications, 300*, 147–159.
77. Jachymski, J. R., & Jóźwik, I. (2007). In *Banach center publications: Vol. 77. Fixed point theory and its applications* (pp. 123–146). Warsaw: Polish Acad. Sci.
78. Jachymski, J. R., & Stein, J. D. Jr. (1999). *Journal of the Australian Mathematical Society. Series A. Pure Mathematics and Statistics, 66*, 224–243.
79. Jachymski, J. R., Schröder, B., & Stein, J. D. Jr. (1999). *Journal of Combinatorial Theory. Series A, 87*, 273–286.
80. Kelley, J. L. (1955). *General topology*. Princeton: Van Nostrand.
81. Kirk, W. A. (1982). *Numerical Functional Analysis and Optimization, 4*, 371–381.
82. Kirk, W. A. (2001). In *Handbook of metric fixed point theory* (pp. 1–34). Dordrecht: Kluwer Academic.
83. Kirk, W. A. (2003). *Journal of Mathematical Analysis and Applications, 277*, 645–650.
84. Konjagin, S. V. (1978). *Soviet Mathematics. Doklady, 19*, 309–312.
85. Krasnosel'skii, M. A., & Zabreiko, P. P. (1984). *Geometrical methods of nonlinear analysis*. Berlin: Springer.
86. Krein, M. G., & Rutman, M. A. (1962). *Translations - American Mathematical Society, 10*, 199–325.
87. Kuratowski, C., & Ulam, S. (1932). *Fundamenta Mathematicae, 19*, 248–251.
88. Lau, K. S. (1978). *Indiana University Mathematics Journal, 27*, 791–795.
89. Lebourg, G. (1979). *Transactions of the American Mathematical Society, 256*, 125–144.
90. Leizarowitz, A. (1985). *SIAM Journal on Control and Optimization, 22*, 514–522.
91. Leizarowitz, A. (1992). *Linear Algebra and Its Applications, 168*, 189–219.
92. Leizarowitz, A. (1994). *Set-Valued Analysis, 2*, 505–527.
93. Li, C. (2000). *Journal of Approximation Theory, 107*, 96–108.
94. Lim, T.-C. (1974). *Bulletin of the American Mathematical Society, 80*, 1123–1126.
95. Lin, P.-K. (1995). *Nonlinear Analysis, 24*, 1103–1108.
96. Lyubich, Y. I. (1995). In *Integral equations and operator theory* (Vol. 23, pp. 232–244).
97. Lyubich, Y. I., & Maistrovskii, G. D. (1970). *Soviet Mathematics. Doklady, 11*, 311–313.
98. Lyubich, Y. I., & Maistrovskii, G. D. (1970). *Russian Mathematical Surveys, 25*, 57–117.
99. Matkowski, J. (1975). *Dissertationes Mathematicae, 127*, 1–68.
100. Merryfield, J., & Stein, J. D. Jr. (2002). *Journal of Mathematical Analysis and Applications, 273*, 112–120.

101. Merryfield, J., Rothschild, B., & Stein, J. D. Jr. (2001). *Proceedings of the American Mathematical Society, 130*, 927–933.
102. Nadler, S. B. Jr. (1969). *Pacific Journal of Mathematics, 30*, 475–488.
103. Neuberger, J. W. (2010). *Lecture notes in mathematics: Vol. 1670. Sobolev gradients and differential equations* (2nd ed.). Berlin: Springer.
104. Nevanlinna, O., & Reich, S. (1979). *Israel Journal of Mathematics, 32*, 44–58.
105. Nikaido, H. (1968). *Convex structures and economic theory*. New York: Academic Press.
106. Nussbaum, R. D. (1988). *Memoirs of the American Mathematical Society, 391*, 1–137.
107. Nussbaum, R. D. (1990). *SIAM Journal on Mathematical Analysis, 21*, 436–460.
108. Ostrowski, A. M. (1967). *Zeitschrift für Angewandte Mathematik und Mechanik, 47*, 77–81.
109. Polyak, B. T. (1979). *Sharp minima. Institute of control sciences lecture notes.*
110. Prager, M. (1960). *Czechoslovak Mathematical Journal, 10*, 271–282.
111. Pustylnik, E., Reich, S., & Zaslavski, A. J. (2008). *Taiwanese Journal of Mathematics, 12*, 1511–1523.
112. Pustylnik, E., Reich, S., & Zaslavski, A. J. (2008). *Fixed Point Theory and Applications, 2008*, 1–10.
113. Radström, H. (1952). *Proceedings of the American Mathematical Society, 3*, 165–169.
114. Rakotch, E. (1962). *Proceedings of the American Mathematical Society, 13*, 459–465.
115. Reem, D., Reich, S., & Zaslavski, A. J. (2007). *Journal of Fixed Point Theory and Applications, 1*, 149–157.
116. Reich, S. (1972). *Bollettino dell'Unione Matematica Italiana, 5*, 26–42.
117. Reich, S. (1973). *Journal of Mathematical Analysis and Applications, 41*, 460–467.
118. Reich, S. (1976). *Journal of Mathematical Analysis and Applications, 54*, 26–36.
119. Reich, S. (1978). *Journal of Mathematical Analysis and Applications, 62*, 104–113.
120. Reich, S. (1979). *Journal of Mathematical Analysis and Applications, 67*, 274–276.
121. Reich, S. (1993). *Dynamic Systems and Applications, 2*, 21–26.
122. Reich, S. (1996). In *Theory and applications of nonlinear operators of accretive and monotone type* (pp. 313–318). New York: Dekker.
123. Reich, S. (2005). In *Proceedings of CMS'05, computer methods and systems* (pp. 9–15). Krakow.
124. Reich, S., & Shafrir, I. (1990). *Nonlinear Analysis, 15*, 537–558.
125. Reich, S., & Zaslavski, A. J. (1999). *Integral Equations and Operator Theory, 35*, 232–252.
126. Reich, S., & Zaslavski, A. J. (1999). *Abstract and Applied Analysis, 4*, 1–19.
127. Reich, S., & Zaslavski, A. J. (1999). *Positivity, 3*, 1–21.
128. Reich, S., & Zaslavski, A. J. (1999). In *Chapman & Hall/CRC research notes in mathematics series: Vol. 410. Calculus of variations and differential equations* (pp. 200–209). Boca Raton: CRC Press.
129. Reich, S., & Zaslavski, A. J. (1999). *Nonlinear Analysis, 36*, 1049–1065.
130. Reich, S., & Zaslavski, A. J. (1999). *Discrete and Continuous Dynamical Systems, 5*, 929–945.
131. Reich, S., & Zaslavski, A. J. (2000). *Comptes Rendus Mathématiques de L'Académie Des Sciences. La Société Royale du Canada, 22*, 118–124.
132. Reich, S., & Zaslavski, A. J. (2000). *Mathematical and Computer Modelling, 32*, 1423–1431.
133. Reich, S., & Zaslavski, A. J. (2000). *Journal of Nonlinear and Convex Analysis, 1*, 107–113.
134. Reich, S., & Zaslavski, A. J. (2000). *Topological Methods in Nonlinear Analysis, 15*, 153–168.
135. Reich, S., & Zaslavski, A. J. (2000). *Mathematics of Operations Research, 25*, 231–242.
136. Reich, S., & Zaslavski, A. J. (2001). *SIAM Journal on Optimization, 11*, 1003–1018.
137. Reich, S., & Zaslavski, A. J. (2001). *Comptes Rendus de L'Académie Des Sciences. Paris, 333*, 539–544.
138. Reich, S., & Zaslavski, A. J. (2001). *Topological Methods in Nonlinear Analysis, 18*, 395–408.

139. Reich, S., & Zaslavski, A. J. (2001). *Far East Journal of Mathematical Sciences, Special Volume*(Part III), 393–401.
140. Reich, S., & Zaslavski, A. J. (2001). *Integral Equations and Operator Theory, 41*, 455–471.
141. Reich, S., & Zaslavski, A. J. (2001). *Nonlinear Analysis, 47*, 3247–3258.
142. Reich, S., & Zaslavski, A. J. (2001). In *Handbook of metric fixed point theory* (pp. 557–575). Dordrecht: Kluwer Academic.
143. Reich, S., & Zaslavski, A. J. (2002). *Nonlinear Functional Analysis and Applications, 7*, 115–128.
144. Reich, S., & Zaslavski, A. J. (2002). In *Set valued mappings with applications in nonlinear analysis* (pp. 411–420). London: Taylor & Francis.
145. Reich, S., & Zaslavski, A. J. (2002). *Set-Valued Analysis, 10*, 287–296.
146. Reich, S., & Zaslavski, A. J. (2003). In *Fixed point theory and applications* (pp. 261–274). Valencia: Yokohama Publishers.
147. Reich, S., & Zaslavski, A. J. (2003). *Journal of Nonlinear and Convex Analysis, 4*, 165–173.
148. Reich, S., & Zaslavski, A. J. (2003). *Electronic Journal of Differential Equations, 2003*, 1–11.
149. Reich, S., & Zaslavski, A. J. (2004). *Fixed Point Theory and Applications, 2004*, 211–220.
150. Reich, S., & Zaslavski, A. J. (2004). In *Abstract and applied analysis* (pp. 305–311). River Edge: World Scientific.
151. Reich, S., & Zaslavski, A. J. (2004). *Nonlinear Analysis Forum, 9*, 135–152.
152. Reich, S., & Zaslavski, A. J. (2005). *Comptes Rendus Mathématiques de L'Académie Des Sciences. La Société Royale du Canada, 27*, 121–128.
153. Reich, S., & Zaslavski, A. J. (2005). *Fixed Point Theory, 6*, 113–118.
154. Reich, S., & Zaslavski, A. J. (2005). *Fixed Point Theory and Applications, 2005*, 207–211.
155. Reich, S., & Zaslavski, A. J. (2006). *Fixed Point Theory, 7*, 323–332.
156. Reich, S., & Zaslavski, A. J. (2007). In *Banach center publications: Vol. 77. Fixed point theory and its applications.* (pp. 215–225). Warsaw: Polish Acad. Sci.
157. Reich, S., & Zaslavski, A. J. (2007). In *Fixed point theory* (Vol. 7, pp. 161–171). New York: Nova Publ.
158. Reich, S., & Zaslavski, A. J. (2007). *Journal of Fixed Point Theory and Applications, 2*, 69–78.
159. Reich, S., & Zaslavski, A. J. (2007). *Fixed Point Theory, 8*, 303–307.
160. Reich, S., & Zaslavski, A. J. (2008). *Fixed Point Theory, 9*, 267–273.
161. Reich, S., & Zaslavski, A. J. (2008). *Bulletin of the Polish Academy of Sciences. Mathematics, 56*, 53–58.
162. Reich, S., & Zaslavski, A. J. (2008). *Journal of Fixed Point Theory and Applications, 3*, 237–244.
163. Reich, S., & Zaslavski, A. J. (2008). *Taiwanese Journal of Mathematics, 12*, 1165–1176.
164. Reich, S., Rubinov, A., & Zaslavski, A. J. (2000). *Nonlinear Analysis, 40*, 537–547.
165. Ricceri, B. (1987). *Atti Della Accademia Nazionale Dei Lincei. Rendiconti Della Classe Di Scienze Fisiche, Matematiche E Naturali, 81*, 283–286.
166. Rockafellar, R. T. (1979). In *Lecture notes in economics and mathematical systems. Convex analysis and mathematical economics* (pp. 122–136). Berlin: Springer.
167. Rockafellar, R. T. (1995). *SIAM Journal on Control and Optimization, 14*, 877–896.
168. Rubinov, A. M. (1984). *Journal of Soviet Mathematics, 26*, 1975–2012.
169. Sach, P. H., & Yen, N. D. (1997). *Set-Valued Analysis, 5*, 37–45.
170. Schaefer, H. H. (1974). *Banach lattices and positive operators*. Berlin: Springer.
171. Schauder, J. (1930). *Studia Mathematica, 2*, 171–180.
172. Spingarn, J. E. (1985). *Linear Algebra and Its Applications, 65*, 45–62.
173. Stechkin, S. B. (1963). *Revue Roumaine de Mathématiques Pures Et Appliquées, 8*, 5–13.
174. Stein, J. D. Jr. (2000). *The Rocky Mountain Journal of Mathematics, 30*, 735–754.
175. von Neumann, J. (1949). *Annals of Mathematics, 50*, 401–485.
176. Zaslavski, A. J. (1981). *Siberian Mathematical Journal, 22*, 63–68.
177. Zaslavski, A. J. (1996). *Numerical Functional Analysis and Optimization, 17*, 215–240.

178. Zaslavski, A. J. (1998). *Journal of Optimization Theory and Applications*, *97*, 731–757.
179. Zaslavski, A. J. (2000). *SIAM Journal on Control and Optimization*, *39*, 250–280.
180. Zaslavski, A. J. (2001). *Calculus of Variations and Partial Differential Equations*, *13*, 265–293.
181. Zaslavski, A. J. (2004). *Abstract and Applied Analysis*, *8*, 691–721.
182. Zaslavski, A. J. (2010). *Optimization on metric and normed spaces*. New York: Springer.

Index

S. Reich, A.J. Zaslavski, *Genericity in Nonlinear Analysis*,
Developments in Mathematics 34, DOI 10.1007/978-1-4614-9533-8,

Printed in the United States
By Bookmasters